OPTICAL SOLITONS
From Fibers to Photonic Crystals

OPTICAL SOLITONS
From Fibers to Photonic Crystals

Yuri S. Kivshar
Research School of Physical Sciences and Engineering
Australian National University
Canberra, Australia

Govind P. Agrawal
The Institute of Optics
University of Rochester
Rochester, New York, USA

ACADEMIC PRESS

An imprint of Elsevier Science

Amsterdam   Boston   London   New York   Oxford   Paris   San Diego
San Francisco   Singapore   Sydney   Tokyo

Academic Press
An imprint of Elsevier Science
525 B Street, Suite 1900, San Diego, California 92101-4495, USA
http://www.academicpress.com

Academic Press
84 Theobald's Road, London WC1X 8RR, UK
http://www.academicpress.com

Library of Congress Catalog Card Number: 2002117795

International Standard Book Number: 0-12-410590-4

PRINTED IN THE UNITED STATES OF AMERICA
03 04 05 06 07 08   9 8 7 6 5 4 3 2 1

For Our Parents and Families

Contents

Preface

Since the word *soliton* was coined in 1965, the field of solitons in general and the field of optical solitons in particular have grown enormously. Most of this growth has occurred over the last 10 years or so, during which many new kinds of optical solitons have been identified. A partial list includes not only the spatial and temporal Kerr solitons described by the already famous nonlinear Schrödinger equation but also many other types of solitons such as spatiotemporal solitons (often called *light bullets*), Bragg and gap solitons in periodically modulated systems, vortex solitons associated with phase singularities in optics, parametric solitons in nonlinear media with quadratic nonlinearities, discrete solitons in waveguide arrays and photonic crystals, and incoherent solitons in photorefractive crystals. Moreover, the whole field of optical solitons has grown to the extent that it requires a comprehensive book that not only covers fundamental aspects, experimental observations, and relevant device applications, but also provides a link to other, related fields, such as the rapidly developing field of Bose–Einstein condensation. In particular, the remarkable similarities between the matter-wave solitons and optical solitons emphasize the intimate connection between classical nonlinear optics and coherent atom optics and may lead to many discoveries in other, seemingly different fields.

The primary objective of this book is to present the entire field of optical solitons in a comprehensive manner so that the widely scattered research material is available in a single source to all the graduate students who want to enter this exciting field. At the same time, the book should serve as a reference for the scientists who are already engaged in the field of optical solitons as well as the researchers from other fields dealing with photonic crystals, Bose-Einstein condensates, and nonlinear matter waves. It is our hope that the book will be able to serve this dual role, also providing links between seemingly different areas.

The first four core chapters of this book describe the temporal and spatial optical solitons, both the bright and dark kinds, in different types of settings, including optical fibers and planar waveguides. The fundamental concepts are then extended in later chapters to cover such solitons in higher dimensions and periodic optical media, such as fiber gratings (Bragg solitons), waveguide arrays (discrete solitons), and photonic crystals. Novel features appear for solitons carrying an angular momentum, such as vortex solitons and soliton clusters. Moreover, optical solitons can form even by partially coherent beams, leading to a novel concept of incoherent solitons. A relatively new area where the soliton concept is becoming more and more important is related to the physics of nonlinear photonic crystals, and we discuss the most recent advances in this field as well.

The final chapter of the book focuses on several important research topics that can be linked to the concept of spatial optical solitons, although they are not always recognized in this context. Topics discussed include the nonlinear effects in liquid crystals, self-written permanent waveguides in photosensitive materials, dissipative and cavity solitons, self-focusing in thin magnetic films resulting in magnetic solitons, and matter-wave solitons in Bose–Einstein condensates. We have included these last topics, even though they do not belong to the field of nonlinear optics, to show that many of the concepts related to optical solitons are quite generic to the entire field of nonlinear physics.

An attempt is made to include as much recent material as possible so that students and researchers are exposed to the most recent advances in this exciting and rapidly growing field. The reference list at the end of each chapter is more elaborate than what is common for a typical textbook. The listing of recent research papers should be useful for researchers from different fields using this book as a reference.

A large number of persons have contributed to this book, either directly or indirectly. It is impossible to mention all of them by name. We thank our graduate students who have worked on topics related to optical solitons and have helped us in developing this emerging branch of nonlinear optics. We are grateful to our colleagues and graduate students at the Australian National University and the Institute of Optics of the University of Rochester for numerous discussions and fruitful collaboration and for providing a cordial and productive atmosphere.

Many friends and colleagues have helped us by reading different chapters and sections of this book, making numerous helpful suggestions. We are very grateful to all of them. Even though the list is too long to mention them all, we are especially indebted to D. Anderson, G. Assanto, Z. Chen, D. Christodoulides, S. Darmanyan, W. Firth, M. Karpierz, R. Kuszelewicz, M. Lisak, L. Lugiato, T. Monro, D. Neshev, E. Ostrovskaya, D. Pelinovsky, S. A. Ponomarenko, N. Rosanov, I. Shadrivov, D. Skryabin, G. Stegeman, A. A. Sukhorukov, V. Taranenko, S. Turitsyn, C. Weiss, and J. Yang for their extremely valuable input.

Last but not the least, we thank our families in Australia and the United States for understanding why we needed to spend many weekends on the book instead of sharing time with them.

Yuri S. Kivshar
Canberra, Australia

Govind P. Agrawal
Rochester, NY, USA

Chapter 1

Introduction

This introductory chapter is intended to provide a general overview of optical solitons, emphasizing the physical principles and the simplest theoretical model based on the cubic nonlinear Schrödinger (NLS) equation and its generalizations. Section 1.1 provides a brief historical account of the discovery of solitons and introduces the concept of temporal and spatial solitons in the context of nonlinear optics. The basic physics behind the spatial and temporal solitons is discussed in Sections 1.2 and 1.3, respectively, in the context of a simple Kerr-type nonlinearity. The concept of modulation instability, the invFerse scattering transform method, and two types of solitons are presented in Section 1.4. More general models of the nonlinearities are considered in Section 1.5. And finally, Section 1.6 provides an overview of how the material in this book is organized.

1.1 Historical Background

Solitary waves—commonly referred to as *solitons*—have been the subject of intense theoretical and experimental studies in many different fields, including hydrodynamics, nonlinear optics, plasma physics, and biology [1]–[8]. The history of solitons, in fact, dates back to 1834, the year in which James Scott Russell observed that a heap of water in a canal propagated undistorted over several kilometers. His report, published in 1844, includes the following text [9]:

> I was observing the motion of a boat which was rapidly drawn along a narrow channel by a pair of horses, when the boat suddenly stopped—not so the mass of water in the channel which it had put in motion; it accumulated round the prow of the vessel in a state of violent agitation, then suddenly leaving it behind, rolled forward with great velocity, assuming the form of a large solitary elevation, a rounded, smooth and well-defined heap of water, which continued its course along the channel apparently without change of form or diminution of speed. I followed it on horseback, and overtook it still rolling on at a rate of some eight or nine miles an hour, preserving its original figure some thirty feet long and a foot to a

1

foot and a half in height. Its height gradually diminished, and after a chase of one or two miles I lost it in the windings of the channel. Such, in the month of August 1834, was my first chance interview with that singular and beautiful phenomenon which I have called the Wave of Translation.

Such waves were later called *solitary waves*. However, their properties were not understood completely until appropriate mathematical models were introduced and the inverse scattering method was developed in the 1960s [10]. The term *soliton* was coined in 1965 to reflect the particle-like nature of solitary waves that remained intact even after mutual collisions [11]. It should be stressed that the distinction between a soliton and a solitary wave is not always made in the optics literature, and it is quite common to refer to all solitary waves as solitons. We adopt the same terminology in this book.

In the context of nonlinear optics, solitons are classified as being either *temporal* or *spatial*, depending on whether the confinement of light occurs in time or space during wave propagation. Temporal solitons represent optical pulses that maintain their shape, whereas spatial solitons represent self-guided beams that remain confined in the transverse directions orthogonal to the direction of propagation. Both types of solitons evolve from a nonlinear change in the refractive index of an optical material induced by the light intensity—a phenomenon known as the *optical Kerr effect* in the field of nonlinear optics [12]–[14]. The intensity dependence of the refractive index leads to spatial self-focusing (or self-defocusing) and temporal self-phase modulation (SPM), the two major nonlinear effects that are responsible for the formation of optical solitons. A spatial soliton is formed when the self-focusing of an optical beam balances its natural diffraction-induced spreading. In contrast, it is the SPM that counteracts the natural dispersion-induced broadening of an optical pulse and leads to the formation of a temporal soliton [15]. In both cases, the pulse or the beam propagates through a medium without change in its shape and is said to be *self-localized* or *self-trapped*.

The earliest example of a spatial soliton corresponds to the 1964 discovery of the nonlinear phenomenon of self-trapping of continuous-wave (CW) optical beams in a bulk nonlinear medium [16]. Self-trapping was not linked to the concept of spatial solitons immediately because of its unstable nature. During the 1980s, stable spatial solitons were observed using nonlinear media in which diffraction spreading was limited to only one transverse dimension [17]. Figure 1.1 shows an example of the spatial soliton formed inside a semiconductor waveguide [18]. The beam diffracts at low input powers but nearly maintains its original shape when the peak power is adjusted to correspond to a spatial soliton.

The earliest example of a temporal soliton is related to the discovery of the nonlinear phenomenon of *self-induced transparency* in a resonant nonlinear medium [19]. In this case, an optical pulse of a specific shape and energy propagates through the nonlinear medium unchanged in spite of large absorption losses. Another example of a temporal soliton was found in 1973, when it was discovered that optical pulses can propagate inside an optical fiber—a dispersive nonlinear medium—without changing their shape if they experience anomalous dispersion [20]. Propagation of such solitons in optical fibers was observed in a 1980 experiment [21]. Since then, fiber solitons have found practical applications in designing long-haul fiber-optic communication systems [22]–[25].

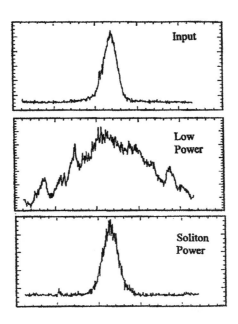

Figure 1.1: Formation of a spatial soliton in a semiconductor waveguide. The input beam (top) diffracts and becomes wider at low powers (middle) but maintains its original shape (bottom) at a certain value of input power. (After Ref. [18]; ©2001 Springer.)

In 1973 it was discovered that optical fibers can support another kind of temporal solitons when the group-velocity dispersion (GVD) is "normal" [26]. Such solitons appear as intensity dips within a CW background and are called *dark solitons*. To make the distinction clear, standard pulse-like solitons are called *bright solitons*. Temporal dark solitons attracted considerable attention during the 1980s [27]–[30]. Spatial dark solitons can also form in optical waveguides and bulk media when the refractive index is lower in the high-intensity region (self-defocusing nonlinearity), and they have been studied extensively [31]–[35]. During the decade of the 1990s, many other kinds of optical solitons were discovered. Examples include spatiotemporal solitons (also called *light bullets*), Bragg solitons, vortex solitons, vector solitons, and quadratic solitons. All of these are covered in this book, intended to provide a comprehensive description of different types of optical solitons in a systematic fashion. In this chapter, we focus on the background material and the basic concepts that would be needed in the later chapters.

1.2 Spatial Optical Solitons

Self-focusing and self-defocusing of CW optical beams in a bulk nonlinear medium has been studied extensively [36]–[38]. The bright or dark spatial solitons form only when the nonlinear effects balance the diffractive effects precisely. Both types of spatial

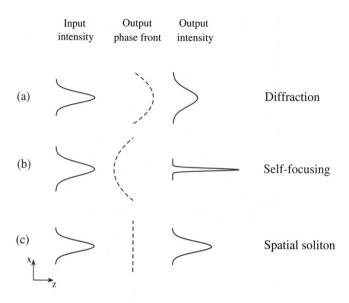

Figure 1.2: Schematic illustration of the lens analogy for spatial solitons. Diffraction acts as a concave lens while the nonlinear medium acts as a convex lens. A soliton forms when the two lenses balance each other such that the phase front remains plane.

solitons have been the focus of extensive research [39]–[41]. This section describes the basic physics and the mathematical tools required for studying spatial solitons.

1.2.1   Basic Concepts

To understand why spatial solitons can form in a self-focusing nonlinear medium, we consider first how light is confined by optical waveguides. Optical beams have an innate tendency to spread (diffract) as they propagate in any homogeneous medium. However, this *diffraction* can be compensated by using *refraction* if the material refractive index is increased in the transverse region occupied by the beam. Such a structure becomes an optical waveguide and confines light to the high-index region by providing a balance between diffraction and refraction. The propagation of the light in an optical waveguide is described by a *linear* but inhomogeneous wave equation whose solution provides a set of *guided modes* that are spatially localized eigenmodes of the optical field in the waveguide that preserve their shape and satisfy all boundary conditions.

It was discovered some time ago [16] that the same effect—suppression of diffraction through a local change of the refractive index—can be produced solely by the nonlinear effects if they lead to a change in the refractive index of the medium in such a way that it is larger in the region where the beam intensity is large. In essence, an optical beam can create its own waveguide and be trapped by this self-induced waveguide. Figure 1.2 shows the basic concept schematically. The input beam diffracts at low power but forms a spatial soliton when its intensity is large enough to create a self-induced waveguide by changing the refractive index. This change is largest at the beam center and gradually reduces to zero near the beam edges, resulting in a graded-index waveguide. The spatial soliton can be thought of as the fundamental mode of this waveguide. Such a nonlinear waveguide can even guide a weak probe beam of a different frequency or polarization [42].

One can also understand the formation of spatial solitons through a lens analogy. Diffraction creates a curved wavefront similar to that produced by a concave lens and spreads the beam to a wider region. The index gradient created by the self-focusing effect, in contrast, acts like a convex lens that tries to focus the beam toward the beam center. In essence, a Kerr medium acts as a convex lens. As seen in Figure 1.2, the beam can become *self-trapped* and propagate without any change in its shape if the two lensing effects cancel each other [16]. Of course, the intensity profile of the beam should have a specific shape for a perfect cancellation of the two effects. These specific beam profiles associated with spatial solitons are the nonlinear analog of the modes of the linear waveguide formed by the self-induced index gradient.

1.2.2 Nonlinear Response

The main equation governing the evolution of optical fields in a nonlinear medium is known as the nonlinear Schrödinger (NLS) equation [1]. In this section we outline the derivation of the NLS equation for a CW beam propagating inside a nonlinear optical medium with Kerr (or cubic) nonlinearity. The Maxwell equations can be used to obtain the following wave equation for the electric field associated with an optical wave propagating in such a medium [15]:

$$\nabla^2 \mathbf{E} - \frac{1}{c^2} \frac{\partial^2 \mathbf{E}}{\partial t^2} = \frac{1}{\varepsilon_0 c^2} \frac{\partial^2 \mathbf{P}}{\partial t^2}, \tag{1.2.1}$$

where c is the speed of light in vacuum and ε_0 is the vacuum permittivity. The induced polarization $\mathbf{P}$ consists of two parts such that

$$\mathbf{P}(\mathbf{r},t) = \mathbf{P}_L(\mathbf{r},t) + \mathbf{P}_{NL}(\mathbf{r},t), \tag{1.2.2}$$

where the linear part $\mathbf{P}_L$ and the nonlinear part $\mathbf{P}_{NL}$ are related to the electric field by the general relations [12]–[14]

$$\mathbf{P}_L(\mathbf{r},t) = \varepsilon_0 \int_{-\infty}^{\infty} \chi^{(1)}(t-t') \cdot \mathbf{E}(\mathbf{r},t') dt', \tag{1.2.3}$$

$$\mathbf{P}_{NL}(\mathbf{r},t) = \varepsilon_0 \iiint_{-\infty}^{\infty} \chi^{(3)}(t-t_1, t-t_2, t-t_3)$$
$$\times \mathbf{E}(\mathbf{r},t_1)\mathbf{E}(\mathbf{r},t_2)\mathbf{E}(\mathbf{r},t_3) dt_1 dt_2 dt_3, \tag{1.2.4}$$

where $\chi^{(1)}$ and $\chi^{(3)}$ are the first- and third-order susceptibility tensors. These relations are valid in the electric-dipole approximation under assumption that the medium response is local. They also neglect the second-order nonlinear effects, assuming that the medium has an inversion symmetry. The second-order nonlinear (quadratic) effects are considered in Chapter 10, which is devoted to the study of quadratic (or parametric) solitons.

Even when only the lowest-order nonlinear effects are included, Eq. (1.2.4) is too complicated to be useful. Considerable simplification occurs if the nonlinear response is assumed to be instantaneous so that the time dependence of $\chi^{(3)}$ is given by the product of three delta functions of the form $\delta(t - t_1)$. Equation (1.2.4) then reduces to

$$\mathbf{P}_{\text{NL}}(\mathbf{r},t) = \varepsilon_0 \chi^{(3)} \, \mathbf{E}(\mathbf{r},t)\mathbf{E}(\mathbf{r},t)\mathbf{E}(\mathbf{r},t). \tag{1.2.5}$$

The assumption of instantaneous nonlinear response amounts to neglecting the contribution of molecular vibrations to $\chi^{(3)}$ (the Raman effect).

Several other simplifying assumptions are necessary. First, $\mathbf{P}_{\text{NL}}$ can be treated as a small perturbation to $\mathbf{P}_L$ because nonlinear changes in the refractive index are $\Delta n/n < 10^{-6}$ in practice. Second, the optical field is assumed to maintain its polarization along the fiber length so that a scalar approach can be used. This is not really the case if the nonlinear medium is birefringent. The birefringence-related effects are considered in Chapter 9 in the context of vector solitons. Third, the optical field is assumed to be quasi-monochromatic. In the slowly varying envelope approximation, it is useful to separate the rapidly varying part of the electric field by writing it in the form

$$\mathbf{E}(\mathbf{r},t) = \tfrac{1}{2}\hat{x}[E(\mathbf{r},t)\exp(-i\omega_0 t) + \text{c.c.}], \tag{1.2.6}$$

where ω_0 is the carrier frequency, $\hat{x}$ is the polarization unit vector, and $E(\mathbf{r},t)$ is a slowly varying function of time (relative to the optical period). The polarization components $\mathbf{P}_L$ and $\mathbf{P}_{\text{NL}}$ can also be expressed in a similar way.

When Eq. (1.2.6) is substituted in Eq. (1.2.5), $\mathbf{P}_{\text{NL}}(\mathbf{r},t)$ is found to have a term oscillating at ω_0 and another term oscillating at the third-harmonic frequency, $3\omega_0$. The latter term requires phase matching and is generally negligible. The slowly varying part $P_{\text{NL}}(\mathbf{r},t)$ of the nonlinear polarization is then given by

$$P_{\text{NL}}(\mathbf{r},t) \approx \varepsilon_0 \varepsilon_{\text{NL}} E(\mathbf{r},t), \tag{1.2.7}$$

where the nonlinear contribution to the dielectric constant is defined as

$$\varepsilon_{\text{NL}} = \tfrac{3}{4}\chi_{xxxx}^{(3)}|E(\mathbf{r},t)|^2. \tag{1.2.8}$$

The linear part of the polarization can be written from Eq. (1.2.3) as $P_L = \varepsilon_0 \chi_{xx}^{(1)} E$. In fact, the linear and nonlinear parts can be combined to provide the following expression for the dielectric constant [15]

$$\tilde{\varepsilon}(\omega) = 1 + \chi_{xx}^{(1)}(\omega) + \varepsilon_{\text{NL}}, \tag{1.2.9}$$

where a tilde denotes the Fourier transform of the quantity under it. The dielectric constant can be used to define the refractive index $\tilde{n}$ and the absorption coefficient $\tilde{\alpha}$. However, both $\tilde{n}$ and $\tilde{\alpha}$ become intensity dependent because of ε_{NL}. It is customary to introduce

$$\tilde{n} = n_0 + n_2|E|^2, \qquad \tilde{\alpha} = \alpha + \alpha_2|E|^2. \tag{1.2.10}$$

The linear index n_0 and the absorption coefficient α are related to the real and imaginary parts of $\tilde{\chi}_{xx}^{(1)}$. Using $\varepsilon = (\tilde{n} + i\tilde{\alpha}c/2\omega_0)^2$ and Eqs. (1.2.8) and (1.2.9), the nonlinear, or Kerr, coefficient n_2 and the two-photon absorption coefficient α_2 are given by

$$n_2 = \frac{3}{8n}\mathrm{Re}(\chi^{(3)}_{xxxx}), \qquad \alpha_2 = \frac{3\omega_0}{4nc}\mathrm{Im}(\chi^{(3)}_{xxxx}), \tag{1.2.11}$$

where Re and Im stand for the real and imaginary parts, respectively.

The nonlinear medium for which the third-order susceptibility dominates and Eq. (1.2.10) describes the nonlinear response accurately is referred to as the *Kerr medium*. Higher-order nonlinear effects can be included in a phenomenological manner by using $\tilde{n} = n_0 + n_{\mathrm{nl}}(I)$ in place of Eq. (1.2.10), where $n_{\mathrm{nl}}(I)$ represents the nonlinear part of the refractive index that depends on the beam intensity $I = |E|^2$. In the case of a Kerr medium, $n_{\mathrm{nl}}(I) = n_2 I$.

1.2.3 Nonlinear Schrödinger Equation

The intensity dependence of the refractive index affects considerably the propagation of electromagnetic waves. In the context of spatial solitons, we can simplify the following analysis by focusing on the case of a CW beam. A general solution of Eq. (1.2.1) can still be written in the form of Eq. (1.2.6) with $E(\mathbf{r},t) = A(\mathbf{r})\exp(i\beta_0 Z)$, where $\beta_0 = k_0 n_0 \equiv 2\pi n_0/\lambda$ is the propagation constant in terms of the optical wavelength $\lambda = 2\pi c/\omega_0$. The beam is assumed to propagate along the Z axis and diffract (or self-focus) along the two transverse directions X and Y, where X, Y, and Z are the spatial coordinates associated with $\mathbf{r}$. The function $A(X,Y,Z)$ describes the evolution of the beam envelope; it would be a constant in the absence of nonlinear and diffractive effects.

When the nonlinear and diffractive effects are included and the envelope A is assumed to vary with z on a scale much longer than the wavelength λ (the paraxial approximation) so that the second derivative $d^2 A/dz^2$ can be neglected, the beam envelope is found to satisfy the following nonlinear parabolic equation:

$$2i\beta_0 \frac{\partial A}{\partial Z} + \left(\frac{\partial^2 A}{\partial X^2} + \frac{\partial^2 A}{\partial Y^2}\right) + 2\beta_0 k_0 n_{\mathrm{nl}}(I)A = 0. \tag{1.2.12}$$

In the absence of the nonlinear effects, this equation reduces to the well-known paraxial equation studied extensively in the context of scalar diffraction theory [43].

We focus on the case of the Kerr (or cubic) nonlinearity and use $n_{\mathrm{nl}}(I) = n_2 I$, n_2 being the Kerr coefficient of the nonlinear material. It is useful to introduce the scaled dimensionless variables as

$$x = X/w_0, \quad y = Y/w_0, \quad z = Z/L_d, \quad u = (k_0|n_2|L_d)^{1/2}A, \tag{1.2.13}$$

where w_0 is a transverse scaling parameter related to the input beam width and $L_d = \beta_0 w_0^2$ is the diffraction length (also called the Rayleigh range). In terms of these dimensionless variables, Eq. (1.2.12) takes the form of a standard $(2+1)$-dimensional NLS equation:

$$i\frac{\partial u}{\partial z} + \frac{1}{2}\left(\frac{\partial^2 u}{\partial x^2} + \frac{\partial^2 u}{\partial y^2}\right) \pm |u|^2 u = 0, \tag{1.2.14}$$

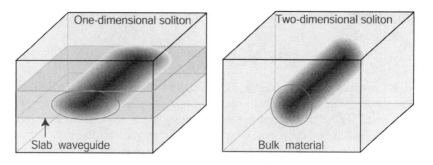

Figure 1.3: Schematics of the planar waveguide (left) and bulk (right) geometries for spatial optical solitons. (Courtesy A. V. Buryak.)

where the choice of the sign depends on the sign of the nonlinear parameter n_2; the minus sign is chosen in the self-defocusing case ($n_2 < 0$). This NLS equation is referred to as being (2+1)-dimensional, where 2 corresponds to the number of the transverse dimensions in the NLS equation and +1 indicates the propagation direction z.

The standard NLS equation has the time variable t in place of z because it has its origin in quantum mechanics. Of course, one can use $Z = (c/n_0)t$ and write the NLS equation in terms of t. However, it is common in optics to use z as the propagation variable. The normalized NLS equation, Eq. (1.2.14), has no free parameters, and even the factor $1/2$ can be removed by renormalizing x and y, as is often done in the soliton literature. We keep this factor for notational simplicity, except for some cases encountered in later chapters.

The dimensionality of the NLS equation can change, depending on the nature of the nonlinear medium. For example, when a nonlinear medium is in the form of a planar waveguide, the optical field is confined in one of the transverse directions, say the vertical y axis, by the waveguide itself, as seen in Figure 1.3. In the absence of the nonlinear effects, the beam will spread only along the x direction. Depending on its size, a planar waveguide supports a finite number of modes. It is often designed to be single-moded, i.e., to support the fundamental mode alone. In that case, the solution of Eq. (1.2.1) can be written in the form of Eq. (1.2.6) with

$$E(\mathbf{r},t) = A(X,Z)B(Y)\exp(i\beta_0 z), \qquad (1.2.15)$$

where the function $B(Y)$ describes the waveguide-mode amplitude and β_0 is the corresponding propagation constant. Using the procedure outlined earlier, we again obtain the normalized NLS equation in the form of Eq. (1.2.14) but without the second-order y derivative, i.e.,

$$i\frac{\partial u}{\partial z} + \frac{1}{2}\frac{\partial^2 u}{\partial x^2} \pm |u|^2 u = 0. \qquad (1.2.16)$$

This NLS equation is referred to as being (1 + 1)-dimensional and constitutes the simplest form of the NLS equation. We focus on this equation because it can be solved exactly using the *inverse scattering method* [44]–[46] for both signs of the nonlinear term. The bright and dark spatial solitons correspond to the choice of + and − signs, respectively.

1.2.4 Bright Spatial Solitons

As a simple example of bright spatial solitons, consider Eq. (1.2.16) with the plus sign for the nonlinear term, assuming that the CW beam is propagating inside a self-focusing Kerr medium. We then need to solve the NLS equation

$$i\frac{\partial u}{\partial z} + \frac{1}{2}\frac{\partial^2 u}{\partial x^2} + |u|^2 u = 0. \tag{1.2.17}$$

Although the inverse scattering method is necessary to find all possible solutions of this equation, the solution corresponding to the fundamental soliton can be obtained by solving the NLS equation directly without using that technique. Since this approach is applicable even when the inverse scattering method cannot be used, we discuss it in some detail.

The approach consists of assuming that a shape-preserving solution of the NLS equation exists and has the form

$$u(z,x) = V(x)\exp[i\phi(z,x)], \tag{1.2.18}$$

where V is independent of z to represent a soliton that maintains its shape during propagation. The phase ϕ can depend on both z and x. If Eq. (1.2.18) is substituted in Eq. (1.2.17) and the real and imaginary parts are separated, we obtain two equations for V and ϕ. The phase equation shows that ϕ should be of the form $\phi(z,x) = Kz + px$, where K and p are constants. Physically p is related to the angle that the soliton trajectory forms with the z axis. Choosing $p = 0$, $V(x)$ is found to satisfy

$$\frac{d^2 V}{dx^2} = 2V(K - V^2). \tag{1.2.19}$$

This nonlinear equation can be solved by multiplying it by $2(dV/dx)$ and integrating over x. The result is

$$(dV/dx)^2 = 2KV^2 - V^4 + C, \tag{1.2.20}$$

where C is a constant of integration. Using the boundary condition that both V and dV/dx vanish as $|x| \to \infty$, C is found to be 0. The conditions that $V = a$ and $dV/dx = 0$, at the soliton peak, which is assumed to occur at $x = 0$, define the constant $K = a^2/2$, and hence $\phi = a^2 z/2$. Equation (1.2.20) is easily integrated to obtain $V(x) = a\,\mathrm{sech}(ax)$, where a is the soliton amplitude.

We have thus found that the $(1+1)$-dimensional NLS equation (1.2.17) has the following simple shape-preserving solution:

$$u(z,x) = a\,\mathrm{sech}(ax)\exp(ia^2 z/2). \tag{1.2.21}$$

As discussed earlier, it represents the fundamental mode of the optical waveguide induced by the propagating beam (higher refractive index near the beam center, where the intensity is the largest). If the input beam has the correct shape, all of its energy will be contained in this mode, and the beam will propagate without change in its shape (no diffraction-induced spreading). If the input beam shape does not exactly match the

Table 1.1 Kerr coefficients for several nonlinear materials

Nonlinear Material	Wavelength (μm)	α (cm^{-1})	n_2 (10^{-14} cm^2/W)
SiO$_2$ glass	1.0–1.6	$<10^{-6}$	0.0024
As$_{0.38}$S$_{0.62}$	1.3–1.6	0.02	4.2
AlGaAs	1.55	0.1	20
PTS crystal	1.6	0.8	220

"sech" shape, some energy will be coupled into higher-order bound modes or into radiation modes of the nonlinear waveguide. The higher-order spatial solitons and the radiation modes are found when Eq. (1.2.17) is solved exactly using the inverse scattering method.

It is important to note that the beam of any size can be self-trapped in the form of a spatial soliton, provided its peak intensity is chosen properly. The intensity needed can be found by writing the soliton solution in Eq. (1.2.21) in physical units as

$$A(Z,X) = \sqrt{I_0}\,\text{sech}(X/w_0)\exp(iZ/2L_d), \qquad (1.2.22)$$

where the peak intensity $I_0 = (k_0 n_2 L_d)^{-1}$. The required input intensity scales with the beam width as w_0^{-2}, indicating that the narrower beams require higher intensities. This is understandable by noting that diffraction effects are stronger for narrower beams. As an example, consider a silica waveguide for which $n_2 = 2.4 \times 10^{-16}$ cm^2/W. The required peak intensity exceeds 10 GW/cm^2 for $w_0 = 100\ \mu$m for a beam at a wavelength of 1.06 μm. It can be reduced by a factor of 100 or more by using other semiconductor or organic materials. Table 1.1 lists the values of the Kerr coefficient for several materials together with their absorption coefficient [47]. Notice that n_2 is smallest for silica glasses and increases by a factor of up to 10,000 for organic materials, such as PTS (*p*-toluene sulfonate) crystals.

1.3   Temporal Optical Solitons

One may wonder whether solitons can exist in a waveguide in which an optical beam is confined in both transverse dimensions. The answer is clearly negative as far as spatial solitons are concerned since the *x*-derivative term in Eq. (1.2.16) disappears in this case if we follow the procedure outlined earlier. It turns out a new kind of solitons can still form in such waveguides if the incident light is in the form of an optical pulse. Such temporal solitons represent optical pulses that maintain their shape during propagation. Their existence was predicted in 1973 in the context of optical fibers [20]. Since then, fiber solitons have been studied extensively and have even found applications in the field of fiber-optic communications [22]–[25]

1.3.1 Pulse Propagation in Optical Fibers

To discuss pulse propagation in optical fibers, we should start from Eq. (1.2.1). The main difference from the CW case discussed in the context of spatial solitons is that the pulse envelope is now time dependent and can be written as

$$E(\mathbf{r},t) = A(Z,t)F(X,Y)\exp(i\beta_0 Z), \qquad (1.3.1)$$

where $F(X,Y)$ is the transverse field distribution associated with the fundamental mode of a single-mode fiber. The time dependence of $A(Z,t)$ implies that all spectral components of the pulse may not propagate at the same speed inside an optical fiber because of the chromatic dispersion. This effect is included by modifying the refractive index in Eq. (1.2.10) as

$$\tilde{n} = n(\omega) + n_2|E|^2. \qquad (1.3.2)$$

The frequency dependence of $n(\omega)$ plays an important role in the formation of temporal solitons. It leads to broadening of optical pulses in the absence of the nonlinear effects and plays the role analogous to that of diffraction in the context of spatial solitons.

Our aim is to obtain an equation satisfied by the pulse amplitude $A(Z,t)$. It is useful to work in the Fourier domain for including the effects of chromatic dispersion and to treat the nonlinear term as a small perturbation. The Fourier transform $\tilde{A}(Z,\omega)$ is found to satisfy [15]

$$\frac{\partial \tilde{A}}{\partial Z} = i[\beta(\omega) + \Delta\beta - \beta_0]\tilde{A}, \qquad (1.3.3)$$

where $\beta(\omega) = k_0 n(\omega)$ and $\Delta\beta$ is the nonlinear part, defined as

$$\Delta\beta = k_0 n_2 |A|^2 \frac{\iint_{-\infty}^{\infty} |F(x,y)|^4\, dx\, dy}{\iint_{-\infty}^{\infty} |F(x,y)|^2\, dx\, dy} \equiv \gamma |A|^2. \qquad (1.3.4)$$

The physical meaning of Eq. (1.3.3) is clear. Each spectral component within the pulse envelope acquires, as it propagates down the fiber, a phase shift whose magnitude is both frequency and intensity dependent.

We can go back to the time domain by taking the inverse Fourier transform of Eq. (1.3.3) and obtain the propagation equation for $A(Z,t)$. However, because an exact functional form of $\beta(\omega)$ is rarely known, it is useful to expand $\beta(\omega)$ in a Taylor series around the carrier frequency ω_0 as

$$\beta(\omega) = \beta_0 + (\omega - \omega_0)\beta_1 + \tfrac{1}{2}(\omega - \omega_0)^2\beta_2 + \tfrac{1}{6}(\omega - \omega_0)^3\beta_3 + \cdots, \qquad (1.3.5)$$

where

$$\beta_m = \left(\frac{d^m\beta}{d\omega^m}\right)_{\omega=\omega_0} \qquad (m = 1, 2, \ldots). \qquad (1.3.6)$$

The cubic and higher-order terms in this expansion are generally negligible if the pulse spectral width $\Delta\omega \ll \omega_0$. Their neglect is consistent with the quasi-monochromatic approximation used earlier. If $\beta_2 \approx 0$ for some specific values of ω_0 (in the vicinity of the zero-dispersion wavelength of the fiber, for example), it may be necessary to include the cubic term.

We substitute Eq. (1.3.5) in Eq. (1.3.3) and take the inverse Fourier transform. During the Fourier-transform operation, $\omega - \omega_0$ is replaced by the differential operator $i(\partial/\partial t)$. The resulting equation for $A(Z,t)$ becomes

$$\frac{\partial A}{\partial Z} + \beta_1 \frac{\partial A}{\partial t} + \frac{i\beta_2}{2} \frac{\partial^2 A}{\partial t^2} = i\gamma |A|^2 A. \tag{1.3.7}$$

The parameters β_1 and β_2 include the effects of dispersion to first and second orders, respectively. Physically, $\beta_1 = 1/v_g$, where v_g is the group velocity associated with the pulse and β_2 takes into account the dispersion of group velocity. For this reason, β_2 is called the group-velocity dispersion (GVD) parameter.

Equation (1.3.7) can be reduced to the $(1 + 1)$-dimensional NLS equation by making the following transformation of variables

$$\tau = (t - \beta_1 Z)/T_0, \qquad z = Z/L_D, \qquad u = \sqrt{|\gamma| L_D} A, \tag{1.3.8}$$

where T_0 is a temporal scaling parameter (often taken to be the input pulse width) and $L_D = T_0^2/|\beta_2|$ is the dispersion length. The variable x now measures time in the reference frame of the moving pulse. In terms of these new variables, Eq. (1.3.7) takes the form

$$i\frac{\partial u}{\partial z} - \frac{s}{2} \frac{\partial^2 u}{\partial \tau^2} \pm |u|^2 u = 0, \tag{1.3.9}$$

where $s = \mathrm{sgn}(\beta_2) = \pm 1$ stands for the sign of the GVD parameter. The GVD parameter β_2 can be positive or negative, depending on the wavelength. The nonlinear term is positive for silica fibers but may become negative for waveguides made of semiconductor materials.

1.3.2 Dispersion Parameter

Because the sign of the GVD parameter plays an important role in determining the soliton solutions of Eq. (1.2.16), it is useful to consider how this parameter varies in optical fibers. In the fiber literature, dispersion is often measured using another parameter, defined as [48]

$$D = \frac{d}{d\lambda}\left(\frac{1}{v_g}\right) = -\frac{2\pi c}{\lambda^2}\beta_2. \tag{1.3.10}$$

D is called the *dispersion parameter* and is expressed in units of ps/(km-nm). Figure 1.4 shows the wavelength dependence of D for three kinds of fibers. For standard fibers, D vanishes at the wavelength of about 1.3 μm. This wavelength λ_{ZD} is called the zero-dispersion wavelength.

The dispersion parameter D is positive at wavelengths such that $\lambda > \lambda_{ZD}$ but becomes negative at shorter wavelengths. From Eq. (1.3.10), β_2 is negative for $\lambda > \lambda_{ZD}$ for optical fibers; i.e, GVD is *anomalous* and we should choose $s = -1$ in Eq. (1.2.16). In contrast, $s = +1$ is chosen when GVD is *normal* in the wavelength region $\lambda < \lambda_{ZD}$. The zero-dispersion wavelength λ_{ZD} can be shifted to longer wavelengths by designing

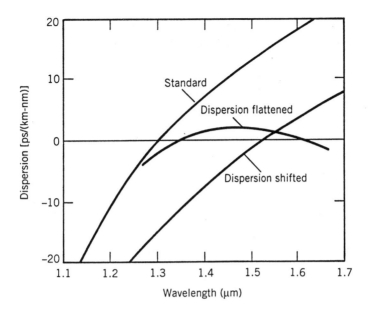

Figure 1.4: Typical wavelength dependence of the dispersion parameter D for standard, dispersion-shifted, and dispersion-flattened fibers.

the fiber suitably. It is in the wavelength region near 1.5 μm in the case of dispersion-shifted fibers. There can even be two such wavelengths in more advanced fibers known as dispersion-flattened fibers.

Because of the two different signs of the GVD parameter, optical fibers, for which $n_2 > 0$, can support two different types of solitons. More specifically, Eq. (1.3.9) has solutions in the form of *dark* temporal solitons in the case of normal GVD ($s = +1$) and *bright* temporal solitons in the case of anomalous GVD ($s = -1$). It is not obvious from Eq. (1.3.9) why dark solitons should exist for $s = 1$ in a self-focusing medium because they were associated with a self-defocusing medium in Section 1.2. However, if one writes this equation as

$$-i\frac{\partial u}{\partial z} + \frac{s}{2}\frac{\partial^2 u}{\partial \tau^2} \mp |u|^2 u = 0, \qquad (1.3.11)$$

it becomes clear that it should support dark solitons even in the self-focusing case. In fact, the solution is the same, provided we replace z by $-z$. In a fiber with self-defocusing nonlinearity ($n_2 < 0$), the situation is reversed, and bright solitons exist in the case of normal GVD. For this reason, even though Eq. (1.3.9) has four possible combinations corresponding to $\pm$ signs for the dispersive and nonlinear terms, it supports only two types of solitons—bright and dark.

1.3.3 Spatiotemporal Dynamics

As seen from Eqs. (1.2.16) and (1.3.9), the same $(1+1)$-dimensional NLS equation describes spatial and temporal solitons. Indeed, from a mathematical standpoint, CW beam propagation in planar waveguides is identical to the phenomenon of pulse propagation in fibers because of the well-known *spatiotemporal analogy* [36]–[38]. This analogy has been exploited extensively and allows one to develop a physical intuition. It is based on the notion that both the beam and pulse propagation can be described by the same cubic NLS equation [49]. However, a crucial difference exists between these two physical phenomena. This difference is responsible for different mathematical tools developed for the analysis of optical solitons—the use of integrable models and their perturbations for temporal solitons and the use of nonintegrable models and soliton instabilities for spatial solitons.

In the case of pulse propagation in fibers, the operating wavelength is usually selected near the zero-dispersion wavelength. As a result, the absolute value of the GVD is small enough to be compensated by a weak nonlinearity produced by the Kerr effect in optical fibers (the nonlinearity-induced change in the refractive index is typically $<10^{-9}$). Therefore, the nonlinear effects in fibers are *always weak*, and pulse propagation in fibers is well modeled by the cubic NLS equation, which is known to be integrable by the inverse scattering method [44]–[46]. However, for ultrashort (femtosecond) pulses, the cubic NLS equation should be corrected to include some additional (but still weak) effects, such as higher-order dispersion, self-steepening, and intrapulse Raman scattering [15].

The physics underlying the propagation of CW beams in planar waveguides and bulk media is quite different. In this case, the nonlinear change in the refractive index should compensate for the beam spreading caused by diffraction, which is not a small effect. That is why much larger nonlinearities are usually required for observing spatial solitons, and very often such nonlinearities are not of the Kerr type (i.e., they saturate at higher intensities). The resulting generalized NLS equation is obtained from Eq. (1.2.14) by replacing the nonlinear term $|u|^2u$ with $F(|u|^2)u$ and takes the form

$$i\frac{\partial u}{\partial z} + \frac{1}{2}\left(\frac{\partial^2 u}{\partial x^2} + \frac{\partial^2 u}{\partial y^2}\right) \pm F(|u|^2)u = 0, \tag{1.3.12}$$

where the functional form of $F(|u|^2)$ is related to $n_{\mathrm{nl}}(I)$, introduced earlier. This equation is not integrable by the inverse scattering method except when $F(|u|^2) = |u|^2$. It can still have solitary-wave solutions, but their properties are quite different from the solutions of the integrable cubic NLS equation. For example, unlike the solitons of the integrable cubic NLS equation, solitary waves associated with a saturable nonlinearity can become unstable and exhibit interesting interaction properties, such as fusion of two colliding solitons.

The second difference between the spatial and temporal solitons is related to the characteristic length scale associated with them. In the case of temporal solitons, nonlinear effects become important over large distances of the order of hundreds of meters or even kilometers. In contrast, propagation distances involved in the case of spatial solitons are of the order of millimeters or centimeters. This difference stems from quite dissimilar values of the dispersion and diffraction lengths. Even for a 1-ps pulse, dispersion length, $L_D = T_0^2/|\beta_2|$, is 1 km if we use a typical value $|\beta_2| = 1$ ps^2/km. The diffraction length, $L_d = \beta_0 w_0^2$, is less than 1 mm for an optical beam with 20-μm spot size at a wavelength near 1 μm.

Under appropriate conditions, the two concepts of spatial and temporal solitons merge into a single concept of the *spatiotemporal soliton*. Indeed, Eq. (1.3.12) can be generalized further to include both the temporal and spatial effects for describing the propagation of optical pulses in a bulk nonlinear medium whose transverse dimensions remain much larger than the beam size (no waveguide effects). In this case, one should include the diffractive, dispersive, and nonlinear effects simultaneously. Following the approach discussed earlier, the general form of the $(3+1)$-dimensional NLS equation (within the scalar approximation) is given by

$$i\frac{\partial u}{\partial z} + \frac{1}{2}\left(\frac{\partial^2 u}{\partial x^2} + \frac{\partial^2 u}{\partial y^2}\right) - \frac{s}{2}\frac{\partial^2 u}{\partial \tau^2} \pm F(|u|^2)u = 0. \qquad (1.3.13)$$

Soliton-like solutions of this equation are called *spatiotemporal* solitons, since their optical field is confined in both space and time. Such solitons, discussed in Chapter 7, are also referred to as *light bullets*. Solutions of the generalized NLS equation (1.3.13) have been obtained in many different contexts, and many interesting properties of bright and dark solitons have been discovered [50]–[62].

1.4 Solutions of the NLS Equation

In this section we consider the simplest $(1+1)$-dimensional NLS equation, describing either the propagation of a CW beam in a planar waveguide or propagation of an optical pulse inside optical fiber, and show that this equation exhibits an instability. This instability is known as the *modulation instability* because it leads to spatial or temporal modulation of a constant-intensity plane wave. We also discuss soliton solutions of the NLS equation under various conditions.

1.4.1 Modulation Instability

The NLS equation (1.2.16) has the simplest solution in the form of a CW plane wave

$$u(z,x) = u_0 \exp(ipz + iqx), \qquad (1.4.1)$$

where u_0 is a constant and p and q satisfy the dispersion relation

$$p = -q^2/2 + \operatorname{sgn}(n_2)u_0^2, \qquad (1.4.2)$$

where $\operatorname{sgn}(n_2) = \pm 1$ depending on the sign of n_2. This solution shows that such a plane wave of amplitude u_0 propagates through the nonlinear medium unchanged except for acquiring an intensity-dependent phase shift.

An important question is whether this plane-wave solution is stable against small perturbations. The answer is provided by a *linear* stability analysis of this solution, an important technique that will be used often in later chapters. We follow the standard procedure and look for solutions of the form

$$u = (u_0 + u_1 + iv_1)\exp(ipz + iqx), \qquad (1.4.3)$$

where u_1 and v_1 represent small perturbations. Substituting Eq. (1.4.3) into the NLS equation (1.2.16) and linearizing the resulting equations, we obtain a system of two coupled linear equations for u_1 and v_1. Looking for solutions to these functions in the form of plane waves $\exp(iKz + iQx)$, we obtain the dispersion relation

$$K = -qQ \pm Q[Q^2/4 - \mathrm{sgn}(n_2)u_0^2]^{1/2}. \tag{1.4.4}$$

The plane-wave solution is stable if perturbations at any wave number Q do not grow with propagation. This is the case as long as K is real.

Equation (1.4.4) shows that the CW plane-wave solution (1.4.1) is absolutely stable only in the case of a self-defocusing medium ($n_2 < 0$). Physically, small-amplitude waves can propagate along with the background intense plane wave in the case of self-defocusing, although their propagation constant K depends on the plane-wave intensity u_0^2. The solution (1.4.1) becomes unstable in a self-focusing medium whenever the beam intensity is such that $u_0^2 > Q^2/4$. This is the modulation instability. Its presence in a self-focusing medium is closely connected with the existence of solitary-wave solutions of the NLS equation. More specifically, spatially localized bright solitons with vanishing amplitude as $|x| \to \infty$ are possible only when the plane-wave solution is unstable. In contrast, dark solitons form in a self-defocusing medium with $n_2 < 0$.

A similar analysis can be carried out for the NLS equation (1.3.9) governing pulse propagation in a dispersive nonlinear medium, such as optical fibers. The dispersion relation in this case is given by

$$K = -qQ \pm Q\sqrt{sQ^2/4 + u_0^2}, \tag{1.4.5}$$

where $s = \mathrm{sgn}(\beta_2)$. Modulation instability is possible only in the case of anomalous dispersion ($\beta_2 < 0$). Once again, bright solitons exist only in this case, while dark solitons occur in the case of normal dispersion. It is easy to see that the defocusing nonlinearity in the spatial problem corresponds to the normal GVD ($s = +1$) in the temporal problem.

In both the spatial and temporal cases, perturbations with certain values of Q become unstable when the last two terms in the NLS equation (1.3.9) have the same sign. Their growth rate g is related to the imaginary part of K and is given by

$$g(Q) = \mathrm{Im}[K(Q)] = |Q|\sqrt{u_0^2 - Q^2/4}. \tag{1.4.6}$$

This growth rate is also called the *modulation-instability gain*. Figure 1.5 shows the instability gain as a function of the perturbation frequency Q (spatial or temporal) for $u_0 = 1$ and 2. The gain exists for both positive and negative values of Q in the range $|Q| < 2u_0$. The maximum gain occurs for $Q = \sqrt{2}u_0$ and has the value $g_{max} = u_0^2$. In normal units, the peak gain is related to the CW intensity of the plane wave as $g_{max} = k_0 n_2 I_0$ and remains relatively small until the intensity becomes large. For example, I_0 should exceed 100 MW/cm^2, even for AlGaAs waveguides, known to have a relatively large value of n_2 (see Table 1.1).

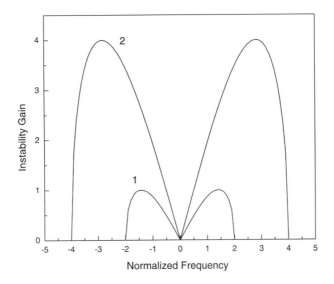

Figure 1.5: Gain spectrum $g(Q)$ of modulation instability for $u_0 = 1$ and 2.

1.4.2 Inverse Scattering Transform Method

In the absence of the nonlinear term in Eq. (1.2.16), the basic mathematical method to analyze any kind of wave propagation in linear media is the standard *Fourier transform method*. By decomposing the input beam $u(z = 0, x)$ into a set of linear Fourier modes, we first reduce the problem of beam propagation to a trivial oscillatory evolution of each Fourier component and then construct the beam shape at any propagation distance z by combining all frequency components. In the case of a homogeneous linear medium, the only possible Fourier modes are the plane waves of the form $\exp(ipx)$, where p labels the Fourier component and is also referred to as the *modal eigenvalue*. Since all values of p are possible in a linear homogeneous medium, the eigenvalue spectrum is continuous, and the solution of Eq. (1.2.16) for low-power beams (negligible nonlinear effects) is given by

$$u(z,x) = \frac{1}{2\pi} \int_{-\infty}^{\infty} \tilde{u}(0,p) \exp(ipx + ip^2 z/2) \, dp, \qquad (1.4.7)$$

where $\tilde{u}(0,p)$ is the Fourier transform of the input field $u(0,x)$.

It should be stressed that the Fourier-transform method is rather general, and it can also be applied to analyze wave propagation in *inhomogeneous* linear media with a spatially varying refractive index. In this case, in addition to the plane-wave modes with a continuous spectrum, there also appear the so-called spatially guided modes, whose existence is due solely to the inhomogeneous nature of the linear medium. The localized or guided modes, however, exist only for certain discrete eigenvalues. Therefore, a complete set of linear eigenstates, which can be obtained by the Fourier method, includes both discrete and continuous modes.

As is well known, the Fourier-transform method cannot be used for nonlinear systems for which the superposition principle does not hold. However, one can try to invent some kind of nonlinear decomposition and obtain *nonlinear modes* whose evolution can be reduced to a simple form similar to the Fourier-transform method. Such a decomposition is known to exist only for some special nonlinear equations (the so-called exactly integrable equations) and is called the *inverse scattering transform*. In some sense, the inverse scattering transform provides an analogy between the modes of *linear inhomogeneous* systems and those of *nonlinear homogeneous* systems. Both types of localized modes—guided waves in the linear case and solitary waves in the nonlinear case—correspond to a specific set of discrete eigenvalues in the linear system.

The main idea behind the nonlinear decomposition used in the inverse scattering transform is to find an appropriate *linear* scattering problem such that it includes the input field $u(z = 0, x)$ as an effective potential. The scattering problem can be solved as an eigenvalue problem owing to its linear nature. Of course, the field and, hence, the potential and the scattering problem change with propagation. However, the eigenvalues of the scattering problem are conserved quantities provided the wave field $u(z, x)$ satisfies the primary nonlinear equation. Finding a scattering problem whose eigenvalue spectrum remains invariant as the potential evolves is the most critical step in applying the inverse scattering method.

The inverse scattering method thus consists of the following steps. For a given $u(z = 0, x)$, solve the linear scattering problem for an auxiliary eigenfunction $\Psi(x; \lambda)$, where λ represents the eigenvalue. The scattering data consist of the amplitude $a(\lambda)$ of a transmitted wave, the amplitude $b(\lambda)$ of a reflected wave, a set of discrete eigenvalues $\{\lambda_n\}$, and the normalized coefficients b_n for the corresponding eigenfunctions. Similar to the case of the standard Fourier-transform method, the evolution of the scattering data $\{a(\lambda), b(\lambda), \lambda_n, b_n\}$ with z is trivial. The solution of the primary nonlinear equation is then found using the inverse scattering method, which allows one to construct the potential $u(z, x)$ from the z-dependent scattering data. A similar method is well known in quantum mechanics.

Each of the discrete eigenvalues of the scattering problem provides a localized solution, which corresponds to a soliton. Solitons maintain their shape because the eigenvalues do not change with z. Moreover, the invariant nature of the eigenvalues results in the important property that two or more solitons maintain their shape even after undergoing collisions. Hence, solitons are important not only as a shape-preserving solution of the nonlinear equation, but as a unique solution whose stability is guaranteed by the invariant property of the corresponding eigenvalue problem. Furthermore, using the scattering data, one can decompose any localized input beam into a set of normal (nonlinear) modes, analogous to the linear Fourier modes. The soliton modes are dominant nonlinear modes in such a decomposition. Thus, any input beam is transformed inside a nonlinear medium asymptotically into a set of solitons associated with the discrete eigenvalues while coupling some of its energy into radiation modes associated with the continuous set of eigenvalues.

It turns out that the NLS equation (1.2.16) is exactly integrable by the inverse scattering method for both signs of the nonlinear term [44]–[46]. The scattering problem associated with this equation involves two auxiliary functions Ψ_1 and Ψ_2 satisfying the following two linear equations [44]:

$$\frac{\partial \Psi_1}{\partial x} = i\lambda \Psi_1 - iu(0,x)\Psi_2, \tag{1.4.8}$$

$$\frac{\partial \Psi_2}{\partial x} = -i\lambda \Psi_2 - iu^*(0,x)\Psi_1, \tag{1.4.9}$$

where the eigenvalues λ are found by solving these equations for a given input field $u(0,x)$. In the case of a self-focusing nonlinearity, the boundary conditions are such that $|u(0,x)| \to 0$ as $|x| \to \infty$, and the solutions are in the form of bright solitons. In contrast, the self-defocusing nonlinearity corresponding to for the minus sign in Eq. (1.2.16) requires the boundary conditions such that $u(0,x) \to u_0$ for $x \to +\infty$, while $u(0,x) \to u_0 e^{i\theta}$ for $x \to -\infty$. Here, u_0 is the background amplitude and θ is a constant phase. The resulting solitons are known as dark solitons because they exhibit an intensity dip near $x = 0$.

1.4.3 Bright and Dark Solitons

Since the CW plane-wave solution is unstable in a self-focusing Kerr medium, bright solitons should exist in this case. Indeed, Zakharov and Shabat showed in 1971 that Eq. (1.2.16) is exactly integrable through the inverse scattering transform [44]. In the case of self-focusing, the fundamental bright soliton has the following general form:

$$u(z,x) = a \operatorname{sech}[a(x-vz)] \exp[ivx + i(a^2 - v^2)z/2], \tag{1.4.10}$$

where a is the soliton amplitude and v is its velocity. This soliton reduces to that obtained earlier in Eq. (1.2.21) in the limit $v = 0$. For spatial solitons, v represents the transverse velocity of solitons propagating at an angle to the z axis. Since this solution exists for all values of a and v, bright solitons form a two-parameter family.

In the case of a self-defocusing Kerr medium, the CW plane-wave solution $|u| = u_0$ is always stable against small modulations. As a result, the soliton solutions exist only in the form of localized "dark" holes created on the CW background. The NLS equation is still exactly integrable by the inverse scattering method [45] if we use the boundary condition $|u| = u_0$ as $|x| \to \infty$. The resulting dark soliton has the following most general form:

$$u(z,x) = u_0\{B \tanh[u_0 B(x - Au_0 z)] + iA\} \exp(-iu_0^2 z), \tag{1.4.11}$$

where the parameters A and B are connected by the relation $A^2 + B^2 = 1$. It is useful to employ a single parameter ϕ introduced using $A = \sin\phi$ and $B = \cos\phi$. The angle ϕ is related to the total phase shift of 2ϕ across the dark soliton. Unlike the bright soliton (1.4.10), the dark-soliton solution (1.4.11) is characterized by a single parameter ϕ (u_0 represents the background amplitude). The magnitude of the dip at the center is governed by $\cos^2\phi$. This can be seen more clearly by noting that

$$|u|^2 = u_0^2\{1 - \cos^2\phi \operatorname{sech}^2[u_0 \cos\phi(x - u_0 \sin\phi z)]\}. \tag{1.4.12}$$

A comparison of Eqs. (1.4.10) and (1.4.11) shows that the quantity $u_0 \sin\phi$ plays the role of the soliton velocity in the x direction; i.e., it represents the relative velocity of

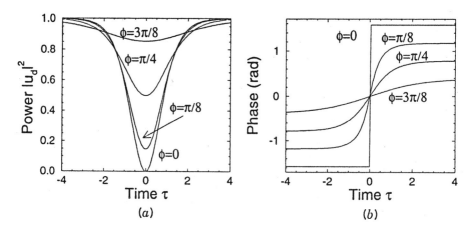

Figure 1.6: (a) Intensity and (b) phase profiles of dark solitons for several values of the internal phase ϕ. The intensity drops to zero at the center for black solitons.

the dark soliton against the background. An important difference between the bright and dark solitons is that the speed of a dark soliton depends on its amplitude through the parameter ϕ.

In the special case of $\phi = 0$, Eq. (1.4.11) reduces to

$$u(z,x) = u_0 \tanh(u_0 x) \exp(-i u_0^2 z).\qquad(1.4.13)$$

This dark soliton is stationary because it does not move against the background. Because of the antisymmetric nature of the "tanh" function, the soliton undergoes a π phase shift at $x = 0$, where the intensity also drops to zero, and the soliton is referred to as being *black*. In contrast, when $\phi \neq 0$, the minimum intensity does not drop to zero at the dip center; such solitons are called *gray* solitons. Another interesting feature of dark solitons is related to their phase. In contrast with bright solitons, which have a constant phase, the phase of a dark soliton changes across its width. Figure 1.6 shows the intensity and phase profiles for several values of ϕ. For a black soliton ($\phi = 0$), a phase shift of π occurs exactly at the center of the dip. For other values of ϕ, the phase changes by an amount $\pi - 2\phi$ in a more gradual fashion. In the case of temporal solitons, the time-dependent phase represents the chirped nature of the dark soliton.

1.5 Solitons of Non-Kerr Media

In the case of the generalized NLS equation (1.3.12), the soliton solutions are not easy to find because this equation is not integrable. If we limit our consideration to one transverse dimension, shape-preserving solutions can be found in an analytical form for some specific forms of the function $F(|u|^2)$. In this section we discuss several models of non-Kerr nonlinearities that will be used in later chapters.

1.5.1 Generalized Nonlinearities

Several different models have been used for the functional form of the nonlinear index $n_{nl}(I)$. They can be divided, generally speaking, into three classes, referred to as *competing*, *saturable*, and *transitive* nonlinearities. We consider each class separately.

Competing nonlinearities

The nonlinear refractive index of any optical material begins to deviate from the n_2I dependence for large enough intensities. Such deviations are observed experimentally for nonlinear materials, such as semiconductor (e.g., AlGaAs, CdS, CdS$_{1-x}$Se$_x$) waveguides, semiconductor-doped glasses, and organic polymers. For example, the measurements for a PTS crystal in the wavelength region near 1600 nm reveal that variations of the nonlinear refractive index with input intensity can be modeled by a cubic-quintic form of the nonlinearity [63]

$$n_{nl}(I) = n_2I + n_3I^2, \tag{1.5.1}$$

where n_2 and n_3 have opposite signs. For $n_2 > 0$ but $n_3 < 0$ this form describes a competition between self-focusing occurring at low intensities and self-defocusing taking over at high intensities. The opposite occurs for $n_2 < 0$ but $n_3 > 0$. A similar model is employed for stabilization of the beam-collapse phenomenon associated with the $(2+1)$-dimensional NLS equation [64]. The nonlinear form in Eq. (1.5.1) can be further generalized by writing it as

$$n_{nl}(I) = n_pI^p + n_{2p}I^{2p}, \tag{1.5.2}$$

where p is a positive constant and the coefficients n_p and n_{2p} have opposite signs so that $n_pn_{2p} < 0$. This model reduces to Eq. (1.5.1) when $p = 1$.

Saturable nonlinearities

Saturable nonlinearities are quite common in nonlinear optics, and saturation of the nonlinearity at high powers has been observed in many nonlinear materials [65]. The most common model for the saturable nonlinearity corresponds to a two-level atomic system, for which the refractive index saturates as $n = n_0(1 + I/I_s)^{-1}$. Such models were introduced more than 25 years ago [66]–[68]. Photorefractive materials such as LiNbO$_3$ also fall in this category, and dark spatial solitons in such materials were observed during the 1990s [59]. Unlike the phenomenological nature of most models used to describe saturation of nonlinearity, this model can be justified rigorously for photovoltaic solitons [60].

From the mathematical point of view, the function $n_{nl}(I)$ describing the saturable nonlinearity is characterized by three independent parameters: the saturation intensity, I_{sat}, the maximum change in the refractive index, n_∞, and the Kerr coefficient n_2 for small I. In a phenomenological approach, the nonlinear part of the refractive index is assumed to vary with intensity I as

$$n_{nl}(I) = n_\infty \left[1 - \frac{1}{(1 + I/I_{sat})^p} \right], \tag{1.5.3}$$

where p is a constant. This form reduces to the Kerr nonlinearity for $I \ll I_s$ with $n_2 = n_\infty p/I_{\text{sat}}$. At the same time, this model describes the nonlinearity of a two-level atomic system for $p = 1$. In the case $p = 2$, the model (1.5.3) yields localized solutions for bright and dark solitons in an explicit analytical form [58].

Transitive Nonlinearities

Bistable solitons introduced during the 1980s require a special form of the intensity-dependent refractive index exhibiting transition from one functional form to another as intensity increases [52]. We call this *transitive nonlinearity* because of the transition in the value of the Kerr coefficient n_2 as the intensity changes from low to high. This type of nonlinearity also supports bistable dark solitons [53]. A simple model for the transitive nonlinearity has the form

$$n_{\text{nl}}(I) = \begin{cases} n_{21}I, & I < I_{\text{cr}}, \\ n_{22}I, & I > I_{\text{cr}}. \end{cases} \tag{1.5.4}$$

The jump occurring at $I = I_{\text{cr}}$ can be avoided by using a smooth transition described by the function [54]

$$n_{\text{nl}}(I) = n_2 I \left\{ 1 + \alpha \tanh \left[\gamma (I^2 - I_{\text{cr}}^2) \right] \right\}. \tag{1.5.5}$$

For $I \ll I_{\text{cr}}$, $n_{\text{nl}}(I) \simeq n_{21}I$ with $n_{21} = n_2[1 - \alpha \tanh^2(\gamma I_{\text{cr}}^2)]$, where α is the model parameter. On the other hand, when $I \gg I_{\text{cr}}$, $n_{\text{nl}}(I) \simeq n_{22}I$ with $n_{22} = (1 + \alpha)$. Thus, n_2 changes from one value to another in a smooth manner in the vicinity of $I = I_{\text{cr}}$. As one may expect, examples of nonlinear optical materials with such a form of the intensity dependence are not yet known. However, bistable solitons possess attractive properties that may be useful for their applications in all-optical logic and switching devices.

1.5.2   Spatial Solitons and Linear Guided Waves

In some cases, one can exploit the concept of a self-induced waveguide discussed in Section 1.2 for finding the solutions of the generalized NLS equation in non-Kerr media [69]–[72]. The basic idea is quite simple and is based on the analogy between the guided modes of the waveguide optics and the solitary waves associated with a nonlinear equation.

As a simple example, consider the problem of light propagation in a homogeneous medium whose refractive index depends on the intensity. If we consider diffraction in only one transverse dimension, the envelope function introduced using $E = A(x) \exp(i\beta Z - i\omega_0 t)$ is a solution of the Helmholtz equation,

$$\left[\frac{d^2}{dx^2} + k_0^2 n^2(|A|^2) - \beta^2 \right] A(x) = 0, \tag{1.5.6}$$

where the propagation constant β corresponds to the discrete eigenvalue associated with each self-induced localized mode. This equation coincides with the stationary form of the NLS equation, as expected.

Each spatially localized solitary wave solution $A(x)$ of Eq. (1.5.6) can be treated as a guided mode of the effective linear waveguide that it induces. Indeed, let us define the *linear* waveguide using the so-called self-consistency relation [71]

$$n_{\text{lin}}^2(x) \equiv n^2(|A(x)|^2). \tag{1.5.7}$$

It is evident that if the solution $A(x)$ of the nonlinear equation (1.5.6) is known, then $A(x)$ is also a solution of the *linear* eigenvalue problem (1.5.6) with the spatially dependent refractive index defined by Eq. (1.5.7). This simple notion allows one to borrow from the literature on linear optical waveguides [73] and to find analytical solutions for the nonlinear media corresponding to them. For example, a slab waveguide is found to correspond to the threshold nonlinearity [69]. Moreover, the interpretation of solitary waves as modes of the self-induced waveguides sometimes helps to provide physical insight for understanding why some kinds of solitary waves are possible while others are not [72].

As an example, consider a linear waveguide with the refractive index profile

$$n_{\text{lin}}^2(x) = n_\infty^2 + (n_0^2 - n_\infty^2)\text{sech}^2(x/a), \tag{1.5.8}$$

where a is related to the width of the index profile. All modes of this waveguide can be expressed with the help of Legendre functions [73]. In particular, the fundamental bound mode is given by

$$A(x) = A_0 \text{sech}^b(x/a), \qquad \beta^2 = (k_0 n_\infty)^2 + (s/a)^2, \tag{1.5.9}$$

where b is an arbitrary constant. The self-consistency condition in Eq. (1.5.7) allows us to come to the conclusion that the index profile in Eq. (1.5.8) corresponds to a bright soliton of the Kerr medium, provided $b = 1$ at a fixed value of the waveguide parameter V given by

$$V^2 \equiv (k_0 a)^2 (n_0^2 - n_\infty^2) = 2. \tag{1.5.10}$$

The preceding example shows that a bright soliton can be treated as the bound mode of the linear waveguide induced by it through the nonlinearity. One may ask if a similar interpretation holds for dark solitons. The answer is yes, provided we realize that the unbounded nature of a dark soliton indicates that it cannot be a bound mode of the self-induced linear waveguide but must be composed of the radiation modes with a continuous spectrum. Indeed, dark solitary waves can be constructed from the radiation modes [71]. This approach provides a simple physical picture of dark solitons through reflectionless plane-wave scattering from a linear dielectric waveguide. For the sech-type waveguide with the index profile (1.5.8), the fields of radiation modes consist of an incident plane wave and its reflections, except for integer values of the parameter s determined from $V^2 = b(b+1)$; for these particular values of b, no reflection occurs. For example, for $b = 1$ ($V = \sqrt{2}$), the reflectionless radiation mode of the sech-type waveguide has the form of a dark soliton of the cubic (or Kerr) medium:

$$A(x) = A_0 \frac{\tanh(x/a) - iaq}{(1 - iaq)} e^{iqx}, \tag{1.5.11}$$

where q is a real continuous variable $(0 < q < k_0 n_\infty)$ that is related to the direction of the incident plane wave. It also defines the propagation constant through $\beta^2 = (k_0 n_\infty)^2 - q^2$. The important point is that the elementary physics of reflectionless plane-wave scattering from a linear waveguide provides a useful insight into the theory of dark solitons.

1.5.3   Limitations of the NLS Equation

The scalar NLS equation (1.3.13) applies for both temporal and spatial aspects of wave propagation and is considered to be a rather universal model. It is derived on the basis of quite general assumptions about the dispersive (and diffractive) effects and the nonlinear properties of physical systems. However, the NLS equation fails in a number of cases, and therefore one should be aware of its limitations. Here we discuss two circumstances under which the scalar NLS equation may fail.

First, the standard derivation of the NLS equation is based on a multiscale asymptotic technique, sometimes called the *reductive perturbation method* (see Refs. [74] and [75]). It assumes that the nonlinear effects are of nonresonant nature and that the most important effects are described by the envelope of the optical field at the fundamental frequency ω propagating with the wave number $k = n(\omega)\omega/c$. All higher-order harmonics are assumed to be too weak to modify the field evolution at the fundamental frequency governed by the NLS equation. However, when several frequencies are generated within the medium, they can affect the propagation of the fundamental harmonic if the so-called *phase-matching condition* is satisfied [14]. For example, strong interaction between the main frequency ω and two other frequencies ω_1 and ω_2 occurs, provided $\omega = \omega_1 + \omega_2$ and the phase mismatch $\Delta k = k - (k_1 + k_2)$ vanishes. This kind of three-wave mixing is possible in a medium for which the lowest-order nonlinearity is quadratic and is governed by the second-order susceptibility $\chi^{(2)}$. In the case of the cubic nonlinearity governed by $\chi^{(3)}$, coupling is possible in the form of a four-wave mixing process. When any such resonance condition is satisfied, the envelope of the fundamental field becomes strongly coupled to one or more secondary fields, and the single NLS equation becomes invalid. In some cases, the nonlinear coupling among multiple waves supports multicomponent solitary waves, which differ considerably from the conventional solitons of the scalar NLS equation (see Chapter 10). Even for a single wave, its orthogonally polarized components can become coupled and form a soliton pair that support each other in a birefringent medium. Such vector solitons are discussed in Chapter 9.

The second class of problems for which the NLS model fails is related to spatial solitons formed in the presence of non-Kerr nonlinearities. Indeed, it is well known that the NLS equation with nonlinearity stronger than cubic (e.g., a power-law nonlinearity $|u|^{2q}u$) has localized solutions that blow up with propagation such that a singularity appears at a finite distance z. This phenomenon occurs for negative values of the system Hamiltonian under the condition $qD \geq 2$, where q is the power of the nonlinearity and D represents th dimension of the $(D+1)$-dimensional NLS equation [76]. This blow-up (or *collapse*) at finite z indicates that the $(D+1)$-dimensional NLS model fails as an envelope equation since it breaks the scale on which it was derived in the framework of the multiscale asymptotic technique. For spatial solitons this condition means that if $D = 2$, the cubic nonlinearity $|u|^2u$ is already sufficient to induce collapse. If $D = 1$, then one needs the quintic (or higher-order) nonlinearity to induce collapse. The occurrence of the beam collapse indicates that the primary NLS model should be corrected by taking into account the higher-order effects related to higher-order dispersion, intrapulse Raman scattering, or the nonparaxial nature of the beam propagation [77]–[81].

We thus conclude that although the NLS equation is quite useful, its applicability must be treated with care. In nonlinear optics, it is a generic model for describing (i) self-guided beams in waveguides and bulk nonlinear media as spatial solitons, (ii) optical pulses in optical fibers as temporal solitons, and (iii) self-focusing with the possibility of beam collapse. When the generalized NLS equation is valid as the main approximation, all corrections to it can be treated by perturbation theory. However, the inclusion of higher-order corrections is very often meaningless near resonances and requires another approach. See Ref. [82] for an example of the failure of the envelope approximation for ultrashort pulses propagating in an an absorbing dispersive medium.

1.6   Book Overview

Over the last decade, the field of optical solitons has grown to the extent that it requires a comprehensive book that not only covers fundamental aspects, experimental observations, and some device applications, but also provides a link to other related fields, such as the rapidly developing field of Bose–Einstein condensation. This book is intended to provide such a comprehensive account of optical solitons. Chapter 1 is devoted to a review of the background material and the basic mathematical tools needed for understanding the various nonlinear effects described in later chapters.

Chapter 2 focuses on *bright spatial solitons* associated with the $(1 + 1)$-dimensional NLS equation. In some cases, these solitons can be described by the exact analytical solutions of the generalized NLS equation. One of the most important feature of such spatially localized solutions of the nonintegrable nonlinear equations is their stability. The stability criterion is discussed together with the concept of the soliton internal modes by considering the properties of linear eigenvalue problems associated with the spatial solitons. This chapter also discusses several aspects of the inelastic soliton collisions in realistic physical models, emphasizing the difference between the soliton interaction in integrable models, such as the cubic NLS equation and more realistic models of non-Kerr nonlinearities.

Chapter 3 is devoted to the study of *temporal solitons* in optical fibers, a topic that has drawn considerable attention because of its fundamental nature as well as potential applications for optical fiber communications. The modulation instability is considered first, to emphasize the importance of the interplay between the dispersive and nonlinear effects that can occur in the anomalous-dispersion regime of optical fibers. The fundamental and higher-order solitons are then introduced, together with the basics of the inverse scattering method used to solve the cubic NLS equation. The last section treats higher-order nonlinear and dispersive effects.

Dark solitons associated with the $(1 + 1)$-dimensional NLS equation are considered in Chapter 4. We begin by studying the case of a Kerr medium for which the cubic

NLS equation is integrable and then focus on the generalized NLS equation needed for describing non-Kerr media. The stability criterion for dark solitons is discussed, using the concept of the renormalized invariants, and applied to several different nonlinear models. Specific effects for temporal dark solitons are also discussed in a separate section of this chapter.

Chapter 5 covers *Bragg solitons* that can form in a nonlinear medium whose linear refractive index varies periodically along its length. The focus is on one-dimensional weakly periodic media. An example of such a medium is provided by fiber Bragg gratings. The coupled-mode theory is used to discuss the concept of photonic bandgap, together with the dispersive effects resulting from such a bandgap. The chapter describes the nonlinear effects, such as the modulation instability, and then focuses on the properties of Bragg solitons. The phenomenon of nonlinear optical switching is also discussed, as an example of the potential applications of Bragg solitons.

Chapter 6 is devoted to *two-dimensional spatial solitons* forming in a bulk nonlinear medium in which diffraction can occur in both transverse dimensions. The chapter begins with a discussion of transverse instability, which breaks a wide optical beam into arrays of stable two-dimensional solitons that maintain their shape in both transverse dimensions. The theoretical and experimental results on two-dimensional spatial solitons are summarized in this chapter. The phenomenon of soliton interaction in a bulk medium is discussed, together with a novel type of two-dimensional self-trapped beam associated with the trapping of the angular momentum.

Chapter 7 treats the simultaneous action of *spatial and temporal effects* using the $(3 + 1)$-dimensional NLS equation that describes self-trapping and self-focusing of optical pulses in a bulk dispersive nonlinear medium. The focus of this chapter is on beam collapse and transverse instabilities associated with multidimensional solitons. The novel effects related to the spatiotemporal dynamics and formation of light bullets are discussed as well.

Chapter 8 is devoted to the physics of *optical vortices* and *optical vortex solitons*. Optical vortices are probably the most interesting structure that can be created by self-focused light, since they carry an orbital angular momentum of light. In some sense, optical vortex solitons can be regarded as a two-dimensional generalization of spatial dark solitons of Chapter 4. Properties of optical vortices are discussed in the framework of the $(2 + 1)$-dimensional NLS equation, with the emphasis on the common features found for optical vortices and the vortices in other fields. Experimental results on optical vortices include the generation of vortices in nonlinear defocusing media, vortex rotation and steering, vortex generation via the transverse modulation instability, and the optical analog of the Aharonov–Bohm effect.

Chapter 9 generalizes many of the important concepts introduced in the previous chapters to the case of *vector solitons* in one and two spatial dimensions. Vector solitons consist of several components with different frequencies or different polarizations. Each of the soliton constituents cannot survive in isolation, but the combination of all components creates a robust entity. In a bulk nonlinear medium, many interesting effects, such as soliton interaction and spiraling, and the formation of dipole- and multipole-vector solitons can occur. Many of these effects has been observed in experiments.

Chapter 10 is devoted to *parametric optical solitons*, also called *quadratic solitons*, when the nonlinear parametric interaction has its origin in the second-order susceptibility. The chapter begins with a discussion of the parametric processes and presents the basic nonlinear equations. It then discusses the properties of one-dimensional quadratic solitons in the waveguide geometry, together with the stability and interaction issues. The experimental results on generation and collision of such solitons are also presented. The chapter then focuses on the solitons supported by competing (quadratic and cubic) nonlinearities, parametric vortex solitons and their instability, and other advanced topics.

Discrete solitons are the topic of Chapter 11. A periodic array of optical waveguides creates a novel device in which new types of spatial solitons can form. The properties of spatially localized modes are first analyzed using a coupled set of equations referred to as the discrete NLS equations. We then present an alternative approach in which the waveguide array is modeled as a sequence of thin-film nonlinear waveguides embedded in an otherwise linear dielectric medium. We discuss modulation instability occurring in such a nonlinear medium and then focus on bright and dark spatial solitons of different types associated with this instability. Experimental results on the formation and steering of spatial solitons in waveguide arrays are also presented.

The novel concepts related to *nonlinear photonic crystals* are introduced in Chapter 12. Photonic crystals can be viewed as an optical analog of semiconductors, in the sense that they modify the propagation characteristics of light just as an atomic lattice modifies the properties of electrons through a bandgap structure. Photonic crystals with embedded nonlinear impurities create an ideal environment for the observation of localized modes in the form of solitons. This chapter focuses on such solitons; they can be viewed as an extension of the gap solitons of Chapter 5 and the discrete solitons of Chapter 11 to two and three spatial dimensions. We use a simple model in the form of a two-dimensional lattice of dielectric rods and show that an isolated defect can support a linear localized mode and that an array of such defects creates a waveguide. We then introduce an effective discrete equation capable of describing quasi-one-dimensional spatial solitons in such photonic waveguides.

The topic of *incoherent solitons* is introduced in Chapter 13. It was discovered in 1996 that optical solitons, normally thought to require a coherent optical field, can form even when the incident beam is only partially coherent, provided the nonlinear medium responds on a time scale much slower than the coherence time. The different theoretical methods developed for analyzing incoherent solitons are covered first. The chapter then focuses on the experimental results obtained using photorefractive crystals because of their relatively slow nonlinear response.

Chapter 14 focuses on several important concepts that can be linked to the concept of spatial optical solitons, although they are not always recognized in this context. Topics covered include the nonlinear effects in *liquid crystals*, *self-written permanent waveguides* in photosensitive materials, solitons in the presence of gain and loss and solitons forming in a cavity containing a nonlinear medium (*cavity solitons*), self-focusing in thin magnetic films resulting in the formation of *magnetic solitons and bullets*, and matter-wave solitons in *Bose–Einstein condensates*. The last two sections describe effects that have little in common with optics. We have included these topics to demonstrate that many of the concepts related to optical solitons are quite generic to the entire field of nonlinear physics. For example, the Bose–Einstein condensates are analyzed using the Gross–Pitaevsky equation, which is closely related to the NLS equation.

References

[1] M. J. Ablowitz and P. A. Clarkson, *Solitons, Nonlinear Evolution Equations, and Inverse Scattering* (Cambridge University Press, New York, 1991).

[2] J. T. Taylor, Ed., *Optical Solitons—Theory and Experiment* (Cambridge University Press, New York, 1992).

[3] F. K. Abdullaev, S. Darmanyan, and P. Khabibulaev, *Optical Solitons* (Springer, Berlin, 1993).

[4] P. G. Drazin, *Solitons: An Introduction* (Cambridge University Press, New York, 1993).

[5] G. L. Lamb, Jr., *Elements of Soliton Theory* (Dover, New York, 1994).

[6] C. H. Gu, *Soliton Theory and its Applications* (Springer, New York, 1995).

[7] N. N. Akhmediev and A. A. Ankiewicz, *Solitons: Nonlinear Pulses and Beams* (Chapman and Hall, London, 1997).

[8] T. Miwa, *Mathematics of Solitons* (Cambridge University Press, New York, 1999).

[9] J. Scott Russell, Report of 14th Meeting of the British Association for Advancement of Science, York, September 1844, pp. 311–390.

[10] C. S. Gardner, J. M. Green, M. D. Kruskal, and R. M. Miura, *Phys. Rev. Lett.* **19**, 1095 (1967); *Commun. Pure Appl. Math.* **27**, 97 (1974).

[11] N. J. Zabusky and M. D. Kruskal, *Phys. Rev. Lett.* **15**, 240 (1965).

[12] Y. R. Shen, *Principles of Nonlinear Optics* (Wiley, New York, 1984).

[13] P. N. Butcher and D. N. Cotter, *The Elements of Nonlinear Optics* (Cambridge University Press, Cambridge, UK, 1990).

[14] R. W. Boyd, *Nonlinear Optics* (Academic Press, San Diego, CA, 1992).

[15] G. P. Agrawal, *Nonlinear Fiber Optics*, 3rd ed. (Academic, San Diego, CA, 2001).

[16] R. Y. Chiao, E. Garmire, and C. H. Townes, *Phys. Rev. Lett.* **13**, 479 (1964).

[17] A. Barthelemy, S. Maneuf, and G. Froehly, *Opt. Commun.* **55**, 201 (1985).

[18] Y. Silberberg and G. I. Stegman, in *Spatial Solitons*, S. Trillo and W. Torruellas, Eds. (Springer, New York, 2001), Chap. 1

[19] S. L. McCall and E. L. Hahn, *Phys. Rev. Lett.* **18**, 908 (1967).

[20] A. Hasegawa and F. Tappert, *Appl. Phys. Lett.* **23**, 142 (1973).

[21] L. F. Mollenauer, R. H. Stolen, and J. P. Gordon, *Phys. Rev. Lett.* **45**, 1095 (1980).

[22] H. Hasegawa and Y. Kodama, *Solitons in Optical Communications* (Oxford University Press, New York, 1995).

[23] H. A. Haus and W. S. Wong, *Rev. Mod. Phys.* **68**, 423 (1996).

[24] L. F. Mollenauer, J. P. Gordon, and P. V. Mamyshev, *Optical Fiber Telecommunications III*, I. P. Kaminow and T. L. Koch, Eds. (Academic Press, San Diego, CA, 1997), Chap. 12.

[25] G. P. Agrawal, *Applications of Nonlinear Fiber Optics* (Academic, San Diego, CA, 2001).

[26] A. Hasegawa and F. Tappert, *Appl. Phys. Lett.* **23**, 171 (1973).

[27] P. Emplit, J. P. Hamaide, F. Reynaud, G. Froehly, and A. Barthelemy, *Opt. Commun.* **62**, 374 (1987).

[28] D. Krökel, N. J. Halas, G. Giuliani, and D. Grischkowsky, *Phys. Rev. Lett.* **60**, 29 (1988).

[29] A. M. Weiner, J. P. Heritage, R. J. Hawkins, R. N. Thurston, E. M. Kirschner, D. E. Learid, and W. J. Tomlinson, *Phys. Rev. Lett.* **61**, 2445 (1988).

[30] A. M. Weiner, R. N. Thurston. W. J. Tomlinson, J. P. Heritage, D. E. Leaird, and E. M. Kirschner, and R. J. Hawkins, *Opt. Lett.* **14**, 868 (1989).

[31] D. R. Andersen, D. E. Hooton, G. A. Swartzlander, Jr., and A. E. Kaplan, *Opt. Lett.* **15**, 783 (1990).

[32] G. A. Swartzlander, Jr., D. R. Andersen, J. J. Regan, H. Yin, and A. E. Kaplan, *Phys. Rev. Lett.* **66**, 1583 (1991).

[33] G. R. Allan, S. R. Skinner, D. R. Andersen, and A. L. Smirl, *Opt. Lett.* **16**, 156 (1991).

[34] S. R. Skinner, G. R. Allan, D. R. Andersen, and A. L. Smirl, *IEEE J. Quantum Electron.* **27**, 2211 (1991).

[35] B. Luther-Davies and X. Yang, *Opt. Lett.* **17**, 496 (1992).

[36] S. A. Akhmanov, A. P. Sukhorukov, and R. V. Khokhlov, *Usp. Fiz. Nauk* **93**, 19 (1967) [*Sov. Phys. Uspekhi* **10**, 609 (1968)].

[37] O. Svelto, in *Progress in Optics*, Vol XII, E. Wolf, Ed. (North-Holland, Amsterdam, 1974).

[38] L. Berge, *Phys. Rep.* **303**, 259 (1998).

[39] P. L. Kelley, *IEEE J. Sel. Topics Quantum Electron.* **6**, 1259 (2000).

[40] G. I. Stegeman, D. N. Christodoulides, and M. Segev, *IEEE J. Sel. Topics Quantum Electron.* **6**, 1419 (2000).

[41] S. Trillo and W. Torruellas, Eds., *Spatial Solitons* (Springer, New York, 2001).

[42] F. Reynaud and A. Barthelemy, *Europhy. Lett.* **12**, 401 (1990).

[43] M. Born and E. Wolf, *Principles of Optics*, 7th ed. (Cambridge University Press, New York, 1999).

[44] V. E. Zakharov and A. B. Shabat, *Zh. Eksp. Teor. Fiz.* **61**, 118 (1971) [*Sov. Phys. JETP* **34**, 62 (1972)].

[45] V. E. Zakharov and A. B. Shabat, *Zh. Eksp. Teor. Fiz.* **64**, 1627 (1973) [*Sov. Phys. JETP* **37**, 823 (1973)].

[46] V. E. Zakharov, S. V. Manakov, S. P. Novikov, and L. P. Pitaevskii, *Theory of Solitons: The Inverse Scattering Transform* (Nauka, Moscow, 1980) [English Translation: Consultant Bureau, New York, 1984].

[47] B. Luther-Davies and G. I. Stegman, in *Spatial Solitons*, S. Trillo and W. Torruellas, Eds. (Springer, New York, 2001).

[48] G. P. Agrawal, *Fiber-Optic Communication Systems*, 3rd ed. (Wiley, New York, 2002).

[49] A. D. Boardman and K. Xie, *Radio Science* **28**, 891 (1993).

[50] V. E. Zakharov, V. V. Sobolev, and V. S. Synakh, *Zh. Eksp. Teor. Fiz.* **60**, 136 (1971) [*Sov. Phys. JETP* **33**, 77 (1971)].

[51] V. E. Zakharov and V. S. Synakh, *Zh. Eksp. Teor. Fiz.* **68**, 940 (1975) [*Sov. Phys. JETP* **41**, 465 (1975)].

[52] A. E. Kaplan, *IEEE J. Quantum Electron.* **21**, 1538 (1985).

[53] L. D. Mulder and R. H. Enns, *IEEE J. Quantum Electron.* **25**, 2205 (1989).

[54] R. H. Enns and L. D. Mulder, *Opt. Lett.* **14**, 509 (1989).

[55] S. Gatz and J. Herrmann, *J. Opt. Soc. Am. B* **8**, 2296 (1991); *Opt. Lett.* **17**, 484 (1992).

[56] J. Herrmann, *Opt. Commun.* **91**, 337 (1992).

[57] A. W. Snyder and A. P. Sheppard, *Opt. Lett.* **18**, 499 (1993).

[58] W. Królikowski and B. Luther-Davies, *Opt. Lett.* **17**, 1414 (1992); *Opt. Lett.* **18**, 188 (1993); *Phys. Rev. E* **48**, 3980 (1993).

[59] G. C. Valley, M. Segev, B. Crosignani, A. Yariv, M. M. Fejer, and M. C. Bashaw, *Phys. Rev. A* **50**, R4457 (1994).

[60] D. N. Christodoulides and M. I. Carvalho, *J. Opt. Soc. Am. B* **12**, 1628 (1995).

[61] D. E. Pelinovsky, V. V. Afanasjev, and Yu. S. Kivshar, *Phys. Rev. E* **53**, 1940 (1996).

[62] R. W. Micallef, V. V. Afanasjev, Yu. S. Kivshar, and J. D. Love, 1996, *Phys. Rev. E* **54**, 2936 (1996).

[63] B. Lawrence, W. E. Torruellas, M. Cha, M. L. Sundheimer, G. I. Stegeman, J. Meth, S. Eteman, and G. Baker, *Phys. Rev. Lett.* **73**, 597 (1994).

[64] C. Josserand and S. Rica, *Phys. Rev. Lett.* **78**, 1215 (1997).

[65] J. L. Coutaz and M. Kull, *J. Opt. Soc. Am. B* **8**, 95 (1991).

[66] T. K. Gustafson, P. L. Kelley, R. Y. Chiao, and R. G. Brewer, *Appl. Phys. Lett.* **12**, 165 (1968).

[67] J. D. Reichert and W. G. Wagner, *IEEE J. Quantum Electron.* **QE-4**, 221 (1968).

[68] J. H. Marburger and E. Dawes, *Phys. Rev. Lett.* **21**, 556 (1968).

[69] A. W. Snyder, D. J. Mitchell, L. Poladian, and F. Ladouceur, *Opt. Lett.* **16**, 21 (1991).

[70] A. W. Snyder, L. Poladian, and D. J. Mitchell, *Opt. Lett.* **17**, 789 (1992).

[71] A. W. Snyder, D. J. Mitchell, and B. Luther-Davies, *J. Opt. Soc. Am. B* **10**, 2341 (1993).

[72] A. W. Snyder, D. J. Mitchell, and Yu. S. Kivshar, *Mod. Phys. Lett. B* **9**, 875 (1995).

[73] A. W. Snyder and D. J. Love, *Optical Waveguide Theory* (Chapman and Hall, London, 1973).

[74] A. Jeffrey and T. Kawahara, *Asymptotic Methods in Nonlinear Wave Theory* (Pitman, London, 1982).

[75] T. Taniuti and K. Nishihara, *Nonlinear Waves* (Pitman, Boston, 1983).

[76] V. E. Zakharov, *Zh. Eksp. Teor. Fiz.* **62**, 1745 (1972) [*Sov. Phys. JETP* **62**, 908 (1972).]

[77] M. Feit and J. Fleck, *J. Opt. Soc. Am. B* **5**, 633 (1988).

[78] G. Fibich, *Phys. Rev. Lett.* **76**, 4356 (1996).

[79] T. Brabec and F. Krauszm, *Phys. Rev. Lett.* **78**, 3282 (1997).

[80] J. K. Ranka and A. L. Gaeta, *Opt. Lett.* **23**, 534 (1998).

[81] Q. Lin and E. Wintner, *Opt. Commun.* **150**, 185 (1998).

[82] K. E. Oughstun and H. Xiao, *Phys. Rev. Lett.* **78**, 642 (1997).

Chapter 2

Spatial Solitons

Spatial solitons in non-Kerr nonlinear materials exhibit many features that differ dramatically from those associated with the exactly integrable NLS equation. This chapter is devoted to the discussion of such features using a nonintegrable generalized NLS equation. Section 2.1 presents a few examples of the explicit analytical solutions for bright spatial solitons in the self-focusing case. Stability of non-Kerr solitons is discussed in Sections 2.2 and 2.3. In particular, Section 2.2 focuses on the important role played by the soliton internal modes, whereas the linear stability analysis of fundamental solitons is outlined in Section 2.3. The concept of *embedded solitons* is introduced in Section 2.4. In Sections 2.5 and 2.6 we discuss the effects of soliton interactions in nonintegrable nonlinear models with examples related to the cubic-quintic and other perturbed NLS equations. Section 2.7 focuses on the experimental results related to the generation, interaction, and steering of spatial solitons in bulk and waveguide nonlinear media. Even though our focus in this chapter is on bright spatial solitons, the discussion in Sections 2.4–2.6 also applies to temporal solitons.

2.1 Analytical Solutions

The properties of optical solitons in a non-Kerr media are governed by the generalized NLS equation (1.2.12) encountered in Section 1.2.3. Using $n_{\mathrm{nl}}(I) = n_2 F(I)$ with $n_2 > 0$ and dimensionless variables of Eq. (1.2.13) and focusing on the waveguide geometry, the $(1+1)$-dimensional NLS equation takes the form

$$i\frac{\partial u}{\partial z} + \frac{1}{2}\frac{\partial^2 u}{\partial x^2} + F(I)u = 0, \tag{2.1.1}$$

where $I = |u(x,z)|^2$ is the beam intensity and the function $F(I)$ characterizes the nonlinear properties of the medium with the condition $F(0) = 0$.

As discussed in Chapter 1, Eq. (2.1.1) can be integrated exactly in the special case of the Kerr nonlinearity [$F(I) = I$] using the inverse scattering transform method. The one-soliton solution of the cubic NLS equation has the following most general form [1]:

$$u(x,z) = a\,\mathrm{sech}[a(x - Vz)]\,\exp[iVx + i(V^2 - a^2)z/2 + i\varphi], \tag{2.1.2}$$

where φ is an arbitrary phase. For a spatial soliton, the parameters a and V are related, respectively, to the amplitude and the transverse velocity of the soliton. When $V \neq 0$, the soliton propagates at an angle to the z axis, and V provides a measure of the transverse displacement of the soliton beam.

The integrable cubic NLS equation is associated with an infinite number of conserved quantities, called the *integrals of motion* [1, 2]. The first three integrals of motion govern the power P, the momentum M, and the Hamiltonian H of the soliton. They are defined as

$$P = \int_{-\infty}^{\infty} |u|^2 dx, \quad M = i \int_{-\infty}^{\infty} (u_x^* u - u_x u^*) \, dx, \tag{2.1.3}$$

$$H = \frac{1}{2} \int_{-\infty}^{\infty} (|u_x|^2 - |u|^4) \, dx, \tag{2.1.4}$$

where u_x stands for the partial derivative $u_x = \partial u / \partial x$.

The generalized NLS equation (2.1.1) cannot be integrated by the inverse scattering transform method. However, it can still have spatially localized solutions that preserve their shape during propagation. To find such soliton-like solutions of Eq. (2.1.1), we assume that a shape-preserving solution has the form

$$u(x, z) = \Phi(x; \beta) e^{i\beta z}, \tag{2.1.5}$$

where β is the soliton propagation constant ($\beta > 0$) and the function $\Phi(x; \beta)$ vanishes for $|x| \to \infty$. The most important conserved quantity for such a soliton is its *power P*, defined as

$$P(\beta) = \int_{-\infty}^{\infty} |u(x, z)|^2 dx = \int_{-\infty}^{\infty} \Phi^2(x; \beta) \, dx. \tag{2.1.6}$$

The spatially localized solutions of Eq. (2.1.1) can be found in an explicit analytical form in a few specific cases. These include the power-law nonlinearity [3], the competing nonlinearity [4] (which includes the case of cubic-quintic nonlinearity [5]), a special case of saturable nonlinearity [6], and the threshold nonlinearity [7]. In the case of the dual-power competing nonlinearity, the nonlinear term in Eq. (2.1.1) is of the form

$$F(I) = \alpha |u|^\sigma + \gamma |u|^{2\sigma}, \tag{2.1.7}$$

where α and γ are constants chosen such that $\alpha \gamma < 0$, to ensure the saturation of nonlinearity with increasing intensity. The parameter σ can be varied to change the form of the nonlinearity. For example, $\sigma = 2$ for the *cubic-quintic nonlinearity* [8]–[12], whereas $\sigma = 1$ for the *quadratic-cubic nonlinearity* [13]. Other integer values of σ describe more exotic type of nonlinear media [14].

Using Eqs. (2.1.5) and (2.1.7) in Eq. (2.1.1) we obtain an ordinary second-order differential equation for Φ, which can be integrated once to yield (assuming $\Phi > 0$)

$$\frac{d\Phi}{dx} = \left[\beta \Phi^2 - \frac{2\alpha}{(\sigma + 2)} \Phi^{\sigma + 2} - \frac{\gamma}{(\sigma + 1)} \Phi^{2(\sigma + 1)} + C \right]^{1/2}, \tag{2.1.8}$$

where C is a constant. Since a bright-soliton solution must satisfy the conditions $\Phi = 0$ and $d\Phi/dx = 0$ as $|x| \to \infty$, we find that $C = 0$. Consequently, Eq. (2.1.8) can be integrated easily with the change of variable $\psi = \Phi^{-\sigma}$. The final soliton solution $u_s(x,z)$ has the form

$$u_s(x,z) = \left[\frac{A}{\cosh(Dx)+B}\right]^{1/\sigma} e^{i\beta z}, \tag{2.1.9}$$

where the real parameters A, D, and B are defined as

$$A = (2+\sigma)B\beta/\alpha, \qquad D = \sigma\sqrt{2\beta}, \tag{2.1.10}$$

$$B \equiv B_{\pm} = \pm\left[1+\frac{(2+\sigma)^2\gamma}{(1+\sigma)\alpha^2}\beta\right]^{-1/2}, \tag{2.1.11}$$

and β can be treated as an arbitrary parameter. In the special case $\sigma = 2$, Eq. (2.1.9) describes the bright solitons associated with the cubic-quintic NLS equation that has been studied as a physically relevant extension of the cubic NLS equation [5],[8]–[12]. The quintic term appears when the contribution of the fifth-order susceptibility is included.

Solution (2.1.9) is useful in many different contexts. For example, when one of the nonlinear terms in Eq. (2.1.7) vanishes, this solution describes the solitons associated with the power-law nonlinearity [3]. For example, when $\alpha = 1$, $\gamma = 0$, the soliton solution can be written in the form (using $B = 1$)

$$u(x,z) = (A/2)^{1/\sigma}\text{sech}^{2/\sigma}(Dx/2)e^{i\beta z}. \tag{2.1.12}$$

It reduces to the bright soliton associated with a Kerr medium for $\sigma = 2$. An algebraic form of the soliton can be obtained in the limit $\beta \rightarrow +0$. For $\alpha < 0$, Solution (2.1.9) reduces in this limit to

$$u_{al}(x) = \left[\frac{2(2+\sigma)(1+\sigma)/|\alpha|}{2\sigma^2(1+\sigma)x^2+(2+\sigma)^2(\gamma/\alpha^2)}\right]^{1/\sigma}. \tag{2.1.13}$$

Such a soliton is called *algebraic* because its amplitude decays as a power law of the form $u_{al}(x) \sim |x|^{-2/\sigma}$ for $|x| \rightarrow \infty$. This solution is a generalization of two particular solutions found in Ref. [13] for $\sigma = 1$ and $\sigma = 2$.

2.2 Soliton Stability and Internal Modes

The solitary waves associated with non-Kerr nonlinear media preserve their shape, but their stability is not guaranteed, because of the nonintegrable nature of the underlying generalized NLS equation. In fact, their stability against small perturbations is a crucial issue because only stable (or weakly unstable) self-trapped beams can be observed experimentally. The stability issue has been studied in several different context, including optics, plasmas, and fluids. The stability of one-parameter solitary waves is well understood for the case of the fundamental (single-hump) solitons [15]–[27]. The main result, known as the *Vakhitov–Kolokolov criterion*, has found its rigorous justification in the mathematical theory developed later [17]. The stability and instability theorems derived there for the scalar NLS equation can be extended to the case of multiparameter solitons [28].

In a *linear* stability analysis, one considers the evolution of a small perturbation of the soliton by modifying the soliton solution as

$$u(x,z) = \left\{ \Phi(x;\beta) + [v(x) - w(x)]\, e^{i\lambda z} + [v^*(x) + w^*(x)]e^{-i\lambda^* z} \right\} e^{i\beta z}, \qquad (2.2.1)$$

where $v(x)$ and $w(x)$ represent small perturbations. Substituting Eq. (2.2.1) in the original NLS equation (2.1.1) and linearizing the resulting equation, $v(x)$ and $w(x)$ are found to satisfy the linear eigenvalue equations

$$L_0 w = \lambda v, \qquad L_1 v = \lambda w, \qquad (2.2.2)$$

$$L_j = -\frac{d^2}{dx^2} + \beta - U_j \quad (j = 0, 1), \qquad (2.2.3)$$

where $U_0 = F(I)$ and $U_1 = F(I) + 2I(\partial F/\partial I)$ are functions of x through the intensity $I(x) = |\Phi(x)|^2$.

A soliton solution of Eq. (2.1.1) is stable if none of the eigenmodes of the linear eigenvalue problem (2.2.2) grows exponentially. This is possible if all eigenvalues are real, i.e., if $\mathrm{Im}(\lambda) = 0$. It turns out that the continuum part of the eigenvalue spectrum associated with Eq. (2.2.2) consists of two symmetric branches such that the eigenvalues are real with the absolute value $|\lambda| > \beta$. The discrete eigenmodes fall into the following three categories:

- neutrally stable *internal modes* with real eigenvalues;
- *instability modes* with imaginary eigenvalues;
- *oscillatory instability modes* with complex eigenvalues.

Since soliton instabilities always occur in nonintegrable models, one may ask what distinct features of such solitary waves are responsible for their instabilities. It is commonly believed that solitons of nonintegrable nonlinear models differ from solitons of integrable models only in their interaction characteristics, i.e, unlike "proper" solitons, the interaction of non-Kerr solitons is accompanied by emission of radiation [23]. However, the stability analysis presented here reveals that the differences are much more substantial. More specifically, non-Kerr solitons exhibit features that are generic and occur whenever Eq. (2.2.2) has discrete eigenvalues such that $|\lambda| < \beta$. For example, even a relatively small perturbation of the integrable NLS equation can create an *internal mode* of the soliton [26]. The analysis of such internal modes is beyond the methods of a regular perturbation theory because solitons of integrable NLS equations do not possess them. In the nonintegrable case, such modes can introduce *qualitatively* new features into the system dynamics and may lead, in particular, to the appearance of new soliton instabilities.

To show that the internal modes constitute a *generic* feature for nonintegrable NLS equations, we consider a slightly perturbed cubic NLS equation and assume that the nonlinear term in Eq. (2.1.1) has the form

$$F(I) = I + \varepsilon f(I), \qquad (2.2.4)$$

where ε is a small parameter and $f(I)$ represents the deviation from the Kerr nonlinearity. The perturbed soliton solution can be expressed as

$$\Phi(x;\beta) = \Phi_0(x) + \varepsilon \Phi_1(x) + O(\varepsilon^2), \qquad (2.2.5)$$

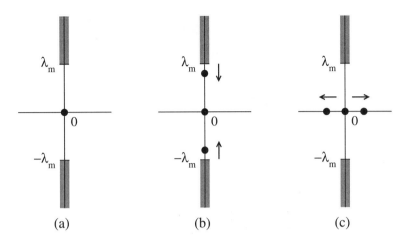

Figure 2.1: Schematic illustration of instabilities associated with non-Kerr solitons: (a) eigenvalue spectrum of Kerr solitons with a neutral mode (real λ), (b) appearance of an internal mode, and (c) collision of the internal mode with the neutral mode. In each case, solid circles show discrete eigenvalues and the hatched region corresponds to continuous real eigenvalues.

where $\Phi_0(x) = \sqrt{2\beta}\,\text{sech}(\sqrt{2\beta}x)$ is the soliton of the cubic NLS equation and $\Phi_1(x)$ is a localized correction derived using Eqs. (2.1.1) and (2.2.4). To the first order in ε, the potentials U_0 and U_1 appearing in the linearized eigenvalue problem (2.2.2) are given as

$$U_0 = \Phi_0^2 + \varepsilon\widetilde{U}_0, \qquad U_1 = 3\Phi_0^2 + \varepsilon\widetilde{U}_1, \qquad (2.2.6)$$

$$\widetilde{U}_0 = f(\Phi_0^2) + 4\Phi_0\Phi_1, \qquad \widetilde{U}_1 = f(\Phi_0^2) + 2\Phi_0^2 f'(\Phi_0^2) + 12\Phi_0\Phi_1, \qquad (2.2.7)$$

where the prime denotes differentiation with respect to x.

The linear eigenvalue problem (2.2.2) can be solved exactly for $\varepsilon = 0$ [29]. Figure 2.1(a) shows the location of eigenvalues in the complex $\lambda = 0$ plane (real values along the vertical axis). The discrete spectrum in the Kerr case contains only one degenerate eigenvalue at $\lambda = 0$, which corresponds to the so-called *neutral mode*. As seen in part (b), the presence of a small perturbation creates an internal mode through the appearance of two symmetric discrete eigenvalues. These eigenvalues bifurcate from the continuous spectral band. If we assume that the cutoff frequencies, $\lambda_m = \pm\beta$, are not affected by the perturbation, the internal-mode eigenvalue on the upper branch can be written as $\lambda = \beta - \varepsilon^2\kappa^2$, where κ is given by [26]

$$\kappa = \frac{\varepsilon}{4|\varepsilon|}\int_{-\infty}^{\infty}\left[V(x,\beta)\widetilde{U}_1 V(x;\beta) + W(x,\beta)\widetilde{U}_0 W(x;\beta)\right]dx. \qquad (2.2.8)$$

Here $V(x;\beta)$ and $W(x;\beta)$ are the eigenfunctions of the cubic NLS equation calculated at the edge of the continuous spectrum and are found to be $V(x;\beta) = 1 - 2\,\text{sech}^2(\sqrt{2\beta}x)$ and $W(x;\beta) = 1$.

As an important example, consider the case of the NLS equation (2.1.1) with

$$F(I) = I + \varepsilon I^3. \qquad (2.2.9)$$

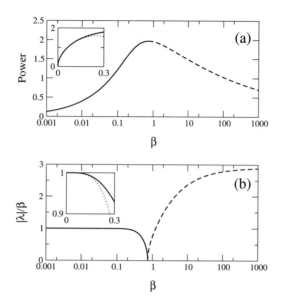

Figure 2.2: (a) Power $P(\beta)$ and (b) the discrete eigenvalues associated with the soliton internal mode (solid) and the instability mode (dashed). Dotted lines in the inset shows the analytical asymptotic dependence. (After Ref. [27]; ©2001 Springer.)

The first-order correction to the soliton profile is then found to be

$$\Phi_1(x) = -\frac{\sqrt{2}\beta^{5/2}[2\cosh(2\sqrt{2\beta}x) + \cosh(4\sqrt{2\beta}x)]}{3\cosh^5(\sqrt{2\beta}x)}. \tag{2.2.10}$$

It is easy to see with the help of Eq. (2.2.8) that the perturbed NLS soliton possesses for $\varepsilon > 0$ an internal mode whose eigenvalue on the upper spectral branch (see Figure 2.1) is given by

$$\lambda = \beta\left[1 - \left(\frac{64\varepsilon}{15}\right)^2\beta^4 + O(\varepsilon^4)\right]. \tag{2.2.11}$$

At high intensities, the last term in Eq. (2.2.9) becomes so large that both the soliton solution and the associated linear spectrum should be calculated numerically. Figure 2.2(a) shows the dependence of soliton power $P(\beta)$ on the propagation constant β calculated using Eq. (2.1.6). The corresponding discrete eigenvalues of the linearized problem (2.2.2) are shown in Figure 2.2(b) (see also Figure 2.1). Both the numerical results and the approximate analytic results are shown. Several conclusions can be drawn from Figure 2.2. First, the approximate theory (the dotted curves in the insets) provides accurate results for small-intensity solitons for which $\beta < 0.1$. Second, the slope of the power dependence changes from positive to negative at the point $\beta = \beta_{cr}$, where the soliton internal modes merge with the neutral modes and split into instability modes, as depicted in Figure 2.1(c). At that point, the nature of the soliton and its stability change because of the appearance of a pair of *purely imaginary* eigenvalues, whose magnitude

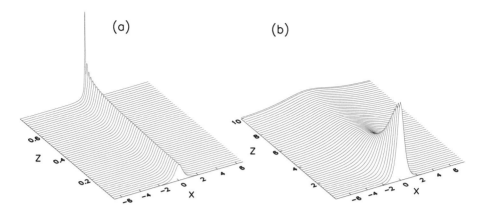

Figure 2.3: Evolution of a perturbed unstable soliton for $\varepsilon = 1$ and $\beta = 2$ showing (a) collapse and (b) decay of the soliton. The initial power was different by 1% from that of the unperturbed soliton. (After Ref. [27]; ©2001 Springer.)

is shown by the dashed curve in Figure 2.2(b). It is shown in Section 2.3 that a direct link exists between the soliton stability and the slope of $P(\beta)$.

Near $\beta = \beta_{\mathrm{cr}}$, the dynamics of an unstable soliton can be described by approximate equations derived using a multiple-scale asymptotic technique (see Section 2.3.3). However, in general, one should perform numerical simulations to study the evolution of linearly unstable solitons. As an example, Figure 2.3 shows two different types of the evolution scenario for a linearly unstable soliton. In case (a), perturbation was chosen such that it *increases* the soliton power, resulting in an unbounded growth of the soliton amplitude and the beam collapse. In case (b), perturbation decreases the soliton power by a small amount and leads to its diffraction and decay. At last, if the model under consideration supports other types of solitons (e.g., for smaller amplitude), the unstable soliton may converge to another soliton if it is stable. Thus, depending on the nature of the nonlinearity, *three distinct scenarios* can exist for the instability-induced soliton dynamics [25].

2.3 Stability Criterion

It is difficult to solve the eigenvalue problem (2.2.2) analytically. However, the analysis can be simplified considerably for fundamental solitons with a single hump (no nodes). In this section we focus on this case and develop a stability criterion for bright solitons.

2.3.1 Linear Stability Analysis

System (2.2.2), consisting of two eigenvalue equations, can be reduced to a single equation of the form

$$L_0 L_1 v = \lambda^2 v. \qquad (2.3.1)$$

The stability condition requires λ^2 to be positive. It is easy to conclude that λ^2 has to be real. Noting that $L_0 = L^+L^-$ with

$$L^{\pm} = \pm\frac{d}{dx} + \Phi^{-1}\left(\frac{d\Phi}{dx}\right), \tag{2.3.2}$$

we can consider the auxiliary eigenvalue problem,

$$L^-L_1L^+\tilde{v} = \lambda^2\tilde{v}, \tag{2.3.3}$$

which reduces to Eq. (2.3.1) after the substitution $v = L^+\tilde{v}$. Since the operator $L^-L_1L^+$ is Hermitian, all eigenvalues λ^2 of Eqs. (2.3.1) and (2.3.3) must be real. This rules out the possibility of *oscillatory instabilities* for fundamental solitons.

The properties of the operators L_0 and L_1 are well known from spectral theory of second-order differential operators [30]). We use two *general* mathematical results of this theory. First, for any linear operator L satisfying the eigenvalue equation $L\varphi_n = \lambda_n\varphi_n$, the eigenvalues always can be ordered as $\lambda_{n+1} > \lambda_n$, where the integer $n \geq 0$ determines the number of zeros in the corresponding eigenfunction φ_n. Second, for a "deeper" potential well, the eigenvalues become smaller (shift downward).

Let us first discuss the properties of the operator L_0 and denote its eigenvalues by $\lambda_{0,n}$. The lowest eigenvalue is zero for the neutral mode; i.e., $L_0\Phi(x;\beta) = 0$. Because $\Phi(x;\beta) > 0$ is the ground-state solution with no nodes, it follows that $\lambda_{0,n} > \lambda_{0,0} = 0$ for $n > 0$. Thus, the operator L_0 is positive definite on the subspace of the functions orthogonal to $\Phi(x;\beta)$. This property allows us to use several general theorems to link the soliton stability with the number of negative eigenvalues of the operator L_1, as discussed in Refs. [20]–[22]. Let us denote the eigenvalues of L_1 as $\lambda_{1,n}$. Two conclusions are: (i) Solitons become unstable if there are two (or more) negative eigenvalues, i.e., if $\lambda_{1,1} < 0$; (ii) solitons are always stable when the operator L_1 is positive definite. In the intermediate case, soliton stability depends on the slope of the $P(\beta)$ curve such that the soliton is stable if $\partial P/\partial\beta > 0$ and *unstable* otherwise. This condition is known as the Vakhitov–Kolokolov criterion [15]. Its derivation is provided in the following subsection.

2.3.2   Vakhitov–Kolokolov Criterion

To derive the Vakhitov–Kolokolov criterion, we follow Ref. [15]. First, we note that, for fundamental solitons with no nodes, the operator L_0 is positive definite for any function orthogonal to $\Phi(x;\beta)$ such that its inverse, L_0^{-1}, exists in a function space orthogonal to $\Phi(x;\beta)$. By applying this inverse operator L_0^{-1} to Eq. (2.3.1), we obtain another eigenvalue problem,

$$L_1v = \lambda^2L_0^{-1}v, \tag{2.3.4}$$

where $v(x)$ satisfies the orthogonality condition

$$\langle v|\Phi\rangle \equiv \int_{-\infty}^{\infty} v^*(x)\Phi(x;\beta)dx = 0 \tag{2.3.5}$$

and is assumed to be normalized such that $\langle v|v \rangle = 1$. Second, we multiply both sides of Eq. (2.3.4) with $v^*(x)$, integrate over x, and obtain

$$\lambda^2 = \frac{\langle v|L_1 v \rangle}{\langle v|L_0^{-1} v \rangle}. \tag{2.3.6}$$

Because the denominator in this equation is positive definite for any v satisfying Eq. (2.3.5), the sign of this ratio depends only on the numerator. Because the instability will appear only if $\lambda^2 < 0$ (so that λ is imaginary), the instability can occur when

$$\min(\langle v|L_1 v \rangle) < 0. \tag{2.3.7}$$

To find the minimum in Eq. (2.3.7) under the constraint in Eq. (2.3.5) and $\langle v|v \rangle = 1$, we use the method of Lagrange multipliers and look for a minimum of the functional

$$\mathcal{L} = \langle v|L_1 v \rangle - \kappa \langle v|v \rangle - \mu \langle v|\Phi \rangle, \tag{2.3.8}$$

where the real parameters κ and μ are unknown. With no lack of generality, we assume that $\mu \geq 0$ (otherwise the sign of the function $v(x)$ can be inverted). The extrema of the functional $\mathcal{L}$ can be found from the condition $\delta \mathcal{L}/\delta v^* = 0$, where δ denotes the variational derivative. As a result, we obtain

$$L_1 v = \kappa v + \mu \Phi, \tag{2.3.9}$$

where the values of κ and μ should be chosen in such a way that Eq. (2.3.5) and the condition $\langle v|v \rangle = 1$ are satisfied. It follows from this equation that $\langle v|L_1 v \rangle = \kappa \langle v|v \rangle$. Thus, the soliton is unstable if and only if there exists a solution with $\kappa < 0$.

The operator L_1 has a full set of orthogonal eigenfunctions such that $\langle \varphi_n|\varphi_m \rangle = 0$ if $n \neq m$ [30]. The eigenvalue spectrum of L_1 consists of discrete ($\lambda_{1,n} < \beta$) and continuous ($\lambda_{1,n} \geq \beta$) parts. Assuming that the eigenmodes are properly normalized, we can expand $v(x)$ as

$$v(x) = \sum_n D_n \varphi_n(x) + \int_\beta^\infty D_n \varphi_n(x) d\lambda_{1,n}, \tag{2.3.10}$$

where the sum extends over the discrete eigenvalues of the operator L_1. The coefficients in Eq. (2.3.10) are given by $D_n = \langle \varphi_n|v \rangle$. The function $\Phi(x; \beta)$ can be decomposed in a similar way with the coefficients $C_n = \langle \varphi_n|\Phi \rangle$. Using Eq. (2.3.9), the coefficients D_n satisfy

$$D_n = \frac{\mu C_n}{(\lambda_{1,n} - \kappa)} \tag{2.3.11}$$

if $\kappa \neq \lambda_{1,n}$.

To find the Lagrange multiplier κ, we substitute Eqs. (2.3.10) and (2.3.11) into the orthogonality condition (2.3.5) and obtain the following relation:

$$Q(\kappa) \equiv \langle v|\Phi \rangle = \sum_n C_n D_n^* + \int_\beta^\infty C_n D_n^* d\lambda_{1,n} = 0. \tag{2.3.12}$$

As mentioned earlier, the instability occurs if this equation has a solution such that $\kappa < 0$. Because the lowest-order modes of the operators L_0 and L_1, $\Phi(x; \beta)$ and φ_0, respectively, do not contain zeros, the coefficient $C_0 \neq 0$. From the structure of Eq. (2.3.12) it follows that $Q(\kappa) > 0$ if $\kappa < \lambda_{1,0}$. Thus, the solution of Eq. (2.3.12) is possible only for $\kappa > \lambda_{1,0}$. This in turn indicates that if $\lambda_{1,0} \geq 0$, the stationary state $\Phi(x; \beta)$ is stable.

From Eqs. (2.3.11) and (2.3.12), $Q(\kappa)$ is monotonic in the interval $(-\infty, \infty)$ for $\mu > 0$ and $\lambda_{1,0} < \kappa < \lambda_{1,n}$, where $n \geq 1$ corresponds to the smallest eigenvalue with $C_n \neq 0$. It follows immediately that instability appears if $\lambda_{1,1} < 0$ and $C_1 \neq 0$. On the other hand, if $C_1 = 0$, the corresponding eigenmode φ_1 satisfies Eq. (2.3.9) and the constraint (2.3.5) with $\kappa = \lambda_{1,1}$ and $\mu = 0$. Therefore, an instability is always present if $\lambda_{1,1} < 0$.

The last possible scenario corresponds to $\lambda_{1,0} < 0$ but $\lambda_{1,1} \geq 0$. Since the eigenmodes with $\kappa = \lambda_{1,n}$ do not lead to instability, we search for solutions with $\mu > 0$. Then, because $\lambda_{1,n} > \lambda_{1,1} \geq 0$, the sign of the solution κ is determined by the value of $Q(0)$. If $Q(0) > 0$, the function $Q(\kappa)$ vanishes at some $\kappa < 0$, which indicates *instability*, and vice versa. From Eqs. (2.3.9) and (2.3.12), it follows that $Q(0) = \langle L_1^{-1} \mu \Phi | \Phi \rangle$. To calculate this value, we differentiate the relation $L_0 \Phi = 0$ with respect to β and obtain

$$L_1 \frac{\partial \Phi}{\partial \beta} = -\Phi. \tag{2.3.13}$$

This equation shows that the sign of $Q(0)$ is opposite to the sign of the derivative $dP/d\beta$. Thus, a soliton solution is stable whenever $dP/d\beta > 0$. This is the Vakhitov–Kolokolov criterion for the stability of solitons.

The foregoing analysis applies for a general form of the operator L_1, e.g. when the NLS equation involves an explicit dependence on x, as in the case of waveguides. An important case concerns the stability of solitons in a homogeneous medium for which $F(I)$ does not depend on x. In this case, a fundamental soliton has a symmetric profile with a single maximum, and $d\Phi/dx$ is the first-order neutral mode of the operator L_1; i.e., $\lambda_{1,1} = 0$. Clearly, soliton stability again follows directly from the sign of the slope $dP/d\beta$.

The preceding stability analysis should be compared with the more general *Lyapunov stability theorem*, which states that a stable solution (in the Lyapunov sense) corresponds to an extremum of the system Hamiltonian for a conservative system provided it is bounded from below. For the NLS equation, a soliton solution is a stationary point of the Hamiltonian H for a fixed power P, and it can be found from the variational problem $\delta(H + \beta P) = 0$. To prove the Lyapunov stability, we need to show that the Hamiltonian has a minimum for a fixed value of P. This can be shown rigorously for a Kerr medium for which $F(I) = I$ [18]. In fact, one can show that

$$H > H_s + (\sqrt{P} - \sqrt{P_s}), \tag{2.3.14}$$

where the subscript s denotes values calculated for the NLS soliton. Condition (2.3.14) proves the soliton stability for both small and finite-amplitude perturbations. A similar relation can be founded for a non-Kerr nonlinearity and is consistent with the Vakhitov–Kolokolov criterion [18].

2.3.3 Marginal Stability Point: Asymptotic Analysis

The preceding linear stability analysis shows that solitons in a homogeneous medium are unstable when the slope of the $P(\beta)$ curve is negative, i.e., when $dP/d\beta < 0$. Near the marginal stability point $\beta = \beta_{cr}$, defined by the condition $(dP/d\beta)_{\beta=\beta_{cr}} = 0$, the instability growth rate is small. Near this point, we can use an *analytical* asymptotic method that describes not only linear instabilities but also the long-term nonlinear evolution of unstable solitons. Such an approach is based on a nontrivial modification of the soliton perturbation theory [23] that is usually applied to analyze soliton dynamics under the action of external perturbations. Here we deal with a qualitatively different physical situation in which an unstable bright soliton evolves under the action of its "own" perturbations.

The main idea behind the asymptotic method is that near a marginal stability point, the propagation constant β associated with the soliton varies slowly along the propagation direction. Because the instability growth rate is small near $\beta = \beta_{cr}$, we can assume that the shape of the perturbed soliton evolves almost adiabatically with z (i.e., it remains self-similar). Thus, the soliton solution of Eq. (2.1.1) can be written in the form

$$u = \phi(x;\beta;Z)\exp\left[i\beta_0 z + i\varepsilon\int_0^Z \beta(Z')dZ'\right], \qquad (2.3.15)$$

where $\beta = \beta_0 + \varepsilon^2\Omega(Z)$, $Z = \varepsilon z$, and $\varepsilon \ll 1$. The constant value β_0 is chosen in the vicinity of the marginal stability point β_{cr}. Using an asymptotic multiscale expansion in the form

$$\phi(x;\beta;Z) = \Phi(x;\beta) + \varepsilon^3\phi_3(x;\beta;Z) + O(\varepsilon^4), \qquad (2.3.16)$$

we obtain the following equation for the small change Ω in the propagation constant β (details can be found in Ref. [31]):

$$\mathcal{M}(\beta_{cr})\frac{d^2\Omega}{dZ^2} + \frac{1}{\varepsilon^2}\left(\frac{dP}{d\beta}\right)_{\beta=\beta_0}\Omega + \frac{1}{2}\left(\frac{d^2P}{d\beta^2}\right)_{\beta=\beta_{cr}}\Omega^2 = 0, \qquad (2.3.17)$$

where $P(\beta)$ is calculated using the stationary soliton solution (2.1.5) and $\mathcal{M}(\beta)$ is defined as

$$\mathcal{M}(\beta) = \int_{-\infty}^{+\infty}\left[\frac{1}{\Phi(x;\beta)}\int_0^x \Phi(x';\beta)\frac{\partial\Phi(x';\beta)}{\partial\beta}dx'\right]^2 dx > 0. \qquad (2.3.18)$$

If we interpret this equation as describing the dynamics of a particle of effective mass $\mathcal{M}$, we can conclude that the dynamics of solitons associated with the generalized NLS equation (2.1.1) can be described near the marginal stability point $\beta = \beta_{cr}$ using a simple collective-coordinate approach. More specifically, the dynamics corresponds to that of an effective (*inertial and conservative*) particle of mass $\mathcal{M}(\beta_{cr})$ whose position Ω shifts under the action of a potential force proportional to the difference $P_0 - P(\beta)$, where $P_0 = P(\beta_0)$. The first two terms in Eq. (2.3.17) give the result of the linear stability theory, according to which a soliton is *linearly unstable* if $dP/d\beta < 0$. The last nonlinear term in Eq. (2.3.17) allows one to describe the long-term nonlinear dynamics of an unstable soliton. They can be used to identify qualitatively different scenarios of the instability-induced soliton dynamics near the marginal stability point [25, 31].

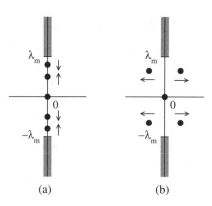

(a)   (b)

Figure 2.4: Schematic presentation of the soliton oscillatory instability born by merging two internal modes. The solid circles show the eigenvalues in the complex plane, whereas the vertical line corresponds to real λ. The hatched region denotes the continuous spectrum band.

2.3.4 Oscillatory Instabilities

When the conditions of the applicability of the Vakhitov–Kolokolov criterion [15] are not satisfied for solitons with nodes, the study of soliton stability requires a numerical approach. No simple stability criterion that involves the system invariants can be suggested in this case. When one or more discrete eigenvalues become *complex*, the corresponding instability is called the *oscillatory instability*, because the growth of perturbation follows an oscillatory pattern. The oscillatory instabilities can often be associated with a resonance between two (or more) soliton internal modes or with a resonance between the soliton internal mode and the mode located at the edge of the continuous spectrum band. The former case is shown schematically in Figure 2.4.

The first example of the oscillatory instability for solitons was found in a 1991 study of a parametrically driven, damped NLS equation of the form [32]

$$i\frac{\partial u}{\partial z} + \frac{1}{2}\frac{\partial^2 u}{\partial x^2} + |u|^2 u = h e^{2iz} u^* - i\gamma u, \qquad (2.3.19)$$

where h is the driving force and γ accounts for losses. This equation has an exact soliton solution given by

$$u(x,z) = A_+ \operatorname{sech}(A_+ x) \exp[i(z - \theta_+)/2], \qquad (2.3.20)$$

where

$$A_+^2 = 1 + \sqrt{h^2 - \gamma^2}, \qquad \theta_+ = \sin^{-1}(\gamma/h). \qquad (2.3.21)$$

The oscillatory instability appears for $\varepsilon > \varepsilon_{\mathrm{cr}} \approx 0.1196$, where $\varepsilon = 2\sqrt{h^2 - \gamma^2}/A_+^2$. Since this instability survives in the limit $\gamma \to 0$ [32], it provides an example of the oscillatory instability in nonlinear homogeneous Hamiltonian systems. Oscillatory instabilities have also been found for nonlinear guided waves in layered media [33, 34]. They are in fact quite common for solitons associated with the coherently coupled NLS equations (see Chapter 9) and parametric solitons (Chapter 10).

A systematic analysis of oscillatory instabilities was carried out in 1998 in the case of *gap solitons* [35] using coupled-mode theory (see Chapter 5). The fact that the system is integrable in a specific limit was used to study *bifurcations* of new eigenvalues from the edge of the continuous spectrum, similar to the case of a scalar weakly perturbed cubic NLS equation [26]. The bifurcating eigenvalues correspond to internal modes of a gap soliton, and the collision of two such eigenvalues results in an oscillatory instability characterized by a pair of complex eigenvalues with a positive real part, similar to the scenario shown in Figure 2.4.

The existence of oscillatory instabilities is usually associated with *multiparameter solitary waves* described by the coupled NLS-type equations. In the case of a gap soliton, the two independent parameters are its frequency and velocity. Another example of the oscillatory instability has been found for vector solitons propagating inside a birefringent optical fiber in the presence of walk-off, self-, and cross-phase modulations and the four-wave mixing effects [36]. Oscillatory instabilities can also occur for dark solitons governed by a discrete NLS equation [37]. In the latter case, even a weak inherent discreteness can lead to oscillatory instabilities. In discrete systems, oscillatory instabilities can appear as a result of either a resonance between a radiation mode and a single internal mode or a resonance between two soliton internal modes.

2.4 Embedded Solitons

Many properties of solitons can be understood from the analysis of the linear-wave spectrum of the underlying nonlinear equation. Since solitons are governed by an NLS equation containing one or more second-order derivatives, the corresponding differential operator has eigenvalues that must necessarily lie outside the continuous spectrum. The situation becomes different for solitons described by a single NLS equation containing higher-order derivatives or by two or more coupled NLS equations. In this case, it can happen that the soliton frequency lies inside the continuous spectrum band of linear waves. The existence of localized solutions then should be studied as the problem of interaction of a soliton with the radiation composed of linear waves [38].

Spatially localized solutions *coexisting with linear waves* are known as *embedded solitons* [39]. Examples of such solitons have been found mostly in the case of nonlinear problems described by a coupled set of NLS-like equations or the Korteweg-de Vries equation with higher-order derivatives [40]–[49]. The majority of the effects associated with the embedded solitons are connected with the fundamental properties of the corresponding linear equations [50]. The linear equations not only determine the asymptotic behavior of the field far from the soliton but also provide considerable information about the possible types of soliton solutions. In this section we discuss embedded solitons using a simple extension of the NLS equation.

To keep the following discussion as general as possible, we consider a generalized NLS equation in the form

$$i\frac{\partial u}{\partial z} + \frac{1}{2}\frac{\partial^2 u}{\partial x^2} + \varepsilon\frac{\partial^4 u}{\partial x^4} + F(|u|^2)u = 0, \qquad (2.4.1)$$

where the nonlinear term can have a general form such as

$$F(|u|^2)u = |u|^2 u + \gamma_1 \left|\frac{\partial u}{\partial x}\right|^2 u + \gamma_2 |u|^2 \frac{\partial^2 u}{\partial x^2} - \gamma_3 |u|^4 u. \tag{2.4.2}$$

The fourth-order derivative in Eq. (2.4.1) can appear as the continuous expansion of a discrete model used for the waveguide arrays (see Chapter 11), or it can result from the effect of higher-order dispersion in the case of temporal solitons.

Consider first the linear solutions of Eq. (2.4.1) in the form of a plane wave $u(z,x) = u_0 \exp(i\beta z + ikx)$ after setting $F = 0$. The propagation constant of these plane waves must satisfy the dispersion relation

$$\beta = -\tfrac{1}{2}k^2 + \varepsilon k^4. \tag{2.4.3}$$

It reaches its lowest value, $\beta_m = -(16\varepsilon)^{-1}$, at the point where $k^2 = k_m^2 = 1/(4\varepsilon)$. Thus, the continuous spectrum occupies a semi-infinite band such that $\beta_m < \beta < \infty$.

The linear equation obtained after setting $F = 0$ in Eq. (2.4.1) has the following Green function $G_\beta(x)$ describing the waves outgoing at infinity:

$$G_\beta(x) = \frac{1}{4\sqrt{\beta - \beta_m}} \left(\frac{i}{k} e^{-ik|x|} + \frac{1}{\kappa} e^{-\kappa|x|}\right). \tag{2.4.4}$$

The constants k and κ are obtained from the dispersion relation (2.4.3) and are given by

$$k^2 = \frac{1}{4\varepsilon}(\sqrt{1 + 16\varepsilon\beta} + 1); \qquad \kappa^2 = \frac{1}{4\varepsilon}(\sqrt{1 + 16\varepsilon\beta} - 1). \tag{2.4.5}$$

In the interval $\beta_m < \beta < 0$, the Green function does not have the exponentially decaying term because κ becomes purely imaginary. For $\beta < \beta_m = -1/(16\varepsilon)$, it changes its form to

$$G_\beta(x) = \frac{1}{4\kappa k} e^{-\kappa|x|} \sin(kx + \varphi), \tag{2.4.6}$$

where $k^2 - \kappa^2 = 1$ and $2\kappa k = \sqrt{\beta_m + \beta}$. This form of the Green function shows that although a localized soliton-like solution of Eq. (2.4.1) can exist for $\beta < \beta_m$, it will decay asymptotically as it losses its energy to the linear waves (continuum radiation). The following conclusions can be drawn from the Green-function analysis:

- Solitons with the propagation constant $\beta < \beta_m$ have exponentially decaying oscillatory "tails" as they loose energy to linear waves.
- The range $\beta_m < \beta < 0$ cannot, in principle, correspond to soliton solutions because there are no asymptotically decaying terms in the Green function.
- The case $\beta > 0$ corresponds to quasi-localized oscillations such that one of the Green-function components is localized in space while the other one describes a standing wave of constant amplitude.

The preceding discussion shows that the soliton solutions of Eq. (2.4.1) with $\beta > 0$ are, as a rule, accompanied by the radiation of linear waves. However, some of such solutions may become *nonradiating*, and they correspond to the embedded solitons. To find the conditions under which radiation will be absent, we consider solutions of the linearized equation in the presence of a distributed force $f(x)$ applied during a small interval near $x = 0$. The general solution of such a linear equation, outside the region of the applied force, can be found using the Green function and is given by

$$u(x) = (1 + 16\varepsilon\beta)^{-1/2}[iQ(k)e^{-ik|x|} + P(k)e^{-\kappa|x|}], \qquad (2.4.7)$$

where

$$kQ(k) = \int f(x)e^{ikx}dx, \qquad \kappa P(\kappa) = \int f(x)e^{\kappa x}dx. \qquad (2.4.8)$$

The oscillating part of the solution vanishes asymptotically when the condition $Q(k) = 0$ is satisfied. This condition yields the specific values of the parameters for which a nonradiating soliton is possible.

As a simple example, consider the generalized NLS equation (2.4.1) with $\gamma_2 = \gamma_3 = 0$ in Eq. (2.4.2). The "sech" soliton solution of the standard NLS equation,

$$u(x,t) = B \operatorname{sech}(\kappa x)e^{i\beta z}, \qquad (2.4.9)$$

becomes a nonradiating soliton solution for Eq. (2.4.1) when $Q(k) = 0$. Using $f(x) = |u(x)|$, this conditions becomes $\int |u(x)| \cos(kx)dx = 0$. It can be satisfied for the following choice of parameters:

$$k^2 = \frac{3\gamma_1 - 12}{2\gamma_1}, \quad \kappa^2 = \frac{\gamma_1 - 42}{2\gamma_1}, \quad \beta = \left(\frac{\gamma_1 - 6}{\gamma_1}\right)^2 - \frac{1}{4}, \qquad (2.4.10)$$

with $B^2 = (24/\gamma_1)\kappa^2 = (12/\gamma_1^2)(\gamma_1 - 12)$.

As another example, the asymmetric two-hump localized structure of the form

$$u(x,z) = A \sinh(\kappa x)\operatorname{sech}^2(\kappa x)\,e^{i\beta_0 z} \qquad (2.4.11)$$

becomes an exact nonradiative localized solution when $k^2 = 11\kappa^2$, $\kappa^2 = 0.1$, $\beta_0 = 0.11$, and $A = \sqrt{6/5}$ [51]. Any deviation from these exact parameter values leads to the emission of radiation. In fact, the solution (2.4.11) is one of many localized solutions of a similar form, consisting of two out-of-phase NLS-like solitons arranged in the way that the linear radiation is trapped in between the soliton pulses but is completely suppressed in the outside region. Figure 2.5 shows the power curve $P(\beta)$ for the discrete set of the stationary localized solutions of Eqs. (2.4.1), (2.4.2) when $\gamma_j = 0$. All these solutions are in the form of two-soliton nonradiating states and can be classified as *two-hump embedded solitons* (see also Ref. [49]). For small values of β these solutions look like two NLS solitons with a standing wave of radiation trapped between them.

It should be stressed that the conditions for the existence of embedded solitons and their analytic form found in a few specific cases do not constitute a definite proof that an analytic form of such solitons can be found in all cases. However, they do show that the soliton structure is determined as much by the properties of the dispersion relation of the linearized equation as by the form of the nonlinear terms in the underlying generalized NLS equation. Thus, many of the localized soliton solutions found

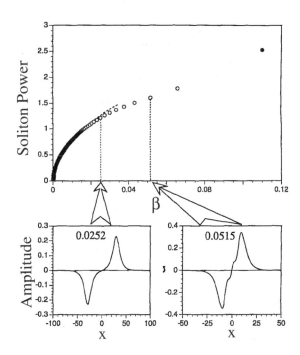

Figure 2.5: $P(\beta)$ for two-hump embedded solitons (circles). The solid circle indicates the exact solution while the dashed line shows the power for two NLS solitons. Two examples are shown for values of β marked by arrows. (After Ref. [40]; ©1995 APS.)

in Refs. [52]–[56] for the NLS-like equations with higher-order dispersive terms are, in fact, embedded solitons. Consequently, they are likely to emit radiation when the specific relations between the soliton parameters are not exactly satisfied.

The most intriguing property of embedded solitons is their dynamics under the action of perturbations, so far studied for the generalized second-harmonic-generation system and the Korteweg de Vries model with the firth-order dispersion [39, 47, 48]. Both numerical and analytical studies indicate that embedded solitons are *semistable*. This means that when the perturbation increases the power (or momentum) of the embedded soliton, the perturbed state approaches asymptotically the embedded soliton, while when the perturbation reduces the power (or momentum) of the embedded soliton, the perturbed state decays into radiation. Moreover, when an embedded soliton is perturbed, it sheds a one-directional continuous-wave radiation, and, generally speaking, the radiation amplitude is finite for any value of perturbation.

2.5   Soliton Collisions

As seen from Eq. (2.1.2), spatial solitons can propagate at an angle to the propagation direction z. Such solitons are of considerable practical interest because they allow steering of light by changing the angle that the soliton makes with the z axis. One can introduce the concept of the soliton transverse velocity V in the framework of the generalized NLS equation (2.1.1) by employing the Galilean transformation, because its use allows us to transform any stationary soliton into a moving soliton according to the rule

$$u(x,z;V) \rightarrow u(x - Vz,z)\exp(iVx - iV^2 z/2), \tag{2.5.1}$$

where the velocity V is also known as the *soliton steering velocity*. Thus, two transversely moving spatial solitons can collide with each other. In this section we discuss the properties of such collisions in nonlinear systems governed by Eq. (2.1.1).

2.5.1 Collisions of Kerr Solitons

Among the solitons governed by the $(1 + 1)$-dimensional generalized NLS equation, only the special case of Kerr nonlinearity is integrable by the inverse scattering transform method. Among the many properties of integrable equations is the existence of exact analytical solutions describing elastic interactions of any number of solitons, the so-called N-soliton solutions. In this case, the normally complicated nonlinear interaction of spatially localized waves reduces asymptotically to a simple linear superposition such that solitons remain unaffected by the collision except for acquiring a phase shift. The shift of the soliton positions after collision can be employed in different concepts of optical switching, as shown in Ref. [57].

The Kerr-soliton solution of the cubic NLS equation is given in Eq. (2.1.2). The parameter V represents the soliton velocity in the transverse direction x. From a physical standpoint, the soliton propagates in a direction that makes an angle, called the *steering angle*, from the z axis whose value depends on the numerical value of V. The integrability of the cubic NLS equation also provides us with explicit analytical solutions that describe the *elastic interaction* of any number of solitons of the form of Eq. (2.1.2). The amplitude a_j and the velocity V_j of the jth soliton maintain their values after the interaction is over.

The elastic nature of the interaction among identical solitons is related the infinite number of the conserved integrals of motion for the cubic NLS equation. During the interaction of N solitons with the parameters a_j and V_j, the three conserved quantities P, M, and H defined in Eqs. (2.1.3) and (2.1.4) remain constants and are given by

$$P = 2\sum_{j=1}^{N} a_j \qquad M = 4\sum_{j=1}^{N} a_j V_j, \qquad H = \sum_{j=1}^{N} \left(a_j V_j^2 - \frac{1}{3}a_j^3\right). \tag{2.5.2}$$

During the collision, the two colliding solitons merge with each other before separating out. The inverse scattering method can be used to find the total field before, during, and after the collision. As an example, we present an analytical form of the two-soliton solution that describes two identical solitons propagating along the z axis such that $a_1 = a_2 = a$ and $V_1 = V_2 = 0$, i.e., with the velocities vanishing at the infinity. The soliton interaction is governed by the analytic solution [58],

$$u(x,z) = \frac{8ia[2ax\sinh(2ax) - \cosh(2ax) - 4ia^2 z\cosh(2ax)]}{\cosh(4ax) + 1 + 8a^2x^2 + 32a^4z^2}\exp(2ia^2 z), \tag{2.5.3}$$

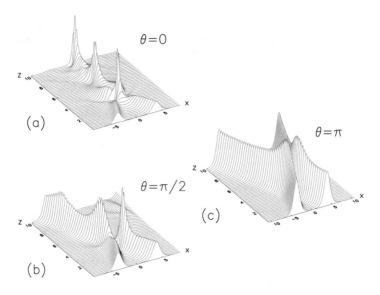

Figure 2.6: Collisions of two solitons as predicted by the the cubic-quintic NLS equation (2.5.4) at $\varepsilon = -0.2$ for three values of the soliton relative phase θ.

which has a single maximum $|u| = 2a$ at $x = z = 0$, indicating that the two solitons exactly overlap at that point.

2.5.2 Collisions of non-Kerr Solitons

In the case of non-Kerr nonlinear media, soliton interaction is not elastic, because Eq. (2.1.1) is not integrable by the inverse scattering transform method and the N-soliton solutions do not exist. The nonintegrability property produces a number of interesting effects that are also observed experimentally. Many of the effects produced by the absence of integrability can be seen through a simple example of the cubic-quintic NLS equation that includes, in addition to the cubic nonlinearity $|u|^2u$, a quintic nonlinear term of the form $|u|^4u$ and can be written as

$$i\frac{\partial u}{\partial z} + \frac{1}{2}\frac{\partial^2 u}{\partial x^2} + |u|^2u = \varepsilon|u|^4u, \qquad (2.5.4)$$

where ε governs the strength of the perturbation term.

When ε is relatively small, the inelastic effects associated with the nonintegrability can be summarized as follows. To the first order in ε, the result of a collision between two solitons of equal amplitudes depends crucially on their *relative phase* θ. Figure 2.6 shows the collision of two solitons of the same amplitude for three values of θ using $\varepsilon = 0.2$. Two solitons attract each other when they are in phase ($\theta = 0$), and fuse together when the collision angle is below a certain critical value. In contrast with the case of Kerr solitons, they do not follow a periodic pattern because the solitons do not emerge unchanged after the collision. Two out-of-phase solitons ($\theta = \pi$) repel each other, as shown in part (c). In the intermediate case, shown in part (b), soliton interaction is accompanied by a strong energy exchange; in the most dramatic case, one of the solitons can even disappear. This behavior is quite different than that of Kerr solitons: The presence of the quintic term in Eq. (2.5.4) makes the soliton interaction inelastic.

To the second order in ε, additional inelastic effects are produced by the continuum radiation that is emitted during soliton collision to the first order in ε. Such radiation produces changes in the soliton amplitudes that are independent of the relative phase of the two solitons. However, in the case of inelastic collisions among three (or more) solitons, an energy exchange between the colliding solitary waves can occur without involving radiation. Similar inelastic effects are produced with the excitation of soliton internal modes.

In the case of two solitons of equal amplitudes but opposite velocities colliding inside a non-Kerr medium, the collision becomes strongly inelastic in a narrow range of the soliton relative phase, even when the perturbation produced by the ε term is almost negligible. A comparatively large exchange of the power, momentum, and energy between the two solitons can occur with practically no radiation escaped, so these three quantities remain conserved with a high accuracy. This effect occurs for solitons in many types of nonlinear media, and its main features can be deduced by solving the cubic-quintic NLS equation (2.5.4) numerically, though the actual form of perturbation is not really that important [59]. The results are shown in Figure 2.7 for three different values of the soliton relative phase. In each case, the power P, the momentum M, and the energy H are calculated after collision and their values compared with the corresponding values before collision.

The trajectories of two colliding solitons are shown in Figure 2.7 in the x–z plane by plotting the constant-intensity contours for $|\text{Re}(u)| > 0.3$. The solitons are assumed to be identical but to have opposite velocities ($V_1 = -V_2 = 0.05$). In the out-of-phase case (a), when the two solitons repel each other and do not overlap during collision, the collision is practically elastic. In cases (b) and (c), for which solitons are nearly in phase and attract each other, the collision is inelastic. When the relative soliton phase is zero, the collision is symmetric, as seen in Figure 2.7(b). A small change in the soliton momentum M is observed because of the radiation emitted, but the power P and energy H do not change much. In contrast, a significant exchange of all three conserved quantities takes place in Figure 2.7(c), where the relative phase is small but not zero.

The power of the soliton changes by as much as 20% for $|\varepsilon| = 0.02$. Its momentum M also increases considerably, indicating that its velocity becomes larger after the collision. Such a strong sensitivity to ε is observed in a narrow region of the relative soliton phase close to zero. Outside that region, the influence of perturbations is by two (or even three) orders of magnitude weaker [59], and the main effects are due to the radiation emitted by colliding solitons.

2.5.3 Chaotic and Fractal Soliton Scattering

Since the inelastic effects between two colliding solitons increase with a decrease in their relative velocity, it is interesting to study the limiting case $V_1 = V_2 = 0$. Such

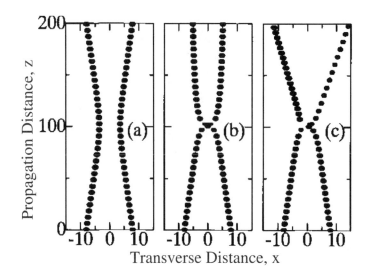

Figure 2.7: Trajectories of two spatial solitons for $\varepsilon = -0.02$ and three values of the relative soliton phase: (a) $\theta = \pi$, (b) $\theta = 0$, and (c) $\theta = -0.08$. Two solitons have equal amplitudes but opposite velocities ($V_1 = -V_2 = 0.05$). The intensity contours were obtained numerically by solving a cubic-quintic NLS equation. (Courtesy S. V. Dmitriev.)

two solitons propagate along the z axis, but their separation can still change if their tails overlap because of the nonlinear coupling induced by cross-phase modulation. It turns out that the most interesting effects in this case occur when solitons have unequal amplitudes [60].

Figure 2.8(a) shows the evolution of two colliding solitons with an amplitude ratio $a_2/a_1 = \sqrt{9/8}$ using the integrable cubic NLS equation ($\varepsilon = 0$) and plotting the region where $\text{Re}(u) > 0.35$. The two solitons attract and repel each other in a periodic fashion. However, because of their different amplitudes, the soliton periods T_1 and T_2 are different for the two solitons. Noting that the soliton period scales inversely with the amplitude, $T_1/T_2 = 9/8$. This mismatch results in an oscillatory motion with the period $T = 8T_1 = 9T_2$. Thus, two overlapping solitons with commensurable periods $nT_1 = mT_2$, where n and m are positive integers, can be regarded as a two-soliton composite state with the period $T = nT_1 = mT_2$. If the ratio n/m or m/n is an irreducible fraction with large values of n and m, the period T is also large. For an irrational ratio, the period T becomes infinitely large, and the soliton dynamics appear *chaotic*.

Figure 2.8(b) shows how the soliton trajectories change when a weak discreteness of the NLS equation is used as a perturbation [60]. In this case, the attraction between the two solitons occurring when they are in phase is not fully compensated by the repulsion when they become out of phase. As a result, the mean distance between the solitons becomes smaller and smaller, and eventually the two solitons collide. Another important effect of perturbation is the energy and momentum exchange occurring during the collision. After the collision, solitons acquire opposite velocities, and the two-soliton state splits into two independent solitons.

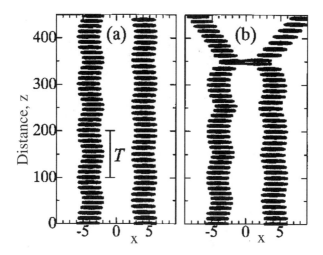

Figure 2.8: Trajectories of two colliding solitons with different amplitudes ($a_2/a_1 = \sqrt{9/8}$) and zero velocity for (a) the cubic NLS equation and (b) the weakly perturbed NLS equation (Courtesy S. V. Dmitriev.)

To study in more details the breakup of the two-soliton state into two separate solitons, the initial phase difference between the two solitons was varied, and the velocities of the solitons after splitting were calculated numerically. The total momentum of the two-soliton system should be conserved (that is why the solitons move in opposite directions after splitting. The velocities V_1 and V_2 are plotted as a function of the relative phase θ in the top panel of Figure 2.9. The structure seen there exhibits *self-similarity* at different scales, a property usually associated with *fractal scattering*. This feature is shown in the remaining four panels. Each subsequent panel shows an expanded view of the interval marked with a thick horizontal bar. The successive expansion coefficients in Figure 2.9 are 17.0, 9.45, 10.1, and 9.96. At each scale, the function $V_j(\theta)$ ($j = 1, 2$) contains alternating smooth and chaotic domains. However, at a larger magnification, each chaotic domain contains again chaotic regions and smooth peaks. In some regions, the width of the peaks vanishes and the density of the peaks goes to infinity. At the same time, the height of the peaks remains the same at each scale. This means that the sensitivity of $V_j(\theta)$ to the phase difference θ becomes infinitely large. The fractal structure of soliton velocity shows the chaotic nature of soliton interaction in a weakly perturbed NLS equation.

The fractal nature of soliton scattering has a simple physical explanation. As can be seen from Figure 2.9, the chaotic regions appear when an extrapolation of the smooth peaks gives nearly zero velocity. In these regions, solitons gain such a small velocity after collision that they cannot overcome their mutual attraction and thus they collide again. During the second collision, solitons may acquire an amount of kinetic energy sufficient to escape each other (because of momentum exchange), but there exists a finite probability that the acquired kinetic energy is not large enough for them to escape. In the latter case, the solitons will collide for a third time, and so on. Such multiple

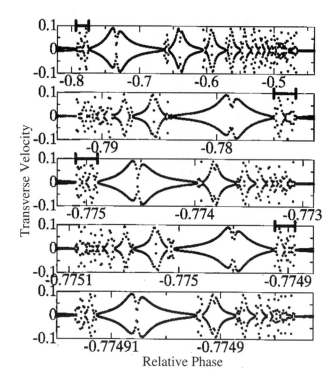

Figure 2.9: Fractal nature of soliton interaction. Each panel shows soliton velocities after split-ting as a function of the soliton relative phase, at five different scales. Thick horizontal bars show the region expanded. (After Ref. [60]; ©2002 APS.)

collisions lead to a resonant energy exchange between solitons and produce the fractal structure seen in Figure 2.9.

2.5.4 Multisoliton Interactions

As already discussed, the elastic collision of two solitons in a Kerr medium results only in a shift of their phases and positions. In the case of several interacting solitons, the shift is equal to the sum of partial shifts resulting from separate collisions with each soliton [1]. This property implies that the so-called many-particle effects are absent in the case of an ideal integrable cubic NLS equation. However, even a relatively small perturbation changes this property. One can again use the cubic-quintic NLS equation (2.5.4) for studying multisoliton interactions while treating the parameter ε as a small perturbation. Even the three-soliton case exhibits the nontrivial effects that occur (to the first order in ε) for any value of relative phases among the solitons. In general, all three solitons change their velocities after the collision [61].

Analytical results can be obtained by considering the collision of a fast-moving soliton, with amplitude a_f and velocity V_f, and a symmetric pair composed of two slow-moving solitons, with equal amplitudes a_s and opposite velocities V_s. Even in this special case, it is necessary to assume that $V_f \gg V_s \gg a_s$. Writing the three-soliton solution in the approximate form

$$u \approx u_f(V_f) + u_{sl}(V_s) + u_{sl}(-V_s),$$ (2.5.5)

we can calculate changes in the soliton parameters during the collision using the soliton perturbation theory based on the inverse scattering transform method [61]. When the initial amplitudes of the slow solitons are equal, the soliton velocities after the collision can be written in the form $V_f' = V_f + \Delta V_f$ and $V_s' = V_s - \Delta V_s$, with the changes given by

$$\Delta V_f = -24\varepsilon(V_s a_s^4 / V_f^2) G(\delta), \qquad \Delta V_s = 48\varepsilon(a_f a_s^3 / V_f) G(\delta).$$ (2.5.6)

The parameter $\delta = a_s(x_{s2}^{(0)} - x_{s1}^{(0)})$ characterizes the initial separation between two slow solitons. The function $G(\delta)$ is defined as

$$G(\delta) = \frac{1}{\sinh^2 \delta} \left[\frac{3(\delta - \tanh \delta)}{\tanh^2 \delta} - \delta \right],$$ (2.5.7)

and it vanishes in the limits $\delta \to 0$ and $\delta \to \infty$. To the first order in ε, there is no change in the soliton amplitudes. It is easy to verify that Eq. (2.5.6) is consistent with the conservation laws.

The analytical result in Eq. (2.5.6) is valid in a relatively narrow region of the soliton parameters. Nevertheless, the same features are observed in numerical simulations shown in Figure 2.10 for $\varepsilon = 0.01$ [62]. Clearly, changes in the velocities of the slow-moving soliton pair are due to an energy exchange among the three solitons during the collision. As one may expect, the energy exchange depends strongly on the separation between the colliding solitons at the moment of collision, and it vanishes for large separations. Interestingly enough, this energy exchange also vanishes when the centers of the slow solitons almost coincide.

2.6 Breathers and Soliton Bound States

The exactly integrable cubic NLS equation does not support bound states formed by multiple solitons. However, two (or more) NLS solitons located at the same position can form a composite state known as the *breather*. A breather constitutes a localized entity whose shape oscillates along z in a periodic fashion as it transforms periodically from a two-hump shape into one with a single hump. In general, such a solution is periodic, with the period $T = \pi/(a_1^2 - a_2^2)$. In the degenerate case $a_1 = a_2 = a$, it reduces to Eq. (2.5.3) describing the interaction of two solitons of equal amplitudes.

Integrability of the NLS equation allows one to obtain different types of the breather solutions as a special case of more general N-soliton solutions when the soliton velocities vanish and their positions coincide. A specific form of the two-soliton breather corresponds to the choice of amplitudes $a_1 = 3/2$ and $a_2 = 1/2$. This breather is interesting because it takes the simple form $u(0,x) = 2\,\text{sech}(x)$ at $z = 0$. Such a solution of the cubic NLS equation is given by [63]

$$u(x,z) = \frac{4(\cosh 3x + 3e^{4iz} \cosh x)e^{iz/2}}{(\cosh 4x + 4\cosh 2x + 3\cos 4z)}.$$ (2.6.1)

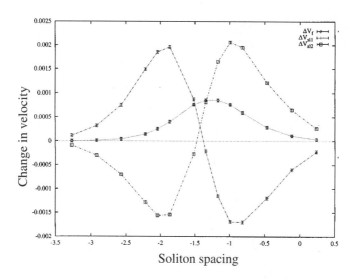

Figure 2.10: Changes in the soliton velocities as a function of the spacing x_D between the slow-moving soliton pair that collides with a fast-moving soliton. Three types of symbols indicate the numerical results for three solitons for $\varepsilon = 0.01$. (After Ref. [62]; ©1996 APS.)

This breather is often referred to as a *second-order soliton* because it corresponds to the case $N = 2$ of the general Nth-order soliton generated by the input field $u(0,z) = N \operatorname{sech}(x)$.

In the presence of perturbations, periodic solutions similar to Eq. (2.6.1) do not exist. Dissipative perturbations always lead to a decay of the breather; the decay may be accompanied by the splitting of a breather into its constituent solitons [64]–[67]. However, the most interesting dynamics are observed for conservative perturbations, such as higher-order nonlinearity or dispersion terms. If the input field is of the form $u(0,z) = 2\operatorname{sech}(x)$, a breather is still formed initially, but it does not evolve in a periodic fashion. The dynamics of the initially excited breather depends on the sign of perturbation term [68, 69]. In the case of a cubic-quintic NLS equation (2.5.4), the breather survives for $\varepsilon < 0$ (self-focusing quintic nonlinearity), but it radiates some energy away. The radiation process is accompanied by a transformation of the internal structure of the breather that can be viewed as the growth of the larger-amplitude soliton ($a_1 = 3/2$) at the expense of the smaller-amplitude one ($a_2 = 1/2$). The decay of the smaller-amplitude soliton is governed by [70]

$$a_2(z) = [a_2^{-3}(0) + \Gamma z]^{-1/3}, \qquad (2.6.2)$$

where $\Gamma = (3/4)\varepsilon^2 b_0^2 [2a_1(0) + a_2(0)]^3$ and $b_0 \approx 1.987$, as found numerically [71]. In the opposite case, $\varepsilon > 0$ (self-defocusing quintic nonlinearity), the breather splits asymmetrically into two solitons of different amplitudes [68, 69]. The splitting can be enhanced by the interaction of the breather with radiation such as a weak beam [72].

The effects of breather splitting can be understood with the help of simple physics. In the self-focusing case, the individual NLS solitons contained in the breather attract

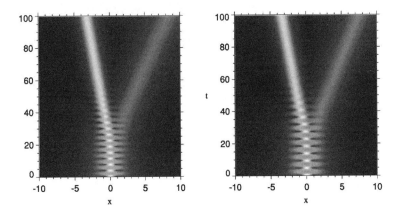

Figure 2.11: A two-soliton bound state (the NLS breather) affected by a repulsive quintic perturbation with $\varepsilon \approx 0.08$. Left: results of perturbation theory. Right: numerical results. Both solitons escape with a nonzero velocity. (After Ref. [69]; ©2000 APS.)

each other and propagate together. Even a small self-defocusing quintic nonlinearity leads to a repulsive force between the solitons because of which the two solitons separate from each other [68]. Moreover, the separating solitons are ejected from the original breather with a finite velocity, as evident from the results of numerical simulations shown in Figure 2.11 (right). This effect is captured accurately by perturbation theory [69], as seen in Figure 2.11 (left). The work done by the perturbation force in moving the soliton determines the asymptotic velocity of an initially stationary particle upon ejection.

As discussed in Section 2.4, many properties of the solitons associated with a non-integrable NLS equation depend on the asymptotic properties of the corresponding linear equations. In particular, the contribution of higher-order derivatives can lead to the solitons with decaying oscillating tails whose overlap provides a mechanism for a novel type of interaction between them. Indeed, the oscillating tails can lead to an effective trapping of two neighboring solitons such that they are located in potential minima created by the tail of the neighboring soliton being separated by a barrier. This scenario of soliton interaction leads to the formation of soliton bound states [73]. As an example, consider again the generalized NLS equation with an additional fourth-order derivative term [74]

$$i\frac{\partial u}{\partial z} + \frac{\partial^2 u}{\partial x^2} - \frac{\partial^4 u}{\partial x^4} + |u|^2 u = 0, \tag{2.6.3}$$

where the fourth-derivative term is taken to be negative. Looking for stationary localized solutions of Eq. (2.6.3) in the form $u(z,x) = U(x)\exp(i\beta z)$, we obtain the following ordinary differential equation for the soliton amplitude U:

$$-\frac{d^4 U}{dx^4} + \frac{d^2 U}{dx^2} - \beta U + U^3 = 0. \tag{2.6.4}$$

Equation (2.6.4) corresponds to a Hamiltonian system in the two-dimensional phase space formed by U and dU/dx. Its localized solutions can only be found numerically. However, Eq. (2.6.4) has an exact solution when $\beta = 0.16$, in the form [53, 54]

$$U(x) = \sqrt{a}\,\text{sech}^2(bx), \qquad (2.6.5)$$

where $a = 0.3$ and $b = \sqrt{20}$. Although other soliton solutions of Eq. (2.6.4) can only be found numerically, we can determine the asymptotic behavior by neglecting the nonlinear term. Substituting $u \sim u_0 \exp(\lambda x)$ into the linearized equation, we obtain a fourth-order polynomial in λ whose roots are given by

$$\lambda = \pm \frac{1}{\sqrt{2}}[1 \pm \sqrt{1-4\beta}]^{1/2}. \qquad (2.6.6)$$

This equation indicates that all localized solutions should have oscillating decaying tails for $\beta = \beta_{cr} > 0.25$, because the parameter λ then has a nonzero imaginary part. In this case, the solitons can bind together, forming multisoliton bound states. A simple analysis indicates that such bound states can be symmetric (in phase) or antisymmetric (out of phase), and they are unstable [38].

2.7   Experimental Results

The experiments on spatial solitons actually precede the 1980 observation of temporal solitons in optical fibers. As indicated in Chapter 1 (see also discussions in Ref. [75]), the idea that an optical beam can induce a waveguide and guide itself in it was first suggested in 1962 [76]. One of the earliest observations in the field of nonlinear optics that is closely related to the soliton concept was the self-focusing of optical beams inside a Kerr medium [77]. To investigate this effect theoretically, the wave equation for a Kerr medium—the cubic NLS equation—was analyzed in 1964 in both one and two transverse dimensions [78]–[80]. In particular, it was discovered that the two-dimensional self-trapped solutions to the cubic NLS equation exhibit the phenomenon of *catastrophic collapse*, in the sense that the beam width shrinks to zero at a finite distance, because the two-dimensional solitons are dynamically unstable [80]. Even one-dimensional solitons in a bulk nonlinear medium are unstable, and they break up into filaments (which are, in fact, solitons of higher dimensions) because of a *transverse* modulation instability. As a result, spatial solitons in the Kerr media can only be observed experimentally in configurations in which one of the two transverse dimensions is redundant, i.e., when diffraction is suppressed in one dimension through a suitable scheme (for example, using a planar waveguide).

The earliest observation of spatial solitons dates back to a 1974 experiment in which self-trapping of an optical beam in a bulk medium was found to occur [81]. It took more than 10 years before soliton experiments were performed in optical waveguides [82] using the reorientational nonlinearity liquid CS_2. Two different approaches were used for suppressing beam diffraction in one transverse dimension. In one approach, an interference pattern was produced in one dimension, creating a parallel sequence of planar waveguides orthogonal to that dimension; light could not diffract across the dark zone

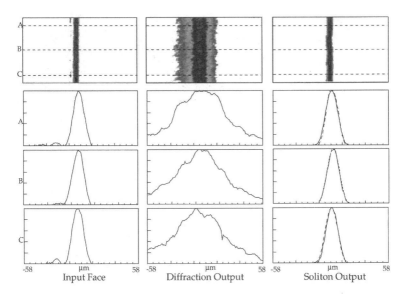

Figure 2.12: Observation of spatial solitons in a photorefractive crystal. Photographs show the input spatial profiles of a 14.5-μm beam (left), the normally diffracting output (middle), and the soliton output beam (right) for a 6-mm-long SBN crystal. The three sets of traces correspond to the locations marked by the dashed lines. (After Ref. [90]; ©1996 APS.)

of the interference pattern. In the second approach, liquid CS_2 was sandwiched between two glass plates, forming effectively a planar waveguide. This 1985 experiment opened the way to numerous demonstrations of one-dimensional bright spatial solitons during the 1990s using materials as diverse as glasses, semiconductors, and polymers [83]–[87].

The discovery of photorefractive solitons was a radical turning point in the field of spatial solitons because it turned out to be important for many reasons [88]–[90]. First, optical power required to generate photorefractive solitons can be as small as 1 μW. As a result, the soliton experiments can be carried out with CW laser beams using simple and inexpensive equipment. Second, photorefractive solitons can be imaged during propagation using scattering impurities. Third, because the response time of photorefractive materials is relatively long (>1 ms), one can work with either coherent or incoherent light, creating new objects such as vector, multimode, and incoherent solitons. Photorefractive solitons are also useful for steering applications.

Figure 2.12 shows the spatial soliton formed inside a 6-mm-long, strontium-barium niobate (SBN), photorefractive crystal using a 14.5-μm-wide input beam [90]. The left column shows the input beam, which diffracts to a size of 56 μm when no voltage is applied to the crystal (middle column). When a voltage of 1.1 kV is applied along the c axis (between two electrodes spaced 5 mm apart), the output beam narrows because of the formation of a spatial soliton. The intensity profiles at three different locations (dashed lines) across the stripe-shaped beam are shown in the bottom three rows in Figure 2.12. Though the soliton profile does not change much along the crystal length

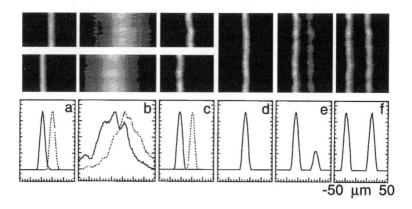

Figure 2.13: Photographs and beam profiles showing the interaction of two spatial solitons in a photorefractive crystal: (a) input beams launched separately; (b) output beams at zero voltage, (c) output beams with voltage applied; (d)–(f) output produced when two solitons propagate simultaneously with a relative input phase of 0, $\pi/2$, and π, respectively. (After Ref. [99]; ©1997 OSA.)

(right column), the diffracted output (middle column) varies significantly because of the roughness caused by material defects and inhomogeneities. The applied voltage not only traps the beam but also reshapes it to a smooth intensity profile that is largely unaffected by the crystal inhomogeneities. Spatial modulation instability and the resulting generation of an array of solitons have also been observed experimentally using a $Bi_{12}TiO_{20}$ photorefractive crystal [91].

The collision properties of spatial solitons have been studied in a number of experiments [92]–[99]. As expected from theory, two in-phase solitons attracted each other but out-of-phase solitons repelled one another. The situation becomes more complex for other values of the soliton relative phase because of the possibility of an energy exchange during inelastic collisions. This feature was clearly observed in a 1992 experiment [96]. When the phase difference between the interacting solitons was $\pi/2$, one soliton gained energy at the expense of the other soliton. The direction of energy exchange was reversed when the phase was increased to $3\pi/2$. Fusion, or "trapping," of two initially overlapping spatial solitons moving in different directions has also been observed [95].

The interaction among solitons is even easier to observe in photorefractive materials [98, 99]. The experimental results obtained using a $Bi_{12}TiO_{20}$ photorefractive crystal with an applied voltage of 1.8 kV were reported in Ref. [98]. At the output end, two solitons fuse together when $\theta = 0$ but acquire different amplitudes when $\theta = 0.65\pi$, because of the energy exchange related to the non-Kerr nature of the photorefractive nonlinearity.

Figure 2.13 shows the experimental results on the collision of two solitons observed by Meng et al. [99]. First two solitons were launched, as shown in part (a). With the zero applied voltage, both beams diffracted and became almost indistinguishable, as seen in part (b). When a voltage of 1250 V was applied, two spatial solitons formed

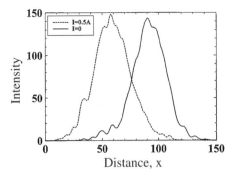

Figure 2.14: Beam profiles showing deflection of of a weak probe beam guided by a spatial soliton that is steered electrically by injecting current. (After Ref. [100]; ©1998 OSA.)

[part (c)]. The solid and dashed profiles of parts (a)–(c) indicate that the two beams were launched separately and did not interact inside the medium. The collision is observed by launching the two solitons simultaneously. When the relative input phase is zero, the solitons merge (fuse) and form an output beam of the same width as each of the input beams, as seen in part (d). When the relative input phase was π, the solitons repelled each other and their separation increased to 46 μm as they diverged [part (f)]. In the case of a $\pi/2$ relative phase, shown in part (e), two solitons were separated by 35 μm and had unequal amplitudes.

This behavior can be understood in terms simple waveguide physics. When two colliding solitons are in phase, the intensity in the overlapping region between the beams increases because of the constructive interference. This, in turn, results in a local increase of the refractive index, which effectively attracts both beams. Exactly the opposite situation occurs when the solitons are out of phase. Then the light intensity drops in the overlap region, and so does the refractive index change. This results in the beams' moving away from each other, which is interpreted as a repulsive force. The spacing between the out-of-phase solitons at the output end depends on the magnitude of the interaction force, which in turn depends on their initial separation.

A practical application of spatial solitons makes use of the fact that the soliton transverse velocity V in Eq. (2.1.2) depends on the spatial phase. Thus, if the phase of an input beam is modulated as $\exp(iVx)$ before launching it inside the nonlinear medium, the generated spatial soliton would travel at an angle to the z axis. By changing the magnitude of V electrically or optically, such a soliton can be steered in any direction. Electrically controlled steering of spatial solitons was demonstrated in a 1998 experiment using a AlGaAs planar waveguide [100]. The phase modulation was produced through index changes that occurred when electrons were injected into the waveguide by applying an external voltage. A weak probe beam of orthogonal polarization, guided by the effective waveguide created by the soliton beam (see Chapter 9), was also deflected with the soliton. Figure 2.14 shows how the probe beam is deflected when a current of 0.5 A is injected (a polarizer was used to filter the spatial soliton). These results show the feasibility of a dynamically reconfigurable optical interconnect based on spatial solitons.

References

[1] M. J. Ablowitz and H. Segur, *Solitons and the Inverse interaction Transform* (SIAM, Philadelphia, 1981).

[2] V. E. Zakharov and A. B. Shabat, *Zh. Eksp. Teor. Fiz.* **61**, 118 (1971) [*Sov. Phys. JETP* **34**, 62 (1972)].

[3] Yu. V. Katyshev, N. V. Makhaldiani, and V. G. Makhankov, *Phys. Lett. A* **66**, 456 (1978).

[4] R. W. Micallef, V. V. Afanasjev, Yu. S. Kivshar, and J. D. Love, *Phys. Rev. E* **54**, 2936 (1996).

[5] K. I. Pushkarov, D. I. Pushkarov, and I. V. Tomov, *Opt. Quantum Electron.* **11**, 471 (1979).

[6] W. Krolikowski and B. Luther-Davies, *Opt. Lett.* **17**, 1414 (1992).

[7] A. W. Snyder, D. J. Mitchell, L. Poladian, and F. Ladouceur, *Opt. Lett.* **16**, 21 (1991).

[8] M. M. Bogdan and A. S. Kovalev, *Pisma Zh. Eksp. Teor. Fiz.* **31**, 213 (1980) [*JETP Lett.* **31**, 195 (1980)].

[9] S. Cowan, R. H. Enns, S. S. Rangnekar, and S. S. Sanghera, *Can. J. Phys.* **64**, 311 (1986).

[10] L. J. Mulder and R. H. Enns, *IEEE J. Quantum Electron.* **25**, 2205 (1989).

[11] L. Gagnon, *J. Opt. Soc. Am. B* **6**, 1477 (1989).

[12] J. Herrman, *Opt. Commun.* **91**, 337 (1992).

[13] K. Hayata and M. Koshiba, *Phys. Rev. E* **51**, 1499 (1995).

[14] Kh. I. Pushkarov and D. I. Pushkarov, *Rep. Math. Phys.* **17**, 37 (1980).

[15] N. G. Vakhitov and A. A. Kolokolov, *Izv. Vyssh. Uchebn. Zaved. Radiofiz.* **16**, 1020 (1973) [*Radiophys. Quantum Electron.* **16**, 783 (1973)].

[16] A. A. Kolokolov, *Izv. Vyssh. Uchebn. Zaved. Radiofiz.* **17**, 1332 (1974) [*Radiophys. Quantum Electron.* **17**, 1016 (1974)].

[17] J. Shatah and W. Strauss, *Comm. Math. Phys.* **100**, 173 (1985); M. I. Weinstein, *Comm. Pure Appl. Math.* **39**, 5 (1986).

[18] E. A. Kuznetsov, A. M. Rubenchik, and V. E. Zakharov, *Phys. Rep.* **142**, 103 (1986).

[19] V. G. Makhankov, Yu. P. Rybakov, and V. I. Sanyuk, *Usp. Fiz. Nauk* **164**, 121 (1994) [*Sov. Phys. Uspekhi* **62**, 113 (1994)].

[20] C. K. R. T. Jones, *J. Diff. Eq.* **71**, 34 (1988); *Ergod. Theor. Dynam. Sys.* **8**, 119 (1988).

[21] M. Grillakis, *Comm. Pure Appl. Math.* **41**, 747 (1988); *Comm. Pure Appl. Math.* **43**, 299 (1990).

[22] M. Grillakis, J. Shatah, and W. Strauss, *J. Funct. Anal.* **74**, 160 (1987); *J. Funct. Anal.* **94**, 308 (1990).

[23] Yu. S. Kivshar and B. A. Malomed, *Rev. Mod. Phys.* **63**, 761 (1989).

[24] D. E. Pelinovsky, A. V. Buryak, and Yu. S. Kivshar, *Phys. Rev. Lett.* **75**, 591 (1995).

[25] D. E. Pelinovsky, V. V. Afanasjev, and Yu. S. Kivshar, *Phys. Rev. E* **53**, 1940 (1996).

[26] Yu. S. Kivshar, D. E. Pelinovsky, T. Cretegny, and M. Peyrard, *Phys. Rev. Lett.* **80**, 5032 (1998).

[27] Yu. S. Kivshar and A. A. Sukhorukov, in *Spatial Solitons*, S. Trillo and W. Torruellas, Eds. (Springer, New York, 2001), pp. 211–246.

[28] D. E. Pelinovsky and Yu.S. Kivshar, *Phys. Rev. E* **62**, 8668 (2000).

[29] D. J. Kaup, *Phys. Rev. A* **42**, 5689 (1990).

[30] E. C. Titchmarsh, *Eigenfunction Expansions Associated with Second-Order Differential Equations* (Oxford University Press, London, 1958).

[31] Yu. S. Kivshar, V. V. Afanasjev, A. V. Buryak, and D. E. Pelinovsky, in *Physics and Applications of Optical Solitons in Fibers*, A. Hasegawa, Ed. (Kluwer, Dordrecht, Netherlands, 1996), pp. 75-88.

[32] I. V. Barashenkov, M. M. Bogdan, and V. I. Korobov, *Europhys. Lett.* **15**, 113 (1991).

[33] H. T. Tran, J. D. Mitchell, N. N. Akhmediev, and A. Ankiewicz, *Opt. Commun.* **93**, 227 (1992).

[34] N. N. Akhmediev, A. Ankiewicz, and H. T. Tran, *J. Opt. Soc. Am.* B **10**, 230 (1993).

[35] I. V. Barashenkov, D. E. Pelinovsky, and E. V. Zemlyanaya, *Phys. Rev. Lett.* **80**, 5117 (1998).

[36] D. Mihalache, D. Mazilu, and L. Torner, *Phys. Rev. Lett.* **81**, 4353 (1998).

[37] M. Johansson and Yu. S. Kivshar, *Phys. Rev. Lett.* **82**, 85 (1999).

[38] A. V. Buryak and N. N. Akhmediev, *Phys. Rev.* E **51**, 3572 (1995).

[39] J. Yang, B. A. Malomed, and D. J. Kaup, *Phys. Rev. Lett.* **83**, 1958 (1999).

[40] A. V. Buryak, *Phys. Rev.* E **52**, 1156 (1995).

[41] Yu. S. Kivshar, A. R. Champneys, D. Cai, and A. R. Bishop, *Phys. Rev.* B **58**, 5423 (1998).

[42] A. R. Champneys and B. A. Malomed, *J. Phys.* A **32**, L547 (1999).

[43] A. R. Champneys and B. A. Malomed, *Phys. Rev.* E **61**, 463 (2000).

[44] A. R. Champneys, B. A. Malomed, J. Yang, and D. J. Kaup, *Physica D* **152**, 340 (2001).

[45] J. Yang, B. A. Malomed, D. J. Kaup, and A. R. Champneys, *Math. Comp. Simulation* **56**, 585 (2001).

[46] J. Yang, *Stud. Appl. Math.* **106**, 337 (2001).

[47] D. E. Pelinovsky and J. Yang, *Proc. Roy. Soc. London A* **458**, 1469 (2002).

[48] Y. Tan, J. Yang, and D. E. Pelinovsky, *Wave Motion* **36**, 241 (2002).

[49] K. Kolossovski, A. R. Champneys, A. V. Buryak, and R.A. Sammut, *Physica D* **171**, 153 (2002).

[50] A. M. Kosevich, *Low Temp. Phys.* **26**, 453 (2000) [*Fiz. Nizk. Temp.* **26**, 620 (2000)].

[51] A. Höök and M. Karlsson, *Opt. Lett.* **18**, 1388 (1993).

[52] E. M. Gromov and V. I. Talanov, *Sov. Phys. JETP* **83**, 73 (1996) [*Zh. Eksp. Teor. Fiz.* **110**, 137 (1996)].

[53] M. Karlsson and A. Höök, *Opt. Commun.* **104**, 303 (1994).

[54] M. Piché, J. F. Cormier, and X. Zhu, *Opt. Lett.* **21**, 845 (1996).

[55] J. Fujioka and A. Espinosa, *J. Phys. Soc. Jpn.* **66**, 2601 (1997).

[56] M. Gedalin, T. C. Scott, and Y. B. Band, *Phys. Rev. Lett.* **78**, 448 (1997).

[57] T. T. Shi and S. Chi, *Opt. Lett.* **15**, 1123(1990).

[58] N. N. Akhmediev and A. Ankiewicz, *Solitons: Nonlinear Pulses and Beams* (Chapman and Hall, London, 1997), Chap. 3.

[59] D. A. Semagin, S. V. Dmitriev, T. Shigenari, Yu. S. Kivshar, and A. A. Sukhorukov, *Physica B* **316**, 136 (2002).

[60] S.V. Dmitriev and T. Shigenari, *Chaos* **12**, 324 (2002).

[61] Yu. S. Kivshar and B. A. Malomed, *Phys. Lett.* A **115**, 377 (1986).

[62] H. Frauenkron, Yu. S. Kivshar, and B. A. Malomed, *Phys. Rev.* E **54**, R2244 (1996).

[63] J. Satsuma and N. Yajima, *Prog. Theor. Phys. Suppl.* **55**, 284 (1974).

[64] N. R. Pereira and F. Y. F. Chu, *Phys. Fluids* **22**, 874 (1979).

[65] T. Yamada and K. Nozaki, *J. Phys. Soc. Jpn.* **58**, 1944 (1989).

[66] S. R. Friberg and K. W. DeLong, *Opt. Lett.* **17**, 979 (1992).

[67] V. V. Afanasjev, J. S. Aitchison, and Yu. S. Kivshar, *Opt. Commun.* **116**, 331 (1995).

[68] D. Artigas, L. Torner, J.P. Torres, and N.N. Akhmediev, *Opt. Commun.* **143**, 322 (1997).

[69] J. A. Besley, P. D. Miller, and N. N. Akhmediev, *Phys. Rev. E* **61**, 7121 (2000).

[70] B. A. Malomed, *Phys. Lett. A* **154**, 441 (1991).

[71] A. V. Buryak and N. N. Akhmediev, *Phys. Rev. E* **50**, 3126 (1994).

[72] A. W. Snyder, A. V. Buryak, and D. J. Mitchell, *Opt. Lett.* **23**, 4 (1998).

[73] K. A. Gorshkov and L. A. Ostrovsky, *Physica D* **3**, 428 (1981).

[74] S. V. Cavalcanti, J. C. Cressoni, H. R. da Cruz, and A. S. Gouveia-Neto, *Phys. Rev. A* **43**, 6162 (1991).

[75] G. I. Stegeman, D. N. Christodoulides, and M. Segev, *IEEE J. Sel. Topics Quantum Electron.* **6**, 1419 (2000).

[76] G. A. Askaryan, *Sov. Phys. JETP* **15**, 1088 (1962).

[77] M. Hercher, *J. Opt. Soc. Amer.* **54** 563 (1964).

[78] R. Y. Chiao, E. Garmire, and C. H. Townes, *Phys. Rev. Lett.* **13**, 479 (1964).

[79] I. Talanov, *Radio Phys.* **7**, 254 (1964).

[80] P. L. Kelley, *Phys. Rev. Lett.* **15**, 1005 (1965).

[81] J. E. Bjorkholm and A. Ashkin, *Phys. Rev. Lett.* **32**, 129 (1974).

[82] A. Barthelemy, S. Maneuf, and C. Froehly, *Opt. Commun.* **55**, 201 (1985).

[83] J. S. Aitchison, A. M. Weiner, Y. Silberberg, M. K. Oliver, J. L. Jackel, D. E. Leaird, E. M. Vogel, and P. W. Smith, *Opt. Lett.* **15**, 471 (1990).

[84] J. S. Aitchison, K. Al-Hemyari, C. N. Ironside, R. S. Grant, and W. Sibbett, *Electron. Lett.* **28**, 1879 (1992).

[85] J. U. Kang, G. I. Stegeman, and J. S. Aitchison, *Opt. Lett.* **20**, 2069 (1995).

[86] U. Bartuch, U. Peschel, Th. Gabler, R. Waldhaus, and H.-H. Horhold, *Opt. Commun.* **134**, 49 (1997).

[87] J. U. Kang, C. J. Hamilton, J. S. Aitchison, and G. I. Stegeman, *Appl. Phys. Lett.* **70**, 1363 (1997).

[88] M. Segev, B. Crosignani, A. Yariv, and B. Fischer, *Phys. Rev. Lett.* **68**, 923 (1992).

[89] G. Duree, J. L. Shultz, G. Salamo, M. Segev, A. Yariv, B. Crosignani, P. DiPorto, E. Sharp, and R. R. Neurgaonkar, *Phys. Rev. Lett.* **71**, 533 (1993).

[90] K. Kos, H. Meng, G. Salamo, M. Shih, M. Segev, and G. C. Valley, *Phys. Rev. E* **53**, R4330 (1996).

[91] M. D. Iturbe-Castillo, M. Torres-Cisneros, J. J. Sánchez-Mondragón, S. Chávez-Cerda, S. I. Srtepanov, V. A. Vysloukh, and G. E. Torres-Cisneros, *Opt. Lett.* **20**, 1853 (1995).

[92] F. Reynaud and A. Barthelemy, *Europhys. Lett.* **12**, 401 (1990).

[93] J. S. Aitchison, A. M. Weiner, Y. Silberberg, D. E. Leaird, M. K. Oliver, J. L. Jackel, and P. W. Smith, *Opt. Lett.* **16**, 15 (1991).

[94] J. S. Aitchison, A. M. Weiner, Y. Silberberg, D. E. Leaird, M. K. Oliver, J. L. Jackel, and P. W. Smith, *J. Opt. Soc. Am. B* **8**, 1290 (1991).

[95] M. Shalaby and A. Barthelemy, *Opt. Lett.* **16**, 1472 (1991).

[96] M. Shalaby, F. Reynaud, and A. Barthelemy, *Opt. Lett.* **17**, 778 (1992).

[97] M. Shih, Z. Chen, M. Segev, T. H. Coskun, and D. N. Christodoulides, *Appl. Phys. Lett.* **69**, 4151 (1996).

[98] G. S. Carcía-Quirino, M. D. Iturbe-Castillo, V. A. Vysloukh, J. J. Sánchez-Mondragón, S. I. Stepanov, G. Lugo-Matínez, and G. E. Torres-Cisneros, *Opt. Lett.* **22**, 154 (1997).

[99] H. Meng, G. Salamo, M. Shih, and M. Segev, *Opt. Lett.* **22**, 448 (1997).

[100] L. Friedrich, G. I. Stegeman, P. Millar, C. J. Hamilton, and J. S. Aitchison, *Opt. Lett.* **23**, 1438 (1998).

Chapter 3

Temporal Solitons

The existence of temporal solitons in optical fibers and their use for optical communications were suggested in 1973 [1], and by 1980 such solitons had been observed experimentally [2]. Since then, rapid progress has converted temporal solitons into a practical candidate for designing modern lightwave systems [3]–[8]. In this chapter we focus on the properties of temporal bright solitons. The basic concepts behind fiber solitons are introduced in Section 3.1, where we also discuss the properties of bright solitons. Section 3.2 shows how temporal solitons can be used for optical communications and how the design of such lightwave systems differs from that of conventional systems. The loss-managed and dispersion-managed solitons are considered in Sections 3.3 and 3.4, respectively. The effect of amplifier noise on such solitons is discussed in Section 3.5, with emphasis on the timing-jitter issue. Section 3.6 focuses on the impact of higher-order dispersive and nonlinear effects on temporal solitons. All issues related to the physics of temporal dark solitons are discussed in Chapter 4.

3.1 Fiber Solitons

As discussed in Section 1.3, temporal solitons form inside optical fibers because of a balance between the group-velocity dispersion (GVD) and self-phase modulation (SPM) induced by the Kerr nonlinearity. One can develop an intuitive understanding of how such a balance is possible. The GVD broadens optical pulses during their propagation inside an optical fiber, except when the pulse is initially chirped in the right way. More specifically, a chirped pulse can be compressed during the early stage of propagation whenever the GVD parameter β_2 and the chirp parameter C happen to have opposite signs such that $\beta_2 C$ is negative [9]. The nonlinear phenomenon of SPM imposes a chirp on the optical pulse such that $C > 0$. Since $\beta_2 < 0$ in the 1.55-μm wavelength region of silica fibers, the condition $\beta_2 C < 0$ is readily satisfied. Moreover, because the SPM-induced chirp is power dependent, it is not difficult to imagine that, under certain conditions, the SPM and GVD may cooperate in such a way that the SPM-induced chirp is just right to cancel the GVD-induced broadening of the pulse. The optical pulse would then propagate undistorted in the form of a soliton.

3.1.1   Nonlinear Schrödinger Equation

The NLS equation governing pulse propagation inside optical fibers was derived in Section 1.3 and is given by Eq. (1.3.9). The bright solitons exist for $s = -1$ (anomalous GVD), for which this equation becomes

$$i\frac{\partial u}{\partial z} + \frac{1}{2}\frac{\partial^2 u}{\partial \tau^2} + |u|^2 u = 0. \tag{3.1.1}$$

In this chapter, we have replaced x with τ to emphasize the temporal nature of the soliton. Physically, τ is a measure of time from the pulse center and is normalized to the input pulse width T_0. The main difference from the spatial solitons discussed in Chapter 2 is that the nonlinear term has the simple form suitable for a Kerr medium. The reason behind this is that the nonlinear effects in optical fibers are weak enough that the third-order nonlinearity does not saturate and can be assumed to increase linearly with the optical intensity $|u(t)|^2$.

As discussed in Chapters 1 and 2, the inverse scattering method applies to Eq. (3.1.1) because this equation is in the form of the standard cubic NLS equation [10]. For this reason, temporal solitons exhibit features that are identical to those for spatial Kerr solitons. The main features associated with temporal solitons can be summarized as follows. When an input pulse having an initial amplitude

$$u(0, \tau) = N \operatorname{sech}(\tau) \tag{3.1.2}$$

is launched into the fiber, its shape remains unchanged during propagation when $N = 1$ but follows a periodic pattern for integer values of $N > 1$ such that the input shape is recovered at $z = m\pi/2$, where m is an integer. The parameter N is related to the input pulse parameters as (see Section 1.3)

$$N^2 = \gamma P_0 L_D = \gamma P_0 T_0^2 / |\beta_2|, \tag{3.1.3}$$

Where L_D is the dispersion length; N represents a dimensionless combination of two pulse parameters (peak power P_0 and width T_0) and two fiber parameters, β_2 and $\gamma = 2\pi n_2(\lambda a_{\text{eff}})$, where a_{eff} is the effective core area of the fiber [9].

An optical pulse whose parameters satisfy the condition $N = 1$ is called the *fundamental* temporal soliton. Pulses corresponding to other integer values of N represent *higher-order* temporal solitons. Noting that $z = Z/L_D$, the soliton period Z_0, defined as the distance over which higher-order solitons recover their original shape, is given by

$$Z_0 = \frac{\pi}{2}L_D = \frac{\pi}{2}\frac{T_0^2}{|\beta_2|}. \tag{3.1.4}$$

The *soliton period* Z_0 and *soliton order* N play an important role for quantifying temporal solitons. Figure 3.1 shows the pulse evolution for the first-order ($N = 1$) and third-order ($N = 3$) solitons over one soliton period by plotting the pulse intensity $|u(z, \tau)|^2$ (top row) and the frequency chirp (bottom row), defined as the time derivative of the soliton phase. Only a fundamental soliton maintains its shape and remains chirp-free during propagation inside optical fibers.

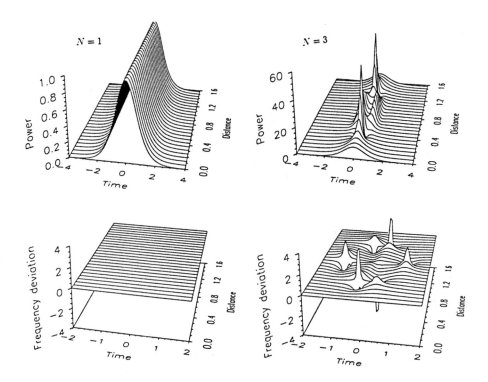

Figure 3.1: Evolution of the first-order (left column) and third-order (right column) temporal solitons over one soliton period. The bottom row shows the chirp profile.

As discussed in Section 1.3, the analytic form of the fundamental temporal soliton for $N = 1$ is given by

$$u(z, \tau) = \text{sech}(\tau) \exp(iz/2). \tag{3.1.5}$$

It shows that the input pulse acquires a phase shift $z/2$ as it propagates inside the fiber but that its amplitude remains unchanged. This property makes temporal solitons an ideal candidate for optical communications. In essence, the effects of fiber dispersion are exactly compensated by the fiber nonlinearity when the input pulse has a "sech" shape and its width and peak power are related by Eq. (3.1.3) in such a way that $N = 1$.

3.1.2 Temporal Soliton Dynamics

An important property of optical solitons is that they are remarkably stable against perturbations. Thus, even though the fundamental soliton requires a specific shape and a certain peak power corresponding to $N = 1$ in Eq. (3.1.3), it can be created even when the pulse shape and the peak power deviate from the ideal conditions. Figure 3.2 shows the numerically simulated evolution of a Gaussian input pulse for which $N = 1$ but $u(0, \tau) = \exp(-\tau^2/2)$. As seen there, the pulse adjusts its shape and width in an attempt to become a fundamental soliton and attains a "sech" profile for $z \gg 1$. A similar behavior is observed when N deviates from 1. It turns out that the Nth-order

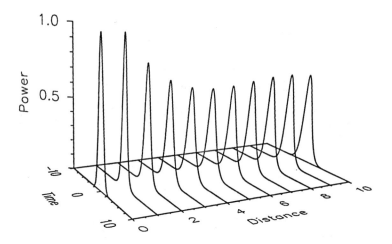

Figure 3.2: Evolution of a Gaussian pulse with $N = 1$ over the range $z = 0$–10. The pulse evolves toward the fundamental soliton by changing its shape, width, and peak power.

soliton can be formed when the input value of N is in the range $N - \frac{1}{2}$ to $N + \frac{1}{2}$ [11]. In particular, the fundamental soliton can be excited for values of N in the range 0.5–1.5. Figure 3.3 shows the pulse evolution for $N = 1.2$ over the range $z = 0$–10 by solving the NLS equation numerically with the initial condition $u(0, \tau) = 1.2 \operatorname{sech}(\tau)$. The pulse width and the peak power oscillate initially but eventually become constant after the input pulse has adjusted itself to satisfy the condition $N = 1$ in Eq. (3.1.3).

It may seem mysterious that an optical fiber can force any input pulse to evolve toward a soliton. A simple way to understand this behavior is to think of optical solitons as the temporal modes of a nonlinear waveguide. Higher intensities in the pulse center create a temporal waveguide by increasing the refractive index only in the central part of the pulse. Such a waveguide supports temporal modes just as the core-cladding index difference leads to spatial modes of fibers. When an input pulse does not match a temporal mode precisely but is close to it, most of the pulse energy can still be coupled into that temporal mode. The rest of the energy spreads in the form of *dispersive waves*. It will be seen later that such dispersive waves affect the system performance and should be minimized by matching the input conditions as close to the ideal requirements as possible. When solitons adapt to perturbations adiabatically, perturbation theory developed specifically for solitons can be used to study how the soliton amplitude, width, frequency, velocity, and phase evolve along the fiber.

3.2   Soliton-Based Communications

Temporal solitons are attractive for optical communications because they are able to maintain their width even in the presence of fiber dispersion. However, their use requires substantial changes in the system design, compared with conventional nonsoliton systems. In this section we focus on several such issues.

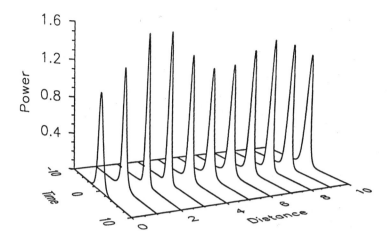

Figure 3.3: Pulse evolution for a "sech" pulse with $N = 1.2$ over the range $z = 0$–10. The pulse evolves toward the fundamental soliton ($N = 1$) by adjusting its width and peak power.

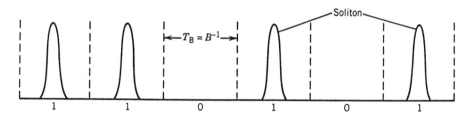

Figure 3.4: Soliton bit stream in RZ format. Each soliton occupies a small fraction of the bit slot, ensuring that neighboring soliton are spaced far apart.

3.2.1 Information Transmission with Solitons

The basic idea is to use a soliton in each bit slot representing 1 in a bit stream. Figure 3.4 shows a soliton bit stream schematically. Typically, the spacing between two solitons exceeds a few times their full width at half maximum (FWHM). The reason behind such a large spacing is easily understood by noting that the pulse width must be a small fraction of the bit slot to ensure that the neighboring solitons are well separated. Mathematically, the soliton solution in Eq. (3.1.5) is valid only when it occupies the entire time window ($-\infty < \tau < \infty$). It remains approximately valid for a train of solitons only when individual solitons are well isolated. This requirement can be used to relate the soliton width T_0 to the bit rate B as

$$B = \frac{1}{T_B} = \frac{1}{2q_0 T_0},\qquad(3.2.1)$$

where T_B is the duration of the bit slot and $2q_0 = T_B/T_0$ is the separation between neighboring solitons in normalized units.

The input pulse characteristics needed to excite the fundamental soliton can be obtained by setting $z = 0$ in Eq. (3.1.5). In physical units, the power across the pulse varies as

$$P(t) = |A(0,t)|^2 = P_0 \text{sech}^2(t/T_0). \tag{3.2.2}$$

The required peak power P_0 is obtained from Eq. (3.1.3) by setting $N = 1$ and is related to the width T_0 and the fiber parameters as

$$P_0 = |\beta_2|/(\gamma T_0^2). \tag{3.2.3}$$

The width parameter T_0 is related to the FWHM of the soliton as

$$T_s = 2T_0 \ln(1 + \sqrt{2}) \simeq 1.763 T_0. \tag{3.2.4}$$

The pulse energy for the fundamental soliton is obtained using

$$E_s = \int_{-\infty}^{\infty} P(t)\, dt = 2P_0 T_0. \tag{3.2.5}$$

Assuming that 1 and 0 bits are equally likely to occur, the average power of the RZ signal becomes $\bar{P}_s = E_s(B/2) = P_0/2q_0$. As a simple example, $T_0 = 10$ ps for a 10-Gb/s soliton system if we choose $q_0 = 5$. The pulse FWHM is about 17.6 ps for $T_0 = 10$ ps. The peak power of the input pulse is 5 mW using $\beta_2 = -1$ ps^2/km and $\gamma = 2$ W^{-1}/km as typical values for dispersion-shifted fibers. This value of peak power corresponds to a pulse energy of 0.1 pJ and an average power level of only 0.5 mW.

3.2.2   Soliton Interaction

The presence of pulses in the neighboring bits perturbs a temporal soliton simply because the combined optical field is not a solution of the NLS equation. Neighboring solitons either come closer or move apart because of a nonlinear interaction between them. This phenomenon of *soliton interaction* has been studied extensively [12]–[16].

One can understand the implications of soliton interaction by solving the NLS equation numerically, with the input amplitude consisting of a soliton pair so that

$$u(0, \tau) = \text{sech}(\tau - q_0) + r\,\text{sech}[r(\tau + q_0)]\exp(i\theta), \tag{3.2.6}$$

where r is the relative amplitude of the two solitons, θ is the relative phase, and $2q_0$ is the initial (normalized) separation. Figure 3.5 shows the evolution of a soliton pair with $q_0 = 3.5$ for several values of the parameters r and θ. Clearly, soliton interaction depends strongly on both the relative phase θ and the amplitude ratio r.

Consider first the case of equal-amplitude solitons ($r = 1$). The two solitons attract each other in the in-phase case ($\theta = 0$) such that they collide periodically along the fiber length. However, for $\theta = \pi/4$, the solitons separate from each other after an initial attraction stage. For $\theta = \pi/2$, the solitons repel each other even more strongly, and their spacing increases with distance. From the standpoint of system design, such behavior is not acceptable. It would lead to jitter in the arrival time of solitons because the relative phase of neighboring solitons is not likely to remain well controlled. One way to avoid soliton interaction is to increase q_0 because the strength of interaction depends on soliton spacing. For sufficiently large q_0, deviations in the soliton position

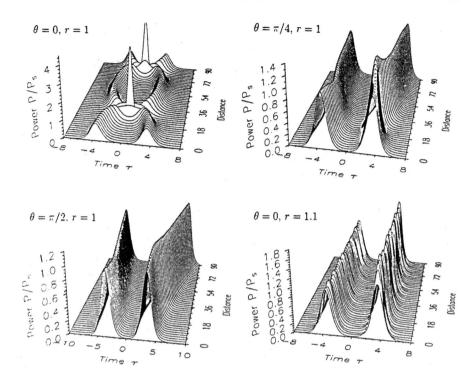

Figure 3.5: Evolution of a soliton pair over 90 dispersion lengths showing the effects of soliton interaction for four different choices of amplitude ratio r and relative phase θ. Initial spacing $q_0 = 3.5$ in all four cases.

are expected to be small enough that the soliton remains at its initial position within the bit slot over the entire transmission distance.

The dependence of soliton separation on q_0 can be studied analytically by using the inverse scattering method [12]. A perturbative approach can be used for $q_0 \gg 1$. In the specific case of $r = 1$ and $\theta = 0$, the soliton separation $2q_s$ at any distance z is given by [13]

$$2\exp[2(q_s - q_0)] = 1 + \cos[4z\exp(-q_0)]. \tag{3.2.7}$$

This relation shows that the spacing $q_s(z)$ between two neighboring solitons oscillates periodically with the period

$$z_p = (\pi/2)\exp(q_0). \tag{3.2.8}$$

A more accurate expression, valid for arbitrary values of q_0, is given by [15]

$$z_p = \frac{\pi \sinh(2q_0)\cosh(q_0)}{2q_0 + \sinh(2q_0)}. \tag{3.2.9}$$

Equation (3.2.8) is quite accurate for $q_0 > 3$. Its predictions are in agreement with the numerical results shown in Figure 3.5, where $q_0 = 3.5$. It can be used for system design as follows. If $z_p L_D$ is much greater than the total transmission distance L_T, soliton interaction can be neglected since soliton spacing would deviate little from its initial value. For $q_0 = 6$, $z_p \approx 634$. Using $L_D = 100$ km for the dispersion length, $L_T \ll z_p L_D$ can be realized even for $L_T = 10,000$ km. If we use $L_D = T_0^2 / |\beta_2|$ and $T_0 = (2Bq_0)^{-1}$ from Eq. (3.2.1), the condition $L_T \ll z_p L_D$ can be written in the form of a simple design criterion,

$$B^2 L_T \ll \frac{\pi \exp(q_0)}{8q_0^2 |\beta_2|}. \tag{3.2.10}$$

For the purpose of illustration, let us choose $\beta_2 = -1$ ps^2/km. Equation (3.2.10) then implies that $B^2 L_T \ll 4.4$ (Tb/s)^2-km if we use $q_0 = 6$ to minimize soliton interactions. The pulse width at a given bit rate B is determined from Eq. (3.2.1). For example, $T_s = 14.7$ ps at $B = 10$ Gb/s when $q_0 = 6$.

A relatively large soliton spacing, necessary to avoid soliton interaction, limits the bit rate of soliton communication systems. The spacing can be reduced by up to a factor of 2 by using unequal amplitudes for the neighboring solitons. As seen in Figure 3.5, the separation for two in-phase solitons does not change by more than 10% for an initial soliton spacing as small as $q_0 = 3.5$ if their initial amplitudes differ by 10% ($r = 1.1$). Note that the peak powers or the energies of the two solitons deviate by only 1%. As discussed earlier, such small changes in the peak power are not detrimental for maintaining solitons. Thus, this scheme is feasible in practice and can be useful for increasing the system capacity. The design of such systems would, however, require attention to many details. Soliton interaction can also be modified by other factors, such as the initial frequency chirp imposed on input pulses.

3.3   Loss-Managed Solitons

Temporal solitons use the nonlinear phenomenon of SPM to maintain their width even in the presence of fiber dispersion. However, this property holds only if fiber losses were negligible. It is not difficult to see that a decrease in soliton energy because of fiber losses would produce soliton broadening simply because a reduced peak power weakens the SPM effect necessary to counteract the GVD. Optical amplifiers can be used for compensating fiber losses. This section focuses on the management of losses through amplification of solitons by considering the lumped- and distributed-amplification techniques shown schematically in Figure 3.6 and studied extensively in the 1980s [17]–[21].

3.3.1   Lumped Amplification

In the lumped-amplification scheme, optical amplifiers are placed periodically along the fiber link such that fiber losses between two amplifiers are exactly compensated by the amplifier gain. An important design parameter is the spacing L_A between amplifiers—it should be as large as possible to minimize the overall cost. For non-soliton systems, L_A is typically 80–100 km. For soliton systems, L_A is restricted to much smaller values because of the soliton nature of signal propagation [18].

The physical reason behind smaller values of L_A is that optical amplifiers boost soliton energy to the input level over a length of a few meters without allowing for

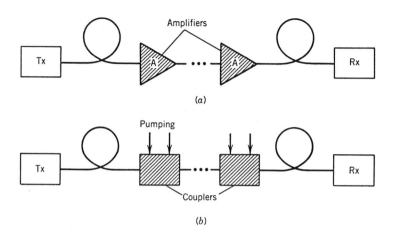

Figure 3.6: (a) Lumped- and (b) distributed-amplification schemes for compensation of fiber losses in soliton communication systems.

gradual recovery of the fundamental soliton. The amplified soliton adjusts its width dynamically in the fiber section following the amplifier. However, it also sheds a part of its energy as dispersive waves during this adjustment phase. The dispersive part can accumulate to significant levels over a large number of amplification stages and must be avoided. One way to reduce the dispersive part is to reduce the amplifier spacing L_A such that the soliton is not perturbed much over this short length. Numerical simulations show [18] that this is the case when L_A is a small fraction of the dispersion length ($L_A \ll L_D$). The dispersion length L_D depends on both the pulse width T_0 and the GVD parameter β_2 and can vary from 10 to 1000 km, depending on their values.

Periodic amplification of solitons can be treated mathematically by writing the NLS equation in the form [22]

$$i\frac{\partial u}{\partial z} + \frac{1}{2}\frac{\partial^2 u}{\partial \tau^2} + |u|^2 u = -\frac{i}{2}\Gamma u + \frac{i}{2}g(z)L_D u, \qquad (3.3.1)$$

where $\Gamma = \alpha L_D$ accounts for fiber losses, $g(z) = \sum_{m=1}^{N_A} g_m \delta(z - z_m)$, N_A is the total number of amplifiers, and g_m is the gain of a lumped amplifier located at z_m. If we assume that amplifiers are spaced uniformly, then $z_m = m z_A$, where $z_A = L_A/L_D$ is the normalized amplifier spacing.

Because of rapid variations in the soliton energy introduced by periodic gain–loss changes, it is useful to make the transformation

$$u(z, \tau) = \sqrt{p(z)}\, v(z, \tau), \qquad (3.3.2)$$

where $p(z)$ is a rapidly varying and $v(z, \tau)$ is a slowly varying function of z. Substituting Eq. (3.3.2) in Eq. (3.3.1), $v(z, \tau)$ is found to satisfy

$$i\frac{\partial v}{\partial z} + \frac{1}{2}\frac{\partial^2 v}{\partial \tau^2} + p(z)|v|^2 v = 0, \qquad (3.3.3)$$

where $p(z)$ is obtained by solving the ordinary differential equation

$$\frac{dp}{dz} = [g(z)L_D - \Gamma]p. \tag{3.3.4}$$

The preceding equations can be solved analytically by noting that the amplifier gain is just large enough that $p(z)$ is a periodic function; it decreases exponentially in each period as $p(z) = \exp(-\Gamma z)$ but jumps to its initial value $p(0) = 1$ at the end of each period. Physically, $p(z)$ governs variations in the peak power (or the energy) of a soliton between two amplifiers. For a fiber with losses of 0.2 dB/km, $p(z)$ varies by a factor of 100 when $L_A = 100$ km.

In general, changes in soliton energy are accompanied by changes in the soliton width. Large, rapid variations in $p(z)$ can destroy a soliton if its width changes rapidly through emission of dispersive waves. The concept of the *path-averaged*, or *guiding-center*, soliton makes use of the fact that solitons evolve little over a distance that is short compared with the dispersion length (or soliton period). Thus, when $z_A \ll 1$, the soliton width remains virtually unchanged even though its peak power $p(z)$ varies considerably in each section between two neighboring amplifiers. In effect, we can replace $p(z)$ by its average value $\bar{p}$ in Eq. (3.3.3) when $z_A \ll 1$. Introducing $u = \sqrt{\bar{p}}v$ as a new variable, this equation reduces to the standard NLS equation obtained for a lossless fiber.

From a practical viewpoint, a fundamental soliton can be excited if the input peak power P_s (or energy) of the path-averaged soliton is chosen to be larger by a factor $1/\bar{p}$. Introducing the amplifier gain as $G = \exp(\Gamma z_A)$ and using $\bar{p} = z_A^{-1} \int_0^{z_A} e^{-\Gamma z} dz$, the energy enhancement factor for loss-managed solitons is given by

$$f_{LM} = \frac{P_s}{P_0} = \frac{1}{\bar{p}} = \frac{\Gamma z_A}{1 - \exp(-\Gamma z_A)} = \frac{G \ln G}{G - 1}, \tag{3.3.5}$$

where P_0 is the peak power in lossless fibers. Thus, soliton evolution in lossy fibers with periodic lumped amplification is identical to that in lossless fibers, provided that amplifiers are spaced such that $L_A \ll L_D$ and that the launched peak power is larger by a factor f_{LM}. As an example, $G = 10$ and $f_{LM} \approx 2.56$ for 50-km amplifier spacing and fiber losses of 0.2 dB/km.

Figure 3.7 shows the evolution of a loss-managed soliton over a distance of 10 Mm assuming that solitons are amplified every 50 km. When the input pulse width corresponds to a dispersion length of 200 km, the soliton is preserved quite well, even after 10 Mm, because the condition $z_A \ll 1$ is reasonably well satisfied. However, if the dispersion length is reduced to 25 km ($z_A = 2$), the soliton is unable to sustain itself because of excessive emission of dispersive waves. The condition $z_A \ll 1$ or $L_A \ll L_D$, required to operate within the average-soliton regime, can be related to the width T_0 by using $L_D = T_0^2/|\beta_2|$. The resulting condition is

$$T_0 \gg \sqrt{|\beta_2|L_A}. \tag{3.3.6}$$

Since the bit rate B is related to T_0 through Eq. (3.2.1), condition (3.3.6) can be written in the form of the following design criterion:

$$B^2 L_A \ll (4q_0^2|\beta_2|)^{-1}. \tag{3.3.7}$$

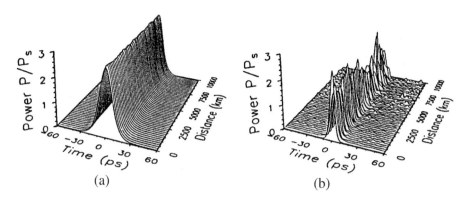

Figure 3.7: Evolution of loss-managed solitons over 10,000 km for (a) $L_D = 200$ km and (b) 25 km with $L_A = 50$ km, $\alpha = 0.22$ dB/km, and $\beta_2 = -0.5$ ps²/km.

Choosing typical values $\beta_2 = -0.5$ ps²/km, $L_A = 50$ km, and $q_0 = 5$, we obtain $T_0 \gg$ 5 ps and $B \ll 20$ GHz. Clearly, the use of path-averaged solitons imposes a severe limitation on both the bit rate and the amplifier spacing for soliton communication systems.

3.3.2 Distributed Amplification

The condition $L_A \ll L_D$, imposed on loss-managed solitons when lumped amplifiers are used, becomes increasingly difficult to satisfy in practice as bit rates exceed 10 Gb/s. This condition can be relaxed considerably when distributed amplification is used. The distributed-amplification scheme is inherently superior to lumped amplification since its use provides a nearly lossless fiber by compensating losses locally at every point along the fiber link. In fact, this scheme was used as early as 1985 using the distributed gain provided by Raman amplification when the fiber carrying the signal was pumped at a wavelength of about 1.46 μm using a color-center laser [20]. Alternatively, the transmission fiber can be doped lightly with erbium ions and pumped periodically to provide distributed gain. Several experiments have demonstrated that solitons can be propagated in such active fibers over relatively long distances [23]–[27].

The advantage of distributed amplification can be seen from Eq. (3.3.4), which can be written in physical units as

$$\frac{dp}{dZ} = [g(Z) - \alpha]p. \tag{3.3.8}$$

If $g(Z)$ is constant and equal to α for all Z, the peak power or energy of a soliton remains constant along the fiber link. This is the ideal situation in which the fiber is effectively lossless. In practice, distributed gain is realized by injecting pump power periodically into the fiber link. Since pump power does not remain constant because of fiber losses and pump depletion (e.g., absorption by dopants), $g(Z)$ cannot be kept constant along the fiber. However, even though fiber losses cannot be compensated

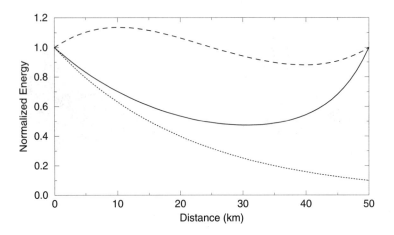

Figure 3.8: Variations in soliton energy for backward (solid line) and bidirectional (dashed line) pumping schemes with $L_A = 50$ km. The lumped-amplifier case is shown by the dotted line.

everywhere locally, they can be compensated fully over a distance L_A provided that

$$\int_0^{L_A} g(Z)\,dZ = \alpha L_A. \tag{3.3.9}$$

A distributed-amplification scheme is designed to satisfy Eq. (3.3.9). The distance L_A is referred to as the *pump-station spacing*.

The important question is how much soliton energy varies during each gain–loss cycle. The extent of peak-power variations depends on L_A and on the pumping scheme adopted. Backward pumping is commonly used for distributed Raman amplification because such a configuration provides high gain where the signal is relatively weak. If we ignore pump depletion, the gain coefficient in Eq. (3.3.8) is given by $g(Z) = g_0 \exp[-\alpha_p(L_A - Z)]$, where α_p accounts for fiber losses at the pump wavelength. The resulting equation can be integrated analytically to obtain

$$p(Z) = \exp\left\{\alpha L_A \left[\frac{\exp(\alpha_p Z) - 1}{\exp(\alpha_p L_A) - 1}\right] - \alpha Z\right\}, \tag{3.3.10}$$

where g_0 was chosen to ensure that $p(L_A) = 1$. Figure 3.8 shows how $p(Z)$ varies along the fiber for $L_A = 50$ km using $\alpha = 0.2$ dB/km and $\alpha_p = 0.25$ dB/km. The case of lumped amplification is also shown for comparison. Whereas soliton energy varies by a factor of 10 in the lumped case, it varies by less than a factor of 2 in the case of distributed amplification.

The range of energy variations can be reduced further using a bidirectional pumping scheme. The gain coefficient $g(Z)$ in this case can be approximated (neglecting pump depletion) as

$$g(Z) = g_1 \exp(-\alpha_p Z) + g_2 \exp[-\alpha_p(L_A - Z)]. \tag{3.3.11}$$

The constants g_1 and g_2 are related to the pump powers injected at both ends. Assuming equal pump powers and integrating Eq. (3.3.8), the soliton energy is found to vary as

$$p(Z) = \exp\left[\alpha L_A \left(\frac{\sinh[\alpha_p(Z - L_A/2)] + \sinh(\alpha_p L_A/2)}{2\sinh(\alpha_p L_A/2)}\right) - \alpha Z\right]. \qquad (3.3.12)$$

This case is shown in Figure 3.8 by a dashed line. Clearly, a bidirectional pumping scheme is the best because it reduces energy variations to below 15%. The range over which $p(Z)$ varies increases with L_A. Nevertheless, it remains much smaller than that occurring in the lumped-amplification case. As an example, soliton energy varies by a factor of 100 or more when $L_A = 100$ km if lumped amplification is used but by less than a factor of 2 when the bidirectional pumping scheme is used for distributed amplification.

The effect of energy excursion on solitons depends on the ratio $z_A = L_A/L_D$. When $z_A < 1$, little soliton reshaping occurs. For $z_A \gg 1$, solitons evolve adiabatically, with some emission of dispersive waves (the quasi-adiabatic regime). For intermediate values of z_A, a more complicated behavior occurs. In particular, dispersive waves and solitons are resonantly amplified when $z_A \simeq 4\pi$; such a resonance can lead to unstable and chaotic behavior [21]. For this reason, distributed amplification is used with $z_A < 4\pi$ in practice [23]–[27].

3.3.3 Experimental Progress

Early experiments on loss-managed solitons concentrated on the Raman-amplification scheme. An experiment in 1985 demonstrated that fiber losses can be compensated over 10 km by the Raman gain while maintaining the soliton width [20]. Two color-center lasers were used in this experiment. One laser produced 10-ps pulses at 1.56 μm, which were launched as fundamental solitons. The other laser operated continuously at 1.46 μm and acted as a pump for amplifying 1.56-μm solitons. In the absence of the Raman gain, the soliton broadened by about 50% because of loss-induced broadening. This amount of broadening was in agreement with the theoretical prediction $T_1/T_0 = 1.51$ for $z = 10$ km and $\alpha = 0.18$ dB/km, the values used in the experiment. When the pump power was about 125 mW, the 1.8-dB Raman gain compensated the fiber losses and the output pulse was nearly identical with the input pulse.

A 1988 experiment transmitted solitons over 4000 km using the Raman-amplification scheme [3]. This experiment used a 42-km fiber loop whose loss was exactly compensated by injecting the CW pump light from a 1.46-μm color-center laser. The solitons were allowed to circulate many times along the fiber loop, and their width was monitored after each roundtrip. The 55-ps solitons could be circulated along the loop up to 96 times without a significant increase in their pulse width, indicating soliton recovery over 4000 km. The distance could be increased to 6000 km with further optimization. This experiment was the first to demonstrate that solitons could, in principle, be transmitted over transoceanic distances. The main drawback was that Raman amplification required pump lasers emitting more than 500 mW of CW power near 1.46 μm. It was not possible to obtain such high powers from semiconductor lasers in 1988, and the color-center lasers used in the experiment were too bulky to be useful for practical lightwave systems.

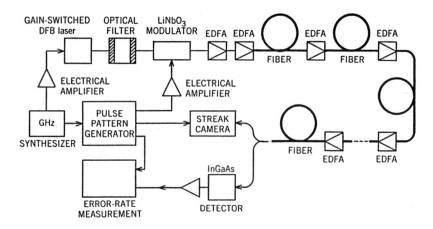

Figure 3.9: Setup used for soliton transmission in a 1990 experiment. Two EDFAs after the LiNbO$_3$ modulator boost pulse peak power to the level of fundamental solitons. (After Ref. [28]; ©1990 IEEE.)

The situation changed with the advent of erbium-doped fiber amplifiers (EDFAs) around 1989, when several experiments used them for loss-managed soliton systems. These experiments can be divided into two categories, depending on whether a linear fiber link or a recirculating fiber loop is used for the experiment. The experiments using fiber link are more realistic because they mimic the actual field conditions. Several 1990 experiments demonstrated soliton transmission over fiber lengths of ~100 km at bit rates of up to 5 Gb/s [28]–[30]. Figure 3.9 shows one such experimental setup in which a gain-switched laser is used for generating input pulses. The pulse train is filtered to reduce the frequency chirp and passed through a LiNbO$_3$ modulator to impose the RZ format on it. The resulting coded bit stream of solitons is transmitted through several fiber sections, and losses of each section are compensated by using an EDFA. The amplifier spacing is chosen to satisfy the criterion $L_A \ll L_D$ and is typically in the range 25–40 km. In a 1991 experiment, solitons were transmitted over 1000 km at 10 Gb/s [31]. The 45-ps-wide solitons permitted an amplifier spacing of 50 km in the average-soliton regime.

Since 1991, most soliton transmission experiments have used a recirculating fiber-loop configuration because of cost considerations. A bit stream of solitons is launched into the loop and forced to circulate many times using optical switches. The quality of the signal is monitored after each round trip to ensure that the solitons maintain their width during transmission. In a 1991 experiment, 2.5-Gb/s solitons were transmitted over 12,000 km by using a 75-km fiber loop containing three EDFAs, spaced 25 km apart [32]. In this experiment, the bit rate–distance product of $BL = 30$ (Tb/s)-km was limited mainly by the timing jitter induced by EDFAs. The use of amplifiers degrades the signal-to-noise ratio (SNR) and shifts the position of solitons in a random fashion. This issue is discussed in Section 3.5.

Because of the problems associated with the lumped amplifiers, several schemes were studied for reducing the timing jitter and improving the performance of soliton systems. Even the technique of Raman amplification was revived in 1999 and has become quite common for both the soliton and nonsoliton systems. Its revival was possible because of the technological advances in the fields of semiconductor and fiber lasers, both of which can provide power levels in excess of 500 mW. The use of dispersion management also helps in reducing the timing jitter. We turn to dispersion-managed solitons next.

3.4 Dispersion-Managed Solitons

Dispersion management is employed commonly for modern wavelength-division multiplexed (WDM) systems. It turns out that soliton systems benefit considerably if the GVD parameter β_2 varies along the fiber length. This section is devoted to such dispersion-managed solitons. We first consider dispersion-decreasing fibers and then focus on dispersion maps that consist of multiple sections of constant-dispersion fibers.

3.4.1 Dispersion-Decreasing Fibers

An interesting scheme proposed in 1987 relaxes completely the restriction $L_A \ll L_D$, imposed normally on loss-managed solitons, by decreasing the GVD along the fiber length [33]. Such fibers are called *dispersion-decreasing* fibers (DDFs) and are designed such that the decreasing GVD counteracts the reduced SPM experienced by solitons weakened from fiber losses.

Since dispersion management is used in combination with loss management, soliton evolution in a DDF is governed by Eq. (3.3.3) except that the second-derivative term has a new parameter, d, that is a function of z because of GVD variations along the fiber length. The modified NLS equation takes the form

$$i\frac{\partial v}{\partial z} + \frac{d(z)}{2}\frac{\partial^2 v}{\partial \tau^2} + p(z)|v|^2 v = 0, \qquad (3.4.1)$$

where $v = u/\sqrt{p}$, $d(z) = \beta_2(z)/\beta_2(0)$, and $p(z)$ takes into account peak-power variations introduced by loss management. The distance z is normalized to the dispersion length, $L_D = T_0^2/|\beta_2(0)|$, defined using the GVD value at the fiber input.

Because of the z dependence of the second and third terms, Eq. (3.4.1) is not a standard NLS equation. However, it can be reduced to one if we introduce a new propagation variable as

$$z' = \int_0^z d(z)\,dz. \qquad (3.4.2)$$

This transformation renormalizes the distance scale to the local value of GVD. In terms of z', Eq. (3.4.1) becomes

$$i\frac{\partial v}{\partial z'} + \frac{1}{2}\frac{\partial^2 v}{\partial \tau^2} + \frac{p(z)}{d(z)}|v|^2 v = 0. \qquad (3.4.3)$$

If the GVD profile is chosen so that $d(z) = p(z) \equiv \exp(-\Gamma z)$, Eq. (3.4.3) reduces the standard NLS equation obtained in the absence of fiber losses. As a result, fiber losses have no effect on a soliton, in spite of its reduced energy, when DDFs are used. Lumped amplifiers can be placed at any distance and are not limited by the condition $L_A \ll L_D$.

The preceding analysis shows that fundamental solitons can be maintained in a lossy fiber provided its GVD decreases exponentially as

$$|\beta_2(z)| = |\beta_2(0)| \exp(-\alpha z). \tag{3.4.4}$$

This result can be understood qualitatively by noting that the soliton peak power P_0 decreases exponentially in a lossy fiber in exactly the same fashion. It is easy to deduce from Eq. (3.1.3) that the requirement $N = 1$ can be maintained, in spite of power losses, if both $|\beta_2|$ and γ decrease exponentially at the same rate. The fundamental soliton then maintains its shape and width even in a lossy fiber.

Fibers with a nearly exponential GVD profile have been fabricated [34]. A practical technique for making such DDFs consists of reducing the core diameter along the fiber length in a controlled manner during the fiber-drawing process. Variations in the fiber diameter change the waveguide contribution to β_2 and reduce its magnitude. Typically, GVD can be varied by a factor of 10 over a length of 20–40 km. The accuracy realized by the use of this technique is estimated to be better than 0.1 ps^2/km [35]. Propagation of solitons in DDFs has been demonstrated in several experiments [35]–[37]. In a 40-km DDF, solitons preserved their width and shape in spite of energy losses of more than 8 dB [36]. In a recirculating loop made using DDFs, a 6.5-ps soliton train at 10 Gb/s could be transmitted over 300 km [37].

Fibers with continuously varying GVD are not readily available. As an alternative, the exponential GVD profile of a DDF can be approximated with a staircase profile by splicing together several constant-dispersion fibers with different β_2 values. This approach was studied during the 1990s, and it was found that most of the benefits of DDFs can be realized using as few as four fiber segments [38]–[42]. How should one select the length and the GVD of each fiber used for emulating a DDF? The answer is not obvious, and several methods have been proposed. In one approach, power deviations are minimized in each section [38]. In another approach, fibers of different GVD values D_m and different lengths L_m are chosen such that the product $D_m L_m$ is the same for each section. In a third approach, D_m and L_m are selected to minimize shading of dispersive waves [39].

3.4.2   Periodic Dispersion Maps

A disadvantage of the DDF is that the average dispersion along the link is often relatively large. Generally speaking, operation of a soliton in the region of low average GVD improves system performance. Dispersion maps consisting of alternating-GVD fibers are attractive because their use lowers the average GVD of the entire link while keeping the GVD of each section large enough that the effects of four-wave mixing (FWM) and third-order dispersion (TOD) remain negligible.

The use of dispersion management forces each soliton to propagate in the normal-dispersion regime of a fiber during each map period. At first sight, such a scheme should not even work, because the normal-GVD fibers do not support bright solitons and lead to considerable broadening and chirping of the pulse. So why should solitons survive in a dispersion-managed fiber link? Already in 1995 the advantages of using periodic dispersion maps were noticed in both theoretical and experimental studies [43, 44]. An intense theoretical effort devoted to this issue since 1996 has provided considerable insight, with a few surprises [45]–[69]. Physically speaking, if the map period is a fraction of the nonlinear length, the nonlinear effects are relatively small, and the pulse evolves in a linear fashion over one map period. On a longer-length scale, solitons can still form if the SPM effects are balanced by the average dispersion. As a result, solitons can survive in an average sense, even though not only the peak power but also the width and shape of such solitons oscillate periodically. This section describes the properties of dispersion-managed (DM) solitons and the advantages offered by them.

Consider a simple dispersion map consisting of two fibers with positive and negative values of the GVD parameter β_2. Soliton evolution is still governed by Eq. (3.4.1), used earlier for DDFs. However, because the GVD parameter changes its sign along the fiber, it is better to use the physical units in this section and write Eq. (3.4.1) as

$$i\frac{\partial B}{\partial Z} - \frac{\beta_2(Z)}{2}\frac{\partial^2 B}{\partial t^2} + \gamma p(Z)|B|^2 B = 0, \qquad (3.4.5)$$

where $B(Z,t) = A(Z,t)/\sqrt{p}$, Z is the physical distance along the fiber and $p(Z)$ is the solution of Eq. (3.3.8). The GVD parameter takes values β_{2a} and β_{2n} in the anomalous and normal sections of lengths l_a and l_n, respectively. The map period $L_{\mathrm{map}} = l_a + l_n$ can be different from the amplifier spacing L_A. As is evident, the properties of DM solitons will depend on several map parameters even when only two types of fibers are used in each map period.

Equation (3.4.5) can be solved numerically using the split-step Fourier method. Numerical simulations show that a nearly periodic solution can often be found by adjusting input pulse parameters (width, chirp, and peak power) even though these parameters vary considerably in each map period. The shape of such DM solitons is typically closer to a Gaussian profile rather than the "sech" shape associated with standard solitons [45]–[49].

Numerical solutions, although essential, do not lead to much physical insight. Several techniques have been used to solve the NLS equation (3.4.5) approximately. A common approach makes use of the variational method [50]–[54] that was first used by Anderson in the context of the NLS equation as early as 1983 [70]. Another approach expands $B(Z,t)$ in terms of a complete set of the Hermite–Gauss functions that are solutions of the linear problem [55, 56]. A third approach solves an integral equation, derived in the spectral domain using perturbation theory [46], [58]–[62].

To simplify the following discussion, we focus on the variational method, which makes use of the fact that Eq. (3.4.5) can be derived from the Euler–Lagrange equation using the following Lagrangian density:

$$\mathcal{L} = \frac{i}{2}\left(B\frac{\partial B^*}{\partial Z} - B^*\frac{\partial B}{\partial Z}\right) + \frac{1}{2}\left[\gamma p(Z)|B|^4 - \beta_2(Z)\left|\frac{\partial B}{\partial t}\right|^2\right]. \qquad (3.4.6)$$

Because the shape of the DM soliton is close to a Gaussian pulse in numerical simulations, the soliton is assumed to evolve as

$$B(Z,t) = a\exp[-(1+iC)t^2/2T^2 + i\phi], \qquad (3.4.7)$$

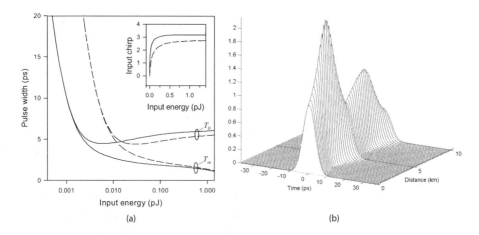

Figure 3.10: (a) Changes in T_0 (upper curve) and T_m (lower curve) with input pulse energy E_0 for $\alpha = 0$ (solid lines) and 0.25 dB/km (dashed lines). The inset shows the input chirp C_0 in the two cases. (b) Evolution of the DM soliton over one map period for $E_0 = 0.1$ pJ and $L_A = 80$ km.

where a is the amplitude, T is the width, C is the chirp, and ϕ is the phase of the soliton. All four parameters vary with Z because of perturbations produced by periodic variations of $\beta_2(Z)$ and $p(Z)$.

Using the variational method, we can obtain four ordinary differential equations for the four soliton parameters. The amplitude equation can be eliminated because $a^2T = a_0^2T_0 = E_0/\sqrt{\pi}$ is independent of Z and is related to the input pulse energy E_0. The phase equation can also be dropped since T and C do not depend on ϕ. The DM soliton then corresponds to a periodic solution of the following two equations for the pulse width T and chirp C:

$$\frac{dT}{dZ} = \beta_2(Z)\frac{C}{T}, \tag{3.4.8}$$

$$\frac{dC}{dZ} = \frac{\gamma E_0 p(Z)}{\sqrt{2\pi}T} + \frac{\beta_2}{T^2}(1+C^2). \tag{3.4.9}$$

These equations should be solved with the periodic boundary conditions

$$T_0 \equiv T(0) = T(L_A), \qquad C_0 \equiv C(0) = C(L_A) \tag{3.4.10}$$

to ensure that the soliton recovers its initial state after each amplifier. The periodic boundary conditions fix the values of the initial width T_0 and the chirp C_0 at $Z = 0$, for which a soliton can propagate in a periodic fashion for a given value of the pulse energy E_0. A new feature of the DM solitons is that the input pulse width depends on the dispersion map and cannot be chosen arbitrarily. In fact, T_0 cannot be below a critical value that is set by the map itself.

Figure 3.10 shows how the pulse width T_0 and the chirp C_0 of allowed periodic solutions vary with input pulse energy for a specific dispersion map. The minimum value T_m of the pulse width occurring in the middle of the anomalous-GVD section of the map is also shown. The map is suitable for 40-Gb/s systems and consists of alternating fibers with GVD of -4 and 4 ps^2/km and lengths $l_a \approx l_n = 5$ km such that the average GVD is -0.01 ps^2/km. The solid lines show the case of ideal distributed amplification, for which $p(Z) = 1$ in Eq. (3.4.9). The lumped-amplification case is shown by the dashed lines in Figure 3.10, assuming 80-km amplifier spacing and 0.25 dB/km losses in each fiber section.

Several conclusions can be drawn from Figure 3.10. First, both T_0 and T_m decrease rapidly as pulse energy is increased. Second, T_0 attains its minimum value at a certain pulse energy E_c while T_m keeps decreasing slowly. Third, T_0 and T_m differ from each other considerably for $E_0 > E_c$. This behavior indicates that the pulse width changes considerably in each fiber section when this regime is approached. An example of pulse breathing is shown in Figure 3.10(b) for $E_0 = 0.1$ pJ in the case of lumped amplification. The input chirp C_0 is relatively large ($C_0 \approx 1.8$) in this case. The most important feature of Figure 3.10 is the existence of a minimum value of T_0 for a specific value of the pulse energy. The input chirp $C_0 = 1$ at that point. It is interesting to note that the minimum value of T_0 does not depend much on fiber losses and is about the same for the solid and dashed curves, although the value of E_c is much larger in the lumped amplification case because of fiber losses.

As seen from Figure 3.10, both the pulse width and the peak power of DM solitons vary considerably within each map period. Figure 3.11(a) shows the width and chirp variations over one map period for the DM soliton of Figure 3.10(b). The pulse width varies by more than a factor of 2 and becomes minimum nearly in the middle of each fiber section, where frequency chirp vanishes. The shortest pulse occurs in the middle of the anomalous-GVD section. For comparison, Figure 3.11(b) shows the width and chirp variations for a DM soliton whose input energy is close to E_c where the input pulse is shortest. Breathing of the pulse is reduced considerably together with the range of chirp variations. In both cases, the DM soliton is quite different from a standard fundamental soliton because it does not maintain its shape, width, or peak power. Nevertheless, its parameters repeat from period to period at any location within the map. For this reason, DM solitons can be used for optical communications, in spite of oscillations in the pulse width. Moreover, such solitons perform better from a system standpoint.

3.4.3 Design Issues

Figures 3.10 and 3.11 show that Eqs. (3.4.8)–(3.4.10) permit periodic propagation of many different DM solitons in the same map by choosing different values of E_0, T_0, and C_0. How should one choose among these solutions when designing a soliton system? Pulse energies much smaller than E_c (corresponding to the minimum value of T_0) should be avoided because a low average power would then lead to rapid degradation of SNR as amplifier noise builds up with propagation. On the other hand, when $E_0 \gg E_c$, large variations in the pulse width in each fiber section would enhance the effects of soliton interaction if two neighboring solitons begin to overlap. Thus, the region near $E_0 = E_c$ is most suited for designing DM soliton systems. Numerical solutions of Eq. (3.4.5) confirm this conclusion.

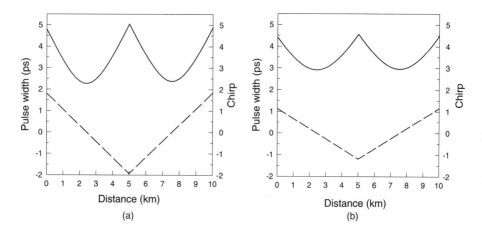

Figure 3.11: (a) Variations of pulse width and chirp over one map period for DM solitons with the input energy (a) $E_0 \gg E_c = 0.1$ pJ and (b) E_0 close to E_c.

The 40-Gb/s system design shown in Figures 3.10 and 3.11 was possible only because the map period L_{map} was chosen to be much smaller than the amplifier spacing of 80 km, a configuration referred to as the *dense* dispersion management. When L_{map} is increased to 80 km using $l_a \approx l_b = 40$ km while keeping the same value of average dispersion, the minimum pulse width supported by the map increases by a factor of three. The bit rate is then limited to about 20 Gb/s. In general, the required map period becomes shorter as the bit rate increases.

It is possible to find the values of T_0 and T_m by solving the variational equations (3.4.8)–(3.4.10) approximately. Equation (3.4.8) can be integrated to relate T and C as

$$T^2(Z) = T_0^2 + 2 \int_0^Z \beta_2(Z) C(Z) \, dZ. \tag{3.4.11}$$

The chirp equation cannot be integrated, but the numerical solutions show that $C(Z)$ varies almost linearly in each fiber section. As seen in Figure 3.11, $C(Z)$ changes from C_0 to $-C_0$ in the first section and then back to C_0 in the second section. Noting that the ratio $(1 + C^2)/T^2$ is related to the spectral width that changes little over one map period when the nonlinear length is much larger than the local dispersion length, we average it over one map period and obtain the following relation between T_0 and C_0:

$$T_0 = T_{\text{map}} \sqrt{\frac{1 + C_0^2}{|C_0|}}, \qquad T_{\text{map}} = \left(\frac{|\beta_{2n} \beta_{2a} l_n l_a|}{\beta_{2n} l_n - \beta_{2a} l_a} \right)^{1/2}, \tag{3.4.12}$$

where T_{map} is a parameter with dimensions of time involving only the four map parameters. It provides a time scale associated with an arbitrary dispersion map in the sense that the stable periodic solutions supported by it have input pulse widths that are close to T_{map} (within a factor of 2 or so). The minimum value of T_0 occurs for $|C_0| = 1$ and is given by $T_0^{\min} = \sqrt{2} T_{\text{map}}$.

Equation (3.4.12) can also be used to find the shortest pulse within the map. Recalling that the shortest pulse occurs at the point at which the propagating pulse becomes unchirped, $T_m = T_0/(1 + C_0^2)^{1/2} = T_{\text{map}}/\sqrt{|C_0|}$. When the input pulse corresponds to its minimum value ($C_0 = 1$), T_m is exactly equal to T_{map}. The optimum value of the pulse stretching factor is equal to $\sqrt{2}$ under such conditions. These conclusions are in agreement with the numerical results shown in Figure 3.11 for a specific map for which $T_{\text{map}} \approx 3.16$ ps. If dense dispersion management is not used for this map and L_{map} equals $L_A = 80$ km, this value of T_{map} increases to 9 ps. Since the FWHM of input pulses then exceeds 21 ps, such a map is unsuitable for 40-Gb/s soliton systems. In general, the required map period becomes shorter and shorter as the bit rate increases, as is evident from the definition of T_{map} in Eq. (3.4.12).

It is useful to look for other combinations of the four map parameters that may play an important role in designing a DM soliton system. Two parameters that help for this purpose are defined as [51]

$$\bar{\beta}_2 = \frac{\beta_{2n}l_n + \beta_{2a}l_a}{l_n + l_a}, \qquad S_m = \frac{\beta_{2n}l_n - \beta_{2a}l_a}{T_{\text{FWHM}}^2}, \qquad (3.4.13)$$

where $T_{\text{FWHM}} \approx 1.665T_m$ is the FWHM at the location where pulse width is minimum in the anomalous-GVD section. Physically, $\bar{\beta}_2$ represents the average GVD of the entire link, while the map strength S_m is a measure of how much GVD varies between two fibers in each map period. The solutions of Eqs. (3.4.8)–(3.4.10) as a function of map strength S for different values of $\bar{\beta}_2$ reveal the surprising feature that DM solitons can exist even when the average GVD is normal, provided the map strength exceeds a critical value S_{cr} [63]–[68]. Moreover, when $S_m > S_{\text{cr}}$ and $\bar{\beta}_2 > 0$, a periodic solution can exist for two different values of the input pulse energy. Numerical solutions of Eqs. (3.4.1) confirm these predictions, but the critical value of the map strength is found to be only 3.9 instead of 4.8 obtained from the variational equations [51].

The existence of DM solitons in maps with normal average GVD is quite intriguing because one can envisage dispersion maps in which a soliton propagates in the normal-GVD regime most of the time. An example is provided by the dispersion map in which a short section of standard fiber ($\beta_{2a} \approx -20$ ps^2/km) is used with a long section of dispersion-shifted fiber ($\beta_{2n} \approx 1$ ps^2/km) such that $\bar{\beta}_2$ is close to zero but positive. The formation of DM solitons under such conditions can be understood by noting that when S_m exceeds 4, input energy of a pulse becomes large enough that its spectral width is considerably larger in the anomalous-GVD section compared with the normal-GVD section [53, 67, 68]. Noting that the phase shift imposed on each spectral component varies as $\beta_2\omega^2$ locally, one can define an effective value of the average GVD as [68]

$$\bar{\beta}_2^{\text{eff}} = \langle\beta_2\Omega^2\rangle/\langle\Omega^2\rangle, \qquad (3.4.14)$$

where Ω is the local value of the spectral width and the angle brackets indicate average over the map period. If $\bar{\beta}_2^{\text{eff}}$ is negative, the DM soliton can exist even if $\bar{\beta}_2$ is positive.

For map strengths below a critical value (about 3.9 numerically), the average GVD is anomalous for DM solitons. In that case, one is tempted to compare them with standard solitons forming in a uniform-GVD fiber link with $\beta_2 = \bar{\beta}_2$. For relatively small values of S_m, variations in the pulse width and chirp are small enough that one can ignore them. The main difference between the average-GVD and DM solitons then stems from the higher peak power required to sustain DM solitons. The energy enhancement factor for DM solitons is defined as [45]

$$f_{DM} = E_0^{DM}/E_0^{av} \qquad (3.4.15)$$

and can exceed 10, depending on the system design. The larger energy of DM solitons benefits a soliton system in several ways. Among other things, it improves the SNR and decreases the timing jitter.

Dispersion-management schemes were used for solitons as early as 1992, although they were referred to by names such as *partial soliton communication* and *dispersion allocation* [71]. In the simplest form of dispersion management, a relatively short segment of dispersion-compensating fiber (DCF) is added periodically to the transmission fiber, resulting in dispersion maps similar to those used for nonsoliton systems. It was found in a 1995 experiment that the use of DCFs reduced the timing jitter considerably [44]. In fact, in this 20-Gb/s experiment, the timing jitter became low enough when the average dispersion was reduced to a value near -0.025 ps^2/km that the 20-Gb/s signal could be transmitted over transoceanic distances.

Since 1996, a large number of experiments have shown the benefits of DM solitons for lightwave systems [72]–[81]. In one experiment, the use of a periodic dispersion map enabled transmission of a 20-Gb/s soliton bit stream over 5520 km of a fiber link containing amplifiers at 40-km intervals [72]. In another 20-Gb/s experiment [73], solitons could be transmitted over 9000 km without using any in-line optical filters since the periodic use of DCFs reduced timing jitter by more than a factor of 3. A 1997 experiment focused on transmission of DM solitons using dispersion maps such that solitons propagated most of the time in the normal-GVD regime [74]. This 10-Gb/s experiment transmitted signals over 28 Mm using a recirculating fiber loop consisting of 100 km of normal-GVD fiber and 8-km of anomalous-GVD fiber such that the average GVD was anomalous (about -0.1 ps^2/km). Periodic variations in the pulse width were also observed in such a fiber loop [75]. In a later experiment, the loop was modified to yield the average-GVD value of zero or slightly positive [76]. Stable transmission of 10-Gb/s solitons over 28 Mm was still observed. In all cases, experimental results were in excellent agreement with numerical simulations [77].

An important application of dispersion management consists of upgrading the existing terrestrial networks designed with standard fibers [78]–[81]. A 1997 experiment used fiber gratings for dispersion compensation and realized 10-Gb/s soliton transmission over 1000 km. Longer transmission distances were realized using a recirculating fiber loop [79] consisting of 102 km of standard fiber with anomalous GVD ($\beta_2 \approx -21$ ps^2/km) and 17.3 km of DCF with normal GVD ($\beta_2 \approx 160$ ps^2/km). The map strength S was quite large in this experiment when 30-ps (FWHM) pulses were launched into the loop. By 1999, 10-Gb/s DM solitons could be transmitted over 16 Mm of standard fiber when soliton interactions were minimized by choosing the location of amplifiers appropriately [80].

3.5 Perturbation of Solitons

The advent of optical amplifiers fueled the development of soliton-based systems. However, with their use came the limitations imposed by the amplifier noise. In this section, we first discuss the method used commonly to analyze the effect of small perturbations on solitons and then apply it to study the impact of fiber losses and amplifier noise. The impact of higher-order dispersive and nonlinear effects is considered in Section 3.6. The fiber is assumed to have uniform dispersion so that the solitons have a "sech" shape and maintain their shape in the absence of perturbations.

3.5.1 Perturbation Methods

Consider the perturbed NLS equation written as

$$i\frac{\partial u}{\partial z} + \frac{1}{2}\frac{\partial^2 u}{\partial \tau^2} + |u|^2 u = i\varepsilon(u), \tag{3.5.1}$$

where $\varepsilon(u)$ is a small perturbation that can depend on u, u^*, and their derivatives. In the absence of perturbation ($\varepsilon = 0$), the soliton solution of the NLS equation is known. The question then becomes what happens to the soliton when $\varepsilon \neq 0$. Several perturbation techniques have been developed for answering this question [82]–[89]. They all assume that the functional form of the soliton remains intact in the presence of a small perturbation, but the soliton parameters change with z as the soliton propagates down the fiber. The most general form of the perturbed soliton can be written as

$$u(z, \tau) = \eta(z)\text{sech}[\eta(z)(\tau - q(z))]\exp[i\phi(z) - i\delta(z)\tau]. \tag{3.5.2}$$

The z dependence of η, δ, q, and ϕ is yet to be determined. In the absence of perturbation ($\varepsilon = 0$), η and δ are constants but $q(z)$ and $\phi(z)$ are obtained by solving the simple ordinary differential equations

$$\frac{dq}{dz} = -\delta, \qquad \frac{d\phi}{dz} = \frac{1}{2}(\eta^2 - \delta^2). \tag{3.5.3}$$

The perturbation techniques developed for solitons include the adiabatic perturbation method, the perturbed inverse scattering method, the Lie-transform method, and the variational method. All of them attempt to obtain a set of the following four ordinary differential equations for the four soliton parameters [4]:

$$\frac{d\eta}{dz} = \text{Re}\int_{-\infty}^{\infty} \varepsilon(u)u^*(\tau)\,d\tau, \tag{3.5.4}$$

$$\frac{d\delta}{dz} = -\text{Im}\int_{-\infty}^{\infty} \varepsilon(u)\tanh[\eta(\tau - q)]u^*(\tau)\,d\tau, \tag{3.5.5}$$

$$\frac{dq}{dz} = -\delta + \frac{1}{\eta^2}\,\text{Re}\int_{-\infty}^{\infty} \varepsilon(u)(\tau - q)u^*(\tau)\,d\tau, \tag{3.5.6}$$

$$\frac{d\phi}{dz} = \text{Im}\int_{-\infty}^{\infty} \varepsilon(u)\{1/\eta - (\tau - q)\tanh[\eta(\tau - q)]\}u^*(\tau)\,d\tau$$

$$+ \frac{1}{2}(\eta^2 - \delta^2) + q\frac{d\delta}{dz}, \tag{3.5.7}$$

where Re and Im stand for the real and imaginary parts, respectively. This set of four equations can also be obtained by using adiabatic perturbation theory or perturbation theory based on the inverse scattering method [82]–[89].

As a simple example, consider the effect of fiber losses on solitons assuming that no amplifiers are used. Fiber losses are detrimental simply because they reduce the peak power of solitons, disturbing the balance between the dispersive and nonlinear effects. As a result, the soliton width begins to increase with propagation. Mathematically, fiber losses are accounted for by adding a loss term to the NLS equation so that it becomes

$$i\frac{\partial u}{\partial z} + \frac{1}{2}\frac{\partial^2 u}{\partial \tau^2} + |u|^2 u = -\frac{i}{2}\Gamma u, \tag{3.5.8}$$

where $\Gamma = \alpha L_D = \alpha T_0^2/|\beta_2|$ and α is the loss parameter.

If $\Gamma \ll 1$, the loss term can be treated as a weak perturbation. Using $\varepsilon(u) = -\Gamma u/2$ in Eqs. (3.5.4)–(3.5.7) and performing the integrations, we find that only soliton amplitude η and phase ϕ are affected by fiber losses and vary along the fiber length as [17]

$$\eta(z) = \exp(-\Gamma z), \qquad \phi(z) = \phi(0) + [1 - \exp(-2\Gamma z)]/(4\Gamma), \tag{3.5.9}$$

where we assumed that $\eta(0) = 1$, $\delta(0) = 0$, and $q(0) = 0$. Both δ and q remain zero along the fiber. Recalling that the amplitude and width of a soliton are related inversely, a decrease in soliton amplitude leads to broadening of the soliton. Indeed, if we write $\eta(\tau - q)$ in Eq. (3.5.2) as T/T_1 and use $\tau = T/T_0$, then T_1 increases along the fiber exponentially as

$$T_1(z) = T_0 \exp(\Gamma z) \equiv T_0 \exp(\alpha Z). \tag{3.5.10}$$

This equation remains valid as long as $\Gamma \ll 1$.

3.5.2 Amplifier Noise

The use of optical amplifiers affects the evolution of solitons considerably. The reason is that amplifiers, although needed to restore the soliton energy, also add noise originating from amplified spontaneous emission (ASE). To study the impact of ASE on soliton evolution, we consider how the four soliton parameters in the NLS solution (3.5.2) are affected by amplifier noise. The effect of ASE is to change randomly the values of η, q, δ, and ϕ at the output of each amplifier. Variances of such fluctuations for the four soliton parameters can be calculated by treating ASE as a perturbation.

In the case of lumped amplification, solitons are perturbed by ASE in a discrete fashion at the location of the amplifiers. However, since the amplifier spacing satisfies $z_A \ll 1$, we assume that noise is distributed all along the fiber length. Such an approach is useful since it can be applied to the case of distributed amplification as well with only minor changes. Assuming fiber losses are fully compensated by amplifiers, adding noise, the NLS equation can be written as

$$i\frac{\partial u}{\partial z} + \frac{1}{2}\frac{\partial^2 u}{\partial \tau^2} + |u|^2 u = in(z,\tau), \tag{3.5.11}$$

where $n(z, \tau)$ is a Markovian stochastic process with Gaussian statistics with $\langle n(z, \tau) \rangle = 0$ and the correlation function

$$\langle n(z, \tau)n^*(z', \tau') \rangle = S_n \delta(\tau - \tau')\delta(z - z'), \tag{3.5.12}$$

where S_n is related to the ASE spectral density $S_{sp} = n_{sp}h\nu_0(G - 1)$ and n_{sp} is the spontaneous-emission factor (typical values in the range 1.3–1.6). In soliton units,

$$S_n = n_{sp}h\nu_0(G - 1)\frac{\gamma L_D^2}{L_A T_0} = \frac{F_n F_G}{N_{ph}z_A}, \tag{3.5.13}$$

where $F_n = 2n_{sp}$ is the amplifier noise figure and $F_G = (G - 1)^2/(G \ln G)$ is related to the amplifier gain $G = \exp(\alpha L_A)$. In Eq. (3.5.13), $N_{ph} = 2P_0 T_0/h\nu_0$ is the average number of photons in the pulse propagating as a fundamental soliton.

Using $\varepsilon(u) = n(z, \tau)$ in Eqs. (3.5.4)–(3.5.7), we can find how the four soliton parameters fluctuate because of amplifier noise. It is useful to define four new noise variables using [7]

$$n_i(z) = \mathrm{Re} \int_{-\infty}^{\infty} n(z, \tau)F_i(\tau) \, d\tau, \tag{3.5.14}$$

where $i = \eta, \delta, q$, and ϕ. Their correlation functions are calculated using Eqs. (3.5.4)–(3.5.7), and the result is

$$\langle n_i(z)n_j(z') \rangle = S_i \delta_{ij}\delta(z - z'). \tag{3.5.15}$$

Assuming $\langle \eta \rangle = 1$, the spectral densities for $i = \eta, \delta, q$, and ϕ are given as [7]

$$S_\eta = S_n, \quad S_\delta = \frac{S_n}{3}, \quad S_q = \frac{\pi^2}{12}S_n, \quad S_\phi = (1 + \pi^2/12)\frac{S_n}{3}. \tag{3.5.16}$$

As a simple example, we calculate the amplitude and frequency fluctuations induced by ASE. Integrating Eqs. (3.5.4) and (3.5.5), we find

$$\eta(z) = 1 + \int_0^z n_\eta(z) \, dz, \qquad \delta(z) = \int_0^z n_\delta(z) \, dz, \tag{3.5.17}$$

where $\eta(0) = 1$ and $\delta(0) = 0$ are assumed for the fundamental soliton launched at $z = 0$. The variance of fluctuations is found to be

$$\sigma_\eta^2 = S_n z, \qquad \sigma_\delta^2 = S_n z/3. \tag{3.5.18}$$

This result shows that variances of both amplitude and frequency fluctuations increase linearly along the fiber link because of the cumulative effects of ASE. Amplitude fluctuations degrade the SNR of the soliton bit stream. The SNR degradation, although undesirable, is not the most limiting factor. In fact, frequency fluctuations affect system performance much more drastically by inducing the timing jitter. We turn to this issue next.

3.5.3   Timing Jitter

A soliton communication system can operate reliably only if all solitons arrive at the receiver within their assigned bit slot. Several physical mechanisms induce deviations in the soliton position from its original location at the bit center. Among them, ASE-induced timing jitter is often dominant in practice. The origin of such jitter can be understood by noting that a change in the soliton frequency by δ affects the speed at which the soliton propagates through the fiber. If δ fluctuates because of amplifier noise, soliton transit time through the fiber link also becomes random. ASE-induced fluctuations in the arrival time of a soliton are referred to as the *Gordon–Haus timing jitter* [90]–[93].

Fluctuations in the soliton position are obtained by integrating Eq. (3.5.6) and calculating the variance of $q(z)$. Following the procedure outlined previously, the variance is found to be

$$\sigma_q^2 = \frac{1}{9}S_n z^3 + \frac{\pi^2}{12}S_n z, \tag{3.5.19}$$

where both δ and q were assumed to be zero at $z = 0$. The second term shows the direct impact of ASE on the soliton position. The first term is due to jitter induced by frequency fluctuations and dominates in practice because of its cubic dependence on the propagation distance. Keeping only the first term, the timing jitter is approximately given by

$$\sigma_q^2 \approx \frac{S_n}{9}z^3 = \frac{F_n F_G}{9N_{\mathrm{ph}}z_A}z^3, \tag{3.5.20}$$

where we used Eqs. (3.5.13) and (3.5.18). Using $\sigma_q = \sigma_t/T_0$, $z = L_T/L_D$, $N_{\mathrm{ph}} = 2P_0 T_0/h\nu_0$, $P_0 = (\gamma L_D)^{-1}$, $F_n = 2n_{\mathrm{sp}}$, and $F_G = (G-1)/f_{\mathrm{LM}}$, the timing-jitter variance in physical units can be written as [91]

$$\sigma_t^2 = \frac{n_{\mathrm{sp}}h\nu_0\gamma|\beta_2|(G-1)L_T^3}{9T_0 L_A f_{\mathrm{LM}}}. \tag{3.5.21}$$

Since a soliton should arrive within its allocated bit slot for its correct identification at the receiver, timing jitter should be a small fraction of the bit slot T_B. This requirement can be written as $\sigma_t/T_B < f_b$, where f_b is the fraction of the bit slot by which a soliton can move without affecting system performance adversely. Using this condition and introducing the bit rate through $B = 1/T_B = (2q_0 T_0)^{-1}$, Eq. (3.5.21) can be written as a design rule:

$$BL_T < \left[\frac{9f_b^2 L_A f_{\mathrm{LM}}}{h\nu_0 F_n (G-1)q_0\gamma|\beta_2|}\right]^{1/3}. \tag{3.5.22}$$

The tolerable value of f_b depends on the acceptable bit-error rate and on details of receiver design; typically, $f_b < 0.1$. To see how amplifier noise limits the total transmission distance, consider a specific soliton communication system operating at 1.55 μm. Using typical parameter values, $q_0 = 5$, $\gamma = 3$ W^{-1}/km, $\beta_2 = -1$ ps^2/km, $F_n = 3$, $L_A = 50$ km, $G = 10$, and $f_b = 0.1$, BL_T must be below 80 (Tb/s)-km. For a 40-Gb/s system, the transmission distance is limited to 2000 km. Similar constraints apply in the case of systems making use of dispersion management [92] and distribution amplification [93]. It turns out that both of these techniques help to reduce the timing jitter. In a 2001 study, the probability density function of the timing jitter has also been derived [94].

3.6 Higher-Order Effects

The properties of optical solitons considered so far are based on the NLS equation (3.1.1). When input pulses are shorter than 4–5 ps, it is necessary to include the higher-order nonlinear and dispersive effects. In terms of the soliton units introduced in Section 5.2, Eq. (2.3.40) takes the form

$$i\frac{\partial u}{\partial z} + \frac{1}{2}\frac{\partial^2 u}{\partial \tau^2} + |u|^2 u = i\delta_3 \frac{\partial^3 u}{\partial \tau^3} - is\frac{\partial}{\partial \tau}(|u|^2 u) + \tau_R u \frac{\partial |u|^2}{\partial \tau}, \tag{3.6.1}$$

where fiber losses are assumed to be compensated through amplification. The parameters δ_3, s, and τ_R govern, respectively, the effects of third-order dispersion (TOD), self-steepening, and intrapulse Raman scattering and are defined as

$$\delta_3 = \frac{\beta_3}{6|\beta_2|T_0}, \qquad s = \frac{1}{\omega_0 T_0}, \qquad \tau_R = \frac{T_R}{T_0}. \tag{3.6.2}$$

All three parameters vary inversely with pulse width and are negligible for $T_0 \gg 1$ ps. They become appreciable for femtosecond pulses. As an example, $\delta_3 \approx 0.03$, $s \approx 0.03$, and $\tau_R \approx 0.1$ for a 50-fs pulse ($T_0 \approx 30$ fs) propagating at 1.55 μm in a standard silica fiber if we take $T_R = 3$ fs. In what follows, we consider each higher-order effect in a separate subsection.

3.6.1 Third-Order Dispersion

When optical pulses propagate relatively far from the zero-dispersion wavelength of an optical fiber, the TOD effects on solitons are small and can be treated perturbatively. To study such effects as simply as possible, let us set $s = 0$ and $\tau_R = 0$ in Eq. (3.6.1) and treat the δ_3 term as a small perturbation. Using Eqs. (3.5.4)–(3.5.7) with $\varepsilon(u) = \delta_3(\partial^3 u/\partial \tau^3)$, it is easy to show that amplitude η, frequency δ, and phase ϕ of the soliton are not affected by TOD. In contrast, the peak position q changes as [4]

$$\frac{dq}{dz} = -\delta + \delta_3 \eta^2. \tag{3.6.3}$$

For a fundamental soliton with $\eta = 1$ and $\delta = 0$, the soliton peak shifts linearly with z as $q(z) = \delta_3 z$. Physically speaking, the TOD slows down the soliton and, as a result, the soliton peak is delayed by an amount that increases linearly with distance. This TOD-induced delay is negligible in most fibers for picosecond pulses for distances as large as $z = 100$ as long as β_2 is not nearly zero.

What happens if an optical pulse propagates so close to the zero-dispersion wavelength of an optical fiber that β_2 is nearly zero? Considerable work has been done to understand propagation behavior in this regime [95]–[101]. Equation (3.6.1) cannot be used in this case because the normalization scheme used for it becomes inappropriate.

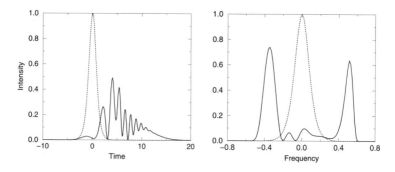

Figure 3.12: Pulse shape and spectrum at $z/L'_D = 3$ of a hyperbolic secant pulse propagating at the zero-dispersion wavelength with a peak power such that $\tilde{N} = 2$. Dotted curves show for comparison the initial profiles at the fiber input.

Normalizing the propagation distance to $L'_D = T_0^3/|\beta_3|$ through $z' = z/L'_D$, we obtain the following equation:

$$i\frac{\partial u}{\partial z'} - \text{sgn}(\beta_3)\frac{i}{6}\frac{\partial^3 u}{\partial \tau^3} + |u|^2 u = 0, \qquad (3.6.4)$$

where $u = \tilde{N}U$, with $\tilde{N}$ defined by

$$\tilde{N}^2 = \frac{L'_D}{L_{NL}} = \frac{\gamma P_0 T_0^3}{|\beta_3|}. \qquad (3.6.5)$$

Figure 3.12 shows the pulse shape and the spectrum at $z' = 3$ for $\tilde{N} = 2$ and compares them with those of the input pulse at $z' = 0$. The most striking feature is splitting of the spectrum into two well-resolved spectral peaks [95]. These peaks correspond to the outermost peaks of the SPM-broadened spectrum. Because the red-shifted peak lies in the anomalous-GVD regime, pulse energy in that spectral band can form a soliton. The energy in the other spectral band disperses away simply because that part of the pulse experiences normal GVD. It is the trailing part of the pulse that disperses away with propagation because SPM generates blue-shifted components near the trailing edge. The pulse shape in Figure 3.12 shows a long trailing edge with oscillations that continues to separate away from the leading part with increasing z'. The important point to note is that, because of SPM-induced spectral broadening, the input pulse does not really propagate at the zero-dispersion wavelength even if $\beta_2 = 0$ initially. In effect, the pulse creates its own $|\beta_2|$ through SPM. The effective value of $|\beta_2|$ is given by Eq. (4.2.7) and is larger for pulses with higher peak powers.

An interesting question is whether soliton-like solutions exist at the zero-dispersion wavelength of an optical fiber. Equation (3.6.4) does not appear to be integrable by the inverse scattering method. Numerical solutions show [97] that for $\tilde{N} > 1$, a "sech" pulse evolves over a length $z' \sim 10/\tilde{N}^2$ into a soliton that contains about half of the pulse energy. The remaining energy is carried by an oscillatory structure near the trailing edge that disperses away with propagation. These features of solitons have also been quantified by solving Eq. (3.6.4) approximately [97]–[101]. In general, solitons at the zero-dispersion wavelength require less power than those occurring in the anomalous-GVD regime. This can be seen by comparing Eqs. (3.1.3) and (3.6.5). To achieve the same values of N and $\tilde{N}$, the required power is smaller by a factor of $T_0|\beta_2/\beta_3|$ for pulses propagating at the zero-dispersion wavelength.

With the advent of the wavelength-division multiplexing (WDM) and dispersion-management techniques, special fibers have been developed in which β_3 is nearly zero over a certain wavelength range while $|\beta_2|$ remains finite. Such fibers are called *dispersion-flattened fibers*. Their use requires consideration of the effects of fourth-order dispersion on solitons. The NLS equation then takes the following form:

$$i\frac{\partial u}{\partial z} + \frac{1}{2}\frac{\partial^2 u}{\partial \tau^2} + |u|^2 u = \delta_4 \frac{\partial^4 u}{\partial \tau^4}, \tag{3.6.6}$$

where $\delta_4 = \beta_4/(24|\beta_2|T_0^2)$.

The parameter δ_4 is relatively small for $T_0 > 1$ ps, and its effect can be treated perturbatively. However, δ_4 may become large enough for ultrashort pulses that a perturbative solution is not appropriate. A shape-preserving, solitary-wave solution of Eq. (3.6.6) can be found by assuming $u(z, \tau) = V(\tau)\exp(iKz)$ and solving the resulting ordinary differential equation for $V(\tau)$. This solution is given by [102]

$$u(z, \tau) = 3b^2 \mathrm{sech}^2(b\tau)\exp(8ib^2 z/5), \tag{3.6.7}$$

where $b = (40\delta_4)^{-1/2}$. Note the sech^2-type form of the pulse amplitude rather than the usual "sech" form required for standard bright solitons. It should be stressed that both the amplitude and the width of the soliton are determined uniquely by the fiber parameters. Such fixed-parameter solitons coexist with radiation, and they are sometimes called *embedded solitons* (see Chapter 2).

3.6.2 Self-Steepening

The phenomenon of self-steepening has been studied extensively [103]–[107]. To isolate the effects of self-steepening governed by the parameter s, it is useful to set $\delta_3 = 0$ and $\tau_R = 0$ in Eq. (3.6.1). Pulse evolution inside fibers is then governed by

$$i\frac{\partial u}{\partial z} + \frac{1}{2}\frac{\partial^2 u}{\partial \tau^2} + |u|^2 u + is\frac{\partial}{\partial \tau}(|u|^2 u) = 0. \tag{3.6.8}$$

Self-steepening creates an optical shock on the trailing edge of the pulse in the absence of the GVD effects. This phenomenon is due to the intensity dependence of the group velocity that makes the peak of the pulse move slower than the wings. The GVD dissipates the shock and smooths the trailing edge considerably. However, self-steepening would still manifest through a shift of the pulse center.

The self-steepening-induced shift is shown in Figure 3.13 where pulse shapes at $z = 0$, 5, and 10 are plotted for $s = 0.2$ and $N = 1$ by solving Eq. (3.6.8) numerically with the input $u(0, \tau) = \mathrm{sech}(\tau)$. Because the peak moves slower than the wings for $s \neq 0$, it is delayed and appears shifted toward the trailing side. The delay is well

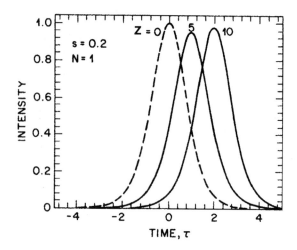

Figure 3.13: Pulse shapes at $z = 5$ and 10 for a fundamental soliton in the presence of self-steepening ($s = 0.2$). Dashed curve shows the initial shape for comparison. The solid curves coincide with the dashed curve when $s = 0$.

approximated by a simple expression $\tau_d = sz$ for $s < 0.3$. It can also be calculated by treating the self-steepening term in Eq. (3.6.8) as a small perturbation. Although the pulse broadens slightly with propagation (by $\sim$20% at $z = 10$), it nonetheless maintains its soliton nature. This feature suggests that Eq. (3.6.8) has a soliton solution toward which the input pulse is evolving asymptotically. Such a solution indeed exists and has the form [70]

$$u(z, \tau) = V(\tau + Mz) \exp[i(Kz - M\tau)], \tag{3.6.9}$$

where M is related to a shift of the carrier frequency. The group velocity changes as a result of the shift. The delay of the peak seen in Figure 3.13 is due to this change in the group velocity. The explicit form of $V(\tau)$ depends on M and s [107]. In the limit $s = 0$, it reduces to the hyperbolic secant form of Eq. (3.1.5). Note also that Eq. (3.6.8) can be transformed into a so-called derivative NLS equation that is integrable by the inverse scattering method and whose solutions have been studied extensively in plasma physics [108]–[111].

The effect of self-steepening on higher-order solitons is remarkable in that it leads to breakup of such solitons into their constituents, a phenomenon referred to as *soliton decay* [104]. Figure 3.14 shows this behavior for a second-order soliton ($N = 2$) using $s = 0.2$. For this relatively large value of s, the two solitons have separated from each other within a distance of two soliton periods and continue to move apart with further propagation inside the fiber. A qualitatively similar behavior occurs for smaller values of s, except that a longer distance is required for the breakup of solitons. The soliton decay can be understood using the inverse scattering method, with the self-steepening term acting as a perturbation. In the absence of self-steepening ($s = 0$), the two solitons form a bound state because both of them propagate at the same speed (the eigenvalues have the same real part). The effect of self-steepening is to break the degeneracy so that the two solitons propagate at different speeds. As a result, they separate from each

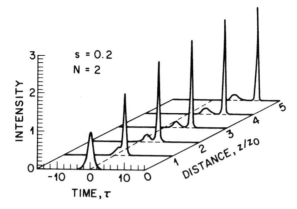

Figure 3.14: Decay of a second-order soliton ($N = 2$) induced by self-steepening ($s = 0.2$). Pulse evolution over five soliton periods is shown.

other, and the separation increases almost linearly with the distance [105]. The ratio of the peak heights in Figure 3.14 is about 9 and is in agreement with the expected ratio $(\eta_2/\eta_1)^2$, where η_1 and η_2 are the imaginary parts of the eigenvalues associated with the inverse scattering transform. The third- and higher-order solitons follow a similar decay pattern. In particular, the third-order soliton ($N = 3$) decays into three solitons whose peak heights are again in agreement with inverse scattering theory.

3.6.3 Intrapulse Raman Scattering

Intrapulse Raman scattering plays the most important role among the higher-order non-linear effects. Its effects on solitons are governed by the last term in Eq. (3.6.1) and were observed experimentally in 1985 [112]. The need to include this term became apparent when a new phenomenon, called the *soliton self-frequency shift*, was observed in 1986 [113] and explained using the delayed nature of the Raman response [114]. Since then, this higher-order nonlinear effect has been studied extensively [115]–[132]. Physically, stimulated Raman scattering leads to a continuous downshift of the soliton carrier frequency when the pulse spectrum becomes so broad that the high-frequency components of a pulse can transfer energy to the low-frequency components of the same pulse through Raman amplification. The Raman-induced frequency shift is negligible for $T_0 > 10$ ps but becomes of considerable importance for short solitons ($T_0 < 5$ ps).

To isolate the effects of intrapulse Raman scattering, it is useful to set $\delta_3 = 0$ and $s = 0$ in Eq. (3.6.1). Pulse evolution inside fibers is then governed by

$$i\frac{\partial u}{\partial z} + \frac{1}{2}\frac{\partial^2 u}{\partial \tau^2} + |u|^2 u = \tau_R u \frac{\partial |u|^2}{\partial \tau}. \tag{3.6.10}$$

Using Eqs. (3.5.4)–(3.5.7) with $\varepsilon(u) = -i\tau_R u(\partial |u|^2/\partial \tau)$, it is easy to see that the amplitude η of the soliton is not affected by the Raman effect, but its frequency δ changes as

$$\frac{d\delta}{dz} = -\frac{8}{15}\tau_R \eta^4. \tag{3.6.11}$$

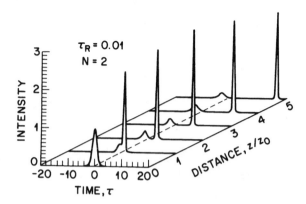

Figure 3.15: Decay of a second-order soliton ($N = 2$) induced by intrapulse Raman scattering ($\tau_R = 0.01$).

This equation is easily integrated for a constant η, with the result $\delta(z) = (8\tau_R/15)\eta^4 z$. Using $\eta = 1$ and $z = z/L_D = |\beta_2|z/T_0^2$, the Raman-induced frequency shift can be written in real units as

$$\Delta\omega_R(z) = -8|\beta_2|T_R z/(15T_0^4). \qquad (3.6.12)$$

The negative sign shows that the carrier frequency is reduced; i.e., the soliton spectrum shifts toward longer wavelengths, or the "red" side.

Physically, the red shift can be understood in terms of stimulated Raman scattering. For pulse widths ~ 1 ps or shorter, the spectral width of the pulse is large enough that the Raman gain can amplify the low-frequency (red) spectral components of the pulse, with high-frequency (blue) components of the same pulse acting as a pump. The process continues along the fiber, and the energy from blue components is continuously transferred to red components. Such an energy transfer appears as a red shift of the soliton spectrum, with shift increasing with distance. As seen from Eq. (3.6.12), the frequency shift increases linearly along the fiber. More importantly, it scales with the pulse width as T_0^{-4}, indicating that it can become quite large for short pulses. As an example, soliton frequency changes at a rate of ~ 50 GHz/km for 1-ps pulses ($T_0 = 0.57$ ps) in standard fibers with $\beta_2 = -20$ ps^2/km and $T_R = 3$ fs. The spectrum of such pulses will shift by 1 THz after 20 km of propagation. This is a large shift if we note that the spectral width (FWHM) of such a soliton is less than 0.5 THz. Typically, the Raman-induced frequency shift cannot be neglected for pulses shorter than 5 ps.

The Raman-induced red shift of solitons was observed in 1986 using 0.5-ps pulses obtained from a passively mode-locked color-center laser [113]. The pulse spectrum was found to shift as much as 8 THz for a fiber length under 0.4 km. The observed spectral shift was called the *soliton self-frequency shift* because it is induced by the soliton itself. In fact, it was in an attempt to explain the observed red shift that the importance of the delayed nature of the Raman response for transmission of ultrashort pulses was first realized [114].

The effect of intrapulse Raman scattering on higher-order solitons is similar to the case of self-steepening. In particular, even relatively small values of τ_R lead to the

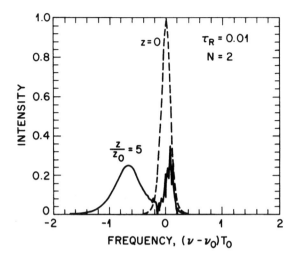

Figure 3.16: Pulse spectrum at $z/z_0 = 5$ for parameter values identical to those of Figure 3.15. Dashed curve shows the spectrum of input pulses.

decay of higher-order solitons into its constituents [120]. Figure 3.15 shows such a decay for a second-order soliton $(N = 2)$ by solving Eq. (3.6.10) numerically with $\tau_R = 0.01$. A comparison of Figures 3.14 and 3.15 shows the similarity and the differences for two different higher-order nonlinear mechanisms. An important difference is that relatively smaller values of τ_R compared with s can induce soliton decay over a given distance. For example, if $s = 0.01$ is chosen in Figure 3.14, the soliton does not split over the distance $z = 5z_0$. This feature indicates that the effects of τ_R are likely to dominate in practice over those of self-steepening.

Another important difference seen in Figs. 3.14 and 3.15 is that both solitons are delayed in the case of self-steepening, while in the Raman case the low-intensity soliton is advanced and appears on the leading side of the incident pulse. This behavior can be understood qualitatively from Figure 3.16, where the pulse spectrum at $z = 5z_0$ is compared with the input spectrum for the second-order soliton (whose evolution is shown in Figure 3.15). The most noteworthy feature is the huge red shift of the soliton spectrum, about four times the input spectral width for $\tau_R = 0.01$ and $z/z_0 = 5$. The red-shifted broad spectral peak corresponds to the intense soliton shifting toward the right in Figure 3.15, whereas the blue-shifted spectral feature corresponds to the other peak moving toward the left in that figure. Because the blue-shifted components travel faster than the red-shifted ones, they are advanced while the others are delayed with respect to the input pulse. This is precisely what is seen in Figure 3.15.

A question one may ask is whether Eq. (3.6.10) has soliton-like solutions. It turns out that pulse-like solutions do not exist when the Raman term is included, mainly because the resulting perturbation is of non-Hamiltonian type [87]. This feature of the Raman term can be understood by noting that the Raman-induced spectral red shift does not preserve pulse energy because a part of the energy is dissipated through the excitation of molecular vibrations. However, a kink-type topological soliton (with in-

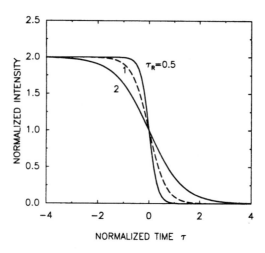

Figure 3.17: Temporal intensity profiles of kink solitons in the form of an optical shock for several values of τ_R. (After Ref. [129]; ©1992 APS.)

finite energy) has been found and is given by [129]

$$u(z, \tau) = \left(\frac{3\tau}{4\tau_R}\right) \left[\exp\left(\frac{3\tau}{\tau_R}\right) + 1\right]^{-1/2} \exp\left(\frac{9iz}{8\tau_R^2}\right). \qquad (3.6.13)$$

Kink solitons appear in many physical systems whose dynamics are governed by a sine–Gordon equation. In the context of optical fibers, the kink soliton represents an optical shock front that preserves its shape when propagating through the fiber. Figure 3.17 shows the shock profiles by plotting $|u(z, \tau)|^2$ for several values of τ_R. The steepness of the shock depends on τ_R such that the shock front becomes increasingly steeper as τ_R is reduced. Even though the parameter N increases as τ_R is reduced, the power level P_0 (defined as the power at $\tau = 0$) remains the same. This can be seen by expressing P_0 in terms of the parameter T_R using Eqs. (3.1.3) and (3.6.2) so that $P_0 = 9|\beta_2|/(16\gamma T_R^2)$. Using typical values for fiber parameters, $P_0 \sim 10$ kW. It is difficult to observe such optical shocks experimentally because of large power requirements.

The kink soliton given in Eq. (3.6.13) is obtained assuming $u(z, \tau) = V(\tau)\exp(iKz)$ and solving the resulting ordinary differential equation for $V(\tau)$. The solution shows that kink solitons form a one-parameter family for various values of K and exist even in the normal-dispersion region of the fiber [130]. They continue to exist even when the self-steepening term in Eq. (3.6.1) is included. The analytic form in Eq. (3.6.13) is obtained only for a specific value $K = 9/(8\tau_R^2)$. When $K < \tau_R^2$, the monotonically decaying tail seen in Figure 3.17 develops an oscillatory structure.

3.6.4 Propagation of Femtosecond Pulses

For femtosecond pulses having widths $T_0 < 1$ ps, it becomes necessary to include all the higher-order terms in Eq. (3.6.1) because all three parameters, δ_3, s, and τ_R, become non-negligible. Evolution of such ultrashort pulses in optical fibers is studied by

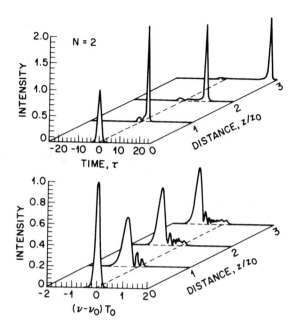

Figure 3.18: Evolution of pulse shapes and spectra for the case $N = 2$. The other parameter values are $\delta_3 = 0.03$, $s = 0.05$, and $\tau_R = 0.1$.

solving Eq. (3.6.1) numerically [133]–[136]. As an example, Figure 3.18 shows the pulse shapes and spectra when a second-order soliton is launched at the input end of a fiber after choosing $\delta_3 = 0.03$, $s = 0.05$, and $\tau_R = 0.1$. These values are appropriate for a 50-fs pulse ($T_0 \approx 30$ fs) propagating in the 1.55-μm region of a standard silica fiber. Soliton decay occurs within a soliton period ($z_0 \approx 5$ cm), and the main peak shifts toward the trailing side at a rapid rate with increasing distance. This temporal shift is due to the decrease in the group velocity $v_g \equiv (d\beta/d\omega)^{-1}$ occurring as a result of the red shift of the soliton spectrum. A shift in the carrier frequency of the soliton changes its speed because v_g is frequency dependent. If we use $T_0 = 30$ fs to convert the results of Figure 3.18 into physical units, the 50-fs pulse has shifted by almost 40 THz, or 20% of the carrier frequency, after propagating a distance of only 15 cm.

When the input peak power is large enough to excite a higher-order soliton such that $N \gg 1$, the pulse spectrum evolves into several bands, each corresponding to splitting of a fundamental soliton from the original pulse. Such an evolution pattern was seen when 830-fs pulses with peak powers up to 530 W were propagated in fibers up to 1 km long [134]. The spectral peak at the extreme red end was associated with a soliton whose width was narrowest ($\approx$55 fs) after 12 m and then increased with a further increase in the fiber length. The experimental results were in agreement with the predictions of Eq. (3.6.1).

The combined effect of TOD, self-steepening, and intrapulse Raman scattering on a higher-order soliton is to split it into its constituents. In fact, the TOD can itself lead to soliton decay even in the absence of higher-order nonlinear effects when the parameter

δ_3 exceeds a threshold value [135]. For a second-order soliton $(N = 2)$, the threshold value is $\delta_3 = 0.022$ but reduces to ≈ 0.006 for $N = 3$. For standard silica fibers, δ_3 exceeds 0.022 at 1.55 μm for pulses shorter than 70 fs. However, the threshold can be reached for pulses wider by a factor of 10 when dispersion-shifted fibers are used.

An interesting question is whether Eq. (3.6.1) permits shape-preserving, solitary-wave solutions under certain conditions. Several such solutions have been found using a variety of techniques [137]–[152]. In most cases, the solution exists only for a specific choice of parameter combinations. For example, fundamental and higher-order solitons have been found when $\tau_R = 0$ with $s = -2\delta_3$ or $s = -6\delta_3$ [144]. From a practical standpoint, such solutions of Eq. (3.6.1) are rarely useful, because it is hard to find fibers whose parameters satisfy the required constraints.

As successful as Eq. (3.6.1) is in modeling the propagation of femtosecond pulses in optical fibers, it is still approximate. A more accurate approach should take into account the time-dependent response of the fiber nonlinearity. In a simple model, the nonlinear term $|u|^2 u$ appearing in Eq. (3.6.1) is replaced with a generalized nonlinearity $F(|u|^2)u$, where

$$F(|u|^2) = \int_{-\infty}^{\tau} |u(t')|^2 R(\tau - t') \, dt', \qquad (3.6.14)$$

and $R(t)$ is the response function associated with the fiber nonlinearity. It can be written as [121]–[127]

$$R(t) = (1 - f_R)\delta(t) + f_R h_R(t), \qquad (3.6.15)$$

where f_R represents the fractional contribution of the delayed Raman response to the nonlinear polarization. The Raman response function $h_R(t)$ is responsible for the Raman gain and can be determined from the experimentally measured spectrum of the Raman gain in silica fibers [121]. An approximate analytic form of the Raman response function is given by [122]

$$h_R(t) = \frac{\tau_1^2 + \tau_2^2}{\tau_1 \tau_2^2} \exp(-t/\tau_2) \sin(t/\tau_1). \qquad (3.6.16)$$

The parameters τ_1 and τ_2 are two adjustable parameters and are chosen to provide a good fit to the actual Raman-gain spectrum. Their appropriate values are $\tau_1 = 12.2$ fs and $\tau_2 = 32$ fs [122]. The fraction f_R is estimated to be about 0.18 [121]–[124].

The delayed nature of the molecular response not only leads to the soliton self-frequency shift but also affects the interaction between neighboring solitons [125]. Equation (2.3.33) has been used to study numerically how intrapulse-stimulated Raman scattering affects evolution of femtosecond optical pulses in optical fibers [126]–[128]. For pulses shorter than 20 fs, even the use of this equation becomes questionable because of the slowly varying envelope approximation made in its derivation (see Section 2.3). Because such short pulses can be generated by modern mode-locked lasers, attempts have been made to improve upon this approximation while still working with the pulse envelope [153]–[155]. For supershort pulses containing only a few optical cycles, it eventually becomes necessary to abandon the concept of the pulse envelope and solve the Maxwell equations directly using an appropriate numerical scheme.

References

[1] A. Hasegawa and F. Tappert, *Appl. Phys. Lett.* **23**, 142 (1973).

[2] L. F. Mollenauer, R. H. Stolen, and J. P. Gordon, *Phys. Rev. Lett.* **45**, 1095 (1980).

[3] L. F. Mollenauer and K. Smith, *Opt. Lett.* **13**, 675 (1988).

[4] A. Hasegawa and Y. Kodama, *Solitons in Optical Communications* (Clarendon Press, Oxford, UK, 1995).

[5] L. F. Mollenauer, J. P. Gordon, and P. V. Mamychev, in *Optical Fiber Telecommunications IIIA*, I. P. Kaminow and T. L. Koch, Eds. (Academic Press, San Diego, 1997), Chap. 12.

[6] R. J. Essiambre and G. P. Agrawal, in *Progress in Optics*, E. Wolf, Ed., Vol. 37 (Elsevier, Amsterdam, 1997), Chap. 4

[7] E. Iannone, F. Matera, A. Mecozzi, and M. Settembre, *Nonlinear Optical Communication Networks* (Wiley, New York, 1998), Chap. 5.

[8] G. P. Agrawal, *Fiber-Optic Communication Systems*, 3rd ed. (Wiley, New York, 2002), Chap. 9.

[9] G. P. Agrawal, *Nonlinear Fiber Optics*, 3rd ed. (Academic Press, San Diego, 2001).

[10] V. E. Zakharov and A. B. Shabat, *Sov. Phys. JETP* **34**, 62 (1972).

[11] J. Satsuma and N. Yajima, *Prog. Theor. Phys.* **55**, 284 (1974).

[12] V. I. Karpman and V. V. Solovev, *Physica* **3D**, 487 (1981).

[13] J. P. Gordon, *Opt. Lett.* **8**, 596 (1983).

[14] F. M. Mitschke and L. F. Mollenauer, *Opt. Lett.* **12**, 355 (1987).

[15] C. Desem and P. L. Chu, *IEE Proc.* **134**, 145 (1987).

[16] Y. Kodama and K. Nozaki, *Opt. Lett.* **12**, 1038 (1987).

[17] A. Hasegawa and Y. Kodama, *Proc. IEEE* **69**, 1145 (1981); *Opt. Lett.* **7**, 285 (1982).

[18] Y. Kodama and A. Hasegawa, *Opt. Lett.* **7**, 339 (1982); **8**, 342 (1983).

[19] A. Hasegawa, *Opt. Lett.* **8**, 650 (1983); *Appl. Opt.* **23**, 3302 (1984).

[20] L. F. Mollenauer, R. H. Stolen, and M. N. Islam, *Opt. Lett.* **10**, 229 (1985).

[21] L. F. Mollenauer, J. P. Gordon, and M. N. Islam, *IEEE J. Quantum Electron.* **22**, 157 (1986).

[22] A. Hasegawa and Y. Kodama, *Phys. Rev. Lett.* **66**, 161 (1991).

[23] D. M. Spirit, I. W. Marshall, P. D. Constantine, D. L. Williams, S. T. Davey, and B. J. Ainslie, *Electron. Lett.* **27**, 222 (1991).

[24] M. Nakazawa, H. Kubota, K. Kurakawa, and E. Yamada, *J. Opt. Soc. Am. B* **8**, 1811 (1991).

[25] K. Kurokawa and M. Nakazawa, *IEEE J. Quantum Electron.* **28**, 1922 (1992).

[26] K. Rottwitt, J. H. Povlsen, S. Gundersen, and A. Bjarklev, *Opt. Lett.* **18**, 867 (1993).

[27] C. Lester, K. Bertilsson, K. Rottwitt, P. A. Andrekson, M. A. Newhouse, and A. J. Antos, *Electron. Lett.* **31**, 219 (1995).

[28] M. Nakazawa, K. Suzuki, and Y. Kimura, *IEEE Photon. Technol. Lett.* **2**, 216 (1990).

[29] N. A. Olsson, P. A. Andrekson, P. C. Becker, J. R. Simpson, T. Tanbun-Ek, R. A. Logan, H. Presby, and K. Wecht, *IEEE Photon. Technol. Lett.* **2**, 358 (1990).

[30] K. Iwatsuki, S. Nishi, and K. Nakagawa, *IEEE Photon. Technol. Lett.* **2**, 355 (1990).

[31] E. Yamada, K. Suzuki, and M. Nakazawa, *Electron. Lett.* **27**, 1289 (1991).

[32] L. F. Mollenauer, B. M. Nyman, M. J. Neubelt, G. Raybon, and S. G. Evangelides, *Electron. Lett.* **27**, 178 (1991).

[33] K. Tajima, *Opt. Lett.* **12**, 54 (1987).

[34] V. A. Bogatyrjov, M. M. Bubnov, E. M. Dianov, and A. A. Sysoliatin, *Pure Appl. Opt.* **4**, 345 (1995).

[35] D. J. Richardson, R. P. Chamberlin, L. Dong, and D. N. Payne, *Electron. Lett.* **31**, 1681 (1995).

[36] A. J. Stentz, R. Boyd, and A. F. Evans, *Opt. Lett.* **20**, 1770 (1995).

[37] D. J. Richardson, L. Dong, R. P. Chamberlin, A. D. Ellis, T. Widdowson, and W. A. Pender, *Electron. Lett.* **32**, 373 (1996).

[38] W. Forysiak, F. M. Knox, and N. J. Doran, *Opt. Lett.* **19**, 174 (1994).

[39] T. Georges and B. Charbonnier, *Opt. Lett.* **21**, 1232 (1996); *IEEE Photon. Technol. Lett.* **9**, 127 (1997).

[40] S. Cardinal, E. Desurvire, J. P. Hamaide, and O. Audouin, *Electron. Lett.* **33**, 77 (1997).

[41] A. Hasegawa, Y. Kodama, and A. Murata, *Opt. Fiber Technol.* **3**, 197 (1997).

[42] S. Kumar, Y. Kodama, and A. Hasegawa, *Electron. Lett.* **33**, 459 (1997).

[43] F. M. Knox, W. Forysiak, and N. J. Doran, *J. Lightwave Technol.* **13**, 1955 (1995).

[44] M. Suzuki, I. Morita, N. Edagawa, S. Yamamoto, H. Taga, and S. Akiba, *Electron. Lett.* **31**, 2027 (1995).

[45] N. J. Smith, F. M. Knox, N. J. Doran, K. J. Blow, and I. Bennion, *Electron. Lett.* **32**, 54 (1996).

[46] I. Gabitov and S. K. Turitsyn, *Opt. Lett.* **21**, 327 (1996).

[47] M. Nakazawa, H. Kubota, and K. Tamura, *IEEE Photon. Technol. Lett.* **8**, 452 (1996).

[48] M. Nakazawa, H. Kubota, A. Sahara, and K. Tamura, *IEEE Photon. Technol. Lett.* **8**, 1088 (1996).

[49] A. B. Grudinin and I. A. Goncharenko, *Electron. Lett.* **32**, 1602 (1996).

[50] I. Gabitov and S. K. Turitsyn, *JETP Lett.* **63**, 861 (1996).

[51] A. Berntson, N. J. Doran, W. Forysiak, and J. H. B. Nijhof, *Opt. Lett.* **23**, 900 (1998).

[52] J. N. Kutz, P. Holmes, S. G. Evangelides, and J. P. Gordon, *J. Opt. Soc. Am. B* **15**, 87 (1998).

[53] S. K. Turitsyn and E. Shapiro, *Opt. Fiber Technol.* **4**, 151 (1998).

[54] S. K. Turitsyn, I. Gabitov, E. W. Laedke, et al., *Opt. Commun.* **151**, 117 (1998).

[55] S. K. Turitsyn and V. K. Mezentsev, *JETP Lett.* **67**, 640 (1998).

[56] T. I. Lakoba and D. J. Kaup, *Electron. Lett.* **34**, 1124 (1998); *Phys. Rev. E* **58**, 6728 (1998).

[57] S. K. Turitsyn and E. G. Shapiro, *J. Opt. Soc. Am. B* **16**, 1321 (1999).

[58] I. R. Gabitov, E. G. Shapiro, and S. K. Turitsyn, *Phys. Rev. E* **55**, 3624 (1997).

[59] M. J. Ablowitz and G. Bioindini, *Opt. Lett.* **23**, 1668 (1998).

[60] V. E. Zakharov and S. V. Manakov, *JETP Lett.* **70**, 573 (1999).

[61] S. K. Turitsyn, T. Schaefer, K.H. Spatschek and V.K. Mezentsev, *Opt. Commun.* **163**, 122 (1999).

[62] C. Paré and P. A. Belangér, *Opt. Lett.* **25**, 881 (2000).

[63] J. H. B. Nijhof, N. J. Doran, W. Forysiak, and F. M. Knox, *Electron. Lett.* **33**, 1726 (1997).

[64] V. S. Grigoryan and C. R. Menyuk, *Opt. Lett.* **23**, 609 (1998).

[65] J. N. Kutz and S. G. Evangelides, *Opt. Lett.* **23**, 685 (1998).

[66] Y. Chen and H. A. Haus, *Opt. Lett.* **23**, 1013 (1998).

[67] S. K. Turitsyn and E. G. Shapiro, *Opt. Lett.* **23**, 683 (1998).

[68] J. H. B. Nijhof, W. Forysiak, and N. J. Doran, *Opt. Lett.* **23**, 1674 (1998).

[69] S. K. Turitsyn, J. H. B. Nijhof, V. K. Mezentsev, and N. J. Doran, *Opt. Lett.* **24**, 1871 (1999).

[70] D. Anderson and M. Lisak, *Phys. Rev. A* **27**, 1393 (1983).

[71] H. Kubota and M. Nakazawa, *Opt. Commun.* **87**, 15 (1992); M. Nakazawa and H. Kubota, *Electron. Lett.* **31**, 216 (1995).

[72] A. Naka, T. Matsuda, and S. Saito, *Electron. Lett.* **32**, 1694 (1996).

[73] I. Morita, M. Suzuki, N. Edagawa, S. Yamamoto, H. Taga, and S. Akiba, *IEEE Photon. Technol. Lett.* **8**, 1573 (1996).

[74] J. M. Jacob, E. A. Golovchenko, A. N. Pilipetskii, G. M. Carter, and C. R. Menyuk, *IEEE Photon. Technol. Lett.* **9**, 130 (1997).

[75] G. M. Carter and J. M. Jacob, *IEEE Photon. Technol. Lett.* **10**, 546 (1998).

[76] V. S. Grigoryan, R. M. Mu, G. M. Carter, and C. R. Menyuk, *IEEE Photon. Technol. Lett.* **10**, 45 (2000).

[77] R. M. Mu, C. R. Menyuk, G. M. Carter, and J. M. Jacob, *IEEE J. Sel. Topics Quantum Electron.* **6**, 248 (2000).

[78] A. B. Grudinin, M. Durkin, M. Isben, R. I. Laming, A. Schiffini, P. Franco, E. Grandi, and M. Romagnoli, *Electron. Lett.* **33**, 1572 (1997).

[79] F. Favre, D. Le Guen, and T. Georges, *J. Lightwave Technol.* **17**, 1032 (1999).

[80] I. S. Penketh, P. Harper, S. B. Aleston, A. M. Niculae, I. Bennion, and N. J. Doran, *Opt. Lett.* **24**, 803 (1999).

[81] M. Zitelli, F. Favre, D. Le Guen, and S. Del Burgo, *IEEE Photon. Technol. Lett.* **9**, 904 (1999).

[82] V. I. Karpman and E. M. Maslov, *Sov. Phys. JETP* **46**, 281 (1977).

[83] D. J. Kaup and A. C. Newell, *Proc. R. Soc. London*, Ser. A **361**, 413 (1978).

[84] V. I. Karpman, *Sov. Phys. JETP* **50**, 58 (1979); *Physica Scripta* **20**, 462 (1979).

[85] Y. S. Kivshar and B. A. Malomed, *Rev. Mod. Phys.* **61**, 761 (1989).

[86] H. Haus, *J. Opt. Soc. Am. B* **8**, 1122 (1991).

[87] C. R. Menyuk, *J. Opt. Soc. Am. B* **10**, 1585 (1993).

[88] T. Georges and F. Favre, *J. Opt. Soc. Am. B* **10**, 1880 (1993).

[89] T. Georges, *Opt. Fiber Technol.* **1**, 97 (1995).

[90] J. P. Gordon and H. A. Haus, *Opt. Lett.* **11**, 665 (1986).

[91] D. Marcuse, *J. Lightwave Technol.* **10**, 273 (1992).

[92] J. Santhanam, C. J. McKinstrie, T. Lakoba, and G. P. Agrawal, *Opt. Lett.* **26**, 1131 (2001).

[93] E. Putrina and G. P. Agrawal, *IEEE Photon. Technol. Lett.* **14**, 39 (2002).

[94] G. Falkovich, V. Lebedev, I. Kolokolov and S. K. Turitsyn, *Phys. Rev. E* **63**, 25601 (2001).

[95] G. P. Agrawal and M. J. Potasek, *Phys. Rev. A* **33**, 1765 (1986).

[96] G. R. Boyer and X. F. Carlotti, *Opt. Commun.* **60**, 18 (1986).

[97] P. K. Wai, C. R. Menyuk, H. H. Chen, and Y. C. Lee, *Opt. Lett.* **12**, 628 (1987); *IEEE J. Quantum Electron.* **24**, 373 (1988).

[98] M. Desaix, D. Anderson, and M. Lisak, *Opt. Lett.* **15**, 18 (1990).

[99] V. K. Mezentsev and S. K. Turitsyn, *Sov. Laser Commun.* **1**, 263 (1991).

[100] Y. S. Kivshar, *Phys. Rev. A* **43**, 1677 (1981); *Opt. Lett.* **16**, 892 (1991).

[101] V. I. Karpman, *Phys. Rev. E* **47**, 2073 (1993); *Phys. Lett. A* **181**, 211 (1993).

[102] M. Karlsson and A. Höök, *Opt. Commun.* **104**, 303 (1994).

[103] N. Tzoar and M. Jain, *Phys. Rev. A* **23**, 1266 (1981).

[104] E. A. Golovchenko, E. M. Dianov, A. M. Prokhorov, and V. N. Serkin, *JETP Lett.* **42**, 87 (1985); *Sov. Phys. Dokl.* **31**, 494 (1986).

[105] K. Ohkuma, Y. H. Ichikawa, and Y. Abe, *Opt. Lett.* **12**, 516 (1987).

[106] A. M. Kamchatnov, S. A. Darmanyan, and F. Lederer, *Phys. Lett. A* **245**, 259 (1998).

[107] W. P. Zhong and H. J. Luo, *Chin. Phys. Lett.* **17**, 577 (2000).

[108] E. Mjolhus, *J. Plasma Phys.* **16**, 321 (1976); **19** 437 (1978).

[109] K. Mio, T. Ogino, K. Minami, and S. Takeda, *J. Phys. Soc. Jpn.* **41**, 265 (1976).

[110] M. Wadati, K. Konno, and Y. H. Ichikawa, *J. Phys. Soc. Jpn.* **46**, 1965 (1979).

[111] Y. H. Ichikawa, K. Konno, M. Wadati, and H. Sanuki, *J. Phys. Soc. Jpn.* **48**, 279 (1980).

[112] E. M. Dianov, A. Y. Karasik, P. V. Mamyshev, A. M. Prokhorov, V. N. Serkin, M. F. Stel'makh, and A. A. Fomichev, *JETP Lett.* **41**, 294 (1985).

[113] F. M. Mitschke and L. F. Mollenauer, *Opt. Lett.* **11**, 659 (1986);

[114] J. P. Gordon, *Opt. Lett.* **11**, 662 (1986).

[115] Y. Kodama and A. Hasegawa, *IEEE J. Quantum Electron.* **QE-23**, 510 (1987).

[116] B. Zysset, P. Beaud, and W. Hodel, *Appl. Phys. Lett.* **50**, 1027 (1987).

[117] V. A. Vysloukh and T. A. Matveeva, *Sov. J. Quantum Electron.* **17**, 498 (1987).

[118] A. B. Grudinin, E. M. Dianov, D. V. Korobkin, A. M. Prokhorov, V. N. Serkin, and D. V. Khaidarov, *JETP Lett.* **46**, 221 (1987).

[119] A. S. Gouveia-Neto, A. S. L. Gomes, and J. R. Taylor, *IEEE J. Quantum Electron.* **24**, 332 (1988).

[120] K. Tai, A. Hasegawa, and N. Bekki, *Opt. Lett.* **13**, 392 (1988).

[121] R. H. Stolen, J. P. Gordon, W. J. Tomlinson, and H. A. Haus, *J. Opt. Soc. Am. B* **6**, 1159 (1989).

[122] K. J. Blow and D. Wood, *IEEE J. Quantum Electron.* **25**, 2665 (1989).

[123] V. V. Afansasyev, V. A. Vysloukh, and V. N. Serkin, *Opt. Lett.* **15**, 489 (1990).

[124] P. V. Mamyshev and S. V. Chernikov, *Opt. Lett.* **15**, 1076 (1990).

[125] B. J. Hong and C. C. Yang, *Opt. Lett.* **15**, 1061 (1990); *J. Opt. Soc. Am. B* **8**, 1114 (1991).

[126] P. V. Mamyshev and S. V. Chernikov, *Sov. Laser Commun.* **2**, 97 (1992).

[127] R. H. Stolen and W. J. Tomlinson, *J. Opt. Soc. Am. B* **9**, 565 (1992).

[128] K. Kurokawa, H. Kubota, and M. Nakazawa, *Electron. Lett.* **28**, 2050 (1992).

[129] G. P. Agrawal and C. Headley III, *Phys. Rev. A* **46**, 1573 (1992).

[130] Y. S. Kivshar and B. A. Malomed, *Opt. Lett.* **18**, 485 (1993).

[131] V. N. Serkin, V. A. Vysloukh, and J. R. Taylor, *Electron. Lett.* **29**, 12 (1993).

[132] S. Liu and W. Wang, *Opt. Lett.* **18**, 1911 (1993).

[133] W. Hodel and H. P. Weber, *Opt. Lett.* **12**, 924 (1987).

[134] P. Beaud, W. Hodel, B. Zysset, and H. P. Weber, *IEEE J. Quantum Electron.* **QE-23**, 1938 (1987).

[135] P. K. A. Wai, C. R. Menyuk, Y. C. Lee, and H. H. Chen, *Opt. Lett.* **11**, 464 (1986).

[136] M. Trippenbach and Y. B. Band, *Phys. Rev. A* **57**, 4791 (1998).

[137] D. N. Christodoulides and R. I. Joseph, *Appl. Phys. Lett.* **47**, 76 (1985).

[138] L. Gagnon, *J. Opt. Soc. Am. B* **9**, 1477 (1989).

[139] A. B. Grudinin, V. N. Men'shov, and T. N. Fursa, *Sov. Phys. JETP* **70**, 249 (1990).

[140] L. Gagnon and P. A. Bélanger, *Opt. Lett.* **9**, 466 (1990).

[141] M. J. Potasek and M. Tabor, *Phys. Lett. A* **154**, 449 (1991).

[142] M. Florjanczyk and L. Gagnon, *Phys. Rev. A* **41**, 4478 (1990); *Phys. Rev. A* **45**, 6881 (1992).

[143] M. J. Potasek, *J. Appl. Phys.* **65**, 941 (1989); *IEEE J. Quantum Electron.* **29**, 281 (1993).

[144] S. Liu and W. Wang, *Phys. Rev. E* **49**, 5726 (1994).

[145] D. J. Frantzeskakis, K. Hizanidis, G. S. Tombrasand, and I. Belia, *IEEE J. Quantum Electron.* **31**, 183 (1995).

[146] K. Porsezian and K. Nakkeeran, *Phys. Rev. Lett.* **76**, 3955 (1996).

[147] G. J. Dong and Z. Z. Liu, *Opt. Commun.* **128**, 8 (1996).

[148] M. Gedalin, T. C. Scott, and Y. B. Band, *Phys. Rev. Lett.* **78**, 448 (1997).

[149] D. Mihalache, N. Truta, and L. C. Crasovan, *Phys. Rev. E* **56**, 1064 (1997).

[150] S. L. Palacios, A. Guinea, J. M. Fernandez-Diaz, and R. D. Crespo, *Phys. Rev. E* **60**, R45 (1999).

[151] C. E. Zaspel, *Phys. Rev. Lett.* **82**, 723 (1999).

[152] Z. Li, L. Li, H. Tian, and G. Zhou, *Phys. Rev. Lett.* **84**, 4096 (2000).

[153] T. Brabec and F. Krauszm, *Phys. Rev. Lett.* **78**, 3282 (1997).

[154] J. K. Ranka and A. L. Gaeta, *Opt. Lett.* **23**, 534 (1998).

[155] Q. Lin and E. Wintner, *Opt. Commun.* **150**, 185 (1998).

Chapter 4

Dark Solitons

This chapter describes the properties of both the spatial and temporal *dark solitons* in the $(1 + 1)$-dimensional geometry. Section 4.1 focuses on dark solitons forming in a Kerr medium, such as an optical fiber. In this case, the cubic NLS equation is exactly integrable by means of the inverse scattering transform method, and the solutions for dark solitons can be found in an analytical form. Sections 4.2 and 4.3 are devoted to the study of the dark solitons in non-Kerr media, such as organic, photorefractive, and semiconductor materials. In particular, Section 4.2 shows that small-amplitude dark solitons are well described by an integrable Korteweg–de Vries (KdV) equation even though the generalized NLS equation itself is not integrable by the inverse scattering transform method. Stability of dark solitons is studied in Section 4.3, where we also discuss the asymptotic analytical results related to the instability-driven dynamics of non-Kerr dark solitons. In Section 4.4 we present the perturbation theory for dark solitons and outline several important features of the perturbation-induced dynamics for both the spatial and temporal dark solitons. The issues that are specific to temporal dark solitons, such as timing jitter, Raman-induced frequency shift, and third-order dispersion, are discussed in Section 4.5. The experimental results related to the observation of temporal and spatial dark solitons and their interaction in various nonlinear media are presented in Section 4.6.

4.1 Kerr Medium

The generalized NLS equation (2.1.1) also applies to dark solitons, provided the leading term in the expansion of the nonlinearity function $F(I)$ is negative, where $I = |u|^2$ is the intensity. In the specific case of a self-defocusing Kerr medium, $F(I) = -I$, and the NLS equation takes the form

$$i\frac{\partial u}{\partial z} + \frac{1}{2}\frac{\partial^2 u}{\partial x^2} - |u|^2 u = 0. \tag{4.1.1}$$

This section focuses on the properties of dark solitons associated with this equation.

4.1.1 Inverse Scattering Transform

As discussed in Section 1.4.2, the NLS equation Eq. (4.1.1) is exactly integrable by the inverse scattering transform method [1]. The linear eigenvalue problem for this equation is the same as that used for bright solitons, and it is given by Eqs. (1.4.8) and (1.4.9). The only specific difference comes through the boundary conditions, which in the case of dark solitons take the form

$$u(0,x) = \begin{cases} u_0 e^{i\theta_1} & \text{for} \quad x \to -\infty, \\ u_0 e^{i\theta_2} & \text{for} \quad x \to +\infty, \end{cases} \tag{4.1.2}$$

where u_0 is the background amplitude and θ_1 and θ_2 are constant phases.

The linear scattering problem for dark solitons possesses a set of N discrete eigenvalues, $|\lambda| < u_0$, and they can be found analytically for some specific input conditions. These eigenvalues remain invariant even when $u(z,x)$ evolves with z. Each real discrete eigenvalue can be represented as $\lambda_n \equiv u_0 \sin\phi_n$, and it corresponds to a dark soliton whose dip has an amplitude given by $u_0 \cos\phi_n$ and that moves against the background with the transverse velocity $V_n = u_0 \sin\phi_n$. Since the asymptotic evolution of any input beam is described by the discrete eigenvalues of the scattering problem, the beam is transformed into a certain set of dark solitons for large enough propagation distances.

Only a single discrete eigenvalue is found for certain input fields. In that case, a single dark soliton will be created. The most general form of a dark soliton has been given in Section 1.5, and it can be written as

$$u(x,z) = u_0 \left\{ \cos\phi \tanh[u_0 \cos\phi (x - zu_0 \sin\phi)] + i \sin\phi \right\} \exp(-iu_0^2 z + i\theta_0), \tag{4.1.3}$$

where ϕ is the angle variable that parameterizes the discrete eigenvalue and θ_0 is an arbitrary phase. This solution of Eq. (4.1.1) describes a dark soliton moving against the background of constant intensity $|u_0|^2$.

The inverse scattering transform method allows one to find more general solutions describing interaction of N dark solitons with arbitrary parameters [1]. In a 1985 study multiple dark-soliton solutions of the cubic NLS equation were obtained using this method [2], and during the 1990s several studies focused on dark-soliton collisions [3]–[5]. The exact, two-parameter, analytical solution of the cubic NLS equation (4.1.1) describing interaction of two dark solitons of equal amplitudes can be written in the form

$$u(x,z) = \left[2\eta \frac{\eta \cosh(2a\eta z) - ia \sinh(2a\eta z)}{a \cosh(2\eta t) + u_0 \cosh(2a\eta z)} - u_0 \right] \exp(-iu_0^2 z), \tag{4.1.4}$$

where η and a are arbitrary parameters. The solution (4.1.4) describes the collision between two dark solitons. For a soliton with constant phase at $z = 0$, this solution also describes the creation of two dark solitons of the opposite velocities from an initial even beam. This case is discussed later in more detail. More general solutions of the multisoliton generation process can be studied numerically in terms of the scattering data [6]–[8].

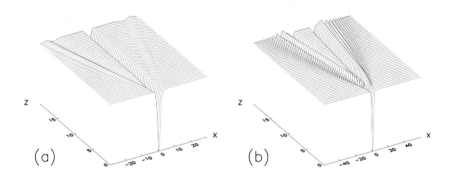

Figure 4.1: Two different scenarios for generating dark solitons using input beams of the form (a) $u(0,x) = u_0 \tanh(ax)$ and (b) $u(0,x) = u_0[1 - b\text{sech}^2(ax)]$ at $b = 0.9$. In both cases, $a = 0.9$ and $u_0 = 1$. (Courtesy A. A. Sukhorukov).

4.1.2   Generation of Dark Solitons

The inverse scattering transform method can be employed to analyze the generation of dark solitons for any shape of the input beam profile within the framework of the cubic NLS equation. From the experimental viewpoint, it is important to know the solution of the linear scattering problem for certain specific input beam profiles. As a simple example, consider an input beam in the form

$$u(0,x) = u_0 \tanh(ax), \tag{4.1.5}$$

where the ratio u_0/a is arbitrary. In this case, the scattering problem (2.3.17) can be solved analytically, and the eigenvalues of the discrete spectrum are given by [9]

$$\lambda_1 = 0, \qquad \lambda_{2n} = -\lambda_{2n+1} = \sqrt{u_0^2 - w_n^2}, \tag{4.1.6}$$

where $n = 1, 2, \ldots, N_0$ and

$$w_n = u_0 \left(1 - \frac{na}{u_0} \right). \tag{4.1.7}$$

The integer N_0 corresponds to the largest integer satisfying the condition $N_0 < u_0/a$. The first eigenvalue with the value zero corresponds to a *black soliton* that has zero intensity at its center. The even number of the secondary eigenvalues in Eq. (4.1.6) correspond to N_0 symmetric pairs of *gray solitons* propagating to the left and right side of the black soliton. Thus, the total number of dark solitons created by the input beam (4.1.5) is $N = 2N_0 + 1$, where N_0 depends on the ratio u_0/a. In the specific case in which $u_0 \leq a$, only the black soliton is created, since $N_0 = 0$.

If the intensity of the input beam does not vanish at any point, the black solitons are not generated. Figure 4.1 shows two characteristic examples of the generation of dark solitons. In the first case, shown in Figure 4.1(a), a single black soliton is always generated, as described earlier. In the second case, shown in Figure 4.1(b), the input beam is selected in the form $u(0,x) = u_0[1 - b\text{sech}^2(ax)]$, with $b = 0.9$ and $u_0 = 1$.

As seen in Figure 4.1(b), only pairs of gray solitons are generated. Both types of problems for the dark-soliton generation can also be analyzed by a variational approach developed for the scattering problem [10].

Another input profile of considerable interest is the step-like profile given in Eq. (4.1.2) which exhibits a phase jump by $\theta_2 - \theta_1$ at $x = 0$ [11]. It turns out that this profile always generates a single dark soliton corresponding to the eigenvalue $\lambda_1 = -u_0 \cos[\frac{1}{2}(\theta_2 - \theta_1)]$. The eigenvalue becomes zero for $\theta_2 - \theta_1 = m\pi$, where m is an integer. A black soliton, with zero intensity at its center, is formed only for these specific values of the phase jump.

To generate several dark solitons by the phase-modulation technique, one can use an input beam with several phase steps. In particular, N phase steps can generate N dark solitons in the asymptotic region $z \gg 1$, provided [12]

$$\Delta x_j > (2u_0)^{-1} |\cot(\theta_{j+1}/2) + \cot(\theta_j/2)|, \qquad (4.1.8)$$

where Δx_j is the distance along x between the two neighboring phase steps and θ_j is the value of the jth phase jump. In particular, two equal steps of opposite signs always produce two dark solitons of opposite velocities [13]. A similar behavior occurs when the amplitude of the background rather than its phase is modulated. The simple case of a box-like input (a square well) was analyzed as early as 1973 assuming that the intensity inside the box drops to zero [1]. This analysis was extended in 1989 to the case of a finite intensity minimum [11]. By 1997, a variational method had been developed for analyzing the eigenvalue problem in the general case of an arbitrary-shape box with a finite minimum intensity [10]. This technique allows one to determine the soliton parameters approximately even when the exact solution of the eigenvalue problem is not available.

The most interesting feature of dark solitons is their *thresholdless* generation [14]. As was seen in Chapter 2, the formation of bright solitons inside a Kerr medium requires a minimum input power that depends on the width of the input profile. More specifically, bright solitons are created when [15]

$$\int_{-\infty}^{+\infty} |u(0,x)| \, dx > \pi/2. \qquad (4.1.9)$$

In contrast, dark solitons can be created by an *arbitrary* small dip on a continuous-wave (CW) background. Consider an input beam with the general form

$$u(0,x) = u_0 e^{i\theta} + u_1(x), \qquad (u_1 \to 0 \quad \text{as} \quad |x| \to \infty), \qquad (4.1.10)$$

where $|u_1(x)|$ is assumed to fall fast enough as $|x| \to \infty$. If we assume that [14]

$$\Delta \equiv \text{Re}\left(e^{-i\theta} \int_{-\infty}^{+\infty} u_1(x) dx \right) < 0, \qquad (4.1.11)$$

then there always exist two discrete eigenvalues in the associated scattering problem such that $\lambda_{1,2} = \pm u_0(1 - \frac{1}{2}\Delta^2)$; they correspond to a pair of dark solitons with equal amplitudes $u_0\Delta$ and opposite velocities. As a consequence of this result, an intensity dip on the input pulse always produces at least one pair of dark solitons.

The formation of dark solitons by an input field of the form $u(x,0) = \exp[iS(x)]$ has been investigated in detail in the context of coherent matter waves in Bose–Einstein condensates (see Chapter 14) by means of phase imprinting [16] (see also Ref. [17]). The inverse scattering eigenvalue problem can be solved in this case by mapping it into the classical problem of a damped driven pendulum. In fact, a formula for the number of dark solitons can be obtained in an analytical form; it describes the odd and even number of generated solitons traveling in both directions. An analytical solution is also possible in the case in which both the amplitude and phase of the input field are piecewise continuous [18]. In this case, the formation of both the even and odd number of dark solitons can be described by a transcendental equation for the eigenvalues. In all cases, the net phase change across the input field plays a key role in the formation of dark solitons, and a small phase variation can change the number of the solitons at the output of the Kerr medium.

4.2   Non-Kerr Media

In the case of a non-Kerr medium, the function $F(I)$ in the generalized NLS equation (2.1.1) does not have the simple form $F(I) = -I$. In the case of dark solitons, we rewrite Eq. (2.1.1) as

$$i\frac{\partial u}{\partial z} + \frac{1}{2}\frac{\partial^2 u}{\partial x^2} + F(|u|^2)u = 0, \tag{4.2.1}$$

where $F(|u|^2)$ is negative in the limit $|u|^2 \to 0$. This equation cannot be solved in general by the inverse scattering transform method. However, it can be reduced to a KdV equation in the small-amplitude approximation, for which the intensity dip associated with the dark soliton is assumed to be small compared with the background intensity [19]–[23]. We focus on this case first.

4.2.1   Small-Amplitude Approximation

To discuss the dark-soliton dynamics in the small-amplitude limit, we look for solutions of Eq. (4.2.1) in the form

$$u(z,x) = [u_0 + a(z,x)]\exp[-iu_0^2 z + i\phi(z,x)], \tag{4.2.2}$$

where u_0 is the amplitude of the background far from the dark soliton of amplitude $a \ll u_0$ and phase ϕ. Substituting Eq. (4.2.2) into the NLS equation (4.2.1), we obtain two equations for the functions a and ϕ. We apply a multiscale perturbation technique for solving these equations and introduce two new variables as

$$\xi = \varepsilon(x - cz), \qquad \zeta = \varepsilon^3 z, \tag{4.2.3}$$

where ε is a small parameter connected with the smallness of the soliton amplitude a and c is the transverse velocity of the soliton (to be determined). We then look for solutions in the form of an asymptotic series in ε using

$$a = \varepsilon^2 a_0 + \varepsilon^4 a_1 + \dots, \qquad \phi = \varepsilon\phi_0 + \varepsilon^3\phi_1 + \dots. \tag{4.2.4}$$

To the lowest-order approximation, the soliton phase is determined from

$$\frac{\partial \phi_0}{\partial \xi} = -\frac{ca_0}{u_0}, \qquad (4.2.5)$$

and the soliton amplitude satisfies the following KdV equation [22]

$$2c\frac{\partial a_0}{\partial \zeta} - 2u_0 a_0 [3F'(u_0^2) + u_0^2 F''(u_0^2)]\frac{\partial a_0}{\partial \xi} - \frac{1}{4}\frac{\partial^3 a_0}{\partial \xi^3} = 0, \qquad (4.2.6)$$

where a prime denotes derivative with respect to the argument and $c \equiv u_0 |F'(u_0^2)|^{1/2}$ is the limiting speed of linear waves associated with Eq. (4.2.1).

In the case of a Kerr medium, $F(I) = -I$. Because $F' = -1$ and $F'' = 0$ in this case, Eq. (4.2.6) reduces to the standard form of the KdV equation,

$$2c\frac{\partial a_0}{\partial \zeta} + 6u_0 a_0 \frac{\partial a_0}{\partial \xi} - \frac{1}{4}\frac{\partial^3 a_0}{\partial \xi^3} = 0. \qquad (4.2.7)$$

Equation (4.2.7) is known to be integrable exactly using the inverse scattering method and has a solution in the form $\text{sech}^2(\xi)$. Thus, the dark-Kerr soliton solution in the small-amplitude limit can be obtained with the help of the KdV equation [21]. This link between the NLS dark solitons and the solitons of the KdV equation (4.2.7) is not as useful as it may first appear because both nonlinear equations are exactly integrable and, in principle, analytical solutions can be obtained for both of them separately. However, the remarkable fact is that this link also exists for the generalized NLS equation and allows us to study its soliton solutions even though the equation itself is not exactly integrable. Additionally, the approach based on the small-amplitude approximation is useful for analyzing the influence of different perturbations on the dark-soliton dynamics using the known analytical results for the KdV solitons [21]–[23]. In the context of temporal solitons, a similar approach can be used for both the normal and anomalous dispersion regimes [24], as well as employed for the study of the dark-soliton collisions near the zero-dispersion point [25].

4.2.2 Integrals of Motion

The dark solitons associated with Eq. (4.2.1) are expected to have a form similar to that found in the Kerr case if the function $F(|u|^2)$ does not deviate too much from its Kerr-medium value. It is thus appropriate to consider the general form of a dark soliton in a non-Kerr medium in the form

$$u(x,z) = u_0\{B\tanh[u_0 B(x - vz)] + iA\}\exp(ikx - i\beta z + i\theta_0), \qquad (4.2.8)$$

where $\beta = \frac{1}{2}k^2 + u_0^2$ characterizes the dispersion relation for the background wave, θ_0 is a constant phase, and the four parameters, u_0, A, B, and v, are connected by the relations

$$u_0 A = v - k, \qquad A^2 + B^2 = 1. \qquad (4.2.9)$$

Thus, the dark-soliton solution (4.2.8) is characterized by three independent parameters; two of them, u_0 and k describe, respectively, the amplitude and wave number of the CW background, and the remaining parameter, A, characterizes the dark soliton itself. The asymptotic form of the solution (4.2.8) coincides with the CW solution of this equation. However, the presence of a dark soliton on the background manifests itself in different phases at $x \rightarrow \pm\infty$; i.e., the plane wave is shifted in phase. The total phase shift across the dark soliton is given by

$$\Delta\theta = 2\left[\tan^{-1}\left(\frac{A}{B}\right) - \frac{\pi}{2}\right] = -2\tan^{-1}\left(\frac{B}{A}\right). \tag{4.2.10}$$

Because the cubic NLS equation (4.1.1) is exactly integrable, it possesses an infinite number of integrals of motion; the first three among them are given in Section 2.1. In the case of dark solitons, they can be written as

$$P = \int_{-\infty}^{\infty} |u|^2 dx, \quad M = i \int_{-\infty}^{\infty} (u_x^* u - u_x u^*)\, dx, \quad H = \frac{1}{2}\int_{-\infty}^{\infty} (|u_x|^2 + |u|^4)\, dx, \tag{4.2.11}$$

where $u_x = \partial u/\partial x$. The only difference appears in H, for which the $|u|^4$ term has the positive sign for dark solitons because of the negative sign of the nonlinear term in Eq. (4.1.1). Another difference between the bright and dark solitons results from different boundary conditions at $|x| = \infty$. It is easy to see that in the case of nonzero boundary conditions suitable for dark solitons, the three integrals in Eq. (4.2.11) *diverge*. This divergence is not surprising because it occurs even for the exact CW solution of Eq. (4.1.1) given by $u = u_0 \exp(ikx - i\beta z)$, where $\beta = \frac{1}{2}k^2 + u_0^2$. In this case, the integrals can easily be performed to obtain

$$P_{\mathrm{cw}} = u_0^2 X_0, \quad M_{\mathrm{cw}} = k u_0^2 X_0 \quad H_{\mathrm{cw}} = \frac{1}{2}u_0^2 (k^2 + u_0^2) X_0, \tag{4.2.12}$$

where X_0 defines the spatial extension of the CW beam in the x direction. Clearly, these values are not finite as $X_0 \rightarrow \infty$, and the same problem persists in the case of dark solitons. However, if we remove the background contribution, it is possible to introduce finite (or renormalized) expressions for the three invariants associated with the dark soliton itself. In what follows we consider this renormalization for the simpler case $k = 0$, for which the CW background consists of a plane wave moving along the z axis.

It is clear that the dark-soliton power should be defined as the difference between the CW power P_{cw} in (4.2.12) and the power P in Eq. (4.2.11) The renormalized power is thus given by

$$P_r = \int_{-\infty}^{\infty} (u_0^2 - |u|^2)\, dx. \tag{4.2.13}$$

The quantity P_r is often called the *complementary power* of a dark soliton. Using $u(x, z)$ from Eq. (4.2.8), we find $P_r = 2u_0 B$. The same procedure can be used to renormalize the Hamiltonian and results in

$$H_r = \frac{4}{3}(c^2 - v^2)^{3/2}, \tag{4.2.14}$$

where $c = u_0$ is the limiting speed of dark solitons.

Renormalization of the field momentum is somewhat tricky. From the definition of M in Eq. (4.2.11) we should subtract the contribution of the background, which is related to the phase difference in Eq. (4.2.10). Because the momentum of the CW background has the form $M = ku_0^2 X_0$ [see Eq.(4.2.12)], this phase difference gives a nonzero contribution even at $k = 0$. Indeed, this contribution is found to be $u_0^2 \int k(x)dx = u_0^2 \Delta\theta$, where $k(x)$ describes a local change in the background wave number. As a result, the renormalized momentum of a dark soliton is defined as $M_r = M - u_0^2 \Delta\theta$. Using $\Delta\theta$ from Eq. (4.2.10) and calculating M from Eqs. (4.2.11) and (4.2.8) with $k = 0$, we find

$$M_r = -2v\sqrt{c^2 - v^2} + 2c^2 \tan^{-1}(\sqrt{c^2 - v^2}/v). \quad (4.2.15)$$

Differentiating H_r and M_r with respect to v, we obtain the simple relation $\partial H_r / \partial M_r = v$. It indicates that the renormalized integrals of motion satisfy the standard expression of classical mechanics, and thus a dark soliton can be treated as an effective classical particle, similar to the case of bright solitons.

In the case of the generalized NLS equation (4.2.1), we can also introduce the renormalized invariants that are free of any divergence and characterize the dark soliton itself using the following general definitions:

$$P_r = \int_{-\infty}^{\infty} (u_0^2 - |u|^2)dx, \quad (4.2.16)$$

$$M_r = \frac{i}{2} \int_{-\infty}^{\infty} (u_x^* u - u_x u^*)(1 - u_0^2/|u|^2)\,dx, \quad (4.2.17)$$

$$H_r = \int_{-\infty}^{\infty} \left\{ \frac{1}{2}\left|\frac{\partial u}{\partial x}\right|^2 + \int_{u_0^2}^{|u|^2} [F(u_0^2) - F(I)]dI \right\} dx. \quad (4.2.18)$$

All three integrals remain finite for any nonlinear function $F(I)$ appearing in the generalized NLS equation.

4.2.3 Examples of Dark Solitons

Consider the generalized NLS equation (4.2.1) and look for stationary solutions on the CW background in the form

$$u(x,z) = \psi(x,z)\exp[iF(q)z], \quad (4.2.19)$$

where $q = u_0^2$ is the background intensity and ψ satisfies the condition $|\psi(x,z)|^2 \to q$ as $|x| \to \infty$. In terms of ψ, the NLS equation (4.2.1) takes the form

$$i\frac{\partial \psi}{\partial z} + \frac{1}{2}\frac{\partial^2 \psi}{\partial x^2} + [F(|\psi|^2) - F(q)]\psi = 0. \quad (4.2.20)$$

A dark soliton ψ_s is a localized traveling-wave solution of Eq. (4.2.20) in the form

$$\psi_s(\xi) = \Phi(\xi)e^{i\theta(\xi)}, \quad (4.2.21)$$

where $\xi = x - vz$ and the *real* functions Φ and θ depend on two parameters, the soliton velocity v and the intensity q of the CW background. These functions satisfy the following ordinary differential equations:

$$\frac{d\theta}{d\xi} = v\left(1 - \frac{q}{\Phi^2}\right),\tag{4.2.22}$$

$$\frac{1}{2}\frac{d^2\Phi}{d\xi^2} + \frac{v^2}{2}\left(\Phi - \frac{q^2}{\Phi^3}\right) + [F(\Phi^2) - F(q)]\Phi = 0.\tag{4.2.23}$$

Solutions of these two equations describe a family of dark solitons in terms of the parameters v and q.

The three renormalized invariants in Eqs. (4.2.16)–(4.2.18) can be written in terms of Φ as

$$P_s(v,q) = \int_{-\infty}^{\infty} (\Phi^2 - q)\,d\xi, \quad M_s(v,q) = -v\int_{-\infty}^{\infty} \frac{(\Phi^2 - q)^2}{\Phi^2}\,d\xi,\tag{4.2.24}$$

$$H_s(v,q) = \int_{-\infty}^{\infty}\left[\frac{1}{2}\left(\frac{d\Phi}{d\xi}\right)^2 + \frac{v^2(\Phi^2 - q)^2}{2\Phi^2} + \int_q^{\Phi^2}[F(q) - F(I)]dI\right]d\xi.\tag{4.2.25}$$

We also find the following analytical expression for the total phase shift S_s of the background wave across the dark soliton:

$$S_s(v,q) = v\int_{-\infty}^{\infty}\left(1 - \frac{q}{\Phi^2}\right)d\xi.\tag{4.2.26}$$

As the most characteristic example of dark solitons, we consider the competing power-law nonlinearity in the form

$$F(I) = \tfrac{1}{2}(\alpha I^p - \beta I^{2p}),\tag{4.2.27}$$

where α and β are both positive. The first term describes the self-focusing occurring at low intensities and prevents the existence of black solitons with a zero value of the minimum intensity. For $p = 1$ the generalized NLS equation (4.2.20) with $F(I)$ of Eq. (4.2.27) describes a self-focusing Kerr medium whose nonlinearity saturates at high intensities. Remarkably, Eq. (4.2.20) possesses in this case an explicit analytical solution in the form of a dark soliton [26]. In particular, the exact dark-soliton solution for the cubic-quintic nonlinearity in Eq. (4.2.27) found for $p = 1$ can be written as

$$\Phi^2(\xi) = 1 - \frac{2k^2}{a + b\cosh(2k\xi)},\tag{4.2.28}$$

where k is a free parameter and the parameters a and b are given by

$$a = (4\beta/3 - 1), \qquad b = (a^2 - 4\beta k^2/3)^{1/2}.\tag{4.2.29}$$

For simplicity, we choose a normalization such that $q = u_0^2 = 1$ and $\alpha = 2$.

In addition to the dark solitons in Eq. (4.2.28), the competing nonlinearity (4.2.27) supports a novel type of localized soliton known as the *kink soliton*. This kink soliton is described by the following exact solution of the generalized NLS equation (4.2.20) with $p = 1$ in Eq. (4.2.27):

$$\psi_k(\xi, z) = \left[\frac{q_c}{1 + \exp(\pm\Delta\xi)} \right]^{1/2} \exp(i\omega_c z), \qquad (4.2.30)$$

where $q_c = 3\alpha/4\beta$, $\omega_c = 2\beta q_c^2$, and $\Delta^2 = 3\alpha^2/4\beta$. The kink solution in Eq. (4.2.30) connects two stable background waves, namely, the CW background of the specific intensity q_c with the zero-intensity background.

In the case of the saturable nonlinearity, the function $F(I)$ varies with intensity as

$$F(I) = \frac{1}{2} \left[\frac{1}{(1 + aI)^p} - 1 \right], \qquad (4.2.31)$$

where p is the saturation index and the parameter a is related inversely to the saturation intensity I_s. This type of nonlinearity is used to analyze the effects of saturation of the nonlinear refractive index at high intensities (see Section 1.4). The case $p = 1$ applies for photorefractive materials [27, 28]. The $p = 2$ case is known to have explicit soliton solutions in the form of bright and dark solitons [29, 30]. These exact solutions reveal that dark solitons supported by the saturable nonlinearity can have a phase shift larger than the limiting value of π realized for the Kerr-type dark solitons for $v = 0$ [31].

Interaction of dark solitons in non-Kerr media does not differ qualitatively from the interaction of dark solitons in a Kerr medium. The new feature in both cases is that the interaction does not depend on the soliton phase. However, the soliton interaction become much more interesting when several different types of dark solitons coexist. For example, in the case of the cubic-quintic nonlinearity, the interaction can involve dark and antidark solitons [32] and may be affected by the presence of a kink soliton [33].

4.3 Instability-Induced Dynamics

In this section we focus on the stability issues associated with the dark solitons governed by generalized NLS equation (4.2.1). We first extend the stability criterion of Section 2.2 to the case of dark solitons. To discuss the dynamics of unstable dark solitons, we develop a multiscale asymptotic analysis and apply it to several specific cases.

4.3.1 Stability Criterion

The stability of bright solitons and their instability-induced dynamics associated with the generalized NLS equation have been studied extensively [34]–[39]. As discussed in Section 2.2, the stability of bright solitons is governed by the simple criterion that states that the solitons are stable as long as $dP/d\beta > 0$, where P is the total soliton power and β is the propagation constant. Moreover, an asymptotic approach has been developed [40] for describing the dynamics of bright solitons undergoing diffraction-induced collapse or switching to a new stable state. One can thus conclude that the instability scenarios are well understood for bright solitons.

In contrast, the stability of dark solitons associated with the generalized NLS equation (4.2.1) has attracted less attention. In fact, the stability issue created some confusion during the 1990s. For example, the concept of the complementary power was used as early as 1989 in this context [41, 42]. This analysis resulted in an incorrect conclusion that a black dark soliton (with zero intensity at its center) is always stable [38]. However, as indicated by numerical simulations, the complementary power does not govern the stability of dark solitons [31].

From a historical point of view, the problem of the stability of dark solitons was posed by the numerical simulations of Ref. [43], where the instability of the localized waves of rarefaction in a Bose-gas condensate—the so-called bubbles— was studied. These nontopological solitary waves are related in the one-dimensional case to the family of dark solitons associated with the cubic-quintic NLS equation, and they survive in higher dimensions [26]. Although the bubbles were found to be always unstable regardless of their dimensionality [44, 45], numerical simulations revealed that moving bubbles can be stabilized at nonzero velocities. Later, this phenomenon was explained through the multivalued dependence of the bubble energy on the bubble momentum [46, 47].

It was believed for many years that dark solitons (in particular, black solitons) are always stable [38]. However, in a 1995 numerical study [48] black solitons were found to be unstable in the case of the saturable nonlinearity with $p = 2$ in Eq. (1.4.3). The variational principle was used to link the instability to the existence of multivalued solutions in terms of the system invariants and to derive the instability condition using an asymptotic expansion.

It has been known for some time that the stability criterion for dark solitons can be expressed in terms of the renormalized soliton momentum as

$$\frac{dM_r}{dv} > 0. \tag{4.3.1}$$

This criterion is consistent with the results of numerical simulations [44, 49] and with the variational principle for bubble- and kink-type dark solitons [46, 48]. A rigorous proof of this stability criterion was provided in 1996 with the help of the Lyapunov function [50]. An asymptotic expansions near the instability threshold results in the same condition [51]. The first approach does not describe the instability itself but only proves the stability if it exists. The second method is only valid in the vicinity of the instability threshold and thus cannot determine the stability domain.

The stability of dark solitons can be formulated using the following alternative definition of the renormalized momentum [52]:

$$\mathbf{M}_r = \frac{i}{2} \int \int [(u - 1)\nabla_T u^* - (u^* - 1)\nabla_T u] \, d\mathbf{r}, \tag{4.3.2}$$

where ∇_T denotes the transverse part of the Laplacian operator in the multidimensional case. This definition was used in 1995 to analyze the transverse instability of dark-soliton stripes and vortex lines in the context of a multidimensional cubic NLS equation [53]. In this chapter we follow the approach based on the asymptotic multiscale analysis [51] because it allows us to describe both the *linear* and *nonlinear* regimes of the instability-induced dynamics of dark solitons. We also discuss the results of numerical simulations describing different scenarios for the evolution of unstable dark solitons.

4.3.2 Asymptotic Multiscale Analysis

The analysis of the instability-induced dynamics of dark solitons can be carried out in the framework of the perturbation theory if all of the soliton parameters change slowly during propagation [51]. This is the case near the instability threshold, where the velocity v of the unstable dark soliton is close to the critical value v_c determined from the condition $\partial M_s/\partial v|_{v=v_c} = 0$. Moreover, if we assume that perturbations remains small, we can introduce a small parameter $\varepsilon \ll 1$ that characterizes the amplitude of perturbations and look for solutions of Eq. (4.2.20) in the form of the following asymptotic multiscale expansion:

$$\psi = [\psi_s(\xi) + \varepsilon\psi_1(\xi;X,T) + \varepsilon^2\psi_2(\xi;X,T) + O(\varepsilon^3)]\exp[iR(X,T)], \qquad (4.3.3)$$

where

$$X = \varepsilon x, \quad T = \varepsilon z, \quad \xi = x - \frac{1}{\varepsilon}X_s(T), \quad X_s(T) = \int_0^T v(T')dT'. \qquad (4.3.4)$$

The quantities $v(T)$ and $R(X,T)$ represent the velocity and local phase of the soliton, respectively, X and T stand for *slow* variables, and $X_s(T)$ represents the position of the soliton center, where the amplitude of the dark soliton is minimum (the intensity dip).

Substituting Eq. (4.3.3) in Eq. (4.2.20), we obtain a set of equations for ψ_1, ψ_2, etc. Generally speaking, solutions of these equations may diverge as $\xi \to \infty$, indicating a breakdown in the asymptotic expansion. However, in the vicinity of the instability threshold, where $\partial M_s/\partial v \sim O(\varepsilon)$, the first-order correction ψ_1 can be expressed in an implicit form, ensuring a bounded solution for ψ_2 [51].

Using the method of matched asymptotics, the soliton velocity v, as derived from the conservation of momentum, is found to satisfy [51]

$$\frac{d}{dT}\left(\mu_s\frac{dv}{dT} + \frac{M_s}{\varepsilon}\right) = K_s\left(\frac{dv}{dT}\right)^2, \qquad (4.3.5)$$

where M_s is the soliton momentum given in Eq. (4.2.24) and the coefficients μ_s and K_s are defined as

$$\mu_s(v,q) = \frac{2c}{q}\left(\frac{\partial P_s}{\partial v}\right)^2 + \frac{q}{2c}\left(\frac{\partial S_s}{\partial v}\right)^2, \qquad (4.3.6)$$

$$K_s(v,q) = \frac{1}{(c^2-v^2)}\left[\frac{2cv}{q}\left(\frac{\partial P_s}{\partial v}\right)^2 + 2c\frac{\partial P_s}{\partial v}\frac{\partial S_s}{\partial v} + \frac{vq}{2c}\left(\frac{\partial S_s}{\partial v}\right)^2\right]. \qquad (4.3.7)$$

In these equations, P_s and S_s are given by Eqs. (4.2.24) and (4.2.26), respectively, and c is the limiting velocity of linear waves, found using $c^2 = qg'(q)$.

Using Eqs. (4.2.24) and (4.2.25) one can show that the first-order variations δM and δH for a perturbed dark soliton are related as $v\,\delta M + \delta H = 0$. However, neither the momentum nor the energy of the perturbed dark soliton is conserved. This feature reveals the essentially dissipative character of the dynamics of unstable dark solitons. It can be understood by noting that the instability is accompanied by the emission of radiation field that propagates away from the perturbed dark soliton in both directions. The radiation field in the $\pm x$ directions at $X = X_s(T)$ can be written as [51]

$$U^\pm(v,q) = W_\pm(v,q)\frac{dv}{dT}, \tag{4.3.8}$$

where the coefficients are given by

$$W_\pm(v,q) = -\frac{1}{c(c \mp v)}\left(c\frac{\partial P_s}{\partial v} \pm \frac{q}{2}\frac{\partial S_s}{\partial v}\right). \tag{4.3.9}$$

Equation (4.3.5) for the soliton velocity cannot always be integrated in an analytic form. Nevertheless, it describes in a simple manner both the dynamics of the dark soliton and the evolution of the radiation field through Eq. (4.3.8). The important features of soliton instability can be understood by solving Eq. (4.3.5) approximately. Consider first the linear approximation and substitute $v = v_0 + v_1 \exp(\lambda T)$ in Eq. (4.3.5), where v_0 is the (initial) velocity of the unperturbed dark soliton and v_1 is a small deviation from it. Linearizing Eq. (4.3.5) in terms of v_1, the instability growth rate is found to be

$$\lambda = -\frac{1}{\varepsilon}[\mu_s(v_0,q)]^{-1}\left(\frac{\partial M_s}{\partial v}\right)_{v=v_0}. \tag{4.3.10}$$

Noting that $\mu_s(v_0,q)$ is always positive, we conclude that dark solitons become unstable whenever $\partial M_s/\partial v|_{v=v_0} < 0$. Therefore, *all* dark solitons with a negative slope of the renormalized momentum $M_s(v)$ are unstable.

Next, we analyze the general conditions under which the instability of dark solitons can occur. Consider the small-amplitude limit in which $|v|$ is close to c. In this limit, dark solitons can be described by an effective KdV equation, and the derivative $\partial M_s/\partial v$ can be calculated explicitly [51]: It is found to be *always positive*. Therefore, small-amplitude dark solitons are *always stable*, and an instability can occur only for dark solitons of larger amplitude or smaller velocity v.

In the limit of small velocities ($v \to 0$), the slope $\partial M_s/\partial v$ can become negative whenever

$$\left.\frac{\partial S_s}{\partial v}\right|_{v=0} < \left.\frac{2P_s}{q}\right|_{v=0} < 0. \tag{4.3.11}$$

For many nonlinear models of dark solitons, the total phase shift S_s is given by a monotonic function rising from the value of $-\pi$ for $v = 0$ (a black soliton) to zero when $v \to c$ (gray solitons). This situation is typical for the the power-law nonlinearity of the form $F(I) \sim I^p$ (which includes the Kerr case for $p = 1$) as well as the nonlinearity with $F(I) = I + \beta I^2$. For these two forms, the slope $\partial S_s/\partial v$ is always positive, and instabilities of dark solitons cannot occur. However, for more complicated but still physically relevant nonlinear models, dark solitons can become unstable. Two examples are shown in Figure 4.2, where the renormalized momentum is plotted as a function of v/c. In both cases, a negative slope (dashed part) indicates unstable dark solitons.

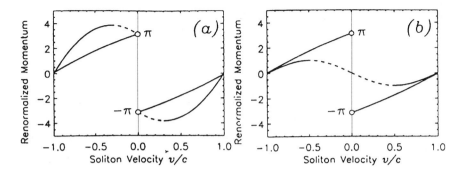

Figure 4.2: Momentum $M_s(v)$ versus v/c when the minimum intensity vanishes as $v \to 0$ (a) or remains finite (b). The straight lines correspond to the stable dark solitons in a Kerr medium. (After Ref. [51]; ©1996 APS.)

The nonlinear dynamics of such unstable dark solitons has been analyzed in Ref. [51]. In the region near the instability threshold, we can use the small-amplitude (but still nonlinear) approximation. Substituting $v = v_0 + \varepsilon V(T)$ in Eq. (4.3.5), we can reduce it to the form

$$\mu_s(v_0,q)\frac{dV}{dT} + \frac{1}{\varepsilon}\left(\frac{\partial M_s}{\partial v}\right)_{v=v_0} V + \frac{1}{2}\left(\frac{\partial^2 M_s}{\partial v^2}\right)_{v=v_0} V^2 = 0. \qquad (4.3.12)$$

This equation resembles the one governing the motion of a classical particle with the effective mass μ_s and velocity V under the action of a nonlinear dissipative force. The nature of instability in this case depends essentially on the sign of initial perturbation and the specific form of $M_s(v)$. In general, dark solitons of larger intensity and smaller velocity are unstable, while small-amplitude solitons with velocities close to the limiting velocity c are stable. Moreover, since the second derivative in the last term of Eq. (4.3.12) is positive, any perturbation with $V(0) \equiv V_0 > 0$ leads to a finite change in the soliton velocity. Thus, the instability process transforms an unstable dark soliton of velocity V_0 into a stable soliton of larger velocity and smaller amplitude. The solution of Eq. (4.3.12) in this case is given by [51]

$$V(T) = \frac{V_0 V_f}{(V_f - V_0)e^{-\lambda T} + V_0}, \qquad (4.3.13)$$

where $\lambda > 0$, as defined in Eq. (4.3.10), and the final velocity of stable solitons is given by

$$V_f = -\frac{2}{\varepsilon}\left(\frac{\partial M_s}{\partial v}\right)_{v=v_0}\left(\frac{\partial^2 M_s}{\partial v^2}\right)_{v=v_0}^{-1}. \qquad (4.3.14)$$

The preceding result is valid only if the renormalized momentum of the perturbed dark soliton is conserved during the soliton transformation. However, this quantity is conserved only up to the order ε^2. The change $\Delta M \equiv M_f - M_0$ can be calculated directly from Eq. (4.3.5) and is found to be

$$\Delta M = \varepsilon \int_{-\infty}^{+\infty} K_s(v,q) \left(\frac{dv}{dT} \right)^2 dT = \frac{\varepsilon^3}{6} \lambda (V_f - V_0)^2 K_s(v_0, q), \qquad (4.3.15)$$

where the coefficient K_s is given in Eq. (4.3.7). Since K_s can be positive or negative, transition from unstable to stable dark solitons can increase or decrease the soliton momentum. The direction of momentum change is determined by a balance between the radiation field U^+ propagating in the same direction as the perturbed dark soliton and the field U^- propagating in the opposite direction. The radiation fields can also be calculated analytically with the help of Eq. (4.3.8) and are given by

$$U^{\pm}(X,T) = \frac{\lambda}{4} |V_f - V_0| W_{\pm}(v_0, q) \operatorname{sech}^2 \left[\frac{\lambda (X \mp cT)}{2(c \mp v_0)} \right]. \qquad (4.3.16)$$

This functional form coincides with the sech^2-type profile of the dark-soliton solution found earlier in the small-amplitude approximation.

The evolution of the radiation field given by Eq. (4.3.16) obeys asymptotically the KdV equation given earlier. It is well known that the sech^2-type initial condition in the KdV equation leads to the formation of solitons only if the field amplitude is negative. In the opposite case, the initial profile (4.3.16) transforms into linear dispersive waves. A simple analysis indicates that in the limit $v_0 \to 0$, the coefficient ζ_+ is *positive* while the coefficient W_- is *negative* [51]. Moreover, the sign of the coefficient W_- remains unchanged throughout the instability region, so the counter-propagating radiation field, described by the function U^-, should always generate a stable (shallow) dark soliton as a result of the transformation of the primary unstable dark soliton. On the other hand, the radiation field, described by the function U^+, can either decay into dispersive waves if $W_+(v_0, q) > 0$ or produce an additional (stable) dark soliton when $W_+(v_0, q) < 0$.

4.3.3 Two Examples

In the case of competing nonlinearities (4.2.27), the dark solitons of Eq. (4.2.20) differ from those of the cubic NLS equation. For example, because of self-focusing at smaller intensities, the minimum amplitude of a dark soliton may not reach zero even for $v = 0$, provided the parameter of the cubic nonlinearity is large enough. As a result, the total phase shift $S_s(v)$ and, therefore, the renormalized momentum $M_s(v)$ tend to zero in the limits $v \to 0$ and $v \to c$. This behavior leads to the appearance of a negative slope of the $M_s(v)$ curve for small v, resulting in an instability of dark solitons, as seen in Figure 4.3.

Although the following analysis is qualitatively valid for any value of p in Eq. (4.2.27), we focus on the case $p = 1$ to be specific. The soliton amplitude k in Eq. (4.2.28) is related to the soliton velocity v as $k^2 + v^2 = \beta - 1$. The condition $k^2 > 0$ yields $|v| < c(\beta) = \sqrt{\beta - 1}$. This condition shows that the dark soliton exists only for $\beta > 1$ [see Figure 4.3(a)]. We now use the instability criterion and evaluate the slope of the function $M_s(v)$. The dark soliton becomes unstable for $\partial M_s(v)/\partial v < 0$. We find that the slope of $M_s(v)$ is negative only for $1 < \beta < 1.5$. In this range, the dark soliton has zero velocity ($v = 0$) and has a finite amplitude at the minimum point such

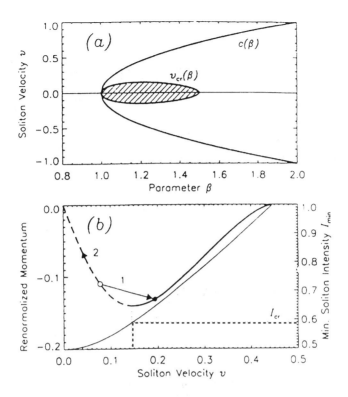

Figure 4.3: (a) The condition $|v| < c$ plotted as a function of β. The instability region $|v| < v_{cr}(\beta)$ is shown hatched. (b) Momentum $M_s(v)$ of the dark soliton for $\beta = 1.2$. The unstable negative-slope region is shown as dashed. The thin, solid curve shows the minimum intensity inside the dip. (After Ref. [51]; ©1996 APS.)

that $\Phi^2(0) = (3 - 2\beta)/\beta$. The function $M_s(v)$ in the specific case $\beta = 1.2$ is shown in Figure 4.3(b), while the instability region is shown in Figure 4.3(a).

To study the evolution of unstable dark solitons, one must perform numerical simulations. In one study [51], the dark soliton in Eq. (4.2.28) was perturbed by adding a symmetric perturbation with the amplitude ε such that

$$\psi_{\text{pert}}(\xi) = \left\{ \Phi(\xi) + \varepsilon[1 - \Phi^2(\xi)] \right\} e^{i\theta(\xi)}, \qquad (4.3.17)$$

and the soliton phase does not change. The initial velocity v_0 of the unstable soliton was chosen in the unstable region, while ε was varied in the interval $0.0001 < |\varepsilon| < 0.02$. The numerical simulations revealed two completely different scenarios for the dynamics of the unstable dark solitons, depending on the sign of ε.

The first scenario is observed for $\varepsilon > 0$, i.e., when initially the soliton amplitude is slightly decreased. Effectively, this corresponds to "pushing" of the unstable soliton toward the stable branch in Figure 4.3 (the arrow marked 1). Figure 4.4 shows an example of the soliton splitting for $v_0 = 0.04$. The initial exponential growth of the perturbation amplitude saturates at approximately $z = 55$, and the unstable dark

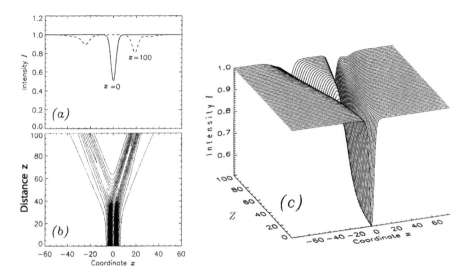

Figure 4.4: Splitting of an unstable dark soliton observed numerically for $\beta = 1.2$, $v_0 = 0.02$, and $\varepsilon = +0.005$. (a) Intensity profiles at $z = 0$ (solid curve) and $z = 100$ (dashed curve); (b) the contour plot and (c) the three-dimensional plot showing propagation dynamics. (After Ref. [51]; ©1996 APS.)

soliton splits into two stable solitons of smaller amplitudes, which move into opposite directions after the splitting. When the initial soliton velocity is selected far from the threshold value v_c, more than two secondary solitons are generated.

In the case $\varepsilon < 0$, the unstable dark soliton is "pushed" deeper into the instability region (the arrow marked 2 in Figure 4.3), resulting in the formation of two kinks that propagate in opposite directions. This scenario of the soliton instability can be referred to as the *collapse* of dark solitons. In this case, the exponential growth of the initial perturbation allows the minimum intensity to reach zero. Then the region of zero intensity starts to spread out while the background intensity increases in the outside region. This process results in the formation two kink-like structures of the form of Eq. (4.2.30).

The other example that we discuss here is the generalized NLS equation (4.2.20) with a saturable nonlinearity of the form (4.2.31). Numerical simulation show that dark solitons of saturable nonlinearities are stable for some parameter values. An instability analysis has also been performed for such solitons [48]. Although we focus on the $p = 2$ case, essentially the same quantitative features are expected for other values of p in Eq. (4.2.27).

Figure 4.5 shows the numerical results obtained by solving Eq. (4.2.20) with the form of $F(I)$ in Eq. (4.2.31) using $p = 2$ and $q = 1$. The parameter a is varied in the range 1–50 in part (a), which shows the curve $v = c(a)$ and the instability region where $v < v_c(a)$. The dashed line depicts the region in which a dark soliton has the total phase shift larger than π across it [31]. The renormalized momentum $M_s(v)$ and the total phase shift $S_s(v)$ are shown for $a = 6$ (curve 1) and $a = 12$ (curve 2) in parts

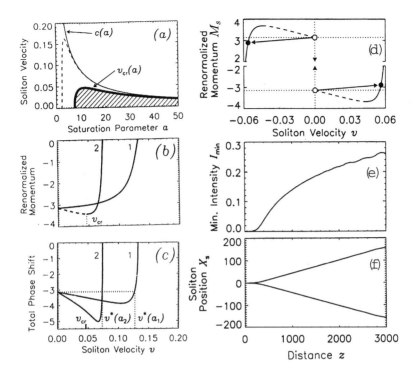

Figure 4.5: Instability of dark solitons in the case of saturable nonlinearity. (a) Instability region with (hatched area) and the curve $v = c$. (b) Soliton momentum $M_s(v)$ and (c) total phase shift $S_s(v)$ for $a = 6$ (curve 1) and $a = 12$ (curve 2). Critical velocity v_c corresponds to the instability threshold. (d) Momentum curves $M_s(v)$ showing transformation of an unstable black soliton into a stable gray soliton. (d) Changes in the dip depth and (e) dip position with z. (After Ref. [51]; ©1996 APS.)

(b) and (c), respectively. The appearance of a large phase shift is indicative of the unstable nature of such dark solitons. However, among all dark solitons with the phase shift larger than π, there exist both stable solitons, with smaller amplitudes and larger velocities, and unstable solitons, with larger amplitudes and smaller velocities.

The right panel of Figure 4.5 shows what happens to unstable solitons. It turns out that the instability-induced soliton dynamics in the case of saturable nonlinearity displays features that are quite different from those found in the case of competing nonlinearities. More precisely, an unstable black soliton transforms itself gradually into a stable gray soliton whose velocity is larger than that of the original soliton. The speed change occurs because the slope of the function $M_s(v)$ must change from negative to positive for stable solitons, as shown in Figure 4.5(d). Despite a relatively small change in the soliton velocity, the minimum intensity within the dip shown in part (e) and the dip position shown in part (f) both exhibit an exponential growth initially and then saturate at the level corresponding to the stable gray soliton. The direction in which the gray soliton moves depends on the sign of initial perturbation of the black soliton, as seen in Figure 4.5(a). The unstable gray solitons develop in a similar manner.

4.4   Perturbation Theory

Renormalized integrals of motion for dark solitons allow us to apply a straightforward technique for describing the perturbation-induced dynamics. Such an approach is simpler and more direct than the technique based on the inverse scattering transform method [54]. In this section we apply the invariant-based perturbation theory for studying several important physical systems.

4.4.1   Constant-Background Case

First, we consider the case of a *constant background*, in which the perturbation does not change the parameters of the CW solution. The basic idea is to write the generalized NLS equation as a perturbed NLS equation in the form

$$i\frac{\partial u}{\partial z} + \frac{1}{2}\frac{\partial^2 u}{\partial x^2} - |u|^2 u = \varepsilon P(u), \tag{4.4.1}$$

where the term $\varepsilon P(u)$ on the right side stands for a small perturbation. Since we assume that the perturbation does not change the CW background, $P(u)$ should vanish as $|x| \to \infty$. It is then useful to eliminate u_0 by introducing new, renormalized variables as

$$\zeta = u_0^2 z, \qquad \xi = u_0 x, \qquad u(x,z) = u_0 v(x,z)\exp(-iu_0^2 z) \tag{4.4.2}$$

and obtain the following equation for the new field $v(\zeta,\xi)$:

$$i\frac{\partial v}{\partial \zeta} + \frac{1}{2}\frac{\partial^2 v}{\partial \xi^2} + (1-|v|^2)v = \varepsilon \tilde{P}(v), \tag{4.4.3}$$

where $\varepsilon \tilde{P}(v)$ is the renormalized perturbation. When $\varepsilon = 0$, the dark-soliton solution of this equation can be written in the form

$$v(\zeta,\xi) = \cos\phi\tan\Theta + i\sin\phi, \qquad \Theta = \eta(\xi - \Omega\zeta), \tag{4.4.4}$$

where $\eta = \cos\phi$ and $\Omega = \sin\phi$ are related to the soliton phase angle ϕ introduced in Section 1.5 ($|\phi| < \pi/2$). As discussed there, ϕ describes the "darkness" of the soliton ($\phi = 0$ for a black soliton).

To treat analytically the influence of a small perturbation $\varepsilon \tilde{P}(v)$ on the parameters of the dark soliton of Eq. (4.4.4), we use the so-called *adiabatic approximation* [54], according to which the soliton maintains its shape but whose parameters evolve slowly in z. Thus, the perturbed soliton is given by Eq. (4.4.4), but the parameter Θ evolves with ζ as

$$\Theta = \cos\phi(\zeta)\left[\xi - \int \sin\phi(\zeta')d\zeta'\right]. \tag{4.4.5}$$

To study how the soliton phase $\phi(\zeta)$ evolves, we start from the renormalized Hamiltonian of the unperturbed system,

$$H_r = \frac{1}{2}\int_{-\infty}^{\infty}\left[\left|\frac{\partial v}{\partial \xi}\right|^2 + (|v|^2-1)^2\right]d\xi. \tag{4.4.6}$$

Note that $H_r = \frac{4}{3}\cos^3\phi$ for the soliton solution (4.4.4) [see also Eq. (4.2.14)]. Calculating the derivative $dH_r/d\zeta$ and using Eq. (4.4.3), we find [55]

$$\frac{d\phi}{d\zeta} = \frac{\varepsilon\sec^2\phi}{8\sin\phi}\text{Re}\left(\int_{-\infty}^{\infty}\tilde{P}(v)\frac{\partial v^*}{\partial\zeta}\,d\xi\right), \tag{4.4.7}$$

where the integral should be calculated in the adiabatic approximation using the solution given in Eq. (4.4.4). The same equation is obtained when an equivalent approach based on the Lagrangian formulation is used [56].

If the perturbation $\varepsilon\tilde{P}(v)$ in Eq. (4.4.7) does not vanish as $|x| \to \infty$, it will certainly affect the background wave. This is what happens for dissipative perturbations, which lead to a slow decay of the background amplitude [57]. Taking the limit $|x| \to \infty$ and focusing on the evolution of the nonpropagating background $u_b(z)$, we obtain the following equation for the background amplitude u_b:

$$i\frac{du_b}{dz} - |u_b|^2 u_b = \varepsilon P(u_b). \tag{4.4.8}$$

Equation (4.4.8) allows us to describe the background evolution in the presence of perturbations. A general solution of Eq. (4.4.8) can be written in the form $u_b(z) = u_0(z)\exp[i\theta(z)]$, where $u_0(z)$ and $\theta(z)$ characterize perturbation-induced changes in the background amplitude and phase, respectively. To describe the evolution of a dark soliton on this varying background we should *remove* the background using the transformation

$$u(z,x) = u_0(z)e^{i\theta(z)}v(x,z) \tag{4.4.9}$$

and find an effective nonlinear equation for the function $v(x,z)$. In many cases the resulting equation can be transformed into a perturbed NLS equation (4.4.3) after a change of the variables, which allows us to apply the result given by Eq. (4.4.7).

In principle, to derive the adiabatic equations for the soliton parameters, we can employ the perturbation theory based on the inverse scattering transform, similar to that developed for bright solitary waves of the NLS equation and other nonlinear models [54]. Development of a comprehensive perturbation theory for dark solitons based on the inverse scattering transform method still remains an open problem. Such a theory should describe not only the evolution of the soliton parameters but also the structure of radiation generated under the action of an external perturbation (see Ref. [58] for discussion of the importance of the radiative effects).

4.4.2   Effect of Gain and Loss

The propagation of spatial solitons is usually associated with two-photon absorption (TPA), a phenomenon that accompanies any third-order nonlinearity and manifests as an intensity-dependent absorption of light [59]. In some cases, TPA is replaced with the intensity-dependent gain. In the presence of either TPA or nonlinear gain, the stationary self-localized states are no longer possible, but stationary solutions in the form of a fundamental dark soliton can exist when TPA is compensated by the gain.

To describe the effects of TPA on dark solitons as simply as possible, we first consider the NLS equation in a Kerr medium and modify it as follows [57, 59]:

$$i\frac{\partial u}{\partial z} + \frac{1}{2}\frac{\partial^2 u}{\partial x^2} - |u|^2 u = -iK|u|^2 u, \tag{4.4.10}$$

where $K = \alpha_2/2k_0 n_2$ is the normalized TPA coefficient, $k_0 = 2\pi/\lambda$, and α_2 and n_2 are the TPA and nonlinear-index coefficients, respectively. The case of nonlinear gain can be treated by allowing K to be negative.

In the absence of TPA ($K = 0$), Eq. (4.4.10) has the CW solution $u = u_0\exp(-iu_0^2 z)$ describing the background wave, and this solution is modulationally stable. Even a small amount of nonlinear absorption or gain leads to the attenuation of the CW background. As a result, its amplitude and phase change with z as

$$u_0(z) = \frac{u_0(0)}{\sqrt{1 + 2Ku_0^2(0)z}}, \tag{4.4.11}$$

$$\theta(z) = \int_0^z u_0^2(z')dz' = \frac{1}{2K}\ln\left[1 + 2Ku_0^2(0)z\right]. \tag{4.4.12}$$

To separate the evolution of the background from the dark soliton, we apply the transformation, $u(z,x) = u_0(z)e^{i\theta(z)}v(z,x)$ and obtain the following equation for v:

$$i\frac{\partial v}{\partial \zeta} + \frac{1}{2}\frac{\partial^2 v}{\partial \xi^2} - (|v|^2 - 1)v = -iK(|v|^2 - 1)v, \tag{4.4.13}$$

where the new coordinates, ζ and ξ, are related to z and x though the relations

$$\zeta = \int_0^z u_0^2(z)dz, \qquad \xi = \int_0^x u_0(z)dx. \tag{4.4.14}$$

After the transformation, the resulting equation has a vanishing perturbation at infinity and can be treated by the preceding perturbation theory developed for dark solitons. Equation for the soliton phase angle ϕ in the primary variables z takes the form

$$\frac{d\phi}{dz} = \frac{1}{3}Ku_0^2(z)\sin(2\phi), \tag{4.4.15}$$

where the background amplitude $u_0(z)$ evolves according to Eq. (4.4.11). Equation (4.4.15) can easily be integrated to obtain the final result

$$\phi(z) = \tan^{-1}\left\{\tan\phi(0)\left[1 + 2Ku_0^2(0)z\right]^{1/3}\right\}. \tag{4.4.16}$$

As an application of this result, we consider the effect of TPA on the steering angle χ of dark solitons [60]. It is easy to show that the total shift x_0 of the dark soliton in the transverse direction is given by the integral

$$x_0(z) = \int_0^z dz' u_0(z')\sin\phi(z'). \tag{4.4.17}$$

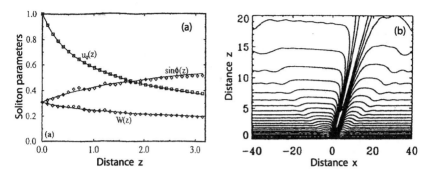

Figure 4.6: (a) Evolution of $u_0(z)$, $\sin\phi(z)$, and $W(z)$ in a Kerr medium with TPA ($K = 0.05$) and $\phi(0) = 0.1\pi$. The results of numerical simulation are shown with the diamond marks. (b) Contour plots showing the evolution of a dark soliton under the same conditions. (After Ref. [55]; ©1994 APS.)

Thus, the steering angle χ can be defined through the local transverse velocity as

$$V(z) = dx_0/dz \equiv \tan\chi(z) = u_0(z)\sin\phi(z). \qquad (4.4.18)$$

We can draw an important conclusion from this equation. When a dark soliton propagates in the presence of TPA on a decaying background $u_0(z)$, the function $\sin\phi(z)$ grows slowly, keeping the product in Eq. (4.4.18) almost constant, at least for initially small values of ϕ [57]. As a consequence, the steering angle for switching devices based on dark soliton is almost preserved in a Kerr nonlinear medium, even in the presence of TPA.

Figure 4.6(a) shows this feature through the evolution of the background $u_0(z)$, $\sin\phi(z)$, and the transverse velocity $W(z) = u_0(z)\sin\phi(z)$. As seen there, the steering angle is almost preserved, provided $\phi(0)$ is small. The analytical results (solid curves) based on Eqs. (4.4.11) and (4.4.16) are in agreement with the results of the numerical simulations (diamond marks). Small deviations between the two are caused by the escaping radiation, which slightly changes the intensity of the background as seen clearly in Figure 4.6(b), where the evolution of the dark soliton is shown in the x–z plane.

It is important to compare the result (4.4.18) with the corresponding result of linear absorption described by the perturbation $\varepsilon P(u) = -i\gamma u$ on the right side of Eq. (4.4.10) instead of the term $-iK|u|^2u$. In this case, the background wave decays exponentially as $u_0(z) = u_0(0)\exp(-\gamma z)$, and the soliton phase angle does not change ($d\phi/dz = 0$). A similar conclusion follows from the analysis presented in Refs. [61] and [62], where a different method that does not allow generalization to the case of nonlinear losses was used.

4.4.3 Stabilization of Dark Solitons

Both the linear and nonlinear absorptions affect the background, but they can be compensated by introducing the gain. If the gain is linear (intensity independent), the background can be stabilized but the dark soliton remains unstable [55]. This result from perturbation theory is in agreement with the existence of the exact solution of the underlying Ginzburg–Landau equation, which is known to be unstable [63, 64].

It turns out that the dark soliton can be stabilized if the gain is also made nonlinear [65]. In this subsection, we consider this case and include the higher-order terms using the following form for the perturbation:

$$\varepsilon P(u) = \delta u + \gamma_1 |u|^2 u + \gamma_2 |u|^4 u. \tag{4.4.19}$$

The dark-soliton angle is then found to evolve as

$$\frac{d\phi}{dz} = \left\{ \frac{\gamma_1}{3} u_0^2(z) - \frac{2\gamma_2}{15} u_0^4(z) [2\cos^2\phi - 5] \right\} \sin(2\phi), \tag{4.4.20}$$

where the background field $u_0(z)$ satisfies

$$\frac{du_0}{dz} = -\left(\delta u_0 + \gamma_1 u_0^2 + \gamma_2 u_0^4 \right) u_0, \tag{4.4.21}$$

The preceding equation shows that $u_0 = 1$ is a stable stationary solution provided $\delta + \gamma_1 + \gamma_2 = 0$ and $\gamma_1 + 2\gamma_2 > 0$. This condition allows one to control dark solitons by varying the nonlinear gain [65]. Moreover, the use of nonlinear gain leads to stable bound states of dark solitons [66]. This conclusion has been verified through direct numerical simulations of the perturbed NLS equation.

Dark solitons can also be stabilized through synchronized phase modulation [67]. In this approach, the perturbation in the NLS equation appears in the form of $\varepsilon P(u) = \mu \cos(\omega x)u$, where μ is the phase-modulation coefficient. The modulation frequency ω is synchronized with the initial pulse separation for a sequence of dark solitons. This perturbation leads to a nontrivial variation of the soliton phase, which can display stable dynamics near the fixed point $\phi = 0$ (corresponding to a black soliton with the zero minimum amplitude at its center) because of phase locking between the soliton and periodically varying perturbation.

An interesting method to compensate for the amplitude variation of a dark soliton in a lossy medium has been suggested in Ref. [68]. In this method, dark solitons are stabilized by using phase-sensitive amplification and spectral filtering together. In a periodically amplified system, spectral filtering inhibits the sideband instabilities that can occur for both the CW waves and solitons [69]. At the same time, the phase-sensitive amplification inhibits the destabilization of the constant-intensity background wave caused by filtering [68].

4.5   Temporal Dark Solitons

The perturbation theory developed in the preceding section can also be applied for analyzing temporal dark solitons. The most common nonlinear medium used for temporal dark solitons is the same used for temporal bright solitons—namely the optical fiber—except that dark solitons require normal GVD ($\beta_2 > 0$). In the subpicosecond regime, it becomes necessary to include several higher-order dispersive and nonlinear effects, similar to those studied in Section 3.6 in the context of bright solitons [70]. In this section we focus on the higher-order effects on temporal dark solitons.

4.5.1 Raman-Induced Frequency Shift

As discussed in Section 3.6, the phenomenon of stimulated Raman scattering (SRS) can lead to the Raman-induced frequency shift for bright solitons [71]–[73]. In the case of dark solitons, the occurrence of such a frequency shift during the initial stages of evolution [74] leads eventually to the decay of dark solitons [75]–[77]. As discussed in Section 3.6.3, the SRS effects can be included using a perturbed NLS equation in the form [78]

$$i\frac{\partial u}{\partial z} - \frac{1}{2}\frac{\partial^2 u}{\partial \tau^2} + |u|^2 u = \tau_R u \frac{\partial |u|^2}{\partial \tau}, \tag{4.5.1}$$

where $\tau_R = T_R/T_0$ is the Raman parameter introduced in Eq. (3.610); typically $T_R = 3$ fs for optical fibers [70]. As was done in Chapter 3, we have replaced x with normalized time τ to emphasize that we are dealing with the temporal solitons. The dispersive term appears with a negative sign in Eq. (4.5.1) because dark solitons form only when the group-velocity dispersion (GVD) is normal. In contrast, the nonlinear Kerr term is positive for optical fibers.

The perturbation theory developed for spatial solitons can be applied to Eq. (4.5.1) if we change first the sign of all terms by multiplying this equation by -1 and replacing z by $-z$. The effect of SRS on dark solitons has been analyzed using both the analytical and numerical techniques [74]–[80]. The small-amplitude limit was considered as early as 1990 using an asymptotic approach based on the perturbed KdV equation [75]. The general formula describing the evolution of dark solitons in the presence of SRS was obtained in a 1993 study [77].

To consider the effects of delayed nonlinear response more accurately than is allowed by the approximated form of the perturbation in Eq. (4.5.1), we can use the general form of the nonlinearity given in Eq. (3.6.14). Using the response function from Eq. (3.6.15), the perturbed NLS equation can be written as

$$-i\frac{\partial u}{\partial z} + \frac{1}{2}\frac{\partial^2 u}{\partial \tau^2} - |u|^2 u = f_R u \int_{-\infty}^{\tau} h_R(\tau - t)|u(t)|^2 dt, \tag{4.5.2}$$

where $h_R(t)$ is the Raman response function and the numerical factor $f_R \approx 0.18$ denotes the Raman contribution to the total nonlinear response. Since the quantity $f_R h(t)$ is relatively small, we can treat the right side of Eq. (4.5.2) as a perturbation. Applying the perturbation theory developed in the preceding section, we obtain the following equation for the soliton phase angle:

$$\frac{d\phi}{dz} = f_R u_0 \cos\phi \int_{-\infty}^{\infty} h(t \sec\phi/u_0) F(t)\, dt, \tag{4.5.3}$$

where

$$F(t) = \frac{2[t(3 - \tanh^2 t) - 3\tanh t]}{\sinh^2 t \tanh^2 t}. \tag{4.5.4}$$

The integral in Eq. (4.5.3) can be evaluated approximately when the Raman response time is short compared with the soliton width $(u_0 \cos\phi)^{-1}$. In this case, we can expand the function $F(t)$ into a Taylor series as $F(t) \simeq F(0) + tF'(0)$ and obtain the result [55, 77]

$$\frac{d\phi}{dz} = \frac{4}{15}\tau_R u_0^3 \cos^3\phi, \tag{4.5.5}$$

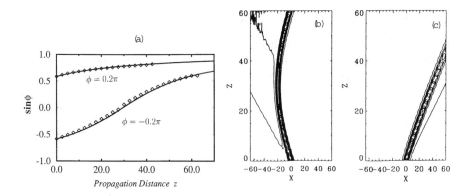

Figure 4.7: (a) Changes in the dark-soliton angle induced by SRS when $\varepsilon = 0.1$ for initial values of $\phi = \pm 0.2\pi$. The diamond symbols show the results of numerical simulations. The contour plots show the evolution of a dark soliton for (b) $\phi = -0.2\pi$ and (c) $\phi = 0.2\pi$. (After Ref. [55]; ©1994 APS.)

where $\tau_R = f_R \int_{-\infty}^{\infty} t h(t)\, dt$. The same equation is obtained if we use the perturbation in the form of Eq. (2.1.9).

Figure 4.7 shows the evolution of a dark soliton in the presence of the Raman contribution using $\tau_R = 0.1$. The solid curves in part (a) are obtained from Eq. (4.5.5) for $\phi(0) = -0.2\pi$ and $\phi(0) = +0.2\pi$. The diamond symbols show for comparison the results of numerical simulations. It is evident that Eq. (4.5.5) describes the soliton dynamics quite well. The evolution of the dark soliton for two initial-phase values is shown in parts (b) and (c) . Notice how the trajectory bends. Physically speaking, the Raman-induced frequency shift transforms a dark soliton with arbitrary parameters into a small-amplitude soliton through a continuous shift in the frequency and position of the soliton.

4.5.2 Dark-Soliton Jitter

For long-distance propagation of temporal dark solitons, fiber losses should be compensated using a periodic amplification scheme. As discussed in Section 3.5.3, an undesirable effect of periodic amplification is the timing jitter for bright solitons due to the random frequency shifts [81]–[83]. This jitter is called the *Gordon–Haus jitter*.

In the case of bright solitons, timing jitter was evaluated in Section 3.5.3 using a perturbation technique. The same result can be obtained using the power and momentum invariants, P and M. We use these two invariants to find the timing jitter of dark solitons. Of course, for dark solitons we need to use the renormalized invariants P_r and H_r, defined earlier in Eqs. (4.2.13) and (4.2.14). Calculating the values of M_r and H_r for the dark-soliton solution, we find $M_r = 2\sin\phi\cos\phi$ and $H_r = (4/3)u_0\cos^3\phi$.

As usual, we first remove the background part by applying the transformation $u = u_0\exp(iu_0^2 z)$ to the NLS equation. The function $v(z, \tau)$ then satisfies the modified NLS equation [84]

$$i\frac{\partial v}{\partial z} - \frac{1}{2}\frac{\partial^2 v}{\partial \tau^2} + u_0^2(|v|^2 - 1)v = 0. \tag{4.5.6}$$

Similar to the case of bright solitons, we follow the approach of Ref. [81] and consider the effect of adding a small perturbation δv to the soliton field v_s so that $u = u_0(v_s + \delta v)\exp(iu_0^2 z)$. Physically, δv represents a fluctuation added by the spontaneous emission at the amplifiers located periodically along the link. The perturbation δv is a Gaussian stochastic process with the statistical properties

$$\langle \delta v(t) \rangle = \langle \delta v^*(t) \rangle = 0, \qquad \langle \delta v(t)\delta v^*(t') \rangle = (D/u_0^2)\delta(t - t'), \tag{4.5.7}$$

where the parameter D is proportional to the mean number of photons per mode at the amplifier output. To the first order, such a perturbation produces a change in the soliton parameters, the radiation corrections being of the second order in δv.

The frequency fluctuation $\delta\Omega$ from its unperturbed value $\Omega = u_0\sin\phi$ depends on the fluctuations of the background amplitude δu_0 and of the internal phase angle $\delta\phi$. From the conservation of the Hamiltonian and the momentum, we find that $\delta\Omega \equiv \alpha(\phi)\delta H_r + \beta(\phi)\delta M_r$ is given by

$$\delta\Omega = 2\,\mathrm{Im}\left\{\int_{-\infty}^{\infty} \delta v\left[\alpha(\phi)\frac{\partial v^*}{\partial z} - \beta(\phi)\frac{\partial v^*}{\partial t}\right]dt\right\}, \tag{4.5.8}$$

$$\alpha(\phi) = \frac{3\sin\phi}{4\cos^3\phi}, \qquad \beta(\phi) = \frac{u_0(3\sin^2\phi + \cos^2\phi)}{4\cos^3\phi}. \tag{4.5.9}$$

We can now calculate the variance of frequency fluctuations by calculating $\langle\delta\Omega^2\rangle$ using Eqs. (4.5.7) and Eq. (4.5.8). The result is found to be

$$\langle\delta\Omega^2\rangle = \frac{2D}{u_0^2}\int_{-\infty}^{\infty}\left|\alpha(\phi)\frac{\partial v}{\partial z} - \beta(\phi)\frac{\partial v}{\partial t}\right|^2 dt. \tag{4.5.10}$$

Substituting the soliton solution into Eq. (4.5.10) and performing the integration, we obtain the final result [84]

$$\langle\delta\Omega^2\rangle = \frac{8D}{3u_0^2}\cos^3\phi[\alpha(\phi)u_0\sin\phi - \beta(\phi)]^2 = \frac{D}{6}\cos\phi. \tag{4.5.11}$$

To make a comparison between the bright and dark solitons, we should consider a dark soliton of the same amplitude as the bright soliton and set $u_0 = 1$ and $\phi = 0$. For this choice of parameters, we obtain the simple relation

$$\langle\delta\Omega^2\rangle_{\mathrm{dark}} = \frac{1}{2}\langle\delta\Omega^2\rangle_{\mathrm{bright}}. \tag{4.5.12}$$

As discussed in Section 3.5.3, frequency fluctuations result in different arrival times of different solitons at the output of the fiber link. This timing jitter has a variance σ_t^2 that is directly proportional to $\langle\delta\Omega^2\rangle$ and is thus reduced by by a factor of 2 for dark solitons. The timing jitter σ_t itself is reduced by a factor of $\sqrt{2}$ when dark solitons are used in place of the bright solitons. This result is in agreement with the numerical simulations based on the NLS equation [85]. The effects of SRS on the timing jitter has also been studied in Ref. [86]. The simultaneous presence of amplifier-induced frequency fluctuations and the Raman-induced frequency shifts can, in some cases, reduce the timing jitter even more. For example, in the case of the background fluctuations, dark-soliton jitter completely vanishes at a specific distance.

4.5.3 Third-Order Dispersion

A 1995 experimental demonstration of long-distance signal transmission with dark solitons [87] has led to renewed efforts to explore this type of signal coding for optical communications. The use of dark solitons requires considerably more power because the signal remains in the "on" state most of the time. The power level can be reduced if the operating wavelength is closer to the zero-dispersion wavelength of the fiber. However, dark solitons are then influenced by the third-order dispersion (TOD).

The effects of the TOD on bright solitons have been studied extensively [88]–[90]. Under the influence of TOD, a bright soliton develops a long tail that extends far from the soliton peak [90]. The tail amplitude depends on the TOD parameter β_3 as $\exp(-1/\beta_3)$ and can be calculated using an asymptotic analysis [89]. This result is consistent with the robustness hypothesis [91], according to which autonomous, homogeneous, Hamiltonian deformations of integrable equations lead to solitary-wave solutions that radiate beyond all orders of perturbation.

The effect of TOD on dark solitons was investigated as early as 1991 in the small-amplitude approximation [92]. The effective KdV equation is valid in this case and shows that the TOD does not affect strongly the existence and properties of gray solitons [21]. However, an interesting feature of dark solitons was found to occur near $\beta_2 = 0$. Somewhat surprisingly, dark solitons can exist in this region as "humps" (instead of "dips") on a constant background [23]. Such *antidark solitons* match exactly the conventional dark solitons at the critical points, where they are described by a modified versions of the KdV equation. Since, the dark and antidark solitons can exist in the same region of the soliton parameters, a head-on collision between them can occur. Such head-on collisions have been investigated analytically using an asymptotic expansion technique to show that the solitons can preserve their identities only to the second order [25]. The effect of the TOD on dark solitons of large or moderate amplitudes has been analyzed analytically as well as numerically [93, 94].

The propagation of optical pulses near the zero-dispersion wavelength of optical fibers is described by the perturbed NLS equation [70]

$$i\frac{\partial u}{\partial z} - \frac{1}{2}\frac{\partial^2 u}{\partial \tau^2} + |u|^2 u = i\delta_3 \frac{\partial^3 u}{\partial \tau^3}, \qquad (4.5.13)$$

where $\delta_3 = \beta_3/(6\beta_2 T_0)$ governs the magnitude of TOD. When $\delta_3 \neq 0$, the existence of localized solutions can be revealed by analyzing the asymptotic behavior. For this purpose, we substitute $u = (u_0 + w)\exp(2iu_0^2 z)$ and linearize Eq. (4.5.13) for small w. Then, for a plane wave moving with the velocity v, we seek a solution in the form $w = (w_r + iw_i)\exp(ik\zeta)$, where $\zeta = \tau - vz$, and obtain the following equation for $\kappa \equiv k^2(v)$:

$$(v - \delta_3 \kappa)^2 = (\kappa/4 + u_0^2). \qquad (4.5.14)$$

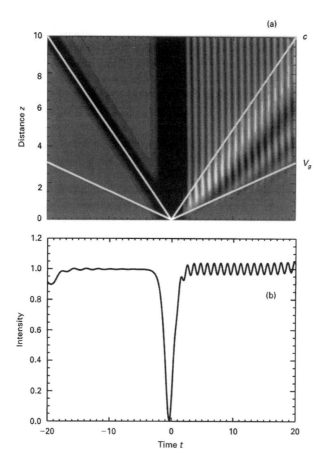

Figure 4.8: (a) Intensity contours showing the formation of an oscillating tail on a black soliton propagating inside an optical fiber with $\delta_3 = 0.18$. The white lines show light propagation with velocities c and V_g. (b) Intensity profile of the dark soliton at $z = 10$. (After Ref. [94]; ©1996 OSA.)

This quadratic equation has two roots. For $|v| < c = u_0$, one root (say, κ_-) is always negative and can be ignored because it leads to an exponentially decaying perturbation. The other positive root, say, κ_+, describes the formation of an oscillating tail on the dark soliton background.

Existence of nonvanishing tails can usefully be viewed as a resonant generation of linear dispersive waves. This process takes place provided the speed v of the dark soliton coincides with the phase velocity V_{ph} of the dispersive waves. Indeed, the condition $V_{ph} = v$ leads immediately to Eq. (4.5.14). From a physical point of view, this results implies that the solitary wave acts as a source of trailing oscillations, with the leading front propagating with the wave group velocity V_g. This process is demonstrated in Figure 4.8 for a black soliton for $\delta_3 = 0.18$. At the early stages of soliton evolution some energy is radiated away because of the TOD-induced perturbation of the dark soliton. This radiation propagates with speed c and quickly separates from the dark soliton, as seen in part (a). The radiation also creates a small-amplitude dark soliton (moving to the left in Figure 4.8) that is stable in the presence of TOD. At the same time, an oscillating tail is formed on the right side of the soliton, and its front propagates with the group velocity V_g, which is different from both the velocity of the dark soliton v and the limiting velocity c. Because of a continuously growing tail, the dark soliton decays such that its amplitude decreases and its velocity increases. A similar behavior is found to occur for initially gray solitons of moderate amplitudes [94].

The calculation of the oscillation amplitude is a delicate task because it requires the inclusion of all orders in the asymptotic expansion [89, 95]. However, a qualitatively correct result can be obtained by considering the linear equation for the soliton perturbations $w = u - u_s$. The corresponding solution is somewhat cumbersome [93], but it has a general structure of the form

$$w = A\theta(\zeta)\theta(-\zeta + V_g z)\sin(\sqrt{\kappa_+}\zeta + \phi), \qquad (4.5.15)$$

where $\zeta = t - vz$ and the step function $\theta(x) = 1$ for $x > 0$ but 0 otherwise. The tail amplitude A depends on the soliton parameters as [94]

$$A \sim CB(\kappa_+)\mathrm{csch}\left(\frac{\pi\sqrt{\alpha}}{u_0\cos\theta}\sqrt{\kappa_+}\right), \qquad (4.5.16)$$

where C is a constant and B is an algebraic function of κ_+, the positive root of Eq. (4.5.14). Using the first-order expansion for small β_3, $k = \sqrt{\kappa_+} \approx \frac{1}{2}\delta_3$, one can easily show that the tail amplitude depends on β_3 exponentially, i.e., in a fashion similar to that for bright solitons [89]. However, a special feature of dark solitons is the presence of the factor $\sqrt{1 - v^2/c^2}$ in the exponent, because of which the radiation amplitude becomes exponentially small for any fixed value of β_3 in the limit $|v| \to c$. This feature underscores the validity of the small-amplitude approximation. It indicates that the oscillation amplitude of the tail becomes small as $|v| \to c$, and in this limit the radiation emitted by the soliton is negligible.

In some specific cases of the generalized NLS equation and for some specific choice of the parameters, the exact analytical tanh-like solutions for dark solitons are known to exist even in the presence of higher-order linear and nonlinear dispersion [96]–[98]. Such dark solitons coexist with the radiation, and they can be understood using the concept of the embedded solitons discussed in Chapter 2. It is likely that such solitons are unstable, and they would emit radiation with any perturbation of the soliton profile or any change in the parameters values.

4.6   Experimental Results

Although the formation of dark solitons in optical fibers was predicted in 1973 [99], dark solitons remained a mathematical curiosity until 1987. In contrast, bright solitons were observed experimentally as early as 1980 [100], and their use for optical communications was pursued vigorously [73]. This progress provided an incentive for studying temporal dark solitons, and they were observed experimentally by 1987. The experiments on spatial dark solitons began around the same time.

4.6.1   Temporal Dark Solitons

Generation of dark solitons

The generation of dark solitons in optical fibers is less straightforward than the case of bright solitons because dark solitons require a localized change in the optical phase as well as a finite amplitude of the background wave. The early experiments used an optical pulse for the background whose width was relatively large compared with the width of the dark soliton, which appeared as a hole in the center of the background pulse [101]–[103]. In later studies, a quasi-continuous train of dark solitons was produced by colliding two optical pulses inside an optical fiber [104]–[106]. It was suggested as early as 1990 that a continuous train of dark solitons can be generated using electro-optic modulators [107]. Such a train can also be generated using temporal shaping with spectral filtering [108] or by beating together two CW signals in a dispersion-tapered fiber [109]. An information-coded stream of dark solitons was produced using a $LiNbO_3$ modulator in a 1995 data-transmission experiment [87].

The technique of spectral filtering for generating dark solitons inside optical fibers was used in a 1987 experiment [101]. The basic idea consisted of creating an optical pulse with a π phase jump at its center. When such a pulse was propagated inside an optical fiber, a dark soliton appeared in the form of a central hole at the pulse center. It was this hole (or the dark pulse) that had properties similar to those of a fundamental dark soliton. This experiment did not demonstrate the propagation of dark solitons over long fiber lengths because of rather large fiber losses at the operating wavelength of 600 nm. In fact, the dispersion length for soliton propagation ($\sim$220 m) was larger than the attenuation length of 140 m.

In a 1988 experiment [103], an even dark pulse (no phase jump) evolved into a symmetric pair of small-amplitude dark pulses that propagated unmistakably as solitons in accordance with the predicted numerical results [2] and a general theory of the pair generation [14]. In this experiment, 0.3-ps dark pulses were produced on a background pulse of 100-ps duration and were launched into a 10-m-long, polarization-maintaining, silica fiber. The output signal was measured for several input power levels using an autocorrelator. At high powers (>9 W), the input pulse evolved into two dark solitons moving relative to the background with opposite velocities. This behavior was different than that observed at low powers, for which the nonlinear effects were negligible. The experimental results were in excellent agreement with the numerical solutions based on the NLS equation. The creation of a pair of dark solitons with opposite velocities and equal amplitudes is related to the fact that the input pulse in this experiment did not contain the phase jump that is needed for creating a single dark soliton.

A pulse-tailoring technique (based on spatial filtering within a standard grating decompressor) was used in another 1988 experiment to produce an optical dark pulse with a π phase jump at its center [102]. Such 185-fs odd-symmetry dark pulses were in the form of a fundamental dark soliton, and they maintained their shape during propagation through a 1.4-m-long optical fiber. Figure 4.9(a) shows the experimental results (dotted curves) and compares them with the numerical predictions (solid curves) based on the NLS equation. The top curve shows the input pulse shape measured using a cross-correlation technique. The duration of the central hole is 185 fs, while the back-

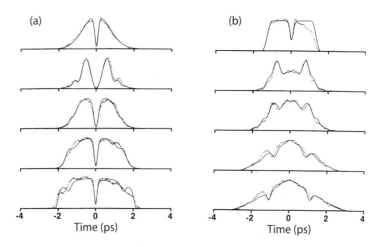

Figure 4.9: Measured (dotted lines) and calculated (solid lines) cross-correlation data showing evolution of dark solitons: (a) odd-symmetry dark pulse with input peak power (from top to bottom) of 0, 1.5, 52.5,150, and 300 W; (b) even-symmetry dark pulses with input peak power of 0, 2.5, 50,150, and 285 W. (After Ref. [102]; ©1998 APS.)

ground pulse is almost 10 times as wide (1.76-ps FWHM). At the lowest power (1.5-W peak input power), the hole widened because the propagation was almost linear. As the power was increased, the background pulse broadened and acquired a square profile because of the combined effects of nonlinearity and dispersion. At the same time, the width of the dark pulse decreased. At the 300-W peak input power, the output dark pulse was of essentially the same duration as the input pulse. The experiment shows clearly that the dark pulse propagates as a dark soliton when the power level is high enough, even though there occurs a significant broadening and chirping of the finite-duration background pulse. Numerical solutions of the NLS equation are in qualitative agreement with the experimental data.

The spectral filtering technique can also produce even-symmetry dark pulses that exhibit no phase jump across them. The experimental results for such pulses are shown in Figure 4.9(b). The top curve again shows the cross-correlation measurements of the input pulse. A comparison of the left and right panels shows the drastic changes induced by the phase jump across the pulse. In the absence of the phase jump, the observed behavior is similar to that of Ref. [103], in the sense that two low-amplitude dark solitons are formed whose separation increases with increasing input power. Again, the experimental data agree closely with numerical solutions of the NLS equation shown as solid lines in Figure 4.9. These experiments confirm the crucial importance of the phase of the input pulse. Only an odd dark pulse propagates undistorted as a black soliton; an even dark pulse splits into a pair of gray solitons.

The nonlinear temporal and spectral shifts of dark solitons were observed in a 1989 experiment [74]. Such effects are well known for bright solitons [73]. In the case of dark solitons, the shifts become increasingly pronounced as the intensity and fiber lengths are increased. The experimental data are in a good agreement with the numerical simulations based on the modified NLS equation that includes the Raman contribution to the nonlinear refractive index. An analytical description of the effect has been given in Section 3.5.1.

Nonlinear propagation of 5.3-ps, odd-symmetry, dark pulses through a 1-km-long single-mode fiber at 850 nm was realized in 1993 [110]. The choice of this wavelength allowed dark-soliton propagation under conditions more compatible with telecommunications. This experiment showed that soliton-like propagation is possible in a fiber 2 5 times longer than the soliton period z_0, despite a finite background and the presence of fiber losses. A quantitative agreement with numerical simulations was found for both temporal and spectral measurements.

The Raman amplification was employed in a 1996 experiment for compensating fiber losses [111]. A 395-m-long, Ge-doped, silica fiber was used as the gain medium. A 2.7-ps dark pulse created on a 39-ps background pulse (at 883 nm) was launched into the fiber. The output pulse was analyzed using a streak camera. A counterpropagating pump laser at 850 nm with an average power of 190 mW was used to produce a Raman gain of up to 3 dB, compensating fully the 6-dB/km fiber losses.

Dark solitons and optical communications

Although these single-pulse propagation studies confirmed the basic properties of dark solitons, the use of a dark-soliton pulse train is essential for optical communication applications. In early experiments, such pulse trains were created by colliding two bright pulses launched into an optical fiber with a time delay between them such that they propagated in the normal-dispersion regime of the fiber [104]–[106]. The process for creating dark-soliton trains involves three stages. In the first stage, the two pulses broaden nonlinearity, forming square-shaped frequency-chirped pulses. The chirped pulses eventually broaden sufficiently that they start to overlap and interfere, forming a pulse with sinusoidal intensity modulation with alternating phases. Finally, the nonlinearity acts on this modulation to produce a train of dark solitons.

In a 1992 experiment [105], a 2-ps pulse from a dye laser was split into a pair of input pulses using an interferometer and launched into a 100-m-long optical fiber. The output was analyzed using a streak camera. By varying the pulse separation and the input power level, the transition from nearly sinusoidal modulation to a dark-soliton-like structure was observed. The later work extended the process to create dark pulse trains at high repetition rates (up to 60 GHz) and propagated these trains for distances of up to 2 km [106]. In a 1992 experiment [109], two distributed feedback (DFB) lasers were temperature tuned to produce two CW beams whose frequencies differed by 100 GHz. The two beams were then launched simultaneously into a 1.5-km-long optical fiber whose dispersion was "normal" and decreased slowly along its length. The output consisted of a train of dark solitons at the 100-GHz repetition rate. The duration of each dark soliton was measured using an autocorrelator and was found to be 1.6 ps. This technique is similar to that first proposed in 1989 for bright solitons [112] and implemented in 1992 [113].

A regular pulse train can be used for information transmission only after it is encoded to produce a pseudo-random sequence of 1 and 0 bits. Nakazawa and Suzuki were successful in 1995 in producing such a pseudo-random data train of dark solitons.

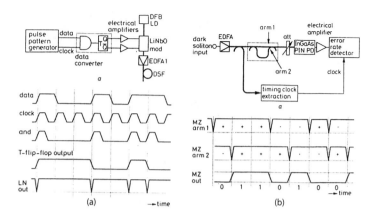

Figure 4.10: (a) Scheme used for (a) generating and (b) detecting a coded train of dark solitons. (After Ref. [87]; ©1995 IEE.)

Figure 4.10 shows their experimental scheme schematically [87]. The electrical data in the nonreturn-to-zero (NRZ) format are combined with a regular clock pattern inside an AND gate. The resulting signal drives a flip-flop circuit to produce a bit stream of electrical dark pulses, which in turn drives a fast LiNbO$_3$ modulator to generate an optically coded dark-soliton train of 50-ps duration.

The coded 10-Gb/s dark-soliton train was transmitted through a 1200-km fiber link exhibiting "normal" dispersion. The output was detected using the scheme shown schematically in Figure 4.10(b). The incoming train of dark solitons was split into two parts, forming the two arms of a Mach–Zehnder interferometer. A one-bit temporal shift was introduced in one arm of the interferometer. Interference between such two optical signals created an inverted version of the original NRZ data train. As a result, it could be processed using a standard receiver. The results showed that the power penalty was below 2.5 dB to maintain a bit-error rate of 10^{-10} or less after 1200 km. This experiment indicates that most of the basic issues involved with the generation, encoding, and detection of dark-soliton pulse trains can be overcome, although higher bit rates will require more sophisticated equipment.

Several interesting features of dark solitons have not yet been experimentally observed. These include the possibility of a reduced timing jitter [84] and a longer collision length compared with that of bright solitons [9]. As a result, dark solitons can be packed more tightly, permitting higher bit rates. The average power needed to transmit dark solitons is generally larger, but it can be made acceptable with a suitable design. However, because the overall phase of a dark soliton rotates at twice the rate of the bright-soliton phase, dark solitons are more sensitive to periodic perturbations over the same transmission distance.

Interaction of dark solitons

The interaction between black and gray solitons was observed in a 1996 experiment by using a 1-km-long fiber with the effective core area of 28 μm^2 and dispersion of 90

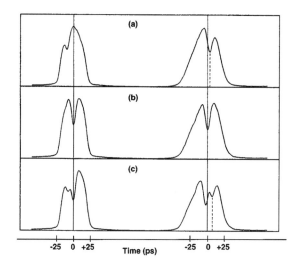

Figure 4.11: Streak camera images showing input (left) and output (right) pulse shapes for a 1-km-long fiber: (a) single gray soliton, (b) single black soliton, and (c) interaction between gray and black solitons. Black solitons are located at the center of the background pulse (vertical solid line), while gray solitons are marked by dashed lines. (After Ref. [114]; ©1996 APS.)

ps/nm/km [114]. The fiber length was long enough to exceed seven soliton periods. A pair of dark solitons with adjustable "blackness" was generated using a pair of phase plates. The power coupled into the fiber slightly exceeded the level of the fundamental dark soliton to compensate for the fiber losses partially. The output pulses were analyzed with a streak camera, and the interaction between the pair of solitons was investigated by measuring the change in their arrival time relative to the interaction-free case.

The streak camera images in Figure 4.11 show input (left) and output (right) temporal profiles in three different situations. In part (a) a single gray soliton ($B^2 = 0.78$) was observed to walk from the leading edge (-10 ps) to the trailing edge ($+4$ ps) of the background pulse. At the same time, the background pulse broadened because of the effects of dispersion and self-phase modulation. The gray soliton also became broader after propagation because of fiber losses that decreased the amplitude of the background pulse. Figure 4.11(b) shows the case of a black soliton positioned initially at the center of the background pulse. Neither its width nor its position changed after propagation, as expected from theory. The case in which both dark and gray solitons were present at the same time is shown in part (c). The gray soliton walked through the black pulse during propagation, but as a result of the interaction of two solitons it was delayed by 3 ps, as evident from the vertical dashed lines. The dark soliton located at the pulse center also changed its position toward the leading edge of the background pulse by 2 ps because of the collision between two solitons. These results provide evidence of the repulsive nature of the interaction between two dark solitons and can be reproduced through numerical simulations based on the cubic NLS equation [114]–[115].

4.6.2   Spatial Dark Solitons

Similar to the case of temporal solitons, bright spatial solitons were observed first using Kerr materials such as silica glasses and semiconductors [116, 117] before the experimental work began on dark spatial solitons. Several experiments were performed almost simultaneously for observing spatial dark solitons in self-defocusing media in a bulk form or in the form of a planar waveguide [118]–[121].

Dark-soliton generation

In the case of bulk nonlinear media, amplitude and phase masks are commonly used in the form of stripes, grids, or crosses on the transverse cross section of a CW beam. In a 1991 experiment [119], a CW beam from a frequency-stabilized dye laser with 100-mW power was passed through a wire mesh and then imaged with a lens into a 18-mm-long cell containing sodium vapor (density $\sim 10^{12}$ cm^{-3}). By tuning the laser beam below the atomic D_2 resonance, a strong defocusing response was obtained with an effective nonlinearity of up to $n_2 \approx -3 \times 10^{-7}$ cm^2/W. The far-field intensity profiles were recorded at a distance of >1 m from the output of the cell. When the laser was detuned far from the resonance, a normal diffraction pattern from the grid was observed because the nonlinear effects were negligible. However, as the D_2 resonance was approached, this pattern underwent a remarkable transformation to form a well-organized array of square dots. Numerical simulations of the NLS equation showed that the formation of regular patterns of dark dots within the background field was a universal phenomenon and was associated with the formation of multiple dark solitons.

To confirm that dark solitons were indeed being created, a further series of experiments was undertaken using a weakly absorbing liquid as a thermally defocusing nonlinear material. The use of either an amplitude mask (a wire cross) or a π phase mask (again in the form of a cross) led to the creation of a single dark soliton or a pair of crossed dark soliton stripes on a uniform background in the nonlinear regime. The transverse velocity of the pair of stripes generated from the amplitude mask varied with the width of the wire in the manner predicted for one-dimensional dark spatial solitons. Similar behavior was observed in an earlier experiment in which a single wire was used to mask an input beam [118].

Although it is well known that such quasi-one-dimensional structures suffer from a transverse modulation instability because they do not exist as stable analytical solutions of the underlying $(2+1)$-dimensional NLS equation (see Chapter 6), the experimental and numerical data presented in these papers provided strong evidence that the observed behavior was indeed due to the creation of spatial dark solitons. A number of experimental factors contributed to the absence of transverse modulation instability in these experiments. First, the use of moderate beam intensities led to stabilization due to a finite size of the laser beam (i.e., the fastest-growing unstable mode had a period much larger than the beam size). Second, the non-Kerr nature of the nonlinear medium helps to eliminate the modulation instability. For example, the diffusive nature of the thermal nonlinearity [119] is known to suppress the beam collapse [122, 123]. The transverse modulation instability cannot always be suppressed, and it was indeed observed in experiments in which a higher nonlinearity with an essentially local response was available [124, 125]. The instability led to a snake-like distortion of the soliton stripe and eventually to its breakup into pairs of optical vortex solitons. This phenomenon is discussed in Chapter 8.

As discussed earlier, gray spatial solitons have a finite transverse velocity relative to the background beam that supports them. A convenient way of understanding this motion is to recognize that a nonzero on-axis intensity can occur only if a traveling wave crosses the soliton. If the background is a plane wave (i.e., it is characterized by a single wave number), the soliton must propagate at some angle to this background wave to generate the traveling-wave component. The direction of the soliton is determined by its grayness, which in turn is determined by the change of phase across the soliton. It follows that adjusting the phase change across the soliton allows it to be scanned relative to the background wave. This feature was first observed in 1992 for single dark solitons [60] and later for an array of dark solitons in both the spatial and temporal domains [126]–[128].

Spatial dark solitons have also been observed using a bulk ZnSe semiconductor crystal as the nonlinear medium [120, 121]. In these experiments, 30-ps pulses from a frequency-doubled Nd:YAG laser (wavelength 532 nm) were passed through phase or amplitude masks and focused onto a 2-mm-long crystal of ZnSe. In this material, defocusing nonlinearity resulted from the dispersive change associated with two-photon absorption. The output from the crystal was imaged onto the slit of a streak camera. When a π phase mask was used, the data showed an intensity-dependent narrowing of the central dark zone, accompanied by defocusing of the background, behavior indicative of the formation of a dark soliton. The interferograms of the output beam for phase and amplitude masks showed the structures expected for black and gray solitons, respectively.

Starting in 1995, photorefractive materials were used for observing spatial solitons [129]–[134]. The nonlinear response in this case results from the photovoltaic fields and can be modeled using a generalized NLS equation with a saturable nonlinearity. The photorefractive dark solitons were first observed in a 1995 experiment [129]. A dark notch was first created through a π phase jump at the center of an argon-ion laser beam and then launched into a strontium–barium niobate (SBN) crystal. An external electric field of 400 V/cm was applied parallel to the c axis of the crystal. In the absence of this field, the 21-μm-wide notch diffracted to a size of 35 μm. With the field turned on, diffraction was fully compensated, and the notch formed a dark soliton, while defocusing of the background still occurred. An important signature of the photorefractive solitons is their insensitivity to the magnitude of the light intensity. Indeed, no change in the shape or the size of the dark soliton was observed when the input power was varied in the range of 3–300 μW.

Screening dark solitons were also first observed in 1995 [130]. The physical process involved is best understood by considering a narrow notch in an otherwise uniform infinite plane wave propagating in a biased photorefractive medium. In the illuminated regions the resistivity decreases. As a result, the voltage drop occurs primarily across the dark region, leading to a local increase in the field and a corresponding local change in the refractive index via the Pockel effect. Dark soliton stripes about 20 μm wide were created when a bismuth–titanium oxide (BTO) crystal was illuminated with a He–Ne laser beam in the presence of an applied field of up to 7.5 kV/cm and an inco-

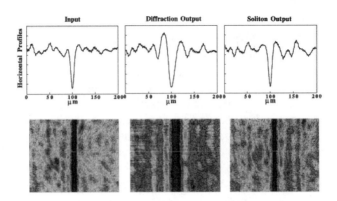

Figure 4.12: Beam profiles (above) and photographs (below) of the input beam (left), the normally diffracting output (middle), and the dark-soliton output beam (right) after propagation along a 5-mm SBN crystal. (After Ref. [131]; ©1996 OSA.)

herent background beam. More recently, the formation of higher-order dark solitons containing odd and even numbers of solitons has been observed without any uniform illumination [135]. Figure 4.12 displays an example of the experimental results [131]. In this experiment, the input notch was 7 μm wide (left), and it diffracted to a size of 12 μm (middle) after 5 mm of propagation in the absence of an external field. When an external voltage of 150 V was applied, the notch got self-trapped to a 7-μm width again, without the background illumination, in accordance with the theory of screening dark solitons.

The photovoltaic effect in a LiNbO$_3$ crystal has also been used to observe dark solitons [136]. A 488-nm CW beam from an argon-ion laser with a π phase step at its center illuminated the LiNbO$_3$ crystal normal to the c axis. At intensities of $\sim$10 W/cm^2, dark solitons with a width of approximately 20 μm were formed after 15 min of exposure. Solitons formed only when the intensity gradient was parallel to the c axis. These photovoltaic waveguides lasted up to 39 hours after the soliton-forming beam was removed. Even a waveguide Y junction can be formed using photovoltaic solitons [132]. More recently, dark photovoltaic spatial solitons were found to form in a slightly Fe-doped LiNbO$_3$ waveguide [137] for both the fundamental and the second-order mode. The formation of dark solitons was followed by the breakup of the initial bright stripe as a result of the transverse modulation instability.

Much of the experimental work on photorefractive materials has concentrated on screening solitons [138]–[141]. Uniform illumination of the crystal to its edges is important for creating the localized screening region, and this can be achieved either by using an incoherent background illumination [130] or by extending the background field to cover the full crystal aperture [131]. Screening solitons have been used to create the Y-shaped waveguide junctions. They can also be used for demonstrating several type of soliton effects, such as higher-order solitons [138], dark-bright soliton pairs [139], and transverse breakup of dark-soliton stripes [140] and their splitting into vortex solitons [141].

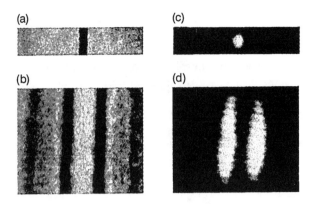

Figure 4.13: Experimental beam profiles showing the formation of light-induced waveguides: (a) input beam; (b) output beam with dark solitons; (c) input probe beam; (d) output probe beam pattern. (After Ref. [144]; ©1992 OSA.)

Dark-soliton Y junctions

From the standpoint of applications, the ability to change the propagation direction of dark spatial solitons is useful when their use is considered for optical switching. Although many concepts for optical switching exist, optical switching based on dark solitons is reconfigurable because it is based on the concept of light-induced waveguides. Spatial solitons can be thought of as self-guided waves; i.e., they "write" a waveguide into the nonlinear material in which they propagate as modes of that waveguide. Bright solitons are bound modes of the waveguide they induce whereas dark solitons are reflectionless radiation modes of the induced waveguide [142, 143].

A common theme of the experiments on photorefractive solitons has been to demonstrate their use as light-induced waveguides. This is in part due to the fact that the photorefractive response is rather slow and leads to quasi-permanent structures. This is particularly true in the case of photovoltaic spatial solitons, where illumination of the nonlinear medium leads to a photovoltaic current that transports charges away from the illuminated region, preferentially along the c axis of a ferroelectric crystal. Photovoltaic solitons exist only when there is a component of the intensity gradient along the c axis. Since the physical mechanism for soliton generation involves the separation and trapping of charge, the index perturbation persists in the dark and may be useful in creating semipermanent waveguides for wavelengths where the material is insensitive to light.

To realize the switching concept, considerable attention has been paid to "writing" optical waveguides into nonlinear materials using dark spatial solitons. The creation of a pair of gray solitons from an intensity dip in a pulse is shown in Figure 4.9(b) for the case of temporal solitons. In the spatial case, a pair of dark solitons emerging from an amplitude jump in the transverse profile of an input beam provides an optical structure that acts as a self-induced Y-junction waveguide [144]–[146]. The input and output field patterns measured experimentally for this structure are shown in Figure 4.13, and they are in good agreement with the corresponding numerical simulations.

The rectangular perturbation evolves into a pair of diverging dark solitons to produce a structure akin to a Y-junction splitter. Figure 4.13(d) shows the experimental result when a 40-μm-wide probe beam is launched along with the soliton beam with amplitude perturbation. The probe beam splits smoothly into a pair of independent guided beams such that more than 98% of its input power is contained in the pair of output beams. Such a device is insensitive to the probe wavelength.

Another interesting property of dark spatial solitons is their behavior during collisions. Numerical simulations show that the soliton-induced X junctions created by such collisions form lossless junctions though which a probe beam can be transmitted [144]. Such X junctions are difficult to realize experimentally in materials with diffusive nonlinearities, but they have been observed in photorefractive crystals [133]. The main impediment to applications of spatial dark solitons is related to the slow switching speed resulting from the slow response time of the photorefractive materials (up to several seconds). In some cases, the response can be improved dramatically by changing the carrier and trap densities.

References

[1] V. E. Zakharov and A. B. Shabat, *Zh. Eksp. Teor. Fiz.* **64**, 1627 (1973) [*Sov. Phys. JETP* **37**, 823 (1973)].

[2] K. J. Blow and N. J. Doran, *Phys. Lett. A* **107**, 55 (1985).

[3] R. N. Thurston and A. M. Weiner, *J. Opt. Soc. Am. B* **8**, 471 (1991).

[4] L. Gagnon, *J. Opt. Soc. Am. B* **10**, 469 (1993).

[5] N. N. Akhmediev and A. Ankiewicz, *Phys. Rev. A* **47**, 3213 (1993).

[6] A. R. Osborne, *Phys. Lett. A* **176**, 75 (1993).

[7] C. S. West and T. A. B. Kennedy, *Phys. Rev. A* **47**, 1252 (1993).

[8] P. B. Lundquist, D. R. Andersen, and G. A. Swartzlander, Jr., *J. Opt. Soc. Am. B* **12**, 698 (1995).

[9] W. Zhao and E. Bourkoff, *Opt. Lett.* **14**, 703 (1989).

[10] F. Kh. Abdullaev, N. K. Nurmanov, and E. N. Tsoy, *Phys. Rev. E* **56**, 3638 (1997).

[11] S. A. Gredeskul and Yu. S. Kivshar, *Opt. Lett.* **14**, 1281 (1989).

[12] Yu. S. Kivshar and S. A. Gredeskul, *Opt. Commun.* **79**, 285 (1990).

[13] S. A. Gredeskul, Yu. S. Kivshar, and M. V. Yanovskaya, *Phys. Rev. A* **41**, 3994 (1990).

[14] S. A. Gredeskul and Yu. S. Kivshar, *Phys. Rev. Lett.* **62**, 977 (1989).

[15] Yu. S. Kivshar, *J. Phys. A* **22**, 337 (1989).

[16] B. Wu, J. Liu, and Q. Niu, *Phys. Rev. Lett.* **88**, 034101(4) (2002).

[17] A. B. Shvartsburg, L. Stenflo, and P. K. Shukla, *Physica Scripta* **65**, 164 (2002).

[18] A. N. Slavin, Yu. S. Kivshar, E. A. Ostrovskaya, and H. Benner, *Phys. Rev. Lett.* **82**, 2583 (1999).

[19] M. S. Cramer and L. T. Watson, *Phys. Fluids* **27**, 821 (1984).

[20] V. G. Makhankov, *Soliton Phenomenology* (Kluwer, Dordrecht, Netherlands, 1990), p. 218.

[21] Yu. S. Kivshar, *IEEE J. Quantum Electron.* **28**, 250 (1993).

[22] Yu. S. Kivshar, D. Anderson, and M. Lisak, *Physica Scripta* **47**, 679 (1993).

[23] Yu. S. Kivshar and V. V. Afanasjev, *Phys. Rev. A* **44**, R1446 (1991).

[24] D. J. Frantzeskakis, *J. Phys. A* **29**, 3631 (1996).

[25] G. Huang and M. G. Velarde, *Phys. Rev. E* **54**, 3048 (1996).

[26] I. V. Barashenkov and V. G. Makhankov, *Phys. Lett. A* **128**, 52 (1988).

[27] G. C. Valley, M. Segev, B. Crosignani, A. Yariv, M. M. Fejer, and M. C. Bashaw, *Phys. Rev. A* **50**, R4457 (1994).

[28] D. N. Christodoulides and M. I. Carvalho, *J. Opt. Soc. Am. B* **12**, 1628 (1995).

[29] W. Królikowski and B. Luther-Davies, *Opt. Lett.* **17**, 1414 (1992).

[30] W. Królikowski and B. Luther-Davies, 1993, *Opt. Lett.* **18**, 188 (1993).

[31] W. Królikowski, N. N. Akhmediev, and B. Luther-Davies, *Phys. Rev. E* **48**, 3980 (1993).

[32] Yu. S. Kivshar, V. V. Afanasjev, and A. W. Snyder, *Opt. Commun.* **126**, 348 (1996).

[33] W.-S. Kim, and H.-T. Moon, *Phys. Lett. A* **266**, 364 (2000).

[34] M. G. Vakhitov and A. A. Kolokolov, *Radiophys. Quantum Electron.* **16**, 783 (1973).

[35] M. I. Weinstein, *SIAM J. Math. Analysis* **16**, 472 (1985).

[36] E. A. Kuznetsov, A. M. Rubenchik, and V. E. Zakharov, *Phys. Rep.* **142**, 113 (1986).

[37] F. V. Kusmartsev, *Phys. Rep.* **183**, 1 (1989).

[38] D. J. Mitchell and A. W. Snyder, *J. Opt. Soc. Am. B* **10**, 1574 (1993).

[39] C. K. R. T. Jones and J. V. Moloney, *Phys. Lett. A* **117**, 175 (1986).

[40] D. E. Pelinovsky, V. V. Afanasjev, and Yu. S. Kivshar, *Phys. Rev. E* **53**, 1940 (1996).

[41] R. H. Enns and L. J. Mulder, *Opt. Lett.* **14**, 509 (1989).

[42] F. G. Bass, V. V. Konotop, and S. A. Puzenko, *Phys. Rev. A* **46**, 4185 (1992).

[43] I. V. Barashenkov and Kh. T. Kholmurodov, JINR Preprint, P17-86-698, Dubno, Russia (1986).

[44] I. V. Barashenkov, A. D. Gosheva, V. G. Makhankov, and I.V. Puzynin, *Physica D* **34**, 240 (1989).

[45] A. De Bouard, *SIAM J. Math. Anal.* **26**, 566 (1995).

[46] M. M. Bogdan, A. S. Kovalev, and A. M. Kosevich, *Fiz. Nizk. Temp.* **15**, 511 (1989).

[47] I. V. Barashenkov and E. Yu. Panova, *Physica D* **69**, 114 (1993).

[48] Yu. S. Kivshar and W. Królikowski, *Opt. Lett.* **20**, 1527 (1995).

[49] Yu. S. Kivshar and V. V. Afanasjev, *Opt. Lett.* **21**, 1135 (1996).

[50] I. V. Barashenkov, *Phys. Rev. Lett.* **77**, 1193 (1996).

[51] D. E. Pelinovsky, Yu. S. Kivshar, and V. V. Afanasjev, *Phys. Rev. E* **54**, 2015 (1996).

[52] C. A. Jones and P. M. Roberts, *J. Phys. A* **15**, 2599 (1982).

[53] E. A. Kuznetsov and J. J. Rasmussen, *Phys. Rev. E* **51**, 4479 (1995).

[54] Yu. S. Kivshar and B. A. Malomed, *Rev. Mod. Phys.* **61**, 763 (1989).

[55] Yu. S. Kivshar and X. Yang, *Phys. Rev. E* **49**, 1657 (1994).

[56] Yu. S. Kivshar and W. Królikowski, *Opt. Commun.* **114**, 353 (1995).

[57] X. Yang, Yu. S. Kivshar, B. Luther-Davies, and D. Andersen, *Opt. Lett.* **19**, 344 (1994).

[58] S. Burtsev and R. Camassa, *J. Opt. Soc. Am. B* **14**, 1782 (1997).

[59] Y. Silberberg, *Opt. Lett.* **15**, 1005 (1990).

[60] B. Luther-Davies and X. Yang, *Opt. Lett.* **17**, 1775 (1992).

[61] M. Lisak, D. Andersen, and B. A. Malomed, *Opt. Lett.* **16**, 1936 (1991).

[62] J. A. Giannini and R. I. Joseph, *IEEE J. Quantum Electron.* **26**, 2109 (1990).

[63] H. Sakaguchi, *Prog. Theor. Phys.* **85**, 417 (1991).

[64] Y. Chen, *Phys. Rev. A* **45**, 6922 (1992).

[65] T. Ikeda, M. Matsumoto, and A. Hasegawa, *Opt. Lett.* **20**, 1113 (1995); *J. Opt. Soc. Am. B* **14**, 136 (1997).

[66] V. V. Afanasjev, P. L. Chu, and B. A. Malomed, *Phys. Rev. E* **57**, 1088 (1998).

[67] A. Maruta and Y. Kodama, *Opt. Lett.* **20**, 1752 (1995).

[68] A. D. Kim, W. L. Kath, and C. G. Goedde, *Opt. Lett.* **21**, 465 (1996).

[69] K. M. Allen, N. J. Doran, and J. A. R. Williams, *Opt. Lett.* **19**, 2086 (1994).

[70] G. P. Agrawal, *Nonlinear Fiber Optics*, 3rd ed. (Academic, San Diego, 2001).

[71] F. M. Mitschke and L. F. Mollenauer, *Opt. Lett.* **11**, 659 (1986).

[72] J. P. Gordon, *Opt. Lett.* **11**, 662 (1986).

[73] A. Hasegawa and Y. Kodama, *Solitons in Optical Communications* (Oxford University Press, Oxford, UK, 1995).

[74] M. J. Weiner, R. N. Thurston. W. J. Tomlinson, J. P. Heritage, D. E. Leaird, and E. M. Kirschner, and R. J. Hawkins, *Opt. Lett.* **14**, 868 (1989).

[75] Yu. S. Kivshar, *Phys. Rev. A* **42**, 1757 (1990).

[76] Yu. S. Kivshar and V. V. Afanasjev, *Opt. Lett.* **16**, 285 (1991).

[77] I. M. Uzunov and V. S. Gerdjikov, *Phys. Rev. A* **47**, 1582 (1993).

[78] R. H. Stolen, J. P. Gordon, W. J. Tomlinson, and H. A. Haus, 1989, *J. Opt. Soc. Am. B* **63**, 1159 (1989).

[79] W. Zhao and E. Bourkoff, *J. Opt. Soc. Am. B* **9**, 1134 (1992).

[80] Yu. S. Kivshar and S. K. Turitsyn, *Phys. Rev. A* **47**, R3502 (1993).

[81] J. P. Gordon and H. A. Haus, *Opt. Lett.* **11**, 665 (1986).

[82] L. F. Mollenauer, J. P. Gordon, and P. V. Mamychev, in *Optical Fiber Telecommunications IIIA*, I. P. Kaminow and T. L. Koch, Eds. (Academic Press, San Diego, 1997), Chap. 12.

[83] G. P. Agrawal, *Fiber-Optic Communication Systems*, 3rd ed. (Wiley, New York, 2002), Chap. 9.

[84] Yu. S. Kivshar, M. Haelterman, Ph. Emplit, and J. P. Hamaide, *Opt. Lett.* **19**, 19 (1994).

[85] J. P. Hamaide, Ph. Emplit, and M. Haelterman, *Opt. Lett.* **16**, 1578 (1991).

[86] N.-C. Panoiu, D. Mihalache, and D.-M. Baboiu, *Phys. Rev. A* **52**, 4182 (1995).

[87] M. Nakazawa and K. Suzuki, *Electron. Lett.* **31**, 1076 (1995); *Electron. Lett.* **31**, 1084 (1995).

[88] P. K. A. Wai, C. R. Menyuk, Y.C. Lee, and H. H. Chen, *Opt. Lett.* **11**, 464 (1986).

[89] P. K. A. Wai, H. H. Chen, and Y. C. Lee, *Phys. Rev. A* **41**, 426 (1990).

[90] C. R. Menyuk and P. K. A. Wai, in *Optical Solitons—Theory and Experiment*, J. R. Taylor, Ed. (Cambridge University Press, Cambridge, UK, 1992), pp. 332–346, 359–369.

[91] C. R. Menyuk, *J. Opt. Soc. Am. B* **10**, 1585 (1993).

[92] Yu. S. Kivshar, *Phys. Rev. A* **43**, 1677 (1991).

[93] V. I. Karpman, *Phys. Lett. A* **181**, 211 (1993).

[94] V. V. Afanasjev, Yu. S. Kivshar, and C. R. Menyuk, 1996, *Opt. Lett.* **21**, 1975 (1996).

[95] Yu. S. Kivshar and B. A. Malomed, *Phys. Rev. Lett.* **60**, 129 (1991).

[96] M. J. Potasek and M. Tabor, *Phys. Lett. A* **154**, 449 (1991).

[97] S. L. Palacios, A. Guinea, J. M. Fernández-Díaz, and R. D. Crespo, *Phys. Rev. E* **60**, R45 (1999).

[98] S. L. Palacios and J. M. Fernández-Díaz, *Opt. Commun.* **178**, 457 (2000).

[99] A. Hasegawa and F. Tappert, *Appl. Phys. Lett.* **23**, 171 (1973).

[100] L. F. Mollenauer, R. H. Stolen, and J. P. Gordon, 1980, *Phys. Rev. Lett.* **45**, 1095 (1980).

[101] Ph. Emplit, J. P. Hamaide, F. Reynaud, G. Froehly, and A. Barthelemy, *Opt. Commun.* **62**, 374 (1987).

[102] A. M. Weiner, J. P. Heritage, R. J. Hawkins, R. N. Thurston, E. M. Kirschner, D. E. Learid, and W. J. Tomlinson, *Phys. Rev. Lett.* **61**, 2445 (1988).

[103] D. Krökel, N. J. Halas, G. Giuliani, and D. Grischkowsky, *Phys. Rev. Lett.* **60**, 29 (1988).

[104] J. E. Rothenberg, *Opt. Commun.* **82**, 107 (1991).

[105] J. E. Rothenberg and H. K. Heinrich, *Opt. Lett.* **17**, 261 (1992).

[106] J. A. R. Williams, K. M. Allen, N. J. Doran, and Ph. Emplit, *Opt. Commun.* **112**, 333 (1994).

[107] W. Zhao and E. Bourkoff, *Opt. Lett.* **15**, 405 (1990).

[108] M. Haelterman and Ph. Emplit, *Electron. Lett.* **29**, 356 (1993).

[109] D. J. Richardson, R. P. Chamberlin, L. Dong, and D. N. Payne, *Electron. Lett.* **30**, 1326 (1994).

[110] Ph. Emplit, M. Haelterman, and J. P. Hamaide, *Opt. Lett.* **18**, 1047 (1993).

[111] D. Foursa and Ph. Emplit, *Electron. Lett.* **32**, 919 (1996).

[112] E. M. Dianov, P. V. Mamyshev, A. M. Prokhorov, and S. V. Chernikov, *Opt. Lett.* **14**, 1008 (1989).

[113] S. V. Chernikov, P. V. Mamyshev, E. M. Dianov, D. J. Richardson, R. I. Laming, and D. N. Payne, *Sov. Lightwave Commun.* **2**, 161 (1992).

[114] D. Foursa and Ph. Emplit, *Phys. Rev. Lett.* **77**, 4011 (1996).

[115] G. L. Diankov, and I. M. Uzunov, *Opt. Commun.* **117**, 424 (1995).

[116] J. S. Aitchinson, A. M. Weiner, Y. Silberberg, M. K. Oliver, J. L. Jackel, D. E. Leaird, E. M. Vogel, and P. W. E. Smith, *Opt. Lett.* **15**, 471 (1990).

[117] F. Reynaud and A. Barthelemy, *Europhys. Lett.* **12**, 401 (1990).

[118] D. R. Andersen, D. E. Hooton, G. A. Swartzlander, Jr., and A. E. Kaplan, *Opt. Lett.* **15**, 783 (1990).

[119] G. A. Swartzlander, Jr., D. R. Andersen, J. J. Regan, H. Yin, and A. E. Kaplan, *Phys. Rev. Lett.* **66**, 1583 (1991).

[120] G. R. Allan, S. R. Skinner, D. R. Andersen, and A. L. Smirl, *Opt. Lett.* **16**, 156 (1991).

[121] S. R. Skinner, G.R. Allan, D. R. Andersen, and A. L. Smirl, *IEEE J. Quantum Electron.* **27**, 2211 (1991).

[122] S. K. Turitsyn, *Teor. Mat. Fiz.* **64**, 226 (1985) [*Theor. Math. Phys.* **64**, 797 (1986)].

[123] D. Suter and T. Blasberg, *Phys. Rev. A* **48**, 4583 (1993).

[124] A. V. Mamaev, M. Saffman, and A. A. Zozulya, *Phys. Rev. Lett.* **76**, 2262 (1996).

[125] V. Tikhonenko, J. Christou, B. Luther-Davies, and Yu. S. Kivshar, *Opt. Lett.* **21**, 1129 (1996).

[126] C. Bosshard, P. V. Mamyshev, and G. I. Stegeman, *Opt. Lett.* **19**, 90 (1994).

[127] P. V. Mamyshev, Ch. Bosshard, and G. I. Stegeman, *J. Opt. Soc. Am. B* **11**, 1254 (1994).

[128] P. V. Mamyshev, P. G. Wigley, J. Wilson, C. Bosshard, and G. I. Stegeman, *Appl. Phys. Lett.* **64**, 3374 (1994).

[129] G. Duree, M. Morin, G. Salamo, M. Segev, B. Crosignani, P. DiPorto, E. Sharp, and A. Yariv, *Phys. Rev. Lett.* **74**, 1978 (1995).

[130] M. D. Iturbe Castillo, J. J. Sánchez-Mondragón, S.I. Stepanov, M.B. Klein, and B.A. Wechsler, *Opt. Commun.* **118**, 515 (1995).

[131] Z. Chen, M. Mitchell, M. Shih, M. Segev, M. H. Garrett, and G. C. Valley, *Opt. Lett.* **21**, 629 (1996).

[132] M. Taya, M. C. Bashaw, M. M. Feier, M. Segev, and G. C. Valley, 1996, *Opt. Lett.* **21**, 943 (1996).

[133] M. Segev, M. Shih, and G. C. Valley, *J. Opt. Soc. Am. B* **13**, 706 (1996).

[134] Z. Chen, M. Segev, S. R. Singh, T. H. Coskun, and D. N. Christodoulides, *J. Opt. Soc. Am. B* **14**, 1407 (1997).

[135] M. M. Méndez-Otero, M. D. Iturbe-Castillo, P. Rodríguez-Montero, E. Martí-Panameño, *Opt. Commun.* **193**, 277 (2001).

[136] M. Taya, M. C. Bashaw, M. M. Fejer, M. Segev, and G. C. Valley, *Phys. Rev. A* **52**, 3095 (1995).

[137] M. Chauvet, S. Chauvin, and H. Mailotte, *Opt. Lett.* **26**, 1344 (2001).

[138] Z. Chen, M. Mitchell, and M. Segev, *Opt. Lett.* **21**, 716 (1996).

[139] Z. Chen, M. Segev, T.H. Coskun, D. N. Christodoulides, Yu. S. Kivshar, and V. V. Afanasjev, *Opt. Lett.* **21**, 1821 (1996).

[140] A. V. Mamaev, M. Saffman, D.Z. Anderson, and A. A. Zozulya, *Phys. Rev. A* **54**, 870 (1996).

[141] A. V. Mamaev, M. Saffman, and A.A. Zozulya, *Phys. Rev. Lett.* **77**, 4544 (1996).

[142] A. W. Snyder, D. J. Mitchell, and B. Luther-Davies, *J. Opt. Soc. Am. B* **10**, 2341 (1993).

[143] P. D. Miller, *Phys. Rev. E* **53**, 4137 (1996).

[144] B. Luther-Davies and X. Yang, *Opt. Lett.* **17**, 496 (1992).

[145] S. Liu, G. Zhang, G. Tian, G. Zhang, and T. Yicheng, *Appl. Opt.* **36**, 8982 (1997).

[146] J. A. Andrade-Lucio, M. M. Méndez-Otero, C.M. Gómez-Sarabia, M. D. Iturbe-Castillo, S. Pérez-Márquez, and G. E. Torres-Cisneros, *Opt. Commun.* **165**, 77 (1999).

Chapter 5

Bragg Solitons

A new kind of soliton, known as the *Bragg soliton* or the *gap soliton*, can form in nonlinear media whose refractive index varies weakly in a periodic fashion along its length. Such a medium is an example of a more general class of materials referred to as the *photonic crystals*. In a photonic crystal, the linear refractive index can be a periodic function in all three spatial dimensions; solitons forming in such materials will be covered in Chapter 12. Here we focus on *one-dimensional periodic media* and assume that the periodicity of the refractive index is limited to the direction in which light is propagating and that this periodicity is *weak*. An example of such a medium is provided by the fiber Bragg gratings, which constitute an essential component of modern lightwave technology and are used routinely for optical telecommunication networks. We discuss in Section 5.1 the history of fiber gratings and introduce the concept of Bragg diffraction that is responsible for creating a backward-propagating wave. *The coupled-mode theory*, which is valid for shallow gratings and weak nonlinearity, is described in Section 5.2, where the concept of the photonic bandgap is also introduced, together with the dispersive effects resulting from such a bandgap. Section 5.3 is devoted to the nonlinear effects, such as the modulation instability that converts a continuous-wave (CW) beam into a pulse train. Section 5.4 then focuses on the properties of Bragg solitons and compares them with the standard temporal solitons observed in optical fibers, also discussing their novel features. The phenomenon of nonlinear optical switching is discussed in Section 5.5 as an example of the potential applications of Bragg solitons.

5.1 Basic Concepts

Silica fibers can change their optical properties permanently when they are exposed to intense radiation from a laser operating in the ultraviolet spectral region. This photosensitive effect can be used to induce periodic changes in the refractive index along the fiber length, resulting in the formation of an intracore Bragg grating. Fiber gratings can be designed to operate over a wide range of wavelengths, extending from the ultraviolet to the infrared region. The wavelength region near 1.5 μm is of particular interest because of its relevance to fiber-optic communication systems.

147

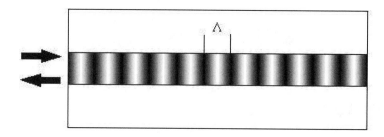

Figure 5.1: Schematic illustration of a fiber grating. Dark and light shaded regions within the fiber core show periodic variations of the refractive index.

Bragg gratings inside optical fibers were first formed in 1978 by irradiating a germanium-doped silica fiber for a few minutes with an intense argon-ion laser beam [1]. The grating period was fixed by the argon-ion laser wavelength, and the grating reflected light only within a narrow region around that wavelength. It was realized that the 4% reflection occurring at the two fiber–air interfaces created a standing-wave pattern and that the laser light was absorbed only in the bright regions. As a result, the glass structure changed in such a way that the refractive index increased permanently in the bright regions. Although this phenomenon attracted some attention during the next 10 years, it was not until 1989 that fiber gratings became a topic of intense investigation, fueled partly by the observation of second-harmonic generation in photosensitive fibers. The impetus for this resurgence of interest was provided by a 1989 paper in which a side-exposed holographic technique was used to make fiber gratings with controllable period [2].

Because of its relevance to fiber-optic communication systems, the holographic technique was quickly adopted to produce fiber gratings in the wavelength region near 1.55 μm [3]. Considerable work was done during the early 1990s to understand the physical mechanism behind photosensitivity of fibers and to develop techniques that were capable of making large changes in the refractive index. By 1995, fiber gratings were available commercially, and by 1999 they became a standard component of lightwave technology [4]–[7].

Figure 5.1 shows schematically how the refractive index varies inside a fiber grating. Mathematically, we need to include both the frequency and intensity dependence of the refractive index in addition to its periodic variation along the fiber length (the z axis) by using

$$\tilde{n}(\omega, z, I) = \bar{n}(\omega) + n_2 I + \delta n_g(z), \qquad (5.1.1)$$

where n_2 is the Kerr coefficient and $\delta n_g(z)$ accounts for periodic index variations inside the grating. Before considering the nonlinear effects governed by n_2, we first set $n_2 = 0$ and ask what happens to light when a CW beam is launched into a one-dimensional weakly periodic medium such as a fiber grating.

Diffraction theory of gratings shows that when light is incident at an angle θ_i (measured with respect to the planes of constant refractive index), it is diffracted at an angle

θ_r such that [8]

$$\sin \theta_i - \sin \theta_r = m\lambda / (\bar{n}\Lambda),\qquad(5.1.2)$$

where Λ is the grating period, $\lambda / \bar{n}$ is the wavelength of light inside the medium with an average refractive index $\bar{n}$, and m is the order of Bragg diffraction. This condition can be thought of as a phase-matching condition, similar to that occurring in the case of four-wave mixing, and can be written as

$$\mathbf{k}_i - \mathbf{k}_d = m\mathbf{k}_g,\qquad(5.1.3)$$

where $\mathbf{k}_i$ and $\mathbf{k}_d$ are the wave vectors associated with the incident and diffracted light, respectively. The grating wave vector $\mathbf{k}_g$ has magnitude $2\pi/\Lambda$ and points in the direction in which the refractive index of the medium is changing in a periodic manner.

In the case of single-mode fibers, all three vectors lie along the fiber axis. As a result, $\mathbf{k}_d = -\mathbf{k}_i$ and the diffracted light propagates backward. Thus, as shown schematically in Figure 5.1, a fiber grating acts as *a reflector for a specific wavelength of light* for which the phase-matching condition is satisfied. In terms of the angles appearing in Eq. (5.1.2), $\theta_i = \pi/2$ and $\theta_r = -\pi/2$. If $m = 1$, the period of the grating is related to the vacuum wavelength as $\lambda = 2\bar{n}\Lambda$. This condition is known as the *Bragg condition*, and a grating satisfying it is referred to as the *Bragg grating*. Physically, the Bragg condition ensures that weak reflections occurring throughout the grating add up in phase to produce a strong reflection. For a fiber grating reflecting light in the wavelength region near 1.5 μm, the grating period $\Lambda \approx 0.5$ μm.

5.2 Photonic Bandgap

Two different approaches have been used to study how the presence of a Bragg grating affects propagation of light. In one approach, the *Bloch-wave formalism*—used commonly for describing motion of electrons in semiconductors—is applied to Bragg gratings [9]. In another, forward- and backward-propagating waves are treated independently, and the Bragg grating provides a coupling between them. This method is known as the *coupled-mode theory* and has been used with considerable success in several different contexts [10]–[12]. In this section we derive the nonlinear coupled-mode equations and use them to discuss propagation of low-intensity CW light through a Bragg grating. We also introduce the concept of *photonic bandgap* and use it to show how a Bragg grating introduces a large amount of chromatic dispersion.

5.2.1 Coupled-Mode Equations

Wave propagation in a linear periodic medium has been studied extensively using coupled-mode theory [10]–[12]. This theory has been applied to distributed-feedback (DFB) semiconductor lasers [13], among other applications. In the case of a dispersive nonlinear medium, the refractive index is given by Eq. (5.1.1). The coupled-mode theory can be generalized to include the nonlinear effects if the nonlinear index change $n_2 I$ in Eq. (5.1.1) is so small that it can be treated as a perturbation [14].

The starting point consists of solving Maxwell's equations with the refractive index given in Eq. (5.1.1). When the nonlinear effects are relatively weak, we can work in the frequency domain and solve the Helmholtz equation

$$\nabla^2 \tilde{E} + \tilde{n}^2(\omega, z)\omega^2/c^2 \tilde{E} = 0, \tag{5.2.1}$$

where $\tilde{E}$ denotes the Fourier transform of the electric field with respect to time.

Noting that $\tilde{n}$ is a periodic function of z, it is useful to expand $\delta n_g(z)$ in a Fourier series as

$$\delta n_g(z) = \sum_{m=-\infty}^{\infty} \delta n_m \exp[2\pi i m(z/\Lambda)]. \tag{5.2.2}$$

Since both the forward- and backward-propagating waves should be included, $\tilde{E}$ in Eq. (5.2.1) is written in the form

$$\tilde{E}(\mathbf{r}, \omega) = F(x, y)[\tilde{A}_f(z, \omega)\exp(i\beta_B z) + \tilde{A}_b(z, \omega)\exp(-i\beta_B z)], \tag{5.2.3}$$

where $\beta_B = \pi/\Lambda$ is the Bragg wave number for a first-order grating. It is related to the Bragg wavelength through the Bragg condition $\lambda_B = 2\tilde{n}\Lambda$ and can be used to define the Bragg frequency as $\omega_B = \pi c/(\tilde{n}\Lambda)$. Transverse variations for the two counterpropagating waves are governed by the same modal distribution $F(x, y)$ in a single-mode fiber.

Using Eqs. (5.1.1)–(5.2.3), assuming that $\tilde{A}_f$ and $\tilde{A}_b$ vary slowly with z and keeping only the nearly phase-matched terms, the frequency-domain coupled-mode equations become [10]–[12]

$$\frac{\partial \tilde{A}_f}{\partial z} = i[\delta(\omega) + \Delta\beta]\tilde{A}_f + i\kappa\tilde{A}_b, \tag{5.2.4}$$

$$-\frac{\partial \tilde{A}_b}{\partial z} = i[\delta(\omega) + \Delta\beta]\tilde{A}_b + i\kappa\tilde{A}_f, \tag{5.2.5}$$

where δ is a measure of detuning from the Bragg frequency and is defined as

$$\delta(\omega) = (\tilde{n}/c)(\omega - \omega_B) \equiv \beta(\omega) - \beta_B. \tag{5.2.6}$$

The nonlinear effects are included through $\Delta\beta$. The coupling coefficient κ governs the grating-induced coupling between the forward and backward waves. For a first-order grating, κ is given by

$$\kappa = \frac{k_0 \iint_{-\infty}^{\infty} \delta n_1 |F(x, y)|^2 \, dx \, dy}{\iint_{-\infty}^{\infty} |F(x, y)|^2 \, dx \, dy}. \tag{5.2.7}$$

In this general form, κ can include transverse variations of δn_g occurring when the photoinduced index change is not uniform over the core area. For a transversely uniform grating $\kappa = 2\pi\delta n_1/\lambda$, as can be inferred from Eq. (5.2.7) by taking δn_1 as constant and using $k_0 = 2\pi/\lambda$. For a sinusoidal grating of the form $\delta n_g = n_a \cos(2\pi z/\Lambda)$, $\delta n_1 = n_a/2$ and the coupling coefficient is given by $\kappa = \pi n_a/\lambda$.

Equations (5.2.4) and (5.2.5) need to be converted to time domain. We assume that the total electric field can be written as

$$E(\mathbf{r}, t) = \tfrac{1}{2}F(x, y)[A_f(z, t)e^{i\beta_B z} + A_b(z, t)e^{-i\beta_B z}]e^{-i\omega_0 t} + \text{c.c.}, \tag{5.2.8}$$

where ω_0 is the frequency at which the pulse spectrum is centered. We expand $\beta(\omega)$ in Eq. (5.2.6) in a Taylor series as [15]

$$\beta(\omega) = \beta_0 + (\omega - \omega_0)\beta_1 + \tfrac{1}{2}(\omega - \omega_0)^2\beta_2 + \tfrac{1}{6}(\omega - \omega_0)^3\beta_3 + \cdots \qquad (5.2.9)$$

and retain terms up to second order in $(\omega - \omega_0)$. The resulting equations can be converted into time domain by replacing $(\omega - \omega_0)$ with the differential operator $i(\partial/\partial t)$. The resulting coupled-mode equations become [6]

$$\frac{\partial A_f}{\partial z} + \beta_1 \frac{\partial A_f}{\partial t} + \frac{i\beta_2}{2}\frac{\partial^2 A_f}{\partial t^2} = i\delta A_f + i\kappa A_b + i\gamma(|A_f|^2 + 2|A_b|^2)A_f, \quad (5.2.10)$$

$$-\frac{\partial A_b}{\partial z} + \beta_1 \frac{\partial A_b}{\partial t} + \frac{i\beta_2}{2}\frac{\partial^2 A_b}{\partial t^2} = i\delta A_b + i\kappa A_f + i\gamma(|A_b|^2 + 2|A_f|^2)A_b, \quad (5.2.11)$$

where δ in Eq. (5.2.6) is evaluated at $\omega = \omega_0$ and becomes $\delta = (\omega_0 - \omega_B)/v_g$. In fact, the δ term can be eliminated from the coupled-mode equations if ω_0 is replaced by ω_B in Eq. (5.2.8). Here $\beta_1 \equiv 1/v_g$ is related inversely to the group velocity, β_2 governs the group-velocity dispersion (GVD), and the nonlinear parameter γ is related to n_2 as $\gamma = n_2\omega_0/(cA_{\mathrm{eff}})$, where A_{eff} is the effective core area.

The nonlinear terms in the time-domain coupled-mode equations contain the contributions of both self-phase modulation (SPM) and cross-phase modulation (XPM). The coupled-mode equations are similar to the set of two coupled NLS equations governing propagation of two copropagating waves inside optical fibers [15]. The two major differences are: (i) the negative sign appearing in front of the $\partial A_b/\partial z$ term in Eq. (5.2.10) because of backward propagation of A_b and (ii) the presence of linear coupling between the counterpropagating waves governed by the parameter κ. Both of these differences change the character of wave propagation profoundly. Before discussing the general case, it is instructive to consider the case in which the nonlinear effects are so weak that the fiber acts as a linear medium.

5.2.2 Continuous-Wave Solution in the Linear Case

First, we focus on the linear case and assume that the input intensity is low enough that the nonlinear effects are negligible. When the SPM and XPM terms are neglected in Eqs. (5.2.10) and (5.2.11), the resulting linear equations can be solved easily in the Fourier domain. In fact, we can use Eqs. (5.2.4) and (5.2.5). These frequency-domain coupled-mode equations include GVD to all orders. After setting the nonlinear contribution $\Delta\beta$ to zero, we obtain

$$\frac{\partial \tilde{A}_f}{\partial z} = i\delta\tilde{A}_f + i\kappa\tilde{A}_b, \qquad (5.2.12)$$

$$-\frac{\partial \tilde{A}_b}{\partial z} = i\delta\tilde{A}_b + i\kappa\tilde{A}_f, \qquad (5.2.13)$$

where $\delta(\omega)$ is given by Eq. (5.2.6).

A general solution of these linear equations takes the form

$$\tilde{A}_f(z) = A_1\exp(iqz) + A_2\exp(-iqz), \qquad (5.2.14)$$

$$\tilde{A}_b(z) = B_1\exp(iqz) + B_2\exp(-iqz), \qquad (5.2.15)$$

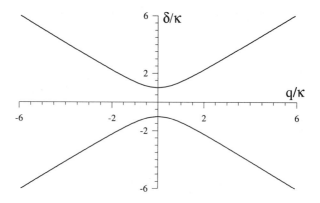

Figure 5.2: Dispersion curves showing variation of δ with q and the existence of the photonic bandgap for a fiber grating.

where q is to be determined. The constants A_1, A_2, B_1, and B_2 are interdependent and satisfy the following four relations:

$$(q - \delta)A_1 = \kappa B_1, \qquad (q + \delta)B_1 = -\kappa A_1, \qquad (5.2.16)$$

$$(q - \delta)B_2 = \kappa A_2, \qquad (q + \delta)A_2 = -\kappa B_2. \qquad (5.2.17)$$

These equations are satisfied for nonzero values of A_1, A_2, B_1, and B_2 if the possible values of q obey the dispersion relation

$$q = \pm\sqrt{\delta^2 - \kappa^2}. \qquad (5.2.18)$$

This equation is of paramount importance for gratings. Its implications will become clear soon.

One can eliminate A_2 and B_1 by using Eqs. (5.2.14)–(5.2.17) and write the general solution in terms of an effective reflection coefficient $r(q)$ as

$$\tilde{A}_f(z) = A_1 \exp(iqz) + r(q)B_2 \exp(-iqz), \qquad (5.2.19)$$

$$\tilde{A}_b(z) = B_2 \exp(-iqz) + r(q)A_1 \exp(iqz), \qquad (5.2.20)$$

where

$$r(q) = \frac{q - \delta}{\kappa} = -\frac{\kappa}{q + \delta}. \qquad (5.2.21)$$

The q dependence of r and the dispersion relation (5.2.18) indicate that both the magnitude and the phase of backward reflection depend on the frequency ω. The sign ambiguity in Eq. (5.2.18) can be resolved by choosing the sign of q such that $|r(q)| < 1$.

The dispersion relation of Bragg gratings exhibits an important property, seen clearly in Figure 5.2, where Eq. (5.2.18) is plotted. If the frequency detuning δ of the incident light falls in the range $-\kappa < \delta < \kappa$, q becomes purely imaginary. Most of the incident field is reflected in that case since the grating does not support a propagating wave. The range $|\delta| \leq \kappa$ is referred to as the *photonic bandgap*, in analogy with the electronic energy bands occurring in crystals. It is also called the *stop band*, since light stops transmitting through the grating when its frequency falls within the photonic bandgap.

5.2.3 Grating-Induced Dispersion

To understand what happens when optical pulses propagate in a fiber grating with their carrier frequency ω_0 outside the stop band but close to its edges, note that the effective propagation constant of the forward- and backward-propagating waves from Eqs. (5.2.3) and (5.2.14) is $\beta_e = \beta_B \pm q$, where q is given by Eq. (5.2.18) and is a function of optical frequency through δ. This frequency dependence of β_e indicates that a grating will exhibit dispersive effects even if it was fabricated in a nondispersive medium. In optical fibers, grating-induced dispersion adds to the material and waveguide dispersion. In fact, the contribution of grating dominates among all sources responsible for dispersion. To see this more clearly, we expand β_e in a Taylor series in a way similar to Eq. (5.2.9) around the carrier frequency ω_0 of the pulse. The result is given by

$$\beta_e(\omega) = \beta_0^g + (\omega - \omega_0)\beta_1^g + \tfrac{1}{2}(\omega - \omega_0)^2\beta_2^g + \tfrac{1}{6}(\omega - \omega_0)^3\beta_3^g + \cdots, \quad (5.2.22)$$

where β_m^g with $m = 1, 2, \ldots$ is defined as

$$\beta_m^g = \frac{d^m q}{d\omega^m} \approx \left(\frac{1}{v_g}\right)^m \frac{d^m q}{d\delta^m}, \quad (5.2.23)$$

where derivatives are evaluated at $\omega = \omega_0$. The superscript g denotes that the dispersive effects have their origin in the grating. In Eq. (5.2.23), v_g is the group velocity of pulse in the absence of the grating ($\kappa = 0$). It occurs naturally when the frequency dependence of $\bar{n}$ is taken into account in Eq. (5.2.6). Dispersion of v_g is neglected in Eq. (5.2.23) but can be included easily.

Consider first the group velocity of the pulse inside the grating. Using $V_G = 1/\beta_1^g$ and Eq. (5.2.23), it is given by

$$V_G = \pm v_g\sqrt{1 - \kappa^2/\delta^2}, \quad (5.2.24)$$

where the choice of the signs $\pm$ depends on whether the pulse is moving in the forward or the backward direction. Far from the band edges ($|\delta| \gg \kappa$), optical pulse is unaffected by the grating, and it travels at the group velocity expected in the absence of the grating. However, as $|\delta|$ approaches κ, the group velocity decreases and becomes zero at the two edges of the stop band where $|\delta| = \kappa$. Thus, close to the photonic bandgap, an optical pulse experiences considerable slowing down inside a fiber grating. As an example, its speed is reduced by 50% when $|\delta|/\kappa \approx 1.18$.

Second- and third-order dispersive properties of the grating are governed by β_2^g and β_3^g, respectively. Using Eq. (5.2.23) together with the dispersion relation, these parameters are given by

$$\beta_2^g = -\frac{\text{sgn}(\delta)\kappa^2/v_g^2}{(\delta^2 - \kappa^2)^{3/2}}, \qquad \beta_3^g = \frac{3|\delta|\kappa^2/v_g^3}{(\delta^2 - \kappa^2)^{5/2}}. \quad (5.2.25)$$

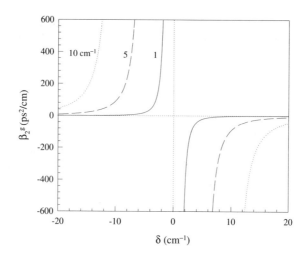

Figure 5.3: Grating-induced group-velocity dispersion plotted as a function of δ for several values of the coupling coefficient κ.

The grating-induced GVD, governed by the parameter β_2^g, depends on the sign of detuning δ. The GVD is anomalous on the upper branch of the dispersion curve in Figure 5.2, where δ is positive and the carrier frequency exceeds the Bragg frequency. In contrast, GVD becomes normal ($\beta_2^g > 0$) on the lower branch of the dispersion curve, where δ is negative and the carrier frequency is smaller than the Bragg frequency. We notice that the third-order dispersion remains positive on both branches of the dispersion curve. Both β_2^g and β_3^g become infinitely large at the two edges of the stop band.

The dispersive properties of a fiber grating are quite different than those of a uniform fiber. First, β_2^g changes its sign on the two sides of the stop band centered at the Bragg wavelength, whose location is easily controlled and can be in any region of the optical spectrum. This is in sharp contrast with β_2 for uniform fibers, which changes sign at the zero-dispersion wavelength, which can be varied only in a range from 1.3 to 1.6 μm. Second, β_2^g is anomalous on the shorter-wavelength side of the stop band, whereas β_2 for fibers becomes anomalous for wavelengths longer than the zero-dispersion wavelength. Third, the magnitude of β_2^g exceeds that of β_2 by a large factor. Figure 5.3 shows how β_2^g varies with detuning δ for several values of κ. As seen there, $|\beta_2^g|$ can exceed 100 ps^2/cm for a fiber grating. This feature can be used for dispersion compensation in the transmission geometry [16]. Typically, a 10-cm-long grating can compensate the GVD acquired over fiber lengths of 50 km or more. Chirped gratings, discussed later in this chapter, can provide even more dispersion when the incident light is inside the stop band, although they reflect the dispersion-compensated signal [7].

5.2.4 Grating as an Optical Filter

What happens to optical pulses incident on a fiber grating depends very much on the location of the pulse spectrum with respect to the stop band introduced by the grating.

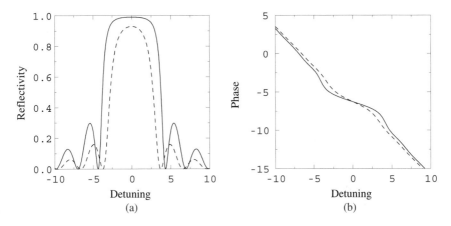

Figure 5.4: (a) The reflectivity $|r_g|^2$ and (b) the phase of the reflection coefficient r_g plotted as a function of detuning δ for two values of κL.

If the pulse spectrum falls entirely within the stop band, the entire pulse is reflected by the grating. On the other hand, if a part of the pulse spectrum is outside the stop band, that part will be transmitted through the grating. Therefore, the shape of the reflected and transmitted pulses will be quite different than that of the incident pulse because of the splitting of the spectrum and the dispersive properties of the fiber grating. If the peak power of pulses is small enough that nonlinear effects are negligible, we can first calculate the reflection and transmission coefficients for each spectral component. The shape of the transmitted and reflected pulses is then obtained by integrating over the spectrum of the incident pulse. Considerable distortion can occur either when the pulse spectrum is wider than the stop band or when it lies in the vicinity of a stop-band edge.

The reflection and transmission coefficients can be calculated by using Eqs. (5.2.19) and (5.2.20) with the appropriate boundary conditions. Consider a grating of length L and assume that light is incident only at the front end, located at $z = 0$. The reflection coefficient is then given by

$$r_g = \frac{\tilde{A}_b(0)}{\tilde{A}_f(0)} = \frac{B_2 + r(q)A_1}{A_1 + r(q)B_2}. \tag{5.2.26}$$

If we impose the boundary condition $\tilde{A}_b(L) = 0$ in Eq. (5.2.20), then

$$B_2 = -r(q)A_1 \exp(2iqL). \tag{5.2.27}$$

Using this value of B_2 and $r(q)$ from Eq. (5.2.21) in Eq. (5.2.26), we obtain

$$r_g = \frac{i\kappa \sin(qL)}{q\cos(qL) - i\delta \sin(qL)}. \tag{5.2.28}$$

The transmission coefficient t_g can be obtained in a similar manner. The frequency dependence of r_g and t_g shows the filter characteristics associated with a fiber grating.

Figure 5.4 shows the reflectivity $|r_g|^2$ and the phase of the reflection coefficient r_g as a function of detuning δ for two values of κL. The grating reflectivity within the stop band approaches 100% for $\kappa L = 3$ or larger. Maximum reflectivity occurs at the center of the stop band and, by setting $\delta = 0$ in Eq. (5.2.28), is given by

$$R_{max} = |r_g|^2 = \tanh^2(\kappa L). \tag{5.2.29}$$

For $\kappa L = 2$, $R_{max} = 0.93$. The condition $\kappa L > 2$ with $\kappa = 2\pi \delta n_1/\lambda$ can be used to estimate the grating length required for high reflectivity inside the stop band. For $\delta n_1 \approx 10^{-4}$ and $\lambda = 1.55$ μm, length L should exceed 5 mm to yield $\kappa L > 2$. These requirements are easily satisfied in practice. Indeed, reflectivities in excess of 99% have been achieved for a grating length of a few centimeters.

An undesirable feature seen in Figure 5.4 from a practical standpoint is the presence of multiple sidebands located on each side of the stop band. These sidebands originate from weak reflections occurring at the two grating ends, where the refractive index changes suddenly compared to its value outside the grating region. Even though the change in refractive index is typically less than 1%, the reflections at the two grating ends form a Fabry–Perot resonator with its own wavelength-dependent transmission. An apodization technique is commonly used to remove the sidebands seen in Figure 5.4 [4]. In this technique, the intensity of the ultraviolet laser beam used to form the grating is made nonuniform in such a way that the intensity drops to zero gradually near the two grating ends.

5.3 Modulation Instability

Wave propagation in a nonlinear, one-dimensional, periodic medium has been studied in several different contexts [17]–[37]. In the case of a fiber grating, the presence of an intensity-dependent term in Eq. (5.1.1) leads to SPM and XPM of counterpropagating waves. These nonlinear effects can be included by solving the nonlinear coupled-mode equations, Eqs. (5.2.10) and (5.2.11). In this section, these equations are used to study the nonlinear CW solution and its destabilization through the modulation instability.

5.3.1 Nonlinear Dispersion Relations

In almost all cases of practical interest, the β_2 term can be neglected in Eqs. (5.2.10) and (5.2.11). For typical grating lengths (< 1 m), the loss term can also be neglected by setting $\alpha = 0$. The nonlinear coupled-mode equations then take the following form:

$$i\frac{\partial A_f}{\partial z} + \frac{i}{v_g}\frac{\partial A_f}{\partial t} + \delta A_f + \kappa A_b + \gamma(|A_f|^2 + 2|A_b|^2)A_f = 0, \tag{5.3.1}$$

$$-i\frac{\partial A_b}{\partial z} + \frac{i}{v_g}\frac{\partial A_b}{\partial t} + \delta A_b + \kappa A_f + \gamma(|A_b|^2 + 2|A_f|^2)A_b = 0, \tag{5.3.2}$$

where v_g is the group velocity far from the stop band associated with the grating. These equations exhibit many interesting nonlinear effects. We begin by considering the CW solution of Eqs. (5.3.1) and (5.3.2) without imposing any boundary conditions. Even though this is unrealistic from a practical standpoint, the resulting dispersion curves provide considerable physical insight. Note that all grating-induced dispersive effects are included in these equations through the κ term.

To solve Eqs. (5.3.1) and (5.3.2) in the CW limit, we neglect the time-derivative term and assume the following form for the solution:

$$A_f = u_f \exp(iqz), \qquad A_b = u_b \exp(iqz), \tag{5.3.3}$$

where u_f and u_b are constant along the grating length. By introducing a parameter $f = u_b/u_f$ that describes how the total power $P_0 = u_f^2 + u_b^2$ is divided between the forward- and backward-propagating waves, u_f and u_b can be written as

$$u_f = \sqrt{\frac{P_0}{1+f^2}}, \qquad u_b = f\sqrt{\frac{P_0}{1+f^2}}. \tag{5.3.4}$$

The parameter f can be positive or negative. For values of $|f| > 1$, the backward wave dominates. By using Eqs. (5.3.1)–(5.3.4), both q and δ are found to depend on f and are given by

$$q = -\frac{\kappa(1-f^2)}{2f} - \frac{\gamma P_0}{2}\left(\frac{1-f^2}{1+f^2}\right), \qquad \delta = -\frac{\kappa(1+f^2)}{2f} - \frac{3\gamma P_0}{2}. \tag{5.3.5}$$

To understand the physical meaning of Eqs. (5.3.5), let us first consider the low-power case so that nonlinear effects are negligible. If we set $\gamma = 0$ in Eqs. (5.3.5), it is easy to show that $q^2 = \delta^2 - \kappa^2$. This is precisely the dispersion relation (5.2.18) obtained previously. As f changes, q and δ trace the dispersion curves shown in Figure 5.2. In fact, $f < 0$ on the upper branch, while positive values of f belong to the lower branch. The two edges of the stop band occur at $f = \pm 1$. From a practical standpoint, the detuning δ of the CW beam from the Bragg frequency determines the value of f, which in turn fixes the values of q from Eqs. (5.3.5). The group velocity inside the grating also depends on f and is given by

$$V_G = v_g \frac{d\delta}{dq} = v_g\left(\frac{1-f^2}{1+f^2}\right). \tag{5.3.6}$$

As expected, V_G becomes zero at the edges of the stop band corresponding to $f = \pm 1$. Note that V_G becomes negative for $|f| > 1$. This is not surprising if we note that the backward-propagating wave is more intense in that case. The speed of light is reduced considerably as the CW-beam frequency approaches an edge of the stop band. As an example, it reduces by 50% when f^2 equals 1/3 or 3.

Equations (5.3.5) can be used to find how the dispersion curves are affected by the fiber nonlinearity. Figure 5.5 shows such curves at two power levels. The nonlinear effects change the upper branch of the dispersion curve qualitatively, leading to the formation of a loop beyond a critical power level. This critical value of P_0 can be found by looking for the value of f at which q becomes zero while $|f| \neq 1$. From Eqs. (5.3.5), we find that this can occur when

$$f \equiv f_c = -(\gamma P_0/2\kappa) + \sqrt{(\gamma P_0/2\kappa)^2 - 1}. \tag{5.3.7}$$

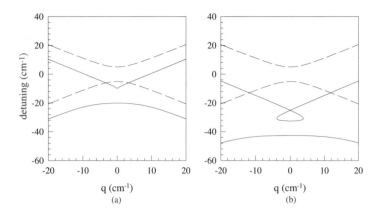

Figure 5.5: Nonlinear dispersion curves showing variation of detuning δ with q for $\gamma P_0/\kappa = 2$ (a) and 5 (b) when $\kappa = 5$ cm^{-1}. Dashed curves show the linear case ($\gamma = 0$).

Thus, a loop is formed only on the upper branch, where $f < 0$. Moreover, it can form only when the total power $P_0 > P_c$, where $P_c = 2\kappa/\gamma$. Physically, an increase in the mode index through the nonlinear term in Eq. (5.1.1) increases the Bragg wavelength and shifts the stop band toward lower frequencies. Since the amount of shift depends on the total power P_0, light at a frequency close to the edge of the upper branch can be shifted out of resonance with changes in its power. If the nonlinear parameter γ were negative (self-defocusing medium with $n_2 < 0$), the loop will form on the lower branch in Figure 5.5, as is also evident from Eq. (5.3.7).

5.3.2   Linear Stability Analysis

The stability issue is of paramount importance and must be addressed for the CW solutions obtained in this section. Similar to the analysis of Section 1.5, modulation instability can destabilize the steady-state solution and produce periodic output even when a CW beam is incident on one end of the fiber grating [38]–[43]. Moreover, the existence of modulation instability is an essential step for the formation of localized states associated with Bragg solitons.

For simplicity, we discuss modulation instability using the CW solution given by Eqs. (5.3.3) and (5.3.4) and obtained without imposing the boundary conditions at the grating ends. Following a standard procedure, we perturb the steady state slightly as

$$A_j = (u_j + a_j)\exp(iqz), \quad (j = f, b), \tag{5.3.8}$$

and linearize Eqs. (5.3.1) and (5.3.2) assuming that the perturbation a_j is small. The resulting equations are [43]

$$i\frac{\partial a_f}{\partial z} + \frac{i}{v_g}\frac{\partial a_f}{\partial t} + \kappa a_b - \kappa f a_f + \Gamma[(a_f + a_f^*) + 2f(a_b + a_b^*)] = 0, \tag{5.3.9}$$

$$-i\frac{\partial a_b}{\partial z} + \frac{i}{v_g}\frac{\partial a_b}{\partial t} + \kappa a_f - \frac{\kappa}{f}a_b + \Gamma[2f(a_f + a_f^*) + f^2(a_b + a_b^*)] = 0, \tag{5.3.10}$$

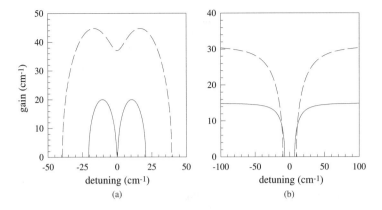

Figure 5.6: Gain spectra of modulation instability in the (a) anomalous- and (b) normal-GVD regions of a fiber grating ($f = \pm 0.5$) at two power levels corresponding to $\Gamma/\kappa = 0.5$ and 2.

where $\Gamma = \gamma P_0/(1 + f^2)$ is an effective nonlinear parameter.

This set of two linear coupled equations can be solved assuming a plane-wave solution of the form

$$a_j = c_j \exp[i(Kz - \Omega t)] + d_j \exp[-i(Kz + \Omega t)], \qquad (5.3.11)$$

where the subscript $j = f$ or b. As a result, we obtain a set of four homogeneous equations satisfied by c_j and d_j. This set has a nontrivial solution only when the (4×4) determinant formed by the coefficients matrix vanishes. This condition leads to the following fourth-order polynomial:

$$(s^2 - K^2)^2 - 2\kappa^2(s^2 - K^2) - \kappa^2 f^2 (s + K)^2$$
$$- \kappa^2 f^{-2}(s - K)^2 - 4\kappa \Gamma f(s^2 - 3K^2) = 0, \qquad (5.3.12)$$

where we have introduced a rescaled spatial frequency as $s = \Omega/v_g$.

The four roots of the polynomial in Eq. (5.3.12) determine the stability of the CW solution. However, a tricky issue must first be resolved. Equation (5.3.12) is a fourth-order polynomial in both s and K. The question is, which one determines the gain associated with modulation instability? In the case of the uniform-index fibers, the gain g can be related to the imaginary part of K when light travels in the forward direction only. In a fiber grating, light travels in both the forward and backward directions simultaneously, and it is the time that moves forward for both of them. As a result, Eq. (5.3.12) should be viewed as a fourth-order polynomial in s whose roots depend on K. The gain of modulation instability is obtained using $g = 2\,\mathrm{Im}(s_m)$, where s_m is the root with the largest imaginary part.

The root analysis of Eq. (5.3.12) leads to several interesting conclusions [43]. Figure 5.6 shows the gain spectra of modulation instability in the anomalous- and normal-GVD regions, corresponding to upper and lower branches of the dispersion curves, for two values of Γ/κ. In the anomalous-GVD case and at relatively low powers ($\Gamma < \kappa$), the gain spectrum is similar to that found for uniform-index fibers. As shown later in this section, the nonlinear coupled-mode equations reduce to a nonlinear Schrödinger (NLS) equation when $\Gamma \ll \kappa$. At high values of P_0 such that $\Gamma > \kappa$, the gain exists even at $s = 0$, as seen in Figure 5.6(a) for $\Gamma/\kappa = 2$. Thus, the CW solution becomes unstable even to zero-frequency (dc) fluctuations at high power levels.

Modulation instability can occur even on the lower branch of the dispersion curve ($f > 0$), where grating-induced GVD is normal. The instability occurs only when P_0 exceeds a certain value such that

$$P_0 > \tfrac{1}{2}\kappa(1 + f^2)^2 f^p, \tag{5.3.13}$$

where $p = 1$ if $f \leq 1$ but $p = -3$ when $f > 1$. The occurrence of modulation instability in the normal-GVD region is solely a grating-induced feature.

The preceding analysis completely ignores boundary conditions. For a finite-length grating, one should examine the stability of the CW solution obtained in terms of the elliptic functions. Such a study is complicated and requires a numerical solution to the nonlinear coupled-mode equations [39]. The results show that the CW solution can still become unstable, resulting in the formation of a pulse train through modulation instability. The resulting pulse train is not necessarily periodic and, under certain conditions, can exhibit period doubling and optical chaos.

5.3.3   Effective NLS Equation

The similarity of the gain spectrum in Figure 5.6 with that occurring in uniform-index fibers (see Section 1.5) indicates that, at not-too-high power levels, the nonlinear coupled-mode equations predict features that coincide with those found for the NLS equation. Indeed, under certain conditions, Eqs. (5.3.3) and (5.3.4) can be reduced formally to an effective NLS equation [44]–[48]. A multiple-scale method is commonly used to prove this equivalence; details can be found in Ref. [29].

The analysis used to reduce the nonlinear coupled-mode equations to an effective NLS equation makes use of the Bloch formalism well known in solid-state physics. Even in the absence of nonlinear effects, the eigenfunctions associated with the photonic bands, corresponding to the dispersion relation $q^2 = \delta^2 - \kappa^2$, are not A_f and A_b but the Bloch waves formed by a linear combination of A_f and A_b. If this basis is used for the nonlinear problem, Eqs. (5.3.3) and (5.3.4) reduce to an effective NLS equation, provided two conditions are met. First, the peak intensity of the pulse is small enough that the nonlinear index change $n_2 I_0$ in Eq. (5.1.1) is much smaller than the maximum value of δn_g. This condition is equivalent to requiring that $\gamma P_0 \ll \kappa$ or $\kappa L_{\mathrm{NL}} \gg 1$, where $L_{\mathrm{NL}} = (\gamma P_0)^{-1}$ is the nonlinear length. This requirement is easy to satisfy in practice even at peak intensity levels as high as 100 GW/cm^2. Second, the third-order dispersion β_3^g induced by the grating should be negligible.

When the foregoing two conditions are satisfied, pulse propagation in a fiber grating is governed by the following NLS equation [43]:

$$\frac{i}{v_g}\frac{\partial U}{\partial t} - \frac{(1 - v^2)^{3/2}}{2\kappa\,\mathrm{sgn}(f)}\frac{\partial^2 U}{\partial \zeta^2} + \frac{1}{2}(3 - v^2)\gamma|U|^2 U = 0, \tag{5.3.14}$$

where $\zeta = z - V_G t$. We have introduced a speed-reduction factor related to the parameter f through Eq. (5.3.6) as

$$v = \frac{V_G}{v_g} = \left(\frac{1 - f^2}{1 + f^2}\right) = \pm\sqrt{1 - \kappa^2/\delta^2}. \tag{5.3.15}$$

The group velocity decreases by the factor v close to an edge of the stop band and vanishes at the two edges ($v = 0$), corresponding to $f = \pm 1$. The reason the first term is a time derivative, rather than the z derivative, was discussed earlier. It can also be understood from a physical standpoint if we note that the variable U does not correspond to the amplitude of the forward- or backward-propagating wave but represents the amplitude of the envelope associated with the Bloch wave formed by a superposition of A_f and A_b.

Equation (5.3.15) has been written for the case in which the contribution of A_f dominates ($|f| < 1$) so that the entire Bloch-wave envelope is propagating forward at the reduced group velocity V_G. With this in mind, we introduce $z = V_G t$ as the distance traveled by the envelope and account for changes in its shape through a local time variable defined as $T = t - z/V_G$. Equation (5.3.15) can then be written in the standard form of the NLS equation as

$$i\frac{\partial U}{\partial z} - \frac{\beta_2^g}{2}\frac{\partial^2 U}{\partial T^2} + \gamma_g |U|^2 U = 0, \tag{5.3.16}$$

where the effective GVD parameter β_2^g and the nonlinear parameter γ_g are defined as

$$\beta_2^g = \frac{(1 - v^2)^{3/2}}{\text{sgn}(f)v_g^2 \kappa v^3}, \qquad \gamma_g = \left(\frac{3 - v^2}{2v}\right)\gamma. \tag{5.3.17}$$

Using Eq. (5.3.15), the GVD parameter β_2^g can be shown to be the same as in Eq. (5.2.25).

Several features of Eq. (5.3.16) are noteworthy when this equation is compared with the standard NLS equation obtained in Chapter 1. First, the variable U represents the amplitude of the envelope associated with the Bloch wave formed by a superposition of A_f and A_b. Second, the parameters β_2^g and γ_g are not constants but depend on the speed-reduction factor v. Both increase as v decreases and become infinite at the edges of the stop band, where $v = 0$. Clearly, Eq. (5.3.16) is not valid at that point. However, it remains valid close to but outside the stop band. Far from the stop band ($v \to 1$), β_2^g becomes quite small (<1 ps^2/km for typical values of κ). One should then include fiber GVD and replace β_2^g by β_2. Noting that $\gamma_g = \gamma$ when $v = 1$, Eq. (5.3.16) reduces to the standard NLS equation, and U corresponds to the forward-wave amplitude since no backward wave is generated under such conditions.

Before we can use Eq. (5.3.16) for predicting the modulation-instability gain and the frequency at which the gain peaks, we need to know the total power P_0 inside the grating when a CW beam with power P_{in} is incident at the input end of the grating, located at $z = 0$. This is a complicated issue for apodized fiber gratings because κ is not constant in the transition zone. However, observing that the nonlinear coupled-mode equations require $|A_f^2| - |A_b^2|$ to remain constant along the grating, one finds that the total power P_0 inside the grating is enhanced by a factor $1/v$ [49]. The predictions of Eq. (5.3.16) are in agreement with the modulation-instability analysis based on the nonlinear coupled-mode equations as long as $\gamma P_0 \ll \kappa$ [43]. The NLS equation provides a shortcut to understanding the temporal dynamics in gratings within its regime of validity.

5.4   Nonlinear Pulse Propagation

As mentioned earlier, modulation instability often indicates the possibility of soliton formation. In the case of Bragg gratings, it is closely related to a new kind of solitons referred to as *Bragg solitons* or *grating solitons*. Such solitons were discovered as early as 1981 [50]. They were called *gap solitons* in 1987 in the context of periodic structures known as *superlattices* [20], because they exist only inside the stop band. Since then, a much larger class of Bragg solitons has been identified [51]–[54] by solving analytically Eqs. (5.3.1) and (5.3.2), or its generalizations [55]).

The advent of fiber gratings provided an incentive during the 1990s for studying propagation of short optical pulses in such gratings [56]–[67]. The peak intensities required to observe the nonlinear effects are quite high (typically >10 GW/cm^2) for Bragg gratings made in silica fibers, because of a short interaction length (typically <10 cm) and a low value of the nonlinear parameter n_2. The use of chalcogenide glass fibers for making gratings can reduce required peak intensities by a factor of 100 or more because of the high values of n_2 in such glasses [68].

5.4.1   Bragg Solitons

It was realized in 1989 that the coupled-mode equations, Eqs. (5.3.1) and (5.3.2), become identical to the well-known massive Thirring model [69] if the SPM term is set to zero. The massive Thirring model of quantum field theory is known to be integrable by the inverse scattering method [70]–[73]. When the SPM term is included, the coupled-mode equations become non-integrable, and solitons do not exist in a strict mathematical sense. However, shape-preserving solitary waves can be obtained through a suitable transformation of the soliton supported by the massive Thirring model. These solitary waves correspond to the following solution [52]:

$$A_f(z,t) = a_+ \text{sech}(\zeta - i\psi/2)e^{i\theta}, \tag{5.4.1}$$

$$A_b(z,t) = a_- \text{sech}(\zeta + i\psi/2)e^{i\theta}, \tag{5.4.2}$$

where

$$a_\pm = \pm\left(\frac{1 \pm v}{1 \mp v}\right)^{1/4}\sqrt{\frac{\kappa(1 - v^2)}{\gamma(3 - v^2)}}\sin\psi, \quad \zeta = \frac{z - V_G t}{\sqrt{1 - v^2}}\kappa\sin\psi, \tag{5.4.3}$$

$$\theta = \frac{v(z - V_G t)}{\sqrt{1 - v^2}}\kappa\cos\psi - \frac{4v}{(3 - v^2)}\tan^{-1}[|\cot(\psi/2)|\coth(\zeta)]. \tag{5.4.4}$$

This solution represents a *two-parameter family* of Bragg solitons. The parameter v is in the range $-1 < v < 1$, while the parameter ψ can be chosen anywhere in the range $0 < \psi < \pi$. The specific case $\psi = \pi/2$ corresponds to the center of the stop band [51]; the soliton solutions of this kind were first obtained analytically in 1981 [50]. Physically, Bragg solitons represent specific combinations of counterpropagating waves that pair in such a way that they move together at the reduced speed $V_G = v v_g$. Since v can be negative, the soliton can move forward or backward. The soliton width T_s is also related to the parameters v and ψ and is given by

$$T_s = \sqrt{1 - v^2}/(\kappa V_G \sin \psi). \tag{5.4.5}$$

One can understand the reduced speed of a Bragg soliton by noting that the counterpropagating waves form a single entity that moves at a common speed. The relative amplitudes of the two waves participating in soliton formation determine the soliton speed. If A_f dominates, the soliton moves in the forward direction but at a reduced speed. The opposite happens when A_b is larger. In the case of equal amplitudes, the soliton does not move at all since V_G becomes zero. This case corresponds to the *stationary gap solitons* predicted in the context of superlattices [20]. In the opposite limit, in which $|v| \to 1$, Bragg solitons cease to exist since the grating becomes ineffective.

Another family of solitary waves is obtained by looking for a general class of the shape-preserving localized solutions of the nonlinear coupled-mode equations [53, 54]. Such solitary waves exist both inside and outside the stop band and for both focusing and defocusing nonlinearities. They reduce to the Bragg solitons described by Eqs. (5.4.2)–(5.4.4) in some specific limits. On the lower branch of the dispersion curve, where the GVD is normal, such solitary waves represent dark solitons similar to those discussed in Chapter 4.

Figure 5.7 shows several different types of gap solitons that can form as the two normalized parameters, δ and v ($|v| < 1$), are varied. The intensity profiles become qualitatively different as these two parameters are changed. First, notice that the two components associated with a gap soliton at rest ($v = 0$) have equal intensities for both the in-gap ($|\delta| < 1$) and out-of-gap ($|\delta| > 1$) solutions as a consequence of the fact that the net photon flux is zero. However, the relative phase ϕ_r between the two components is not the same. As a result, the gap soliton is of the bright kind inside the bandgap ($\phi_r = 0$) but forms a dark–antidark pair outside of it ($\phi_r = \pi$). Moreover, the amplitude of the in-gap (bright) solitons increases and their width decreases as δ is varied in the range -1–1.

As seen in Figure 5.7, the symmetry between the counterpropagating waves is broken for moving gap solitons, which have a stronger component in the direction of motion (the inset marked AS) . More importantly, the bright gap solitons with high amplitudes (HA inset) close to the upper edge of the bandgap first acquire a *Lorentzian* shape (LZ inset) and then bifurcate into the dark–antidark pairs (DK and AK insets) outside the bandgap.

Stability of the gap solitons associated with the nonlinear couple-mode equations has been analyzed recently [74]–[79]. Applying a singular perturbation theory to the generalized Thirring model for a small value of the SPM term, it was discovered that the oscillatory instability of gap solitons is associated with collisions of the soliton internal modes [74]. However, in the limit in which the NLS equation is valid, all gap solitons are found to be stable. When the ratio between the SPM and the XPM coefficients

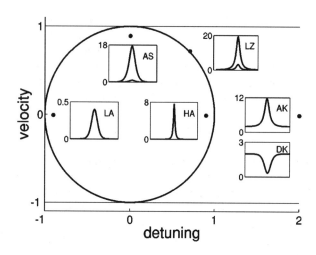

Figure 5.7: Different types of gap solitons forming in the self-defocusing case as v and $delta$ are varied. The insets show the amplitude profiles $|A_f|$ and $|A_b|$ at several points (solid circles). Bright gap solitons of low amplitude (LA), high amplitude (HA), and asymmetric (AS) kinds exist inside the bandgap. The HA solitons become the Lorentzian (LZ) type over the right semicircle and then bifurcate into dark (DK) and antidark (AK) pairs. In the self-focusing case the same picture holds with the reversed symmetry. (After Ref. [54]; ©2001 APS.)

is varied, the instability scenario of the multihump gap solitons changes considerably [78]. In the case in which the XPM term dominates, the gap soliton is highly unstable and often splits into two pulses. When the SPM term dominates, only the oscillatory instabilities with small growth rates remain. In general, a numerical stability analysis is the only tool available for studying such oscillatory instabilities, but the results are strongly influenced by the grid size used and can be affected by numerical instabilities [77], [79]. The resonant mechanism behind the oscillatory instabilities of localized modes in a nonlinear periodic system is discussed in more detail in Chapter 11.

It should be mentioned that the nonlinear coupled-mode equations, which possess the exact analytical solutions in the form of a two-parameter family of gap solitons, have also been discussed in several other contexts. An example is provided by the diatomic nonlinear lattices with an on-site nonlinearity; such lattices have been found to support gap solitons [80]–[83]. The relation of these solitons to the optical gap solitons has been clarified in Ref. [84], where it is shown that the main equations of Ref. [80] represent a generalization of the coupled-mode equations and that they also appear in optics in the case of the so-called *deep gratings* [55].

Similar to the spatial and temporal solitons discussed in Chapters 2 and 3, the gap solitons behave like classical particles in the presence of external perturbations [84]–[86]. In particular, the interaction of gap solitons with the grating defects displays features such as the capture, reflection, and transmission of gap solitons [86]. As an example, for sufficiently low velocities, a gap soliton, incident on a defect, can transfer its energy to a nonlinear defect mode formed at the defect location, provided such a mode exists with the same frequency (resonance) and has an intensity less than that of the gap soliton (energetic accessibility).

5.4.2 Relation to NLS Solitons

As discussed earlier, the nonlinear coupled-mode equations reduce to the NLS equation when $\gamma P_0 \ll \kappa$, where P_0 is the peak power of the pulse propagating inside the grating. Since the NLS equation is integrable by the inverse scattering transform method, the fundamental and higher-order NLS solitons (see Section 1.2) should also exist for a fiber grating. The question then becomes how they are related to the Bragg soliton described by Eqs. (5.4.1) and (5.4.2).

To answer this question, we write the NLS equation (5.3.16) using soliton units in its standard form,

$$i\frac{\partial u}{\partial \xi} + \frac{1}{2}\frac{\partial^2 u}{\partial \tau^2} + |u|^2 u = 0, \tag{5.4.6}$$

where $\xi = z/L_D$, $\tau = T/T_0$, $u = \sqrt{\gamma_g L_D}$, and $L_D = T_0^2/|\beta_2^g|$ is the dispersion length. The fundamental soliton of this equation, in its most general form, is given by

$$u(\xi, \tau) = \eta \operatorname{sech}[\eta(\tau - \tau_s + \varepsilon\xi)]\exp[i(\eta^2 - \varepsilon^2)\xi/2 - i\varepsilon\tau + i\phi_s], \tag{5.4.7}$$

where η, ε, τ_s, and ϕ_s are four arbitrary parameters representing amplitude, frequency, position, and phase of the soliton, respectively. The soliton width is related inversely to the amplitude as $T_s = T_0/\eta$. The physical origin of such solitons is the same as that for conventional solitons, except that the GVD is provided by the grating rather than by material dispersion.

At first sight, Eq. (5.4.7) looks quite different than the Bragg soliton described by Eqs. (5.4.2)–(5.4.4). However, one should remember that u represents the amplitude of the Bloch wave formed by superimposing A_f and A_b. If the total optical field is considered and the low-power limit ($\gamma P_0 \ll \kappa$) is taken, the Bragg soliton indeed reduces to the fundamental NLS soliton [29]. The massive Thirring model also allows for higher-order solitons [87]. One would expect them to be related to higher-order NLS solitons in the appropriate limit. It has been shown that any solution of the NLS equation (5.3.16) can be used to construct an approximate solution of the coupled-mode equations [48].

The observation that Bragg solitons are governed by an effective NLS equation in the limit $\kappa L_{NL} \gg 1$, where L_{NL} is the nonlinear length, allows us to use the concept of soliton order N and the soliton period z_0 developed in Chapter 3. These parameters are defined as

$$N^2 = \frac{L_D}{L_{NL}} \equiv \frac{\gamma_g P_0 T_0^2}{|\beta_2^g|}, \qquad z_0 = \frac{\pi}{2}L_D \equiv \frac{\pi}{2}\frac{T_0^2}{|\beta_2^g|}. \tag{5.4.8}$$

We need to interpret the meaning of the soliton peak power P_0 carefully since the NLS soliton represents the amplitude of the Bloch wave formed by a combination of A_f and A_b. This aspect is discussed later in this section.

An interesting issue is related to the collision of Bragg solitons. Since Bragg solitons described by Eqs. (5.4.1) and (5.4.2) are only solitary waves (because of the nonintegrability of the underlying nonlinear coupled-mode equations), they may not survive collisions. On the other hand, the NLS solitons are guaranteed to remain unaffected by their mutual collisions. Numerical simulations based on Eqs. (5.3.1) and (5.3.2) show that Bragg solitons indeed exhibit features reminiscent of an NLS soliton in the low-power limit $\gamma P_0 \ll \kappa$ [64]. More specifically, two Bragg solitons attract or repel each other depending on their relative phase. The new feature is that the relative phase depends on the initial separation between the two solitons.

5.4.3 Formation of Bragg Solitons

Formation of Bragg solitons in fiber gratings was first observed in a 1996 experiment [56]. Since then, more careful experiments have been performed, and many features of Bragg solitons have been extracted. While comparing the experimental results with the coupled-mode theory, one needs to implement the boundary conditions properly. For example, the peak power P_0 of the Bragg soliton formed inside the grating when a pulse is launched is not the same as the input peak power P_{in}. The reason can be understood by noting that the group velocity of the pulse changes as the input pulse crosses the front end of the grating, located at $z = 0$. As a result, pulse length given by $v_g T_0$ just outside the grating changes to $V_G T_0$ on crossing the interface located at $z = 0$ [9], and the pulse peak power is enhanced by the ratio v_g/V_G. Mathematically, one can use the coupled-mode equations to show that $P_0 = |A_f^2| + |A_b^2| = P_{\text{in}}/v$, where $v = V_G/v_g$ is the speed-reduction factor introduced earlier. The argument becomes more complicated for apodized fiber gratings, used often in practice, because κ is not constant in the transition region [61]. However, the same power enhancement occurs at the end of the transition region.

From a practical standpoint, one needs to know the amount of peak power P_{in} required to excite the fundamental Bragg soliton. The soliton period z_0 is another important parameter relevant for soliton formation since it sets the length scale over which optical solitons evolve. We can use Eq. (5.4.8) with $N = 1$ to estimate both of them. Using the expressions for β_2^g and γ_g from Eq. (5.3.17), the input peak power and the soliton period are given by

$$P_{\text{in}} = \frac{2(1-v^2)^{3/2}}{v(3-v^2)v_g^2 T_0^2 \kappa \gamma}, \qquad z_0 = \frac{\pi v^3 v_g^2 T_0^2 \kappa}{2(1-v^2)^{3/2}}, \qquad (5.4.9)$$

where T_0 is related to the FWHM as $T_{\text{FWHM}} \approx 1.76 T_0$. Both P_{in} and z_0 depend through v on detuning of the laser wavelength from the edge of the stop band located at $\delta = \kappa$. Because $v \to 0$ near the edge, P_{in} becomes infinitely large while z_0 tends toward zero.

Bragg solitons have been formed in a 7.5-cm-long apodized fiber grating by using 80-ps pulses obtained from a Q-switched, mode-locked Nd:YLF laser operating at 1053 nm [65]. Figure 5.8 shows pulse shapes at the output end of the grating when input pulses having a peak intensity of 11 GW/cm^2 are used. The coupling coefficient κ was estimated to be 7 cm^{-1} for this grating, while the detuning parameter δ was varied over the range from 7 to 36 cm^{-1} on the blue side of the stop band (anomalous GVD). The arrival time of the pulse depends on δ because of the reduction in group velocity as δ is reduced and tuned closer to the stop-band edge. This delay occurs even when nonlinear effects are negligible.

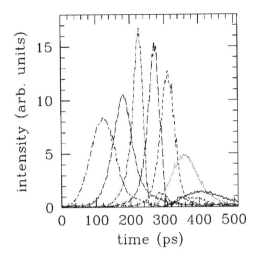

Figure 5.8: Output pulse shapes for different δ when 80-ps pulses with a peak intensity of 11 GW/cm^2 are propagated through a 7.5-cm-long fiber grating. Values of δ from left to right are 3612, 1406, 1053, 935, 847, 788, and 729 m^{-1}. (After Ref. [65]; ©1999 OSA.)

At the high peak intensities used for Figure 5.8, SPM in combination with the grating-induced anomalous GVD leads to formation of Bragg solitons. However, since both β_2^g and γ_g depend on the detuning parameter δ through v, a Bragg soliton can form only in a limited range of δ. With this in mind, we can understand the pulse shapes seen in Figure 5.8. Detuning is so large and β_2^g is so small for the leftmost trace that the pulse acquires some chirping through SPM, but its shape remains nearly unchanged. This feature can also be understood from Eq. (5.4.9), where the soliton period becomes so long as $v \to 1$ that nothing much happens to the pulse over a few-centimeters-long grating. As δ is reduced, the pulse narrows down considerably. A reduction in pulse width by a factor of 3 occurs for $\delta = 1053$ m^{-1} in Figure 5.8. This pulse narrowing is an indication that a Bragg soliton is beginning to form. However, the soliton period is still much longer than the grating length. In other words, the grating is not long enough to observe the final steady-state shape of the Bragg soliton. Finally, as the edge of the stop band is approached and δ becomes comparable to κ (rightmost solid trace), the GVD becomes so large that the pulse cannot form a soliton and becomes broader than the input pulse. This behavior is also deduced from Eq. (5.4.8), which shows that both N and z_0 tend toward zero as β_2^g tends toward infinity. A Bragg soliton can form only if $N > \frac{1}{2}$. Since the dispersion length becomes smaller than the grating length close to the stop-band edge, pulse can experience considerable broadening. This is precisely what is observed for the smallest value of δ in Figure 5.8 (solid curve).

A similar behavior was observed over a large range of pulse energies, with some evidence of the second-order soliton for input peak intensities in excess of 20 GW/cm^2 [65]. A careful comparison of the experimental data with the theory based on the nonlinear coupled-mode equations and the effective NLS equation showed that the NLS equation provides an accurate description within its regime of validity. Figure 5.9

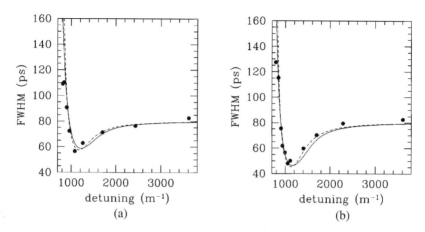

Figure 5.9: Measured pulse widths (circles) as a function of detuning for 80-ps pulses with a peak intensity of (a) 3 GW/cm² and (b) 6 GW/cm². Predictions of the coupled-mode theory (solid line) and the effective NLS equation (dashed line) are shown for comparison. (After Ref. [65]; ©1999 OSA.)

compares the measured values of the pulse width with the two theoretical models for peak intensities of 3 and 6 GW/cm². The NLS equation is valid as long as $\kappa L_{\rm NL} \gg 1$. Using $\kappa = 7$ cm^{-1}, we estimate that the peak intensity can be as high as 50 GW/cm² for the NLS equation to remain valid. This is also what was found in Ref. [65].

Gap solitons that form within the stop band of a fiber grating have not been observed because of a practical difficulty: A Bragg grating reflects light whose wavelength falls inside the stop band. Stimulated Raman scattering may provide a solution to this problem since a pump pulse, launched at a wavelength far from the stop band, can excite a *Raman gap soliton* that is trapped within the grating and propagates much more slowly than the pump pulse itself [67]. The energy of such a gap soliton leaks slowly from the grating ends, but it can survive for durations greater than 10 ns even though it is excited by pump pulses of duration 100 ps or so.

5.5 Nonlinear Switching

A fiber grating can exhibit bistable switching even when a CW beam is incident on it [17]. However, optical pulses should be used in practice because of the high intensities required for observing SPM-induced nonlinear switching. We discuss both the CW and pulse-based optical switching in this section.

5.5.1 Optical Bistability

The simple CW solution given in Eq. (5.3.3) is modified considerably when boundary conditions are introduced at the two grating ends. For a finite-size grating, the simplest manifestation of the nonlinear effects occurs through optical bistability, first predicted in 1979 [17].

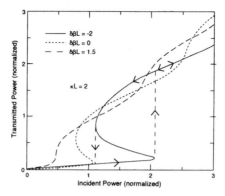

Figure 5.10: Transmitted versus incident power for three values of detuning within the stop band. (After Ref. [17]; ©1979 American Institute of Physics.)

Consider a CW beam incident at one end of the grating and ask how the fiber non-linearity would affect its transmission through the grating. It is clear that both the beam intensity and its wavelength with respect to the stop band will play an important role. Mathematically, we should solve Eqs. (5.3.1) and (5.3.2) after imposing the appropriate boundary conditions at $z = 0$ and $z = L$. These equations can be solved in terms of the elliptic functions; see e.g. Ref. [17]. The analytic solution is somewhat complicated and provides only an implicit relation for the transmitted power at $z = L$. We refer to Ref. [29] for details.

Figure 5.10 shows the transmitted versus incident power [both normalized to a critical power $P_{cr} = 4/(3\gamma L)$] for several detuning values within the stop band by taking $\kappa L = 2$. The S-shaped curves are well known in the context of optical bistability occurring when a nonlinear medium is placed inside a cavity [88]. A linear stability analysis shows that the middle branch of each bistability curve with a negative slope is *unstable*. Clearly, the transmitted power would exhibit bistability with hysteresis, as shown by the arrows on the solid curve. At low powers, transmittivity is small, as expected from the linear theory, since the nonlinear effects are relatively weak. However, above a certain input power, most of the incident power is transmitted. Switching from a low to a high transmission state can be understood qualitatively by noting that the effective detuning δ in Eqs. (5.3.1) and (5.3.2) becomes power dependent because of the nonlinear contribution to the refractive index in Eq. (5.1.1). Thus, light that is mostly reflected at low powers because its wavelength is inside the stop band may tune itself out of the stop band and get transmitted when the nonlinear index change becomes large enough.

The observation of optical bistability in fiber gratings is hampered by the large switching power required ($P_0 > P_{cr} > 1$ kW). Short optical pulses are often used to enhance the peak intensity. Even then, one needs peak-intensity values in excess of 10 GW/cm^2. For this reason, bistable switching was first observed during the 1980s using DFB semiconductor amplifiers for which large carrier-induced nonlinearities reduce the switching threshold to power levels below 1 mW [89]–[91]. Nonlinear switching in a passive grating was observed in a 1992 experiment using a semiconductor waveguide grating [26]. The nonlinear response of such gratings is not governed by

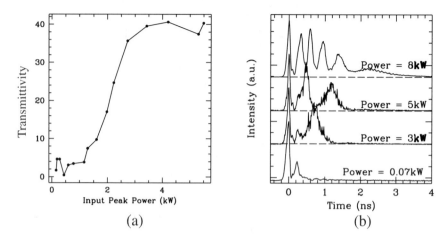

Figure 5.11: (a) Transmittivity as a function of input peak power showing nonlinear switching; (b) output pulse shapes at several peak power levels. (After Ref. [62]; ©1998 OSA.)

the Kerr-type nonlinearity seen in Eq. (5.1.1) because of the presence of free carriers (electrons and holes) whose finite lifetime limits the nonlinear response time. The switching power can also be reduced by introducing a $\pi/2$ phase shift in the middle of the Bragg grating [92]. Such gratings are called $\lambda/4$-shifted or phase-shifted gratings, since a distance of $\lambda/4$ (50% of the grating period) corresponds to a $\pi/2$ phase shift. They are used routinely for making distributed-feedback (DFB) semiconductor lasers [13].

The bistable switching does not always lead to a constant output power when a CW beam is transmitted through a grating. As early as 1982, numerical solutions of Eqs. (5.3.1) and (5.3.2) showed that transmitted power can become not only periodic but also chaotic under certain conditions [18]. In physical terms, portions of the upper branch in Figure 5.10 may become unstable. As a result, the output becomes periodic or chaotic once the beam intensity exceeds the switching threshold.

5.5.2   SPM-Induced Switching

Nonlinear switching in a fiber Bragg grating was observed in 1998 in the 1.55-μm-wavelength region useful for fiber-optic communications [60]. An 8-cm-long grating, with its Bragg wavelength centered near 1536 nm, was used in the experiment. It had a peak reflectivity of 98% and its stop band was only 4 GHz wide. The 3-ns input pulses were obtained by amplifying the output of a pulsed DFB semiconductor laser to power levels as high as 100 kW. Their shape was highly asymmetric because of gain saturation occurring inside the amplifier chain. The laser wavelength was inside the stop band on the short-wavelength side but was set very close to the edge (offset of about 7 pm, or 0.9 GHz).

Figure 5.11(a) shows a sharp rise in the transmittivity from a few percent to 40% when the peak power of input pulses increases beyond 2 kW. Physically, the nonlinear increase in the refractive index at high powers shifts the Bragg wavelength far enough that the pulse finds itself outside the stop band and switches to the upper branch of the bistability curves seen in Figure 5.10. The pulse shapes seen in Figure 5.11(b) show what happens to the transmitted pulse. The initial spike near $t = 0$ in these traces is due to a sharp leading edge of the asymmetric input pulse and should be ignored. Multiple pulses form at the grating output whose number depends on the input power level. At a power level of 3 kW, a single pulse is seen, but the number increases to five at a power level of 8 kW. The pulse width is smallest (about 100 ps) near the leading edge of the pulse train but increases substantially for pulses near the trailing edge.

Several conclusions can be drawn from these results. First, the upper bistability branch in Figure 5.10 is not stable and converts the quasi-CW signal into a pulse train [18]. Second, each pulse evolves toward a constant width. Pulses near the leading edge have had enough propagation time within the grating to stabilize their widths. These pulses can be thought of as a gap soliton since they are formed even though the input signal is inside the photonic bandgap and would be completely reflected in the absence of the nonlinear effects. Third, pulses near the trailing edge are wider simply because the fiber grating is not long enough for them to evolve completely toward a gap soliton. This interpretation was supported by a later experiment in which the grating length was increased to 20 cm [66]. Six gap solitons were found to form in this grating at a peak power level of 1.8 W. The observed data were in agreement with theory based on the nonlinear coupled-mode equations.

The nonlinear switching seen in Figure 5.11 is called *SPM-induced* or *self-induced switching*, since the pulse changes the refractive index to switch itself to the high-transmission state. Clearly, another signal at a different wavelength can also induce switching of the pulse by changing the refractive index through XPM, resulting in XPM-induced switching. This phenomenon was first observed in 1991 as an increase in the transmittivity of a 514-nm signal caused by a 1064-nm pump beam [24]. The increase in transmission was less than 10% in this experiment.

It was suggested later that XPM could be used to form a "push broom" such that a weak CW beam (or a broad pulse) would be swept by a strong pump pulse and its energy piled up at the front end of the pump pulse [93]. The basic idea behind the optical push broom is quite simple. If the wavelength of the pump pulse is far from the stop band while that of the probe is close to but outside the stop band (on the lower branch of the dispersion curve), the pump travels faster than the probe. In the region where pump and probe overlap, the XPM-induced chirp changes the probe frequency such that it moves with the leading edge of the pump pulse. As the pump pulse travels further, it sweeps more and more of the probe energy and piles it up at its leading edge. In effect, the pump acts like a push broom. At the grating output, a significant portion of the probe energy appears at the same time as the pump pulse in the form of a sharp spike because of the XPM-induced increase in the probe speed. Such a push-broom effect was seen in a 1997 experiment [94].

5.5.3 Effects of Birefringence

Fiber birefringence manifests as slightly different refractive indices for the orthogonally polarized components of the optical mode. It plays an important role and affects the nonlinear phenomena considerably. Its effects should be included if Bragg gratings are made inside the core of polarization-maintaining fibers. The coupled-mode theory can easily be extended to account for fiber birefringence [95]–[98]. However, the problem becomes quite complicated, since one needs to solve a set of *four coupled equations* describing the evolution of two orthogonally polarized components, each containing both the forward- and backward-propagating waves. This complexity, however, leads to a rich class of nonlinear phenomena with practical applications, such as optical logic gates.

From a physical standpoint, the two orthogonally polarized components have slightly different mode indices. Since the Bragg wavelength depends on the mode index, the stop bands of the two modes have the same widths but are shifted by a small amount with respect to each other. As a result, even though both polarization components have the same wavelength (or frequency), one of them may fall inside the stop band while the other remains outside it. Moreover, as the two stop bands shift due to nonlinear index changes, the shift can be different for the two orthogonally polarized components because of the combination of the XPM and birefringence effects. It is this feature that leads to a variety of interesting nonlinear effects.

In the case of CW beams, the set of four coupled equations was solved numerically in 1994, and several birefringence-related nonlinear effects were identified [96]. One such effect is related to the onset of polarization instability well known for the vector solitons (see Chapter 9). The critical power at which this instability occurs can be reduced considerably in the presence of a Bragg grating [99], [100]. Nonlinear birefringence also affects Bragg solitons. In the NLS limit ($\gamma P_0 \ll \kappa$), the four equations reduce to a pair of coupled NLS equations, similar to those appearing in Chapter 9. In the case of low-birefringence fibers, the two polarization components have nearly the same group velocity, and the coupled NLS equations take the following form [95]:

$$\frac{\partial U_x}{\partial z} + \frac{i\beta_2^g}{2}\frac{\partial^2 U_x}{\partial T^2} = i\gamma_g \left(|U_x|^2 + \frac{2}{3}|U_y|^2 \right) U_x + \frac{i\gamma_g}{3} U_x^* U_y^2 e^{-2i\Delta\beta z}, \qquad (5.5.1)$$

$$\frac{\partial U_y}{\partial z} + \frac{i\beta_2^g}{2}\frac{\partial^2 U_y}{\partial T^2} = i\gamma_g \left(|U_y|^2 + \frac{2}{3}|U_x|^2 \right) U_y + \frac{i\gamma_g}{3} U_y^* U_x^2 e^{2i\Delta\beta z}, \qquad (5.5.2)$$

where $\Delta\beta \equiv \beta_{0x} - \beta_{0y}$ is related to the beat length L_B as $\Delta\beta = 2\pi/L_B$. These equations support *a vector soliton* with equal amplitudes such that the peak power required for each component is only $\sqrt{3/5}$ of that required when only one component is present. Such a vector soliton is referred to as the *coupled-gap soliton* [95].

A coupled-gap soliton can be used for making an all-optical AND gate. The x and y polarized components of the input light represent bits for the gate, each bit taking a value of 0 or 1, depending on whether the corresponding signal is absent or present. The AND gate requires that a pulse appear at the output only when both components are present simultaneously. This can be achieved by tuning both polarization components inside the stop band but close to the upper branch of the dispersion curve. Their combined intensity can increase the refractive index (through a combination of SPM and XPM) enough that both components are transmitted. However, if one of the components is absent at the input (0 bit), the XPM contribution vanishes and both components are reflected by the grating. This occurs simply because the coupled-gap soliton forms

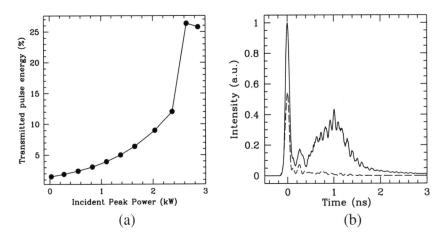

Figure 5.12: (a) Grating transmissivity as a function of input peak power showing the operation of the AND gate and (b) output pulse shapes at a peak power level of 3 kW when only one polarization component (dashed line) or both polarization components (solid line) are incident at the input end. (After Ref. [62]; ©1998 OSA)

at a lower peak power level than the Bragg soliton associated with each individual component [95].

An all-optical AND gate was realized in a 1998 experiment in which a switching contrast of 17 dB was obtained at a peak power level of 2.5 kW [59]. Figure 5.12 shows the fraction of total pulse energy transmitted (a) as a function of input peak power and the transmitted pulse shapes (b) at a peak power of 3 kW. When only one polarization component is incident at the input end, little energy is transmitted by the grating. However, when both polarization components are launched, each having the same peak power, an intense pulse is seen at the output end of the grating, in agreement with the prediction of the coupled NLS equations.

The XPM-induced coupling can be advantageous even when the two polarization components have different wavelengths. For example, it can be used to switch the transmission of a CW probe from low to high by using an orthogonally polarized short pump pulse at a wavelength far from the stop band associated with the probe [101]. In contrast with the self-induced bistable switching discussed earlier, XPM-induced bistable switching can occur for a CW probe too weak to switch itself. Furthermore, the short pump pulse switches the probe beam permanently to the high-transmission state.

References

[1] K. O. Hill, Y. Fujii, D. C. Johnson, and B. S. Kawasaki, *Appl. Phys. Lett.* **32**, 647 (1978).

[2] G. Meltz, W. W. Morey, and W. H. Glen, *Opt. Lett.* **14**, 823 (1989).

[3] R. Kashyap, J. R. Armitage, R. Wyatt, S. T. Davey, and D. L. Williams, *Electron. Lett.* **26**, 730 (1990).

[4] R. Kashyap, *Fiber Bragg Gratings* (Academic Press, San Diego, 1999).

[5] A. Othonos and K. Kalli, *Fiber Bragg Gratings* (Artec House, Boston, 1999).

[6] G. P. Agrawal, *Applications of Nonlinear Fiber Optics* (Academic Press, San Diego, 2001).

[7] G. P. Agrawal, *Fiber-Optic Communication Systems*, 3rd ed. (Wiley, New York, 2002).

[8] M. Born and E. Wolf, *Principles of Optics*, 7th ed. (Cambridge University Press, New York, 1999), Section 8.6.

[9] P. S. J. Russell, *J. Mod. Opt.* **38**, 1599 (1991).

[10] H. A. Haus, *Waves and Fields in Optoelectronics* (Prentice-Hall, Englewood Cliffs, NJ, 1984).

[11] D. Marcuse, *Theory of Dielectric Optical Waveguides* (Academic Press, San Diego, 1991).

[12] A. Yariv, *Optical Electronics in Modern Communications*, 5th ed. (Oxford University Press, New York, 1997).

[13] G. P. Agrawal and N. K. Dutta, *Semiconductor Lasers*, 2nd ed. (Van Nostrand Reinhold, New York, 1993).

[14] B. Crosignani, A. Cutolo, and P. Di Porto, *J. Opt. Soc. Am. B* **72**, 515 (1982).

[15] G. P. Agrawal, *Nonlinear Fiber Optics*, 3rd ed. (Academic Press, San Diego, 2001).

[16] N. M. Litchinitser, B. J. Eggleton, and D. B. Patterson, *J. Lightwave Technol.* **15**, 1303 (1997).

[17] H. G. Winful, J. H. Marburger, and E. Garmire, *Appl. Phys. Lett.* **35**, 379 (1979).

[18] H. G. Winful and G. D. Cooperman, *Appl. Phys. Lett.* **40**, 298 (1982).

[19] H. G. Winful, *Appl. Phys. Lett.* **46**, 527 (1985).

[20] W. Chen and D. L. Mills, *Phys. Rev. Lett.* **58**, 160 (1987); *Phys. Rev. B* **36**, 6269 (1987).

[21] D. L. Mills and S. E. Trullinger, *Phys. Rev. B* **36**, 947 (1987).

[22] C. M. de Sterke and J. E. Sipe, *Phys. Rev. A* **38**, 5149 (1988); *Phys. Rev. A* **39**, 5163 (1989).

[23] C. M. de Sterke and J. E. Sipe, *J. Opt. Soc. Am. B* **6**, 1722 (1989).

[24] S. Larochelle, V. Mizrahi, and G. Stegeman, *Electron. Lett.* **26**, 1459 (1990).

[25] C. M. de Sterke and J. E. Sipe, *Phys. Rev. A* **43**, 2467 (1991).

[26] N. D. Sankey, D. F. Prelewitz, and T. G. Brown, *Appl. Phys. Lett.* **60**, 1427 (1992); *J. Appl. Phys.* **73**, 1 (1993).

[27] J. Feng, *Opt. Lett.* **18**, 1302 (1993).

[28] Y. S. Kivshar, *Phys. Rev. Lett.* **70**, 3055 (1993).

[29] C. M. de Sterke and J. E. Sipe, in *Progress in Optics*, Vol. 33, E. Wolf, Ed. (Elsevier, Amsterdam, 1994), Chap. 3.

[30] P. S. J. Russell and J. L. Archambault, *J. Phys. III France* **4**, 2471 (1994).

[31] M. Scalora, J. P. Dowling, C. M. Bowden, M. J. Bloemer, *Opt. Lett.* **19**, 1789 (1994).

[32] S. Radic, N. George, and G. P. Agrawal, *Opt. Lett.* **19**, 1789 (1994); *J. Opt. Soc. Am. B* **12**, 671 (1995).

[33] S. Radic, N. George, and G. P. Agrawal, *IEEE J. Quantum Electron.* **31**, 1326 (1995).

[34] A. R. Champneys, B. A. Malomed, M. J. Friedman, *Phys. Rev. Lett.* **80**, 4169 (1998).

[35] A. E. Kozhekin, G. Kurizki, and B. Malomed, *Phys. Rev. Lett.* **81**, 3647 (1998).

[36] C. Conti, G. Asanto, and S. Trillo, *Opt. Exp.* **3**, 389 (1998).

[37] Y. A. Logvin and V. M. Volkov, *J. Opt. Soc. Am. B* **16**, 774 (1999).

[38] C. M. de Sterke and J. E. Sipe, *Phys. Rev. A* **42**, 2858 (1990).

[39] H. G. Winful, R. Zamir, and S. Feldman, *Appl. Phys. Lett.* **58**, 1001 (1991).

[40] A. B. Aceves, C. De Angelis, and S. Wabnitz, *Opt. Lett.* **17**, 1566 (1992).

[41] C. M. de Sterke, *Phys. Rev. A* **45**, 8252 (1992).

[42] B. J. Eggleton, C. M. de Sterke, R. E. Slusher, and J. E. Sipe, *Electron. Lett.* **32**, 2341 (1996).

[43] C. M. de Sterke, *J. Opt. Soc. Am. B* **15**, 2660 (1998).

[44] J. E. Sipe and H. G. Winful, *Opt. Lett.* **13**, 132 (1988).

[45] C. M. de Sterke and J. E. Sipe, *Phys. Rev. A* **42**, 550 (1990).

[46] C. M. de Sterke, D. G. Salinas, and J. E. Sipe, *Phys. Rev. E* **54**, 1969 (1996).

[47] T. Iizuka and M. Wadati, *J. Opt. Soc. Am. B* **14**, 2308 (1997).

[48] C. M. de Sterke and B. J. Eggleton, *Phys. Rev. E* **59**, 1267 (1999).

[49] B. J. Eggleton, C. M. de Sterke, A. B. Aceves, J. E. Sipe, T. A. Strasser, and R. E. Slusher, *Opt. Commun.* **149**, 267 (1998).

[50] Yu.I. Voloshchenko, Yu.N. Ryzhov, and V.E. Sotin, *Zh. Tekh. Fiz.* **51**, 902 (1981) [*Sov. Phys. Tech. Phys.* **26**, 541 (1981)].

[51] D. N. Christodoulides and R. I. Joseph, *Phys. Rev. Lett.* **62**, 1746 (1989).

[52] A. B. Aceves and S. Wabnitz, *Phys. Lett. A* **141**, 37 (1989).

[53] J. Feng and F. K. Kneubühl, *IEEE J. Quantum Electron.* **29**, 590 (1993).

[54] C. Conti and S. Trillo, *Phys. Rev. E* **64**, 036617 (2001).

[55] T. Iizuka and C.M. de Sterke, *Phys. Rev. E* **61**, 4491 (2000).

[56] B. J. Eggleton, R. R. Slusher, C. M. de Sterke, P. A. Krug, and J. E. Sipe, *Phys. Rev. Lett.* **76**, 1627 (1996).

[57] C. M. de Sterke, N. G. R. Broderick, B. J. Eggleton, and M. J. Steel. *Opt. Fiber Technol.* **2**, 253 (1996).

[58] B. J. Eggleton, C. M. de Sterke, and R. E. Slusher, *J. Opt. Soc. Am. B* **14**, 2980 (1997).

[59] D. Taverner, N. G. R. Broderick, D. J. Richardson, M. Isben, and R. I. Laming, *Opt. Lett.* **23**, 259 (1998).

[60] D. Taverner, N. G. R. Broderick, D. J. Richardson, R. I. Laming, and M. Isben, *Opt. Lett.* **23**, 328 (1998).

[61] C. M. de Sterke, *Opt. Exp.* **3**, 405 (1998).

[62] N. G. R. Broderick, D. Taverner, and D. J. Richardson, *Opt. Exp.* **3**, 447 (1998).

[63] B. J. Eggleton, G. Lenz, R. E. Slusher, and N. M. Litchinitser, *Appl. Opt.* **37**, 7055 (1998).

[64] N. M. Litchinitser, B. J. Eggleton, C. M. de Sterke, A. B. Aceves, and G. P. Agrawal, *J. Opt. Soc. Am. B* **16**, 18 (1999).

[65] B. J. Eggleton, C. M. de Sterke, and R. E. Slusher, *J. Opt. Soc. Am. B* **16**, 587 (1999).

[66] N. G. R. Broderick, D. J. Richardson, and M. Isben, *Opt. Lett.* **25**, 536 (2000).

[67] H. G. Winful and V. Perlin, *Phys. Rev. Lett.* **84**, 3586 (2000).

[68] M. Asobe, *Opt. Fiber Technol.* **3**, 142 (1997).

[69] W. E. Thirring, *Ann. Phys. (NY)* **3**, 91 (1958).

[70] A. V. Mikhailov, *JETP Lett.* **23**, 320 (1976).

[71] E. A. Kuznetsov and A. V. Mikhailov, *Teor. Mat. Fiz.* **30**, 193 (1977).

[72] D. J. Kaup and A. C. Newell, *Lett. Nuovo Cimento* **20**, 325 (1977).

[73] D. J. Kaup and T. J. Lakoba, J. Math. Phys. **37**, 308 (1996).

[74] I. V. Barashenkov, D. E. Pelinovsky, and E. V. Zemlyanaya, *Phys. Rev. Lett.* **80**, 5117 (1998).

[75] A. De Rossi, C. Conti, and S. Trillo, *Phys. Rev. Lett.* **81**, 85 (1998).

[76] J. Schöllmann, R. Scheibenzuber, A. S. Kovalev, A. P. Mayer, and A.A. Maradudin, *Phys. Rev. E* **59**, 4618 (1999).

[77] I. V. Barashenkov and E. V. Zemlyanaya, *Comp. Phys. Commun.* **126**, 22 (2000).

[78] J. Schöllmann and A. P. Mayer, *Phys. Rev. E* **61**, 5830 (2000).

[79] J. Schöllmann, *Physica A*, **288**, 218 (2000).

[80] Yu. S. Kivshar and N. Flytzanis, *Phys. Rev. A* **46**, 7972 (1993).

[81] G. Huang, *Phys. Rev. E* **49**, 5893 (1994).

[82] Yu. S. Kivshar, *Int. J. Mod. Phys. B* **9**, 2963 (1995).

[83] A. S. Kovalev, O. V. Usatenko, and A. V. Gorbatch, *Phys. Rev. E* **60**, 2309 (1999).

[84] C. M. de Sterke, *Phys. Rev. E* **48**, 4136 (1993).

[85] M. J. Steel and C. M. de Sterke, *Phys. Rev. A* **48**, 1625 (1993).

[86] R. H. Goodman, R. E. Slusher, and M. I. Weinstein, *J. Opt. Soc. Am. B* **19**, 1635 (2002).

[87] D. David, J. Harnad, and S. Shnider, *Lett. Math. Phys.* **8**, 27 (1984).

[88] H. M. Gibbs, *Optical Bistability: Controlling Light with Light* (Academic Press, San Diego, 1985).

[89] H. Kawaguchi, K. Inoue, T. Matsuoka, and K. Otsuka, *IEEE J. Quantum Electron.* **21**, 1314 (1985).

[90] M. J. Adams and R. Wyatt, *IEE Proc.* **134** 35 (1987).

[91] M. J. Adams, *Opt. Quantum Electron.* **19**, S37 (1987).

[92] G. P. Agrawal and S. Radic, *IEEE Photon. Technol. Lett.* **6**, 995 (1994).

[93] C. M. de Sterke, *Opt. Lett.* **17**, 914 (1992).

[94] N. G. R. Broderick, D. Taverner, D. J. Richardson, M. Isben, and R. I. Laming, *Phys. Rev. Lett.* **79**, 4566 (1997); *Opt. Lett.* **22**, 1837 (1997).

[95] S. Lee and S. T. Ho, *Opt. Lett.* **18**, 962 (1993).

[96] W. Samir, S. J. Garth, and C. Pask, *J. Opt. Soc. Am. B* **11**, 64 (1994).

[97] W. Samir, C. Pask, and S. J. Garth, *Opt. Lett.* **19**, 338 (1994).

[98] S. Pereira and J. E. Sipe, *Opt. Exp.* **3**, 418 (1998).

[99] R. E. Slusher, S. Spälter, B. J. Eggleton, S. Pereira, and J. E. Sipe, *Opt. Lett.* **25**, 749 (2000).

[100] S. Pereira, J. E. Sipe, R. E. Slusher, and S. Spälter, *J. Opt. Soc. Am. B* **19**, 1509 (2002).

[101] S. Broderick, *Opt. Commun.* **148**, 90 (1999).

Chapter 6

Two-Dimensional Solitons

The preceding chapters focused on one-dimensional solitons, which maintain their shape either in time or in one spatial dimension. When such a spatial soliton propagates in a bulk nonlinear medium, it takes the form of a planar soliton stripe and experiences a modulation instability, referred to as the *transverse instability* because it produces modulations in the transverse dimensions. In many cases, this instability leads to the formation of an array of stable two-dimensional solitons that maintain their shape in both transverse dimensions. One can even imagine three-dimensional solitons capable of maintaining their shape in the temporal dimension as well. This chapter focuses on two-dimensional solitons and provides a link between the solitons forming inside a planar waveguide and a bulk medium. Section 6.1 is devoted to the transverse instability associated with a soliton stripe. The theoretical and experimental results on two-dimensional solitons are summarized in Sections 6.2 and 6.3, respectively, where we also consider the nonparaxial effects. The phenomenon of soliton interaction in a bulk medium is discussed in Section 6.4. In Section 6.5 we introduce new types of ring-shaped spatial solitons that carry a finite angular momentum.

6.1 Soliton Transverse Instability

The fundamental physical mechanism behind the transverse instability induced by the self-focusing of soliton stripes is similar to the mechanism governing the self-focusing and modulation instabilities of small-amplitude, quasi-harmonic, wave packets [1]. The instability occurs when transverse modulations on the wave front of a planar soliton stripe decrease the local value of the soliton energy. Two analytical methods can be used for studying the transverse instability of solitons—the *ray-optics approach* and *linear stability analysis*. We consider both of them in this section.

6.1.1 Ray-Optics Approach

The geometrical-optics approach is based on the assumption that a transversely modulated plane wave remains locally close to its steady-state profile, so each individual

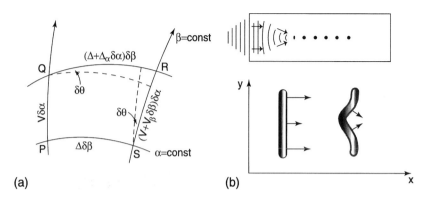

Figure 6.1: (a) Notation and the coordinate system used in the ray-optics approach. (b) Schematic illustration of soliton self-focusing and the instability mechanism. (After Ref. [5]; ©2000 Elsevier.)

segment of the wave evolves along an individual ray [2]. This assumption is valid provided the soliton remains stable against longitudinal perturbations that preserve its symmetry. The ray-optics method for describing transverse self-focusing of solitons was first developed in a 1976 study [3, 4].

To present the essential features of this method, we introduce an orthogonal system of coordinates (α, β) shown in Figure 6.1(a) and consider the evolution of the local soliton velocity V, the width Δ of a ray tube, and the phase angle θ that the ray makes with the x axis [5]. Using the notation $\Delta_\alpha = \partial\Delta/\partial\alpha$ and $V_\beta = \partial V/\partial\beta$, V and Δ are connected through the following kinematic relations:

$$\frac{\partial s}{\partial \alpha} + \frac{1}{\Delta}\frac{\partial V}{\partial \beta} = 0, \qquad \frac{\partial s}{\partial \beta} - \frac{1}{V}\frac{\partial \Delta}{\partial \alpha} = 0, \tag{6.1.1}$$

where s is the distance measured along the ray path. This set of equations can be closed with the help of the energy conservation law,

$$\Delta(V)W(V) = \text{constant}, \tag{6.1.2}$$

where W is the energy density. The resulting closed system of equations is elliptic provided $dV/d\Delta > 0$, and it predicts a transverse instability of the soliton. Since the ray width Δ is inversely proportional to the soliton amplitude, the ray equations provide the following universal criterion for the self-focusing instability of solitons in an isotropic medium: Transverse self-focusing of a planar soliton occurs when *its velocity decreases with an increase in its amplitude*. Figure 6.1(b) shows that the corresponding rays form a converging cylindrical wave.

This simple physical mechanism explains the transverse instability of the dark Kerr solitons [6], for which the dependence of the velocity v on the amplitude a is fixed by the relation $V = (u_0^2 - a^2)^{1/2}$, where u_0 is the background amplitude (see Section 4.1.1). It is straightforward to generalize Eqs. (6.1.1) and (6.1.2) to the case in which solitons are modulated with respect to their frequency rather than their velocity. As a result, the same criterion for soliton self-focusing is valid when the soliton frequency ω decreases with the soliton amplitude a. This is the case for bright Kerr solitons, for which the propagation constant β lies inside an effective gap within the linear spectral band and is given by $\beta = \beta_c - \gamma a^2$, where β_c is the cutoff value of the linear band and γ is positive. Indeed, it has been known since 1973 that bright solitons are unstable with respect to transverse perturbations [7].

6.1.2 Linear Stability Analysis

The linear stability analysis discussed in Section 1.5.1 deals with the instability of a CW beam. However, the same basic approach can be used for analyzing the self-focusing instabilities of a soliton stripe. It makes use of a linear eigenvalue problem that is obtained by linearizing the $(2+1)$-dimensional NLS equation near the exact one-dimensional soliton solution. Because the soliton solution depends only on a few parameters (such as amplitude and phase), one can separate the dynamical variables and reduce the corresponding linear problem to the analysis of the eigenvalue spectrum of a linear operator.

In the case of bright NLS solitons forming inside a Kerr medium, we consider the $(2+1)$-dimensional NLS equation given in Eq. (1.3.12) with $F(|u|^2) = |u|^2$. Introducing $\psi = \sqrt{2}\,u$, we write this equation in the form

$$i\frac{\partial \psi}{\partial z} + \frac{\partial^2 \psi}{\partial x^2} + \sigma_d \frac{\partial^2 \psi}{\partial y^2} + 2|\psi|^2 \psi = 0, \qquad (6.1.3)$$

where the parameter $\sigma_d = \pm 1$ is introduced to include the case in which the variable y corresponds to the temporal dimension ($\sigma_d = -1$ in the case of normal dispersion). A planar bright soliton is described by the y-independent solution of the NLS equation (6.1.3) and has the general form

$$\psi(x,z) = \Phi_s(x - x_0 - 2vz; \beta)\exp[i(vx - v^2 z + \beta z + \theta)], \qquad (6.1.4)$$

where x_0 is the location of the soliton peak, $2v$ represents the transverse velocity of the soliton, β is the propagation constant, θ is the soliton phase, and the shape of the soliton is governed by

$$\Phi_s(x;\beta) = \sqrt{\beta}\,\mathrm{sech}(\sqrt{\beta}x). \qquad (6.1.5)$$

To analyze the stability of this soliton with respect to transverse modulations, we consider a linear perturbation in the form

$$\delta\psi \equiv \psi(x,y,z) - \Phi_s(x;\beta)e^{i\beta z} = [u(x) + iw(x)]\exp(i\beta z + \Gamma z + ipy), \qquad (6.1.6)$$

where p is the transverse wave number along the y axis and $\Gamma(p)$ determines the growth rate of this perturbation along z. After linearizing the NLS equation, the real functions $u(x)$ and $w(x)$ are found to satisfy the following linear eigenvalue problem:

$$\left(\hat{\mathcal{L}}_1 + \sigma_d p^2\right)u = -\Gamma w, \qquad \left(\hat{\mathcal{L}}_0 + \sigma_d p^2\right)w = \Gamma u, \qquad (6.1.7)$$

where the two linear operators are given by

$$\hat{\mathcal{L}}_0 = -\partial^2/\partial_x^2 + \beta - 2\Phi_s^2, \qquad \hat{\mathcal{L}}_1 = \hat{\mathcal{L}}_0 - 4\Phi_s^2. \qquad (6.1.8)$$

No exact analytic solutions have been found for this linear eigenvalue problem. However, several analytical methods allow us to approximate the unstable eigenvalue $\Gamma(p)$ and to find the threshold condition for the onset of the transverse instability. Two of these methods are based on the asymptotic expansions valid for either small wave numbers ($p \to 0$) or finite wave numbers close to a critical value p_c ($p \to p_c$). We call these asymptotic approximations *long-scale* and *short-scale* expansions, respectively, because the spatial period s of modulations depends on p as $s = 2\pi/p$.

The long-scale asymptotic method is based on the Taylor expansion of the unstable eigenvalue for small values of p [7]–[11]. Such a technique is sometimes called the *p-expansion method*. It is often successful in detecting the existence of the self-focusing instability because the transverse instability begins with modulations having large periods (i.e., small p). Using the discrete-spectrum modes of the linear eigenvalue problem for $p = 0$ and $\Gamma = 0$, usually known in an explicit analytical form (the so-called *neutral modes*), one can construct the eigenfunctions and eigenvalues of the corresponding linear problem and find the first dominant term in the asymptotic expansion. This term determines the spatial symmetry of perturbations leading to the transverse instability and also provides an approximate expression for the instability growth rate $\Gamma(p)$ for small p. Since the growth rate depends on the sign of the parameter σ_d, we consider the two cases separately.

Consider first the case $\sigma_d = +1$, for which the NLS equation is elliptic. This case corresponds to two-dimensional bright spatial solitons and was first studied in 1973 [7]. The functions u and w near $p = 0$ are simply $u = 0$ and $w = \Phi_s(x; \beta)$, while the growth rate is given by

$$\Gamma^2(p) = 4\beta p^2 - \frac{4}{3}\left(1 + \frac{\pi^2}{3}\right)p^4 + O(p^6). \qquad (6.1.9)$$

In the opposite case, $\sigma_d = -1$, the NLS equation is hyperbolic. Physically, this situation corresponds to an optical pulse propagating inside a planar waveguide. The functions u and w near $p = 0$ are now found to be $w = 0$ and $u = \partial\Phi_s/\partial x$, while the growth rate is given by [8]

$$\Gamma^2(p) = \frac{4}{3}\beta p^2 - \frac{4}{9}\left(\frac{\pi^2}{3} - 1\right)p^4 + O(p^6). \qquad (6.1.10)$$

The numerical coefficients in Eqs. (6.1.9) and (6.1.10) are slightly different from those found in the original papers [7, 8] because of misprints.

In the case of the short-scale asymptotic method, the Taylor expansion is not around zero but around p_c, where p_c is the critical (or cutoff) value for which Γ vanishes, i.e., $\Gamma(p_c) = 0$. This method uses the eigenmodes of the linear eigenvalue problems at $p = p_c$ and $\Gamma = 0$ [12]–[14]. These eigenmodes can often be found by a direct analysis of the linearized equations. The growth rate $\Gamma(p)$ again depends on the sign of the

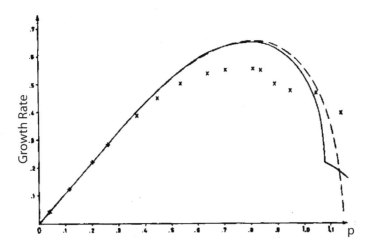

Figure 6.2: Numerically calculated growth rate $\Gamma(p)$ of the soliton transverse instability in a self-focusing medium ($\sigma_d = +1$) as a function of spatial frequency p (solid curve). The dashed curve shows the analytical approximation (6.1.9), while crosses mark the results of another numerical study. (After Ref. [10].)

parameter σ_d. In the case $\sigma_d = +1$, $p_c = \sqrt{3\beta}$, the eigenfunctions are $u = \mathrm{sech}^2(\sqrt{\beta}x)$ and $w = 0$, and the growth rate is given by [13]

$$\Gamma^2(p) = \frac{24\beta p_c}{(\pi^2 - 6)}(p_c - p) + \mathrm{O}[(p_c - p)^2]. \qquad (6.1.11)$$

In the opposite case, $\sigma_d = -1$, $p_c = \sqrt{\beta}$, the eigenfunctions are $u = 0$ and $w = \tanh(\sqrt{\beta}x)$, and the growth rate is given by [14]

$$\Gamma^2(p) = \frac{16\beta p_c}{3\pi^2}\sqrt{p_c^2 - p^2} + \mathrm{O}(p_c - p). \qquad (6.1.12)$$

Note that the eigenfunction w is delocalized in this case and becomes infinitely large as $|x| \to \infty$.

The solid line in Figure 6.2 shows the growth rate Γ as a function of p as calculated numerically. The analytical approximation based on the long-scale asymptotic method is shown by the dashed line. It works quite well for values of p as large as 0.6 but begins to deviate considerably near $p = 1$. Crosses show the results of another numerical study [9]. The main conclusion one can draw is that the growth rate Γ takes its maximum value for a certain value of p. As a result, the soliton solution given in Eq. (6.1.4) becomes modulated at that spatial frequency.

6.1.3 Equations for Soliton Parameters

The preceding analysis of transverse instability is based on two different approximations. The geometrical-optics approach describes the evolution of strongly nonlinear, long-wavelength modulations in the absence of diffraction (or dispersion), whereas linear stability analysis allows us to describe the evolution of perturbations of all scales but within the small-amplitude (linear) approximation.

The important question is: How does the soliton stripe evolve after the onset of transverse instability? This question can be answered only after the nonlinear effects are included, simply because the exponential growth of the perturbation, predicted by the linear analysis, cannot continue once the perturbation grows to a finite size. Several different analytical methods have been employed in the soliton literature for deriving the modulation equations in the nonlinear regime. In this section we review one of these methods and discuss its limitations.

Two basic assumptions are made in the following analysis: (i) A planar soliton is stable against all symmetry-preserving perturbations, and (ii) instability occurs when the period of transverse modulations is much larger than the soliton width (the amplitude of such modulations need not be small). The modulation equations can then by derived by using a variational method based on the concept of the averaged Lagrangian density. These equations result, to the leading order, in an ill-posed (elliptic) problem and correspond to the dynamics of an unstable gas. Their solution displays the development of singularities over a finite medium length. However, it turns out that the gas-dynamics equations represent the dispersionless limit of the NLS equation. One can improve the applicability of the asymptotic equations by extending the perturbation theory to higher orders and including the dispersive and/or dissipative effects [5].

The variational method is a particular case of the general Whitham modulation theory [15]. The basic idea consists of casting the NLS equation in terms of a Lagrangian density $\mathcal{L}$ and introducing the action S as

$$S = \int_0^z dz \int\int_{-\infty}^{\infty} \mathcal{L}(x,y,z)\,dx\,dy. \tag{6.1.13}$$

A trial function is then chosen in the form of a steady-state soliton, but its parameters are allowed to evolve slowly with z. Variations of the soliton parameters then produce the modulation equations responsible for soliton self-focusing [16, 17]. However, a straightforward application of the variational method often encounters difficulties associated with the appearance of diverging integrals that are responsible for the radiation emitted by solitary waves. To solve this problem, the trial function should include a nonlocalized radiation component.

As an example, we consider a bright spatial soliton by choosing $\sigma_d = +1$ in Eq. (6.1.3). The Lagrangian density for this NLS equation is found to be

$$\mathcal{L} = \frac{i}{2}(\psi^*\psi_z - \psi\psi_z^*) - |\psi_x|^2 - |\psi_y|^2 + |\psi|^4, \tag{6.1.14}$$

where a subscript denotes a particle derivative. As a trial function, we take the bright NLS soliton solution in Eq. (6.1.4) and assume for simplicity that $x_0 = 0$ and $v = 0$. However, both β and θ are allowed to vary with y and z to include the modulations produced by transverse instability. We also introduce a perturbation parameter ε by writing the solution in the form $\psi = \Phi_s(x;\beta)e^{i\theta/\varepsilon}$. Integrating $\mathcal{L}$ with respect to x, the average Lagrangian in terms of the soliton parameters β and θ is given by

$$\langle \mathcal{L} \rangle = \frac{1}{2} \int_{-\infty}^{\infty} \mathcal{L}(x,z)\,dx = -\sqrt{\beta}(\theta_z + \theta_y^2) + \frac{1}{3}\beta^{3/2} - \varepsilon^2 \mu \beta^{-3/2} \beta_y^2, \qquad (6.1.15)$$

where $\mu = \frac{1}{12}(1 + \pi^2/12)$ and a subscript denotes a derivative.

The variation of $\langle \mathcal{L} \rangle$ with respect to θ yields the energy conservation law, whereas the variation with respect to β generates an eikonal equation. The soliton modulation equations were first obtained with this method by Makhankov [16], and they were later extended to include dispersive and dissipative effects. The steady-state solution of the resulting equations corresponds to an unperturbed planar soliton stripe. For example, the modulation equations for the Kerr soliton have the solution $\beta = \beta_0$ and $\theta = \omega_0 z + \theta_0$, where ω_0 and θ_0 are constants; this solution corresponds to a bright NLS soliton.

Although the gas-dynamics equations always lead to the formation of singularities for elliptically unstable problems, the higher-order dispersive or dissipative effects can suppress an exponential growth of modulations. Depending on the type of higher-order effects, one can distinguish, in general, among four different *scenarios* of the instability-driven self-focusing dynamics:

- *wave collapse*, or formation of a chain of two-dimensional localized singularities along the front of a planar soliton;
- *monotonic transition* from a planar soliton to a periodic chain of two-dimensional solitons;
- *breakup* of a planar soliton into *localized states* that gradually decay because of dispersion;
- *quasi-recurrence*, i.e., a periodic growth and damping of transverse modulations along the front of a planar soliton.

The extended modulation equations capable of describing these four scenarios usually need to be solved numerically. As an exception, an explicit analytical solution can be found when the original nonlinear equation is integrable by the inverse scattering method [5]. Approximate analytic solutions of the extended modulations equations are often possible by employing small-amplitude asymptotic expansions. Applying such methods, one can show that the bright solitons of the elliptic NLS equation ($\sigma_d = +1$) display the first scenario, while the bright solitons of the hyperbolic NLS equation ($\sigma_d = -1$) follow the third scenario. Dark solitons, depending on their initial parameters, can display either the second or the fourth scenario of the instability-induced dynamics.

6.2 Spatial Solitons in a Bulk Medium

In this section we focus on the propagation of CW beams in a nonlinear bulk medium and choose $\sigma_d = +1$ in the NLS equation (6.1.3). In the absence of any nonlinearity, the beam spreads because of diffraction. The possibility of self-focusing and self-trapping of optical beams in nonlinear media was predicted in the early 1960s [18, 19].

Since then, these phenomena have been studied extensively, both experimentally and theoretically [20]–[23]. A self-trapped CW beam represents a spatial soliton that is confined in both transverse dimensions. Such solitons can be found by analyzing the radially symmetric solutions of the $(2+1)$-dimensional NLS equation. This section focuses on the shape and stability of such solutions.

6.2.1 Structure and Stability of Solitons

As discussed in Section 1.2.3, the propagation of a linearly polarized CW optical beam in a saturable self-focusing medium is described by Eq. (1.2.12). If we adopt the saturable nonlinearity model $n_{nl}(I) = n_2 I/(1 + I/I_s)$, where I_s is the saturation intensity and use a normalization scheme similar to that of Eq. (1.2.13), we obtain the following $(2+1)$-dimensional NLS equation:

$$i\frac{\partial u}{\partial z} + \frac{1}{2}\nabla_\perp^2 u + \frac{|u|^2 u}{1 + |u|^2} = 0. \tag{6.2.1}$$

The transverse Laplacian can be written in the cylindrical coordinates as

$$\nabla_\perp^2 \equiv \frac{\partial^2}{\partial x^2} + \frac{\partial^2}{\partial y^2} = \frac{\partial^2}{\partial r^2} + \frac{1}{r}\frac{\partial}{\partial r} + \frac{1}{r^2}\frac{\partial^2}{\partial \theta^2}. \tag{6.2.2}$$

To find the soliton solutions, we assume a solution in the form $u = Ue^{iqz}$, where U does not depend on z and q plays the role of an eigenvalue representing the shift in the propagation constant $(0 < q < 1)$. Assuming that the CW beam maintains its cylindrical symmetry during propagation so that U is independent of θ, Eq. (6.2.1) becomes a simple ordinary differential equation for $U(r)$:

$$\frac{d^2 U}{dr^2} + \frac{1}{r}\frac{dU}{dr} + \frac{|U|^2 U}{1 + |U|^2} = qU. \tag{6.2.3}$$

Assuming that the beam amplitude peaks at $r = 0$ and vanishes as $r \to \infty$, this equation should be solved with the boundary conditions

$$\left.\frac{dU}{dr}\right|_{r=0} = \left.\frac{dU}{dr}\right|_{r=\infty} = 0. \tag{6.2.4}$$

It is straightforward to show that Eq. (6.2.3) has multiple solutions for each value of q, each of which represents a bound state such that the beam shape remains unchanged during propagation [24]–[26]. It is common to label this family of two-dimensional spatial solitons as $U_n(r; q)$, where the integer n takes integer values $n = 0, 1, 2, \ldots$ and is chosen such that the nth solution has n nodes where the optical field vanishes. Figure 6.3 shows the first three amplitude profiles obtained numerically for $q = 0.5$ and $n = 0, 1, 2$. Each of these profiles corresponds to a two-dimensional spatial soliton. The nodeless solution $(n = 0)$ is called the *fundamental soliton*.

Stability of the fundamental soliton can be determined by applying the general stability theory developed in Chapter 2 in the case of one-dimensional solitons. In the framework of the generalized NLS equation, stability can be determined by calculating

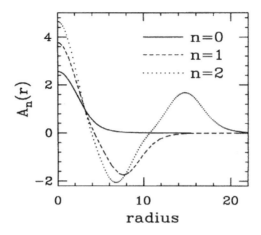

Figure 6.3: Radial distribution of two-dimensional spatial solitons forming in a saturable nonlinear medium with $q = 0.5$. (After Ref. [26]; ©1975 Pergamon Press.)

the power of the self-trapped bound solution using $P_n = 2\pi \int_0^\infty A_n^2(r)\, r\, dr$, and applying the Vakhitov–Kolokolov stability criterion $\partial P_0 / \partial q > 0$. It turns out that the fundamental two-dimensional solitons of the NLS equation (6.2.1) are always stable in a saturable nonlinear medium (but not in a Kerr medium). The stability of the higher-order solitons with $n \geq 1$ does not follow from this criterion and should be analyzed numerically. Numerical simulations indicate that all higher-order solitons are unstable and break up during propagation [26]–[28]. However, the instability growth rate depends on the type of nonlinearity and can be dramatically suppressed in saturable nonlinear media.

Figure 6.4 shows how the transverse instability for the $n = 1$ mode can be suppressed in a saturable nonlinear medium using high input power levels. At low power levels ($P \ll P_s$), the nonlinear medium acts as a Kerr medium, and the bound mode breaks up into five filaments after propagating a distance of 25 diffraction lengths (left column). Such a breakup of the CW beam is referred to as *beam filamentation*. At high power levels ($P \gg P_s$), the filamentation instability does not not occur up to $z = 25$, as seen in the right column of Figure 6.4. It is important to stress that the instability is not eliminated but only suppressed. If the CW beam were to propagate over long lengths, it would eventually break up into filaments, irrespective of the input power level.

Similar to the cases discussed in Chapter 2, the dynamics of stable two-dimensional solitons display long-lived, radially symmetric oscillations of the soliton amplitude [29, 30], a phenomenon that can be understood in terms of the *internal modes* associated with the breaking of the phase symmetry [31]. Internal modes are the discrete eigenfunctions of the linear operator found linearizing the NLS equation around the soliton solution. To determine these internal modes, we write the perturbed soliton solution as

$$U(r, \theta; z) = U_s(r) + \phi_k(r) \exp[i(\lambda z + k\theta)] + \psi_k^*(r) \exp[-i(\lambda^* z + k\theta)], \quad (6.2.5)$$

where $U_s(r)$ represents the fundamental soliton, ϕ_k and ψ_k are small perturbations, λ

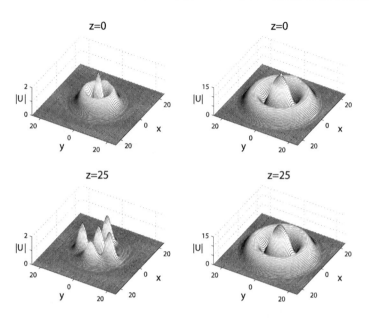

Figure 6.4: Suppression of transverse instability for the first higher-order soliton ($n = 1$) at high input power levels. Input- (top row) and output-beam profiles (bottom row) at low (left column) and high (right column) power levels. (After Ref. [28]; ©2002 APS.)

is the eigenvalue, and k is the integer representing the angle dependence of the perturbation. At least four such internal modes ($k = 0, 1, 2, 3$) can be found in the case of a saturable nonlinearity [28].

The numerical results obtained when the soliton amplitude is perturbed by multiplying it by the factor $(1 + \varepsilon)$, with $\varepsilon = 0.2$, are displayed in Figure 6.5. The left panel shows how the amplitude at the beam center ($r = 0$) oscillates with propagation. The oscillation frequency is close to the frequency associated with the $k = 0$ internal mode (about 0.2187 in normalized units). To examine the radiation damping of these oscillations, the radial profiles $|U(r)|$ at three distances are shown in the right panel of Figure 6.5. Remarkably, radiation emission from these internal oscillations is extremely small, indicating that oscillations will persist for a very long distance.

The oscillations caused by an internal mode with $k \neq 0$ affect not only the amplitude but also the shape of a two-dimensional soliton. If the soliton is perturbed only by the $k = 2$ mode, such oscillations are very robust. The effect of internal-mode perturbation is to prolong the fundamental soliton in one direction, and the soliton visually appears to exhibit *a rotation* in the perturbed state. This is evident in Figure 6.6, where the soliton shape is shown at five distances using the contours of the solution amplitude $|U(r,z)|$ in the x–y plane. The fundamental soliton under the $k = 2$ internal-mode perturbation appears to rotate counterclockwise. This is an interesting and distinctive visual feature of the internal oscillations associated with two-dimensional solitons. If the soliton is perturbed by both the $k = 2$ and $k = -2$ modes, its evolution is visually different.

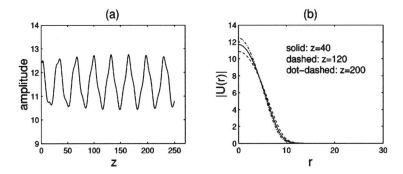

Figure 6.5: Evolution of a fundamental soliton under radially symmetric perturbations: (a) changes in the on-axis soliton amplitude $|u(r = 0, z)|$ with z: (b) soliton profiles at three values of z. (After Ref. [28]; ©2002 APS.)

6.2.2 Nonparaxial Solitons

The $(2+1)$-dimensional NLS equation (6.2.1) used in the preceding analysis is based on the scalar and paraxial approximations. As the beam size decreases because of self-focusing, both of these approximations begin to breakdown. The use of complete Maxwell's equations is then essential to estimate the importance of the effects not included within the scalar NLS equation.

The paraxial approximation or, equivalently, the slowly varying envelope approximation is valid if the typical transverse scale (such as the beam width w) is much longer than the light wavelength λ. Limitations of the scalar NLS equation become clear from the analysis of the self-focusing phenomenon in a Kerr medium. Indeed, the NLS equation in this case predicts catastrophic collapse of a radially symmetric optical beam when the input power exceeds a critical power level (see Chapter 7). However, it is evident that as the beam collapses, its width will approach the optical wavelength λ, and the paraxial theory is not applicable in this regime.

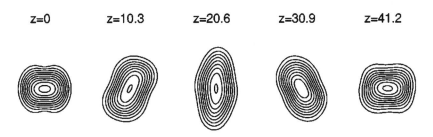

Figure 6.6: Evolution of the fundamental soliton under $k = 2$ internal-mode perturbations. The contours of the solution amplitude $|U(r, z)|$ are shown at five distances in the x–y plane. (After Ref. [28]; ©2002 APS.)

The nonparaxial and all other higher-order effects can be included completely by solving the full Maxwell's equations directly. In one approach, a rigorous solution of these equations is sought in the form

$$\mathbf{E}(x,y,z,t) = \mathbf{A}(r_\perp)\exp(i\beta z - i\omega_0 t), \tag{6.2.6}$$

where z is the direction of beam propagation, $r_\perp = (x^2 + y^2)^{1/2}$ is the radial distance in the transverse plane, and β is the propagation constant at the optical frequency ω_0. This solution corresponds to a spatial soliton because of its z-independent envelope. Such nonparaxial solutions in the form of self-trapped beams (i.e., nonparaxial solitons) were obtained as early as 1964 [32, 33]. In contrast to the paraxial theory based on the NLS equation, the results were shown to be very sensitive to the field polarization state. In the case of transverse-electric (TE) polarization with only one nonzero electric-field component E_x, the nonparaxial soliton is a solution of the scalar wave equation and has the same shape as the NLS soliton in the paraxial approximation [34]). In a planar-waveguide geometry, this soliton has the familiar "sech" shape such that $A_x = a\,\mathrm{sech}(ax)$. The soliton width $w \sim a^{-1}$ can be arbitrarily small for a sufficiently large propagation constant β. In the case of transverse-magnetic (TM) polarization, the electric field has two nonzero components, E_x and E_z, and the soliton shape can be found only numerically [35]–[37]. It turns out that the width of such TM solitons cannot become arbitrarily small, and the minimum possible value is $w_{\min} \sim \lambda/2$ [38]–[40]. Mixed polarization states are also possible [41]. Similar to the one-dimensional scalar solitons of the NLS equation, all one-dimensional nonparaxial solitons are unstable in a bulk self-focusing medium, because of the transverse instability discussed in Section 6.1, and produce filaments in the orthogonal transverse direction (along the y axis).

The two-dimensional nonparaxial solitons were studied beginning in 1977 by allowing the optical beam to spread in both transverse dimensions [42]–[49]. Maxwell's equations permit solutions in the form of cylindrically symmetric solitons. However, it turns out that at least a part of the nonrparaxial soliton is unstable. This feature has been demonstrated in the Kerr and saturable nonlinear media for solitons with only the tangential field component [46]. A new type of stable but *cylindrically asymmetric* spatial soliton was discovered in 1999 for the Kerr nonlinearity [47] and in 2001 for the saturable nonlinearity [49]. The possibility of forming stable, ultranarrow, subwavelength, two-dimensional spatial solitons (the so-called *optical needles*) has also been suggested [48].

6.2.3   Beam Collapse and Filamentation

The use of full vector theory is not essential for including the nonparaxial effects. It is possible to stabilize a two-dimensional spatial soliton, even within the scalar approximation, with a suitable modification of the scalar NLS equation, as discussed as early as 1972 [50]. Since then, many studies based on the scalar Helmholtz equation have shown that the polarization effects are not essential for the stabilization of two-dimensional spatial solitons [51]–[58].

The main idea behind the studies based on the scalar wave equation is to retain the NLS equation but to generalize it to include the nonparaxial corrections. Following the

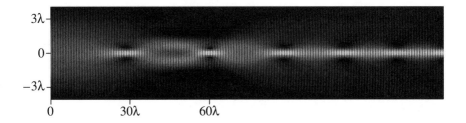

Figure 6.7: Evolution of a self-focusing optical beam in a Kerr medium as described by the NLS equation with the nonparaxial effects included. (After Ref. [57]; ©2002 APS.)

derivation of the NLS equation outlined in Section 1.2.3 but retaining the nonparaxial term $\partial^2 A / \partial z^2$, the generalized NLS equation can be written in the following normalized form:

$$i\frac{\partial u}{\partial z} + \frac{1}{2}\nabla_\perp^2 u + |u|^2 u + \varepsilon\frac{\partial^2 u}{\partial z^2} = 0, \qquad (6.2.7)$$

where the last term accounts for the nonparaxial effects. The small parameter ε is defined as $\varepsilon = \frac{1}{2}(\beta_0 w_0)^{-2}$, where w_0 is a measure of the beam width. Equation (6.2.7) is an ill-posed Cauchy problem, and, as a result, its direct numerical integration often encounters problems [56]. However, these problems can be solved, and as early as 1988 it was discovered that *no singularity occurs* when the nonparaxial term in Eq. (6.2.7) is included [52]. This numerical result is supported by the asymptotic analysis of Fibich [55] showing that the beam collapse is indeed arrested for any nonzero value of ε [59, 60].

Figure 6.7 shows how the nonparaxial effect can arrest the beam collapse. The evolution of a self-focusing beam inside a Kerr medium is shown in the x–z plane by solving Eq. (6.2.7) numerically with an input beam width $w_0 \approx 3\lambda$. Notice that the transverse and longitudinal scales are different in this figure. The beam begins to collapse and its size becomes smaller than the optical wavelength λ at a distance of 30λ. However, the nonparaxial effects become so strong in this region that the beam does not collapse but follows a quasi-periodic evolution pattern and undergoes self-focusing multiple times. At the same time, a part of the beam power escapes as radiation. It turns out that the high-frequency spatial components occurring in the focal regions are responsible for the radiation. The beam dynamics in Figure 6.7 display not only the oscillatory character of the optical field but also a gradual damping of these oscillations, resulting eventually in the formation of a soliton-like beam with a size close to λ. This final state corresponds to the nonparaxial spatial soliton with TE polarization.

More accurate models of nonparaxial self-focusing should include the *vectorial effects* that couple the TE and TM components and also lead to backscattering, both of which produce additional power losses. The vectorial coupling becomes significant when the beam width becomes comparable to the wavelength. The complete vector treatment of the collapse arrest for different states of beam polarization has been carried out in Refs. [61] and [62]. An approximate analytic theory has also been developed [63]

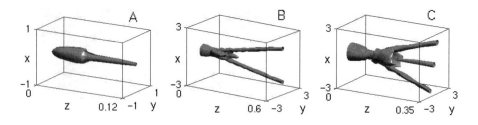

Figure 6.8: Evolution of a circularly polarized Gaussian beam with an elliptical spot size in the self-focusing regime at a power level of (A) $P = 5P_0$, and (B) $P = 8.3P_0$. (C) A cylindrically symmetric beam in the presence of noise at $P = 10P_0$. (After Ref. [65]; ©2002 APS.)

following the two earlier proposals in Refs. [42, 64]. The main conclusions are that the coupling between the TE and TM components can arrest the beam collapse and that the polarization effects are of the same order of magnitude as the nonparaxial effects [63].

Most studies on beam collapse have focused on linearly polarized optical beams The propagation of circularly polarized laser beams in a Kerr medium was considered in a 2002 study [65]. This analysis revealed a principal difference between the two types of polarizations—linear polarization breaks up the cylindrical symmetry of an input beam but the circular polarization does not. As a result, the use of circular polarization is more likely to suppress multiple filaments. Because the creation of multiple filaments limits the power that each filament can carry, the use of circular polarization may lead to other differences between the two polarization states.

Numerical simulations for circularly polarized optical beams reveal an interesting feature of two-dimensional spatial solitons [65]. Even though a cylindrically symmetric beam does not break into multiple filaments, even a small departure from the perfect cylindrical symmetry (e.g., a slightly elliptical spot size) can produce filamentation. This behavior is shown in Figure 6.8 for a Gaussian beam. When the spot size is slightly elliptical, the beam forms a spatial soliton and maintains its shape for $P = 5P_0$ (part A) but breaks into filaments at a higher power level of $P = 8.3P_0$ (part B). Here P_0 is the critical power at which a linearly polarized beam will undergo collapse. Another feature found numerically is that the presence of amplitude noise can lead to the formation of multiple filaments, even for a beam with circular spot size (part C), because such amplitude fluctuations destroy the cylindrical symmetry. Thus, the suppression of filamentation for circularly polarized beams should focus on producing a perfect cylindrically symmetric input beam rather than on producing perfect circular polarization. It is important to note that one cannot suppress the formation of multiple filaments for linearly polarized beams by using a cylindrically symmetric input beam.

Several other studies have considered the nonparaxial effects in both the linear and nonlinear regimes of beam propagation [66]–[68]. In a 1996 study, the vector effects were taken into account but the scalar nonparaxial effects were ignored [69]. The inclusion of the vectorial effects supports the existence of stable self-trapped beams when the coupled transverse and longitudinal fields are of the same order of magnitude [37]. In fact, at the final stage of self-focusing, the self-trapped beam appears in the orm of an "optical needle" [49]. In the case of ultrashort laser pulses, the "transverse" onparaxial effects have also been taken into account [70]. Numerical simulations ased on the time-dependent Maxwell equations are perhaps the most accurate as far s the beam-collapse regime is concerned [71, 72]; these are discussed in Chapter 7 in he context of optical bullets.

5.3 Experimental Results

Historically, starting in 1965 several experiments studied the phenomena of self-focusing nd self-trapping using nonlinear materials such as optical glasses, organic and inoranic liquids, and gas vapors [73]–[78]. The self-trapping was not associated at that ime with the concept of solitons, and its connection to two-dimensional spatial solions become apparent only in the 1980s. In 1965, Pilipetskii and Rustamov were the irst to observe the formation of up to three filaments during self-focusing of a laser eam in organic liquids [73]. One year later, Garmire et al. observed self-trapping of n optical beam inside CS_2 liquid [74]. It was found that an input beam of circular ymmetry collapsed to form a *bright filament* with diameter as small as 50 μm. The hreshold power, the trapping length, and the nonlinearity-induced increase in the reractive index in the trapped region were consistent with the theoretical predictions. In indsight, we might say that this was the first experimental manifestation of the entity ow called a *spatial* soliton.

In the experiment of Garmire et al., a ruby laser beam with 10–100 kW of power vas passed through a 0.5-mm pinhole to produce a diffraction-limited plane wave [74]. Changes in the beam diameter were produced by placing various lenses after the pinole. The development of self-trapping inside a CS_2 cell was studied by immersing microscope slides every 2 cm along the beam to reflect a small fraction of power out of the cell. Figure 6.9 shows a sketch of the experimental setup together with the magified images (8×) of the beam profile at seven different locations within the cell. The op row (a) shows the control beam from a gas laser, while rows (b)–(d) show the evoution of the self-trapped beam with increasing power. Simple diffraction would double he size of such a beam between each glass plate. However, because of the nonlinear effects, the beam propagates without spreading and maintains a 100-μm diameter.

Because the ruby-laser beam used in the experiment had intensities far above the elf-trapping threshold for CS_2 (about 25 kW), it was possible to observe not only the ormation of rings around the self-focused spots but also the growth of multiple filanents at power levels far above the threshold from an apparently homogeneous beam of 1-mm diameter. The former effect can be associated with the higher-order circularly ymmetric soliton modes [24] shown in Figure 6.3, whereas the latter effect is a direct manifestation of the transverse instability capable of producing spatial modulations on a relatively broad beam.

A detailed study of the spatial breakup of a broad optical beam was carried out in a 1973 experiment using a 50-cm CS_2 cell [77]. It was observed that radially symmetric ring patterns created by circular apertures broke up into focal spots having azimuthal ymmetry and a regular spacing. This kind of effect can be associated with the transverse instability of quasi-plane bright rings created by the input beam. Indeed, the

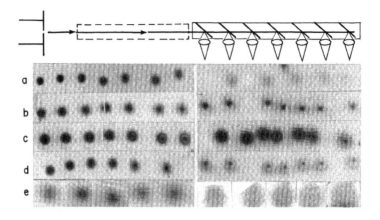

Figure 6.9: Self-trapping of an optical beam inside a CS_2 cell. Spot sizes at seven different locations are shown. An additional CS_2 cell (shown dashed) is inserted for data on the right side to add 25 cm of path length. (a) Gas-laser control; (b)–(d) self-trapping at increasing power levels; (e) output from the 1-mm pinhole. (After Ref. [74]; ©1966 APS.)

number of bright spots and the critical powers are in a good qualitative agreement with the simple theory of transverse instabilities [78]. In a 1975 experiment, an artificial Kerr medium consisting of submicrometer particles in a liquid suspension was used [79]. The self-trapped filaments in this experiment exhibited the smallest diameter (∼2 μm). In a more recent experiment, an organic nonlinear material in the form of a polydiacetylene paratoluene sulfonate (PTS) crystal was used to observe self-trapping in the spectral region near 1600 nm [80]. The nonlinearity of such a crystal is among the largest because of which the threshold power was only 1 kW in this experiment.

Vortex rings were first observed in a 1995 experiment [81]. A vortex ring corresponds to bright ring-shaped region and has a finite angular momentum. It was created by passing a laser beam through a diffracting phase mask and then propagating it inside a a 20-cm-long Rb-vapor cell. A quadratically nonlinear medium (a KTP crystal) was used in a later experiment [82]. The angular momentum associated with the input beam affects strongly the dynamics of bright spots (i.e, spatial solitons) that are created by the transverse instability of vortex rings; they can attract or repel each other, and may even fuse together under certain conditions.

In a 1991 experiment, self-focusing inside sodium vapor produced a variety of spatial patterns of spots, each spot being a bright spatial soliton, as the input beam power was varied from 30 to 460 mW [83]. A series of bifurcations were observed to occur because of spatial instabilities seeded by the intentionally introduced aberrations. In this experiment, the structure of the transverse-instability gain curve was used for encoding the input-beam wave front such that the growth of certain unstable spatial frequencies accelerated. As a result, complicated spatial patterns could be created by varying the beam intensity or tuning the optical wavelength close to an atomic resonance; these patterns had the "complexity and beauty rivaling that of a kaleidoscope" [83].

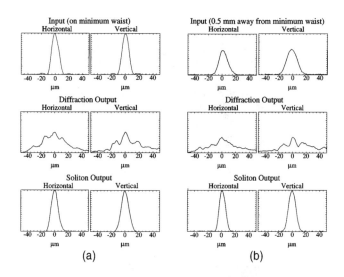

Figure 6.10: Intensity profiles along the two transverse dimensions of the input beam (upper panel), diffracted output (middle panel), and soliton output (lower panel) when the input facet of the crystal is (a) at the minimum waist of the input beam and (b) 0.5 mm away from the minimum waist. (After Ref. [92]; ©1996 OSA.)

The analysis of spatial instabilities, bifurcations, and formation of spatial solitons based on the $(2+1)$-dimensional NLS equation is, strictly speaking, valid for only CW beams. It can be applied, in practice, even for optical pulses, provided they are relatively long. In contrast, short pulses undergoing self-focusing do not collapse to wavelength dimensions. Several experiments have revealed the resistance of short pulses ($\sim$50 fs) to self-trapping [84]. The reason is related to the presence of group-velocity dispersion, which plays an important role and should be included. Chapter 7 considers such spatiotemporal effects using the $(3+1)$-dimensional NLS equation.

Starting in 1993, a photorefractive nonlinear medium was used to study self-trapping and solitons [89]–[97]. Its use permitted the formation of two-dimensional spatial solitons at relatively low input powers. In a 1996 experiment [92], an electric field of 5.8 kV/cm was applied across a strontium–barium niobate (SBN) crystal to form screening solitons, trapped in both transverse dimensions, by launching an extraordinary polarized beam at $\lambda = 524$ nm. The beam profiles were measured throughout the crystal. Self-trapping of the beam produced filaments as small as 9.6 μm at microwatt power levels. The creation of an (almost) axially symmetric soliton required 3400 V along the crystalline c axis between the two electrodes separated by 5.5 mm. A slightly lower voltage generated an elliptical soliton beam that was narrower in the direction parallel to the external field. In contrast, a slightly higher voltage created a soliton narrower in the direction perpendicular to the field.

Figure 6.10 compares the evolution of input beams of different sizes under several different experimental conditions [92]. In part (a) the input facet of the photorefractive crystal is located at the beam waist (where the phase front is planar), whereas in part (b) the input facet of the photorefractive crystal is located 0.5 mm away from the beam

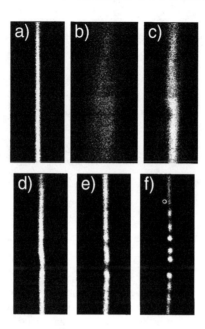

Figure 6.11: Evolution of a narrow soliton stripe inside a photorefractive crystal. The input beam is shown in (a), while photos (b)–(f) show the output at applied voltages of 0, 620, 900, 1290, and 1790 V, respectively. (After Ref. [98]; ©1996 APS.)

waist. In each case, the upper trace shows the intensity profiles of the input beam (12-μm FWHM), the middle trace shows the diffracted output at zero voltage (linear propagation), and the lower trace shows the soliton output when the nonlinear effects are turned on by applying a bias voltage. Although the two beams differ by 40% in width and even more in their transverse phases, the output soliton profiles are nearly identical. The extent of beam reshaping can be seen by comparing the profiles of the diffracted beams in the middle traces. The beam shape is significantly perturbed by material inhomogeneities in the linear case, but the soliton profiles seen in the lower traces are relatively smooth.

The breakup of a quasi-planar bright spatial soliton stripe into a sequence of two-dimensional solitons initiated by the transverse instability was observed in an experiment [98] in which a 10-mW CW beam from a He–Ne laser (λ = 632.8 nm) was used to create a highly elliptical beam (cross section 15 μm × 2 mm). This beam was directed into a photorefractive SBN crystal. A variable external voltage was applied along the crystal c axis to take advantage of the largest component of the electro-optic nonlinearity and to vary its magnitude in a controlled fashion.

Figure 6.11 shows the near-field distributions of the input beam (a) and output beams (b–f) for different values of the applied voltage (i.e., different values of the nonlinearity). Without the applied voltage ($V = 0$), the output beam spreads, owing to diffraction (b). As the nonlinearity is increased, the beam starts to self-focus (c) and forms a self-trapped channel of light in the photo (d). A further increase of nonlinearity leads to transverse instability, and the self-trapped stripe beam breaks into a multiple filaments, as seen in photos (e) and (f). Because no artificial noise was added to the input beam to seed the instability, the instability developed from the natural level of noise present on the laser beam or noise added inside the crystal [99].

It should be stressed that the transverse instability of a planar soliton stripe inside photorefractive media is a much more complicated phenomenon than that occurring inside a Kerr medium. The reason is that the applied electric field makes the problem anisotropic. The theoretical analysis of Section 6.2 can be generalized to include these effects, resulting in the following set of two equations for the normalized beam amplitude u and the normalized electrostatic potential ϕ induced by the beam [99]:

$$i\frac{\partial u}{\partial z} + \frac{1}{2}\nabla_\perp^2 u + \phi_x u = 0, \tag{6.3.1}$$

$$\nabla_\perp^2 \phi + \nabla_\perp \ln(1 + |u|^2)\nabla_\perp \phi = \frac{\partial}{\partial x}\ln(1 + |u|^2). \tag{6.3.2}$$

Numerical simulations based on these equations show that the self-focusing of a soliton stripe can still occur. In fact, each stationary solution of Eqs. (6.3.1) and (6.3.2) satisfies an NLS equation with an effective saturable nonlinearity because Eq. (6.3.2) can be integrated to yield $\partial\phi/\partial x = (|u|^2 - u_0^2)/(1 + |u|^2)$, where u_0 is a constant. The transverse instability of the resulting soliton solution has been analyzed in both the self-focusing and self-defocusing cases using an asymptotic-expansion technique [100].

6.4 Soliton Interaction and Spiraling

The physics behind the *coherent interaction* among spatial solitons does not depend much on the type of nonlinear medium in which the solitons are propagating. As discussed in Chapters 2 and 3, the nature of coherent interaction between two one-dimensional solitons depends crucially on their relative phase θ [101]; the solitons attract each other for $\theta = 0$ and repel each other for $\theta = \pi$. For intermediate values of the soliton phase, $0 < \theta < \pi$, the solitons experience inelastic interaction and exchange energy between them. This behavior persists for two-dimensional solitons. The new features in a bulk medium turns out to be that solitons can change their direction [102], fuse together, [103], and even spiral around each other [104, 105]. We focus on these new features in this section.

6.4.1 Two-Soliton Interaction

Because two-dimensional spatial solitons are unstable in a Kerr medium, the soliton interaction and spiraling can be observed only in non-Kerr media. The interaction can be modeled theoretically either using an NLS equation with saturable nonlinearity [106] or by employing the cubic-quintic NLS equation [107]. If the two-dimensional spatial solitons are treated as "effective particles," their interaction potential can be calculated by using the collective-coordinate approach [107]–[109]. In this case, soliton interaction is described by an effective Hamiltonian,

$$H_{\text{eff}} = \frac{m}{2}\left(\frac{dR}{dz}\right)^2 + U_{\text{eff}}(R), \tag{6.4.1}$$

with the effective potential

$$U_{\text{eff}}(R, \theta) = \frac{M_a^2}{2mR^2} - I(R, \theta),$$ (6.4.2)

where $m = \mu/2$, $\mu = E_1 E_2/(E_1 + E_2)$ is the reduced mass of two solitons and $E_n = \int u_n^2 d\mathbf{r}$ is related to the soliton power ($n = 1, 2$). Further,

$$I(R) = \int u_1^2(|\mathbf{r} - \mathbf{r}_1|) u_2^2(|\mathbf{r} - \mathbf{r}_2|) d\mathbf{r}$$ (6.4.3)

is the overlap integral for the two solitons with amplitudes u_1 and u_2, and $\mathbf{R} = \mathbf{r}_1 - \mathbf{r}$ is the relative distance between them. The parameter

$$M_a = \frac{\mu}{2}\left(\mathbf{R} \times \frac{d\mathbf{R}}{dz}\right) \cdot \mathbf{e}_z,$$ (6.4.4)

where $\mathbf{e}_z$ is a unit vector along the z axis, is a conserved quantity that is related to the initial angular momentum of two solitons; M_a has a nonzero value whenever the soliton trajectories do not belong to a common straight line and the impact parameter is nonzero during the collision.

The feature that makes the soliton interaction rather complicated is that the interaction potential U_{eff} depends on the relative phase θ. For a fixed value of the relative phase, the interaction is similar to the motion of a particle in a central potential. As a result, it should display many types of bounded trajectories, including the spiraling of solitons around each other [104]. However, the dependence of the potential on the relative phase makes the soliton interaction more complicated. In particular, the spiraling phenomenon becomes unstable. Figure 6.12 shows two numerical examples of soliton interaction for the cases in which the collective-particle approach predicts that solitons would spiral around each other. Depending on the initial value of the angular momentum M_a, two solitons eventually either fuse [coumns (a) and (b)] or repel each other [coumns (c) and (d)] because of the phase instability. Notice that the evolution of the two interacting solitons is virtually identical in the two cases until just before they merge or separate. Stable spiraling of two solitons becomes possible when the solitons are incoherent because the phase instability is then suppressed [110, 111]. This case is discussed in Chapter 9 in the context of vector solitons.

6.4.2   Multisoliton Clusters

In this section we consider the possibility that N coherently interacting solitons may form a *stationary configuration*, as a simple extension of the two-soliton interaction problem. It is natural to look for such a stationary structure in a ring-like geometry, similar to the case of two-soliton spiraling. However, such a configuration turns out to be *radially unstable* because of an effective tension induced by the bending of the soliton array. A ring of N solitons either collapses or expands, depending on whether the mutual interaction between the neighboring solitons is attractive or repulsive.

A simple physical mechanism can stabilize the ring-like configuration of N solitons. The idea consists of introducing an *additional phase* that twists by $2\pi m$ along the

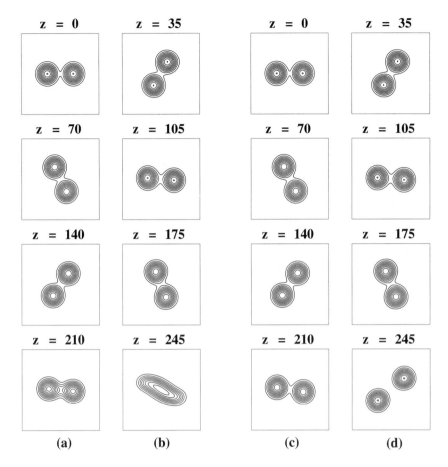

Figure 6.12: Spiraling of two spatial solitons in a non-Kerr medium, leading to fusion [columns (a) and (b)] or separation [columns (c) and (d)], depending on their initial angular momentum. (After Ref. [106]; ©1998 Elsevier.)

soliton ring, where m is an integer. This phase introduces an effective centrifugal force that can balance the tension effect and stabilize the ring-like soliton cluster. Owing to a net angular momentum induced by such a phase distribution, the soliton clusters will rotate with an angular velocity that depends on the number of solitons and the integer m. As a matter of fact, this idea does not work for a small number of solitons in the ring. More specifically. it does not apply to the case of two solitons because the phase difference between the neighboring beams in the ring should be less than $\pi/2$ to make the soliton interaction attractive. Instead, the spiraling of two-soliton beams requires a nonzero input angular momentum, as discussed earlier.

To describe the soliton clusters analytically, we consider a coherent superposition of N solitons with the envelopes $u_n(x, y, z)$, where n varies in the range 1–N. The total field $u = \sum u_n$ satisfies the NLS equation. For a ring of *weakly overlapping* identical solitons launched simultaneously, we can analyze the evolution of the ring by using the following Gaussian ansatz for each two-dimensional spatial soliton:

$$u_n = A \exp\left(-\frac{|\mathbf{r} - \mathbf{r}_n|^2}{2a^2} + i\alpha_n\right), \tag{6.4.5}$$

where $\mathbf{r}_n = (x_n, y_n)$ describes the location of the nth soliton and α_n is its phase.

The three integrals of motion introduced in Section 2.1 for one-dimensional solitons can be extended to include two transverse dimensions. A new feature of two-dimensional solitons is that one must distinguish between the linear and angular momenta, both of which become vector quantities. Indeed, the following three quantities are conserved for the two-dimensional spatial solitons:

$$P = \int |u|^2 \, d\mathbf{r}, \qquad \mathbf{M}_l = \mathrm{Im} \int u^* \nabla u \, d\mathbf{r}, \qquad M_a = \mathrm{Im} \int u^* (\mathbf{r} \times \nabla u) \cdot \mathbf{e}_z \, d\mathbf{r}, \tag{6.4.6}$$

where the integration is over the entire x–y plane. Note that it is the z component of the angular momentum that appears in Eq. (6.4.6). Using $u = \sum u_n$ with u_n as in Eq. (6.4.5), we obtain the following three conservation relations for the soliton cluster:

$$P = \pi a^2 A^2 \sum_{n=1}^{N} \sum_{k=1}^{N} \exp(-Y_{nk}) \cos \theta_{nk}, \tag{6.4.7}$$

$$\mathbf{M}_l = \frac{\pi}{2} A^2 \sum_{n=1}^{N} \sum_{k=1}^{N} \exp(-Y_{nk})(\mathbf{r}_n - \mathbf{r}_k) \sin \theta_{nk}, \tag{6.4.8}$$

$$M_a = \pi A^2 \sum_{n=1}^{N} \sum_{k=1}^{N} \exp(-Y_{nk})(\mathbf{r}_n \times \mathbf{r}_k) \sin \theta_{nk}, \tag{6.4.9}$$

where $Y_{nk} = |\mathbf{r}_n - \mathbf{r}_k|^2 / 4a^2$ and $\theta_{nk} = \alpha_n - \alpha_k$. When the beams are arranged in a ring-shaped array, $\mathbf{r}_n = (R \cos \theta_n, R \sin \theta_n)$ with $\theta_n = 2\pi n/N$, and we can use $Y_{nk} = (R/a)^2 \sin^2[\pi(n-k)/N]$.

For analyzing multiple-soliton clusters, we remove the motion of the center of the mass and put $\mathbf{M}_l = 0$. Applying this constraint to Eqs. (6.4.7)–(6.4.9), we find that the soliton phases should satisfy the condition $\alpha_{i+n} - \alpha_i = \alpha_{k+n} - \alpha_k$. This condition is satisfied provided the phase α_n depends on n linearly, i.e., $\alpha_n = n\theta$, where θ is the relative phase between two neighboring solitons in the ring. Employing the periodicity condition $\alpha_{n+N} = \alpha_n + 2\pi m$ for a ring of N solitons, where m is an integer, the angle θ is found to be

$$\theta = 2\pi m/N. \tag{6.4.10}$$

In the context of field theory, Eq. (6.4.10) corresponds to the condition for vanishing energy flow, because the linear momentum, $\mathbf{M}_l = \int \mathbf{j}(\mathbf{r}) \, d\mathbf{r}$, is related to the local current density $\mathbf{j} = \mathrm{Im}(u^* \nabla u)$. Therefore, Eq. (6.4.10) determines the nontrivial phase distribution for which soliton interaction is effectively elastic along the ring. When the ring consists of just two solitons ($N = 2$), this condition gives only two states with zero energy exchange; for even values of m, the two solitons attract each other ($\theta = 0$). In contrast, they repel each other when m is odd ($\theta = \pi$).

For $N > 2$, Condition (6.4.10) predicts the existence of a discrete set of allowed states, corresponding to a set of values $\theta = \theta^{(m)}$ with $m = 0, \pm 1, \ldots, \pm(N-1)$. Here, the plus and minus signs correspond to the sign of the angular momentum (similar to the case of vortex solitons discussed in Chapter 8). For any value of m such that $\theta^{(m)}$ lies in the domain $\pi < |\theta| < 2\pi$, the corresponding angle for $-m$ has the domain $0 < |\theta| < \pi$. Thus, both m and $-m$ describe the same stationary state. For example, in the case $N = 3$, the three states with zero energy exchange correspond to $\theta^{(0)} = 0$, $\theta^{(1)} = 2\pi/3$, and $\theta^{(2)} = 4\pi/3$ such that $\theta^{(\pm 1)} \leftrightarrow \theta^{(\mp 2)}$. Therefore, it is useful to limit the allowed values of θ to the domain $0 \leq \theta \leq \pi$, keeping in mind that all allowed soliton states are doubly degenerate with respect to the sign of the angular momentum. The absolute value of the angular momentum vanishes at both ends of this domain, when $m = 0$ for any N and when $m = N/2$ for even N. The number m determines the full phase twist around the ring, and it plays the role of the topological charge for the corresponding phase dislocation.

To explain the basic properties of the ring-like soliton clusters through a specific example, we select a saturable nonlinear Kerr medium with $F(I) = I(1 + sI)^{-1}$, where $I = |u|^2$. First, we apply the variational technique and find the parameters of a single soliton described by the ansatz (6.4.5). For $s = 1/2$, we obtain $A = 3.604$ and $a = 1.623$. Then, substituting Eq. (6.4.5) into the Hamiltonian

$$H = \int\int_{-\infty}^{\infty} \left[|\nabla u|^2 - \frac{1}{s}|u|^2 + \frac{1}{s^2} \ln\left(1 + s|u|^2\right) \right] dx\,dy, \qquad (6.4.11)$$

the effective interaction potential $U(R)$ can be calculated using

$$U(R) = H(R)/|H(\infty)|. \qquad (6.4.12)$$

For any value of N, we find *three distinct forms* for the interaction potential (6.4.12). These are shown in Figure 6.13 for the specific case $N = 5$. Noting that only one of these three potentials has a local minimum at a finite distance R, it is clear that a cluster that is stable against collapse or expansion can form only for this potential.

Extensive numerical simulations based on the NLS equation for different values of N confirm the predictions of the effective-particle approach. According to Figure 6.13, the effective potential is always attractive for $m = 0$, and thus a ring of five in-phase solitons should exhibit oscillations and, possibly, fusion. Exactly this behavior is observed in Figure 6.14(a). Although the ring oscillations are well described by the effective potential $U(R)$, the ring dynamics observed numerically are more complicated. The $m = 2$ curve in Figure 6.13 corresponds to a repulsive potential ($\theta = 4\pi/5$). In the numerical simulations corresponding to this case, the ring of soliton array expands, as is shown in Figure 6.14(d). The evolution of the soliton bound state that corresponds to a minimum of the effective potential $U(R)$ in Figure 6.13 (for $m = 1$) is shown in Figure 6.14(b). In this case, the nonzero angular momentum produces a repulsive centrifugal force that balances out an effective attraction among solitons.

To draw an analogy between the soliton cluster and a rigid body, one can calculate the *moment of inertia* $\mathfrak{I}$ of the cluster and its *angular velocity* Ω using

$$\mathfrak{I} = \int |u|^2 \mathbf{r}^2 d\mathbf{r}, \qquad \Omega = M_a/\mathfrak{I}. \qquad (6.4.13)$$

For the case shown in Figure 6.14(b), the numerically calculated value of the angular velocity is found to be $\Omega \simeq \pi/20 \simeq 0.157$, while Eq. (6.4.13) gives $\Omega = 0.154$. Nu-

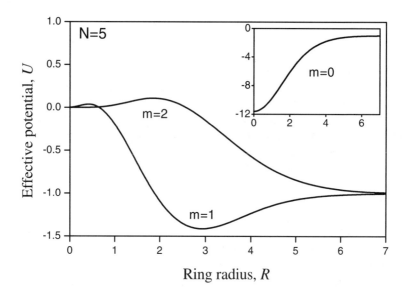

Ring radius, R

Figure 6.13: Examples of three effective potentials for a ring of five solitons. The corresponding topological charge m is shown near each curve. A dynamically stable bound state is possible for $m = 1$ only.

merical simulations reveal the existence of the "excited states" associated with this stationary state. One such state is shown in Figure 6.14(c); it exhibits oscillations around the equilibrium state with a period $z_p = 22$. Such a *vibrational* state of the "N-soliton molecule" demonstrates the radial stability of the bound state, in agreement with the effective-particle approximation.

The preceding analysis is valid for any N, and it allows us to classify all possible scenarios of the soliton interaction in terms of the phase jump θ between the neighboring solitons along the ring. Its main results can be summarized as follows. For $\theta = 0$, a ring of N solitons collapses after several oscillations. If the value of θ lies in the range $0 < \theta \leq \pi/2$, the interaction among the solitons is *attractive*, the value of the induced angular momentum is finite, and a rotating bound state of N solitons is possible. In contrast, if θ lies in the range $\pi/2 < \theta \leq \pi$, the soliton interaction is *repulsive* and the soliton ring expands with or without (for $\theta = \pi$) rotation, similar to the necklace-type beams discussed later [112]–[115]. For example, in the case $N = 3$, two states are possible for $\theta = 0$ and $\theta = 2\pi/3 > \pi/2$, but none of them is *stationary*.

For $N = 4$ and $m = 1$, the value of θ is $\pi/2$, and the stabilization of interacting solitons in the form of a ring-like cluster is indeed possible. In numerical simulations, such clusters are found to be long-lived objects that rotate while propagating over many diffraction lengths. However, similar to the scenario of the vortex breakup discussed in the next section, the soliton clusters are azimuthally unstable and they decay, after relatively long propagation, into several fundamental solitons that fly away in different directions. The rotation of different clusters observed in numerical simulations (seen in Figure 6.14) is always *clockwise* for positive m in Eq. (6.4.10); i.e., the clusters rotate in

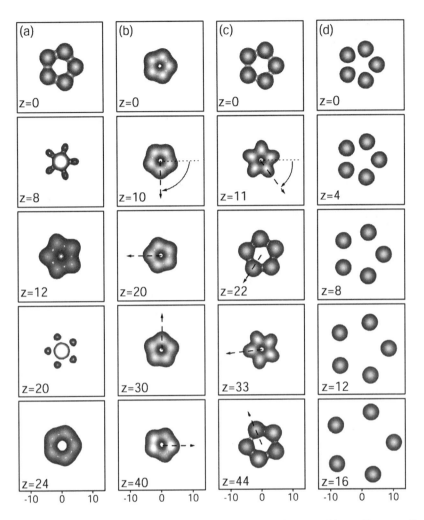

Figure 6.14: Four different scenarios for a ring of five solitons in a saturable nonlinear medium.
(a) $m = 0$, collapse and fusion through oscillations; (b) $m = 1$, a stationary bound state at $R_0 = 3$;
(c) $m = 1$, an excited bound state with an oscillation period $z_p = 22$; (d) $m = 2$, soliton repulsion.
The initial radius of the ring in all cases except (b) is $R = 5$.

the direction *opposite* to their phase gradient, in sharp contrast to the behavior observed
for necklace-type beams [114] and shown later in Figure 6.17.

The repulsion and subsequent diffraction of the individual beamlets in a self-trapped
ring-like structure is the main physical mechanism behind its disintegration. As discussed later in this book, the incoherent interaction between the components of a *vector*
ring-like beam allows one to compensate for the repulsion of beamlets, creating a new
type of quasi-stationary self-trapped structure that exhibits the properties of the necklace beams and ring-vortex solitons. The physical mechanism for creating such composite vector ring-like clusters is somewhat similar to the mechanism responsible for

the formation of the so-called *solitonic gluons* [116] and the multihump vector solitary waves [117]. It can be explained by a balance of the interaction forces acting between the coherent and incoherent components of a composite soliton.

6.5 Solitons with Angular Momentum

The self-trapping phenomenon is usually associated with optical beams that have a single central intense region (optical field without nodes). However, it has been known since 1966 that quasi-stable self-trapping can also occur for optical beams that exhibit radial symmetry but have one or more rings [25] . More recently, "necklace-shaped" beams lacking a radial symmetry have been found to be quasi-stable in the sense that they expanded relatively slowly in a self-focusing Kerr medium [112]. All such self-trapped optical structures possess *zero* angular momentum and, as a result, they do not rotate during propagation.

A novel class of rotating self-trapped optical beams was introduced in 1985 by Kruglov and Vlasov [118]. Such beams have a ring shape, and their intensity vanishes at the beam center as well as outside the ring. They also have a spiral phase structure with a singularity at the origin, representing a *phase dislocation* of the wave front and resembling the structure of an optical vortex [119]. Such an optical beam can be associated with a higher-order spatial soliton that carries a *finite* angular momentum. It is also referred to as a *vortex soliton*. During the 1990s, such ring-like vortex solitons were discovered for several types of nonlinear media [120]–[122], including those with quadratic nonlinearities [123]–[126]. Furthermore, it was shown, both numerically and analytically, that such ring-shaped vortex beams undergo an azimuthal symmetry-breaking instability and eventually decay into multiple fundamental solitons [120]–[126]. This symmetry-breaking instability has been observed experimentally in both Kerr-type and quadratic nonlinear media [81]–[82].

Stabilization of a ring-shaped vortex soliton is possible only in some exceptional cases, one example being the competing quadratic and self-defocusing nonlinearities [127]–[130]). It turns out that ring-shaped structures carrying *integer* or *fractional* angular momentum are quasi-stable when they are in the form of a *modulated* necklace beams [112] or a *necklace-ring vector soliton* [131]. The vectorial-beam interaction can also support self-trapped rotating structures in the form of "propeller solitons" [132]. We focus in this section on the scalar spinning solitons with nonzero angular momentum and discuss the effects of vectorial and parametric interactions in Chapters 9 and 10, respectively.

6.5.1 Spinning Solitons

The starting point is the $(2 + 1)$-dimensional NLS equation,

$$i\frac{\partial u}{\partial z} + \nabla_{\perp}^2 u + F(I)u = 0, \tag{6.5.1}$$

where $I = |u|^2$ represents the local beam intensity. The function $F(I) = I(1+sI)^{-1}$ for a saturable nonlinear medium. In the case of a weakly saturating medium ($s \leq 1$), one can use the expansion $F(I) = I - sI^2$, resulting in the cubic-quintic NLS equation.

We look for spatial solitons that maintain their intensity profile during propagation as well as phase profile by assuming a solution in the form

$$u(x,y,z) = U(x,y)\exp[ikz + i\phi(x,y)], \tag{6.5.2}$$

where U and ϕ represent the soliton amplitude and phase, respectively, and k is the propagation constant. Substituting Eq. (6.5.2) into Eq. (6.5.1), we obtain two coupled equations for the soliton amplitude and phase:

$$\nabla_\perp^2 U - kU - (\nabla\phi)^2 U + F(U^2)U = 0, \tag{6.5.3}$$

$$\nabla_\perp^2 \phi + 2(\nabla\phi)(\nabla \ln U) = 0. \tag{6.5.4}$$

Consider first the constant-phase case by setting $\phi = \phi_0$. This case was considered earlier, in Section 6.2, and it was found that all localized solutions must possess a radial symmetry such that $U(x,y) \equiv U(r)$, where $r = \sqrt{x^2 + y^2}$. Solutions of this type include the fundamental (bell-shaped) soliton and higher-order solitons with several rings surrounding the central peak [24, 25]. The main parameter characterizing such a spatial soliton is its power P, defined as in Eq. (6.4.6). The family of transversely moving solitons can be obtained by applying the Galilean transformation, $\mathbf{r} \rightarrow \mathbf{r} - 2\mathbf{q}z$ and $\phi \rightarrow \phi + \mathbf{q} \cdot (\mathbf{r} - \mathbf{q}z)$, where $\mathbf{v} = 2\mathbf{q}$ is the transverse velocity of the soliton. Such moving solitons are characterized by their linear momentum,

$$\mathbf{M}_l = \text{Im} \int u^* \nabla u \, d\mathbf{r} = \int \nabla\phi \, U^2 d\mathbf{r}. \tag{6.5.5}$$

For the fundamental soliton, $\mathbf{M}_l = \mathbf{q}P$, where P is the soliton power.

Vortex-type bright solitons, first introduced in 1985 [118], correspond to a bright spatial soliton whose optical field depends not only on the radial coordinate r but also on the azimuthal coordinate θ. This feature leads to a *finite angular momentum* for such solitons. Vortex solitons correspond to solutions of Eqs. (6.5.3) and (6.5.4) with a radially symmetric amplitude $U(r)$ such that $U = 0$ at the beam center $r = 0$. They also have a rotating spiral phase, varying *linearly* with the polar angle θ, so that $\phi = m\theta$. The rotation velocity is quantized by the condition that the optical field should not change after θ changes by 2π. This condition forces m to be an integer. Figure 6.15(a) shows the intensity distribution for several values of m for the fundamental soliton ($m = 0$) and vortex solitons ($m \leq 1$) in a saturable nonlinear medium using $s = 0.5$. The radial distribution is different for different values of m. Moreover, the soliton power grows with the saturation parameter s, as shown in Figure 6.15(b) and discussed in Ref. [133].

Physically, the index m stands for a phase twist around the intensity ring and is usually called the *topological index* or *topological charge* of the vortex. The *constant of motion* associated with this type of soliton is the axial component of the *angular momentum*, defined earlier in Eq. (6.4.6). It is related to the soliton phase as

$$M_a = \int \frac{\partial \phi}{\partial \theta} U^2(r) d\mathbf{r}, \tag{6.5.6}$$

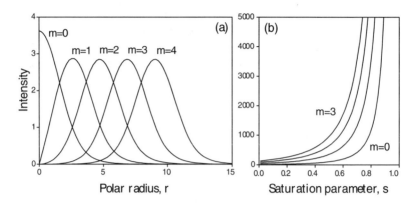

Figure 6.15: (a) Radially symmetric vortex solitons in a saturable nonlinear medium for $s = 0.5$ and several values of the topological charge m; $m = 0$ corresponds to the fundamental soliton. (b) Increase in the soliton power with the saturation parameter s. (After Ref. [133]; ©2002 IOP.)

where Eq. (6.5.2) was used to express M_a in terms of the soliton amplitude and phase. The angular momentum M_a is an intrinsic characteristic of a vortex, and it is usually regarded as its *spin* angular momentum, to distinguish it from the *orbital* angular momentum discussed in the context of a cluster of several interacting solitons. The ratio $S = M_a/P$ can be identified with the *soliton spin*. Using $\phi = m\theta$ in Eq. (6.5.6), the spin of a vortex soliton is simply equal to its topological charge, i.e., $S = m$, and it is zero for a fundamental soliton.

The existence of ring-shaped solitary waves can be understood in terms of simple physics. As discussed in Section 6.1, soliton stripes formed by one-dimensional solitons undergo a transverse instability when propagating inside a bulk Kerr medium, [5]. One way to suppress this instability is to employ a ring structure created from a one-dimensional soliton stripe wrapped around its tail [134]. It turns out that when the optical phase depends on the radial coordinate, the ring exhibits quasi-stabilization, and even an initially expanding beam can shrink and eventually collapse [122]. The spiral beams with a rotating phase provide another example of *nonstationary* ring-shaped solitary waves [120].

Vortex and ring-like bright solitons experience another transverse instability in a self-focusing medium known as the *azimuthal instability*. It breaks up the ring into a number of *moving* fundamental solitons, which fly off the ring [120]–[123]. Figure 6.16 shows a numerical example of the instability-induced decay for two vortex solitons with different topological charges. The initial ring-shaped vortex decays, in the presence of numerical noise, to two or four fundamental solitons, depending on its topological charge ($m = 1$ or 2). Because the total angular momentum should be conserved during the decay, solitons fly off the ring along the tangential trajectories [123]. In other words, the initial *spin* angular momentum of the vortex is transformed into a net *orbital* angular momentum of the multiple flying solitons [108]. This process has been observed experimentally for both the Kerr-type [81] and quadratic [82] nonlinear media.

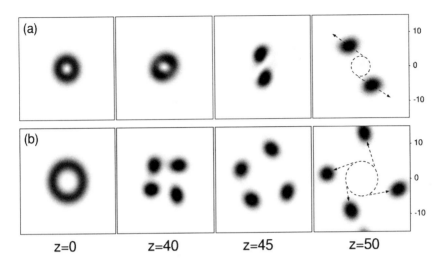

Figure 6.16: Breakup of a vortex solitons with the topological charge (a) $m = 1$ and (b) $m = 2$. Dashed curves at $z = 50$ show the initial rings and the trajectories along which solitons fly away. (After Ref. [133]; ©2002 IOP.)

6.5.2 Necklace Beams

Since the decay of vortex solitons is associated with the azimuthal modulation of their intensity resulting from a symmetry-breaking instability, one can try to stabilize the ring structure by modulating the intensity or the phase of the input beam before it is launched into the nonlinear medium. To understand this idea and to link it to the vortex beams, we assume that the initial field (at $z = 0$) is of the form

$$u(r, \theta, 0) = U(r)[\cos(m\theta) + ip\sin(m\theta)],\qquad(6.5.7)$$

where p is the modulation parameter with values in the range $0 < p < 1$. The structure (6.5.7) corresponds to a vortex with the topological charge m when $p = 1$. The *spin* of this nonstationary structure is given by

$$S = \frac{2mp}{1 + p^2},\qquad(6.5.8)$$

and it vanishes as $p \to 0$.

Figure 6.17 shows the numerical results for such a beam propagating inside a saturable nonlinear medium using three values of p and assuming $s = 0.5$ and $m = 6$. When the ring vortex is only slightly modulated (a), it decays into a complex pattern because of competition among different instability modes. The direction of the rotation associated with the pattern is determined, according to Eqs. (6.5.5) and (4.3.5), by the phase gradient, which grows *anticlockwise* for $m > 0$. The angular velocity diminishes as the ring-like structure expands.

When the modulation becomes deeper, the initial ring-like vortex transforms into a necklace-like structure, seen in Figure 6.17(b). The modulated ring-shape structure

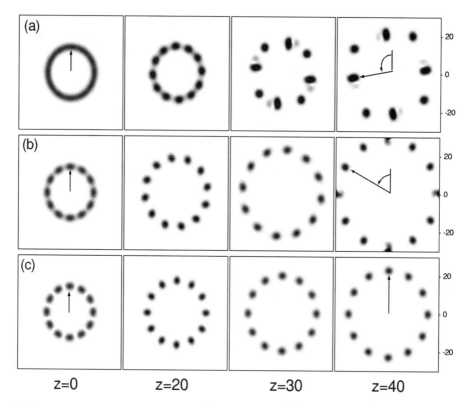

$z=0$ $\qquad$ $z=20$ $\qquad$ $z=30$ $\qquad$ $z=40$

Figure 6.17: Breakup of a ring-shape vortex with $m = 6$ for different values of the modulation parameter p: (a) $p = 0.95$ and $S \simeq 5.992$, (b) $p = 0.5$ and $S = 4.8$, and (c) $p = 0$. (After Ref. [133]; ©2002 IOP.)

does not decay but, instead, expands while rotating slowly. The rotation is much weaker because the initial angular momentum (or spin) is much smaller than the case shown in Figure 6.17(a). Similar ring-like structures were first suggested in Ref. [114], with a *fractional spin* and a rotating intensity pattern. When the initial angular momentum is zero ($p = 0$), the necklace-type beam expands, with no rotation. These necklace beams have been studied in detail for self-focusing Kerr media [112], where they exhibit a quasi-stable expansion. The reason why such necklace beams do not exist as stable stationary structures of constant radius is related to the fact that neighboring "pearls" of the necklace interact and repel each other [113] when their interaction forces are not balanced.

Experimental results on the self-trapping of necklace-type beams were reported as early as 1993 [135], even before the concept of the necklace beams was developed. Figure 6.18 shows the experimental data for the case in which an input beam in the form of a higher-order Laguerre–Gaussian mode shown as (a) was launched at the input end of a CS_2 cell. The beam diameter was 260 μm, with a petal thickness of about 80 μm. The output pattern after 5 cm of propagation is shown in part (b), when the intensity was

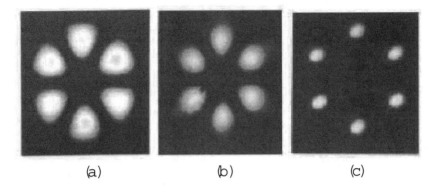

(a)   (b)   (c)

Figure 6.18: Propagation of a Laguerre–Gaussian beam inside a Kerr medium: (a) input beam, (b) diffracted beam at low intensity, and (c) the self-trapped necklace beam at high intensities. (After Ref. [135]; ©1993 Springer.)

low enough that the nonlinear effects were negligible. As expected, the beam diameter increased to 265 μm, because of diffraction, while the petal thickness remained close to 80 μm. As the beam power was gradually increased, the petal thickness decreased, because of self-focusing. The petal size reduced to 30 μm at an intensity level of 5×10^7 W/cm^2 [115]. Such self-trapped structures are remarkably stable and allow one to transport optical beams with powers several times the critical power at which a Gaussian beam would otherwise collapse because of self-focusing; however, they disintegrate for input intensities lower than the self-trapping intensity.

References

[1] B. B. Kadomtsev, *Collective Phenomena in Plasmas* (Nauka, Moscow, 1976) [in Russian].

[2] A. M. Anile, J. K. Hunter, P. Pantano, and G. Russo, *Ray Methods for Nonlinear Waves in Fluids and Plasmas* (Longman Scientific & Technical, Essex, UK, 1993).

[3] L. A. Ostrovsky and V. I. Shrira, *Zh. Eksp. Teor. Fiz.* **71**, 1412 (1976) [*Sov. Phys. JETP* **44**, 738 (1976)].

[4] V. I. Shrira V. I., *Zh. Eksp. Teor. Fiz.* **79**, 87 (1980) [*Sov. Phys. JETP* **52**, 44 (1980)].

[5] Yu. S. Kivshar and D. E. Pelinovsky, *Phys. Rep.* **331**, 117 (2000).

[6] E. A. Kuznetsov and S. K. Turitsyn, *Zh. Eksp. Teor. Fiz.* **94**, 119 (1988) [*Sov. Phys. JETP* **67**, 1583 (1988)].

[7] V.E. Zakharov and A. M. Rubenchik, *Zh. Eksp. Teor. Fiz.* **65**, 997 (1973) [*Sov. Phys. JETP* **38**, 494 (1974)].

[8] E. A. Kuznetsov, A. M. Rubenchik, and V. E. Zakharov, *Phys. Rep.* **142**, 103 (1986).

[9] N. R. Pereira, A. Sen, and A. Bers, *Phys. Fluids* **21**, 117 (1978).

[10] D. Anderson, A. Bondeson, and M. Lisak, *Phys. Scripta* **20**, 343 (1979).

[11] B. B. Kadomtsev and V. I. Petviashvili, *Dokl. Acad. Nauk SSSR* **192**, 753 (1970) [*Sov. Phys. Doklady* **15**, 539 (1970)].

[12] P. A. E. M. Janssen and J. J. Rasmussen, *Phys. Fluids* **26**, 1279 (1983).

[13] E. E. Kuznetsov, S. L. Musher, and A. V. Shafarenko, *Zh. Eksp. Teor. Fiz.* **37**, 204 (1983) [*JETP Lett.* **37**, 241 (1983)].

[14] D.E. Pelinovsky, Math. Comput. Simulations **55**, 585 (2001).

[15] G. B. Whitham, *Linear and Nonlinear Waves* (Wiley, New York, 1974).

[16] V. G. Makhankov, *Phys. Rep.* **35**, 1 (1978).

[17] B. A. Trubnikov and S. K. Zhdanov, *Phys. Rep.* **155**, 137 (1987).

[18] G. A. Askar'yan, *Zh. Eksp. Teor. Fiz.* **42**, 1567 (1962) [*Sov. Phys. JETP* **15**, 1088 (1962)].

[19] R. Y. Chiao, E. Garmire, and C. H. Townes, *Phys. Rev. Lett.* **13**, 479(1964).

[20] Y. R. Shen, *Prog. Quantum Electron.* **4**, 1 (1975).

[21] J. H. Marburger, *Prog. Quantum Electron.* **4**, 35 (1975).

[22] Y. R. Shen, *The Principles of Nonlinear Optics* (Wiley, New York, 1984).

[23] S. N. Vlasov and V. I. Talanov, *Wave Self-Focusing* (Institute of Applied Physics, Nizhnii Novgorod, Russia, 1997) [in Russian].

[24] Z. K. Yankauskas, *Izv. Vuzov Radiofiz.* **9**, 412 (1966) [*Sov. Radiophys.* **9**, 261 (1966)].

[25] H. A. Haus, *Appl. Phys. Lett.* **8**, 128 (1966).

[26] J. M. Soto-Crespo, D. R. Heatley, E. M. Wright, and N. N. Akhmediev, *Phys. Rev. A* **44**, 636 (1991).

[27] D. Edmundson, *Phys. Rev. E* **55**, 7636 (1997).

[28] J. Yang, *Phys. Rev. E* **66**, 026601 (2002).

[29] V. M. Malkin and E.G. Shapiro, *Physica D* **53**, 25 (1991).

[30] F. Vidal and T.W. Johnson, *Phys. Rev. E* **55**, 3571 (1997).

[31] D. V. Skryabin, *J. Opt. Soc. Am. B* **19**, 529 (2002).

[32] V. I. Talanov, *Izv. Vuzov Radiofiz.* **7**, 564 (1964).

[33] A. G. Litvak, *Izv. Vuzov Radiofiz.* **9**, 675 (1966).

[34] T. A. Laine and A. T. Friberg, *J. Opt. Soc. Am. B* **17**, 751 (2000).

[35] V. M. Eleonskii and V.P. Silin, *Sov. Phys. JETP. Lett.* **13**, 167 (1971).

[36] V. M. Eleonskii and V.P. Silin, *Sov. Phys. JETP* **33**, 1039 (1971).

[37] V. M. Eleonskii, L.G. Oganes'yants, and V.P. Silin, *Sov. Phys. JETP* **36** 282 (1973).

[38] E. Granot, S. Stemklar, Y. Isbi, B. Malomed, and A. Lewis. *Opt. Lett.* **22** 1290 (1997).

[39] E. Granot, S. Stemklar, Y. Isbi, B. Malomed, and A. Lewis. *Opt. Commun.* **166**, 121 (1999).

[40] E. Granot, S. Stemklar, Y. Isbi, B. Malomed, and A. Lewis. *Opt. Commun.* **178**, 431 (2000).

[41] A. W. Snyder, D. J. Mitchell, and Y. Chen, *Opt. Lett.* **19** 524 (1994).

[42] D. I. Abakarov, A. A. Alopyan, and S.I. Pekar, *Sov. Phys. JETP* **25**, 303 (1967).

[43] D. Pohl, *Opt. Commun.* **2**, 307 (1970).

[44] D. Pohl, *Phys. Rev. A* **5**, 1908 (1972).

[45] V. M. Eleonskii, L. G. Oganes'yants, and V. P. Silin, *Sov. Phys. JETP* **35**, 44 (1972).

[46] D. A. Kirsanov and N. N. Rosanov, *Opt. Spectr.* **87**, 390 (1999).

[47] V. E. Semenov, N. N. Rosanov, and N. V. Vyssotina, *Sov. Phys. JETP* **89**, 243 (1999) [*Zh. Eksp. Teor. Fiz.* **116**, 458 (1999)].

[48] N. N. Rosanov, N. V. Vyssotina, and A. G. Vladomirov, *Sov. Phys. JETP* **91**, 1130 (2000) [*Zh. Eksp. Teor. Fiz.* **118**, 1307 (2000)].

[49] N. N. Rosanov, V. E. Semenov, and N. V. Vyssotina, *J. Opt. B* **3**, S96 (2001).

[50] S. A. Darznek and A. F. Suchkov, *Sov. J. Quantum Electron.* **1** 400 (1972).

[51] N. A. Tikhonov, *Sov. Phys. Dokl.* **21**, 663 (1976).

[52] M. D. Feit and J. A. Fleck, Jr., *J. Opt. Soc. Am. B* **5**, 633 (1988).

[53] N. Akhmediev, A. Ankiewicz, and J.M. Soto-Crespo, *Opt. Lett.* **18** 411 (1993).

[54] J. M. Soto-Crespo and N. N. Akhmediev, *Opt. Commun.* **101**, 223 (1993).

[55] G. Fibich, *Phys. Rev. Lett.* **76**, 4356 (1996).

[56] A. P. Sheppard and M. Haelterman, *Opt. Lett.* **23**, 1820 (1998).

[57] S. A. Izyurov and S. A. Kolov, *JETP Lett.* **71**, 453 (2000).

[58] P. Chamorro-Posada, G. S. McDonald, and G. H. C. New, *Opt. Commun.* **192**, 1 (2001).

[59] G. Fibich and G.C. Papanicolaou, *Phys. Lett. A* **239**, 167 (1998).

[60] G. Fibich and G.C. Papanicolaou, SIAM J. Appl. Math. (1999)

[61] A. A. Kolokolov and A. I. Sukov, *Prikl. Mekh. Teor. Fiz.* **6**, 77 (1977) (in Russian).

[62] S. N. Vlasov, *Sov. J. Quantum Electron.* **17** 1191 (1987).

[63] S. Chi and Qi Guo, *Opt. Lett.* **20**, 1598 (1995).

[64] M. Lax, W. H. Loisell, and W. B. Khight, *Phys. Rev. A* **11**, 1365 (1975).

[65] G. Fibich and B. Illan, *Phys. Rev. Lett.* **89**, 013901 (2002).

[66] B. Crosignani, P. Di Porto, and A. Yariv, *Opt. Lett.* **22**, 778 (1997).

[67] A. Ciattoni, P. Di Porto, B. Crosignani, and A. Yariv, *J. Opt. Soc. Am. B* **17**, 809 (2000).

[68] A. Ciattoni, B. Crosignani, amd P. Di Porto, *Opt. Commun.* **202**, 17 (2002).

[69] C. S. Milsted and C.D. Cantrell, *Phys. Rev. A* **53**, 3536 (1996).

[70] S. Blair and K. Wagner, *Opt. Quantum Electron.* **30**, 697 (1998).

[71] R. M. Joseph, P. M. Goorjian, and A. Taflone, *Opt. Lett.* **18**, 491 (1993).

[72] P. M. Goorjian and Y. Silberberg, *J. Opt. Soc. Am. B* **14**, 3253 (1997).

[73] N. F. Pilipetskii and A. R. Rustamov, *Zh. Eksp. Teor. Fiz.* **2**, 88 (1965).

[74] E. Garmire, R. Y. Chiao, and C. H. Townes, *Phys. Rev. Lett.* **16**, 347 (1966).

[75] D. Grishkowsky, *Phys. Rev. Lett.* **24**, 866 (1970).

[76] J.E. Bjorkholm and A. Ashkin, *Phys. Rev. Lett.* **32**, 129(1974).

[77] A.J. Campillo, S. L. Shapiro, and B. R. Suydam, *Appl. Phys. Lett.* **23**, 628 (1973).

[78] A.J. Campillo, S. L. Shapiro, and B.R. Suydam, *Appl. Phys. Lett.* **24**, 178 (1974).

[79] A. Ashkin, J. M. Dziedzic, and P. W. Smith, *Opt. Lett.* **7**, 276 (1982).

[80] W. Torruellas, B. Lawrence, and G. I. Stegeman, *Electron. Lett.* **32**, 2092 (1996).

[81] V. Tikhonenko, J. Christou, and B. Luther-Davies, *J. Opt. Soc. Am. B* **12**, 2046 (1995); *Phys. Rev. Lett.* **76**, 2698 (1996).

[82] D. V. Petrov, L. Torner, J. Martorell, R. Vilaseca, J.P. Torres, and C. Cojocaru, *Opt. Lett.* **23**, 1444 (1998).

[83] J. W. Grantham, H. M. Gibbs, G. Khitrova, J. F. Valley, and Xu Jiajin, *Phys. Rev. Lett.* **66**, 1422 (1991).

[84] D. Strickland and P.B. Corkum, *J. Opt. Soc. Am. B* **11**, 492 (1994).

[85] J. K. Ranka, R. W. Schirmer, and A. L. Gaeta, *Phys. Rev. Lett.* **77**, 3783 (1996).

[86] R. R. Alfano and S. L. Shapiro, *Phys. Rev. Lett.* **24**, 584, 592 (1970).

[87] S. A. Diddams, H. K. Eaton, A. A. Zozulya, and T. S. Clement, *Opt. Lett.* **23**, 379 (1998).

[88] A. A. Zozulya, S. A. Diddams, and T. S. Clement, *Phys. Rev. A* **58**, 3303 (1998).

[89] G. C. Duree, Jr., J. L. Shultz, G. J. Salamo, M. Segev, A. Yariv, B. Crosignani, P. Di Porto, E. J. Sharp, and R. R. Neurgaoukar, *Phys. Rev. Lett.* **71**, 533 (1993); *Opt. Lett.* **19**, 1195 (1994).

[90] M. D. Iturbe-Castillo, P. A. Marquez Aquilar, J. J. Sanchez-Mondragon, S. Stepanov, and V. Vysloukh, *Appl. Phys. Lett.* **64**, 408 (1994).

[91] M. F. Shih, M. Segev, G. C. Valley, G. Salamo, B. Crosignani, and P. Di Porto, *Electron. Lett.* **31**, 826 (1995).

[92] M. F. Shih, P. Leach, M. Segev, M. H. Garrett, G. Salamo, and G. C. Valley, *Opt. Lett.* **21**, 324 (1996).

[93] B. Crosignani, A. Degasperis, E. DelRe, P. Di Porto, and A. J. Agranat, *Phys. Rev. Lett.* **82**, 1664 (1999).

[94] W. L. She, K. K. Lee, and W. K. Lee, *Phys. Rev. Lett.* **85**, 2498 (2000).

[95] E. DelRe, A. Ciattoni, and A. J. Agranat, *Opt. Lett.* **26**, 908 (2001).

[96] W. L. She, C. W. Chan, and W.K. Lee, *Opt. Lett.* **26**, 1093 (2001).

[97] J. Petter, C. Denz, A. Stepken, and F. Kaiser, *J. Opt. Soc. Am. B* **19**, 1145 (2002).

[98] A. V. Mamaev, M. Saffman, D. Z. Anderson, and A. A. Zozulya, *Phys. Rev. A* **54**, 870 (1996).

[99] A. V. Mamaev, M. Saffman, and A. A. Zozulya, *Europhys. Lett.* **35**, 25 (1996).

[100] E. Infeld and T. Lenkowska-Czerwińska, *Phys. Rev. E* **55**, 6101 (1997).

[101] G. I. Stegeman and M. Segev, *Science* **286**, 1518 (1999), and references therein.

[102] J. K. Drohm, L. P. Kok, Yu. A. Simonov, J. A. Tjon, and A.I. Veselov, *Phys. Lett. A* **101**, 204 (1981).

[103] L. Bergé, M. R. Schmidt, J. J. Rasmussen, P. L. Christiansen, and K. Ø. Rasmussen, *J. Opt. Soc. Am. B* **14**, 2550 (1997).

[104] L. Poladian, A.W. Snyder, and D. J. Mitchell, *Opt. Commun.* **85**, 59 (1991).

[105] D. E. Edmundson and R. H. Enns, *Opt. Lett.* **18**, 1609 (1993).

[106] J. Schjødt-Eriksen, M. R. Schmidt, J. J. Rasmussen, P. L. Christiansen, Yu. B. Gaididei, and L. Bergé, *Phys. Lett. A* **246**, 423 (1998).

[107] B. A. Malomed, *Phys. Rev. E* **58**, 7928 (1998).

[108] A. S. Desyatnikov and A. I. Maimistov, *Quantum Electron.* **30**, 1009 (2000).

[109] D. J. Mitchell, A. W. Snyder, and L. Poladian, *Electron. Lett.* **27**, 848 (1991).

[110] M. Shih, M. Segev, and G. Salamo, *Phys. Rev. Lett.* **78**, 2551 (1997).

[111] A. V. Buryak, Yu. S. Kivshar, M. Shih, and M. Segev, *Phys. Rev. Lett.* **82**, 81 (1999).

[112] M. Soljaĉić, S. Sears, M. Segev, *Phys. Rev. Lett.* **81**, 4851 (1998).

[113] M. Soljaĉić and M. Segev, *Phys. Rev. E* **62**, 2810 (2000).

[114] M. Soljaĉić and M. Segev, *Phys. Rev. Lett.* **86**, 420 (2001).

[115] A. Barthelemy, C. Froehly, and M. Shalaby, *Proc. SPIE* **2041**, 104 (1994).

[116] E. A. Ostrovskaya, Yu. S. Kivshar, Z. Chen, and M. Segev, *Opt. Lett.* **24**, 327(1999).

[117] E. A. Ostrovskaya, Yu. S Kivshar, D. V. Skryabin, and W. Firth, *Phys. Rev. Lett.* **83**, 296 (1999).

[118] V. I. Kruglov and R. A. Vlasov, *Phys. Rev. A* **111**, 401(1985).

[119] Yu. S. Kivshar and E. A. Ostrovskaya, *Opt. Photon. News* **12**(4) 27 (2001).

[120] V. I. Kruglov, Yu. A. Logvin, and V. M. Volkov, *J. Mod. Opt.* **39**, 2277 (1992).

[121] J. Atai, Y. Chen, and J. M. Soto-Crespo, *Phys. Rev. A* **49**, R3170 (1994).

[122] V. V. Afanasjev, *Phys. Rev. E* **52**, 3153 (1995).

[123] W. J. Firth and D. V. Skryabin, *Phys. Rev. Lett.* **79**, 2450 (1997).

[124] D. V. Skryabin and W.J. Firth, *Phys. Rev. E* **58**, 3916 (1998).

[125] L. Torner and D. V. Petrov, *Electron. Lett.* **33**, 608 (1997).

[126] L. Torner and D. V. Petrov, *J. Opt. Soc. Am. B* **14**, 2017 (1997).

[127] M. Quiroga-Teixeiro and H. Michinel, *J. Opt. Soc. Am. B* **14**, 2004 (1997).

[128] I. Towers, A. V. Buryak, R. A. Sammut, and B. A. Malomed, *Phys. Rev. E* **63**, 055601(R) (2001); *Phys. Lett. A* **288**, 292 (2001).

[129] B. A. Malomed, L.-C. Crasovan, and D. Michalache, *Physica D* **161**, 187 (2002).

[130] H. Michinel, J. Campo-Táboas, R. García-Fernández, J. R. Salqueiro, and M. L. Quiroga-Teixeiro, *Phys. Rev. E* **65** 066604 (2002); *J. Opt. B* **3**, 314 (2001).

[131] A. S. Desyatnikov and Yu. S. Kivshar, *Phys. Rev. Lett.* **87**, 033901 (2001).

[132] T. Carmon, R. Uzdin, C. Pigier, Z. H. Musslimani, M. Segev, and A. Nepomnyashchy, *Phys. Rev. Lett.* **87**, 143901 (2001).

[133] A. S. Desyatnikov and Yu. S. Kivshar, *J. Opt. B* **4**, S58 (2002).

[134] P. S. Lomdahl, O. H. Olsen, and P. L. Christiansen, pla78, 125 (1980).

[135] A. Bathelemy, C. Froehly, M. Shalaby, P. Donnat, J. Paye, and A. Migus, in *Ultrafast Phenomena VIII*, J. J. Martin et al., Eds. (Springer, Berlin, 1993), pp. 299–305.

Chapter 7

Spatiotemporal Solitons

The spatial and temporal solitons discussed in the preceding chapters are the special cases of a much larger class of nonlinear phenomena in which the spatial and temporal effects are coupled and occur simultaneously. When a pulsed optical beam propagates through a bulk nonlinear medium, it is affected by diffraction and dispersion simultaneously, but at the same time the two effects become coupled through the medium's nonlinearity. Such a space–time coupling leads to a plethora of novel nonlinear effects, including the possibility of spatiotemporal collapse or pulse splitting and the formation of light bullets. This chapter is devoted to the study of such effects. Section 7.1 introduces the basic concepts through the phenomena of spatiotemporal modulation instability. The $(3 + 1)$-dimensional nonlinear Schrödinger (NLS) equation is solved in Section 7.2, with emphasis on the phenomenon of beam collapse and the formation of light bullets. Section 7.3 focuses on the pulse splitting occurring when an ultrashort pulse propagates in a normally dispersive Kerr medium. Section 7.4 is devoted to mechanisms that can be used to arrest the pulse collapse in the case of anomalous dispersion. The effects of higher-order effects are discussed in Section 7.5. Section 7.6 summarizes some of the major experimental results.

7.1 Spatiotemporal Modulation Instability

As discussed in Section 1.5.1, modulation instability can be purely spatial (leading to beam filamentation) or purely temporal (leading to pulse-train formation), depending on whether diffraction or dispersion is responsible for it. When both of them are present simultaneously, the same instability acquires a spatiotemporal character such that a continuous-wave (CW) beam becomes modulated both spatially and temporally. Such instabilities have been studied in several different fields, including fluid dynamics, plasma physics, and optics [1]–[9].

The starting point is the $(3 + 1)$-dimensional NLS equation (1.4.1), capable of accounting for the diffractive and dispersive effects occurring simultaneously within the nonlinear medium. For the discussion of modulation instability, it is useful to write this

212

equation in its unnormalized form as

$$i\frac{\partial A}{\partial Z} + \frac{1}{2\beta_0}\left(\frac{\partial^2 A}{\partial X^2} + \frac{\partial^2 A}{\partial Y^2}\right) - \frac{\beta_2}{2}\frac{\partial^2 A}{\partial T^2} + \gamma|A|^2 A = 0. \tag{7.1.1}$$

This equation is identical to Eq. (1.2.12) except that a time-dependent term has been added using Eq. (1.3.7) and introducing $T = t - \beta_1 z$ as the reduced time. All the parameters appearing in this equation were introduced in Chapter 1. Physically, β_0 is the propagation constant, β_2 is the group-velocity dispersion (GVD) parameter, and the nonlinear parameter $\gamma = (2\pi/\lambda)n_2$ is responsible for self-phase modulation (SPM). The last two parameters, β_2 and γ, can be positive or negative, depending on the nature of the nonlinear medium.

In the case of modulation instability, a CW plane wave is launched at the input end of the nonlinear medium, located at $Z = 0$. It is easy to see that Eq. (7.1.1) permits a plane-wave solution of the form

$$\bar{A} = \sqrt{I_0}\exp(i\phi_{NL}), \tag{7.1.2}$$

where I_0 is the incident intensity and $\phi_{NL} = \gamma I_0 Z$ is the nonlinear phase shift induced by SPM inside the nonlinear medium. This solution shows that, in principle, the CW plane wave should remain unchanged except for acquiring a nonlinear phase shift.

The important question is whether the CW plane-wave solution in Eq. (7.1.2) is stable against small perturbations. To answer this question, we should perform a linear stability analysis, perturb the steady state slightly such that

$$A = (\sqrt{I_0} + a)\exp(i\phi_{NL}), \tag{7.1.3}$$

and study the evolution of the weak perturbation $a(X,Y,Z,T)$ inside the nonlinear medium. Substituting Eq. (7.1.3) in Eq. (7.1.1) and linearizing in a, the evolution of a is found to be governed by

$$i\frac{\partial a}{\partial Z} + \frac{1}{2\beta_0}\left(\frac{\partial^2 a}{\partial X^2} + \frac{\partial^2 a}{\partial Y^2}\right) - \frac{\beta_2}{2}\frac{\partial^2 a}{\partial T^2} + \gamma I_0(a + a^*) = 0. \tag{7.1.4}$$

The linear equation (7.1.4) can be solved easily using the Fourier-transform method. Its general solution is of the form

$$a(\mathbf{r},T) = a_1\exp[i(\mathbf{K}\cdot\mathbf{r} - \Omega T)] + a_2\exp[-i(\mathbf{K}\cdot\mathbf{r} - \Omega T)], \tag{7.1.5}$$

where $\mathbf{r} = (X,Y,Z)$ and $\mathbf{K}$ and Ω are the wave vector and frequency of the plane wave, respectively, associated with a specific Fourier component of the perturbation. Equations (7.1.4) and (7.1.5) provide a set of two homogeneous equations for a_1 and a_2. This set has a nontrivial solution only when $\mathbf{K}$ and Ω satisfy the following dispersion relation:

$$4\beta_0^2 K_z^2 = (K_x^2 + K_y^2 - \beta_0\beta_2\Omega^2)(K_x^2 + K_y^2 - \beta_0\beta_2\Omega^2 - 4\gamma P_0). \tag{7.1.6}$$

This dispersion relation reduces to that obtained in Section 1.5.1 in the limit $K_x = K_y = 0$ (no diffraction of perturbation because of waveguiding) or the limit $K_y = \Omega = 0$ (no dispersion and diffraction in only one transverse dimension).

The dispersion relation (7.1.6) shows that the stability of the CW solution (7.1.2) depends critically not only on the spatial and temporal frequencies of the perturbation propagates but also on the signs of the parameters, β_2 (normal versus anomalous GVD) and γ (self-focusing versus self-defocusing). The CW solution (7.1.2) becomes unstable whenever K_z in Eq. (7.1.6) has an imaginary part, since the perturbation then grows exponentially with the intensity gain $g = 2\,\text{Im}(K_z)$. In general, g is a function of K_x, K_y, and Ω. However, g depends on K_x and K_y only through $K_s = (K_x^2 + K_y^2)^{1/2}$. It is useful to introduce a spatial frequency Ω_s as

$$\Omega_s = \left(\frac{K_x^2 + K_y^2}{\beta_0\beta_2}\right)^{1/2}. \tag{7.1.7}$$

In terms of the frequencies Ω and Ω_s, the gain associated with the modulation instability can be written as

$$g(\Omega, \Omega_s) = |\beta_2|(\Omega_s^2 - s_d\Omega^2)^{1/2}(s_n\Omega_c^2 + s_d\Omega^2 - \Omega_s^2)^{1/2}, \tag{7.1.8}$$

where $s_d = \text{sgn}(\beta_2) = \pm 1$ depending on the sign of β_2, $s_n = \text{sgn}(\gamma) = \pm 1$ depending on the sign of γ (or n_2), and Ω_c^2 is defined as

$$\Omega_c^2 = \frac{4\gamma I_0}{|\beta_2|} = \frac{4}{|\beta_2|L_{\text{NL}}}. \tag{7.1.9}$$

The nonlinear length L_{NL} is introduced using $L_{\text{NL}} = (\gamma I_0)^{-1}$; it becomes shorter as the intensity I_0 increases. The solution (7.1.2) becomes unstable if $g(\Omega, \Omega_s)$ is real and positive for some values of Ω and Ω_s.

Equation (7.1.8) shows that depending on the nature of GVD and nonlinearity (signs of s_d and s_n), the CW plane-wave solution of the NLS equation (7.1.1) can become unstable. Such an instability is referred to as the *spatiotemporal instability* since it leads to simultaneous spatial and temporal modulations of the incident CW beam. It is easy to see that no instability occurs when both β_2 and n_2 are negative ($s_d = s_n = -1$). For the three remaining choices of s_d and s_n, there exists some range of Ω and Ω_s over which g is real and positive. Figure 7.1 shows the instability domains in the Ω–Ω_s plane for the three cases. In each case the dashed line displays the contour of maximum gain and the solid line shows the contour of zero gain.

In the absence of diffractive effects ($\Omega_s = 0$), the spatiotemporal instability reduces to the temporal modulation instability that has been observed in optical fibers [10] and occurs only for anomalous GVD ($s_d = -1$) and positive n_2 ($s_n = 1$). In a bulk Kerr medium, the presence of diffraction can lead to spatial oscillations that accompany temporal oscillations. The most interesting feature is that the spatiotemporal instability does not necessarily require anomalous GVD and can occur even in a normally dispersive medium. In a self-focusing medium ($s_n = 1$) with normal GVD ($s_d = 1$), the instability gain occurs for all frequencies such that

$$\Omega_s^2 - \Omega_c^2 \le \Omega^2 \le \Omega_s^2. \tag{7.1.10}$$

A similar situation occurs for a self-defocusing medium ($s_n = -1$) with normal GVD ($s_d = 1$) except that Ω is now bounded in the range

$$\Omega_s^2 \le \Omega^2 \le \Omega_s^2 + \Omega_c^2. \tag{7.1.11}$$

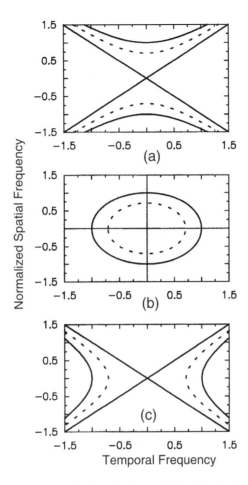

Figure 7.1: Instability domains (region bounded by solid lines) in the Ω–Ω_s plane for (a) $s_d = 1$, $s_n = 1$, (b) $s_d = -1$, $s_n = 1$, and (c) $s_d = 1$, $s_n = -1$. Dashed lines represent the contours of maximum gain. (After Ref. [7]; ©1992 APS.)

In both cases, the spatiotemporal modulation instability can convert a CW beam into a train of ultrashort pulses while simultaneously breaking it into spatial filaments. The self-defocusing case is most interesting, since the repetition rate of the pulse train can be made quite large by increasing Ω_s and Ω_c.

7.2 Optical Bullets and Their Stability

The presence of spatiotemporal modulation instability suggests that the combination of self-focusing, diffraction, and dispersion may lead to the formation of optical wave packets that remain confined in all three spatial dimensions (a finite pulse width corresponds to a finite pulse length along the propagation direction). Such a confined wave packet is often referred to as a *light bullet*, and it represents an extension of self-trapped optical beams into the temporal domain. The existence and stability of self-trapped beams in a nonlinear medium has been investigated since 1964, when this phenomenon was first discovered [11]–[14].

7.2.1 Shape-Preserving Solutions

We have seen in the preceding chapters that the NLS equation has solutions in the one-dimensional case that preserve their shape during propagation. We begin by asking whether such solutions exist in the multidimensional case. To find shape-preserving solutions of Eq. (7.1.1) it is useful to write it in a normalized form by introducing

$$z = Z/L_d, \quad x = X/w_0, \quad y = Y/w_0, \tag{7.2.1}$$

$$\tau = T/(w_0^2 \beta_0 \beta_2)^{1/2}, \quad u = (\gamma L_d)^{1/2} A, \tag{7.2.2}$$

where w_0 is a measure of the spatial width and $L_d = \beta_0 w_0^2$ is the diffraction length. In terms of the normalized amplitude u, Eq. (7.1.1) can be written as

$$i\frac{\partial u}{\partial z} + \frac{1}{2}\left(\frac{\partial^2 u}{\partial x^2} + \frac{\partial^2 u}{\partial y^2} - s_d\frac{\partial^2 u}{\partial \tau^2}\right) + |u|^2 u = 0, \tag{7.2.3}$$

where $s_d = \text{sgn}(\beta_2)$.

The preceding equation shows the total equivalence among the spatial and temporal coordinates when dispersion is anomalous ($\beta_2 < 0$). We focus on the anomalous-GVD case and exploit this symmetry for solving Eq. (7.2.3) by setting $s_d = -1$. It is useful to introduce a three-dimensional vector $\mathbf{R}$ with components x, y, and τ and to write Eq. (7.2.3) in a compact form as

$$i\frac{\partial u}{\partial z} + \frac{1}{2}\nabla_R^2 u + |u|^2 u = 0, \tag{7.2.4}$$

where ∇_R^2 is the Laplacian operator.

The shape-preserving solutions of Eq. (7.2.4) can be found by looking for a solution with the property

$$u(x, y, \tau, z) = U(x, y, \tau)\exp(i\kappa_0 z), \tag{7.2.5}$$

where κ_0 is the propagation constant. Since U does not depend on z, such a pulse would propagate without change in its spatial or temporal shape, resulting in an optical bullet. If we write the Laplacian in Eq. (7.2.4) in spherical coordinates and focus on the radially symmetric solutions, $U(x, y, \tau)$ depends only on $R \equiv (x^2 + y^2 + \tau^2)^{1/2}$ and satisfies an ordinary differential equation,

$$\frac{1}{2}\left[\frac{d^2 U}{dR^2} + \frac{(D-1)}{R}\frac{dU}{dR}\right] - \kappa_0 U + U^3 = 0. \tag{7.2.6}$$

This equation should be solved with the boundary condition $U(\infty) = 0$. The parameter D takes values 1, 2, or 3 depending on the dimensionality of the vector $\mathbf{R}$. The one-dimensional case ($D = 1$) corresponds to the purely spatial or temporal solitons discussed in earlier chapters. The two-dimensional case ($D = 2$) applies to self-focusing of CW beams and was discussed in Chapter 6. It also applies to pulse propagation inside planar waveguides. The three-dimensional case corresponds to short optical pulses propagating inside a bulk nonlinear medium. As discussed in Section 1.2.5, an analytic solution of Eq. (7.2.6) in the form sech(R) can easily be found for $D = 1$ and corresponds to either a spatial soliton ($R = x$) or a temporal soliton ($R = \tau$). For $D > 1$, one can solve Eq. (7.2.6) numerically. Such numerical solutions were obtained for the $D = 2$ case during the 1960s in the context of self-focusing of CW beams [11]. In fact, Eq. (7.2.6) was found to have an infinite number of solutions that are analogous to the fundamental and higher-order solitary-wave solutions found in the $D = 1$ case [12]. The mth-order solution in the $D = 2$ case displays a central spot surrounded by m rings.

Equation (7.2.6) has been solved numerically in the $D = 3$ case. The lowest-order solution is of primary interest because it corresponds to an optical bullet. Figure 7.2 shows this solution by plotting $U(R)/U(0)$ versus R and compares it with the corresponding solutions obtained for $D = 1$ and 2. The propagation constant κ_0 is different in each case and has a value of 0.5 for $D = 1$, 0.2055 for $D = 2$, and 0.05316 for $D = 3$ [15]. The peak amplitude U_0 is also different in the three cases. It is useful to introduce a parameter representing the pulse energy and defined as

$$E = \int |u|^2 \, d^D R, \qquad (7.2.7)$$

where the single integral sign represents integration over the entire region in D dimensions. For the fundamental solutions shown in Figure 7.2, $E = 2$ for $D = 1$, $E = 7.850$ for $D = 2$, and $E = 28.87$ for $D = 3$. Note that in the purely spatial case of a CW beam, E has the physical meaning of being the total beam power.

7.2.2 Spatiotemporal Collapse

The stability of any shape-preserving solution found by solving Eq. (7.2.4) should be examined by performing a linear stability analysis similar to that used for CW beams in Section 7.1. Such an analysis shows [16] that the shape-preserving solution is stable only in the $D = 1$ case. When $D > 1$, small fluctuations in the intensity, beam size, or pulse width can grow with propagation and lead to a phenomenon known as *spatiotemporal collapse* (or *beam collapse* in the context of a CW beam). A consequence of this instability is that, if pulse energy exceeds a critical value E_c, the pulse collapses in such a way that the intensity $|u|^2$ becomes infinitely large at a finite distance as the the size of the beam diminishes and shrinks to zero both spatially and temporally.

The collapse of CW beams has been studied extensively in the context of self-focusing [17]–[25]. The techniques used there can also be applied for studying the spatiotemporal collapse [26]. A variational technique based on the Lagrangian formalism of classical mechanics is quite useful in this context [27]. Here we follow the moment method based on the virial theorem [13]; it was used in 1994 to study the effect of spatial and temporal chirp on the threshold of spatiotemporal collapse [28].

As seen in Chapters 2 and 4, the conservation laws associated with the $(1 + 1)$-dimensional NLS equation are quite useful for analyzing soliton properties. The multidimensional NLS equation (7.1.1) also has several conserved quantities associated

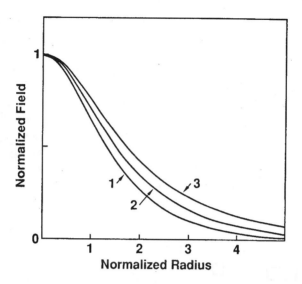

Figure 7.2: Shape-preserving solutions of the NLS equation for $D = 1$–3. The field is normalized to 1 at $R = 0$. (After Ref. [15]; ©1990 OSA.)

with it [14, 21]. The pulse energy E, defined in Eq. (7.2.7), is the three-dimensional analog of the power P introduced in Eq. (2.1.3) as a conserved quantity. The invariance of the NLS equation with respect to space and time translations leads to another two conserved quantities:

$$\mathbf{M} = \frac{1}{2i} \int \left(u^* \nabla_R u - u \nabla_R u^* \right) d^D R, \tag{7.2.8}$$

$$H = \frac{1}{2} \int \left(|\nabla_R u|^2 - |u|^4 \right) d^D R. \tag{7.2.9}$$

Physically, M and H are analogs of the linear momentum and the Hamiltonian, respectively, introduced first in Section 2.1.1.

The moment method uses the three conserved quantities, E, M, and H, to find the first and second moments of $R = (x^2 + y^2 + \tau^2)^{1/2}$, defined using

$$\langle R^n \rangle = \frac{\int R^n |u|^2 d^D R}{\int |u|^2 d^D R}, \tag{7.2.10}$$

where n is an integer. Spatiotemporal collapse is studied using the variance

$$\sigma^2 = \langle R^2 \rangle - \langle R \rangle^2, \tag{7.2.11}$$

where σ is a measure of the spatial and temporal spreads associated with an optical pulse. If σ approaches zero, the beam size diminishes in all three dimensions (x, y, and τ) simultaneously, and the beam undergoes spatiotemporal collapse.

Using Eqs. (7.2.8)–(7.2.11), σ^2 is found to satisfy a second-order ordinary differential equation [28]

$$\frac{d^2\sigma^2}{dz^2} = \frac{1}{E}\left(4H - \frac{2M^2}{E} - \int (D-2)|u|^4 d^D R\right). \tag{7.2.12}$$

If the expression inside the large parentheses is negative, σ^2 decreases monotonically along the medium length, becoming zero at a finite distance associated with the spatiotemporal collapse. Equation (7.2.12) cannot be solved in general because of the last term. The two-dimensional case ($D = 2$) is special, since this term vanishes in that case. We consider the $D = 2$ case first.

7.2.3   Pulse Propagation in Planar Waveguides

Consider the case of an ultrashort pulse propagating inside a planar waveguide such that it diffracts along the x axis but is confined in the y direction. Even in that case, the spatiotemporal collapse depends on input parameters, such as beam and pulse widths and pulse energy. It also depends on whether the input pulse is chirped and whether the phase front is curved. To be specific, we assume that the input field can be written as

$$u(x,\tau,0) = u_0 \exp\left[-(1+iC_s)\frac{x^2}{2W_s^2} - (1+iC_t)\frac{\tau^2}{2W_t^2}\right], \tag{7.2.13}$$

where W_s and W_t are the normalized widths and C_s and C_t are the chirp parameters in the spatial and temporal dimensions, respectively. This field distribution corresponds to a chirped Gaussian pulse in the form of a Gaussian optical beam with a curved wave front. Positive and negative values of C_s correspond to a converging and a diverging wave front, respectively.

The integrations in Eqs. (7.2.7)–(7.2.9) can easily be performed using Eq. (7.2.13), resulting in

$$E = \pi W_s W_t |u_0|^2, \qquad M = 0, \tag{7.2.14}$$

$$H = \frac{E}{4}\left[\frac{1+C_s^2}{W_s^2} + \frac{1+C_t^2}{W_t^2} - \frac{E}{\pi W_t W_s}\right]. \tag{7.2.15}$$

Equation (7.2.12) can be integrated when $D = 2$ to find how σ^2 changes with propagation inside the nonlinear medium. The result is given by

$$\sigma^2(z) = \sigma_0^2 - (C_s + C_t)z + (2H/E)z^2, \tag{7.2.16}$$

where σ_0 is the value of σ at $z = 0$ and is related to W_s and W_t as $\sigma_0^2 = \frac{1}{2}(W_s^2 + W_t^2)$.

Consider first the case of an unchirped pulse such that $C_s = C_t = 0$. From Eq. (7.2.16), spatiotemporal collapse can occur only if $H < 0$. From Eq. (7.2.15), this is possible only when the pulse energy exceeds a critical value such that

$$E > E_c = \pi(W_s^2 + W_t^2)/(W_s W_t). \tag{7.2.17}$$

The distance at which collapse occurs corresponds to the value of z for which $\sigma = 0$ in Eq. (7.2.16) and is given by $z_c = (-E/2H)^{1/2}\sigma_0$. The critical energy E_c and the collapse distance z_c are two important parameters that characterize spatiotemporal collapse. In the special case $W_s = W_t = 1$, these parameters are given by

$$E_c = 2\pi, \qquad z_c = (E/E_c - 1)^{-1/2}. \tag{7.2.18}$$

This case corresponds to the situation in which the input pulse width T_0 and the beam width w_0 are chosen such that the dispersion length, defined as $L_D = T_0^2/|\beta_2|$, equals the diffraction length, $L_d = \beta_0 w_0^2$.

The spatial and temporal chirping parameters affect considerably both E_c and z_c. In fact, the critical energy depends on whether the total chirp $C = C_s + C_t$ is positive or negative. One can show using Eqs. (7.2.14)–(7.2.16) that E_c and z_c are given by

$$E_c = \pi \left[(1 + C_s^2)\frac{W_t}{W_s} + (1 + C_t^2)\frac{W_s}{W_t} - \frac{\theta(C)C^2 W_s W_t}{(W_s^2 + W_t^2)} \right], \tag{7.2.19}$$

$$z_c = \frac{\pi W_s W_t}{E - E'} \left[\left(C^2 + (E - E')\frac{W_s^2 + W_t^2}{\pi W_s W_t} \right)^{1/2} - C \right], \tag{7.2.20}$$

where $\theta(C)$ is the step function defined such that $\theta = 1$ if $C > 0$ but zero otherwise. The quantity E' is defined as

$$E' = \pi \left[(1 + C_s^2)\frac{W_t}{W_s} + (1 + C_t^2)\frac{W_s}{W_t} \right] = E_c + \theta(C)\frac{\pi W_s W_t C^2}{2(W_s^2 + W_t^2)}. \tag{7.2.21}$$

Thus, $E' = E_c$ only when $C = C_s + C_t < 0$.

As an example, consider again the special case $W_s = W_t = 1$. From Eq. (7.2.19), the critical energy is given by

$$\frac{E_c}{E_0} = \begin{cases} 1 + (C_s - C_t)^2/4 & \text{if } C > 0, \\ 1 + (C_s^2 + C_t^2)/2 & \text{if } C < 0, \end{cases} \tag{7.2.22}$$

where $E_0 = 2\pi$ is the critical energy needed for an unchirped pulse. Clearly, chirping—whether spatial or temporal—helps to raise the threshold of spatiotemporal collapse. Effects of spatial chirp are well known in the context of self-focusing of CW beams. Effects of temporal chirp on self-focusing are evident from Eq. (7.2.22). In fact, when C_s and C_t are equal, the collapse threshold does not change when they are positive but increases by a factor of $1 + C_t^2$ when they are negative. This is understandable if we note that temporal chirp induced by SPM and spatial chirp induced by self-focusing are both positive.

The general case in which $W_s \neq W_t$ is shown in Figure 7.3, where the energy enhancement factor E_c/E_0 is plotted as a function of W_t for several values of C_t, assuming $W_s = 1$ and $C_s = 0$. These values are relevant for a chirped Gaussian pulse of arbitrary width in the form of a Gaussian-shaped beam incident at the planar waveguide with a plane wave front. It is evident that chirping of the pulse increases the collapse threshold considerably.

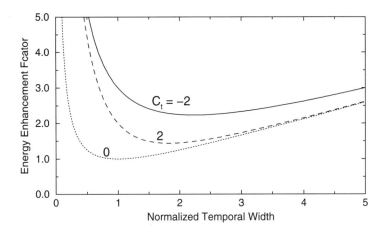

Figure 7.3: Energy enhancement factor for chirped Gaussian pulses. E_c/E_0 is plotted as a function of W_t for several values of C_t.

7.2.4 Pulse Propagation in Bulk Kerr Media

In the case of a bulk Kerr medium, we should include diffraction in both transverse dimensions. Assuming a Gaussian spatial profile but allowing for different beam sizes in the x and y directions (elliptical spot size), Eq. (7.2.13) for the input field takes the form

$$u(x, y, \tau, 0) = u_0 \exp\left[-\frac{(1 + iC_x)}{2W_x^2}x^2 - \frac{(1 + iC_y)}{2W_y^2}y^2 - \frac{(1 + iC_t)}{2W_t^2}\tau^2 \right], \qquad (7.2.23)$$

where W_x and W_y are the spatial widths and C_x and C_y are the initial chirp parameters at the input plane $z = 0$. The input pulse diffracts in x and y spatial dimensions, while it disperses in the temporal dimension τ. The three conserved quantities can be calculated, as before, using Eq. (7.2.23) in Eqs. (7.2.7)–(7.2.9) and are given by

$$E = \pi^{3/2} W_x W_y W_t |u_0|^2, \qquad M = 0 \qquad (7.2.24)$$

$$H = \frac{E}{8}\left[\sum_i \frac{1 + C_i^2}{W_i^2} - \frac{E}{\pi^{3/2} W_x W_y W_t} \right], \qquad (7.2.25)$$

where $i = x$, y, and t in the $D = 3$ case.

Spatiotemporal collapse of the pulse is governed by Eq. (7.2.12) with the choice of $D = 3$. However, this equation cannot be solved analytically because the last term involves a volume integral and contains the field $u(x, y, \tau, z)$, which is not known without solving the NLS equation itself. However, an upper bound on the magnitude of this term can easily be obtained if we assume that the pulse is localized in both space and time [28]. Indeed, it follows from the Cauchy–Schwartz inequality that

$$\int |u|^4 d^D R < E^2/V, \qquad (7.2.26)$$

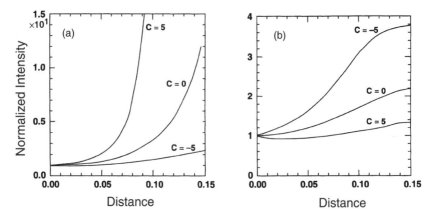

Figure 7.4: Increase in the peak intensity during propagation of chirped Gaussian pulses in a Kerr medium with (a) anomalous and (b) normal GVD. The quantity $|u(z, R = 0)|^2/u_0^2$ is calculated numerically and plotted as a function of z for three values of C using $u_0 = 5$ and $C_x = C_y = 0$. (After Ref. [28]; ©1994 APS.)

where V is a constant representing the volume in which most of the pulse energy is initially located. This inequality allows us to find an upper bound on the critical value of the pulse energy using the method outlined earlier. The result is given by [28]

$$E_c = \frac{\pi\sqrt{2\pi}VW_xW_yW_t}{V + \pi\sqrt{2\pi}W_xW_yW_t} \sum_i \frac{1+C_i^2}{W_i^2} - \frac{C_t(C)(C_x+C_y+C_t)^2}{8(W_x^2+W_y^2+W_t^2)}, \qquad (7.2.27)$$

where $C = C_x + C_y + C_t$ is the total chirp.

Similar to the $D = 2$ case discussed earlier, the net magnitude of total chirp affects the process of spatiotemporal collapse considerably. As a simple example, consider the case of an unchirped pulse whose spot size w_0 is the same in the two transverse dimensions (circular spot) and whose pulse width T_0 is chosen such that $L_d = L_D$. For such a beam, $W_x = W_y = W_t = 1$ and $C_x = C_y = C_t = 0$. The threshold energy for spatiotemporal collapse to occur is given by

$$E_c = \frac{3\pi\sqrt{2\pi}V}{V + \pi\sqrt{2\pi}}. \qquad (7.2.28)$$

Typically $V \gg \pi\sqrt{2\pi}$, and $E_c \approx 3\pi\sqrt{2\pi} \approx 23.6$ provides an upper bound on the value of E_c. This should be compared with the value $E_c = 2\pi$ found in the $D = 2$ case. The effect of spatial or temporal chirp is to increase the value of E_c, similar to the $D = 2$ case. The enhancement factor $E_c/E_0 = 1 + (C_x^2 + C_y^2 + C_t^2)/3$ can be quite large when the total chirp $C = C_x + C_y + C_t$ is negative in magnitude.

Numerical solutions of the multidimensional NLS equation, based on the split-step Fourier method [29], confirm that temporal chirp plays an important role in the process of spatiotemporal collapse. To reduce the computation time, it is common to assume that the beam maintains cylindrical symmetry. A fast Hankel transform can then be used in the spatial dimensions [30]–[32]. Figure 7.4 shows how the peak intensity increases along the medium length for a chirped Gaussian pulse using the initial field given in Eq. (7.2.23) with $u_0 = 5$, $W_x = W_y = W_t = 1$, and $C_x = C_y = 0$. The temporal chirp C_t is varied in the range -5–5. An unchirped pulse ($C_t = 0$) with such parameters collapses after a distance $z \approx 0.15$. However, the same pulse collapses sooner for $C_t = 5$ and much later for $C_t = -5$. The case of normal dispersion ($\beta_2 > 0$), also shown in Figure 7.4, indicates that collapse does not occur for any value of C_t. Since this behavior is quite different qualitatively, we discuss pulse propagation in a normally dispersive nonlinear medium in the next section.

7.3 Normal-Dispersion Regime

The spatiotemporal collapse occurring in the case of anomalous GVD is replaced by the phenomenon of pulse splitting if the GVD associated with a nonlinear medium is normal ($\beta_2 > 0$). The normal-GVD case has been studied extensively and found to have several interesting features [33]–[43]; we focus on them in this section, including the localized waves in the form of X waves.

7.3.1 Shape-Preserving Solutions

Similar to the anomalous-GVD case, consider first the shape-preserving solutions of the multidimensional NLS equation (7.2.3). In the normal-GVD case ($s_d = 1$), this equation becomes

$$i\frac{\partial u}{\partial z} + \frac{1}{2}\left(\frac{\partial^2 u}{\partial x^2} + \frac{\partial^2 u}{\partial y^2} - \frac{\partial^2 u}{\partial \tau^2}\right) + |u|^2 u = 0. \tag{7.3.1}$$

Exact analytic solutions of this equation do not appear to exist. However, an approximate solitary-wave solution was found in 1993 using a factorization approach [35].

Consider the case in which a short pulse propagates inside a planar waveguide by setting the y-derivative term to zero, and assume the following form of the shape-preserving solution:

$$u(z, x, \tau) = f(x)g(\tau)\exp(iK_0 z), \tag{7.3.2}$$

where the form of the spatial and temporal functions is yet to be determined. Substituting Eq. (7.3.2) in Eq. (7.3.1), we obtain

$$\frac{1}{2}\left(g\frac{d^2 f}{dx^2} - f\frac{d^2 g}{d\tau^2}\right) + (|f|^2|g|^2 - K_0)fg = 0. \tag{7.3.3}$$

Multiplying Eq. (7.3.3) by g^* and integrating over the time domain, $f(x)$ is found to satisfy

$$\frac{1}{2}\frac{d^2 f}{dx^2} + (a_f|f|^2 + \tfrac{1}{2}b_f - K_0)f = 0, \tag{7.3.4}$$

where the constants a_f and b_f are defined as

$$a_f = \frac{\int_{-\infty}^{\infty} |g|^4\, d\tau}{\int_{-\infty}^{\infty} |g|^2\, d\tau}, \qquad b_f = \frac{\int_{-\infty}^{\infty} |dg/d\tau|^2\, d\tau}{\int_{-\infty}^{\infty} |g|^2\, d\tau}. \tag{7.3.5}$$

Similarly, multiplying Eq. (7.3.3) by f^* and integrating over the spatial dimension, $g(\tau)$ is found to satisfy

$$-\frac{1}{2}\frac{d^2g}{d\tau^2} + (a_g|f|^2 - \tfrac{1}{2}b_g - K_0)g = 0, \qquad (7.3.6)$$

where the constants a_g and b_g are defined as

$$a_g = \frac{\int_{-\infty}^{\infty}|f|^4\,dx}{\int_{-\infty}^{\infty}|f|^2\,dx}, \qquad b_g = \frac{\int_{-\infty}^{\infty}|df/dx|^2\,dx}{\int_{-\infty}^{\infty}|f|^2\,dx}. \qquad (7.3.7)$$

Equations (7.3.4) and (7.3.6) can easily be integrated in the form of a bright spatial soliton and a dark temporal soliton, respectively, and have the solutions

$$f(x) = \text{sech}(\sqrt{a_f}x), \qquad g(\tau) = \tanh(\sqrt{a_g}\tau), \qquad (7.3.8)$$

provided the propagation constant K_0 satisfies the constraint

$$K_0 = \tfrac{1}{2}(a_f + b_f) = a_g - \tfrac{1}{2}b_g. \qquad (7.3.9)$$

The solution (7.3.8) can be used to evaluate the four parameters, and the result is found to be

$$a_f = 1, \quad b_f = 0, \quad a_g = 2/3, \quad b_g = 1/3. \qquad (7.3.10)$$

Using these values in Eq. (7.3.9), we find a self-consistent solitary-wave solution with $K_0 = 1/2$.

The factorization method can also be used in the three-dimensional case. Assuming a solitary-wave solution of Eq. (7.3.1) in the form

$$u(z,x,y\tau) = f(x)g(y)h(\tau)\exp(iK_0z), \qquad (7.3.11)$$

and following the same approach, we find

$$f(x) = \text{sech}(\sqrt{a}x), \quad g(y) = \text{sech}(\sqrt{a}y), \quad h(\tau) = \tanh(a\tau) \qquad (7.3.12)$$

with $a = 2/3$ and $K_0 = 2/9$. Thus, the solitary-wave solution has the shape of a bright soliton along the two spatial dimensions but that of a dark soliton along the temporal dimension. It should be stressed that this solution is not an exact solution of Eq. (7.3.1). In fact, the error can be estimated by substituting it back into the NLS equation [35]. Stability of this solution is also not guaranteed and should be checked either analytically using a linear stability analysis or numerically. It turns out that this solitary-wave solution is not stable.

7.3.2 Numerical Simulations

The multidimensional NLS equation (7.1.1) has been integrated numerically using the split-step Fourier method or a finite-difference scheme such as the Crank–Nicholson method. The results show that when a pulse propagates in the normal-dispersion regime ($\beta_2 > 0$), it experiences self-focusing in the spatial dimensions and also gets

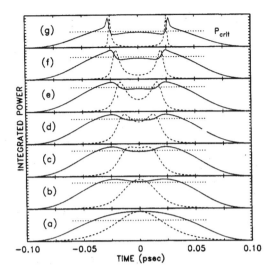

Figure 7.5: Splitting of a 50-fs "sech" pulse with a 0.5-mm Gaussian spot size. Dashed curves show the on-axis intensity, while solid curves show the total integrated power. Propagation distances are (a) 17.53, (b) 18.60, (c) 18.79, (d) 18.86, (e) 18.82, (f) 18.98, and (f) 19.04 cm. The dotted line indicates the critical power level. (After Ref. [34]; ©1992 OSA.)

compressed initially. However, the large increase in the peak power at the pulse center leads to pulse splitting because of chirping and spectral broadening induced by SPM.

Figure 7.5 shows an example of pulse splitting for an input pulse launched into the nonlinear medium with the field amplitude

$$A(x,y,0,T) = \sqrt{I_0}\exp\left[-\left(\frac{x^2+y^2}{2r_0^2}\right)\right]\operatorname{sech}\left(\frac{T}{T_0}\right), \qquad (7.3.13)$$

where T_0 is the temporal width, r_0 is the spatial width, and I_0 is the peak intensity at $z = 0$. Numerical simulations were done for 50-fs pulses (FWHM) with a 0.5-mm spot size (FWHM). The peak input power of the pulse defined as

$$P(z = 0, T) = \iint_{-\infty}^{\infty} |A(x,y,0,T)|^2 \, dx \, dy, \qquad (7.3.14)$$

was three times larger than the 15-MW critical power estimated for the catastrophic collapse of the Gaussian beam. The dashed curves in Figure 7.5 show the on-axis intensity $|A(0,0,z,T)|^2$ while the solid curves show the spatially integrated power $P(z,T)$. Panels (a)–(g) correspond to propagation distances in the range 17.53–19.04 cm. The pulse splitting is clearly seen through the dashed curves, which show the intensity at the beam center. As seen by the dashed curves, at the same time the GVD transfers the pulse energy from the center of the pulse toward its wings, until the peak intensity drops below the critical power level, thereby terminating the collapse of the pulse center [34].

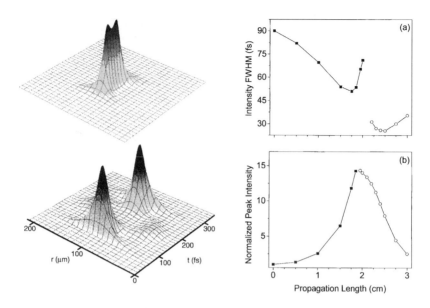

Figure 7.6: Pulse splitting occurring in the normal-GVD regime of fused silica. The spatiotemporal profiles at distances of 2 and 3 cm are shown together with changes in the pulse width and the on-axis intensity occurring with z. The input pulse has a "sech" shape and a Gaussian spatial profile with the peak intensity $I_0 = 87$ GW/cm^2. (After Ref. [43] ©1998 IEEE.)

Figure 7.6 shows another example of pulse splitting for an input pulse with the field amplitude in the form of Eq. (7.3.13). The numerical values used for the pulse parameters correspond to an actual experiment in which mode-locked pulses emitted from a Ti-sapphire laser ($\lambda_0 = 0.8$ μm) propagated inside a bulk sample of fused silica [43]. The spatiotemporal profiles shown in Figure 7.6 were obtained at distances of 2 and 3 cm using $r_0 = 70$ μm, $T_0 = 90$ fs, and $I_0 = 87$ GW/cm^2.

Figure 7.6 also shows how the pulse width and the intensity at beam center change with propagation. Initially, the intensity increases rapidly, because of self-focusing, while the pulse width decreases, features that occur even in the anomalous-GVD region and that lead to spatiotemporal collapse. However, at a distance of about 1.95 cm, the pulse splits into two parts when GVD is normal. After that, the intensity at beam center decreases. The two pulses begin to separate, and their width also increases after acquiring a minimum value near $z = 2.5$ cm. This last feature can be attributed to the higher-order effects, discussed later in Section 7.5.

7.3.3 Nonlinear X Waves

Since the normal-GVD regime does not permit spatiotemporal localization in the form of light bullets, the field evolution is qualitatively different, and it involves complex phenomena, such as temporal splitting and spectral breaking. This is why no attempts have been made to answer the fundamental question as to whether any form of nonlinearity-induced localization could still take place in normally dispersive media. It was only in 2002 that the existence of a novel type of localized wave in the form of *nonlinear X waves*, or X-wave solitons, was noted [44]. The X-shaped waves are well known in the context of the linear propagation of acoustic [45] or electromagnetic [46] waves. They constitute a polychromatic generalization of the diffraction-free Bessel beams [47], which have been observed in both the acoustical [48] and optical [49] experiments using beam-shaping techniques. Conti et al. found through numerical simulations that the the concept of X waves can be extended into the nonlinear regime [44]. A fundamental difference between linear and nonlinear X waves is that the formation of X-shaped nonlinear waves occurs spontaneously at high intensities through self-induced spectral reshaping triggered by a mechanism of conical emission. Thus, nonlinear X waves are expected to have a stronger impact on experiments than linear X waves.

To find nonlinear X waves, we start from Eq. (7.3.1) and look for its soliton-like, radially symmetric solutions in the form $u(x, y, z, \tau) = u_0 U(r, \tau) \exp(-ibz/2)$. The normalized field $U(r, \tau)$ is then found to satisfy the following equation:

$$\frac{\partial^2 U}{\partial r^2} + \frac{1}{r} \frac{\partial U}{\partial r} - \frac{\partial^2 U}{\partial \tau^2} + bU + \gamma U^3 = 0, \tag{7.3.15}$$

where the parameter $\gamma = 2u_0^2$ characterizes the strength of the nonlinearity.

Equation (7.3.15) must be integrated with the boundary conditions $U(r, \tau = \pm\infty) = 0$, $U(r = \infty, \tau) = 0$, and $\partial U(0, \tau)/\partial r = 0$. While the anomalous-GVD regime guarantees that nearly separable light bullets exist with exponentially decaying tails for $b < 0$, the nature of the localized solutions (if any) must change dramatically because the low-intensity tails cannot decay exponentially in the normal-GVD regime. To find the localized solutions in this case, a pseudo-spectral numerical technique was used in Ref. [44] for solving Eq. (7.3.15) as a dynamical evolution problem in r with appropriate discretization in time [50]; this technique is well suited for searching strongly nonseparable objects with slow spatiotemporal decay. The convergence was relatively quick when the following trial function was used during numerical simulations:

$$U(r, \tau) = [(\Delta - i\tau)^2 + r^2]^{-1/2}. \tag{7.3.16}$$

This function represents an X-shaped solution in the *linear* limit ($\gamma = 0$). Here, Δ is a free parameter: The smaller the value of Δ, the stronger the localization.

Two nonlinear X-wave solutions, obtained by solving Eq. (7.3.15) numerically with $\Delta = 1$ in Eq. (7.3.16), are shown in Figure 7.7. For $b = 0$ and moderate nonlinearities ($\gamma = 1$), the "ground-state" mode shown in part (a) has a clear X shape in the x–τ plane (or V shape in the r–τ plane). It represents a tightly confined structure that decays slowly spatially ($\sim 1/r$) and that is accompanied by radially increasing temporal pulse splitting. In the case $b = 1$, while the field maintains its basic X shape, it develops damped oscillations in the spatial dimensions as seen in Figure 7.7(b) for the strongly nonlinear case ($\gamma = 10$).

These results show that a novel type of space–time localization can take place in normally dispersive Kerr media. The resulting spatiotemporal soliton represents a non-diffractive and nondispersive wave packet in the shape of an X, and it decays slowly in the spatial dimensions. Such high-intensity X waves, unlike linear ones, can be formed

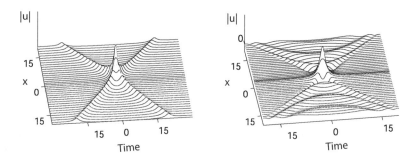

Figure 7.7: Spatiotemporal profile $u(x, y = 0, \tau)$ of X-wave solitons in a normally dispersive Kerr media for (a) $b = 0$, $\gamma = 1$ and (b) $b = 1$, $\gamma = 10$. (After Ref. [44].)

spontaneously through a trigger mechanism of conical emission and are likely to play an important role in experiments.

7.4 Collapse-Arresting Mechanisms

The discussion so far in this chapter has been based on the multidimensional NLS equation (7.1.1) written for a Kerr medium with a refractive index of the form $n = n_0 + n_2|A|^2$. It was realized in the late 1960s that a modification of the Kerr nonlinearity can arrest the collapse of CW beams [51]–[53]. Examples of such nonlinear mechanisms include self-steepening [51], a saturable nonlinearity [52], and nonlinear absorption [53]. Several other effects can also arrest the collapse of a CW beam. For example, if the linear part of the refractive index n_0 is spatially inhomogeneous, it may stop the spatiotemporal collapse of an optical pulse even in the anomalous-GVD regime of the medium. We consider several collapse-arresting mechanisms in this section.

7.4.1 Saturable Nonlinearity

In a saturable nonlinear medium, the refracting index first increases with intensity I at low intensity levels but then begins to saturate when intensity becomes large enough. The refracting index of such a medium can be written as $n = n_0 + n_2 F(I)$, where the form of the function $F(I)$ depends on details of the saturation mechanism. For most nonlinear media, this function has the form

$$F(I) = \frac{I}{1 + I/I_s} \approx I(1 - I/I_s), \tag{7.4.1}$$

where I_s is the saturation intensity; the approximate form of $F(I)$ is valid for $I \ll I_s$. This type of saturable nonlinearity has attracted considerable attention [54]–[58].

The NLS equation for a saturable nonlinear medium remains in the form of Eq. (7.1.1), but the nonlinear term contains $F(|A|^2)$ in place of $|A|^2$. Using the dimensional variables introduced in Section 7.2.1, the normalized NLS equation becomes

$$i\frac{\partial u}{\partial z} + \frac{1}{2}\left(\frac{\partial^2 u}{\partial x^2} + \frac{\partial^2 u}{\partial y^2} - s_d\frac{\partial^2 u}{\partial \tau^2}\right) + F(|u|^2)u = 0. \tag{7.4.2}$$

The bullet-like solutions of this equation exist only in the case of anomalous dispersion. Choosing $s_d = -1$, the shape-preserving solutions of the form

$$u(x,y,\tau,z) = U(x,y,\tau)\exp(i\kappa_0 z) \tag{7.4.3}$$

are obtained by solving the differential equation

$$\frac{1}{2}\left[\frac{\partial^2 U}{\partial R^2} + \frac{(D-1)}{R}\frac{\partial U}{\partial R}\right] - \kappa_0 U + F(U^2)U = 0, \tag{7.4.4}$$

where $R = (x^2 + y^2 + \tau^2)^{1/2}$; $D = 2$ for planar waveguides but equals 3 for a bulk nonlinear medium.

Numerical solutions of Eq. (7.4.4) show that $U(R)$ varies with R qualitatively in the same way as shown in Figure 7.2 in the special case $F(U^2) = U^2$. The lowest-order solution exhibits a single peak at $R = 0$ and corresponds to an optical bullet. The pulse energy E, defined as in Eq. (7.2.7), depends on κ_0, and the derivative $dE/d\kappa_0$ can be used to judge the stability of the solution. More specifically, the solution is stable if $dE/d\kappa_0 > 0$. In contrast with the Kerr case (no saturation of nonlinearity), for which this derivative is always negative, the light bullets are stable for sufficiently intense pulses for which I exceeds the saturation intensity I_s.

Further insight can be gained by using the variational method based on the Lagrangian density [57]

$$\mathcal{L} = -R^{D-1}\left|\frac{\partial u}{\partial R}\right| + \frac{i}{2}R^{D-1}\left(u*\frac{\partial u}{\partial z} - u\frac{\partial u^*}{\partial z}\right) + R^{D-1}F(|u|^2), \tag{7.4.5}$$

where $F(x) = \int_0^x F(x)\,dx$. The $(D+1)$-dimensional NLS equation can be derived from $\mathcal{L}$ using the Euler–Lagrange equation. The variational method approximates the lowest-order solution of Eq. (7.4.4) with a Gaussian function. Allowing for the chirp, the solution $u(R,z)$ takes the form

$$u(R,z) = A\exp\left[-(1+ib)\frac{R^2}{2a^2} + i\phi\right], \tag{7.4.6}$$

where A is the peak amplitude, a is the width, b is the chirp, and ϕ is the phase, all of which are functions of z.

Following the standard method [57], we can derive evolution equations for these four parameters. The amplitude equation is not needed, since $A^2 a^D = c_0$ is related to the pulse energy E. This constant of motion c_0 can be used to find A if a is known. The chirp parameter can also be obtained from a using the simple relation

$$b(z) = \frac{a}{2}\frac{da}{dz}. \tag{7.4.7}$$

The evolution of $a(z)$ is governed by

$$\frac{d^2a}{dz^2} = \frac{4}{a^3} - \frac{2}{a}[h'(A^2) - A^{-2}h(A^2)], \tag{7.4.8}$$

where a prime denotes derivative and the function $h(x)$ is defined as

$$h(x) = \frac{4}{\Gamma(D/2)} \int_0^\infty y^{D-1} F[x\exp(-y^2)]\, dy. \tag{7.4.9}$$

The bullet solution corresponds to the case in which the pulse parameters do not change with propagation. From Eq. (7.4.8) a does not change with z when the right-hand side is zero. This condition provides the steady-state value a_0 of the bullet size (in both spatial and temporal dimensions) as

$$a_0 = \sqrt{2}[h'(A_0^2) - A^{-2}h(A_0^2)]^{-1/2}, \tag{7.4.10}$$

where A_0^2 is itself related to a_0 as $A_0^2 = c_0/a_0$.

The stability of the optical bullet can be checked using a linear stability analysis. If $\delta a = a - a_0$ represents a small fluctuation from a_0, linearization of Eq. (7.4.8) leads to the following equation for δa:

$$\frac{d^2}{dz^2}\delta a + K^2\, \delta a = 0, \tag{7.4.11}$$

where the propagation constant K depends on the bullet parameters as

$$K = a_0^{-2}[4 + 2D - Da_0c_0h''(A_0^2)]^{1/2}, \tag{7.4.12}$$

The bullet solution is stable if K is real or $K^2 > 0$. This condition can be written as $dc_0/dA_0 > 0$ and is equivalent to the stability condition given earlier.

The preceding variational analysis applies to any form of saturable nonlinearity. As an example, consider the $D = 3$ case with a weakly saturable nonlinearity given in Eq. (7.4.1). Using $F(x) = x - \mu x^2$, where μ is related inversely to the saturation intensity, the function $F(x) = x^2/2 - \mu x^3/3$ in Eq. (7.4.5). We can now calculate $h(x)$ by performing the integral in Eq. (7.4.9) and obtain

$$h(x) = 2^{-3/2}x^2 - 3^{-5/2}(2\mu)x^3. \tag{7.4.13}$$

Using Eqs. (7.4.10) and (7.4.12) we can find both a_0 and the condition for this solution to be stable. The stability condition is found to be

$$\mu A_0^2 = I_0/I_s > 9\sqrt{3}/32\sqrt{2} \approx 0.344, \tag{7.4.14}$$

where I_0 is the peak intensity of the bullet. This value is only approximate since it is based on the Gaussian-shaped ansatz. However, one can draw a general conclusion that the input intensity should be comparable to the saturation intensity for optical bullets to be stable in a saturable nonlinear medium.

Numerical results obtained by solving the NLS equation (7.4.2) directly are in qualitative agreement with the predictions of the variational analysis. Figure 7.8 shows

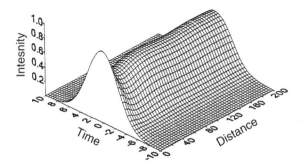

Figure 7.8: Optical bullet formed inside a saturable nonlinear medium when $I_0 = I_s$. (After Ref. [57]; ©1997 APS.)

an optical bullet formed when $I_0 = I_s$ and the input pulse parameters are such that $c_0 = A_0^2 a_0^3 = 50$. In this case, the pulse initially broadens both spatially and temporally but forms a bullet after a few diffraction lengths.

In the normal-dispersion case, $s_d = 1$ in the NLS equation (7.4.2). A spherically symmetric solution does not exist in this case. The variational analysis can still be used with the following ansatz for the solution [58]:

$$u(R,z) = A \exp\left[-(1+ib_s)\frac{x^2+y^2}{2a_s^2} - (1+ib_t)\frac{T^2}{2a_t^2} + i\phi \right], \qquad (7.4.15)$$

where a_s and a_t represent, respectively, spatial and temporal widths of the pulse and b_s and b_t are the corresponding chirp parameters. The analysis is more complicated in the normal-dispersion case because of two additional parameters. Both the numerical and analytic results show the possibility of a new effect. If the pulse intensity exceeds the saturation intensity and the approximation (7.4.1) is used, the pulse can collapse temporally even if it spreads spatially. This effect is due to the fact that the nonlinearity becomes of self-defocusing type under such conditions. In reality, the approximation (7.4.1) cannot be used when $I > I_s$, and one should use the more accurate form of the saturating nonlinearity.

7.4.2 Graded-Index Nonlinear Media

This section considers the possibility of avoiding spatiotemporal collapse using an inhomogeneous Kerr medium in which the linear part of the refractive index depends on the spatial coordinates [59]–[62]. As a simple example, we consider a graded-index nonlinear medium for which the refractive index can be written as

$$n(x,y,\omega) = n_0(\omega) + n_1(x^2+y^2) + n_2 I, \qquad (7.4.16)$$

where n_1 governs changes in the refractive index in the transverse dimensions x and y. An example of such a nonlinear medium is provided by graded-index silica fibers for which $n_1 < 0$.

The NLS equation can easily be generalized to include the spatial dependence of the refractive index in Eq. (7.4.16). In its normalized form, the NLS equation can be written as

$$i\frac{\partial u}{\partial z} + \frac{1}{2}\left(\frac{\partial^2 u}{\partial x^2} + \frac{\partial^2 u}{\partial y^2} - s_d\frac{\partial^2 u}{\partial \tau^2}\right) + s_g(x^2 + y^2)u + |u|^2 u = 0, \qquad (7.4.17)$$

where $s_g = \text{sgn}(n_1)$ and the transverse beam width w_0 in Eq. (7.2.1) has been chosen as $w_0 = (2k_0^2|n_1|)^{-1/4}$. Numerical solutions of this equation govern whether an input pulse will collapse spatially and temporally or form an optical bullet. In the following we focus on the case of anomalous GVD ($s_d = -1$) and a graded-index fiber with high index along the axis ($s_g = -1$).

As before, the variational method can be used to gain considerable physical insight. The Lagrangian density corresponding to Eq. (7.4.17) is given by

$$\mathcal{L} = \frac{i}{2}\left(u\frac{\partial u^*}{\partial z} - u^*\frac{\partial u}{\partial z}\right) + \frac{1}{2}\left(\left|\frac{\partial u}{\partial x}\right|^2 + \left|\frac{\partial u}{\partial y}\right|^2\right)$$

$$+ \frac{1}{2}\left|\frac{\partial u}{\partial \tau}\right|^2 + \frac{1}{2}(x^2 + y^2)|u|^2 - \frac{1}{2}|u|^4. \qquad (7.4.18)$$

The choice of the trial function is crucial. Noting that the one-dimensional NLS equation ($x = y = 0$) supports "sech"-type bright solitons in the temporal dimension while a graded index supports a fundamental Gaussian-shaped spatial mode, an appropriate trial function is of the form

$$u(x, y, z, \tau) = \sqrt{\frac{EW_t^{-1}}{2\pi W_s^2}}\,\exp\left[-(1 + iC_s)\frac{(x^2 + y^2)}{2W_s^2}\right]\text{sech}\left(\frac{\tau}{W_t}\right)\exp(i\phi - iC_t\tau^2/2),$$

$$(7.4.19)$$

where E is the pulse energy as defined in Eq. (7.2.7). The parameters W_t, W_s, C_t, C_s, and ϕ are allowed to vary with z and represent, respectively, the temporal width, spatial width, temporal chirp, spatial chirp, and phase associated with the pulse.

Following the standard method [61, 62], we find the evolution equations for the five pulse parameters. The phase equation is decoupled from the others and can be ignored. The remaining four equations are

$$\frac{dW_s}{dz} = -C_s/W_s, \qquad (7.4.20)$$

$$\frac{dC_s}{dz} = \left(W_s^2 - \frac{1}{W_s^2}\right) + \frac{E}{12\pi W_s^4 W_t}. \qquad (7.4.21)$$

$$\frac{dW_t}{dz} = -C_t W_t, \qquad (7.4.22)$$

$$\frac{dC_t}{dz} = \left(C_t^2 - \frac{4}{\pi^2 W_T^4}\right) + \frac{E}{\pi^3 W_s^2 W_t^3}. \qquad (7.4.23)$$

The stationary states corresponding to an optical bullet are found by setting the z derivatives to zero in the preceding four equations. All such solutions are unchirped;

that is, $C_s = C_t = 0$. From Eq. (7.4.23), the temporal width of bullet-like solutions is related to the spatial width as $W_t = 4\pi W_s^2/E$. From Eq. (7.4.21), W_s^2 satisfies the cubic polynomial

$$W_s^6 - W_s^2 + E^2/(48\pi^2) = 0. \tag{7.4.24}$$

Physically meaningful roots ($W_s^2 > 0$) of this polynomial exist when the normalized pulse energy E is below a critical value E_c such that

$$E < E_c = 4\pi(4/3)^{1/4} \approx 13.5. \tag{7.4.25}$$

Thus, a spatiotemporal soliton can form in a graded-index Kerr medium if the pulse energy is below the critical threshold.

Stability of the optical-bullet soliton can be examined by linearizing Eqs. (7.4.20)–(7.4.23) around the steady-state solution. Such a linear stability analysis shows that two bullet-like solutions exist for a given value of the pulse energy E, but only one solution for which $dW_s/dE < 0$ is stable [61]. An approximate value of the spatial width can be obtained from Eq. (7.4.24) by noting that W_s is close to 1 for $E \ll E_c$ and is given by

$$W_s \approx [1 - E^2/(48\pi^2)]^{1/4}. \tag{7.4.26}$$

The corresponding pulse width is obtained using $W_t = 4\pi W_s^2/E$. As an example, consider a graded-index fiber for which n_1 is chosen such that the scale factor $w_0 = 50 \ \mu$m. For a pulse energy such that $E = 2\pi$, the spatial and temporal widths of the optical bullet are about 45 μm and 28 fs, respectively, if we use $\beta_2 = -20 \ ps^2/km$ at the 1.55-μm operating wavelength. The formation of a stable optical bullet in such a Kerr medium is due solely to the inhomogeneous nature of the nonlinear medium.

A graded-index nonlinear medium also allows the possibility of bullet formation in the case of self-defocusing and normal dispersion [62]. This case corresponds to $s_d = 1$, $s_g = -1$, and a negative sign for the nonlinear term in Eq. (7.4.17). Without the index gradient, an optical bullet cannot form since self-defocusing leads to a rapid increase in W_s. However, the focusing provided by the index gradient can counteract this spatial broadening, resulting in the formation of an optical bullet.

7.5 Higher-Order Effects

For ultrashort pulses having widths below 1 ps, the multidimensional NLS equation should be generalized to include several new effects, such as higher-order dispersion and diffraction, intrapulse Raman scattering, plasma generation through multiphoton absorption, and self-steepening [63]–[77]. In this section, we consider several higher-order effects and discuss their impact on spatiotemporal collapse.

7.5.1 Generalized NLS Equation

Various higher-order effects can be grouped into two categories, referred to here as linear and nonlinear. Linear higher-order effects have their origin in the dispersion relation

$$k_z = [\beta^2(\omega) - (k_x^2 + k_y^2)]^{1/2}, \tag{7.5.1}$$

where k_x, k_y, and k_z are the components of the propagation vector inside the nonlinear medium. Expanding $\beta(\omega)$, as in Eq. (1.3.5), around the carrier frequency ω_0, k_z can be written as [64]

$$k_z = \beta_0 + \frac{\Omega}{v_g} + \frac{1}{2}\beta_2\Omega^2 + \frac{1}{6}\beta_3\Omega^3 - \frac{k_s^2}{2\beta_0}\left(1 - \frac{\Omega}{\beta_0 v_g}\right) - \frac{k_s^4}{8\beta_0^3} + \cdots, \qquad (7.5.2)$$

where $v_g = 1/\beta_1$ is the group velocity, $\Omega = \omega - \omega_0$, and $k_s^2 = k_x^2 + k_y^2$. When this equation is converted into the space–time domain, k_x, k_y, k_z, and Ω are replaced by the corresponding spatial and temporal derivatives. Following this approach, the NLS equation should be generalized to include new higher-order terms resulting from the terms, such as $k_s^2\Omega$ and k_s^4. These terms become important only when the spatial and temporal widths become so small that the slowly varying envelope approximation and the paraxial approximation begin to break down because of a large reduction in the beam size and the pulse width. However, such a reduction can easily occur in the beam-collapse regime, and one must include higher-order linear effects in this regime.

The dominant higher-order nonlinear effects have their origin in the fact that both electronic and nuclear motions contribute to nonlinear polarization as

$$P_{NL}(\mathbf{r},t) = \varepsilon_0\chi^{(3)}E(\mathbf{r},t)\int_{-\infty}^{t} R(t-t_1)|E(\mathbf{r},t_1)|^2\,dt_1. \qquad (7.5.3)$$

The response function is given by

$$R(t) = (1 - f_R)\delta(t) + f_R h_R(t), \qquad (7.5.4)$$

where f_R represents the fraction of Raman contribution resulting from nuclear motion and $h_R(t)$ is the Raman-response function. The electronic response is nearly instantaneous on the time scale of the Raman contribution. In silica glasses, $f_R \approx 0.18$ [78].

Several other nonlinear effects should be considered for highly energetic ultrashort optical pulses, and they become especially relevant when such pulses are propagated through air or water [73]–[77]. The new physical phenomenon one must consider is the generation of plasma through multiphoton absorption. This process affects the pulse evolution in two ways. First, multiphoton absorption is a nonlinear process, in contrast with the one-photon absorption process. Second, the generated plasma itself absorbs light and may even change the refractive index in a nonlinear fashion.

When Eqs. (7.5.2)–(7.5.4) are used in the Maxwell equations, we obtain a generalized NLS equation. In its normalized form, this equation takes the following form:

$$i\frac{\partial u}{\partial z} + \frac{1}{2}\nabla_T^2 u - \frac{1}{2}s_d\frac{\partial^2 u}{\partial\tau^2} + H_d(u) + H_p(u) + \left(1 + \frac{i}{\omega_0 T_0}\frac{\partial}{\partial\tau}\right)$$
$$\times \left[(1 - f_R)|u|^2 + f_R\int_{-\infty}^{\tau} h_R(\tau-t)|u(t)|^2 dt\right]u = 0, \qquad (7.5.5)$$

where Δ_T^2 is the transverse Laplacian and T_0 is the pulse-width scale introduced in Eq. (7.2.2) and defined as $T_0 = (w_0^2\beta_0\beta_2)^{1/2}$. The function $H_d(u)$ takes into account higher-order dispersive and diffractive effects and contains third- and higher-order derivatives of u. Its functional form can be written from Eq. (7.5.2) as

$$H_d(u) = -i\delta_t \frac{\partial^3 u}{\partial \tau^3} - i\delta_{st} \frac{\partial}{\partial \tau}(\nabla_T^2 u) + \delta_s \nabla_T^4 u + \cdots, \qquad (7.5.6)$$

where the three small parameters are defined as

$$\delta_t = \frac{\beta_3}{6\beta_2 T_0}, \qquad \delta_{st} = \frac{w_0^2}{2L_d v_g T_0}, \qquad \delta_s = \frac{1}{8\beta_0 L_d}. \qquad (7.5.7)$$

The function $H_p(u)$ in Eq. (7.5.5) takes into account the nonlinear effects produced by the generated plasma. Assuming that the plasma is produced through absorption of m photons simultaneously, this term can be written as [75]

$$H_p(u) = \alpha_m |u|^{2m-2} + i\delta_p \int_{-\infty}^{\tau} |u|^{2m} d\tau, \qquad (7.5.8)$$

where α_m is related to the m-photon absorption coefficient and the parameter δ_p accounts for the absorption by plasma.

7.5.2 Space–Time Focusing

The relative importance of the higher-order linear terms is governed by the magnitudes of the five parameters in Eq. (7.5.7). For a 100-fs pulse propagating at 1.55 μm inside silica glass and having a spatial size such that $w_0 = 50$ μm, the parameters values are about $\delta_t = 0.05$, $\delta_{st} = 0.009$, and $\delta_s = 1.5 \times 10^{-6}$. The plasma generation is of concern only at extremely high values of pulse energies. In many experiments, pulse energy is well below the threshold of plasma generation. If we focus on this case and assume pulse widths of 100 fs or more, we can neglect during the initial stage of propagation of such pulses all terms in Eq. (7.5.6) except for the first term responsible for the third-order dispersion. However, if the spatial width decreases during propagation because of self-focusing, higher-order diffractive effects begin to become important. As an example, when the spatial width is reduced to to 10 μm, $\delta_{st} \approx 0.04$ and $\delta_s \approx 0.003$. It was shown in 1991 that even the inclusion of the smallest δ_s term can eliminate the collapse of a CW beam in a self-focusing Kerr medium [63].

The effect of the δ_{st} term containing mixed space–time derivatives in Eq. (7.5.6) was first studied in 1992 numerically [64]. This term is responsible for a phenomenon known as *space–time focusing* resulting from variations in the group velocity associated with the off-axis spatial frequency components of a pulse. Figure 7.9 shows the effects of space–time focusing on the spatiotemporal collapse of a 50-fs "sech" pulse, with 80-μm Gaussian spot size, propagating in the anomalous GVD regime of silica glass ($\beta_2 = -22$ ps^2/km at $\lambda = 1.55$ μm). The input peak power of 29 MW corresponds to 1.9 times the critical power associated with a CW Gaussian beam. The dashed curves show how the pulse compresses rapidly to a width of only a few femtoseconds when the higher-order effects are ignored. This pulse compression, occurring in the anomalous-dispersion regime, is due to the phenomenon of the spatiotemporal collapse discussed earlier. When the effects of space–time focusing are included (solid curves), the pulse is still compressed but the details are quite different, as is evident from Figure 7.9. More

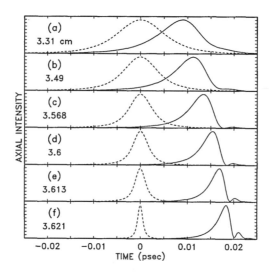

Figure 7.9: Effect of space–time focusing on the catastrophic collapse of a 50-fs "sech" pulse with 80-μm Gaussian spot size. On-axis intensity versus time is shown at several distances with (solid curves) and without (dashed curves) the term responsible for space–time focusing. The vertical scale is different in all cases for ease of comparison. (After Ref. [64]; ©1992 OSA.)

specifically, the pulse is delayed, acquires an asymmetric shape, and compresses much less. Moreover, the compressed pulse appears to form a shock front near its training edge and develops small satellite peaks. These features show that the spatiotemporal collapse is affected considerably by the higher-order linear effects.

7.5.3 Intrapulse Raman Scattering

As seen in Eq. (7.5.5), the relative importance of the higher-order nonlinear effects is governed by the parameter f_R, the Raman Response function $h_R(t)$, and the pulse width T_0. The term proportional to $1/\omega_0 T_0$ lead to the phenomenon of self-steepening. Since this quantity is quite small unless pulse width reduces to just a few optical cycles, self-steepening can often be ignored. However, the Raman term is responsible for an effect known as *intrapulse Raman scattering* and becomes important for pulses as wide as a few picoseconds. Physically speaking, high-frequency components of an optical pulse pump the low-frequency components of the same pulse through stimulated Raman scattering. As a result, the pulse spectrum shifts toward lower frequencies as the pulse propagates inside the nonlinear medium. This spectral shift manifests as changes in the pulse position, since the speed of light is frequency dependent in a dispersive medium.

For studying the effects of intrapulse Raman scattering we need the functional form of the Raman Response function $h_R(t)$. On physical grounds, $h_R(t)$ is expected to exhibit damped oscillations and can be approximated as [79]

$$h_R(t) = \frac{1 + (\Omega_r T_r)^2}{\Omega_r T_r^2} \exp(-t/T_r) \sin(\Omega_r t), \qquad (7.5.9)$$

where Ω_r and τ_r represent, respectively, the frequency and damping time of oscillations and are chosen to provide a good fit to the actual Raman-gain spectrum.

For pulses shorter than 5 ps but wide enough to contain many optical cycles (width $\gg 10$ fs), one can simplify Eq. (7.5.5) by using

$$u(t - t')|^2 \approx |u(t)|^2 - t' \frac{\partial |u|^2}{\partial t}. \tag{7.5.10}$$

Defining the first moment of the nonlinear response function as

$$T_R = \int_{-\infty}^{\infty} t R(t) \, dt \equiv f_R \int_{-\infty}^{\infty} t \, h_R(t) \, dt, \tag{7.5.11}$$

and retaining only the dominant terms in Eq. (7.5.5), the generalized NLS equation can be written as

$$i \frac{\partial u}{\partial z} + \frac{1}{2} \nabla_T^2 u - i \delta_{st} \frac{\partial}{\partial \tau} (\nabla_T^2 u) - \frac{1}{2} s_d \frac{\partial^2 u}{\partial \tau^2} - i \delta_t \frac{\partial^3 u}{\partial \tau^3}$$
$$+ (1 - f_R) \left(1 + \frac{i}{\omega_0 T_0} \frac{\partial}{\partial \tau} \right) |u|^2 u - \tau_R u \frac{\partial |u|^2}{\partial \tau} = 0. \tag{7.5.12}$$

This equation includes most of the higher-order effects and is appropriate for pulses containing even a few optical cycles.

Numerical simulations show that the inclusion of intrapulse Raman scattering affects the pulse evolution in the cases of both normal and anomalous dispersion. Figure 7.10 shows how the pulse-splitting phenomenon occurring in the normal-dispersion regime is affected by this Raman effect by using $f_R = 0.18$, $T_r = 50$ fs, and $\Omega_r T_r = 4.2$ [43]. It should be compared with Figure 7.6, for which Raman effects were ignored. The input field amplitude is in the form of Eq. (7.3.13) with $r_0 = 70$ μm, $T_0 = 90$ fs, and $I_0 = 87$ GW/cm^2. The spatiotemporal profiles at distances of 2 and 3 cm are shown on left. The right side shows how the pulse width and the intensity at beam center change with propagation.

The main effect of intrapulse Raman scattering in Figure 7.10 is is to introduce an asymmetry in the two subpulses created after the splitting of the original pulse. At $z = 3$ cm, the leading subpulse has almost twice the peak intensity of the trailing subpulse. This behavior can be understood as a combination of the normal GVD and the Raman gain. The intrapulse Raman scattering amplifies the red-shifted frequency components of the pulse at the expense of the blue-shifted components, and the GVD moves the red components ahead of the blue components. The net result is that energy from the trailing subpulse is transferred to the leading subpulse during propagation. This interpretation is supported by the changes in the pulse width and peak intensity. Before splitting, the pulse is compressed and experiences a 10-fold increase in the peak intensity after 2 cm of propagation. After splitting starts, the peak intensities of both subpulses fall rapidly and their widths begin to increase. Note that when the pulse first begins to split, the trailing subpulse has a higher peak intensity than the leading one. However, this trend reverses soon because of intrapulse Raman scattering, and the leading subpulse becomes more intense.

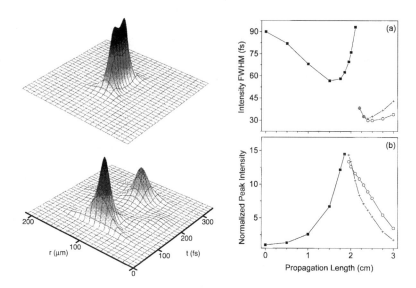

Figure 7.10: Effect of intrapulse Raman scattering on pulse splitting in the normal-GVD regime of fused silica. The spatiotemporal profiles at distances of 2 and 3 cm are shown together with changes in the pulse width and the on-axis intensity occurring with z. Pulse parameters are identical to those used for Figure 7.6. (After Ref. [43]; ©1998 IEEE.)

In some practical situations, all higher-order effects can be ignored, with the exception of the Raman term. In such cases, one can reduce Eq. (7.5.12) to the following simpler NLS equation:

$$i\frac{\partial u}{\partial z} + \frac{1}{2}\left(\nabla_T^2 u - s_d\frac{\partial^2 u}{\partial \tau^2}\right) + |u|^2 u - \tau_R u\frac{\partial |u|^2}{\partial \tau} = 0, \qquad (7.5.13)$$

where the coefficient $1 - f_R$ in the $|u|^2 u$ term has been absorbed in the normalization of u. The important question is whether Eq. (7.5.13) allows for the formation of stable, shape-preserving spatiotemporal solitons in the case of anomalous dispersion.

It turns out that although bullet-like solutions do not exist when the Raman term is included, Eq. (7.5.13) has an approximate analytic solution in the case of anomalous GVD ($s_d = -1$) in the form of a kink soliton [35]. An exact kink-soliton solution of Eq. (7.5.13) was first found in 1993 [80] for the one-dimensional case ($\Delta_T^2 u = 0$). When the diffraction term is included, the temporal kink soliton is coupled with a spatial soliton in such a way that the pair supports each other in a symbiotic fashion. An approximate analytic form of this solution can be found for both a planar waveguide (diffraction in only one transverse dimension) and bulk nonlinear media [35]. The method for finding such a soliton pair is identical to that used in Section 7.3. In the waveguide case, the solution is in the form of Eq. (7.3.2), but the function $f(x)$ and $g(\tau)$ are given as

$$f(x) = \text{sech}[3\sqrt{3}x/(4\tau_R)], \qquad g(\tau) = \exp(3\tau/4\tau_R)[\text{sech}(3\tau/2\tau_R)]^{1/2}. \qquad (7.5.14)$$

The preceding discussion shows that the Raman effects should in general be considered whenever spatiotemporal effects are studied for femtosecond pulses.

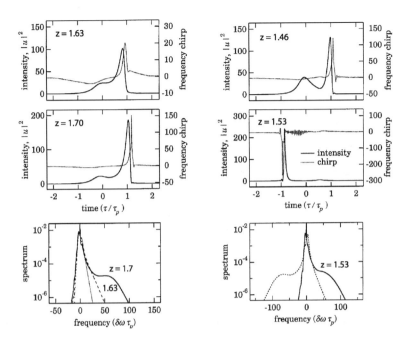

Figure 7.11: On-axis intensity (solid curve) and chirp (dotted curve) profiles together with pulse spectra at several distances without (left column) and with (right column) with the plasma effects. The input peak power of 70-fs pulses is such that $P/P_{cr} = 1.8$ on the left and 2 for curves on the right. Dashed lines show the total pulse spectrum. (After Ref. [72]; ©2000 APS.)

7.5.4 Multiphoton Absorption

As mentioned earlier, one must consider the phenomenon of plasma generation through multiphoton absorption for high-energy ultrashort pulses. When such pulses are propagated through air, they can propagate over considerable distances (~ 100 m) because of the formation of filaments after the plasma has been generated through multiphoton ionization of air [73]–[77]. The dispersive effects are less important in this case. The main mechanism behind the filament formation is related to a dynamic balance between the Kerr self-focusing and the defocusing induced by the plasma [75].

Even the phenomenon of catastrophic beam collapse is considerably affected by the multiphoton absorption and the plasma created by it. It was found in a 2000 numerical study that a steep edge (an optical shock wave) is formed at the back of an ultrashort pulse propagating in the anomalous-GVD regime of a Kerr medium as the pulse approaches the collapse point [72]. At the same time, the pulse spectrum develops a broad blue-shifted pedestal, resulting in a wide spectrum known as a *supercontinuum*. Figure 7.11 shows the effect of plasma on the pulse shapes and pulse spectra under several different conditions. Although the optical shock and supercontinuum are generated even in the absence of the plasma effects, both of them are affected considerably by the formation of plasma. These calculations were done for 70-fs pulses propagating inside a sapphire sample, but similar features are expected for other nonlinear materials.

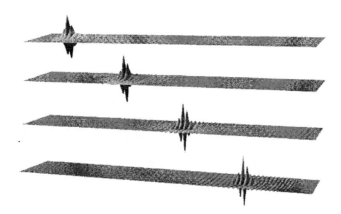

Figure 7.12: (a) Electric field associated with the spatiotemporal soliton formed in a planar glass waveguide. The four frames show the electric field after a delay of 155, 310, 465, and 620 fs. (After Ref. [88]; ©1997 OSA.)

7.5.5 Direct Integration of Maxwell Equations

There are several limitations inherent in the use of even a generalized NLS equation whenever ultrashort optical pulses propagate inside a nonlinear medium. The NLS equation is based on the slowly varying envelope approximation and is valid only if the spatial and temporal variations occur on a scale much larger than the optical wavelength and the optical cycle, respectively. The other major limitation is related to the neglect of the vector nature of the electromagnetic fields. In essence, the polarization effects are completely ignored whenever a scalar NLS equation is used. All of these limitations can be removed if the Maxwell equations are solved directly.

In the case of a linear medium, the algorithms that solve the Maxwell equations directly in the time domain by using a finite-difference method have been known for many years [81]. This method, known as the *finite-difference time-domain* (FDTD) method and extended during the 1990s to the case of nonlinear media, has been used to study the evolution of femtosecond optical pulses [82]–[89]. Conceptually, the main difference between the FDTD method and the split-step Fourier method commonly employed for solving the NLS equation is that the former deals with all electromagnetic components without eliminating the carrier frequency and thus automatically included all higher-order linear and nonlinear effects once a suitable method has been adopted for calculating the polarization induced inside the optical medium.

Numerical simulations carried out using the FDTD method have revealed that catastrophic collapse indeed does not occur when the response of the nonlinear medium follows Eq. (7.5.3), even though it is dominated by Kerr-type nonlinearity. Even more promising is the result that spatiotemporal solitons in the form of optical bullets can form for pulses as short as 10 fs [88]. Figure 7.12 shows the results obtained by solving the Maxwell equations inside a planar glass waveguide with the input

$$E_y(X,0,t) = E_0 \operatorname{sech}[(t - T_d)/T_0]\operatorname{sech}(X/w_0), \qquad (7.5.15)$$

where $T_0 = 10.3$ fs, $T_d = 61.8$ fs, $w_0 = 0.2\ \mu$m, and $E_0 = 29$ GV/m. The four frames

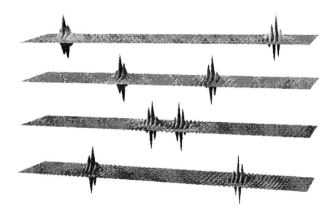

Figure 7.13: (a) Collision of two spatiotemporal solitons inside a planar glass waveguide. The four frames show the electric field after a delay of 155, 310, 465, and 620 fs. (After Ref. [88]; ©1997 OSA.)

show the electric field after a delay of 155, 310, 465, and 620 fs. After some initial adjustment, the electric field was found to remain nearly unchanged over a propagation distance of 28 diffraction lengths. This feature indicates the robust nature of the optical bullet formed inside the nonlinear medium.

An important question is whether such solitary waves are true solitons that can survive mutual collisions. To answer this question, two identical pulses were launched such that they propagated in counter-propagating directions inside the nonlinear medium. Figure 7.13 shows the results obtained. It is evident that solitons remain nearly intact after the collision. In fact, the comparison of the final pulse shapes in Figs. 7.12 and 7.13 after 620 fs shows that they are virtually identical except for a slight retardation occurring in the case of collision. This is a well-known feature of one-dimensional solitons associated with the standard $(1 + 1)$-dimensional NLS equation. A new feature of the two colliding multidimensional solitons is that their trajectories are also spatially displaced slightly. This feature can form the basis of a nonlinear optical switch [88].

The FDTD method is certainly more accurate than the NLS equation because it solves the Maxwell equations directly with a minimum number of approximations. However, this improvement in accuracy is achieved only at the expense of a considerable increase in computational effort. This can be understood by noting that the time step needed to resolve the optical carrier is of necessity a fraction of the optical period and should be less than 1 fs in practice. The step size along the medium length is also required to be a fraction of the optical wavelength. As a result, the use of the FDTD method in practice is limited to optical pulses with $T_0 < 1$ ps. At the same time, the total length of the nonlinear medium should be 1 cm or less. Such distances are still long enough for studying optical bullets, provided the diffraction length is reduced to 10 μm or less by focusing the beam to a spot size comparable to the optical wavelength. It should be stressed that a supercomputer is not required for such simulations. For example, the results shown in Figure 7.13 took only twelve minutes on a workstation with one processor operating at a 250-MHz clock rate.

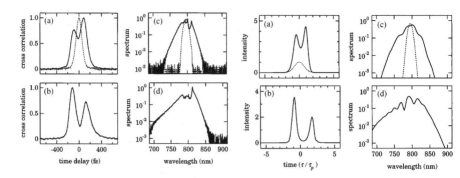

Figure 7.14: Comparison of measured (left) and calculated (right) pulse shapes and spectra for a 78-fs pulse transmitted through 3 cm of fused silica. The on-axis intensity profiles (left column) and power spectra (right column) are shown at an input peak power level of 5.1 MW (top) and 5.9 MW (bottom). Input pulse shape and spectrum are shown by dotted curves. (After Ref. [69]; ©1998 OSA.)

7.6   Experimental Results

The spatiotemporal effects on ultrashort pulses have been observed since 1995 in a number of experiments using several different types of nonlinear media [90]–[101]. Most of these experiments employ pulses shorter than 100 fs and focus the input pulsed beam tightly onto the nonlinear medium to ensure that both the diffractive and dispersive effects play an important role during pulse propagation. In this section we discuss separately the major experimental results observed in the normal- and anomalous-GVD regimes of the nonlinear medium. The nonlinear medium itself can range from a glass waveguide to plane air.

7.6.1   Normal-Dispersion Regime

The evidence of pulse splitting in the normal-GVD regime of a nonlinear medium was first seen in a 1996 experiment [42]. A 85-fs mode-locked pulse at 800 nm with a 55-μm Gaussian spot was transmitted through a 2.54-cm-long BK7-glass sample and was found to split into two after the peak power exceeded 3 MW (70% above the critical threshold of self-focusing). The evidence of splitting was inferred from the 3-peak autocorrelation trace. In a later experiment, the pulse-splitting regime was explored in more detail using a cross-correlation technique that allowed an accurate measurement of the pulse shape directly [69]. In this experiment, 78-fs pulses from a Ti:sapphire laser operating at 795 nm were focused to a 75-μm spot size at the front face of a 3-cm-long fused silica glass. Figure 7.14 shows the measured pulse shapes and spectra at the input peak power levels $P_0 = 5.1$ and 5.9 MW. The critical power for the beam collapse (in the absence of dispersive effects) is estimated to be 3 MW for this experiment. The presence of normal dispersion ($\beta_2 = 385$ fs^2/cm) eliminates the collapse through pulse splitting.

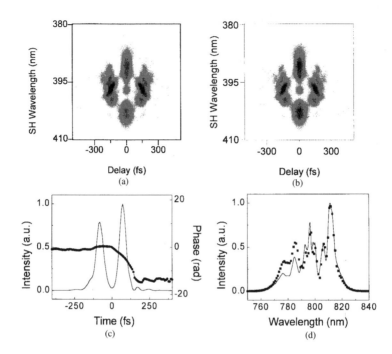

Figure 7.15: (a) Measured and (b) recovered FROG traces after a 92-fs pulse passes through 2.54-cm of fused silica. (c) Intensity (thin line) and phase (solid dots) profiles of the output pulse. (d) Measured (thin line) and FROG-inferred (solid squares) pulse spectra of the output pulse. (After Ref. [43]; ©1998 IEEE.)

Two features of Figure 7.14 are most noteworthy. First, the splitting is asymmetrical, in the sense that the two subpulses have different heights. Second, even though the trailing pulse is more intense at $P_0 = 5.1$ MW, the reverse occurs for $P_0 = 5.9$ MW. All experimentally observed features can be explained by using a generalized NLS equation that includes the effects of both the space–time focusing and self-steepening and is written in the normalized form as

$$i\frac{\partial u}{\partial z} + \frac{1}{2}\left(1 + i\delta_{st}\frac{\partial}{\partial \tau}\right)^{-1}\nabla_T^2 u - s_d\frac{\partial^2 u}{\partial \tau^2} + \left(1 + i\delta_{st}\frac{\partial}{\partial \tau}\right)|u|^2 u = 0, \qquad (7.6.1)$$

where δ_{st} is a small parameter introduced earlier and responsible for both space–time focusing and self-steepening. The numerical simulations based on this equation provide an excellent fit to the experimental data, as is evident from Figure 7.14. This indicates that intrapulse Raman scattering did not play a major role in the experiment. The plasma effects were also found to be negligible at the pulse energies used in this experiment.

In another 1998 experiment on pulse splitting, the technique of frequency-resolved optical gating (FROG) was used to measure both the pulse shape and phase profiles directly [43]. The experimental parameters used in this experiment were close to those used for Figure 7.6. Figure 7.15 shows the FROG traces on top and the shape, phase,

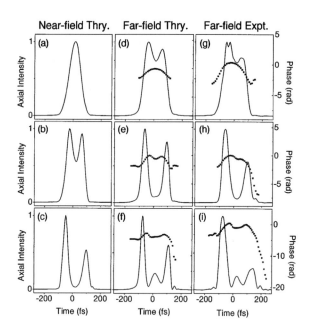

Figure 7.16: Theoretical and experimental traces of the axial intensity $(x = y = 0)$ in the near and far fields as a function of time at input power levels of 3.9 (upper row), 4.6 (middle row) and 5.4 MW (bottom row). Solid circles represent the phase profile. (After Ref. [95]; ©1999 APS.)

and spectrum of the pulse at the bottom after the 92-fs pulse has propagated through a 2.54-cm-long fused silica sample. The peak intensity of input pulse was 88 GW/cm^2 (90% above the critical threshold).

Several features of Figure 7.15 are noteworthy. First, the agreement between the measured and FROG-inferred pulse spectra is reasonably good. Second, the measured phase profile indicates that the pulse becomes chirped as it propagates inside the nonlinear medium. Third, although pulse splitting is evident in Figure 7.15(c), the two sub-pulses exhibit asymmetry, in contrast with the numerical results shown in Figure 7.6. As discussed earlier, this asymmetry is due to the higher-order effects that must be included for ultrashort pulses of widths less than 100 fs.

Depending on the experimental values of the parameters, different higher-order effects can become important. In a 1999 experiment, performed by transmitting 90-fs pulses with 70-μm spot size (both FWHM) through 3 cm of fused silica [95], it was necessary to include all of the higher-order terms in Eq. (7.5.12) for explaining the experimentally observed features. At the operating wavelength of 800-nm, silica exhibits normal GVD with $\beta_2 = 360$ fs^2/cm and $\beta_3 = 275$ fs^3/cm. As a result, pulse splitting is expected to occur at high peak intensities in place of spatiotemporal collapse. Figure 7.16 shows the theoretical and experimental results by plotting the axial intensity $(x = y = 0)$ in the near and far fields as a function of time at input power levels of 3.9 (upper row), 4.6 (middle row), and 7.4 MW (bottom row). The far-field experimental data are in remarkable agreement with the predictions based on Eq. (7.5.12)

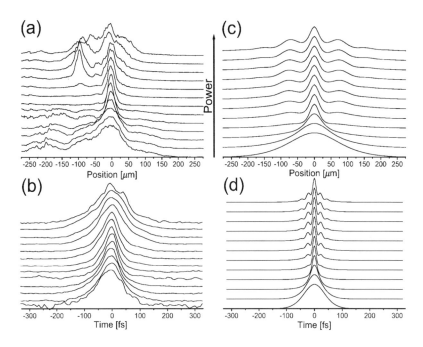

Figure 7.17: (a) Spatial profiles and (b) temporal autocorrelation traces at several input peak power levels after a 60-fs pulse has propagated through a 3-cm-long glass waveguide. Corresponding numerical results are shown in parts (c) and (d). In each case, input peak power increases form bottom to top in the range 0–1 MW. (After Ref. [101]; ©2001 APS.)

using $f_R = 0.15$, $T_r = 50$ fs, and $\Omega_r T_r = 4.2$, values that are quite reasonable for silica glasses. The asymmetry seen in the near- and far-field pulse profiles is due solely to the higher-order effects.

7.6.2 Anomalous-Dispersion Regime

The anomalous-dispersion regime of a nonlinear medium is of considerable experimental interest because it offers the possibility of observing spatiotemporal solitons in the form of optical bullets. Unfortunately, most nonlinear materials exhibit normal dispersion in the 800-nm spectral region in which Ti:sapphire lasers can produce optical pulses as short as 5 fs. For this reason, the cascaded quadratic nonlinearity was used in a 1999 experiment through second-harmonic generation in a nonlinear crystal [97]. This case is discussed in Chapter 10, devoted to quadratic solitons. The use of optical parametric amplifiers and oscillators has solved the dispersion problem to some extent for such oscillators can produce pulses shorter than 100 fs in the spectral region beyond 1.3 μm, where silica glass exhibits anomalous dispersion.

In a 2001 experiment, 60-fs pulses were propagated through a 3-cm-long planar glass waveguide to study the spatiotemporal effects [101]. The anomalous dispersion at the pulse wavelength of 1.52 μm was large enough and the spot size of the launched beam small enough that the dispersion and diffraction lengths were nearly the same in this experiment. Figure 7.17 shows the spatial and temporal autocorrelation profiles of the output pulse at several different input power levels together with the results of numerical simulations. As the input power is increased, the beam shrinks in both the spatial and temporal dimensions, as expected for a pulse propagating inside a self-focusing Kerr medium. However, after a peak power level of 600 kW, further increase in power does not reduce the spatial and temporal widths. The two widths remain nearly constant over a power range of 600–800 kW and begin to increase when the input power is increased beyond that. These power levels exceed considerably the threshold power of 160 kW predicted for the beam collapse. The generalized NLS equation is able to explain the observed features qualitatively if the effects of both intrapulse Raman scattering and multiphoton absorption are included. The main conclusion that one can draw is that beam collapse does not occur in nonlinear Kerr-type materials because of the higher-order effects and that a spatiotemporal soliton resembling a light bullet may form in a certain power range.

References

[1] A. G. Litvak and V. I. Talanov, *Radiophys. Quantum Electron.* **10**, 296 (1967).

[2] D. U. Martin and H. C. Yuen, *Phys. Fluids* **21**, 881 (1981).

[3] A. G. Litvak, T. A. Petrova, A. M. Sergeev, and A. D. Yunakovskii, *Sov. J. Plasma Phys* **9**, 287 (1983).

[4] P. K. Newton and J. B. Keller, *SIAM J, Appl. Math.* **47**, 959 (1987).

[5] N. N. Akhmediev, V. I. Korneev, and R. F. Nabiev, *Opt. Lett.* **17**, 393 (1992).

[6] A. B. Aceves, C. De Angelis, and S. Wabnitz, *Opt. Lett.* **17**, 1758 (1992).

[7] L. W. Liou, X. D. Cao, C. J. McKinstrie, and G. P. Agrawal, *Phys. Rev. A* **46**, 4202 (1992).

[8] M. F. Shih, C. C. Jeng, F.W. Sheu, and C. Y. Lin, *Phys. Rev. Lett.* **88**, 133902(2002).

[9] S. C. Wen and D. Y. Fan, *J. Opt. Soc. Am. B* **19**, 1653 (2002).

[10] K. Tai, A. Hasegawa, and A. Tomita, *Phys. Rev. Lett.* **56**, 135 (1986).

[11] R. Y. Chiao, E. Garmire, and C. H. Townes, *Phys. Rev. Lett.* **13**, 479 (1964).

[12] Z. K. Yankauskas, *Sov. Radiophys.* **9**, 261 (1966).

[13] S. N. Vlasov, V. A. Petrishchev, and V. I. Talanov, *Radiophys. Quantum Electron.* **14**, 1353 (1971).

[14] V. E. Zakharov, V. V. Sobolev, and V. S. Synakh, *JETP Lett.* **14**, 390 (1971).

[15] Y. Silberberg, *Opt. Lett.* **15**, 1282 (1990).

[16] N. G. Vakhitov and A. A. Kolokolov, *Radiophys. Quantum Electron.* **16**, 1020 (1974).

[17] V. E. Zakharov and V. S. Synakh, *Sov. Phys. JETP* **41**, 465 (1975).

[18] Y. R. Shen, *Prog. Quantum Electron.* **4**, 1 (1975).

[19] J. H. Marburger, *Prog. Quantum Electron.* **4**, 35 (1975).

[20] F. H. Berkshire and J. D. Gibbon, *Stud. Appl. Math.* **69**, 229 (1983).

[21] J. J. Rasmussen and K. Rypdal, *Phys. Scr.* **33**, 481 (1986).

[22] K. Rypdal and J. J. Rasmussen, *Phys. Scr.* **33**, 498 (1986).

[23] V. M. Malkin, *JETP Lett.* **48**, 653 (1988).

[24] S. N. Vlasov, L. V. Piskunova and V. I. Talanov, *Sov. Phys. JETP* **68**, 1125 (1989).

[25] V. E. Zakharov, N. E. Kosmatov, and V. F. Shvets, *JETP Lett.* **49**, 492 (1989).

[26] L. Berge, *Phys. Rep.* **303**, 259 (1998).

[27] M. Desaix, D. Anderson, and M. Lisak, *J. Opt. Soc. Am. B* **8**, 2082 (1991).

[28] X. D. Cao, C. J. McKinstrie, and G. P. Agrawal, *Phys. Rev. A* **49**, 4085 (1994).

[29] G. P. Agrawal, *Nonlinear Fiber Optics*, 3rd ed. (Academic, San Diego, 2001).

[30] A. E. Siegman, *Opt. Lett.* **1**, 13 (1977).

[31] G. P. Agrawal and M. Lax, *Opt. Lett.* **6**, 171 (1981).

[32] M. Lax, J. H. Batteh, and G. P. Agrawal, *J. Appl. Phys.* **52**, 109 (1981).

[33] P. Chernev and V. Petrov, *Opt. Lett.* **17**, 172 (1992); *Opt. Commun.* **87**, 28 (1992).

[34] J. E. Rothenberg, *Opt. Lett.* **17**, 583 (1992).

[35] K. Hayata and M. Koshiba, *Phys. Rev. E* **48**, 2312 (1993).

[36] G. G. Luther, J. V. Moloney, A. C. Newell, and E. M. Wright, *Opt. Lett.* **19**, 862 (1994).

[37] D. Strickland and P. B. Corkum, *J. Opt. Soc. Am. B* **11**,492 (1994).

[38] G. Fibich, V. M. Malkin, and G. C. Papanicolaou, *Phys. Rev. A* **52**, 4218 (1995).

[39] L. Berge and J. J. Rasmussen, *Phys. Rev. A* **53**, 4476 (1996).

[40] L. Berge, J. J. Rasmussen, E. A. Kuznetsov, E. G. Shapiro, and S. K. Turitsyn, *J. Opt. Soc. Am. B* **13**, 1879 (1996).

[41] M. Trippenbach and Y. B. Band, *Phys. Rev. A* **56**, 4242 (1997).

[42] J. K. Ranka, R. W. Schirmer, and A. L. Gaeta, *Phys. Rev. Lett.* **77**, 3783 (1996).

[43] S. A. Diddams, H. K. Eaton, A. A. Zozulya, and T. S. Clement, *IEEE J. Sel. Topics Quantum Electron.* **4**, 306 (1998).

[44] C. Conti, S. Trillo, P. Di Trapani, G. Valiulis, O. Jedrkiewicz, and J. Trull, preprint webpage arXiv.org/physics/0204066.

[45] P. R. Stepanishen and J. Sun, *J. Acoust. Soc. Am.* **102**, 3308 (1997); J. Salo, J. Fagerholm, A. T. Friberg, and M. M. Salomaa, *Phys. Rev. Lett.* **83**, 1171 (1999).

[46] J. Salo, J. Fagerholm, A. T. Friberg, and M. M. Salomaa, *Phys. Rev. E* **62**, 4261 (2000); K. Reivelt and P. Saari, *J. Opt. Soc. Am. A* **17**, 1785 (2000).

[47] J. Durnin, J.J. Miceli, and J.H. Eberly, *Phys. Rev. Lett.* **58**, 1499 (1987).

[48] J. Lu and J. F. Greenleaf, *IEEE Trans. Ultrason. Ferrelec. Freq. contr.* **39**, 441 (1992).

[49] P. Saari and K. Reivelt, *Phys. Rev. Lett.* **79**, 4135 (1997); H. Sönajalg, M. Rtsep, and P. Saari, *Opt. Lett.* **22**, 310 (1997).

[50] C. Canuto, M. Y. Hussaini, A. Quaternoni, and M. A. Zang, *Spectral Methods in Fluidodynamics* (Springer, New York, 1988).

[51] F. DeMartini, C. H. Townes, T. K. Gustafson, and P. L. Kelly, *Phys. Rev.* **164**, 312 (1967).

[52] E. L. Dawes and J. H. Marburger, *Phys. Rev.* **179**, 862 (1969).

[53] A. L. Dyshko, V. N. Lugovi, and A. M. Prokhorov, *Sov. Phys. JETP* **34**, 1235 (1972).

[54] J. M. Soto-Crespo, D. R. Heatley, E. Wright, and N. N. Akhmediev, *Phys. Rev. A* **44**, 636 (1991).

[55] N. Akhmediev and J.M. Soto-Crespo, *Phys. Rev. A* **47**, 1358 (1993).

[56] R. McLeod, K. Wagner, and S. Blair, *Phys. Rev. A* **52**, 3254 (1995).

[57] V. Skarka, V.I. Berezhiani, and R. Miklaszewski, *Phys. Rev. E* **56**, 1080 (1997).

[58] V. Skarka, V.I. Berezhiani, and R. Miklaszewski, *Phys. Rev. E* **60**, 7622 (1999).

[59] M. Karlsson, D. Anderson, and M. Desaix, *Opt. Lett.* **17**, 22 (1992).

[60] J. T. Manassah, *Opt. Lett.* **17**, 1259 (1992).

[61] S. S. Yu, C. H. Chien, Y. Lai, and J. Wang, *Opt. Commun.* **119**, 167 (1995).

[62] S. Raghvan and G. P. Agrawal, *Opt. Commun.* **180**, 377 (2000).

[63] V. I. Karpman, *Phys. Lett. A* **160**, 531 (1991).

[64] J. E. Rothenberg, *Opt. Lett.* **17**, 1340 (1992).

[65] J. T. Manassah and B. Gross, *Laser Phys.* **6**, 363 (1996).

[66] G. Fibich and G. C. Papanicolaou, *Opt. Lett.* **22**, 1379 (1996).

[67] T. Brabec and F. Krausz, *Phys. Rev. Lett.* **78**, 3282 (1997).

[68] M. Mlenjek, E. M. Wright, and J. V. Moloney, *Opt. Lett.* **23**, 382 (1998).

[69] J. K. Ranka and A. L. Gaeta, *Opt. Lett.* **23**, 534 (1998).

[70] A. A. Zozulya, S. A. Diddams, and T. S. Clement, *Phys. Rev. A* **58**, 3303 (1998).

[71] Y. B. Band and M. Trippenbach, *Phys. Rev. A* **57**, 4791 (1998).

[72] A. L. Gaeta, *Phys. Rev. Lett.* **84**, 3582 (2000).

[73] N. Mlejnek, M. Kolesik, J. V. Moloney, and E. M. Wright, *Phys. Rev. Lett.* **83**, 2938 (1999).

[74] N. Mlejnek, E. M. Wright, and J. V. Moloney, *IEEE J. Quantum Electron.* **35**, 1771 (1999).

[75] N. Aközbek, C. M. Bowden, A. Talebpour, and S. L. Chin, *Phys. Rev. E* **61** 4540 (2000)

[76] N. Aközbek, M. Scalora, C. M. Bowden, and S. L. Chin, *Opt. Commun.* **191** 353 (2001)

[77] N. Aközbek, C. M. Bowden, and S. L. Chin, *J. Mod. Opt.* **49** 475 (2002).

[78] R. H. Stolen and W. J. Tomlinson, *J. Opt. Soc. Am. B* **9**, 565 (1992).

[79] K. J. Blow and D. Wood, *IEEE J. Quantum Electron.* **25**, 2665 (1989).

[80] G. P. Agrawal and C. Headley III, *Phys. Rev. A* **46**, 1573 (1991).

[81] A. Taflove and S. C. Hagness, *Computational Electrodynamics: The Finite-Difference Time-Domain Method*, 2nd ed. (Artech House, Norwood, MA, 2000).

[82] P. M. Goorjian and A. Taflove, *Opt. Lett.* **16**, 180 (1992).

[83] P. M. Goorjian, A. Taflove, R. M. Joseph, and S. C. Hagness, *IEEE J. Quantum Electron.* **28**, 2416 (1992).

[84] R. M. Joseph, P. M. Goorjian, and A. Taflove, *Opt. Lett.* **17**, 491 (1993).

[85] R. W. Ziolkowski and J. B. Judkins, *J. Opt. Soc. Am. B* **10**, 186 (1993).

[86] M. Zoboli, F. Di Pasquale, and S. Selleri, *Opt. Commun.* **97**, 11 (1993).

[87] C. V. Hile and W. L. Kath, *J. Opt. Soc. Am. B* **13**, 1135 (1996).

[88] P. M. Goorjian and Y. Silberberg, *J. Opt. Soc. Am. B* **14**, 3523 (1997).

[89] H. S. Eisenberg, and Y. Silberberg, *Phys. Rev. Lett.* **83**, 540 (1999).

[90] A. Braun, G. Korn, X. Liu, D. Du, J. Squier, and G. Mourou, *Opt. Lett.* **20**, 73 (1995).

[91] E. T. J. Nibbering, P. F. Curley, G. Grillon, B. S. Prade, M. A. Franco, F. Salin, and A. Mysyrowicz, *Opt. Lett.* **21**, 62 (1996).

[92] J. K. Ranka, R. W. Schirmer, and A. L. Gaeta, *Phys. Rev. Lett.* **77**, 3783 (1996).

[93] A. Brodeur , C. Y. Chien, F. A.Ilkov, S. L. Chin, O. G. Kosareva, and V. P. Kandidov, *Opt. Lett.* **22**, 304 (1997).

[94] S. A. Diddams, H. K. Eaton, A. A. Zozulya, and T. S. Clement, *Opt. Lett.* **23**, 379 (1998).

[95] A. A. Zozulya, S. A. Diddams, A. G. van Engen, and T. S. Clement, *Phys. Rev. Lett.* **82**, 1430 (1999).

[96] A. Brodeur and S. L. Chin, *J. Opt. Soc. Am. B* **16**, 637 (1999).

[97] X. Liu, L. J. Qian, and F. W. Wise, *Phys. Rev. Lett.* **82**, 4631 (1999).

[98] I. G. Koprinkov, A. Suda, P. Wang, and K. Midorikawa, *Phys. Rev. Lett.* **84**, 3847 (2000).

[99] A. L. Gaeta and F. W. Wise, *Phys. Rev. Lett.* **87**, 229401 (2001).

[100] F. W. Wise, *Pramana* **57**, 1129 (2001).

[101] H. S. Eisenberg, R. Morandotti, Y. Silberberg, S. Bar-Ad, D. Ross, and J. S. Aitchison, *Phys. Rev. Lett.* **87**, 43902 (2001).

Chapter 8

Vortex Solitons

Optical vortex solitons represent a generalization of the concept of spatial dark solitons discussed in Chapter 4 to two transverse dimensions. A vortex soliton is associated with the self-trapping of a phase singularity embedded within a broad optical beam in a self-defocusing medium. Experimentally, vortex solitons appear in a natural way through the transverse instability of dark-soliton stripes in a nonlinear bulk medium (see Chapter 6). Unlike the spinning solitons of Section 6.5, which carry an angular momentum and require a self-focusing medium, the vortex solitons discussed in this chapter exist only in self-defocusing nonlinear media ($n_2 < 0$). Section 8.1 provides an introduction and links optical vortex solitons to similar solitons appearing in other fields. In Section 8.2, we discuss the transverse instability of dark-soliton stripes and how it leads to the formation of vortex solitons. The basic properties of vortex solitons are discussed in Section 8.3 together with the experimental methods used for their generation. Section 8.4 is devoted to the Aharonov–Bohm effect in the context of vortex scattering. Vortex arrays and lattices are discussed in Section 8.5. Section 8.6 is devoted to the concept of ring-shaped dark solitons.

8.1 Introduction

Spatial solitons, discussed in Chapters 2 and 4, form inside a planar optical waveguide and can be regarded as one-dimensional objects whose evolution is governed by the $(1 + 1)$-dimensional NLS equation. However, an additional transverse dimension must be considered in the case of a bulk nonlinear medium. As discussed in Chapter 6, the resulting $(2 + 1)$-dimensional NLS equation supports two-dimensional bright solitons that are confined by the nonlinear effects in both transverse dimensions in a self-focusing medium.

A natural question is: What is the two-dimensional extension of a spatial dark soliton in the case of a self-defocusing medium? The simplest object one may consider would be a *dark stripe* that propagates unchanged as a quasi-one-dimensional dark

249

soliton. This idea can be extended, and it leads to different kinds of objects, such as a *dark cross* (two stripes perpendicular to each other), a *dark grid* formed by superposing multiple dark stripes in the two transverse dimensions [1], and a superposition of bright and dark spatial solitons [2]. However, it has turned out that all such extended objects are unstable to the transverse modulational instability [3] because it leads to spatial modulations in the transverse dimensions. In the case of dark solitons, linear stability analysis was first carried out in 1988 for self-defocusing Kerr media using the cubic NLS equation [4]. It was found that dark-soliton stripes are always unstable to long-wavelength modulations (or low spatial frequencies) in the transverse dimensions. This instability can be suppressed when the dark stripe is bent to form a loop such that its radius is less than the smallest unstable wavelength [5]. However, such ring-shaped dark solitons do not exist as a stationary object; they either expand or shrink, depending on the initial conditions.

The important physical question then is: What kind of *stable stationary structures* can exist in the $(2 + 1)$-dimensional geometry? In the self-focusing case, transverse instability of a bright-soliton stripe leads to the creation of two-dimensional bright solitons of circular symmetry that can be stable in a non-Kerr bulk medium [3]. A similar scenario is expected for dark solitons. Indeed, numerical simulations, an asymptotic analytical theory, and experimental results all show that the transverse instability of dark-stripe solitons leads to the creation of multiple pairs of optical vortices with alternate polarities [6]–[12].

Importantly, optical vortices appear even in linear optics as phase singularities of optical beams [13]. However, like all linear beams of finite dimensions they diffract and the singular "holes" expand. It is only in a nonlinear medium that the singularity can be self-trapped to form the stable stationary structure associated with a vortex soliton. In fact, vortex solitons are the only stable two-dimensional stationary structure in the case of a bulk self-defocusing medium. Optical vortex solitons are sometimes described as the *stable, black, self-guided beams* of circular symmetry [14]. Such objects were encountered as early as 1958 in the form of topological excitations of an imperfect Bose gas in the context of superfluids [15, 16]).

Vortex solitons can be generalized to include the polarization properties of light, and two such generalizations are possible. First, two optical fields of different polarizations can have a vortex-like structure. This kind of *double-vortex soliton* is known to exist in other fields [17]–[19] and was introduced in 1994 to nonlinear optics [20, 21]. Second, a vortex of one polarization can guide the optical field of another polarization, giving rise to the two-dimensional generalization of the bright-dark soliton pairs [19, 22] (such objects are discussed in Chapter 9). Vortex solitons can also be generalized to include the temporal dimension, as discussed in Chapter 7. Such an extension leads to the concept of optical bullets in the context of bright solitons [23]. In the case of dark solitons, a vortex soliton in $(3 + 1)$-dimensional space (i.e., two transverse dimensions plus time and the propagation dimension) forms the so-called vortex line, which itself is unstable to transverse modulations [24]. This instability should lead to the formation of *dark optical bullets* in the anomalous-dispersion regime of the material. The situation is far from clear in the case of normal GVD, even for a self-focusing medium. In this case, the soliton-like objects are associated with a nontrivial phase structure and have not yet been fully explored.

8.2 Transverse Instability

n the case of a CW beam propagating inside a bulk nonlinear medium, dark solitons can be observed experimentally as dark stripes or grids, with properties similar to those of one-dimensional dark solitons [1]. However, a linear stability analysis shows that a dark-soliton stripe is unstable to transverse spatial perturbations [4]. Numerical simulations reveal that, as a result of the development of this instability, each dark stripe decays into multiple vortex solitons of alternative polarities [6]–[12]. This section focuses on the basic features of this transverse instability.

8.2.1 Linear Stability Analysis

The propagation of a CW beam in a bulk self-defocusing medium is governed by a $2 + 1$)-dimensional NLS equation. In the specific case of the Kerr nonlinearity, this equation can be written in its normalized form as

$$i\frac{\partial u}{\partial z} + \frac{1}{2}\left(\frac{\partial^2 u}{\partial x^2} + \frac{\partial^2 u}{\partial y^2}\right) - |u|^2 u = 0. \tag{8.2.1}$$

Similar to the analysis of Section 4.4, we can eliminate the background of constant amplitude u_0 through the transformation

$$z' = u_0^2 z, \quad x' = u_0 x, \quad y' = u_0 y, \quad u = u_0 \psi e^{-iu_0^2 z}, \tag{8.2.2}$$

and obtain the following equation for the new field ψ:

$$i\frac{\partial \psi}{\partial z} + \frac{1}{2}\nabla^2 \psi + (1 - |\psi|^2)\psi = 0, \tag{8.2.3}$$

where we have dropped the primes for simplicity of notation. The dimensionless field variable is defined such that $|\psi| \to 1$ as x and $y \to \pm\infty$. The NLS equation (8.2.3) has an exact solution describing a dark-soliton stripe, e.g., parallel to the x or y axis. For the dark stripe extending along the y axis, this solution can be written as

$$\psi_s(x,y,z) = a \tanh[a(x - vz)] + iv, \tag{8.2.4}$$

where the amplitude parameter a $(0 < a < 1)$ and the transverse velocity v $(v^2 \le 1)$ are coupled through the relation $a^2 + v^2 = 1$.

The linear stability analysis of this solution can be performed following the method outlined in Section 6.1. If we consider a perturbation in the form $\exp(py)$, the dark stripe is found to be unstable for all values of p such that $p < p_{cr}(k)$, where the critical value of p is given by [4]

$$p_{cr}^2(k) = k^2 - 2 + 2\sqrt{k^4 - k^2 + 1}. \tag{8.2.5}$$

The instability domain is bounded by the curve $p = p_{cr}(k)$, and when we apply a periodic perturbation with any wave number $p < p_{cr}(k)$, the amplitude of the dark-soliton stripe will grow exponentially in the z direction. Of course, the exponential growth is

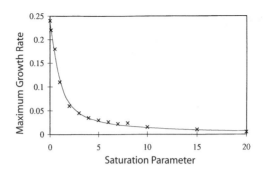

Figure 8.1: Maximum growth rate of the transverse instability in a saturable nonlinear mediu as a function of the saturation parameter s. Crosses indicate numerical results. (After Ref. [25] ©1997 OSA.)

halted once the perturbation becomes large enough for the linear stability analysis to become invalid. Nonlinear regimes of such an instability have been investigated nu merically [6, 7] as well as analytically using an asymptotic technique valid near th marginal stability point [9]. The results show that the dark-soliton stripe decays int an array of optical vortices such that two neighboring vortices have opposite polari ties. This kind of instability-induced evolution always occurs for a black stripe (zer intensity at the center). A gray-soliton stripe does not decay into vortices but under goes oscillations that persist over long lengths and are accompanied by the emission o radiation propagating along the background.

The transverse instability of a dark-soliton stripe in a strongly saturable optica medium has also been studied using a generalized NLS equation, both analytically an numerically [25]. By employing an asymptotic-expansion technique for perturbation with small wave numbers ($p \ll 1$), the growth rate $\lambda_\perp$ of transverse modulations i found to be

$$\lambda_\perp = p(\partial P_s/\partial v)^{1/2}\sqrt{H_s}, \tag{8.2.6}$$

where $P_s(v)$ and $H_s(v)$ are, respectively, the renormalized momentum and Hamiltonia introduced in Chapter 4 for dark solitons. The subscript s is a reminder that thes invariants are calculated using the stationary solution given in Eq. (8.2.4). Figure 8. shows the maximum growth rate $\lambda_\perp^{max}$ of the transverse instability as a function of th saturation parameter s when the nonlinearity saturates with intensity I as $(1+sI)^{-}$ Clearly, saturation of the nonlinearity helps to suppress the transverse instability.

8.2.2 Experimental Observations

Any experiment on spatial dark solitons must use a finite-size background beam eve though the theoretical analysis of the transverse instability assumes a background ex tending to infinity. However, at high enough nonlinearities, the transverse instabilit has been observed even with finite-size beams in several experiments performed us ing using nonlinear media as diverse as rubidium vapor and photorefractive crystal [10]–[12].

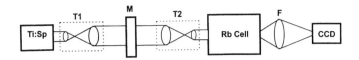

Figure 8.2: Schematic of the experimental setup used for observing vortex solitons. (After Ref. [10]; ©1996 OSA.)

In the 1996 experiment involving rubidium vapor [10], the CW beam was linearly polarized and its Gaussian-beam shape was slightly elliptical. Its wavelength was tuned close to the atomic resonance line near 780 nm. A π phase jump was imposed across the beam center using a mask, and the resulting beam was focused onto the nonlinear medium. The vapor density could be increased up to 10^{13} cm^{-3} by changing the cell temperature. The near-field images of the beam at the output of the cell were recorded using a CCD camera. Figure 8.2 shows schematically the experimental setup used to observe the vortex solitons. The important steps that enabled the observation of transverse instability were (i) the resonant enhancement of the nonlinearity by tuning the laser frequency close to the D_2 line of rubidium atoms (detuning < 1 GHz) and (ii) the use of maximum vapor pressure consistent with tolerable absorption. The input-beam power was 240 mW, and the beam waist at the $1/e^2$ point was 0.3 mm. The maximum nonlinear change in the refractive index was estimated to be $\sim 10^{-4}$ in this experiment.

Figure 8.3 shows a series of output intensity patterns calculated numerically (left column) and observed experimentally (right column) as the vapor density was increased by increasing cell temperature while the detuning was kept fixed at 0.85 GHz. For the smallest vapor density (a), the nonlinear effects were negligible. As a result, both the dark stripe and the background beam spread. With increasing temperature (i.e., higher nonlinearity), the background beam spread even more because of self-defocusing, but it developed a narrow dark-soliton stripe at its center, as seen in frame (b). Further increase in the temperature led to a periodic modulation of the the dark stripe, seen in frame (c). At higher temperatures, the breakup of the stripe began, initially appearing as "snake-type" bending, seen in frame (d), and then breaking into dark spots at the inflection points in the bends [frame (e)]. At the highest nonlinearity, the dark spots acquired a circular symmetry, seen in frame (f). These dark spots represent the predicted formation of pairs of optical vortex solitons.

The preceding breakup scenario was found to be sensitive to the size of the phase step imposed on the input beam. Even a slight misalignment of the phase mask from the optimum location (corresponding to a phase jump close to π) stopped the formation of optical vortex pairs at the final stage. Under such conditions, only the snake-type bending of the soliton stripe was observed. This observation supports the theoretical prediction that the growth rate of the instability is reduced for gray-soliton stripes.

Numerical simulations based on a generalized NLS equation that included the saturation and dissipative effects for comparison with the experimental results [10] are shown in the left column of Figure 8.3. The agreement between theory and experiment is remarkably good, in the sense that most of the experimental features are reproduced quite well. Even the instability growth rate seen in the experiment is well approximated

Simulation   Experiment

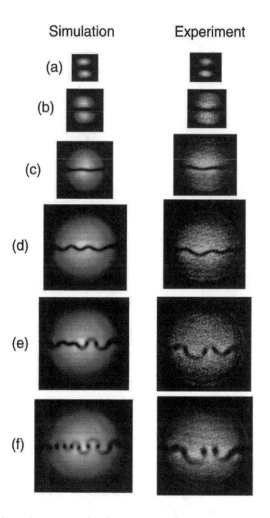

Figure 8.3: Output intensity patterns showing the onset of transverse instability of a dark-soliton stripe as the nonlinearity is increased by varying the vapor density through temperature changes. The cell temperature was (a) 40, (b) 72, (c) 82, (d) 90, (e) 112, and (f) 125°C. Laser detuning was held constant at 0.85 GHz. (After Ref. [10]; ©1996 OSA.)

by the simulations. However, the period of the transverse perturbation corresponding to the maximum growth rate appears to be *smaller* than that seen in the experiment, by a factor of about 1.5. This discrepancy is most likely due to a physically more complicated nonlinear response of rubidium atoms than that used in the simulations. Difficulty in accurately characterizing the input field may also be responsible for this disagreement, because the numerical simulations are found to be quite sensitive to the exact form of the input field. It should be noted that this sensitivity to input field was not observed in the experiments, suggesting that the breakup process may have been partially stabilized by some physical mechanisms not included in the model.

A similar behavior was observed in the experiments in which a photorefractive crystal was used as the nonlinear medium [11, 12]. A biased SBN crystal was irradiated with a 10-mW CW beam (from a He–Ne laser) containing the phase step. The effective nonlinearity was varied by increasing the bias voltage. When no voltage was applied, the dark stripe spread due to diffraction, as did the background beam. As the applied voltage was increased, a dark-soliton stripe formed. A further increase in the voltage to 990 V led to snake-like bending of the stripe, similar to that seen in Figure 8.3(c). However, further evolution at higher voltages was quite different from that seen in Figure 8.3, in the sense that the vortex solitons were not clearly visible. Interferometric measurements on the output beam indicated that the optical field did vanish periodically at intervals of about 40 μm.

8.3 Properties of Vortex Solitons

As is well known in the fields of both optics and acoustics, vortices can appear as the specific modes of a linear wave equation, and they are associated with a phase singularity of the linearly diffracting field [26]–[29]. A phase singularity, also called *wave front screw dislocation*, can be generated, for example, through the scattering of light or sound from a rough surface [30, 31]. In a self-defocusing nonlinear medium, such screw dislocations can create a stationary beam structure with a phase singularity, resulting in a vortex soliton. This section focuses on the study of the properties and dynamics of vortex solitons.

8.3.1 Stationary Solutions

The existence of vortex solutions for the $(2+1)$-dimensional cubic NLS equation can be established using an analogy between the fields of optics and fluid dynamics. Employing the Madelung transformation [32]–[34],

$$\psi(\mathbf{r},z) = \chi(\mathbf{r},z)\exp[\varphi(\mathbf{r},z)], \qquad (8.3.1)$$

where $\mathbf{r}$ is a two-dimensional vector with coordinates x and y, we can transform the NLS equation (8.2.3) into the following set of two coupled equations:

$$\frac{\partial \chi^2}{\partial z} + \nabla \cdot (\chi^2 \nabla \varphi) = 0, \qquad (8.3.2)$$

$$\frac{\partial \varphi}{\partial z} + \frac{1}{2}(\nabla \varphi)^2 = 1 - \chi^2 + \frac{\nabla^2 \chi}{2\chi}. \qquad (8.3.3)$$

These equations can be viewed as those governing the conservation of mass and momentum for a compressible inviscid fluid of density $\rho = \chi^2$ and velocity $\mathbf{V} = \nabla \varphi$, with the pressure defined as $p = \rho^2/2$. More importantly, this kind of analogy between optics and fluid mechanics remains valid even for the generalized NLS equation (4.2.1) with an arbitrary form of $g(|u|^2)$, provided the effective pressure is defined as

$$p(\rho) = \int \rho \frac{dg(\rho)}{d\rho} d\rho. \qquad (8.3.4)$$

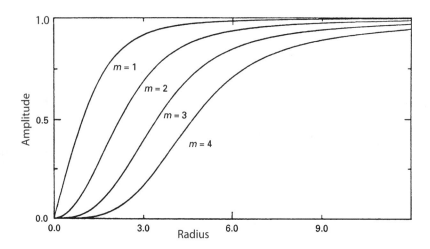

Figure 8.4: Vortex profiles in a self-defocusing Kerr medium for four values of the integer vortex charge m. (After Ref. [35]; ©1990 Elsevier.)

The analogy, however, is not exact because, in addition to the standard pressure, Eq. (8.3.3) includes a second term that has no analog in fluid mechanics. This term results from the so-called *quantum-mechanical pressure* in the context of superfluids.

The Madelung transformation is singular at the points where $\chi = 0$. Around such points located on the plane (x, y), the circulation of $\mathbf{V}$ is not zero but equals 2π. These points are the *topological defects* of the scalar field and are called *vortices*. To find the stationary solution corresponding to a vortex soliton (a dark soliton with circular symmetry), we look for solutions of the cubic NLS equation in the polar coordinates r and θ and assume that the solution of Eq. (8.2.3) can be written as

$$\psi(r, \theta; z) = U(r)e^{im\theta}, \tag{8.3.5}$$

where the integer m is the so-called *winding number*, also called the *vortex charge*, and the real function $U(r)$ satisfies

$$\frac{d^2U}{dr^2} + \frac{1}{r}\frac{dU}{dr} - \frac{m^2}{r^2}U + (1 - U^2)U = 0, \tag{8.3.6}$$

with the boundary conditions

$$U(0) = 0, \qquad U(\infty) = 1. \tag{8.3.7}$$

The continuity of u at $r = 0$ forces the first condition, while $U(\infty) = 1$ is consistent with a uniform background of intensity U_0^2 as $r \to \infty$.

Equation (8.3.6) can be solved numerically to find the shape of the vortex soliton for different values of m. Figure 8.4 depicts $U(r)$ for four different values of m [35, 36]. The region in the vicinity of $r = 0$, where $U(r)$ is significantly less than 1, is called the *vortex core*. The functional form of $U(r)$ near $r = 0$ and $r = \infty$ can be established directly from Eq. (8.3.6) by taking the appropriate limit and is found to be

$$U(r) \sim \begin{cases} ar^{|m|} + O(r^{|m|+2}) & \text{as } r \to 0, \\ 1 - \frac{m^2}{2r^2} + O(1/r^4) & \text{as } r \to \infty. \end{cases} \qquad (8.3.8)$$

The structure of the vortex soliton for an arbitrary form of the nonlinearity can be found using the same method and solving numerically for the amplitude function $U(r)$. No qualitatively new features are found when the nonlinearity is allowed to saturate [37]. However, the effective diameter of the vortex core increases almost linearly with the saturation parameter $s = I_0/I_s$, where I_s and I_0 are the saturation and background intensities, respectively [38].

Stability of vortex solitons associated with the generalized NLS equation has not yet been fully addressed. However, it is believed that vortices with the winding numbers $m = \pm 1$ are *topologically stable* but that those with larger values of $|m|$ are unstable and decay into $|m|$ single-charge vortices. In the context of a superfluid, the multicharged vortex solitons are found to survive for a relatively long time [39]. A similar behavior is found to occur in optics [40]. For this reason, multicharged vortices are classified as being metastable. As an example, an intentional perturbation of a triply charged vortex leads to its incomplete decay to a long-lived doubly charged vortex and a singly charged vortex [41], confirming that saturation of nonlinearity can effectively suppress the instability. We should, however, stress that multicharged vortices are strongly unstable in anisotropic nonlinear media [42].

8.3.2 Vortex Rotation and Drift

Experimentally, a vortex soliton appears as a dark region that maintains its shape on a diffracting background beam and displays a nontrivial dynamical behavior. Phenomena such as the rotation and radial drift of the vortex relative to the background CW beam are often observed experimentally, even though they cannot be predicted from a casual analysis of the stationary solution of the NLS equation. A proper theoretical description of these effects requires special analytical techniques capable of analyzing the vortex motion [43, 44]. Physically, a dynamical feature such as rotation or drift of a vortex soliton results from a nonuniform intensity profile of the background field—typically a Gaussian beam. The stationary solutions shown in Figure 8.4 were obtained assuming a background beam of infinite extent and constant intensity.

As discussed in Section 1.2, the propagation of a CW beam in a bulk nonlinear medium with the intensity-dependent refractive index, $n = n_0 + n_{nl}(I)$, is governed by Eq. (1.2.12) in the paraxial and scalar approximations. Rather than using the normalization scheme adopted in preceding chapters, it is useful to introduce the dimensionless variables by using the value of the background field at the vortex center, $I_0 \equiv I_b(\mathbf{r}_0)$. If we define the nonlinear length as $L_{nl} = (k_0|n_2|I_0)^{-1}$ and introduce the transformation

$$z = Z/L_{nl}, \qquad \mathbf{r} = \mathbf{R}\sqrt{2\beta_0/L_{nl}}, \qquad A = Bu, \qquad (8.3.9)$$

where $B = \sqrt{I_b}\, \exp(i\theta_b)$ is the background field, the vortex field $u(\mathbf{r}, z)$ satisfies the following equation [45]:

$$i\frac{\partial u}{\partial z} + \nabla^2 u + [g(I_b|u|^2) - g(I_b)]u = -\nabla u \cdot \mathbf{f}, \qquad (8.3.10)$$

where $g(I) = (|n_2|I_0)^{-1}n_{nl}(I)$ and the complex vector $\mathbf{f}$ depends on the gradient of the background field as

$$\mathbf{f} \equiv \mathbf{f}_r + i\mathbf{f}_i = \nabla B \equiv \tfrac{1}{2}\nabla \ln I_b + i\nabla \theta_b. \qquad (8.3.11)$$

The boundary condition $|u| \to 1$ applies for large values of $\mathbf{r}$. The background field $B(\mathbf{r}, z)$ itself satisfies the generalized NLS equation

$$i\frac{\partial B}{\partial z} + \nabla^2 B + g(|B|^2)B = 0. \qquad (8.3.12)$$

The physical meaning of Eqs. (8.3.10)–(8.3.12) is as follows. The nonuniform background field evolves as dictated by Eq. (8.3.12). However, the radial dependence of the background intensity I_b and the phase θ_b creates a force $\mathbf{f}$ under which the vortex field evolves. The main effect of this force is that the vortex drifts from its original position and may even rotate. The drift is quantified by the vortex velocity V, defined as

$$\frac{d\mathbf{r_0}}{dz} = \mathbf{V}, \qquad (8.3.13)$$

where r_0 denotes the location of the vortex.

Consider first the simpler case of a self-defocusing Kerr medium by choosing $n_{nl}(I) = -|n_2|I$. In this case, Eq. (8.3.10) takes the form of the following perturbed NLS equation [cf. Eq. (8.2.3)]:

$$i\frac{\partial u}{\partial z} + \nabla^2 u + \frac{I_b}{I_0}(1 - |u|^2)u = -\nabla u \cdot \mathbf{f}. \qquad (8.3.14)$$

This equation can be used to analyze the motion of a slow-moving vortex within a shallow-gradient background field using the method of *matched asymptotic expansion* [44]. More specifically, the asymptotic expansions of the vortex field in the regions near and far from the vortex core are matched to find the field at an intermediate distance, a technique that has been used with success in several other contexts [46]–[48]. In the present context, it is based on the assumption that the background intensity and phase, I_b and θ_b, vary slowly in comparison with the vortex scale. Then the problem is to describe changes in the position $\mathbf{r_0}(z)$ of the vortex (the vortex drift) under the action of a slowly varying background field. To understand the notion of a background field, one may picture it as the field that would exist if the vortices were somehow removed.

The basic idea consists of transforming Eq. (8.3.14) to a reference frame moving with the vortex-drift velocity $\mathbf{V}(z)$ so that the vortex appears stationary in the new reference frame. For this purpose, we make the transformation $\mathbf{r} - \mathbf{V}z \to \mathbf{r}$. Moreover, since the background field does not change significantly on the length scale associated with the vortex, the term I_b/I_0 in Eq. (8.3.14) can be expanded around $\mathbf{r} = \mathbf{r_0}$ as

$$I_b/I_0 \approx 1 + \mathbf{r} \cdot \nabla \ln I_b|_{\mathbf{r}=\mathbf{r_0}} \equiv 1 + \mathbf{r} \cdot \mathbf{f}_{0r}, \qquad (8.3.15)$$

where f_0 is is the value of f at the vortex position $\mathbf{r} = \mathbf{r_0}$ and the subscript r denotes the real part of this quantity. Since u does not change with z in the new reference frame, Eq. (8.3.14) reduces to the following equation:

$$\nabla^2 u + (1 - |u|^2)u = \varepsilon F_p(u), \tag{8.3.16}$$

where ε denotes that $F_p(u)$ is a small perturbation given by

$$F_p(u) = (i\mathbf{V} - \mathbf{f}_0) \cdot \nabla u - \mathbf{r} \cdot \mathbf{f}_{0r}(1 - |u|^2)u. \tag{8.3.17}$$

Equation (8.3.16) can be solved using perturbation theory. For this purpose, we expand the vortex field u in terms of the perturbation parameter as $u = u_0 + \varepsilon u_1 + \dots$ and substitute the expansion into Eq. (8.3.16). In the zeroth-order approximation in ε, we obtain the standard stationary NLS equation for the vortex in the form

$$\nabla^2 u_0 + (1 - |u_0|^2)u_0 = 0. \tag{8.3.18}$$

In the polar coordinates of the moving frame, its solution can be written as $u_0 = U(r)\exp(im\phi)$, where m is the vortex charge. The radial part $U(r)$ satisfies the following ordinary differential equation:

$$\frac{d^2 U}{dr^2} + \frac{1}{r}\frac{dU}{dr} + \left(1 - \frac{m^2}{r^2} - U^2\right)U = 0. \tag{8.3.19}$$

The solution of this equation describes the vortex formed within a constant-intensity background. It was first studied in the context of superfluids [15].

The effects of a nonuniform background appear in the the first-order approximation in ε and are governed by the following linear but inhomogeneous equation:

$$\nabla^2 u_1 + u_1 - 2|u_0|^2 u_1 - u_0^2 u_1^* = F_p(u_0). \tag{8.3.20}$$

The solvability condition for Eq. (8.3.20) can be used to find the equation of motion for the vortex core. These conditions follow from the orthogonality of the inhomogeneous part, $F_p(u_0)$, to the two components of the translational eigenfunction, ∇u_0^*, of the adjoint homogeneous equation. The method of matched asymptotics then allows us to write the solvability condition in the following vector form [44]:

$$\frac{d\mathbf{r_0}}{dz} \equiv V(z) = \left(-\nabla \theta_b + \frac{m}{2}C\mathbf{J}\nabla(\ln I_b)\right)\Bigg|_{\mathbf{r}=\mathbf{r_0}}, \tag{8.3.21}$$

where C is slowly varying function of the background intensity I_b and is given by

$$C(I_b) = -\ln\left[c\exp(\gamma_e)|\nabla I_b|/4I_b\right]. \tag{8.3.22}$$

In these equations, c is found numerically ($c \approx 1.126$ for the cubic nonlinearity), γ_e is the Euler constant ($\gamma_e \approx 0.577$), and the matrix

$$\mathbf{J} = \begin{pmatrix} 0 & -1 \\ 1 & 0 \end{pmatrix}. \tag{8.3.23}$$

Equation (8.3.21) governs the motion of the vortex core for a nonuniform background field in a self-defocusing Kerr medium.

In the case of a non-Kerr medium with the nonlinearity $g(I)$ in Eq. (8.3.10), we can follow the same approach to find that Eq. (8.3.14) should be replaced with

$$i\frac{\partial u}{\partial z} + \nabla^2 u + \frac{g(I_b)}{g(I_0)}\left[1 - \frac{g(|u|^2 I_b)}{g(I_b)}\right]u = -\nabla u \cdot \mathbf{f}, \tag{8.3.24}$$

The zeroth-order equation for the stationary vortex profile is now given by

$$\nabla^2 u_0 + (1 - G(|u_0|)^2)u_0 = 0, \tag{8.3.25}$$

where $G(x) = g(I_0 x)/g(I_0)$. In the first-order approximation, we recover Eq. (8.3.20) but with a modified form of the perturbation $F_p(u_0)$. As a result, the further analysis remains unchanged. More specifically, the vortex motion is still governed by Eq. (8.3.21). Even the coefficient C in Eq. (8.3.22) has the same form. The only difference appears in the normalization used: The nonlinear index change $n_2 I_0$ should be replaced with its non-Kerr value $n_{nl}(I_0)$. In the case of a saturable nonlinearity, one can use $g(I) = I/(1 + sI)$, where larger values of s correspond to a stronger saturation of the nonlinearity. The parameter c in Eq. (8.3.22) changes from 1.126 at $s = 0$ to 1.412 when $s = 1$ and to 1.639 when $s = 2$ [44].

To describe the drift and rotation of a vortex induced by a diffracting background field, one should know the evolution of the background field a priori so that the radial and angular velocity components associated with the vortex motion can be calculated from Eq. (8.3.21). It is often assumed that the background field in the absence of the vortex evolves in approximately the same manner as it would when hosting a vortex. Even a qualitative knowledge of the propagation behavior of the background field can be used, along with Eq. (8.3.21), to predict the motion of the vortex nested in that field. The following example of a vortex nested in a Gaussian beam serves to illustrate how vortex motion can be predicted using this technique [43].

The velocity of a vortex in the transverse x–y plane is governed by Eq. (8.3.21) and has two components arising from the phase and intensity gradients of the background field at the vortex position. The first component, $-\nabla\theta_b$, is directed normal to the wave front of the background (in the direction of energy flow) and gives rise to the *radial motion* of the vortex. The second component, $\frac{1}{2}mC\mathbf{J}\nabla(\ln I_b)$, is directed along the intensity contours (called *isophotes*) of the background, with the sense of direction given by the vortex charge $m = \pm 1$. For a Gaussian-shape background, the isophote in any transverse plane is a circle, and the second component of velocity describes the *angular motion* of the vortex [49]. A continuous flattening of the intensity profile induced by self-defocusing of the background reduces the intensity gradient and slows down the rotation experienced by a vortex in the case of linear propagation. For flatter intensity profiles, the motion of the vortex depends mostly on the background wave front, and we recover the results obtained in this limit [50, 51]. A qualitatively similar behavior was observed in numerical simulations based on the complex Ginzburg–Landau equation [52].

Equation (8.3.21) can be integrated analytically in the case of a Gaussian background beam diffracting linearly [43]. Although such an approximation neglects the nonlinear evolution of the background field, it is nonetheless quite useful for gaining physical insight. If we neglect the nonlinear term in Eq. (8.3.12), a Gaussian beam is found to diffract as

$$B(r,z) = b\exp(-br^2/2) \tag{8.3.26}$$

where $b = (1 + izL_{nl}/L_d)^{-1}$ and $L_d = \beta_0 w_0^2$ is the diffraction length for an input Gaussian beam of width w_0. Integrating Eq. (8.3.21) and using the polar coordinates (r_0, ϕ) for the position vector $\mathbf{r}_0$, we obtain the simple relations

$$\frac{r_0(z)}{r_0(0)} = \frac{w(z)}{w_0}, \qquad \phi(z) = \phi(0) + mC\int_0^z \frac{d\zeta}{w^2(\zeta)}, \tag{8.3.27}$$

where the beam width at a distance z is given by [53]

$$w(z) = w_0[1 + (zL_{nl}/L_d)^2]^{1/2}. \tag{8.3.28}$$

These equations have been found useful in the context of vortex steering [44].

The problem of *vortex interaction* can also be studied using the preceding approach if the background field in Eq. (8.3.10) is assumed to include not only the background but also the remaining vortices, while the field u corresponds to the vortex under study. A single vortex has circular isophotes centered on the vortex core. Thus, one vortex interacting with the background field generated by the other vortex will move in the direction normal to the line connecting its core with the background vortex. The situation is exactly the same for the other vortex forming the pair. The resultant motion of the vortex pair can therefore only be circular or parallel, depending on the vortex chirality. It is also possible to include the effects of a nonplanar background for estimating its influence on the interaction between two vortices [49].

A simple physical argument can clarify the physical mechanism underlying the vortex motion. Consider the "momentum" of a small element of the background field surrounding the vortex core and defined as $\int I\nabla\theta \, d\mathbf{r}$. First assume that the intensity is uniform. Then the momentum is proportional to the sum of the vortex-phase gradient and the background-phase gradient around the core. Since the former contribution is zero, the element around the core has a momentum proportional to the background-phase gradient at the vortex position. In the second case, assume that the background phase is uniform. Then the background-phase gradient around the core is zero. The momentum of the element is then proportional to the sum of the vortex-phase gradient weighted by the background intensity around the core. Using the concept of vector summation, it is clear that any imbalance in the background intensity over a small region around the vortex core gives rises to a net momentum component in the direction normal to the intensity gradient (i.e., along the isophote). That the origin of vortex dynamics can be justified using such a simple physical picture suggests that the equation of motion may remain valid even beyond the specific approximations made in its formal derivation.

8.3.3 Experimental Results

Optical vortex solitons were first observed in a 1992 experiment using a self-defocusing thermal nonlinearity and a phase mask that imposed an approximately helical phase structure of an optical vortex on the input beam [55]. The mask contained regions of 0, π, and 2π phase thickness surrounding a single point in the plane of the mask. At the output of the nonlinear medium, a dark spot on the CW beam was observed localized at that point. The interferometric measurements indicated the presence of a phase dislocation at that point, supporting the idea that an optical vortex soliton had formed. By copropagating a He–Ne beam as a guided mode through the induced structure, it was deduced that the a two-dimensional waveguide existed in the vicinity of the vortex.

A more practical method of imposing a helical phase structure on the input beam has been suggested [28]. In this method, the phase mask is computer generated by numerically calculating the interference pattern for an on-axis spherical wave (or an off-axis plane wave). The first-order diffracted beam from such a mask contains the phase structure needed to form an optical vortex. This method has been used in several experiments [56]–[58]. Its main advantage is that both the number and the chirality (clockwise or anticlockwise) of the vortices induced on the input beam can be precisely controlled. Another method of creating vortices is based on the fact that the far field of a single vortex is identical to the first-order Gauss–Laguerre mode of an optical resonator [59].

To observe rotating vortex solitons, a similar phase-mask technique was used in a 1994 experiment to create a pair of vortices, shifted symmetrically from the beam center and having the same charge [49]. Each off-axis vortex rotated around the axis of the Gaussian background beam by $90°$ as it propagated from the beam waist to infinity [54]. This rotation matched exactly the well-known Guoy shift, which characterizes the net change in the on-axis phase for a Gaussian beam relative to that of a plane wave. The defocusing action of the nonlinear medium acts to flatten the phase front of the background beam, reducing the net Guoy shift at a given distance. Hence, the effect of the nonlinearity is to reduce the natural rotation of a vortex experienced during linear propagation (rotation does not occur for a uniform plane wave). Using rubidium vapor pumped resonantly near 780 nm together with a focused Gaussian beam with its confocal parameter much shorter than the cell length, the rotation of a vortex by about $90°$ could be observed with increasing nonlinearity with little change in the radial position of the vortex. This motion may be useful for creating an optical rotary switch based on the concept of vortex solitons.

Several experiments have used the nonlocal response of photorefractive crystals for studying vortex solitons [60]–[62]. An SBN crystal is often used for this purpose. In a 1994 experiment, either the input beam was the coherent 'donut' mode of a laser or such a beam shape was created artificially by combining two independent beams with vertical and horizontal notches and a $\pi/2$ relative phase between them [60]. Both methods produced a Gauss–Laguerre beam with the desired azimuthal phase dependence and were able to create a vortex soliton at the output of the nonlinear medium. In spite of the anisotropy of the nonlinear response in photorefractive materials, it was possible to generate nearly circular vortex solitons under appropriate input conditions.

Figure 8.5 shows the photographs and beam profiles of the optical beam at the input (column a) and output faces (columns b and c) of the crystal. The vortex size is 18 μm (FWHM) at the input (a), and it diffracts to roughly 43 μm at the output (b) when no bias voltage is applied to the 11.7-mm-long crystal. When a bias voltage is applied (relative to the crystal c axis), the crystal turns into a self-defocusing medium. Figure

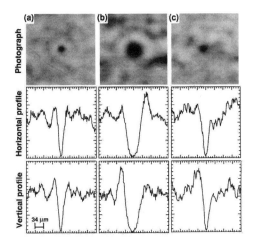

Figure 8.5: Creation of a vortex soliton in a 11.7-mm-long SBN crystal: (a) input beam, (b) output beam at $V = 0$, and (c) output beam at $V = -450$ V. Horizontal and vertical intensity profiles are shown in the three cases in the middle and bottom rows, respectively. (After Ref. [61]; ©1997 OSA.)

8.5(c) is obtained when the applied voltage (between the electrodes separated by 5.3 mm) is set to -450 V and the intensity ratio (the ratio between the peak intensity of the vortex beam and the sum of the background illumination and the dark irradiance) is set to 0.95. It is apparent that the vortex is self-trapped in both transverse dimensions and maintains a nearly circular shape. Because the vortex is self-trapped to its initial size and because the self-trapping persists in the steady state, such a vortex is called a *screening vortex soliton* [61]. The right experimental conditions are critical for generating such vortex solitons. If the values of applied voltage or the intensity ratio, both of which control the nonlinearity, deviate considerably from those supporting a circular vortex soliton, the vortex shape becomes elliptical rather than circular.

Several other experiments have shown the extreme sensitivity of vortex solitons to the input conditions. In a 1996 study of the spatial dynamics of vortices in an anisotropic photorefractive medium [62], the experimental and numerical results indicated a stretching of the vortex and its subsequent decay. Sometimes the decay of a single vortex soliton resulted in the bound state of two counterrotating vortices. These results clearly indicate a high sensitivity to the initial conditions for generating vortex solitons in photorefractive media. They also show the importance of a finite-width background that can lead to a chance of the vortex charge. Further experimental studies will help in resolving these issues.

Many other experiments have studied the creation and dynamics of optical vortex solitons in saturating nonlinear media [6, 10]. In a 1996 experiment [63], single optical vortices were generated in rubidium vapor (for experimental details, see Ref. [25]). A comparison of numerical simulations with the experimental data showed that both saturation and absorption must be taken into account to obtain a reasonable agreement. The interaction of the vortex soliton with a weak coherent beam can be used for steer-

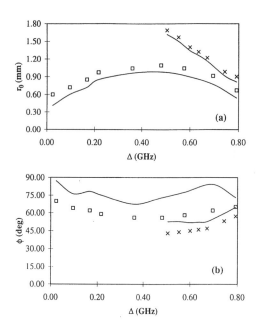

Figure 8.6: Radial (a) and angular (b) positions of the vortex as a function of the detuning below the resonance at a Rb-cell temperature of 88° (squares) and 108°C (crosses). Solid curves show the analytical results. (After Ref. [44]; ©1998 Elsevier.)

ing the soliton. This idea is closely related to the steering phenomenon occurring for vortices nested inside a linearly propagating beam [29].

Steering of vortex solitons was realized in a 1996 experiment [43] using a coherent background wave whose intensity was ∼20% of that of the background beam containing the off-axis vortex. By adjusting the relative phase of the background wave, the position of the vortex could be moved to any selectable angular position in the output beam. The analytic model developed in Section 8.3.2. can be used to describe the vortex motion. It shows that the position of the vortex could be described to result from the combination of a radial drift proportional to the gradient of the phase of the background beam and an azimuthal drift proportional to the gradient of the intensity (this effect was observed earlier in numerical simulations [64]). Using the Gaussian approximation for the background wave, it is found that the radial displacement of the vortex is magnified by the ratio of the beam radii with and without self-defocusing. Experimental results confirm this prediction even when the background deviates from the exact Gaussian form. Figure 8.6 shows the radial (a) and angular (b) positions of the vortex at the output of the Rb cell as a function of the detuning below the atomic resonance (which changes the strength of self-defocusing). To assess the affect of a non-Gaussian background, numerical simulations were carried out with the actual form of the background and yielded an excellent fit with the experimental data. Interestingly, a small but definite saturation of the transfer function can be observed even when the vortex is launched very close to an edge of the background beam.

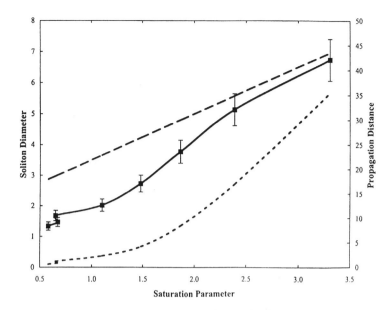

Figure 8.7: Experimentally measured values of the vortex diameter (solid curve with error bars) in a saturable medium as a function of the saturation parameter s. The corresponding theoretical values in the stationary regime are shown as a dashed curve. The dotted curve displays the effective propagation distance. (After Ref. [38]; ©1998 OSA.)

A detailed comparison between the theoretical and experimentally measured vortex diameters has been carried out for a saturable self-defocusing medium [38]. Saturation was characterized by a dimensionless saturation parameter $s = I_0/I_{sat}$, where I_0 and I_{sat} are the background and saturation intensities, respectively. It was noticed that the vortex profile and its diameter depended strongly on the degree of saturation and that the FWHM diameter of the vortex increased almost linearly with s. To link the analytic theory, which deals with stationary solutions on an infinite uniform background, with the experiments where input beams with a somewhat arbitrary intensity profile and helical phase are used, numerical simulations were used to study the formation of a vortex soliton for several typical input profiles [38]. At the same time, the dynamics of the vortex propagation was investigated experimentally using Rb vapor as a medium with a variable saturating nonlinearity. Rotation of the initially elliptical vortex core was observed as the soliton formed. Measurements of the vortex diameter as a function of the saturation parameter are shown in Figure 8.7. A nearly linear growth of the vortex size with s, as predicted by theory, could be observed only in the region of high saturation, where the effective propagation distance (shown in Figure 8.7 by a dotted curve) was long enough to realize the stationary case. At lower saturation levels, the measured vortex diameter was less than the theoretically predicted value. These results indicate that the experimentally observed vortex was far from being in the steady-state regime.

As a potential application of vortex solitons, we mention the three-dimensional

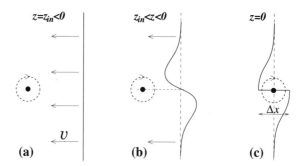

Figure 8.8: Schematic illustration of the wave-front deformation and splitting during scattering of a plane wave with a vortex. (After Ref. [71]; ©2001 APS.)

trapping of low-index particles (20-μm-diameter hollow glass spheres in water) by using a single, strongly focused Gaussian beam containing an optical vortex [65]. Transverse trapping was attributed to the gradient force directed toward the vortex core; it allowed to trap the high-index particles in a ring pattern. The computer-based holographic technique was used in this experiment to generate the optical vortex.

8.4   Aharonov–Bohm Effect

The magnetic vector potential is known to influence the dynamics of a charged particle, even if the magnetic field vanishes (e.g., when the magnetic field is confined to a cylinder into which the charged particle cannot penetrate). This phenomenon is known as the *Aharonov–Bohm effect* [66]. It has been linked to a more general phenomenon of the *geometrical phase* induced by wave-front dislocations [67]. It has also been shown to have a classical analog in the scattering of a linear wave by a vortex [68, 69]. In a 1999 study, the interaction of a water wave with a vortex [70] revealed both similarities with and differences from the Aharonov–Bohm effect, allowing one to observe directly the macroscopic aspects of the geometrical phases. A similar phenomenon occurs for optical vortices in the framework of the NLS equation—scattering of an optical vortex in a bulk nonlinear medium—and results in an optical analog of the Aharonov–Bohm effect [71].

Figure 8.8 shows the scattering problem schematically. Consider a plane-wave front moving with velocity v and encountering a vortex. As a result of the scattering, the wave front is deformed, as shown in part (b). After the scattering process is complete, the wave front is split into two parts by an amount $\Delta x = -2\pi/v$. In the linear limit, the velocity v coincides with the sound speed, and the result is valid in the long-wavelength approximation in which the wave dispersion is neglected and all linear waves propagate with the same velocity. The phenomenon of the wave-front splitting after the scattering by a vortex is related directly to the Aharonov–Bohm effect [72, 73].

In the nonlinear regime of this scattering process, the plane wave is replaced with a spatial dark-soliton stripe forming in a self-defocusing nonlinear medium [74]. The corresponding nonlinear scattering problem is rather complicated. It has been studied only approximately at large distances from the vortex core using a multiscale asymptotic analysis [71]. The deformation of a dark-soliton stripe because of its interaction with the vortex field is found to be

$$h(y,z) = \frac{v^2}{(v^2 - \lambda^2)} h_L(y,z),$$ (8.4.1)

where $h_L(y,z)$ is is the deformation in the linear limit, v is the velocity of the soliton stripe, and λ is an eigenvalue of the scattering problem.

Equation (8.4.1) describes the result of the nonlinear Aharonov-Bohm scattering from a vortex. The effect is stronger than its linear analog by the factor $v^2/(v^2 - \lambda^2)$, provided $\lambda < v$. The general analytical solution found in Ref. [71] describes an exponential growth for the instability of the dark-soliton stripe. The nonlinear regime of this instability is hard to study analytically, but numerical simulations provide a useful way to explore it. In the numerical simulations, the contrast or the amplitude of the dark-soliton stripe (i.e., its transverse velocity) is varied to monitor the stripe deformation for different propagation distances. Scattering of a small-amplitude dark-soliton stripe by an optical vortex corresponds to the scattering of a linear wave packet associated with the Aharonov–Bohm effect. The snapshots of the scattering process in Figure 8.9 show the details of the stripe deformation (first symmetric, then becoming asymmetric) during the scattering. The vortex itself shifts during the scattering process. The transverse instability of the stripe starts developing for not-too-low values of the amplitudes, and it is accompanied by the subsequent formation of mixed (edge screw) and, later, screw phase dislocations. Consequently, the stripe breaks up into vortices of opposite charges in a way that resembles the formation of vortices in the field of hydrodynamics. This phenomenon can be interpreted as an effective "unzipping" of the dark-soliton stripe by the vortex in the strongly nonlinear regime of the Aharonov-Bohm scattering—a remarkable nonlinear effect.

The primary vortex itself undergoes a large shift from its initial position after the scattering. This shift is a consequence of a novel type of interaction associated with the creation–annihilation process of a vortex pair [71]. Physically, during the interaction, a part of the stripe closest to the vortex breaks up, creating a pair of vortices with opposite charges. One of the new vortices annihilates with the primary vortex, in effect replacing it with a new vortex of the generated pair. This change in the vortex identity appears as a large shift in the position of the original vortex.

Interaction of a vortex soliton with a dark-soliton stripe in a bulk nonlinear defocusing medium was studied in a 2000 experiment [75], and both the vortex-induced bending and the breakup of the stripe were observed. A CW beam from a Ti:sapphire laser was focused onto a rubidium vapor cell after its phase was modified such that it produced a vortex and a soliton stripe next to each other. The experiment was similar to those performed earlier for observing stripe breakup and vortex steering [43]. The laser output was linearly polarized with a slightly elliptical Gaussian beam and was tuned close to the atomic resonance line near 780 nm (detuning 0.8 GHz). The dark-soliton stripe was created using a π phase jump imposed across the beam center by a mask. Without a vortex, this phase jump evolves into an almost straight stripe-shape dark soliton, as seen in part (b) of Figure 8.10. A vortex was created by imaging of the waist of the beam onto a computer-generated phase mask [43]. The beam was then imaged

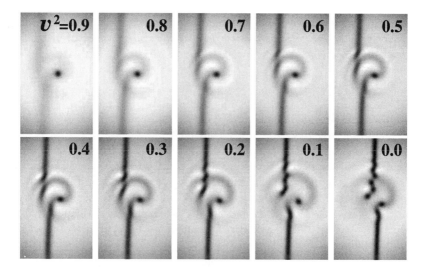

Figure 8.9: Snapshots of a dark-soliton stripe scattered by an optical vortex soliton for its different initial velocities v. The propagation distance $z = 16$ (in units of diffraction length). The initial
offset x_0 of the stripe is chosen such that it reaches the center ($x = 0$) at $z = 8$ in the absence of
the vortex. In the case $v = 0$, $x_0 = 2.5$. (After Ref. [71]; ©2001 APS.)

onto the input end of a 20-cm-long rubidium vapor cell (at 101 °C). A vortex soliton can
be formed at any position in the Gaussian beam; its experimentally observed image in
the absence of a stripe is shown in part (a). When both the vortex and the stripe are
created by combining the two phase masks, they interact inside the nonlinear medium.
As a result of this interaction, the stripe bends and the vortex itself shifts slightly, as
seen in part (c). When the nonlinearity is increased by tuning the frequency closer to
the atomic resonance, the stripe breaks up into pairs of component vortices because of
the transverse instability as seen in part (d). Transverse instability can also occur for
an isolated stripe but requires much higher powers than those used here (60 mW).

8.5   Vortex Arrays and Lattices

When several vortices are generated, they may exhibit an interesting dynamical behavior often associated with the fluid-like motion. The propagation of the simplest vortex
ensemble—a pair of vortices—has been investigated in several experiments [76, 77].
The rotation of a pair of vortices with equal topological charge formed on a focused
Gaussian beam was controlled by changing the beam intensity, which changes the position of the beam waist inside the medium [49]. A comparison between the degree of
rotation of a vortex pair in the linear and nonlinear regimes shows that the effect of the
rotation in the nonlinear regime can be larger by more than three times [79]. The enhancement is due to to the nonlinear confinement of the vortex core, which allows the
vortices to propagate as vortex filaments. Similar effects are observed for more complicated configurations, such as a linear array of vortices [80, 81]. In the low-intensity

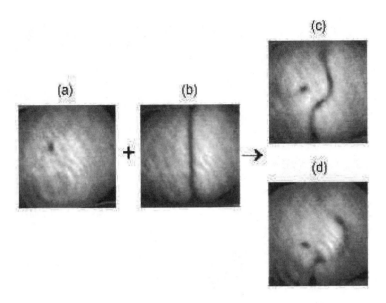

Figure 8.10: Experimental photographs of a vortex soliton (a) interacting with a dark-soliton stripe (b). During interaction, the stripe first bends (c) and then breaks up (d) into multiple vortices. (After Ref. [75]; ©2000 OSA.)

regime, a vortex array was produced by a bent glass plate. At larger intensities, the nonlinear rotation of the vortices was observed; it was found to depend on the vortex density and was not uniform for each vortex pair because of the effect associated with the vortex–vortex interaction.

When highly charged vortices are generated in a configuration that is stable for singly charged ones ($m = \pm 1$), not only do they decay into multiple vortices with $m = \pm 1$ but the new vortices form a pattern similar to a hexagonal crystal lattice [78]. When propagating inside a self-defocusing nonlinear medium, such a pattern presents a two-dimensional periodic mode that induces a periodic modulation of the medium refractive index. The induced change in the refractive index is sufficient to diffract a low-intensity probe beam propagating through the nonlinear medium index. Such diffraction could be controlled by controlling the degree of rotation of the entire lattice [78].

Several different lattice patterns have been studied numerically. In each case, two neighboring vortices can have the same topological charge, or the charge alternates between $+1$ and -1. Depending on the topological charges, the lattice may or may not rotate [78]. Lattices also exhibit elasticity against displacement of one or more vortices from their equilibrium position. These features has been observed in a 2002 experiment in which lattice structures of optical vortices were propagated in a saturable nonlinear medium [82]. We should emphasize the close link between the fields of nonlinear optics and coherent matter waves in the form of Bose–Einstein condensates. In the latter field, several experiments have focused on vortex ensembles, vortex arrays, and vortex lattices [83]–[85].

Figure 8.11 shows the numerical results obtained by solving a generalized NLS

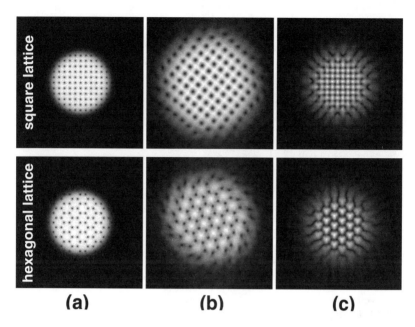

(a)   **(b)**   **(c)**

Figure 8.11: (a) Input optical beams containing vortices in the form of square and hexagonal lattices; Output obtained numerically at $z = 10$ for vortices of the (b) same charge and (c) alternating charge. (After Ref. [82]; ©2002 OSA.)

equation [82] for two lattices of different geometries and different topological charges. No qualitative differences are observed in the propagation of vortex structures with respect to the lattice geometry (square or hexagonal). In contrast. large differences appear for lattices formed with the same or alternating vortex charges. Two main differences are evident. In the case of equal topological charges, the superposition of the phase of all vortices generates a phase gradient, which causes rotation of the whole structure. At the same time, this phase structure causes increased broadening of the background beam seen in Figure 8.11(b). In the case of alternating topological charges, rotation does not occur, because phases cancel on average. For the same reason, the background beam spreads less. We should stress that both effects are topological. In fact, the dependence on the intensity of the background beam is negligible in numerical simulations. However, the broadening of the background beam in the case of alternating topological charges [Figure 8.11(c)] is due to the combined action of diffraction and self-defocusing nonlinearity and does depend on the background-beam intensity. The effect of the lattice structure in Figure 8.11 can also be understood from the preceding discussion. Since the rotation in both cases is caused by the phase gradient, which is larger for the denser lattice, the square lattice rotates faster than the hexagonal one.

The experimental setup used to observe these effects was similar to that used in the earlier work [86]. The 488-nm beam from an argon-ion laser was used to reproduce a computer-generated phase mask with the lattice structure imposed on it. The ± 1 diffraction orders were separated by a diaphragm and focused onto the input end of the nonlinear medium containing an organic dye. The output was imaged by a CCD camera, and neutral density filters were used to avoid its saturation [82]. The nonlinear parameters of the dye were measured at two concentrations. At the lower concentration, the power necessary to form a dark-soliton stripe was found to be only $P_{sol} \simeq 22$ mW, while the saturation power was $P_{sat} \simeq 60$ mW [86]. At the higher concentration, these values were $P_{sol} \simeq 20$ mW and $P_{sat} \simeq 16$ mW. The intensity distributions for two hexagonal lattices with alternating and equal charges are shown in Figure 8.12 (at the lower concentration of the dye). The number of vortices within the lattice is smaller in the case of equal charges because of technical difficulties in synthesizing the hologram. However, the main geometrical feature—the elementary cell—is the same in both cases. The propagation behavior is clearly different for the two lattices shown in Figure 8.12. While the lattice in (a), with alternating charges, exhibits steady propagation, the one with equal charges (b) tends to rotate (about $28°$ counterclockwise). The background beam spreads more in the case of alternating charges, although this feature is not obvious from Figure 8.12 because of the different number of vortices in each lattice. A comparison between the elementary cells of the two lattices shows 18% bigger size for the one with equal charges.

The nonlinear effects are apparent in Figure 8.12 when the images for each lattice are compared at 10- and 50-mW power levels. The increase of the beam power does not influence the degree of rotation of the lattice. The higher power vmainly to enhanced beam broadening. Comparing the size of the elementary cell of the lattice at both powers, it was estimated that 15% and 12% broadening occurred in the two cases. The difference in the two cases is attributed to the increased beam size because of initial topological broadening in the case of equal charges.

8.6 Ring Dark Solitons

As discussed in Section 8.4, dark-soliton stripes are unstable to long-wavelength transverse modulations. The instability region is characterized by a critical wave number p_{cr} such that a soliton stripe is stable to the transverse perturbations of wavelengths $\lambda_\perp < \lambda_{cr}$, where $\lambda_{cr} \equiv 2\pi/p_{cr}$. Consider a *dark-soliton loop* of radius R formed by a quasi-two-dimensional dark soliton. From a physical point of view, it is clear that transverse instabilities would be suppressed when the condition $R < \lambda_{cr}$ holds. The loop of the lowest energy is expected to have a circular symmetry. It is thus expected that *ring-shaped* dark solitons will be stable in self-defocusing nonlinear media [5].

To study the features of ring dark solitons [87], we consider the radially symmetric solutions of the $(2+1)$-dimensional NLS equation on the unit-intensity background $(u_0 = 1)$ in the form

$$u(z,r) = \{\cos\phi\tanh[\cos\phi(r-R)] + i\sin\phi\}\, e^{-iz}, \qquad (8.6.1)$$

where $\phi(z)$ and $R(z)$ are two slowly varying soliton parameters representing the angle and the center of the soliton, respectively, such that $|\phi| < \pi/2$. Physically, the soliton angle ϕ describes the contrast of a ring dark soliton $(B^2 = \cos^2\phi)$ and is connected with the phase jump of 2ϕ occurring across the ring, if we calculate the phase difference

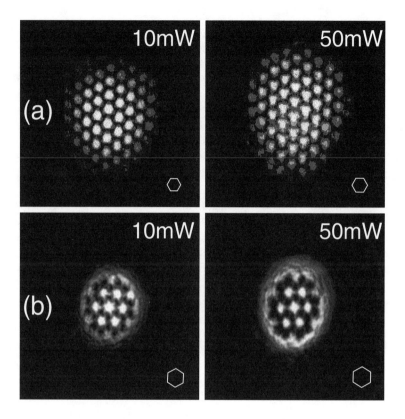

Figure 8.12: Experimental images of the vortex lattices at the output of a 10-cm-long dye cell. (a) Hexagonal lattices with alternative charges at 10 and 50 mW. (b) Hexagonal lattices with equal charges at the same two power levels. Insets in each image show the orientation and size of the elementary cell. (After Ref. [82]; ©2002 OSA.)

between the outer and inner regions separated by the ring. The variable $R(z)$ represents the radius of the ring at a distance z.

For large values of the soliton radius, the ring soliton can be regarded as a quasi-one-dimensional object, and its curvature can be treated as a perturbation. This feature allows us to study the evolution of soliton parameters using perturbation theory developed for dark solitons (see Section 4.1), with r^{-1} acting as an effective perturbation parameter. The resulting evolution equations take the form [5, 87]

$$\frac{d\phi}{dz} = \frac{(D-1)}{3R}\cos\phi, \qquad \frac{dR}{dz} = \sin\phi. \qquad (8.6.2)$$

Solving these coupled equations, the radial velocity of the ring dark soliton is found to vary with its radius R as

$$W \equiv \frac{dR}{dz} = \kappa \left[1 - \cos^2\phi_0 \left(R_0/R\right)^{\frac{2}{3}(D-1)}\right]^{1/2}, \qquad (8.6.3)$$

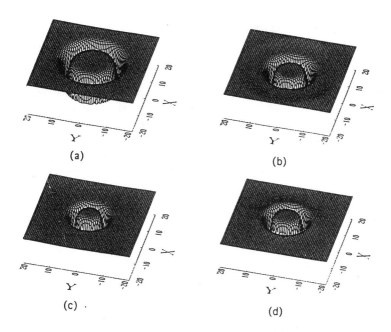

(a)

(b)

(c)

(d)

Figure 8.13: Evolution of a ring dark soliton at (a) $z = 0$, (b) $z = 3$, (c) $z = 6$, and (d) $z = 9$. The ring collapses in (b) and (c) and begins to expand after that. (After Ref. [5]; ©1994 APS.)

where $\kappa = \text{sgn}[\sin \phi_0] = \pm 1$ and ϕ_0 and R_0 represent the initial values of the parameters. Equation (8.6.3) shows that the minimum radius of the collapsing ring dark soliton is defined by the initial conditions as

$$R_{\min} = R_0 \left[\cos \phi_0\right]^{3/(D-1)}, \qquad (8.6.4)$$

and at $R = R_{\min}$ the dark soliton has the maximum contrast. Depending on the initial value ϕ_0 of the soliton phase ϕ, the dark soliton can collapse to reach $R_{\min}$, or it diverges decreasing its contrast.

The linear stability analysis predicts that the dark-soliton stripe is stable when the condition [4]

$$p_\perp > p_{\text{cr}}(\phi) = \left[2\sqrt{\sin^4 \phi + \cos^2 \phi} - (1 + \sin^2 \phi)\right]^{1/2} \qquad (8.6.5)$$

is satisfied. This result shows that the instability band vanishes for small-amplitude dark solitons, for which $\cos \phi \to 0$. Thus, when the length of the ring $2\pi R_{\min}$ is smaller than the minimum wavelength $2\pi/p_{\text{cr}}(0)$ associated with the the transverse instability, we expect the ring dark soliton to be stable because it gets "grayer" with expansion and even more stabilized. This feature leads to the following stability condition:

$$R_{\min} p_{\text{cr}}(0) < 1. \qquad (8.6.6)$$

Numerical simulations carried out using the NLS equation [5] show an excellent agreement with the preceding analytical results. Figure 8.13 shows the evolution of a ring dark soliton over the range $z = 0$–9 (normalized to the diffraction length). The dark ring first collapses, reaching its minimum radius at $z = 6$ (c), and then begins to expand (d). At the turning point, the validity of the adiabatic approximation breaks down, and the dark ring expands along a trajectory slightly different from that predicted by theory. Nevertheless, the solitary wave is robust and it perfectly conserves its radial symmetry, as is seen in Figure 8.13. In the small-amplitude approximation discussed in Section 4.1, ring dark solitons can be described by the cylindrical Korteweg–de Vries equation known to be exactly integrable [5].

In an interesting idea, a ring dark soliton is used to guide multiple bright-soliton beams [88]. In effect, the dark ring acts as an all-optical cable for transmitting $(2 + 1)$-dimensional bright solitons. The idea applies equally well to the transmission of light bullets through waveguides "written" by dark solitons. However, since a dark-soliton stripe suffers from a transverse modulation instability, a ring dark soliton forms a stable curved waveguide. This idea is a nontrivial generalization of the concept of the soliton-induced waveguide discussed earlier for one-dimensional dark solitons. It allows us to use more than one signal beam and also offers stability against misalignments, similar to the inherent stability of a bright-dark soliton pair [89].

The existence of the ring dark solitons has been verified experimentally [90] by placing an amplitude mask consisting of opaque dots ranging from 50 to 250 μm in diameter in front of a thermally nonlinear medium (ethanol containing a red dye). A copper-vapor laser (average power 4 W) was used to produce the background beam. A single dark ring was observed to form at an optimum power level. At higher power levels, a double dark-ring structure was found to form, separating regions of roughly uniform intensity. The dependence of the transverse velocity of the ring on the intensity was also measured for a fixed value of the dark-beam diameter. The results were in qualitative agreement with the theory and numerical simulations [5, 91]. The phase profile of ring dark solitons has also been measured [92, 93], confirming that the dark rings correspond to regions in which a phase jump of nearly π occurs across the background beam.

References

[1] G. A. Swartzlander, Jr., D. R. Andersen, J. J. Regan, H. Yin, and A. E. Kaplan, *Phys. Rev. Lett.* **66**, 1583 (1991).

[2] K. Hayata and M. Koshiba, *Phys. Rev. E* **48**, 2312 (1993).

[3] E. A. Kuznetsov, A. M. Rubenchik, and V. E. Zakharov, *Phys. Rep.* **142**, 113 (1986).

[4] E. A. Kuznetsov and S. K. Turitsyn, *Zh. Eksp. Teor. Fiz.* **94**, 119 (1988) [*Sov. Phys. JETP* **67**, 1583 (1988)].

[5] Yu. S. Kivshar and X. Yang, *Phys. Rev. E* **50**, R40 (1994).

[6] C. T. Law and G. A. Swartzlander, Jr., *Opt. Lett.* **18**, 586 (1993).

[7] G. S. McDonald, K. S. Syed, and W. J. Firth, *Opt. Commun.* **95**, 281 (1993).

[8] C. Josserand and Y. Pomeau, *Europhys. Lett.* **30**, 43 (1995).

[9] D. E. Pelinovsky, Yu. A. Stepanyants, and Yu. S. Kivshar, *Phys. Rev. E* **51**, 5016 (1995).

[10] V. Tikhonenko, J. Christou, B. Luther-Davies, and Yu. S. Kivshar, *Opt. Lett.* **21**, 1129 (1996).

[11] A. V. Mamaev, M. Saffman, and A. A. Zozulya, *Phys. Rev. Lett.* **76**, 2262 (1996)

[12] A. V. Mamaev, M. Saffman, D.Z. Anderson, and A. A. Zozulya, *Phys. Rev. A* **54**, 870 (1996).

[13] M. S. Soskin and M. V. Vasnetsov, in *Progress in Optics*, E. Wolf, Ed., Vol. 42 (Elsevier, Amsterdam, 2001).

[14] A. W. Snyder, L. Poladian, and D. J. Mitchell, *Opt. Lett.* **17**, 789 (1992).

[15] L. P. Pitaevsky, *Zh. Eksp. Teor. Fiz.* **40**, 646 (1961) [*Sov. Phys. JETP* **13**, 451 (1961)].

[16] V. L. Ginzburg and L. P. Pitaevski, *Zh. Eksp. Teor. Fiz.* **34**, 1240 (1958) [*Sov. Phys. JETP* **7**, 858 (1959)].

[17] L. Perivolaropoulos, *Phys. Lett. B* **316**, 528 (1993).

[18] L. M. Pismen, *Physica D* **73**, 244 (1994).

[19] L. M. Pisman, *Phys. Rev. Lett.* **72**, 2557 (1994).

[20] C.T. Law and G. A. Swartzlander, Jr., *Chaos, Solitons Fractals* **4**, 1759 (1994).

[21] I. Velchev, A. Dreischuh, D. Neshev, and S. Dinev, *Opt. Commun.* **130**, 385 (1996).

[22] A.P. Sheppard and M. Haelterman, *Opt. Lett.* **19**, 859 (1994).

[23] Y. Silberberg, *Opt. Lett.* **15**, 1282 (1990).

[24] E. A. Kuznetsov and J. J. Rasmussen, *Phys. Rev. E* **51**, 4479 (1995).

[25] B. Luther-Davies, J. Christou, V.V. Tikhonenko, and Yu. S. Kivshar, *J. Opt. Soc. Am. B* **14**, 3045 (1997).

[26] J.F. Nye and M. V. Berry, *Proc. R. Soc. Lond. A* **336**, 165 (1974).

[27] V. Yu. Bazhenov, V. Yu., M. S. Soskin, and M. V. Vasnetsov, *J. Mod. Opt.* **39**, 985 (1992).

[28] N. R. Heckenberg, R. McDuff, C. P. Smith, and A. G. White, *Opt. Lett.* **17**, 221 (1992).

[29] I. V. Basistiy, V. Yu. Bazhenov, M. S. Soskin, and M. V. Vasnetsov, *Opt. Commun.* **103**, 422 (1993).

[30] I. Freund, *J. Opt. Soc. Am. A* **11**, 1644 (1980).

[31] N. B. Baranova, B. Ya. Zel'dovich, A. V. Mamaev, N. F. Pilipetsky, and V. V. Shkunov, *Zh. Eksp. Teor. Fiz.* **33**, 206 (1981) [*JETP Lett.* **33**, 195 (1981)].

[32] E. A. Spiegel, *Physica D* **1**, 236 (1980).

[33] R. J. Donnelly, *Quantized Vortices in Helium II* (Cambridge University Press, Cambridge, UK, 1991).

[34] C. Nore, M. E. Brachet, and S. Fauve, *Physica D* **65**, 154 (1993).

[35] J. C. Neu, *Physica D* **43**, 385 (1990).

[36] I. Velchev, A. Dreischuh, D. Neshev, and S. Dinev, *Opt. Commun.* **140**, 77 (1997).

[37] Y. Chen and J. Atai, *J. Opt. Soc. Am. B* **9**, 2252 (1992).

[38] V. Tikhonenko, Yu. S. Kivshar, V.V. Steblina, and A. A. Zozulya, *J. Opt. Soc. Am. B* **15**, 79 (1998).

[39] I. Aranson and V. Steinberg, *Phys. Rev. B* **53**, 75 (1996).

[40] A. Dreischuh, G. G. Paulus, F. Zacher, F. Grabson, D. Neshev, and H. Walther, *Phys. Rev. E* **60**, 7518 (1999).

[41] A. Dreischuh, G. G. Paulus, F. Zacher, F. Grasbou, and H. Walter, *Phys. Rev. E* **60**, 6111 (1999).

[42] A. V. Mamaev, M. Saffman, and A. A. Zozulya, *Phys. Rev. Lett.* **78**, 2108 (1997).

[43] J. Christou, V. Tikhonenko. Yu. S. Kivshar, and B. Luther-Davies, *Opt. Lett.* **21**, 1649 (1996).

[44] Yu. S. Kivshar, J. Christou, V. Tikhonenko, B. Luther-Davies, and L. Pismen, *Opt. Commun.* **152**, 198 (1998).

[45] Yu. S. Kivshar and X. Yang, *Opt. Commun.* **107**, 93 (1994).

[46] J. Rubinstein and L. M. Pismen, *Physica D* **78**, 1 (1994).

[47] L. M. Pismen and J.D. Rodriquez, *Phys. Rev. A* **42**, 2471 (1990).

[48] L. M. Pismen and J. Rubinstein, *Physica D* **47**, 353 (1991).

[49] B. Luther-Davies, R. Powles, and V. Tikhonenko, *Opt. Lett.* **19**, 1816 (1994).

[50] K. Staliunas, *Chaos, Solitons, and Fractals*, **4**, 1783 (1994).

[51] F.S. Roux, *J. Opt. Soc. Am. B* **12**, 1215 (1995).

[52] K. Staliunas, *Opt. Commun.* **90**, 123 (1994).

[53] V. S. Butylkin, A. E. Kaplan, Yu. G. Khronopulo, and E. I. Yakubovich, *Resonant Non linear Interactions of Light with Matter* (Springer, Berlin, 1989).

[54] G. Indebetouw, *J. Mod. Opt.* **40**, 73 (1993).

[55] G. A. Swartzlander, Jr., and C. Law, *Phys. Rev. Lett.* **69**, 2503 (1992).

[56] G. A. Swartzlander, Jr., and C. Law, *Opt. Phon. News* 10(12) (1993).

[57] V. Tikhonenko, J. Christou, and B. Luther-Davies, *J. Opt. Soc. Am. B* **12**, 2046 (1995).

[58] V. Tikhonenko, J. Christou, and B. Luther-Davies, *Phys. Rev. Lett.* **76**, 2698 (1996).

[59] G. Duree, M. Morin, G. Salamo, M. Segev, B. Crosignani, P. DiPorto, E. Sharp, and A Yariv, *Phys. Rev. Lett.* **74**, 1978 (1995).

[60] M. Segev, G. C. Valley, B. Crosignani, P. DiPorto, and A. Yariv, *Phys. Rev. Lett.* **73**, 3211 (1994).

[61] Z. Chen, M. Shih, M. Segev, D. W. Wilson, R. E. Muller, and P. D. Maker, *Opt. Lett.* **22**, 1751 (1997).

[62] A. V. Mamaev, M. Saffman, and A. A. Zozulya, *Phys. Rev. Lett.* **77**, 4544 (1996).

[63] V. Tikhonenko and N. N. Akhmediev, *Opt. Commun.* **126**, 108 (1996).

[64] G. S. McDonald, K. S. Syed, and W. J. Firth, *Opt. Commun.* **94**, 469 (1992).

[65] K. T. Gahagan and G. A. Swartzlander, Jr., *Opt. Lett.* **21**, 827 (1996).

[66] Y. Aharonov and D. Bohm, *Phys. Rev.* **115**, 485 (1959).

[67] M. V. Berry, *Proc. R. Soc. London A* **392**, 45 (1984).

[68] E. B. Sonin, *Sov. Phys. JETP* **42**, 469 (1976).

[69] M. V. Berry et al., *Eur. J. Phys.* **1**, 154 (1980).

[70] F. Vivanco et al., *Phys. Rev. Lett.* **83**, 1966 (1999).

[71] D. Neshev, A. Nepomnyashchy, and Yu. S. Kivshar, *Phys. Rev. Lett.* **87**, 043901 (2001).

[72] L. P. Pitaevskii, *Sov. Phys. JETP* **8**, 888 (1959).

[73] L. M. Pismen, *Vortices in Nonlinear Fields* (Clarendon, Oxford, UK, 1999), Sec. 4.3.4.

[74] Yu. S. Kivshar and B. Luther-Davies, *Phys. Rep.* **298**, 81 (1998).

[75] Yu. S. Kivshar, A. Nepomnyashchy, V. Tikhonenko, J. Christou, and B. Luther-Davies, *Opt. Lett.* **25**, 123 (2000).

[76] D. Rozas, Z. S. Sacks, and G. A. Swartzlander, Jr., *Phys. Rev. Lett.* **79**, 3399 (1997).

[77] D. Rozas, C.T. Law, and G. A. Swartzlander Jr., *J. Opt. Soc. Am. B* **14**, 3054 (1997).

[78] D. Neshev, A. Dreischuh, M. Assa, and S. Dinev, *Opt. Commun.* **151**, 413 (1998).

[79] D. Rozas and G. A. Swartzlander, Jr., *Opt. Lett.* **25**, 126 (2000).

[80] G. H. Kim, J. H. Jeon, Y. C. Noh, K. H. Ko, H. J. Moon, J. H. Lee, and J. S. Chang, *Opt. Commun.* **147**, 131 (1998).

[81] G. H. Kim, J. H. Jeon, Y. C. Noh, J. H. Lee, and J. S. Chang, K. H. Ko, H. J. Moon, *J. Korean Phys. Soc.* **33**, 308 (1998).

[82] A. Dreischuh, S. Chervenkov, D. Neshev, G. G. Paulus, and H. Walther, *J. Opt. Soc. Am. B* **19**, 550 (2002).

[83] K.W. Madison, F. Chevy, W. Wohlleben, and J. Dalibard, *Phys. Rev. Lett.* **84**, 806 (2000).

[84] B. P. Anderson, P. C. Haljan, C. A. Regal, D. L. Feder, L. A. Collins, C. W. Clark, and E. A. Cornell, *Phys. Rev. Lett.* **86**, 2926 (2001).

[85] J. R. Abo-Shaeer, C. Raman, J. M. Vogels, W. Ketterle, *Science* **292**, 476 (2001).

[86] A. Dreischuh, D. Neshev, G. G. Paulus, and H. Walther, *J. Opt. Soc. Am. B* **17**, 2011 (2000).

[87] Yu. S. Kivshar and X. Yang, *Chaos, Solitons Fractals* **4**, 1745 (1994).

[88] A. Dreischuh, V. Kamenov, and S. Dinev, *Appl. Phys. B* **63**, 145 (1996).

[89] A. P. Sheppard and Yu. S. Kivshar, *Phys. Rev. E* **55**, 4773 (1997).

[90] S. Baluschev, A. Dreischuh, I. Velchev, S. Dinev, and O. Marazov, *Phys. Rev. E* **52**, 5517 (1995); *Appl. Phys. B* **61**, 121 (1995).

[91] V. Kamenov, A. Dreischuh, and S. Dinev, *Phys. Scripta* **55**, 68 (1997).

[92] A. Dreischuh, W. Fliesser, I. Velchev, S. Dinev, and L. Windholz, *Appl. Phys. B* **62**, 139 (1996).

[93] D. Neshev, A. Dreischuh, V. Kamenov, I. Stefanov, S. Dinev, W. Fliesser, and L. Windholz, *Appl. Phys. B* **64**, 429 (1997).

Chapter 9

Vector Solitons

Solitons discussed so far are described by a single NLS equation for a scalar field. Such scalar solitons form when a single wave propagates inside a nonlinear medium in such a way that it maintains its polarization state. When these conditions are not satisfied, one must consider interaction of several field components at different frequencies or polarizations and solve simultaneously a set of coupled NLS equations. A shape-preserving solution of such equations is called a *vector soliton* because of its multicomponent nature. In some cases, the soliton constituents correspond to the components of the vector field associated with the soliton, but, generally speaking, the term *vector soliton* is rather broad and covers different types of multicomponent solitons. This chapter is devoted to vector solitons. Section 9.1 introduces the simplest physical model of a two-component field governed by two coupled NLS equations that describe the coupling between two pulses propagating inside optical fibers. This model is extended in Section 9.2 to the case of N-component vector solitons governed by a set of N incoherently coupled NLS equations. Section 9.3 focuses on the stability of such vector solitons. Section 9.4 is devoted to the properties of spatial vector solitons associated with the *coherently coupled NLS equations*. It uses the example of spatial vector solitons forming inside a planar semiconductor waveguide because of nonlinear coupling between the transverse-electric (TE) and transverse-magnetic (TM) modes. In Section 9.5 we consider the structure and stability of multihump vector solitons. The generalization of vector solitons to two spatial dimensions is discussed in Section 9.6, where we consider both the radially symmetric and asymmetric cases. In Section 9.7 we focus on the transverse instability of vector soliton stripes. In the final Section we discuss several types of vector solitons associated with dark solitons, such as dark-bright vector solitons, polarization-domain walls, and vector vortex solitons.

9.1 Incoherently Coupled Solitons

In this section we focus on two-component vector solitons. Such solitons can be spatial or temporal and form using two orthogonally polarized components of a single optical field or two fields of different frequencies but the same polarization. We derive the underlying incoherently coupled NLS equation using the example of *a temporal vector soliton* with two different frequency components and explain how these equations are modified in the case of spatial solitons. As discussed in Section 9.4, similar equations appear for spatial vector solitons forming in planar waveguides.

9.1.1 Coupled Nonlinear Schrödinger Equations

In the case of a single pulse propagating inside a multimode optical fiber, vector solitons may form because of the coupling among different guided modes induced by the fiber nonlinearity. Mathematically, pulse propagation in a multimode fiber is described by a set of incoherently coupled NLS equations [1]. Even in a single-mode fiber, a single pulse can form a vector soliton if the birefringence effects lead to a coupling between its two orthogonally polarized components [2]. In this case, however, one must take into account the coherent nature of the coupling, as discussed in Section 9.4.

In this section we focus on the case of two pulses of two distinct carrier frequencies propagating along the Z direction inside a single-mode fiber [3]. The total optical field at any point inside the fiber can be written as

$$\mathbf{E}(\mathbf{r},t) = \mathrm{Re}\{\hat{x}F(X,Y)[A_1\exp(i\beta_{10}Z - i\omega_1 t) + A_2\exp(i\beta_{20}Z - i\omega_2 t)]\}, \quad (9.1.1)$$

where $\hat{x}$ is the polarization unit vector, $F(X,Y)$ governs the spatial mode distribution, ω_1 and ω_2 are the carrier frequencies associated with the two pulses, and β_{j0} is the zeroth-order term in the Taylor expansion of the propagation constant $\beta_j(\omega)$ $(j = 1, 2)$ given by

$$\beta_j(\omega) = \beta_{j0} + (\omega - \omega_j)\beta_{j1} + \tfrac{1}{2}(\omega - \omega_j)^2\beta_{j2} + \cdots, \quad (9.1.2)$$

where $\beta_{jk} = (d^k\beta_j/d\omega^k)_{\omega=\omega_j}$.

Following the method outlined in Section 1.3, the slowly varying pulse envelopes A_1 and A_2 are found to satisfy the following set of two equations [3]:

$$i\left(\frac{\partial A_1}{\partial Z} + \beta_{11}\frac{\partial A_1}{\partial t}\right) - \frac{\beta_{12}}{2}\frac{\partial^2 A_1}{\partial t^2} + \gamma_1(|A_1|^2 + \sigma|A_2|^2)A_1 = 0, \quad (9.1.3)$$

$$i\left(\frac{\partial A_2}{\partial Z} + \beta_{21}\frac{\partial A_2}{\partial t}\right) - \frac{\beta_{22}}{2}\frac{\partial^2 A_2}{\partial t^2} + \gamma_2(|A_2|^2 + \sigma|A_1|^2)A_2 = 0, \quad (9.1.4)$$

where the nonlinear parameter $\gamma_j = n_2\omega_j/(ca_{\mathrm{eff}})$ and a_{eff} is the effective core area. The nonlinear coupling between the two fields is governed by the cross-phase modulation (XPM) term containing the parameter σ. Its value σ equals 2 in the case under study but may change depending on the physical problem. For example, $\sigma = 2/3$ when the two fields have the same frequency but different polarizations [3]. This difference has its origin in the tensor nature of the third-order nonlinear susceptibility $\chi^{(3)}$.

As in previous chapters, it is useful to normalize Eqs. (9.1.3) and (9.1.4) using

$$x = (t - \beta_{1r}Z)/T_0, \qquad z = Z/L_D, \qquad u_j = \sqrt{\gamma_r L_D}A_j, \quad (9.1.5)$$

where $L_D = T_0^2/|\beta_{2r}|$, T_0 is a measure of pulse width, and β_{1r}, β_{2r}, and γ_r are the reference values used for normalization. In terms of the normalized variables, Eqs.

(9.1.3) and (9.1.4) become

$$i\left(\frac{\partial u_1}{\partial z} + \delta_1 \frac{\partial u_1}{\partial x}\right) + \frac{d_1}{2}\frac{\partial^2 u_1}{\partial x^2} + \gamma_1'(|u_1|^2 + \sigma|u_2|^2)u_1 = 0, \qquad (9.1.6)$$

$$i\left(\frac{\partial u_2}{\partial z} + \delta_2 \frac{\partial u_2}{\partial x}\right) + \frac{d_2}{2}\frac{\partial^2 u_2}{\partial x^2} + \gamma_2'(|u_2|^2 + \sigma|u_1|^2)u_2 = 0, \qquad (9.1.7)$$

where the new dimensionless parameters are defined as

$$\delta_j = (\beta_{j1} - \beta_{1r})L_d/T_0, \quad d_j = -\beta_{j2}/\beta_{2r}, \quad \gamma_j' = \gamma_j/\gamma_r, \qquad (9.1.8)$$

with $j = 1$ or 2. These six parameters govern the properties of two-component vector solitons. They can be reduced to five by choosing $\beta_{1r} = \frac{1}{2}(\beta_{11} + \beta_{21})$ as the average value so that the group velocity of pulses is measured in a reference frame moving at the speed

$$v_{gr} = 1/\beta_{1r} = v_{g1}v_{g2}/(v_{g1} + v_{g2}). \qquad (9.1.9)$$

With this choice, $\delta = \delta_1 = -\delta_2$ is a measure of the group-velocity mismatch between the two pulses.

A major hurdle for the formation of vector solitons is the group-velocity mismatch δ_j appearing in Eqs. (9.1.6) and (9.1.7). In the absence of the nonlinear effects, the two pulses will separate or walk away from each other after propagating a distance known as the *walk-off length* and defined as [3]

$$L_W = T_0/|\beta_{11} - \beta_{21}| = L_D/|\delta_1 - \delta_2|. \qquad (9.1.10)$$

However, solitons can shift their frequencies in such a way that the faster-moving pulse slows down while the slower-moving pulse speeds up so that the two pulses continue to overlap indefinitely. This is the main mechanism behind the formation of vector solitons in spite of a group-velocity mismatch. Mathematically, we can eliminate the δ_j term in Eqs. (9.1.6) and (9.1.7) by making the following transformation:

$$u_j = U_j \exp(iK_j z - i\Omega_j x). \qquad (9.1.11)$$

With this transformation, Eqs. (9.1.6) and (9.1.7) reduce to

$$i\frac{\partial U_1}{\partial z} + \frac{d_1}{2}\frac{\partial^2 U_1}{\partial x^2} + \gamma_1(|U_1|^2 + \sigma|U_2|^2)U_1 = 0, \qquad (9.1.12)$$

$$i\frac{\partial U_2}{\partial z} + \frac{d_2}{2}\frac{\partial^2 U_2}{\partial x^2} + \gamma_2(|U_2|^2 + \sigma|U_1|^2)U_2 = 0, \qquad (9.1.13)$$

provided that $\Omega_j = \delta_j/d_j$ and $K_j = \delta_j^2/(2d_j)$. For notational simplicity, the prime over γ is not shown explicitly here and in what follows.

Equations (9.1.12) and (9.1.13) are the simplest coupled NLS equations. They are said to be incoherently coupled because the coupling depends only on the local intensities and is therefore phase insensitive. The same set of equations applies in the case of two CW beams propagating inside a planar optical waveguide such that the light diffracts along the x direction only. The only difference is that the parameters d

($j = 1, 2$) are related not to dispersion but to diffraction and are defined as $d_j = \beta_{0r}/\beta_{j0}$, where β_{0r} is a reference value. Of course, as discussed in Chapter 1, d_j takes only positive values in the spatial case but can be positive or negative in the temporal case, depending on the nature of the dispersion (anomalous versus normal). The nonlinear parameter γ_j is positive in the case of self-focusing nonlinearity but takes negative values for a self-defocusing nonlinear medium. For a non-Kerr medium, Eqs. (9.1.12) and (9.1.13) should be generalized by replacing $|U_j|^2$ in the nonlinear terms with a function $F(|U_j|^2)$, as was done in Chapter 2.

The coupled NLS equations, such as Eqs. (9.1.12) and (9.1.13), have been studied extensively for finding solutions that represent two spatial (or temporal) solitons whose shape does not change during propagation. Such vector solitons are sometimes referred to as *coupled solitons* or *symbiotic solitons* to emphasize the dependence of various components of a vector soliton on each other. In the case of a self-focusing nonlinearity, temporal vector solitons can be grouped into the following five different classes.

- Both components of a vector soliton are bright solitons, each experiencing anomalous dispersion ($d_1 > 0$ and $d_2 > 0$). This case corresponds to a two-component *bright vector soliton* [4]–[10].

- A bright soliton experiencing anomalous dispersion ($d_1 > 0$) is coupled to a dark soliton experiencing normal dispersion ($d_2 < 0$). This case corresponds to a "normal" dark-bright soliton pair [11]–[17].

- A bright soliton experiencing normal dispersion ($d_1 < 0$) is incoherently coupled to a dark soliton experiencing anomalous dispersion ($d_2 > 0$). This case corresponds to an "inverted" dark-bright pair [11, 18].

- Two coupled dark solitons propagate in the normal-dispersion regime ($d_1 < 0$ and $d_2 < 0$). This is the case of a two-component dark vector soliton [19]–[25].

- A bright pulse is coupled to a dark soliton, and both of them propagate in the normal-dispersion region ($d_1 < 0$ and $d_2 < 0$), in which the bright pulse cannot exist by itself [24], [26]–[30]. In this case, the bright pulse is guided by a dark-soliton waveguide.

An important generalization includes the case of *coherent* nonlinear coupling between the two components, and it is discussed later using the example of spatial vector solitons. We should also mention that there exist other types of the mode coupling and, correspondingly, other types of coupled NLS equations. A well-known example is provided by a nonlinear directional coupler in which the modes in two physically separated fiber cores become coupled. Such a nonlinear coupler can be used for all-optical switching [31].

9.1.2 Soliton-Induced Waveguiding

To understand the physics behind the existence of two-component vector solitons, it is useful to discuss first the phenomenon of soliton-induced waveguiding. The concept was first suggested by Manassah [32] for temporal solitons. It was was later applied to spatial solitons, whose use also allowed it to be observed experimentally [33]–[35]. It should, however, be stressed that the basic idea behind this phenomenon was discussed as early as 1962 by Askar'yan [36] while analyzing the effect of a light intensity gradient on electrons and atoms. We discuss the basic underlying idea using the coupled NLS equations (9.1.12) and (9.1.13).

Consider the case in which one of the fields is so weak compared with the other that it does not affect the propagation of the intense field. However, the intense field can still affect the weak field through the XPM-induced coupling. Under appropriate conditions, the soliton can trap the weak field completely through a phenomenon called *soliton-induced waveguiding*. Physically, the intense field propagating as a a soliton changes the refractive index of a nonlinear medium and creates an effective waveguide. Under appropriate conditions, a weak field is trapped by this waveguide and propagates as a guided mode, creating the simplest example of a vector soliton.

Assume that $|U_1|^2 \gg |U_2|^2$ in the coupled NLS equations (9.1.12) and (9.1.13). If we choose $d_1 = 1$ and $\gamma_1 = 1$ (using the field U_1 as a reference for normalization), these equations become

$$i\frac{\partial U_1}{\partial z} + \frac{1}{2}\frac{\partial^2 U_1}{\partial x^2} + |U_1|^2 U_1 = 0, \qquad (9.1.14)$$

$$i\frac{\partial U_2}{\partial z} + \frac{d_2}{2}\frac{\partial^2 U_2}{\partial x^2} + \gamma_2\sigma|U_1|^2 U_2 = 0. \qquad (9.1.15)$$

Equation (9.1.14) is the standard NLS equation encountered in Chapter 1, and it has the fundamental-soliton solution $U_1(x,z) = \text{sech}(x)e^{iz/2}$. If we substitute this solution in Eq. (9.1.15) and look for shape-preserving solutions in the form $U_2(x,z) = f(x)e^{i\Lambda z}$, $f(x)$ is found to satisfy the following eigenvalue equation:

$$\frac{d_2}{2}\frac{d^2 f}{dx^2} + \frac{\gamma_2\sigma}{\cosh^2 x}f = \Lambda f, \qquad (9.1.16)$$

where the eigenvalue Λ is yet to be determined. The physical meaning of Eq. (9.1.16) is clear: The bright soliton increases the refractive index, depending on its intensity, and creates a graded-index waveguide that can then guide the weak field U_2.

The eigenvalue problem (9.1.16) appears in many branches of physics. In particular, it has been studied in the context of linear graded-index optical waveguides with an index distribution in the form of $\text{sech}^2 x$ [37]. For $d_2 > 0$, this eigenvalue problem has bounded solutions such that $f(x) \to 0$ as $|x| \to \infty$ only if $\Lambda < \gamma_2\sigma$ (the guided modes). The number of such bound modes depends on σ. In particular, for $\sigma = 2$ and $\gamma_2 \approx 1$, two localized solutions exist [37]. The lowest-order (symmetric) solution has the form $f(x) = \text{sech}^\nu(x)$, with $\nu = 1.56$. For $d_2 < 0$, no localized eigenmodes exist for Eq. (9.1.16) because this case corresponds to a so-called "antiwaveguide."

The preceding discussion shows that, under certain conditions, an intense optical field can guide a weaker field through XPM-induced coupling. This soliton-induced waveguiding phenomenon is observed experimentally by launching a CW probe beam into a self-focusing Kerr medium together with an intense CW beam whose power is chosen such that it propagates as a bright spatial soliton [35]. Such a trapping of a weak beam and its guidance by a strong soliton beam has been seen in many experiments using several different types of self-focusing media [38]–[47] and is often exploited for practical applications using soliton-based Y junctions, splitters, and directional couplers [45]–[41]. A soliton-induced waveguide has also led to efficient second-harmonic generation [42]. The use of dark solitons for inducing waveguides has also been suggested and verified experimentally [49].

9.1.3 Exact Solutions

The soliton-induced waveguiding phenomenon shows that a soliton at one wavelength can be used to guide a weak beam at another wavelength. Even if the second beam becomes intense enough that the assumption $|U_1|^2 \gg |U_2|^2$ no longer holds in the coupled NLS equations, one expects such a mutual guiding to continue under some conditions. Indeed, Eqs. (9.1.12) and (9.1.13) have exact analytical solutions in the form of soliton pairs that preserve their shape through the mutual XPM interaction [4]–[24]. Because such vector solitons specify intensity profiles of both beams (or pulses) and always occur in pairs [50, 51], they are also referred to as *XPM-paired solitons* or *symbiotic solitons*.

An interesting example is provided by the inverted dark-bright soliton pair formed when $\beta_{12} < 0$ and $\beta_{22} > 0$. This solution of Eqs. (9.1.3) and (9.1.4) with $\sigma = 2$ was first obtained in 1988 and can be written as [18]

$$A_1(Z,t) = B_1 \tanh[W(t - Z/V)] \exp[i(K_1 Z - \Omega_1 t)], \tag{9.1.17}$$

$$A_2(Z,t) = B_2 \mathrm{sech}[W(t - Z/V)] \exp[i(K_2 Z - \Omega_2 t)], \tag{9.1.18}$$

where V is the common group velocity. The soliton amplitudes are determined from

$$B_1^2 = (2\gamma_1 \beta_{22} + \gamma_2 |\beta_{12}|)W^2/(3\gamma_1 \gamma_2), \tag{9.1.19}$$

$$B_2^2 = (2\gamma_2 |\beta_{12}| + \gamma_1 \beta_{22})W^2/(3\gamma_1 \gamma_2), \tag{9.1.20}$$

and the wave numbers K_1 and K_2 are given by

$$K_1 = \gamma_1 B_1^2 - |\beta_{12}|\Omega_1^2/2, \qquad K_2 = \beta_{22}(\Omega_2^2 - W^2)/2. \tag{9.1.21}$$

The effective group velocity of the soliton pair is obtained from

$$V^{-1} = v_g^{-1} - |\beta_{12}|\Omega_1 = v_g^{-1} + \beta_{22}\Omega_2, \tag{9.1.22}$$

where $v_g = 1/\beta_{11}$ is the group velocity of the pulse in the absence of the nonlinear effects (assumed to be the same for both pulses).

As seen from Eq. (9.1.22), the frequency shifts Ω_1 and Ω_2 must have opposite signs and cannot be chosen independently. The parameter W governs the pulse width and determines the soliton amplitudes through Eqs. (9.1.19) and (9.1.20). Thus, two members of the soliton pair have the same width, the same group velocity, but different shapes and amplitudes such that they support each other through the XPM coupling. The most striking feature of this soliton pair is that the dark soliton propagates in the anomalous-GVD regime, whereas the bright soliton propagates in the normal-GVD regime, which is exactly the opposite of the behavior expected in the absence of XPM. The physical mechanism behind such an unusual pairing can be understood as follows.

Because XPM is twice as strong as SPM, it can counteract the temporal spreading of an optical pulse induced by the combination of SPM and normal GVD, provided the XPM-induced chirp is of the opposite kind than that produced by SPM. A dark soliton can generate this kind of chirp. At the same time, the XPM-induced chirp on the dark soliton is such that the pair of bright and dark solitons can support each other in a symbiotic manner.

The dark soliton within the pair does not have to be black. Even a gray soliton can pair with a bright soliton. Such a soliton pair can be obtained by solving Eqs. (9.1.3) and (9.1.4) with the ansatz

$$A_j(Z,t) = Q_j(t - Z/V)\exp[i(K_jZ - \Omega_jt + \phi_j)], \tag{9.1.23}$$

where V is the common velocity of the soliton pair, Q_j governs the soliton shape, and ϕ_j is the phase ($j = 1,2$). The resulting solution has the form [13]

$$Q_1(\tau) = B_1[1 - b^2\mathrm{sech}^2(W\tau)], \qquad Q_2(\tau) = B_2\mathrm{sech}(W\tau), \tag{9.1.24}$$

where $\tau = t - Z/V$. The parameters W and b depend on the soliton amplitudes, B_1 and B_2, and on fiber parameters through the relations

$$W = \left(\frac{3\gamma_1\gamma_2}{2\gamma_1\beta_{22} - 4\gamma_2\beta_{21}}\right)^{1/2} B_2, \qquad b = \left(\frac{2\gamma_1\beta_{22} - \gamma_2\beta_{21}}{\gamma_1\beta_{22} - 2\gamma_2\beta_{21}}\right)^{1/2}\frac{B_2}{B_1}. \tag{9.1.25}$$

The constants K_1 and K_2 are also fixed by various fiber parameters and soliton amplitudes. The phase of the bright soliton is constant, but the dark-soliton phase ϕ_1 is time dependent. The frequency shifts Ω_1 and Ω_2 are related to the soliton-pair velocity, as in Eq. (9.1.22).

The new feature of the XPM-coupled soliton pair in Eq. (9.1.24) is that the dark soliton is of a "gray" type. The parameter b controls the depth of the intensity dip associated with a gray soliton. Both solitons have the same width W but different amplitudes. In addition, the two GVD parameters can be positive or negative. However, the soliton pair exists only under certain conditions. The soliton solution is always possible if $\beta_{12} < 0$ and $\beta_{22} > 0$, but it does not exist when $\beta_{12} > 0$ and $\beta_{22} < 0$. As discussed before, this behavior is opposite to what would normally be expected and is due solely to XPM. If both solitons experience normal GVD, the bright-gray soliton pair can exist if $\gamma_1\beta_{22} > 2\gamma_2\beta_{12}$. Similarly, if both solitons experience anomalous GVD, the soliton pair can exist if $2\gamma_1|\beta_{22}| < \gamma_2|\beta_{12}|$.

The soliton-pair solutions given in the preceding are not the only possible exact solutions of Eqs. (9.1.12) and (9.1.13). These equations also support pairs with two bright or two dark solitons, depending on various parameter values [11]. A simple way to find the conditions under which XPM-paired solitons can exist is to postulate an appropriate solution, substitute it in Eqs. (9.1.12) and (9.1.13), and then investigate whether soliton parameters (amplitude, width, group velocity, frequency shift, and wave number) can be determined with physically possible values. Stability of the XPM-paired solitons is not always guaranteed. We discuss this issue later in Section 9.3.

The coupled NLS equations (9.1.12) and (9.1.13) also have periodic solutions that represent two pulse trains that propagate undistorted through an optical fiber because of the XPM-induced coupling between them. One such periodic solution in terms of the elliptic functions was found in 1989 in the specific case in which both pulse trains had the same group velocity and has experienced anomalous GVD inside the fiber [52]. By 1998, nine periodic solutions, written as different combinations of the elliptic functions, had been found [53]. All of these solutions assume equal group velocities and anomalous GVD for both pulse trains.

9.2   Multicomponent Vector Solitons

The concept of two-component vector solitons can be easily generalized to multicomponent vector solitons. Such temporal solitons can form, for example, when multifrequency composite optical pulses are transmitted over the same fiber in wavelength-division-multiplexed (WDM) channels of optical communication systems [54, 55] or a bit-parallel-wavelength scheme is used for high-speed interconnect applications [56]–[61].

9.2.1   Multiple Coupled NLS Equations

To describe the simultaneous transmission of N pulses of different wavelengths in a single-mode fiber, we extend the analysis of Section 9.1 and obtain the following set of N incoherently coupled NLS equations [1]:

$$ i\left(\frac{\partial u_j}{\partial z} + \delta_j \frac{\partial u_j}{\partial x}\right) + \frac{d_j}{2}\frac{\partial^2 u_j}{\partial x^2} + \left(\gamma_j |u_j|^2 + \sigma \sum_{m\neq j} \gamma_m |u_m|^2\right) u_j = 0, \qquad (9.2.1) $$

where the reference value $\beta_{1r} \equiv 1/v_{\mathrm{gr}}$ is chosen to correspond to the middle channel so that $\delta_j = 0$ for that channel and takes positive and negative values for channels located on both sides of it. Typically, fiber dispersion for the middle channel is $\beta_{2r} \sim -5$ ps^2/km. The dispersion length is then ~ 10 km for pulses widths ~ 10 ps. We assume that the design parameters are such that the group-velocity-mismatch parameter $|\delta_j|$ does not exceed 5 for any channel. As $|\delta_j|$ increases, it becomes harder for all components to propagate at a common group velocity. Because channel spacing is typically less than 1 nm, the parameters d_j and γ_j differ only slightly and have values close to unity [56]–[60].

The concept of soliton-induced waveguiding discussed earlier can be generalized to the case of multiple optical beams or pulses. In the temporal case, a single intense pulse may create a waveguide in which several weaker pulses propagate as linear guided modes of the effective waveguide induced by the primary soliton [33]. This phenomenon is sometimes referred to as *pulse shepherding*. Mathematically, one can show that the set of equations (9.2.1) admits stationary solutions in the form of a multicomponent vector soliton such that an intense shepherd pulse forces all other lower-amplitude pulses to propagate at its own speed and remain synchronized.

To find the soliton solutions of Eqs. (9.2.1), we assume a solution in the form of Eq. (9.1.11) so that U_j satisfies an ordinary differential equation

$$\frac{d_j}{2}\frac{d^2 U_j}{dx^2} + \left(\gamma_j |U_j|^2 + \sigma \sum_{m \neq j} \gamma_m |U_m|^2\right) U_j = \lambda_j U_j, \tag{9.2.2}$$

where, as before, the frequency shift $\Omega_j = \delta_j / d_j$ and $\lambda_j = K_j - \delta_j^2 / (2d_j)$. This transformation removes the walk-off term from the evolution equations and transfers it into a frequency shift of each individual pulse. In a more rigorous analysis, the moving solitons are found directly in the presence of walk-off. Of course, the two treatments produce the same results.

The set of equations (9.2.2) has *exact* analytical solutions under certain conditions [62, 63]. To find these conditions, we denote the central channel by $j = 0$ and use it as a reference channel so that $d_0 = \gamma_0 = 1$. Renormalizing x using $\sqrt{2\lambda_0}x \rightarrow x$, we rewrite system (9.2.2) in the form

$$\frac{1}{2}\frac{d^2 U_0}{dx^2} + \left(|U_0|^2 + \sigma \sum_{n \neq 0} \gamma_n |U_n|^2\right) U_0 = \tfrac{1}{2} U_0, \tag{9.2.3}$$

$$\frac{d_n}{2}\frac{d^2 U_n}{dx^2} + \left(\gamma_n |U_n|^2 + \sigma \sum_{m \neq n} \gamma_m |U_m|^2\right) U_n = \lambda_n U_n, \tag{9.2.4}$$

where both d_n and γ_n are rescaled.

In the absence of XPM terms, Eq. (9.2.3) has an analytic solution: $U_0(x) = \text{sech}(x)$. We assume that in the presence of XPM all components of the vector soliton have the same "sech" shape but different amplitudes, so $U_n(x) = a_n \text{sech}(x)$. Substituting this solution in Eq. (9.2.4), the amplitudes a_n are found to satisfy the following set of N coupled algebraic equations:

$$a_0^2 + \sigma \sum_{n=1} \gamma_n a_n^2 = 1, \qquad \gamma_n a_n^2 + \sigma \sum_{m \neq n} \gamma_m a_m^2 = d_n. \tag{9.2.5}$$

As long as all parameters d_n and γ_n are close to 1, this solution describes a vector soliton with N components of nearly equal intensity. In the degenerate case, $d_n = \gamma_n = 1$, the amplitudes can be found exactly [58] in the form $U_n = [1 + \sigma(N-1)]^{-1/2}$.

Approximate analytical solutions of Eq. (9.2.4) can be obtained in the *waveguiding limit*, in which the central channel amplitude U_0 is much larger than all other channels [33]. The linearization of Eqs. (9.2.4) for $U_n \ll U_0$ yields $(N-1)$ decoupled linear equations for U_n ($n \neq 0$). Each of these equations possesses a localized solution, provided (assuming $\sigma = 2$)

$$\lambda_n = \Lambda_n \equiv (d_n/8)[1 - \sqrt{1 + 16d_n}]^2. \tag{9.2.6}$$

Near this point in the parameter space, called the *bifurcation point*, the central soliton (the shepherd pulse) can be thought of as inducing an effective waveguide that supports a fundamental mode u_n, with the corresponding cutoff Λ_n. It can be shown that, since d_n and γ_n are close to 1, the soliton-induced waveguide always supports no more than

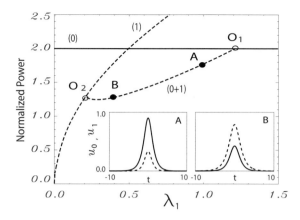

Figure 9.1: Bifurcation diagram for a vector soliton with two components. Curves marked (0) and (1) show powers of the two individual components in the absence of coupling. The curve (0 + 1) shows the total power of the vector soliton. Two insets show the two components u_0 (solid) and u_1 (dashed) at the points A and B marked by solid dots. (After Ref. [61]; ©2002 IEEE.)

two different modes (fundamental and the first excited one) of the *same wavelength* but with largely separated eigenvalues λ_n. As a result, the effective waveguide induced by the shepherd pulse stays predominantly *single-moded* for all operating wavelengths.

We should mention that the special case in which the SPM and XPM nonlinear terms in the set (9.2.2) of N coupled NLS equations are equal ($\gamma_m = \gamma_j = 1$) corresponds to an N-component generalization of the Manakov model for the vector solitons [5]. In this special case, Eqs. (9.2.2) are known to possess exact N-soliton solutions that describe the structure and interaction of multicomponent vector solitons [64]–[67]. Such solitons have been studied in the context of incoherent solitons (see Chapter 13).

9.2.2 Bifurcation Diagram for Vector Solitons

In this subsection we discuss the physical mechanism behind pulse shepherding leading to the formation of vector solitons. This mechanism is closely linked to the soliton-waveguiding phenomenon discussed earlier, and it serves as a mathematical tool for describing the families of different types of vector solitons. From the mathematical point of view, this mechanism is associated with bifurcations of a scalar soliton [68, 69].

Consider the simplest case of $N = 2$ first to get a better physical insight. If pulses in the two channels ($n = 0, 1$) do not interact through XPM, the *uncoupled* Eqs. (9.2.3) and (9.2.4) possess the scalar soliton solutions

$$U_0(x) = \mathrm{sech}(x), \qquad U_1 = (2\lambda_1\gamma_1/d_1)^{1/2}\mathrm{sech}(\sqrt{2\lambda_1/d_1}x). \qquad (9.2.7)$$

The corresponding normalized powers obtained using $P_n = \int_{-\infty}^{\infty} |U_n|^2(x)\,dx$ are given by

$$P_0 = 2, \qquad P_1 = 2(2\lambda_1\gamma_1)^{1/2}d_1^{-3/2}. \qquad (9.2.8)$$

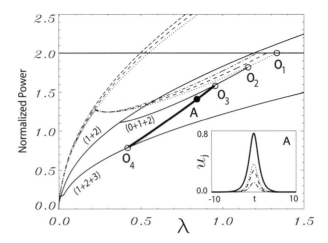

Figure 9.2: Bifurcation diagram for vector solitons with four components shown by plotting $P(\lambda)$ for noninteracting pulses (uppermost dotted, dashed, and dash-dotted curves and the horizontal line) and pulses interacting in pairs or triplets (thin solid lines). The thick solid line shows the case of four interacting pulses. The circles show the bifurcation points and the inset shows the amplitudes at point A, where U_0 (solid line) dominates. (After Ref. [61]; ©2002 IEEE.)

These solutions are shown as curves marked (0) and (1) in Figure 9.1, where the soliton powers are plotted as a function of λ_1. When the two copropagating pulses interact through XPM, a new branch of the vector soliton appears [branch marked $(0+1)$]. This branch is characterized by the total power $P(\lambda_1) = P_0 + P_1$ and joins the two branches $P_0(\lambda_1)$ and $P_1(\lambda_1)$ at the bifurcation points O_1 and O_2, respectively. Near point O_1, the vector soliton consists of an intense component that guides a much weaker component of the second field U_1. As an example, inset A shows the two components at point A close to O_1. Point O_1 coincides with the cutoff condition $\lambda_1 = \Lambda_1$ for the mode U_1 guided by the waveguide created by the shepherd pulse U_0. The shapes and amplitudes of the two components evolve as λ_1 changes. For example, the U_1 component becomes dominant at point B as seen from inset B in Figure 9.1. Component U_0 disappears at the bifurcation point marked O_2.

Next, we consider a more complicated case ($N = 4$) in which a shepherd pulse guides three pulses ($n = 1$–3) and forms a vector soliton with four components. Figure 9.2 shows the bifurcation diagram in this case using $\alpha_1 = \gamma_1 = 0.65$, $\alpha_2 = \gamma_2 = 0.81$, and $\alpha_3 = \gamma_3 = 1$. Such four-component vector solitons can be found numerically as localized solutions of Eqs. (9.2.3) and 9.2.4). The situation is quite complex in this case because of the existence of three parameters, λ_1, λ_2, and λ_3. Figure 9.2 shows the case in which $\lambda_1 = \lambda_2 = \lambda_3 \equiv \lambda$. The uppermost dotted, dashed, and dash-dotted curves in Figure 9.2 correspond to the case of three pulses with no XPM interaction. The second set of dotted, dashed, and dash-dotted curves correspond to pairwise interaction of the three pulses with the central pulse but not with each other; as expected, the bifurcation pattern for each one of them is similar to that shown in Figure 9.1. The thin solid lines show the other combinations of interacting pulses. As one might guess, the three-pulse interaction curves bifurcate from the curves corresponding to pairwise interaction. When all four pulses interact with each other, we obtain the thick solid line connecting bifurcation points O_3 and O_4. This line corresponds to the four-component $(0+1+2+3)$ vector solitons and represents the total power in all four components of such solitons. Pulse shapes for such a soliton at point A are shown in the inset of Figure 9.2. This solution corresponds to the shepherding regime, in which an intense central pulse traps and guides three weaker pulses at different wavelengths.

It is evident from Figure 9.2 that, starting from the central pulse branch (horizontal line), the solution undergoes a *cascade of bifurcations*: $O_1 \rightarrow O_2 \rightarrow O_3 \rightarrow O_4$. Vector solitons with different components form after each bifurcation. After O_1, a $(0+1)$ vector soliton forms and it connects with the $(0+1+2)$ soliton at O_2, which in turn connects with the $(0+1+2+3)$ soliton at O_3. At point O_4, this four-components soliton meets the $(1+2+3)$ soliton branch with only three components. The values of the parameters in Figure 9.2 were chosen to provide a clear bifurcation picture, although they correspond to a wavelength spacing that is much larger than that likely to be used in an experiment [57]–[60]; typically, $\gamma_n/\gamma_{n+1} \approx 0.997$. If the modal parameters are tuned closer to each other, the first two links of the bifurcation cascade tend to disappear. Ultimately, for equal parameters, bifurcation points O_2 and O_3 merge at point O_1, and the four-mode soliton family (thick line in Figure 9.2) branches off directly from the central pulse branch. This qualitative picture of the *bifurcation cascade* found for $N = 4$ is preserved for other values of N.

9.3 Stability of Vector Solitons

As discussed in Section 2.2, the Vakhitov–Kolokolov stability criterion governs the stability of scalar solitons. One would like to find a similar criterion for vector solitons. However, the linear stability analysis becomes much more complicated for vector solitons. Although a direct numerical solution of the corresponding eigenvalue problem is the only possible approach in many cases, it is possible in some cases to extend the scalar stability analysis to vector solitons. We follow Ref. [70] to show how the Vakhitov–Kolokolov stability criterion can be generalized to the case of N coupled NLS equations. Such an analysis includes, as a special limiting case, the bifurcation analysis near the marginal stability curve.

9.3.1 General Stability Analysis

One can extend the method outlined in Section 9.2 to the case of N optical fields interacting nonlinearly and derive the following set of N incoherently coupled NLS equations:

$$i\frac{\partial U_n}{\partial z} + \frac{d_n}{2}\nabla_x^2 U_n + \left(\sum_{m=1}^{N} \gamma_{nm}|U_m|^2\right) U_n = 0, \qquad (9.3.1)$$

where the SPM and XPM terms have been combined by introducing a nonlinear matrix γ whose diagonal components account for SPM. We have also generalized the

NLS equations to multiple spatial dimensions through the Laplacian ∇_x^2 in the D-dimensional space. This generalization allows one to consider vector solitons in two spatial dimensions by choosing $D = 2$. Even the spatiotemporal case can be handled using $D = 3$ if one of the coordinates stands for time. Equation (9.3.1) then describes the spatiotemporal dynamics of multicomponent optical bullets.

Similar to the scalar case discussed in Chapter 2, Eq. (9.3.1) has several conserved quantities associated with it. Provided the symmetry condition $\gamma_{nm} = \gamma_{mn}$ is satisfied, this set of the NLS equations conserves the Hamiltonian

$$H = \int \cdots \int_{-\infty}^{\infty} \left(\sum_{n=1}^{N} \frac{d_n}{2} |\nabla_x U_n|^2 - \frac{1}{2} \sum_{n=1}^{N} \sum_{m=1}^{N} \gamma_{nm} |U_n|^2 |U_m|^2 \right) d^D x. \qquad (9.3.2)$$

Moreover, the individual mode powers, $P_n = \int |U_n|^2 d^D x$, where $d^D x = dx_1 \ldots dx_N$, as well as the total momentum are conserved.

Localized solutions of Eq. (9.3.1) in the form of a multicomponent vector soliton can be obtained in a standard form using $U_n = \Phi_n(x) e^{i\beta_n z}$, where $\Phi_n(x)$ is a real function *with no nodes* and β_n is the corresponding *positive* propagation constant. Vector solitons correspond to the stationary points of the Lyapunov functional

$$\Lambda[U] = H[U] + \sum_{n=1}^{N} \beta_n P_n[U], \qquad (9.3.3)$$

where $[U]$ stands for the set $(U_1, U_2, \ldots, U_N)$. At each stationary point, the first variation, $\delta \Lambda[U]$, vanishes, whereas the second variation, $\delta^2 \Lambda[U]$, governs the stability properties. More specifically, the negative directions of the second variation correspond to unstable eigenvalues in the soliton stability problem.

The stability problem is solved by minimizing the second variation of the Lyapunov functional given by

$$\delta^2 \Lambda = \int_{-\infty}^{\infty} dx [\langle v | L_1 v \rangle + \langle w | L_0 w \rangle], \qquad (9.3.4)$$

where $v(x)$ and $w(x)$ are perturbations of the vector soliton in the form

$$U' = U + [v + iw] e^{\lambda z}, \qquad (9.3.5)$$

and the scalar product is defined as $\langle f | g \rangle = \sum_{n=1}^{N} f_n^* g_n$. The operator L_0 has a diagonal form with the elements

$$(L_0)_{nn} = -\frac{d_n}{2} \nabla_x^2 + \beta_n - \sum_{m=1}^{N} \gamma_{nm} \Phi_m^2, \qquad (9.3.6)$$

whereas the elements of the operator L_1 are given by

$$(L_1)_{nn} = -\frac{d_n}{2} \nabla_x^2 + \beta_n - \sum_{m=1}^{N} \gamma_{nm} \Phi_m^2 - 2\gamma_{nn} \Phi_n^2, \qquad (L_1)_{nm} = -2\gamma_{nm} \Phi_n \Phi_m. \qquad (9.3.7)$$

Similar to the scalar case discussed in Chapter 2, the matrix operators L_0 and L_1 satisfy the linear eigenvalue problem

$$L_1 v = -\lambda w, \qquad L_0 w = \lambda v. \qquad (9.3.8)$$

Both the eigenvalue problem (9.3.8) and the minimization problem (9.3.4) should also satisfy N additional constraints,

$$F_n = \int_{-\infty}^{\infty} d\mathbf{x} \langle \Phi_n \mathbf{e}_n | \mathbf{v} \rangle = 0, \qquad (9.3.9)$$

where $\mathbf{e}_n$ is a unit vector. These constraints correspond to the conservation of individual powers P_n in the presence of a vector perturbation $(\mathbf{v}, \mathbf{w})$.

We recall from Section 2.3 that the scalar solitons ($N = 1$) are stable in the framework of the constrained variational problem governed by Eqs. (9.3.4)–(9.3.9) when

$$\frac{d^2 \Lambda_s}{d\beta_1^2} = \frac{dP_1}{d\beta_1} > 0. \qquad (9.3.10)$$

When this condition is satisfied, the eigenvalue problem (9.3.8) has no eigenvalues with a *positive* real part. The stability criterion for scalar NLS solitons holds when the self-adjoint operator $\mathbf{L}_1$ has a single negative eigenvalue, i.e., when the second variation in Eq. (9.3.4) has a single negative direction without constraint (9.3.9) imposed. If the latter condition is not satisfied, as happens for multihump solitons with nodes, the scalar criterion for soliton instability cannot be extended because of more generic mechanisms of oscillatory instabilities associated with complex eigenvalues for the linear eigenvalue problem.

Fortunately, the scalar stability criterion can be extended in the case of single-hump vector solitons associated with the incoherently coupled NLS equations. We assume that the number of negative directions (eigenvalues) of the second variation, $\delta^2 \Lambda$, is fixed and denote it by $n(\Lambda)$. The unstable eigenvalues λ of the eigenvalue problem (9.3.8) are connected with the negative eigenvalues of the *Hessian matrix* $\mathbf{U}$ with the elements

$$U_{nm} = \frac{\partial^2 \Lambda_s}{\partial \beta_n \partial \beta_m} = \frac{\partial P_n}{\partial \beta_m} = \frac{\partial P_m}{\partial \beta_n}. \qquad (9.3.11)$$

We denote the number of positive eigenvalues of the matrix $\mathbf{U}$ as $p(U)$ and the number of its negative eigenvalues as $n(U)$. Clearly, $p(U) + n(U) \leq N$, since some eigenvalues may be zero in a degenerate (bifurcation) case. As discussed in Ref. [70], both $p(U)$ and $n(U)$ satisfy an additional constraint of the form

$$p(U) \leq \min\{N, n(\Lambda)\}, \qquad n(U) \geq \max\{0, N - n(\Lambda)\}. \qquad (9.3.12)$$

With these notations, one can deduce the following results on the stability of multicomponent vector solitons [70]:

- The eigenvalue problem (9.3.8) can have at most $n(\Lambda)$ unstable eigenvalues λ such that they are all *real* and *positive*.
- A vector soliton is *linearly unstable* provided $p(U) < n(\Lambda)$. The eigenvalue problem (9.3.8) then has $n(\Lambda) - p(U)$ real eigenvalues.
- A vector soliton is *linearly stable* provided $p(U) = n(\Lambda) \leq N$. In the case $n(\Lambda) = N$, this criterion implies that the energy surface $\Lambda_s(\boldsymbol{\beta})$ has a concave shape in the $\boldsymbol{\beta}$ space.

- A single eigenvalue λ crosses *a marginal stability curve* when the matrix $\mathbf{U}$ possesses a zero eigenvalue that becomes negative under perturbation.

The instability-induced dynamics of vector solitons is governed by an equation that resembles the equation of motion for a classical particle subjected to an N-dimensional potential field such that the total energy is given by

$$E = \frac{1}{2} \sum_{n=1}^{N} \sum_{m=1}^{N} M_{nm} \frac{dv_n}{dz} \frac{dv_m}{dz} + W(\boldsymbol{\beta}, \mathbf{v}), \qquad (9.3.13)$$

where $\mathbf{v}$ is the vector describing perturbation of the soliton parameters $\boldsymbol{\beta}$ and the elements of the positive-definite "mass matrix" have the form

$$M_{nm} = \int_{-\infty}^{\infty} dx \sum_{k=1}^{N} G_{kn}(\mathbf{x}) G_{km}(\mathbf{x}), \qquad (9.3.14)$$

with

$$G_{kn}(\mathbf{x}) = \frac{1}{\Phi_k(\mathbf{x})} \left(\int_0^{\mathbf{x}} d\mathbf{x}' \Phi_k(\mathbf{x}') \frac{\partial \Phi_k(\mathbf{x}')}{\partial \beta_n} \right). \qquad (9.3.15)$$

The effective potential energy in Eq. (9.3.13) is given by

$$W(\boldsymbol{\beta}, \mathbf{v}) = \Delta H(\boldsymbol{\beta}) + \sum_{n=1}^{N} (\beta_n + v_n) \Delta P_n(\boldsymbol{\beta}), \qquad (9.3.16)$$

where

$$\Delta H(\boldsymbol{\beta}) = H(\boldsymbol{\beta} + v) - H(\boldsymbol{\beta}), \qquad \Delta P_n(\boldsymbol{\beta}) = P_n(\boldsymbol{\beta} + v) - P_n(\boldsymbol{\beta}). \qquad (9.3.17)$$

We refer to Ref. [70] for details on derivation of these results and their comparison with the Grillakis theorems [71]–[74].

9.3.2 Two Coupled Modes

To demonstrate how the preceding general theory can be applied to a specific physical problem, we consider the case of two simplest incoherently coupled NLS equations of the form

$$i \frac{\partial U_1}{\partial z} + \frac{\partial^2 U_1}{\partial x^2} + (|U_1|^2 + \sigma |U_2|^2) U_1 = 0, \qquad (9.3.18)$$

$$i \frac{\partial U_2}{\partial z} + \frac{\partial^2 U_2}{\partial x^2} + (|U_2|^2 + \sigma |U_1|^2) U_2 = 0. \qquad (9.3.19)$$

These equations are obtained from Eq. (9.3.1) using $N = 2$, $d_1 = d_2 = 2$, $\gamma_{11} = \gamma_{22} = 1$, and $\gamma_{12} = \gamma_{21} = \sigma$. An explicit soliton solution of them can easily be found for $\beta_1 = \beta_2 = \beta$ and $\sigma > -1$ in the form

$$\Phi_1(x) = \Phi_2(x) = \sqrt{\frac{2\beta}{1 + \sigma}} \operatorname{sech}\left(\sqrt{\beta} x\right). \qquad (9.3.20)$$

This solution describes a vector soliton whose two components have equal amplitudes. It corresponds to a straight line $\beta_1 = \beta_2$ in the parameter plane (β_1, β_2) of a general *two-parameter family* of vector solitons associated with Eqs. (9.3.18) and (9.3.19). For $-1 < \sigma \leq 0$, two-parameter solitons exist everywhere in the plane (β_1, β_2), while for $\sigma > 0$, the existence region is restricted by the two bifurcation curves determined from

$$\beta_2 = \omega_\pm(\sigma)\beta_1, \qquad \omega_\pm(\sigma) = \left(\frac{\sqrt{1+8\sigma}-1}{2} \right)^{\pm 2}. \tag{9.3.21}$$

Approximate analytical expressions for the two-component vector solitons can also be obtained in the vicinity of the bifurcation curves (9.3.21) because one of the components is relatively small in this region while the other one is described by a scalar NLS equation. From the physical point of view, this region corresponds to the situation in which the stronger component creates an effective waveguide that guides the weaker component (the *shepherding effect* described in Section 9.2). Assuming that the component Φ_1 corresponds to the stronger component, the solution can be written in terms of a small parameter ε as [70]

$$\Phi_1(x) = R_0(x) + \varepsilon^2 R_2(x) + O(\varepsilon^4), \qquad \Phi_2(x) = \varepsilon S_1(x) + O(\varepsilon^3), \tag{9.3.22}$$

where the dominant terms are given by

$$R_0(x) = \sqrt{2\beta_1}\, \mathrm{sech}\left(\sqrt{\beta_1} x \right), \qquad S_1(x) = \sqrt{\beta_1}\, \mathrm{sech}^{\sqrt{\omega_+}}\left(\sqrt{\beta_1} x \right). \tag{9.3.23}$$

This solution is valid in the vicinity of the bifurcation curve

$$\beta_2 = \omega_+(\sigma)\beta_1 + \varepsilon^2 \omega_{2+}(\sigma)\beta_1 + O(\varepsilon^4), \tag{9.3.24}$$

where ω_{2+} is given by

$$\omega_{2+} = \frac{\int_{-\infty}^{\infty} (S_1^4 + 2\sigma R_0 R_2 S_1^2)\, dx}{\int_{-\infty}^{\infty} S_1^2(x)\, dx}. \tag{9.3.25}$$

The second-order correction $R_2(x)$ can also be obtained by solving

$$\left[-\frac{\partial}{\partial x^2} + \beta_1 - 6\beta_1 \,\mathrm{sech}^2\left(\sqrt{\beta_1} x \right) \right] R_2 = \sigma R_0 S_1^2. \tag{9.3.26}$$

In the region $\sigma > 0$, where the two-component soliton exists, it follows from Eq. (9.3.25) that $\omega_{2+}(\sigma) > 0$ for $0 < \sigma < 1$ but it becomes negative for $\sigma > 1$. When $\sigma = 1$, this family of two-parameter vector solitons becomes degenerate in the sense that $\beta_1 = \beta_2$, but the resulting vector soliton is different from the solution given in Eq. (9.3.20). The vector soliton is known to be stable in the integrable case $\sigma = 1$, as first shown by Manakov [5] . Here we apply stability theory to show that all two-component vector solitons are *stable* for $\sigma \geq 0$ but become *unstable* for $\sigma < 0$.

For this purpose, we first evaluate the number of positive eigenvalues $p(U)$ using the explicit solution given in Eq. (9.3.20). The elements of the Hessian matrix **U** from Eq. (9.3.11) are found to be [70]

$$U_{11} = U_{22} = \frac{1/\sqrt{\beta}}{(1+\sigma)} \qquad U_{12} = U_{21} = -\frac{\sigma/\sqrt{\beta}}{(1+\sigma)}. \tag{9.3.27}$$

It is easy to show that this Hessian matrix has two positive eigenvalues [$p(U) = 2$] for $-1 < \sigma < 1$ but that only one positive eigenvalue exists for $\sigma > 1$.

Consider now the number $n(\Lambda)$ of negative eigenvalues for the linear matrix operator $\mathbf{L}_1$ given in Eq. (9.3.7). It can be diagonalized using linear combinations of the eigenfunctions $v_1 = u_1 + u_2$ and $v_2 = u_1 - u_2$ such that

$$\left[-\frac{\partial}{\partial x^2} + \beta - 6\beta \operatorname{sech}^2 \left(\sqrt{\beta} x \right) \right] v_1 = \mu v_1, \qquad (9.3.28)$$

$$\left[-\frac{\partial}{\partial x^2} + \beta - 2\beta \frac{(3 - \sigma)}{(1 + \sigma)} \operatorname{sech}^2 \left(\sqrt{\beta} x \right) \right] v_2 = \mu v_2. \qquad (9.3.29)$$

Both of these eigenvalue equations involve sech-type potentials and can easily be solved to find the eigenvalues. The first equation always has a single negative eigenvalue $\mu = -3\beta$. The second equation has no negative eigenvalues for $\sigma > 1$, has a single negative eigenvalue for $0 < \sigma < 1$, and has more than one negative eigenvalues for $-1 < \sigma < 0$. Thus, $n(\Lambda)$ exceeds 2 for $-1 < \sigma < 0$, equals 2 for $0 < \sigma < 1$, and is 1 for $\sigma > 1$. Applying stability analysis [70], we come to the conclusion that the soliton solution (9.3.20) with equal amplitudes is *stable* for $\sigma > 0$, since in this domain $p(U) = n(\Lambda)$ but that it becomes *unstable* for $-1 < \sigma < 0$, because $p(U) < n(\Lambda)$ in that case.

The other limiting case related to the shepherding effect can be analyzed in the same way. The elements (9.3.11) of the Hessian matrix U can again be obtained in an explicit analytical form. The Hessian matrix U is found to have two [$p(U) = 2$] positive eigenvalues for $0 < \sigma < 1$ but only one for $\sigma > 1$. The linear matrix operator $\mathbf{L}_1$ cannot be diagonalized for the shepherding soliton (9.3.22) unless $\varepsilon = 0$. In this decoupled case, it has a single negative eigenvalue $\mu = -3\beta_1$ and a doubly degenerate zero eigenvalue. When $\varepsilon \neq 0$, the zero eigenvalue shifts to become $\mu = -2\omega_{2+}(\sigma)\beta_1 \varepsilon^2$. Therefore, the matrix operator $\mathbf{L}_1$ for the shepherding soliton (9.3.22) has $n(\Lambda) = 2$ for $0 < \sigma < 1$ and $n(\Lambda) = 1$ for $\sigma > 1$. Since $p(U) = n(\Lambda)$ when $\sigma > 0$, we come to the conclusion that the shepherding soliton is stable for $\sigma > 0$.

9.3.3 Effect of Walk-off

It follows from the preceding discussion that stationary vector solitons in which a strong component shepherds several weaker ones are *linearly stable*. However, the walk-off effect resulting from the group-velocity mismatch would endanger the integrity of such a composite soliton. In the case of two components of comparable amplitudes, the nonlinearity can provide an effective trapping mechanism that keeps the pulses together [2], [75]–[77]. In the shepherding regime, one or more strong pulses create an effective attractive potential, which traps a weaker component if its group velocity is less than the *escape velocity* associated with this potential.

In reality, all components of a vector soliton have nonzero values of the walk-off parameters. It is difficult to treat this situation analytically. Figure 9.3 shows the results of numerical simulations for $N = 4$ obtained by launching the exact four-component vector soliton shown in the inset of Figure 9.2. In the absence of walk-off, all components would evolve without changing their shape. In the presence of small

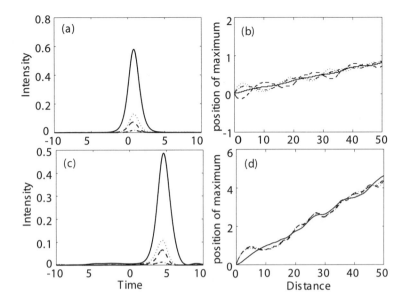

Figure 9.3: Effect of group-velocity mismatch in the shepherding regime for the vector soliton corresponding to point A in Figure 9.2. (a) Four components at $z = 50$ and (b) shift in the position of the four components with z for $\delta_1 = 0.45, \delta_2 = -0.35, \delta_3 = 0.25$. Parts (c) and (d) show the behavior when $\delta_1 = \delta_2 = \delta_3 = 0.9$. In all cases, the dominant component is shown by a solid line. (After Ref. [61]; ©2002 IEEE.)

to moderate walk-off, the four components remain localized and mutually trapped except for shifting their position slightly, as seen in the top row of Figure 9.3 for $\delta_1 = 0.45, \delta_2 = -0.35, \delta_3 = 0.25$. However, the vector soliton begins to shed considerable energy into radiation for large values of the relative walk-off, as seen in the bottom row for $\delta_1 = \delta_2 = \delta_3 = 0.9$. The former situation is more likely to be realized experimentally because the group-velocity mismatch for pulses of different wavelengths is expected to be different [58]. In this case, the estimate for the threshold values of δ_j can be given only if the shepherd pulse is much stronger than the trapped weaker pulses, which are treated as noninteracting fundamental modes of the effective waveguide induced by the shepherd pulse. In all other regimes, a numerical linear stability analysis is needed.

9.4 Coherently Coupled Solitons

Vector solitons discussed so far are said to be incoherently coupled, in the sense that the coupling is phase insensitive. Another important class of vector solitons is associated with the coherent coupling among the optical fields; i.e., the coupling depends on relative phases of the interacting fields. Coherent interaction occurs when the nonlinear medium is weakly anisotropic or weakly birefringent. Coherently coupled vector solitons possess many properties that are different from the case of incoherent coupling.

We focus on these differences by considering two-component vector solitons formed in planar waveguides through nonlinear coupling between the TE and TM spatial modes. Such planar waveguides have the potential of producing compact devices when a semiconducting AlGaAs layer is used for making the waveguide. The reason is that the detrimental effects of two-photon absorption can be completely eliminated by operating at a wavelength in the 1.5-μm region, where the photon energy is then below the half-bandgap. At the same time, the Kerr coefficient n_2 is almost *three* orders of magnitude larger than that of silica [78, 79]. Even though a GaAs crystal has no material birefringence in the bulk form, a waveguide formed using this material acquires weak birefringence from the presence of an unavoidable growth-related stress in the guiding region.

9.4.1   Coherently Coupled NLS Equations

The basic equations governing the propagation of the orthogonally polarized TE and TM waveguide modes can be obtained using the standard coupled-mode theory [80]–[82]. The total field inside the waveguide at a frequency ω can be written in the form

$$\mathbf{E}(\omega) = \mathrm{Re}\left[\hat{x}A_x(\omega)e^{ik_xz-i\omega_0t} + \hat{y}A_y(\omega)e^{ik_yz-i\omega_0t}\right], \tag{9.4.1}$$

where k_α is the propagation constant with $\alpha = x$ and y for the TE and TM modes, respectively. The third-order nonlinear polarization can also be separated into TE and TM parts as $\mathbf{P} = \mathrm{Re}[(\hat{x}P_x + \hat{y}P_y)e^{-i\omega_0t}]$. Using the third-order susceptibility for a Kerr medium, the polarization components are given by

$$P_\alpha(\omega) = \sum_{\beta\gamma\sigma} \chi^{(3)}_{\alpha\beta\gamma\sigma}[A_\beta A_\gamma A_\sigma^* e^{i(k_\beta+k_\gamma-k_\sigma)} + A_\beta^* A_\gamma A_\sigma e^{i(-k_\beta+k_\gamma+k_\sigma)}$$

$$+ A_\beta A_\gamma^* A_\sigma e^{i(k_\beta-k_\gamma+k_\sigma)}]. \tag{9.4.2}$$

To proceed further, we assume that Kleinman symmetry holds and also notice that the zinc-blend-type semiconductor crystals, such as GaAs, possess cubic symmetry. The third-order susceptibility tensor $\chi^{(3)}$ then has only *four* independent components: $\chi_{xxxx}, \chi_{xyyx}, \chi_{xyxy}$, and χ_{xxyy} [83]. Retaining only the nonzero components in Eq. (9.4.2), we obtain

$$P_x = \frac{3a}{8}\left(A_x|A_x|^2 + \frac{2b}{3a}A_x|A_y|^2 + \frac{b}{3a}A_x^*A_y^2e^{-i\Delta kz}\right)e^{ik_xz}, \tag{9.4.3}$$

$$P_y = \frac{3a}{8}\left(A_y|A_y|^2 + \frac{2b}{3a}A_y|A_x|^2 + \frac{b}{3a}A_y^*A_x^2e^{i\Delta kz}\right)e^{ik_yz}, \tag{9.4.4}$$

where $a = \chi_{xxxx}$, $b = \chi_{xxyy} + \chi_{xyxy} + \chi_{xyyx}$, and $\Delta k = 2(k_x - k_y)$. For the cubic symmetry classes 432, $\bar{4}3m$, and $m3m$, $\chi_{xxyy} = \chi_{xyxy}$ far from resonances [84]. Deviations from the isotropic properties are usually quantified by a parameter $\eta = 1 - b/a$ [85]; $\eta = 0$ in isotropic crystals.

The last term in Eqs. (9.4.3) and (9.4.4) is phase sensitive and is responsible for coherent coupling. If the wave-vector mismatch Δk is sufficiently large (the strong-birefringence limit), this term oscillates rapidly and can be eliminated after averaging over the phase [2]. The resulting equations are then coupled incoherently, and the situation reduces to the one discussed in preceding sections. In the case of *weak* birefringence, the coherent-coupling terms cannot be neglected, and they play an important role.

The evolution equations for the two orthogonally polarized components in the limit of *weak birefringence* take the form [86]

$$i\frac{\partial A_x}{\partial z} + \frac{1}{2k}\frac{\partial^2 A_x}{\partial x^2} - \frac{(k^2 - k_x^2)}{2k}A_x + \gamma\left[(|A_x|^2 + A|A_y|^2)A_x + BA_x^*A_y^2\right] = 0, \quad (9.4.5)$$

$$i\frac{\partial A_y}{\partial z} + \frac{1}{2k}\frac{\partial^2 A_y}{\partial x^2} - \frac{(k^2 - k_y^2)}{2k}A_y + \gamma\left[(A|A_x|^2 + |A_y|^2)A_y + BA_y^*A_x^2\right] = 0, \quad (9.4.6)$$

where $k = (k_y + k_x)/2 = k_0\bar{n}$, $k_0 = \omega/c$, $\bar{n}$ is the average refractive index, $\gamma = n_2 k_0/d_e$, and d_e is the effective waveguide thickness. The incoherent coupling is governed by the parameter

$$A = 2b/(3a) = 2(2\chi_{xxyy} + \chi_{xyyx})/(3\chi_{1111}), \quad (9.4.7)$$

and the coherent coupling is governed by the parameter $B = b/3a$. When the TE mode is polarized along the x axis, $B = A/2$. However, in general, A, B, and the SPM coefficient (1 in our case) depend on the orientation of the wave vector with respect to the principal directions of the nonlinear crystal [84]. For some orientations of the waveguide, these coefficients can become different for the TE and TM modes [85, 87].

The coupled equations (9.4.5) and (9.4.6) describe the dynamics of two coherently coupled fields in a Kerr-like medium, and it is a special case of a more general four-wave mixing (FWM) process. Solitons in such generalized FWM models have also been studied [88]–[90], but many issues still remain open. However, the model of third-harmonic generation has been studied in more detail [91].

As usual, it is helpful to normalize Eqs. (9.4.5) and (9.4.6). Using the transformations

$$x' = kx, \quad z' = kz, \quad U = \sqrt{\gamma/k}A_x, \quad V = \sqrt{\gamma/k}A_y, \quad (9.4.8)$$

we obtain the following set of two coupled NLS equations:

$$i\frac{\partial U}{\partial z} - \beta U + \frac{1}{2}\frac{\partial^2 U}{\partial x^2} + (|U|^2 + A|V|^2)U + BU^*V^2 = 0, \quad (9.4.9)$$

$$i\frac{\partial V}{\partial z} + \beta V + \frac{1}{2}\frac{\partial^2 V}{\partial x^2} + (A|U|^2 + |V|^2)V + BV^*U^2 = 0, \quad (9.4.10)$$

where the parameter $\beta = (k_y - k_x)/2k$ defines the degree of birefringence. The prime over x and z has been dropped for notational simplicity. The TM component with amplitude V is assumed to have a larger propagation constant than the TE component, as observed under typical experimental conditions. In this case β is positive, although it can be negative under some conditions [80, 92]. The same set of equations (9.4.9) and (9.4.10) is also obtained in the temporal case, in which a short pulse propagates inside a weakly birefringent fiber, provided the coordinate x is interpreted as a time variable. The temporal case has been extensively studied in the field of nonlinear fiber optics [93]–[98].

Despite the obvious homology of the underlying equations in the spatial and temporal cases, neither the results nor the conclusions can be transferred directly to the spatial domain from the corresponding studies in the temporal domain. The reason is related to the *anisotropic* nature of the nonlinear medium in the spatial case. In isotropic materials, such as optical fibers, parameters A and B are *fixed* by the symmetry properties of the susceptibility tensor and have values $A = 2/3$, $B = 1/3$. In contrast, in an anisotropic material, only the ratio of the coefficients is fixed, but their absolute value can vary, depending on the nature of the third-order susceptibility tensor [78]. For example, it was the closeness of the A coefficient to unity in GaAs in the 1.5-μm-wavelength region [86] that enabled the first experimental observation of the *Manakov soliton* [99]—an exact solution of Eqs. (9.4.9) and (9.4.10) for $A = 1$ and $B = 0$. The exact solutions of these equations for $B = 0$ have been found even when $A \neq 1$ [100, 101].

It should be noticed that Eqs. (9.4.9) and (9.4.10) cover the case of incoherently coupled NLS equations (at $B = 0$ and $\beta = 0$) discussed earlier, and the analysis of their stationary localized solutions is similar. The main difference between the incoherently and coherently coupled NLS equations and the vector solitons they describe is the dynamic energy exchange between the components that occurs for $B \neq 0$.

The spatial walk-off term has been neglected in the coherently coupled NLS equations, Eqs. (9.4.9) and (9.4.10). This can often be justified in the weak-birefringence regime. The effect of this term is analogous to that of the group-velocity mismatch for the orthogonally polarized pulses propagating in birefringent fibers [97]. The walk-off effects on the stability of spatial solitons have been considered in Ref. [103].

9.4.2   Shape-Preserving Solutions

In this section we focus on the shape-preserving solutions of Eqs. (9.4.9) and (9.4.10) that qualify as vector solitons. Similar to the scalar case discussed in section 2.1, these equations have two quantities that remain invariant during propagation. One of them is the total power of both components, given by

$$P = \int_{-\infty}^{\infty} (|U|^2 + |V|^2)\, dx. \tag{9.4.11}$$

The second invariant is the system Hamiltonian, defined as

$$H = \int_{-\infty}^{\infty} \left[\beta(|U|^2 - |V|^2) + \frac{1}{2}\left(\left|\frac{\partial U}{\partial x}\right|^2 + \left|\frac{\partial V}{\partial x}\right|^2 \right) \right.$$
$$\left. - \frac{1}{2}(|U|^4 + |V|^4) - A|U|^2|V|^2 - \frac{B}{2}(U^2 V^{*2} + U^{*2} V^2) \right] dx. \tag{9.4.12}$$

Any stationary soliton solution of Eqs. (9.4.9) and (9.4.10) can be represented by a point in the two-dimensional phase space formed using P and H. Similar to the scalar case, the functional dependence of the Hamiltonian H on power P can be used for judging the stability of stationary solutions (provided no complex eigenvalues are found to occur for the corresponding linear problem).

The β term in Eqs. (9.4.9) and (9.4.10) resulting from the birefringence leads to a phase shift, in addition to the nonlinearity-induced phase shift. Because of these phase shifts, the coherent-coupling term responsible for energy exchange becomes crucially important because of its phase-sensitive nature. It turns out that the energy exchange leads to a phase-locking effect: Both components of the field can be *locked in phase* and form a vector soliton [97].

To describe the *phase-locking* effect, we look for stationary solutions of Eqs. (9.4.9) and (9.4.10) in the form

$$U = u(x,q)\exp(iqz + i\varphi_1), \qquad V = v(x,q)\exp(iqz + i\varphi_2), \qquad (9.4.13)$$

where $u(x,q)$ and $v(x,q)$ govern the shape of the two orthogonally polarized components. The parameter q represents the shift of the wave number for both components. The shape-preserving solutions exist only when the phase difference $\Delta\varphi \equiv \varphi_1 - \varphi_2$ between the two components takes the value 0, π, or $\pm\pi/2$. Substituting Eq. (9.4.13) in Eqs. (9.4.9) and (9.4.10), u and v are found to satisfy the following two coupled differential equations:

$$\frac{1}{2}\frac{d^2u}{dx^2} - (q+\beta)u + \left[|u|^2 + (A\pm B)|v|^2\right]u = 0, \qquad (9.4.14)$$

$$\frac{1}{2}\frac{d^2v}{dx^2} - (q-\beta)v + \left[|v|^2 + (A\pm B)|u|^2\right]v = 0, \qquad (9.4.15)$$

where the plus sign is chosen for $\Delta\varphi = 0$ or π and the minus is chosen for for $\Delta\varphi = \pm\pi/2$. The former case corresponds to linearly polarized vector solitons, while the soliton is elliptically polarized in the latter case. To find such vector solitons, the preceding two equations should be solved with the boundary conditions that u, v, and their first derivatives vanish as $x \to \pm\infty$.

Equations (9.4.14) and (9.4.15) have simple exact solutions in the form of TE- and TM-polarized bright *scalar* solitons. In the TE case, $v = 0$ and u is given by

$$u_0(x,q) = \sqrt{2(q+\beta)}\,\text{sech}(\sqrt{2(q+\beta)}x). \qquad (9.4.16)$$

In contrast, in the TM case, $u = 0$ and v is given by

$$v_0(x,q) = \sqrt{2(q-\beta)}\,\text{sech}(\sqrt{2(q-\beta)}x). \qquad (9.4.17)$$

It can seen from these solutions that the width of the beam (or the pulse width in the temporal case) is determined by the birefringence parameter β.

Vector solitons are formed when both u and v are nonzero. Such solutions of Eqs. (9.4.14) and (9.4.15) can be found numerically and form a continuous family with respect to the parameter q [104]. We quantify this family in Figure 9.4 by plotting the soliton power $P(q)$ in the left column and the invariant $H(P)$ in the right column. The three rows correspond to different values of A, with $B = A/2$ and $\beta = 1$. These results were obtained by using a shooting technique in combination with the relaxation method [105].

For values of q below a critical value q_{cr} or values of P below P_{cr}, only the TE and TM solutions shown by the solid lines exist. However, when q exceeds q_{cr} or

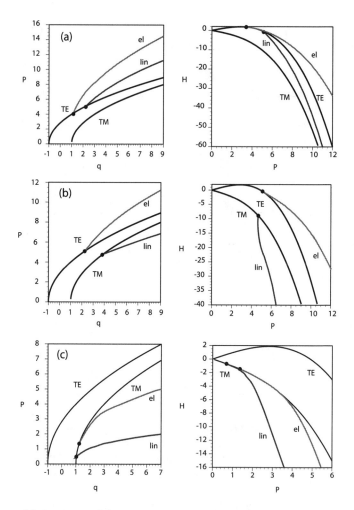

Figure 9.4: $P(q)$ (left) and $H(P)$ (right) curves for (a) $A = 1/3$, (b) $A = 1$, and (c) $A = 4$. In all cases, $B = A/2$ and $\beta = 1$. Black curves correspond to scalar TE and TM solitons. Vector solitons form after the bifurcation points (solid dots) and are linearly or elliptically polarized on the gray branches, marked "lin" and "el," respectively. (After Ref. [104]; ©1997 OSA.)

$P \geq P_{cr}$, two new branches emerge corresponding to two different families of vector solitons. The bifurcation points indicated by solid dots in Figure 9.4 can be found using a perturbation approach similar to that used earlier in the case of incoherently coupled NLS equations. Consider a bifurcation from the TE mode, i.e., branching of the solution $(u_0, 0)$. In the vicinity of the bifurcation point, the condition $v/u \sim \varepsilon \ll 1$ holds. Therefore, we can expand u and v in a series in powers of ε as

$$u = u_0 + \varepsilon^2 u_2, \qquad v = \varepsilon v_1. \tag{9.4.18}$$

Substituting these expansions into Eqs. (9.4.14) and (9.4.15) and linearizing around the exact solution u_0, we obtain the following equation for v_1:

$$\frac{1}{2}\frac{d^2v_1}{dx^2} + \left[-(q-\beta) + \frac{(B\pm A)(q+\beta)}{\cosh^2[\sqrt{2(q+\beta)}x]}\right]v_1 = 0. \qquad (9.4.19)$$

This equation represents a linear eigenvalue problem that can be solved analytically. The eigenfunctions have the general form

$$v_1 = (1-\xi^2)^{\nu/2}F\left(\nu-s,\nu+s+1,\nu+1,\frac{1-\xi}{2}\right), \qquad (9.4.20)$$

where F is a hypergeometric function [106] and

$$\xi = \tanh[\sqrt{2(q+\beta)}x], \quad \nu = \sqrt{\frac{q-\beta}{q+\beta}}, \quad s = \tfrac{1}{2}\left(-1+\sqrt{1+4(B\pm A)}\right). \qquad (9.4.21)$$

Localized eigenfunctions ($v_1 \to 0$ as $|x| \to \infty$) require $s - \nu = n$, where n is an integer. In this case, the hypergeometric function reduces to the nth-order Jacobi polynomial [106]. Depending on the numerical value of the parameters A and B, N discrete eigenvalues $\lambda_n \equiv q - \beta$ can exist for $1 \leq N \leq s - \frac{1}{2}$. Thus the parameter of the medium, A, defines the total number of localized solutions v_{1n}. Each eigenvalue λ_n corresponds to a unique value of the soliton parameter q_n such that

$$q_n^{\pm} = \frac{\beta(1+v_{\pm}^2)}{|1-v_{\pm}^2|}, \quad v_{\pm} = \tfrac{1}{2}[-(1+2n)+\sqrt{1+4(B\pm A)}]. \qquad (9.4.22)$$

Each $q_n^{\pm}$ for $n = 0, 1, 2, \ldots N$ corresponds to the *branching* of a new *vector* solution with the components (u_0, v_{1n}), where $v_{1n} \equiv v_1(x, q_n)$. The relative phase of these components is 0 or π for a $+$ sign and leads to linearly polarized vector solitons. In contrast, the two components differ in phase by $\pi/2$ and form an elliptically polarized vector soliton for the choice of a $-$ sign. A similar bifurcation analysis can be carried out for the TM scalar soliton and indicates that both linearly and elliptically polarized vector solitons can again form through multiple bifurcations.

The *fundamental* vector soliton corresponds to the lowest-order eigenvalue ($n = 0$). Near the bifurcation point, this soliton has two symmetric sech-like components. Two examples of the transverse profiles are shown in Figure 9.5 for linearly and elliptically polarized vector solitons. The shape of two components can be quite different for higher-order vector solitons. For example, in the neighborhood of the second bifurcation point ($n = 1$), an *antisymmetric* solution with $u \approx u_0$ and

$$v \approx \sinh[\sqrt{2(q+\beta)}x]\cosh^{-(\nu+1)}[\sqrt{2(q+\beta)}x] \qquad (9.4.23)$$

emerges. Higher-order bifurcations ($n \geq 1$) can form multihump vector solitons. Such bifurcations can occur even in the absence of the energy-exchange term ($B = 0$), as discussed in Refs. [107] and [108].

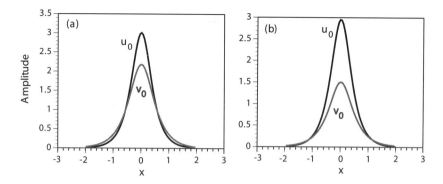

Figure 9.5: Transverse profiles for vector solitons with (a) elliptical and (b) linear polarization at the beam center for $A = 1/3$, $B = A/2$, $\beta = 1$, and $q = 4$. (After Ref. [104]; ©1997 OSA.)

9.4.3 Polarization Dynamics

The polarization state of a vector soliton may change with propagation. Strictly speaking, the polarization is not uniform across the entire beam for a vector soliton, as evident from Figure 9.4, where the ratio v_0/u_0 is different at different points across the beam profile. However, one can often employ an *average profile approximation* for calculating the polarization state associated with the beam as a whole. This approximation has proven to be quite accurate in numerical simulations [109]. However, one should be aware that it does not take into account the radiation emitted by a vector soliton or reshaping of the soliton components [95].

To follow the polarization dynamics, we consider the evolution of the Stokes vector on the Poincaré sphere. The state of polarization of a vector soliton governed by Eqs. (9.4.9) and (9.4.10) can be described by the following four Stokes parameters [110]:

$$s_0 = |U|^2 + |V|^2, \quad s_1 = |U|^2 - |V|^2, \quad s_2 = \mathrm{Re}(UV^*), \quad s_3 = \mathrm{Im}(UV^*), \quad (9.4.24)$$

which satisfy the condition $s_0^2 = s_1^2 + s_2^2 + s_3^2$. This allows the state of polarization of the soliton to be represented by the tip of the Stokes vector $\mathbf{s} = \{s_1, s_2, s_3\}$ on the Poincaré sphere [110]. All trajectories representing the evolution of the polarization state are located on this sphere of fixed radius s_0.

The Stokes vector $\mathbf{s}$ describes the state of polarization at one point of the soliton profile and varies with both x and z. To discuss the polarization of the whole soliton, we assume that the shape of both components is identical and does not change during propagation and write $U = f(x)X(z)$ and $V = f(x)Y(z)$, where $f(x)$ represents the shape of the vector soliton and X and Y are the evolving amplitudes of the soliton components. With this assumption, we can introduce new Stokes parameters after integrating over the soliton shape as [109]

$$S_i(z) = \frac{\int_{-\infty}^{\infty} s_i(x,z)\,dx}{\int_{-\infty}^{\infty} f^2(x)\,dx}. \quad (9.4.25)$$

Again, $S_0^2 = S_1^2 + S_2^2 + S_3^2$, and the polarization state of the whole soliton can be represented by the motion of the Stokes vector $\mathbf{S} = \{S_1, S_2, S_3\}$ on the Poincaré sphere.

Using Eqs. (9.4.9) and (9.4.10), the three components of the Stokes vector are found to satisfy the following set of first-order differential equations:

$$\frac{dS_1}{dz} = AgS_2S_3, \tag{9.4.26}$$

$$\frac{dS_2}{dz} = 2\beta S_3 - (1 - A/2)gS_1S_3, \tag{9.4.27}$$

$$\frac{dS_3}{dz} = -2\beta S_2 + (1 - 3A/2)gS_1S_2, \tag{9.4.28}$$

where we have used $B = A/2$ and the parameter g is defined as

$$g = \int_{-\infty}^{\infty} f^4(x)dx \Big/ \int_{-\infty}^{\infty} f^2(x)dx. \tag{9.4.29}$$

The only parameter describing trajectories on the Poincaré sphere is the parameter g, and it depends implicitly on the beam power P. Taking the beam profile as $f(x) = \sqrt{2q}\,\text{sech}(\sqrt{2q}x)$, which corresponds to the linearly polarized solution (9.4.16) at $\beta = 0$, we obtain $g = 4q/3$. This relation allows us to establish a *correspondence* between the bifurcations patterns shown in Figure 9.4 using $P(q)$ and $H(P)$ curves and the trajectories on the Poincaré sphere governing the evolution of the state of polarization.

Before analyzing the trajectories of the Stokes vector on the Poincaré sphere, we first identify the so-called fixed points of Eqs. (9.4.26)–(9.4.28) after setting $dS_i/dz = 0$ for $i = 1,\ 2,\ 3$. These points correspond to the stationary soliton solutions of Eqs. (9.4.9) and (9.4.10). Since the system is Hamiltonian, they can only be *focus*- or *saddle*-type points, corresponding to the stable and unstable dynamics of the stationary solutions, respectively. The main feature of these solutions is that the axes of the polarization ellipse *do not* change their length or orientation with propagation.

Two stationary points on the sphere, $\{-S_0, 0, 0\}$ and $\{S_0, 0, 0\}$, correspond to the scalar TM and TE solutions of Eqs. (9.4.9) and (9.4.10). They exist for any value of beam power. In addition, at high powers, *bifurcation* of the fixed points occurs, and we find *four stationary solutions* that can be either elliptically or linearly polarized. Their existence depends on the parameter g, which has two critical values. Normalizing the total power such that $S_0 = 1$, these critical values are given by

$$g_{\text{el}} = 2\beta/(1 - A/2), \qquad g_{\text{lin}} = 2\beta/(1 - 3A/2). \tag{9.4.30}$$

A direct correspondence exists between the bifurcation of the polarization eigenstates and the bifurcations of the soliton families shown in Figure 9.4.

Two fixed points on the Poincaré sphere corresponding to vector solitons with *elliptical polarization* appear when $g \geq |g_{\text{el}}|$ (and $q \geq q_{\text{el}}$). The stokes vector at these fixed points has the components

$$S_1 = g_{\text{el}}/g, \quad S_2 = 0, \quad S_3 = \pm\sqrt{1 - (g_{\text{el}}/g)^2}. \tag{9.4.31}$$

The points located above and below the equatorial line ($S_3 = 0$) correspond to the right-handed and left-handed elliptical polarization, respectively. The two components of the vector soliton are out of phase by $+\pi/2$ and $-\pi/2$, respectively, in the two cases. Another two fixed points represent *linearly polarized* vector solitons whose components differ in phase by either 0 or π and that exist when $g \geq |g_{\text{lin}}|$. They are located on the equatorial line and have the components

$$S_1 = g_{\text{lin}}/g, \quad S_2 = \pm\sqrt{1 - (g_{\text{lin}}/g)^2}, \quad S_3 = 0. \tag{9.4.32}$$

The plane of polarization has different orientation for these two fixed points.

As seen in Figure 9.4, the number and the location of bifurcation points depend on the value of the material parameter $A > 0$. In general, there are two bifurcation points where two pairs of vector solitons with linear and elliptical polarization split off from the scalar TE- and TM-mode solitons. These mixed-mode solitons exist only when the beam power P is above a certain threshold given by Eq. (9.4.22). The new stationary solutions of Eqs. (9.4.9) and (9.4.10) can start from the same or different branches, depending on the value of parameter A. If $A > 2$, then both branches of the new stationary solutions split off from the TM branch (see Figure 9.4). If $A < 2/3$, the pairs of new stationary solutions branch off from the TE branch. For $2/3 < A < 2$, elliptically polarized vector solitons bifurcate from the TE branch, whereas linearly polarized solitons bifurcate from the TM branch as seen in Figure 9.4(b).

Although the bifurcation pattern for the eigenpolarizations is identical for spatial solitons and polarizationally unstable CW waves, the preceding evolution scenario for spatial solitons has *no analogy* in the CW case [93, 94]. In the CW case a nonstationary solution cannot converge to a stable eigenpolarization because of the conservation of total power S_0. As a result, the representative point on the Poincaré sphere follows the trajectories obtained by solving Eqs. (9.4.26)–(9.4.28).

Correspondence between the dynamics based on Eqs. (9.4.26)–(9.4.28) and the original Eqs. (9.4.9) and (9.4.9) can be established through numerical simulations of the later equations. The beam profiles $U(x,z)$ and $V(x,z)$ at each point z are calculated by using the standard split-step method [3], and the components of the Stokes vector are calculated using

$$S_i(z) = \int_{-\infty}^{\infty} s_i(x,z)\,dx \Big/ \int_{-\infty}^{\infty} s_0(x,0)\,dx. \tag{9.4.33}$$

Figure 9.6 compares the results in the two cases and shows that the simulations based on Eqs. (9.4.9) and (9.4.9) have qualitatively the same features as those based on the average-profile approximation. Convergence of the corresponding trajectories to the fixed points on the Poincaré sphere is clearly seen in Figure 9.6. If the input beam is launched into the strongly unstable TM mode with a power level above the bifurcation threshold, then this TM soliton transforms into a new, polarizationally stable, linearly polarized vector soliton after emitting some radiation.

9.4.4   Experimental Results

As early as 1996, dragging and trapping of two initially overlapping, orthogonally polarized spatial solitons was observed in an AlGaAs slab waveguide [111]. The mutual trapping was due both to XPM-induced and coherent-mode coupling between the TE

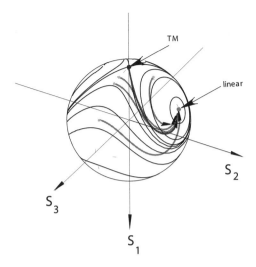

Figure 9.6: Comparison of trajectories of the Stokes vector on the Poincaré sphere. Thin curves are calculated using the approximate analysis, whereas thick curves are found by solving the coupled NLS equations numerically. Arrows indicate the convergence of a polarizationally unstable TM soliton to a stable, linearly polarized vector soliton. (Courtesy E. Ostrovskaya).

and TM polarization components and it provided an indirect proof of the existence of mixed-mode vector solitons.

It is difficult to investigate experimentally the variation of beam parameters with propagation distance within the waveguide, but the behavior of nonstationary vector beams as a function of the input peak power can easily be studied for a *fixed* length of the waveguide. Since the shape and the length of the Stokes trajectory for a fixed sample length depend on the input power, the behavior of the mixed-mode vector solitons can be observed using power levels exceeding the critical power (i.e., beyond the bifurcation point). Using typical parameter values for an AlGaAs waveguide, $n_2 \approx 1.5 \times 10^{-13}$ cm^2/W, $d_e \approx 1.8$ μm, $\lambda_0 = 1.55$ μm, $\bar{n} = 3.3$, $\beta = 2.5 \times 10^{-5}$, and assuming $A = 0.95$, a value close to the experimentally measured value [112], the threshold power levels is estimated to be 300 and 330 W for the bifurcation of TE and TM solitons, respectively. Such power levels can easily be realized in practice.

The experimental setup for observing mixed-mode vector solitons is shown in Figure 9.7 [112]. A color-center laser operating near 1.55 μm was used to ensure that two-photon absorption did not occur. The mode-locked laser produced 670-fs pulses at a repetition rate of 76 MHz. The 15-mm-long waveguide consisted of a 1.5-μm-thick guiding layer of Al$_{0.18}$Ga$_{0.82}$As, a 4-μm-thick lower cladding region, and a 1.5-μm-thick upper cladding region (both made using Al$_{0.24}$Ga$_{0.76}$As). The loss of this waveguide, measured to be 0.16 cm^{-1}, was negligible at the operating wavelength. The waveguide was designed to minimize birefringence, and the TE–TM index difference of about 0.0004 was due to residual stress-induced birefringence. The combination of a polarizer and a quarter-wave plate was used to create the desirable polarization state for the input beam. This beam was elliptically shaped to a cross section of 5×35 μm

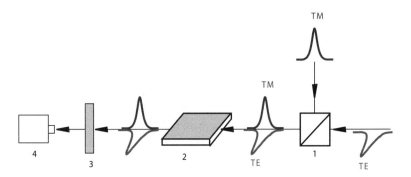

Figure 9.7: Schematic of the experimental setup used for investigating TE–TM vector solitons formed inside an AlGaAs waveguide (marked as 2). Other optical elements are: (1) polarizing beam splitter, (3) polarizer, and (4) camera. (After Ref. [99]; ©1996 APS.)

$(1/e^2$ radius) using a cylindrical lens and coupled into the waveguide using a microscope objective. At the output of the waveguide, a quarter-wave wave plate, a polarizer, a polarizing beam splitter, and an infrared camera were used to characterize the output beam.

The conventional scalar solitons could be formed by orienting the polarization state of the input beam along the TE or TM axis. When the input power was near 600 W for either the TE or TM mode launched *separately*, the beam width with was roughly equal to the input width in both cases, as expected for the scalar (fundamental) solitons. To detect the formation of a mixed-mode vector soliton, the input beam was *linearly polarized* at a 45° angle to the TE-mode axis so that both components were launched with equal powers. The output-beam profiles of both TE and TM components waves were measured using the polarizer and recorded by a camera.

Figure 9.8 shows the fraction of power in the TE and TM components. Clearly, power transfer between the two components begins to occur at input power levels beyond 50 W because of nonlinear rotation of the polarization ellipse. However, the direction of power transfer reverses around 330 W (the region shown shaded in Figure 9.8). These input power levels are close to the values of the critical powers associated with the waveguide. The divergence of the data points at very low input powers ($<$50 W) is due to experimental noise (output polarization state fluctuates widely in this region because of diffraction). Recent experiments show the evidence of a polarization instability associated with the spatial vector solitons [114, 115]. The observed instability is related to the coupling of the faster-traveling component to the radiation modes of the slow-moving component because of phase matching.

9.5   Multihump Vector Solitons

So far, we have focused on vector solitons forming inside a self-focusing Kerr medium. However, vector solitons can form in other types of non-Kerr media, and such solitons have been studied using photorefractive crystals [116, 117]. The nonlinear response of such a medium is so slow that the nonlinear index $\Delta n = n_2 I$ depends on the total beam

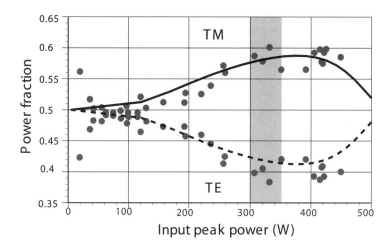

Figure 9.8: Fraction of the output power in the TE (black circles) and TM (gray circles) polarization components for a linearly polarized input beam oriented at a 45° to the TE direction. Solid and dashed lines show the results of numerical simulations under the same conditions. (After Ref. [113]; ©1997 OSA.)

intensity I such that (see also Chapter 13)

$$I(x,z) = \left| \sum_{j=1}^{N} E_j(x,z) \right|^2 \approx \sum_{j=1}^{N} \left| E_j(x,z) \right|^2, \qquad (9.5.1)$$

where E_j with $j = 1$–N is the electric field associated with the jth component of a vector soliton with N components. The formation of solitons in this case can be interpreted in terms of linear waveguide theory [118]. The dominant component of the vector soliton creates a waveguide in which the soliton itself propagates as the fundamental mode. The other components of the vector soliton can propagate as higher-order modes of the *same* waveguide. The simplest two-mode soliton of this type consists of the fundamental (nodeless) and first-order (single-node) waveguide modes, propagating as a multihump structure. The shape of the vector soliton can become quite complicated when more than two waveguide modes are involved [119]. It turns out that two-peak structures can be dynamically stable in a certain region of their existence domain, but all three-peak structures are unstable [120].

To study the stability of multihump solitons, first we consider the simplest model of two optical beams interacting incoherently in a photorefractive medium [117]. Such a two-component vector soliton is described by the following set of two coupled normalized NLS equations:

$$i\frac{\partial u}{\partial z} + \frac{1}{2}\frac{\partial^2 u}{\partial x^2} + \frac{u(|u|^2 + |w|^2)}{1 + s(|u|^2 + |w|^2)} - u = 0, \qquad (9.5.2)$$

$$i\frac{\partial w}{\partial z} + \frac{1}{2}\frac{\partial^2 w}{\partial x^2} + \frac{w(|u|^2 + |w|^2)}{1 + s(|u|^2 + |w|^2)} - \lambda w = 0, \qquad (9.5.3)$$

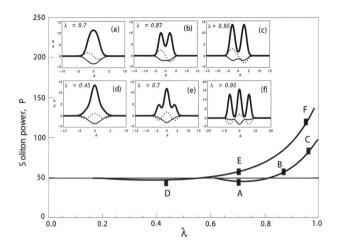

Figure 9.9: Soliton bifurcation diagram for $s = 0.8$. Curves ABC and DEF correspond to $|0,1\rangle$ and $|0,2\rangle$ solitons, respectively. The horizontal line shows the fundamental u soliton. Six insets show u (thin line), w (dashed line), and total intensity (thick line) at the six marked points. (After Ref. [120]; ©1999 APS.)

where the parameter λ represents the ratio of the nonlinear propagation constants β_1 and β_2 of the two components and s is an effective *saturation parameter*.

We look for z-independent, localized solutions of Eqs. (9.5.2) and (9.5.3) such that both $u(x)$ and $w(x)$ vanish as $|x| \to \infty$. Different types of such two-component vector solitons can be characterized by their total power, $P(\lambda, s) = P_u + P_w$, where $P_u = \int_{-\infty}^{\infty} |u|^2 dx$ and $P_w = \int_{-\infty}^{\infty} |w|^2 dx$. If one of the components is small (e.g., $w/u \sim \varepsilon \ll 1$), Eqs. (9.5.2) and (9.5.3) become decoupled. Equation (9.5.2) for the u component reduces to the standard NLS equation to the leading order and has a solution $u_0(x)$ in the form of a sech-shape soliton with no nodes. The w equation then becomes

$$\frac{1}{2}\frac{d^2w}{dx^2} + \frac{w|u_0|^2}{1+s|u_0|^2} - \lambda w = 0. \tag{9.5.4}$$

It can be considered as an eigenvalue problem for the "modes" $w_n(x)$ of a waveguide created by the soliton $u_0(x)$. The parameter s determines the total number of guided modes and the cutoff value $\lambda_n(s)$ for each mode. Such a vector soliton has two components and consists of a fundamental soliton with amplitude u_0 and the nth-order mode of the soliton-induced waveguide, which has n nodes in its amplitude profile. For notational simplicity, we denote this soliton by its "state vector" $|0,n\rangle$.

Figure 9.9 shows the $P(\lambda)$ curve by plotting the total power of both components as a function of the parameter λ using $s = 0.8$. The thin horizontal line shows the power of the fundamental soliton alone. The thick solid lines show two branches representing $|0,1\rangle$ (branch ABC) and $|0,2\rangle$ solitons (branch DEF) that emerge at the two bifurcation points. The shapes of the two components associated with the vector solitons at the six points marked by solid squares are also shown [120]. The modal description is valid only near the bifurcation points. For larger values of λ, the amplitude of

w component grows and the self-induced waveguide deforms. Two- and three-hump solitons are members of the soliton families $|0,1\rangle$ (branch *ABC*) and $|0,2\rangle$ (branch *DEF*), respectively. Close to the bifurcation point, the w component is so small that all $|0,n\rangle$ solitons remain *single-humped*, as seen in insets (a) and (d) in Figure 9.9. As the amplitude of the w component grows with increasing λ, the total intensity profile, $I(x) = u^2(x) + w^2(x)$, develops two or three humps, as seen in insets (b) and (e). For sufficiently large λ, the u component itself becomes *multihumped*, as seen in insets (c) and (f). Spacing among the soliton humps tends to infinity as $\lambda \to 1$.

To analyze the linear stability of these multihump solitons, we seek solutions of Eqs. (9.5.2) and (9.5.3) in the form of

$$u(x,z) = u_0(x) + \varepsilon[F_u(x,z) + iG_u(x,z)], \tag{9.5.5}$$

$$w(x,z) = w_n(x) + \varepsilon[F_w(x,z) + iG_w(x,z)], \tag{9.5.6}$$

where $\varepsilon \ll 1$. Setting $F_q \sim f_q(x)e^{\mu z}$ and $G_q \sim g_q(x)e^{\mu z}$ with $q = u$ and w, we obtain the following two eigenvalue equations:

$$\mathcal{L}_1\mathcal{L}_0\vec{g} = -\Lambda\vec{g}, \qquad \mathcal{L}_0\mathcal{L}_1\vec{f} = -\Lambda\vec{f}, \tag{9.5.7}$$

where $\Lambda = \mu^2$, $\vec{g} \equiv (g_u, g_w)^T$, $\vec{f} \equiv (f_u, f_w)^T$, and two matrix operators are defined as

$$\mathcal{L}_m = \begin{pmatrix} -\frac{1}{2}\frac{d^2}{dx^2} + 1 - a_m & b_m \\ b_m & -\frac{1}{2}\frac{d^2}{dx^2} + \lambda - c_m \end{pmatrix}, \tag{9.5.8}$$

where $m = 0$ or 1, $a_0 = c_0 = I/(1+sI)$, $a_1 = a_0 + 2u_0^2/(1+sI)^2$, $b_1 = -2u_0w_n/(1+sI)^2$, $b_0 = 0$, and $c_1 = c_0 + 2w_n^2/(1+sI)^2$.

Since $\mathcal{L}_1\mathcal{L}_0$ and $\mathcal{L}_0\mathcal{L}_1$ are adjoint operators with identical spectra, we can consider the spectrum of only one of them, say, $\mathcal{L}_1\mathcal{L}_0$. It is straightforward to show that the continuum part of the spectrum lies in the range $-\infty < \Lambda < -\mu^2$. Stable bounded eigenmodes of the discrete spectrum, the so-called *soliton internal modes*, can have eigenvalues only inside the range $-\mu^2 < \text{Re}\Lambda < 0$. Similar to the general theory discussed in Chapter 2, a vector soliton is stable if all eigenvalues Λ have a negative real part.

Numerical solutions of the eigenvalue problem (9.5.7) show that multihump vector solitons are stable only in a certain region of their existence domain. Figure 9.10 shows the stability domain for the two- and three-hump vector solitons of Figure 9.9. The curves marked $\lambda_1(s)$ and $\lambda_2(s)$ show the bifurcation curves for two- and three-hump vector solitons, respectively. The shaded region shows where the two-hump solitons are unstable. The instability appears after the total intensity I becomes two-humped (the dashed line) and a pair of the soliton internal modes splits from the continuum spectrum into the gap. Squares and circles show the numerically obtained instability threshold for the two- and three-hump solitons, respectively.

With the help of the analytical asymptotic technique, it is possible to show that a vector soliton with components u and w becomes unstable when the determinant

$$J = \frac{\partial P_u}{\partial s}\frac{\partial P_w}{\partial \lambda} - \frac{\partial P_w}{\partial s}\frac{\partial P_u}{\partial \lambda} \tag{9.5.9}$$

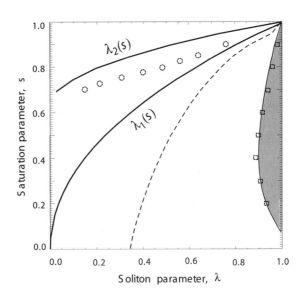

Figure 9.10: The existence and stability domains for two- and three-hump solitons. Solid curve show the thresholds $\lambda_1(s)$ and $\lambda_2(s)$ for $|0,1\rangle$ and $|0,2\rangle$ solitons, respectively. Dashed lin shows the boundary where the total intensity develops humps, and the shaded region show the instability domain for two-hump solitons. Squares and circles show the numerical results fc two- and three-hump solitons, respectively. (After Ref. [120]; ©1999 APS.)

becomes negative. This is the generalization of the Vakhitov–Kolokolov stability cri terion of Chapter 2 to the case of two-component vector solitons. However, as we discussed in Section 9.3, such a generalization is rigorous for fundamental component: only (no nodes), whereas here we deal with higher-order solitons that may contair nodes. Therefore, other instabilities (which are not associated with the bifurcation con dition $J = 0$ and can even have larger growth rates) are still possible. This is indeec what happens for three-hump solitons. The stability condition of two-hump solitons is fully governed by the determinant condition, as verified numerically by the results shown in Figure 9.10.

Since the concept of vector solitons is broad and multicomponent solitons are pre dicted theoretically for several different types of nonlinear media [121]–[127], the exis tence of multimode vector solitons is intimately linked to their stability. The preceding stability analysis focused on the case of a saturable nonlinearity. A number of theo retical studies has shown that the multimode vector solitons are unstable in the case o a pure Kerr nonlinearity [9, 122, 126]. More precisely, when the strength of the XPM term is larger than that of the SPM term (which is the case for a Kerr medium), the multimode soliton undergoes a space-inversion *symmetry-breaking instability*.

As seen in Figure 9.11, the multimode soliton exhibits, in general, a two-hump intensity distribution, a feature that makes it analogous to a directional coupler basec on a two-core waveguide. Since the odd mode is unstable in a nonlinear directiona coupler due to a symmetry-breaking instability (the so-called fast-mode instability) one should expect the multimode soliton of Figure 9.11(a) to be unstable for the same

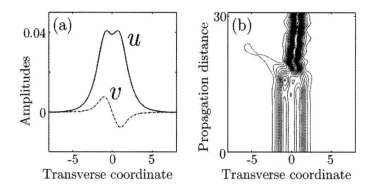

Figure 9.11: (a) Envelopes of the circular polarization components of the two-component vector soliton. (b) Contour plot showing the evolution of the vector soliton in (a) when slightly perturbed by random noise. (After Ref. [128]; ©2002 APS.)

reason. This can be checked by solving numerically a set of two coupled NLS equations. Figure 9.11(b) shows the propagation of the multimode soliton seen in part (a). As can be seen, the energy that is initially evenly distributed between the two cores of the induced waveguide goes abruptly from one core to the other, resulting eventually in the destruction of the waveguide. Interestingly, after the transient regime, the field becomes confined to a single-hump distribution (i.e., both fields exhibit even parity), as expected for a fundamental elliptically polarized vector soliton.

The space-inversion symmetry-breaking instability was observed in a Kerr medium in a 2002 experiment [128]. The experiment was performed using a CS_2 planar waveguide, and the orthogonal circular polarization states of the input beam were used for the two components of a vector soliton. The 10-Hz repetition rate of the laser permitted the researchers to record a large number of events separately and to analyze their statistics. It was found that about 62% of the laser shots were characterized by a strongly asymmetric output in which the beam was displaced either to the left or to the right from the initial beam axis. This observation is consistent with the interpretation that the two-component spatial vector solitons undergo a space-inversion symmetry-breaking instability inside Kerr media.

9.6 Two-Dimensional Vector Solitons

The concept of spatial vector solitons can be extended to the $(2 + 1)$-dimensional geometry in which each component diffracts in two transverse dimensions. Such vector solitons can develop a complex internal topology, a process that is somewhat analogous to the formation of a "solitonic molecule" from the elementary constituents. An example is provided by the *vortex vector soliton* [129], whose one component has the shape of a vortex. Such a radially symmetric, ring-like vector soliton (analogous in shape to the Laguerre–Gaussian modes of a cylindrical waveguide) can undergo a *symmetry-breaking* instability [130] that transforms it into a radially asymmetric *dipole vector soliton*, even in a perfectly isotropic nonlinear medium. The dipole vector soliton originates from the trapping of a Hermite–Gaussian HG_{01} mode associated with the waveguide created by the stronger component. Although many other topologically complex structures can be created, only the dipole vector solitons appear to be dynamically robust. In particular, when launched with a nonzero angular momentum, such a dipole structure survives in the form of a "propeller soliton" [131].

In this section, we extend the analysis of incoherently coupled vector solitons of Section 9.2 to the case of two-dimensional vector solitons formed in a bulk Kerr medium. As discussed in Chapter 6, two-dimensional scalar solitons collapse in a Kerr-type medium, but they are stable in a saturable nonlinear medium. The interesting question is whether vector solitons remain stable in a Kerr medium even when they spread in both transverse dimensions. To answer this question, we first focus on the simple case of Kerr nonlinearity and consider vector solitons formed thorough an incoherent coupling between two fundamental nodeless solitons. We then analyze the vortex- and dipole-type vector solitons in the non-Kerr media with a saturable nonlinearity.

9.6.1 Radially Symmetric Vector Solitons

Consider two incoherently interacting beams propagating along the z direction in a bulk Kerr medium. Vector solitons in this case are governed by the coupled NLS equations (9.1.12) and (9.1.13), with the difference that the diffraction term should include both x and y derivatives. We assume that the wavelengths of the two beams are close enough that we can set $d_1 = d_2 = 2$ and $\gamma_1 = \gamma_2 = 1$. The resulting equations become

$$i\frac{\partial U_1}{\partial z} + \left(\frac{\partial^2 U_1}{\partial x^2} + \frac{\partial^2 U_1}{\partial y^2}\right) + (|U_1|^2 + \sigma|U_2|^2)U_1 = 0, \qquad (9.6.1)$$

$$i\frac{\partial U_2}{\partial z} + \left(\frac{\partial^2 U_2}{\partial x^2} + \frac{\partial^2 U_2}{\partial y^2}\right) + (|U_2|^2 + \sigma|U_1|^2)U_2 = 0. \qquad (9.6.2)$$

Depending on the state of polarization of the two beams, the nature of nonlinearity, and anisotropy of the material, σ varies over a wide range. For a Kerr-type electronic nonlinearity $\sigma \geq 2/3$, whereas $\sigma \leq 7$ for the nonlinearity resulting from changes in the molecular orientation [132].

We use the cylindrical coordinates, r and φ, in place of x and y and look for solutions of Eqs. (9.6.1) and (9.6.2) in the form

$$U_1(r,\varphi,z) = \sqrt{\beta_1}\,u(r)\,e^{im_1\varphi}e^{i\beta_1 z}, \qquad U_2(r,\varphi,z) = \sqrt{\beta_1}\,v(r)\,e^{im_2\varphi}e^{i\beta_2 z}, \qquad (9.6.3)$$

where β_1 and β_2 are two independent propagation constants. The integers m_1 and m_2 represent topological charges. Measuring the radial coordinate in units of $\sqrt{\beta_1}$ and introducing the ratio of the propagation constants, $\lambda = \beta_2/\beta_1$, we obtain from Eqs. (9.6.1) and (9.6.2) the following set of two ordinary differential equations governing the radially symmetric envelopes u and v:

$$\frac{d^2u}{dr^2} + \frac{1}{r}\frac{du}{dr} - \frac{m_1^2}{r^2}u - u + (u^2 + \sigma v^2)u = 0, \qquad (9.6.4)$$

$$\frac{d^2v}{dr^2} + \frac{1}{r}\frac{dv}{dr} - \frac{m_2^2}{r^2}v - \lambda v + (v^2 + \sigma u^2)v = 0. \qquad (9.6.5)$$

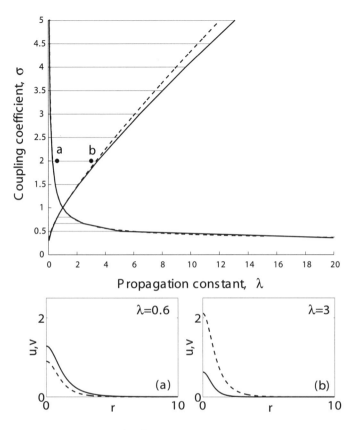

Figure 9.12: Existence region for $|0,0\rangle$ vector solitons (hatched area). Solid lines show the numerically calculated cutoff values λ_1 and λ_2 as a function of σ. Dashed lines show the results of a variational analysis. Amplitudes u (solid) and v (dashed) of the vector soliton at points a and b are also shown. (After Ref. [130]; ©2000 OSA.)

Following the notation introduced in Ref. [120], we denote such vector solitons with their "state vector" $|m_1, m_2\rangle$.

Consider first the case $m_1 = m_2 = 0$ so that the resulting $|0,0\rangle$ vector soliton is radially symmetric. This family of vector solitons is characterized by a single parameter λ and exists in the range between the cutoff values λ_1 and λ_2, which depend on the numerical value of the coupling parameter σ. Figure 9.12 shows the cutoff values λ_1 and λ_2 as a function of σ. Near the two cutoff points ($\lambda \approx \lambda_1$ or $\lambda \approx \lambda_2$), the formation of vector solitons can be interpreted in terms of a waveguide created by the dominant field component, in which the weaker component propagates as a higher-order mode of this self-induced waveguide. Two examples of $|0,0\rangle$ solitons are shown in Figure 9.12 for $\sigma = 2$ at the points marked by solid dots.

The crossover of the existence domains $\lambda_1(\sigma)$ and $\lambda_2(\sigma)$ occurs at the point $\lambda = \sigma = 1$. As expected, when $\sigma = 1$, the $|0,0\rangle$ vector solitons exist only for $\lambda = 1$, and their properties resemble those of the Manakov solitons [5]. Such solitons can be found

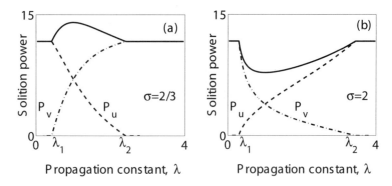

Figure 9.13: Total power of vector solitons as a function of λ for (a) $\sigma = 2/3$ and (b) $\sigma = 2$. Powers of individual components are shown by dashed and dot-dashed curves. (After Ref. [130]; ©2000 OSA.)

by using the transformation $u = U\cos\theta$ and $v = U\sin\theta$ in Eqs. (9.6.4) and (9.6.5), where θ is an arbitrary angle and U satisfies the scalar equation

$$\frac{d^2U}{dr^2} + \frac{1}{r}\frac{dU}{dr} - U + U^3 = 0. \tag{9.6.6}$$

To describe the existence domain of vector solitons analytically, one can employ the variational technique [133]. We look for the stationary solutions of Eqs. (9.6.4) (9.6.5) in the form

$$u(r) = A\exp(-r^2/a^2), \qquad v(r) = B\exp(-r^2/b^2), \tag{9.6.7}$$

where the parameters A, B, a, and b depend on z. An effective Lagrangian is used to find how these parameters vary with z. This approach allows one to find the following analytical approximation to the two solid curves shown in Figure 9.12:

$$\sigma(\lambda) = (1 + \sqrt{\lambda_2})^2/4, \qquad \sigma(\lambda) = (1 + \sqrt{\lambda_1})^2/(4\lambda_1). \tag{9.6.8}$$

The dashed curves in Figure 9.12 show that the variational analysis is reasonably accurate in determining the existence domain for the vector solitons, in spite of its use of a Gaussian ansatz, and it provides an alternative to the numerical approach that is often much more time-consuming.

An important physical characteristic of vector solitons is their total power, found using $P = 2\pi\int_0^\infty (u^2 + v^2)r\,dr$. Figure 9.13 shows the total power of two-dimensional vector solitons as a function of λ for (a) $\sigma = 2/3$ and (b) $\sigma = 2$. As seen there, the cases $\sigma < 1$ and $\sigma > 1$ are qualitatively different. In the former case, the total power of the vector soliton is more than the power of the scalar soliton ($P_0 \approx 11.7$) forming when $\sigma = \lambda = 1$ [see Figure 9.13(a)]. This is an important result because it opposes the commonly held view that the formation of vector solitary waves always requires less input power in comparison with scalar solitons. For $\sigma > 1$, the situation is the opposite and vector solitons exist at lower powers than that required for scalar solitons. This

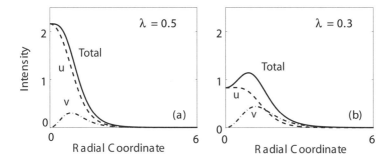

Figure 9.14: Intensity profiles of $|0,\pm1\rangle$ solitons for two values of λ. Solid curve shows the total intensity; dashed and dash-dotted curves show the shape of individual components. (After Ref. [130]; ©2000 OSA.)

feature explains the effective suppression of the beam collapse observed numerically for $\sigma = 2$ in the special case of $\lambda = 1$, for which an analytical form of the vector soliton can be found by using a Hartree-type ansatz [134].

9.6.2 Ring-Shaped Vector Solitons

We now focus on higher-order vector solitons $|m_1, m_2\rangle$ with topological charges. In the simplest case of $|0,\pm1\rangle$ solitons, $m_1 = 0$ but $m_2 = \pm1$. Such vector solitons can be found by solving Eqs. (9.6.4) and (9.6.5), and two examples are shown in Figure 9.14, where we show the total intensity $I(r) = |u|^2 + |v|^2$ as well as the individual mode intensities. Similar to the case of $|0,0\rangle$ solitons, near the cut-off, the weaker component (dash-dotted line) appears as a higher-order mode of the waveguide created by the stronger component (dashed curve). The total intensity has a maximum at the beam center in this case. However, far from cutoff, the total intensity can develop a ring shape, as seen in Figure 9.14(b). In that case the v component becomes strong enough to deform the waveguide and reduce the intensity of the u component near the beam center.

The important question is whether the vector solitons of types $|0,0\rangle$ and $|0,1\rangle$ are stable. A numerical stability analysis and the Vakhitov–Kolokolov stability criterion developed for vector solitons [120] reveal that all such vector solitons are *linearly unstable*, and the presence of the second component cannot arrest the collapse of the scalar two-dimensional solitons in a Kerr medium.

Given the stability of one-dimensional vector solitons in a saturable medium [120], one is tempted to consider $|0,1\rangle$-type, ring-like, two-dimensional vector solitons in such a medium [129]. In photorefractive materials, the saturable nonlinearity is of the form such that the nonlinear term in Eqs. (9.6.4) and (9.6.5) can be replaced with $(1 + |u|^2 + |v|^2)^{-1}$. Seeking stationary solutions in the form (9.6.3) and introducing the relative propagation constant $\lambda = (1 - \beta_2)/(1 - \beta_1)$, we obtain the following set of two equations [116, 135]:

$$\frac{d^2u}{dr^2} + \frac{1}{r}\frac{du}{dr} - \frac{m_1^2}{r^2}u - u + \frac{Iu}{1+sI} = 0, \tag{9.6.9}$$

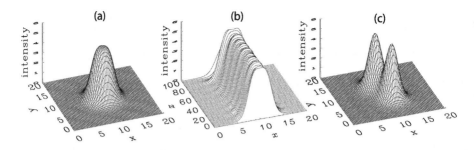

Figure 9.15: Evolution of a $|0,1\rangle$ soliton in a saturable medium for $s = 0.65$ and $\lambda = 0.6$: (a) Intensity at $z = 0$; (b) evolution of the intensity profile at $y = 10$; (c) intensity at $z = 100$. (After Ref. [130]; ©2000 OSA.)

$$\frac{d^2 v}{dr^2} + \frac{1}{r}\frac{dv}{dr} - \frac{m_2^2}{r^2}v - \lambda v + \frac{Iv}{1+sI} = 0, \tag{9.6.10}$$

where $\lambda = (1 - \beta_2)/(1 - \beta_1)$, $I = u^2 + v^2$, and $s = 1 - \beta_1$ plays the role of a saturation parameter. For $s = 0$, this set of equations describes the case of Kerr nonlinearity with $\sigma = 1$. In this case, the lowest-order $|0,0\rangle$ solitons exist only for $\lambda = 1$. In the remaining region of the parameter plane (s, λ), vector solitons of the type $|0,1\rangle$ exist and exhibit a ring structure far from the cutoff, similar to the Kerr-case discussed earlier. Figure 9.15 shows the evolution of such a soliton for $s = 0.65$ and $\lambda = 0.6$. Numerical simulations indicate that although the saturation has a strong *stabilizing* effect on the $|0,1\rangle$ soliton, vector solitons of this type appear to be linearly unstable. The instability, although largely suppressed by saturation, triggers the decay of the soliton into a dipole structure, discussed in the next section.

9.6.3 Multipole Vector Solitons

As the name implies, multipole vector solitons are not radially symmetric. To find them, we consider again the case of a saturable photorefractive medium and write the two coupled NLS equations, Eqs. (9.6.1) and (9.6.2), in the form

$$i\frac{\partial U_1}{\partial z} + \left(\frac{\partial^2 U_1}{\partial x^2} + \frac{\partial^2 U_1}{\partial y^2}\right) - \frac{U_1}{1 + |U_1|^2 + |U_2|^2} = 0, \tag{9.6.11}$$

$$i\frac{\partial U_2}{\partial z} + \left(\frac{\partial^2 U_2}{\partial x^2} + \frac{\partial^2 U_2}{\partial y^2}\right) - \frac{U_2}{1 + |U_1|^2 + |U_2|^2} = 0, \tag{9.6.12}$$

To find their localized shape-preserving solutions, we substitute

$$U_1 = \sqrt{\beta_1}\, u(x,y) \exp(i\beta_1 z), \qquad U_2 = \sqrt{\beta_1}\, w(x,y) \exp(i\beta_2 z), \tag{9.6.13}$$

where β_1 and β_2 are two independent propagation constants. Measuring the transverse coordinates in the units of $\sqrt{\beta_1}$ and introducing parameter $\lambda = (1 - \beta_2)/(1 - \beta_1)$, we obtain the following two equations for u and w:

$$\frac{\partial^2 u}{\partial x^2} + \frac{\partial^2 u}{\partial y^2} - u + \frac{Iu}{1 + sI} = 0, \tag{9.6.14}$$

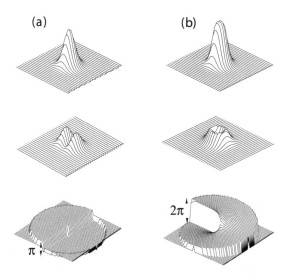

Figure 9.16: Examples of (a) dipole ($s = 0.3$) and (b) vortex ($s = 0.65$) vector solitons. The top and middle rows show the intensity profiles for u and w components, while the bottom row displays the phase profile of the w component. (After Ref. [139]; ©2002 OSA.)

$$\frac{\partial^2 w}{\partial x^2} + \frac{\partial^2 w}{\partial y^2} - \lambda w + \frac{Iw}{1 + sI} = 0, \qquad (9.6.15)$$

where $I = u^2 + w^2$ and $s = 1 - \beta_1$ is an effective saturation parameter. The limit $s \to 0$ corresponds to the case of a Kerr medium. These equations should be solved with the boundary condition that both u and w vanish as $r \to \infty$.

When one of the components, say, w, is weak, we can use the concept of soliton-induced waveguiding. If the waveguide is induced by a sech-shape fundamental soliton in the u component, the guided modes form a set analogous to the Hermite–Gaussian (HG) or Laguerre–Gaussian (LG) modes supported by a radially symmetric waveguide [136]. Depending upon which mode of the induced waveguide is excited by the weaker component, a fundamental soliton can trap and guide beams with various topologies. If the excited mode corresponds to LG_{01} mode, the resulting structure is radially symmetric, and such a vector soliton is referred to as the *vortex vector soliton* [137]. In contrast, if the weaker component corresponds to an HG_{01} mode, the vector soliton becomes radially asymmetric and is referred to as a *dipole vector soliton* because of its dipole-like structure [138, 139]. This concept can be extended to a larger number of components, and, in particular, for the case of three components the stable structure is formed with two perpendicular dipoles trapped by a fundamental mode of the third component [140, 141].

Figure 9.16 shows an example of the dipole-type (left column) and vortex-type (right column) vector solitons. The top and middle rows show the intensity distribution of the u and w components, respectively. The bottom row shows the phase profile of the w competent that is responsible for the different properties of the two types of vector solitons. For a given value of s, the solutions are characterized by a certain cutoff value of λ, above which the dipole mode appears. Near the cutoff, vector solitons can be approximately described by linear waveguide theory. With increasing λ, the w component grows and deforms the effective waveguide generated by the fundamental u mode such that at large intensities the u-component elongates and loses its radial symmetry. *Linear* and *dynamical* stability analyses of the dipole vector solitons have revealed their robustness with respect to both small and large perturbations [138]. The dynamical stability was checked numerically for propagation distances up to several hundreds of diffraction lengths.

The phase distribution of the radially symmetric vortex soliton shown in Figure 9.16 corresponds to a singly charged vortex in the w component. Dynamical and linear stability analyses reveal that all such vector solitons become unstable on some distance from the bifurcation point. Similar to the case of a scalar vortex beam, which is unstable in a self-focusing medium [142]–[144], the vortex-mode soliton exhibits a symmetry-breaking instability and breaks up into several fragments during propagation. It turns out that it can decay into a radially asymmetric dipole-mode soliton that has a nonzero angular momentum and can survive for *very long* propagation distances [138].

In contrast to asymmetric dipole-type solitons, radially symmetric solutions of Eqs. (9.6.14) and (9.6.15) can be found numerically using a simple shooting or relaxation technique. Via this method, one can find an entire family of vortex-mode solitons with charge $m = 1$ in the w component. Figure 9.17(a) shows the power curve $P(\lambda)$ for this family of solitons using $s = 0.5$. The total power as well as the individual mode powers are shown in this figure. The dotted and dashed lines show the power levels associated with the sech-shape and vortex-shape scalar solitons. Point A marks the location of the *bifurcation* where a scalar soliton transforms into a vector soliton as λ increases. At bifurcation point B, the vector soliton transforms back into a scalar vortex soliton. The amplitude profiles of the two components of a vector soliton in the vicinity of bifurcation points A and B are shown in parts (c) and (d).

Near bifurcation point B, the vortex component w becomes so strong that it induces an effective waveguide that traps the u component. The shape of the u component becomes somewhat complex, as seen in part (d) of Figure 9.17, in the sense that intensity has a shallow hole near the beam center. This is expected for an induced waveguide with an index dip. The range over which vector solitons can exist depends on the saturation parameter s. This dependence is shown in Figure 9.17(b), where the locations of bifurcation points A and B are plotted in the (λ, s) parameter space. For a given value of the saturation parameter s, the vector structure can exist only for λ in the region limited by the bifurcation points.

Extrapolating the results of linear waveguide theory to the nonlinear regime, in which w and u are comparable in magnitude, one may expect that the mutual trapping of two beams can exhibit a structure that mimics the shape of higher-order Hermite–Gaussian or Laguerre–Gaussian modes. Indeed, Eqs. (9.6.14) and (9.6.15) are found to possess a family of solutions that can be identified as *multipole vector solitons*. However, their modal structure is more complex than those of HG_{mn} and LG_{mn} modes in the nonlinear regime because it is is determined by the nonlinear interaction of both components and thus cannot be predicted by a linear theory. Direct numerical methods for solving stationary coupled NLS equations are often the only available tool for analyzing the nonlinear regime, especially for solitons with no radial symmetry [138].

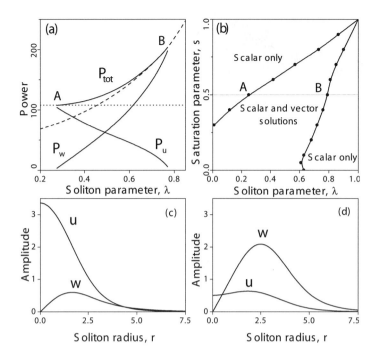

Figure 9.17: Characteristics of vortex-mode vector solitons for $s = 0.5$. (a) Total and individual mode powers as a function of λ. The dotted and dashed lines show the power levels associated with the sech-shape and vortex-shape scalar solitons. (b) Existence domain in the (λ, s) parameter space. Points A and B represent two bifurcations points. Amplitude profiles near these points are shown using (c) $\lambda = 0.3$ and (d) $\lambda = 0.75$. (After Ref. [139]; ©2002 OSA.)

Some physical insight into the behavior of multipole vector solitons can be gained by using a variational approach [145] for describing the structure of the localized solutions of Eqs. (9.6.14) and (9.6.15). Using the cylindrical coordinates (r, φ), the Lagrangian associated with these equations can be written in the form

$$
\mathcal{L} = r \left(\left| \frac{\partial u}{\partial r} \right|^2 + \left| \frac{\partial w}{\partial r} \right|^2 \right) + \frac{1}{r} \left| \frac{\partial w}{\partial \varphi} \right|^2 + \frac{1}{r} \left| \frac{\partial u}{\partial \varphi} \right|^2
$$
$$
+ r \left(|u|^2 + \lambda |w|^2 - \frac{I}{s} + \frac{1}{s^2} \ln(1 + sI) \right), \tag{9.6.16}
$$

where $I = |u|^2 + |w|^2$. Following the analogy with linear theory, we assume that one component is radially symmetric, $u(x, y) = \tilde{u}(r)$ but that the other component has the general form

$$
w(x, y) = \tilde{w}(r)[\cos(m\varphi) + ip \sin(m\varphi)]. \tag{9.6.17}
$$

The integer p is used to distinguish between the multipole ($p = 0$) and vortex ($p = \pm 1$) vector solitons. The other integer m determines the charge of the vortex.

Applying the variational technique, we integrate the Lagrangian (9.6.16) over φ using $\langle L \rangle = \int_0^{2\pi} L\, d\varphi$. We then obtain the following Euler–Lagrange equations for the

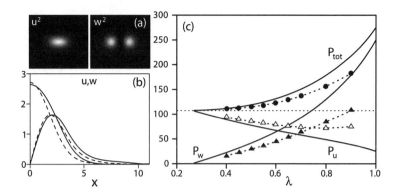

Figure 9.18: (b) Comparison of the variational (dashed) and numerical (solid) results for the dipole-mode soliton components found for $s = 0.5$ and $\lambda = 0.5$; (a) the corresponding intensity patterns. (c) Total and partial powers of the two components as a function of λ obtained using variational (solid curves) and numerical (symbols) methods for $s = 0.5$. (After Ref. [145]; ©2001 OSA.)

radially symmetric functions $\tilde{u}(r)$ and $\tilde{w}(r)$:

$$\frac{d^2\tilde{u}}{dr^2} + \frac{1}{r}\frac{d\tilde{u}}{dr} + \frac{1}{s}[1 - s - f_1(\tilde{u},\tilde{w})]\tilde{u} = 0, \tag{9.6.18}$$

$$\frac{d^2\tilde{w}}{dr^2} + \frac{1}{r}\frac{d\tilde{w}}{dr} + \frac{1}{s}\left[1 - s\lambda - \frac{m^2 s}{r^2} - f_2(\tilde{u},\tilde{w})\right]\tilde{w} = 0, \tag{9.6.19}$$

where the functions f_1 and f_2 are defined as

$$f_1(\tilde{u},\tilde{w}) = [(1 + s\tilde{u}^2 + p^2 s\tilde{w}^2)(1 + s\tilde{u}^2 + s\tilde{w}^2)]^{-1/2}, \tag{9.6.20}$$

$$f_2(\tilde{u},\tilde{w}) = \frac{qf_1^2(1 + s\tilde{u}^2) + p^2 sf_1^2\tilde{w}^2 + qf_1}{qf_1(1 + s\tilde{u}^2) + q^2 f_1 s\tilde{w}^2 + q}, \tag{9.6.21}$$

and $q = (p^2 + 1)/2$.

In the case $p = 0$ and $m = 1$, Eqs. (9.6.18) and (9.6.19) describe the dipole-mode vector soliton discussed earlier. The variational results are quite close to those obtained by solving Eqs. (9.6.14) and (9.6.15) numerically. The variational technique has an advantage over the direct numerical solution, in the sense that Eqs. (9.6.18) and (9.6.19) are much easier to solve owing to their radial symmetric nature. Figure 9.18 compares the variational solution with the numerical solution in the specific case of $s = 0.5$. Even though there exist quantitative differences, especially for large values of λ, the variational analysis captures all qualitative features. More specifically, for any value of s, the soliton solutions are characterized by a certain cutoff value of λ above which the dipole-mode solitons can form. It also turns out that the asymmetric multipolar vector soliton always has the lowest possible power among the solutions found for various values of p and m.

For $p = 0$ and values of $m > 1$, the variational results suggest the existence of two-dimensional vector solitons with a much more complicated spatial structure. For example, a quadrupole-like structure consisting of four beamlets is expected for $m = 2$. In the case $m = 3$, a hexapole structure should form. As shown in Figure 9.19, numerical simulations confirm this expected behavior. Parts (a) and (b) of this figure show the intensity patterns ($|u|^2$ and $|w|^2$) for $m = 2$ and 3, respectively. The multipetal intensity pattern of the w component is related to the π phase jump that occurs between two neighboring beamlets. Even more complicated patterns have been discovered by integrating Eqs. (9.6.14) and (9.6.15) numerically. Parts (c) and (d) of Figure 9.19 show two examples. In part (d), the dodecagon-type structure resembles the "necklace-type" solitons discussed earlier, in Chapter 6, as expanding scalar structures in a self-focusing Kerr medium [146, 147]. The patterns shown in Figure 9.19 were obtained by using the approximate solution obtained variationally as the initial condition for the numerical solution. Even though the variation solution is not exact and changes initially with propagation, it rapidly converges to the final stationary solution. In the examples shown in Figure 9.19, numerical solutions converged after only a few diffraction lengths to the final stationary state, which then remained unchanged for tens of diffraction lengths. It should be stressed that if the w component of these solitons is propagated separately (without the presence of u component), it disintegrates rapidly because of a repulsive force between the out-of-phase neighboring lobes.

9.6.4 Effect of Anisotropy and Nonlocality

So far we have discussed multipole vector solitons in the context of an isotropic non-linearity. However, the experiments often employ a photorefractive crystal as a nonlinear medium known to exhibit a strongly *anisotropic and nonlocal* nonlinear response [148]–[150]. Given the strong anisotropy of photorefractive nonlinearities, it is important to consider whether the theoretical predictions on multipole vector solitons made for isotropic nonlinear media would survive for anisotropic, nonlocal, self-focusing media. To answer this question, we show using a simple model that stable dipole-mode vector solitons can exist even in an anisotropic medium with nonlocal nonlinear response and exhibit a number of new anisotropy-driven features. More specifically, all such vector solitons with broken radial symmetry exhibit orientation-dependent dynamics. Several such features have been already observed experimentally [151].

The nonlinear interaction of two optical beams propagating inside an anisotropic, nonlocal, nonlinear medium, such as a photorefractive crystal, depends on an externally applied voltage. When the characteristic spatial scales (e.g., diffraction length) are longer than the Debye length, the propagation along the z axis of a photorefractive crystal, with an externally applied electric field along the y axis, is described by the following set of equations [148]–[150]:

$$i\frac{\partial U_j}{\partial z} + \frac{1}{2}\nabla^2 U_j = -\frac{g}{2}\frac{\partial \varphi}{\partial y}U_j \quad (j = 1, 2), \tag{9.6.22}$$

$$\nabla^2 \Phi + \nabla\Phi\nabla\ln(1+I) = E_0 x_0 \frac{\partial}{\partial y}\ln(1+I), \tag{9.6.23}$$

where g is the effective nonlinear parameter, E_0 is the external field, $I \equiv |U_1|^2 + |U_2|^2$ is the total intensity, $\nabla = \hat{x}(\partial/\partial x) + \hat{y}(\partial/\partial y)$, and Φ is the electrostatic potential induced

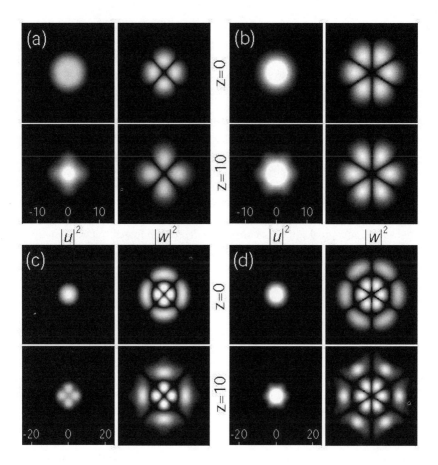

Figure 9.19: Examples of higher-order multipole vector solitons. (a) Quadrupole soliton ($m = 2$) for $s = 0.5$ and $\lambda = 0.5$. (b) Hexapole soliton ($m = 3$) for $s = 0.7$ and $\lambda = 0.3$. Parts (c) and (d) examples of more complicated octopole and dodecagon structures. (After Ref. [139]; ©2002 OSA.)

by optical fields. As usual, z is measured in units of the diffraction length, and the transverse coordinates are normalized to the input beam size x_0.

It is evident from the presence of y derivatives in Eqs. (9.6.22) and (9.6.23) that these equations are highly anisotropic and do not allow radially symmetric soliton solutions. Therefore, we look for stationary solutions in the form

$$U_1(x,y) = u(x,y)\exp(i\lambda_1 z), \qquad U_2(x,y) = w(x,y)\exp(i\lambda_2 z), \qquad (9.6.24)$$

and solve numerically the resulting two-dimensional eigenvalue problem for the mode amplitudes u and w using an iterative relaxation scheme. We search for a dipole-mode vector soliton created when the stronger component, say, U_1, propagates as a standard spatial soliton (single peak in the center) and induces a waveguide that traps and guides the weaker component U_2 in the form of the HG_{01} mode with two lobes of opposite

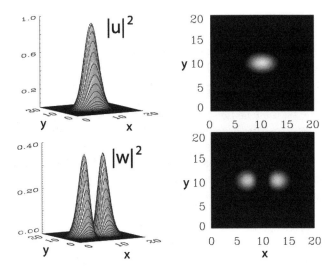

Figure 9.20: Example of a dipole vector soliton forming in a photorefractive crystal with an electric field applied along the y axis for $g = 0.5$, $E_0 = 2.5$, and $\lambda_2/\lambda_1 = 0.5$. (After Ref. [151]; ©2001 OSA.)

phase [138]. In the case of isotropic nonlinearity, such a dipole-like structure has an arbitrary orientation in the x-y plane. However, in the anisotropic case the stationary solutions of this kind exist only for a specific orientation of the dipole axis, more specifically, perpendicular to the direction of the external field. Figure 9.20 shows an example of such a dipole-mode vector soliton found numerically. Except for the dipole orientation fixed by the applied external field, other features of the soliton are qualitatively similar to the isotropic case.

A more detailed investigation of the stability of dipole-mode vector solitons shows that such a soliton can also form with the orientation of its dipole axis along the direction of the applied field, but it is only quasi-stable. In contrast, the vector soliton whose dipole axis is perpendicular to the field corresponds to a stable state. It appears from numerical simulations that when we launch the initial beams according to the "correct" orientation of the dipole (stable direction), they create a very robust dipole-mode soliton, in the sense that increasing the input intensity of the radially symmetric component leads to only small oscillations of the beam intensities near the stationary solution. In contrast, when the initial conditions excite a vector soliton whose dipole axis is oriented along the direction of the applied field, the weaker dipole component begins to diffract when the intensity of the stronger component is increased. However, the diffraction spreading is slow enough that the vector soliton can be observed over many diffraction lengths before it is destroyed. For this reason, this soliton is referred to as being *quasi-stable*. When the dipole axis is tilted around the stationary state, its dynamics becomes much more complicated. More specifically, the dipole starts wobbling along the vertical axis while diffracting slowly with propagation. A similar behavior occurs for other types of multipole solitons [151].

9.6.5   Experimental Results

The multipole vector solitons have been observed using a strontium–barium niobate
(SBN) photorefractive crystal [151]–[153]. The experimental setup is shown schematically in Figure 9.21. An intense laser beam at a wavelength of 532 nm was split into
two parts. One of the beams was transmitted through a phase mask (a set of glass
slides) to impose on it the phase structure required to create a dipole-like structure (a
phase jump of π across the beam along one of the transverse directions). Different
phase masks can be used for generating a higher-order structure consisting of an even
number of symmetrically distributed (and out-of-phase) beamlets. The second beam
was combined with the first one without any modification, and both beams were then
focused using a set of spherical and cylindrical lenses onto the input facet of the photorefractive crystal.

Two SBN crystals doped with cerium (0.002% by weight) were used in the experiment. It is well known that photorefractive crystals biased with a strong static electric
field exhibit strong positive or negative nonlinearities, depending on the polarity of the
field [154]–[156]. In the experiment, the crystal was biased by applying 1.5–2.5 kV of
external voltage along the optical axis, resulting in saturable self-focusing nonlinearity.
The degree of saturation was controlled by illuminating the crystal with a wide beam
from a white light source. Since the both components forming a vector soliton have to
be mutually incoherent, one of the beams was reflected from a vibrating mirror (M2).
This imposed a fast phase variation onto the beam and made both components *effectively incoherent* inside the crystal because of the slow photorefractive response of the
crystal.

Typical experimental results are shown in Figure 9.22 for dipole (a), quadrupole
(b), and hexapole (c) vector solitons. The first column of this figure shows the input intensities of both (u and w) components, the second column displays the output
intensities when each beam propagates individually in isolation, and the last column
shows the results when both beams propagate simultaneously. When the two beams
propagate alone, the u component (without a phase jump) always forms a fundamental
soliton, while the w component forms multiple fundamental solitons that repel each
other because of the initially imposed π relative phase shifts among them. However,
when the two beams propagate together, as shown in the right column of Figure 9.22,

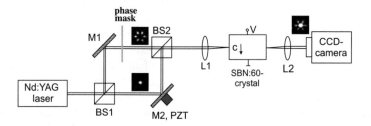

Figure 9.21: Experimental setup used for observing multipolar vector solitons; BS and PZT
stand for beam splitter and piezoelectric transducer, respectively. (After Ref. [151]; ©2001
OSA.)

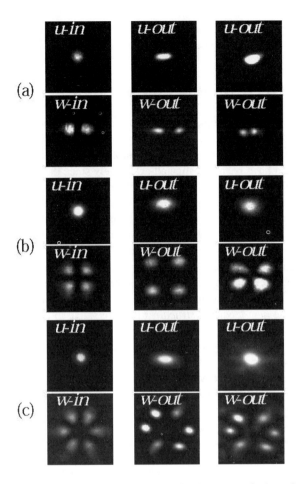

Figure 9.22: Two-component vector solitons observed at the output of a 1-cm-long SBN crystal. (a) Dipole soliton ($V_e = 2$ kV); (b) quadrupole soliton ($V_e = 2.3$ kV); and (c) hexapole soliton ($V_e = 2$ kV). Left column: input intensities of two components launched with 0.3-μW power. Middle column: output intensities when each component travels in isolation. Right column: output intensities when they propagate simultaneously. (After Ref. [151]; ©2001 OSA.)

the nonlinear coupling between them creates an effective waveguide that traps all lobes of the w component, leading to the formation of a multipole vector soliton.

Although the observed multipolar vector solitons appear to be quite stable, they could easily be destabilized by a small relative spatial shift of the two components. The only exception occurs for the dipole vector soliton, a result that confirms the stable nature of this vector soliton, in agreement with the numerical simulations [138]. All other multipole solitons break up into a set of fundamental and dipole-type solitons; the breakup process is found to be strongly influenced by the anisotropy of the photorefractive nonlinearity. Numerical simulations also indicate a strong influence of the anisotropy on the formation and propagation of the dipole and multipole vector solitons. In agreement with numerical predictions, multipole vector solitons are observed experimentally only for a specific orientation of the beam lobes.

9.7 Transverse Modulation Instability

In this section, we discuss another aspect of two-dimensional vector solitons, namely, the transverse modulation instability of one-dimensional vector-soliton stripes. As was discussed in Chapter 6, transverse instabilities of scalar solitons were first predicted in 1973 [157] and have been investigated extensively since then [70]. Such instabilities have been observed experimentally for both bright and dark scalar solitons [158]–[163]. It turns out that the nonlinear interaction among multiple components of a vector soliton adds new features to the associated transverse instability [120, 164, 165]. In particular, an incoherent coupling between the components of a composite dark-bright soliton can suppress transverse instabilities [166, 167].

As discussed in Chapter 6, instabilities developing under the action of higher-order spatial or temporal perturbations can initiate the breakup of a soliton stripe into multiple fragments. Several different scenarios of such a breakup are known [70], and some of them can generate *new localized structures* that are stable in two dimensions. In the scalar case, this situation corresponds to the breakup of a bright-soliton stripe into multiple two-dimensional bright solitons in a self-focusing medium. In the case of a dark-soliton stripe, the same instability produces an array of vortices. In the case of a vector-soliton stripe, the transverse instability turns out to be even more interesting. Indeed, if the vector-soliton stripe is created by the fundamental modes, it decays into an array of two-dimensional vector solitons (see Ref. [165]). However, if the stripe is composed of mutually coupled fundamental and first-order modes, it can force the decay of a quasi-one-dimensional vector soliton into an array of dipole-type vector solitons [168]. This effect provides additional evidence of the robust nature of the dipole-mode vector soliton; one can think of them as resembling "molecules of light" as far as their dynamics are concerned.

Any investigation of the modulation instability makes use of linear stability analysis. In the case of an isotropic Kerr medium, we need to linearize Eqs. (9.6.1) and (9.6.2) and solve the resulting set of linear equations using an asymptotic analysis. This approach shows that vector-soliton stripes are indeed transversally unstable, and the growth rate (or the gain coefficient) g of this instability is given by $g(p) = p\sqrt{2/\mu_+}$, where p is the unstable transverse wave number and μ_+ is the positive eigenvalue of the Hessian matrix. This matrix was introduced earlier in Eq. (9.3.11) and has components $U_{ij} = \partial P_i/\partial \beta_j$, where $i, j = 1$ and 2, $P_i = \int_{-\infty}^{\infty} |U_i|^2 dx$ is the conserved power of the ith component, and β_i is the corresponding propagation constant. Figure 9.23 shows the growth rate $g(p)$ calculated numerically (circles). The dashed line shows the results of an asymptotic analysis valid for small values of p. The solid line shows a simple fit to the numerics obtained using $g = \alpha p(1 - p/p_c)^{1/2}$ with $\alpha = 0.89$ and $p_c = 1.36$.

To investigate the transverse instability of two incoherently coupled beams in a photorefractive medium, one must solve Eqs. (9.6.22) and (9.6.23) numerically using the split-step method. First, these equations are solved without the y-derivative term to find the vector-soliton stripe with components $U_1(x)$ and $U_1(x)$. Then the y derivatives

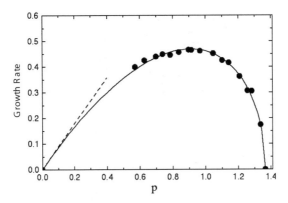

Figure 9.23: Growth rate $g(p)$ of the transverse instability found numerically (circles) and analytically using an asymptotic method (dashed line). Solid curve shows a simple fit to numerics. (After Ref. [168]; ©2001 APS.)

are retained and the same equations are solved with an input that represents a vector-soliton stripe perturbed transversely as

$$U_1'(x,y;0) = [1 + \varepsilon q(x,y)]U_1(x), \qquad U_2'(x,y;0) = U_2(x), \qquad (9.7.1)$$

where $q(x,y)$ represents a noise term with amplitude $\varepsilon \sim 10^{-5}$.

The same experimental setup shown in Figure 9.21 was used for observing the transverse instability after replacing the spherical lenses with a set of cylindrical lenses so that the two input beams were so wide in one transverse dimension that they formed a vector-soliton stripe during propagation inside the crystal. Figure 9.24 shows an example of the behavior observed at the crystal output. The CCD-camera images display the intensity patterns observed for the fundamental (top) and the dipole (bottom) components after propagating through a 15-mm-long SBN crystal. Notice how both beams break up in two-dimensional vector solitons. Moreover, the weaker component breaks up into an array of dipole-type vector solitons. These dipole solitons are generated vertically along the direction of the applied electric field. Both the initial decay of the stripe and the formation of the dipole array seem to be only weakly affected by the anisotropy. As a result, the main qualitative features are well reproduced, even by an isotropic model.

9.8 Dark Vector Solitons

So far this chapter has focused on bright vector solitons. As discussed in Section 9.1, it is possible to realize vector solitons whose one or more components correspond to a dark soliton. In the temporal case, the possibility of different signs for the dispersive term (normal versus anomalous dispersion) leads to several different combinations of the bright- and dark-soliton pairs, forming a vector soliton. In this section we consider several types of vector solitons associated with dark solitons.

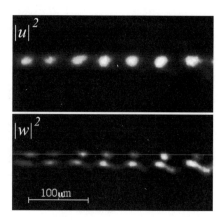

Figure 9.24: Experimental observation of the transverse modulation instability of a vector-soliton stripe. Input powers for the upper and lower components are 20 and 8 μW, respectively. (After Ref. [168]; ©2001 APS.)

9.8.1 Polarization-Domain Walls

An interesting class of vector solitons in the form of two coupled kink solitons can form when the nonlinear interaction of two orthogonally polarized components is considered in an isotropic medium [169]–[175]. A kink soliton represents a wave front whose amplitude changes rapidly in a narrow region but the amplitude does not vanish as $x \rightarrow \pm\infty$. Such a soliton is sometimes referred to as a *domain wall*, because it separates two spatially distinct regions.

Mathematically, domain-wall vector solitons are also described by Eqs. (9.4.9) and (9.4.10). If we use the circularly polarized components in place of the linearly polarized ones with $U_\pm = (U_1 \pm iU_2)/\sqrt{2}$, we obtain the following set of two coupled NLS equations [171]:

$$i\frac{\partial U_+}{\partial z} \pm \frac{1}{2}\frac{\partial^2 U_+}{\partial x^2} + (|U_+|^2 + \sigma|U_-|^2)U_+ = 0, \qquad (9.8.1)$$

$$i\frac{\partial U_-}{\partial z} \pm \frac{1}{2}\frac{\partial^2 U_-}{\partial x^2} + (|U_-|^2 + \sigma|U_+|^2)U_- = 0, \qquad (9.8.2)$$

where $\sigma = (1 + B)/(1 - B)$ is the coupling coefficient. These equations can also describe the temporal case of pulses propagating inside a highly birefringent optical fiber [2]. The choice of minus sign corresponds to normal dispersion, in the temporal case, or to a self-defocusing medium, in the spatial case.

The simplest solution of Eqs. (9.8.1) and (9.8.2) is in the form of a linearly polarized CW beam such that

$$U_+ = U_- = \sqrt{P_0}\exp[i(1 + \sigma)P_0 z], \qquad (9.8.3)$$

where P_0 is the initial input power of the beam. The linear stability analysis of this solution was carried out as early as 1970, and the CW solution was found to be unstable [4]. If we perturb the solution as

$$U_\pm = (\sqrt{P_0} + a_\pm)\exp[i(1 + \sigma)P_0 z], \qquad (9.8.4)$$

linearize Eqs. (9.8.1) and (9.8.2) for small perturbations $a_\pm$, and look for their solution in the form $a_\pm = b_\pm \exp(\lambda z)\cos(\Omega x)$, we obtain an eigenvalue equation as a fourth-degree polynomial in λ. One of the eigenvalues of this polynomial can become negative at high input powers in the case of a self-defocusing medium. This eigenvalue is given by

$$\lambda_1 = \Omega\sqrt{(\sigma-1)P_0 - \Omega^2/4}, \tag{9.8.5}$$

The maximum gain of this modulation instability is found to be $\lambda_{\max} = (\sigma-1)P_0$ and occurs at a spatial frequency $\Omega_m = [2(\sigma-1)P_0]^{1/2}$. This kind of modulation instability occurs in the normal-dispersion regime of a self-focusing medium and is referred to as the *polarization instability*. Note that it ceases to occur for $\sigma < 1$.

As discussed in Chapter 1, modulation instability is always associated with the existence of localized soliton-like solutions. In fact, scalar bright solitons are associated with the modulation instability of a scalar NLS equation. Using this analogy, one expects to find vector solitons associated with Eqs. (9.8.1) and (9.8.2) when the second-derivative term is negative [3]. Indeed, Haelterman and Sheppard found in 1994 the so-called *polarization-domain walls* as the vector solitons associated with this kind of polarization instability [21].

As usual, to find the vector solitons, we look for localized solutions of Eqs. (9.8.1) and (9.8.2) in the form [21]

$$U_+(x,z) = u(x)e^{i\beta z}, \qquad U_-(x,z) = v(x)e^{i\beta z}, \tag{9.8.6}$$

where the functions $u(x)$ and $v(x)$ are real and β is the propagation constant. With this substitution, Eqs. (9.8.1) and (9.8.2) reduce to the following coupled ordinary differential equations:

$$\frac{1}{2}\frac{d^2u}{dx^2} = -\beta u + u^3 + \sigma v^2 u, \tag{9.8.7}$$

$$\frac{1}{2}\frac{d^2v}{dx^2} = -\beta v + v^3 + \sigma u^2 v. \tag{9.8.8}$$

In general, one must solve these two equations numerically. We can gain considerable physical insight by using a mechanical analogy if we note that the same equations describe the motion of a particle (of unit mass) in the (u,v) plane when it is subjected to the potential

$$U(u,v) = \beta(u^2+v^2) - \frac{1}{2}(u^4+v^4) - \sigma u^2 v^2. \tag{9.8.9}$$

The soliton solutions correspond to the separatrix trajectories of this potential. The separatrix trajectories connecting a pair of maxima correspond to left and right circularly polarized dark solitons with amplitudes

$$u(x) = \sqrt{\beta}\tanh(\sqrt{\beta}x), \quad v(x) = 0, \tag{9.8.10}$$

$$v(x) = \sqrt{\beta}\tanh(\sqrt{\beta}x), \quad u(x) = 0. \tag{9.8.11}$$

The separatrix connecting a pair of opposite saddle points correspond to the linearly polarized dark solitons with amplitudes

$$u(x) = \pm v(x) = \sqrt{\beta/(1+\sigma)}\tanh(\sqrt{\beta}x). \tag{9.8.12}$$

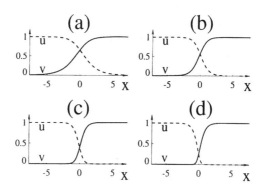

Figure 9.25: Envelopes $u(x)$ and $v(x)$ of the polarization domain wall found numerically for (a) $\sigma = 1.2$, (b) $\sigma = 2$, (c) $\sigma = 7$, and (d) $\sigma = 40$. In all cases, $\beta = 1$.

The separatrix trajectories connecting adjacent maxima of the potential $U(u,v)$ can be found only numerically for each value of σ. They correspond to the kink-shaped vector solitons whose examples are shown in Figure 9.25 using four different values of σ. A solution of this kind connects two domains of orthogonal stable polarization states of the Kerr medium; for this reason it is referred to as a *polarization-domain wall*. It describes a change in the ellipticity of the state of polarization of the optical field from $q = +1$ to $q = -1$, where q is defined as $q = (u-v)/(u+v)$. Note that $q = 0$ corresponds to a linearly polarized optical field.

The predictions based on Eqs. (9.8.7) and (9.8.8) have been checked by solving Eqs. (9.8.1) and (9.8.2) numerically [21]. The results show that such solutions indeed exist and are stable. A polarization-domain wall can be treated as the limiting case of a periodic solution, and it can be interpreted as a solitary wave associated with the polarization instability. The domain walls are also known in other fields. Two examples are provided by (i) the domain walls that separate convection patterns of different symmetry and are governed by two coupled Ginzburg–Landau equations [176, 177] and (ii) the gap solitons that separate different standing waves in a discrete lattice [178, 179].

Experimental observation of polarization domain walls was achieved by mixing two intense counterpropagating laser beams in a nonlinear isotropic dielectric [180] and as a result of modulation instability in normal dispersive bimodal fiber [181].

9.8.2   Vector Solitons Created by Optical Vortices

As described in Chapter 8, vortex solitons are dark solitons, self-trapped in two spatial dimensions and carrying a phase singularity [182]. Vortex solitons can be generalized to vector solitons [183]–[185], but the most interesting generalization is associated with the waveguiding properties of the vortex solitons. Indeed, similar to bright solitons, vortex solitons create waveguides that can guide and steer another beam, resulting in a configurable all-optical circuit. Dark solitons and vortices are more attractive for waveguiding applications because of their greater stability and steerability. In particular, it was demonstrated that optical vortices may be useful for trapping small particles [186]. The numerical and analytical results indicate that although most waveguiding properties of vortices are similar to those of planar dark solitons, some new features are also possible.

Consider two incoherently coupled CW beams at different frequencies ω_1 and ω_2 propagating in a self-defocusing Kerr medium. Their evolution is governed by Eqs. (9.1.12) and (9.1.13). These equations can be written in the following form after noting that both d_j and γ_j ($j = 1, 2$) depend on the carrier frequency of the beams [187]:

$$i\frac{\partial U_1}{\partial z} + \left(\frac{\partial^2 U_1}{\partial x^2} + \frac{\partial^2 U_1}{\partial y^2}\right) - (|\eta^{-1}|U_1|^2 + \sigma|U_2|^2)U_1 = 0, \qquad (9.8.13)$$

$$i\kappa\frac{\partial U_2}{\partial z} + \left(\frac{\partial^2 U_2}{\partial x^2} + \frac{\partial^2 U_2}{\partial y^2}\right) \pm (\eta|U_2|^2 + \sigma|U_1|^2)U_2 = 0, \qquad (9.8.14)$$

where $\eta = \omega_2^2/\omega_1^2$ and $\kappa = \omega_2/\omega_1$. The nonlinear terms are negative for a self-defocusing medium. The same equations describe the interaction of two orthogonally polarized beams of the same frequency if we set $\eta = \kappa = 1$. In the case of a photorefractive medium, the nonlinear terms can have opposite signs for the two polarization components [188]. We allow for this possibility through the choice of the sign so that U_2 can experience either defocusing or focusing.

To find vector solitons created by optical vortices, we consider the situation in which U_1 corresponds to a vortex soliton but the field U_2 describes either the fundamental or the first-order mode of the waveguide induced by the vortex. Thus, we look for radially symmetric solutions of Eqs. (9.8.13) and (9.8.14) in the form

$$U_1(R, \varphi; z) = \eta^{1/2}B_0 u(R)e^{-iB_0^2 z}e^{in\varphi}, \qquad (9.8.15)$$

$$U_2(R, \varphi; z) = \eta^{-1/2}B_0 v(R)e^{-i(\lambda/\kappa)B_0^2 z}e^{im\varphi}, \qquad (9.8.16)$$

where B_0 is the background field associated with the vortex, $R = \sqrt{x^2 + y^2}$ and n and m are topological charges. Using cylindrical coordinates in Eqs. (9.8.13) and (9.8.14), the coupled equations for the real functions u and v become

$$\frac{d^2u}{dr^2} + \frac{1}{r}\frac{du}{dr} - \frac{n^2}{r^2}u + u - [u^2 + (\sigma/\eta)v^2]u = 0, \qquad (9.8.17)$$

$$\frac{d^2v}{dr^2} + \frac{1}{r}\frac{dv}{dr} - \frac{m^2}{r^2}v + \lambda v \pm [v^2 + \eta\sigma u^2]v = 0, \qquad (9.8.18)$$

where $r = RB_0$. We set $n = \pm 1$ assuming a singly charged vortex.

For any σ and λ, Eqs. (9.8.17) admit a solution in the form of a scalar vortex soliton with $u = u_0(r)$ and $v = 0$. For fixed values of σ and η, the family of vector solitons for which $v \neq 0$ is described by a single parameter λ. For $|v|/|u| \ll 1$, Eq. (9.8.18) for $v(r)$ becomes a linear eigenvalue problem with an effective potential created by the vortex. When the nonlinear term is negative, the potential is *attractive*, and it can support spatially localized solutions as *guided modes* of the vortex-induced waveguide, each of them appearing above a certain cutoff value of λ. When the nonlinear term is positive, the potential is *repulsive*, and no guiding is possible in the linear limit.

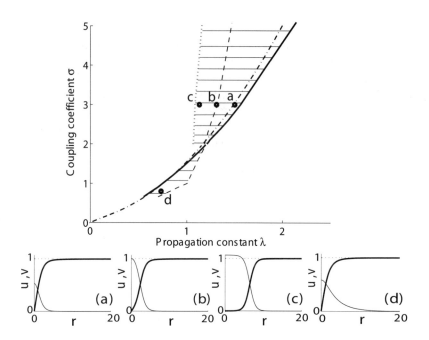

Figure 9.26: Existence domain (hatched region) for the fundamental mode guided by a vortex. Filled circles correspond to the four examples shown at bottom. The solid and dot-dashed curves show the cutoff for the mode guiding calculated numerically and by using a variational approach, respectively. Dashed curve shows the parameters where the amplitude of the guided mode coincides with the background. (After Ref. [137]; ©2000 OSA.)

Let us focus on the attractive-potential case, for which guided modes are possible. As the amplitude of the guided mode grows, the linear waveguide approximation becomes invalid. In this nonlinear regime, the guided mode is strong enough that it deforms the vortex waveguide, and together they form a composite vector soliton [189]. To find such solitons, one must solve Eqs. (9.8.17) and (9.8.18) numerically with the boundary conditions that $u \to 1$ and $v \to 0$ as $r \to \infty$. Multiple solutions have been found and they can be classified by the order of the guided mode [190]. Figure 9.26 shows several examples for $m = 0$, i.e., when the vortex guides the fundamental mode of the waveguide. The existence region of such solutions is also shown on the parameter plane (λ, σ), where the points (a) to (d) correspond to the four examples shown.

In the (λ, σ) plane, the cutoff of the fundamental mode is characterized by the $\sigma(\lambda)$ curve shown by the thick solid line in Figure 9.26. Near this curve, the amplitude of the guided mode is small, and the linear waveguiding regime is valid. Linearizing Eqs. (9.8.17) and (9.8.18) around the vortex solution, one can obtain the following two decoupled equations:

$$\frac{d^2 u_0}{dr^2} + \frac{1}{r}\frac{du_0}{dr} - \frac{n^2}{r^2}u_0 - u_0 + u_0^3 = 0, \qquad (9.8.19)$$

$$\frac{d^2 v}{dr^2} + \frac{1}{r}\frac{dv}{dr} - \frac{m^2}{r^2}v - \lambda v + u_0^2(r)v = 0, \qquad (9.8.20)$$

The effective guiding potential for the v components depends on the vortex shape u_0, which is not known an an explicit form. For this reason, the eigenvalue problem for mode $v(r)$ cannot be solved analytically. However, an estimate of the cutoff can be obtained by using a standard variational technique.

In the variational method, we fix the vortex shape using a simple trial function $u_0(r) = \tanh^n(ar)$. Noting that Eq. (9.8.19) has the Lagrangian

$$\mathcal{L}_u = r\left(\frac{du_0}{dr}\right)^2 + \frac{n^2}{r^2}u_0^2 + \frac{r}{2}(1+u_0^2)^2, \tag{9.8.21}$$

we find the averaged Lagrangian using $\langle L \rangle = a^{-1}\int_0^\infty L_u(x)\,dx$, where $x = ar$. Variation of $\langle L_u \rangle$ with respect to a yields the following expression for the parameter a:

$$a = \frac{1}{n}\left\{\int_0^\infty x[1 - \tanh^{2n}(x)]^2 dx\right\}. \tag{9.8.22}$$

This variational result provides a reasonable fit to the numerical solution for $n \leq 3$. Using it, the shape of the effective potential created by a single-charge vortex ($n = 1$) is found to be $u_0^2(r) = \tanh^2(0.554r) \approx 1 - \exp(-r^2/4)$. Using this expression in Eq. (9.8.20), the following simple analytical expression for the cutoff can be obtained:

$$\sigma = \frac{1}{32}\left[1 + 24\lambda + 16\lambda^2 - (1 - 4\lambda)\sqrt{1 - 4\lambda)^2}\right]. \tag{9.8.23}$$

This prediction is in good agreement with the numerically calculated cutoff value, as seen in Fig 9.26.

The properties of the vortex-induced waveguides and the corresponding vector solitons are different, depending on whether $\sigma < 2$ or > 2. For $\sigma > 2$, the amplitude of the guided mode grows with decreasing λ. As a result, the amplitude of the guided mode reaches the vortex background far from the cutoff point (dashed curve in Figure 9.26) and goes above the background value before the solution disappears (dotted curve in Figure 9.26). Close to this limit, the solution corresponds to a type of polarization-domain wall [190]. For $\sigma < 2$, the scenario is similar as λ increases, but the mode amplitude never reaches the background level before the solution disappears on the other border of the dashed curve in Figure 9.26. In the case of a self-focusing non-linearity, no bound modes of the vortex antiwaveguide can exist in the linear regime ($|v| \ll |u|$). However, the numerical results suggest that for large intensities of the bright component, mutual trapping is still possible [137]. This effect is similar to that occurring in nonlinear antiwaveguides assisted by an external potential [191].

References

[1] B. Crosignani and P. DiPorto, *Opt. Lett.* **6**, 329 (1981); *J. Opt. Soc. Am. B* **72**, 1136 (1982).

[2] C. R. Menyuk, *IEEE J. Quantum Electron.* **25**, 2674 (1989).

[3] G. P. Agrawal, *Nonlinear Fiber Optics*, 3rd ed. (Academic, San Diego, 2001).

[4] A. L. Berkhoer and V. E. Zakharov, *Zh. Eksp. Teor. Fiz.* **58**, 903 (1970) [*Sov. Phys. JETP* **31**, 486 (1970)].

[5] S. V. Manakov, *Sov. Phys. JETP* **38**, 248 (1974).

[6] M. V. Tratnik and J. E. Sipe, *Phys. Rev. A* **38**, 2011 (1988).

[7] D. N. Christodoulides and R. I. Joseph, *Opt. Lett.* **13**, 53 (1988).

[8] M. Haelterman, A.P. Sheppard, and A.W. Snyder, *Opt. Lett.* **18**, 1406 (1993).

[9] M. Haelterman and A. P. Sheppard, *Phys. Rev. E* **49**, 3376 (1994); Phys. Lett. A **194**, 191 (1994).

[10] M. Segev, G. C. Valley, S. R. Singh, M. I. Carvalho, and D. N. Christodoulides, *Opt. Lett.* **20**, 1764 (1995).

[11] V. V. Afanasjev, Yu. S. Kivshar, V. V. Konotop, and V.N. Serkin, *Opt. Lett.* **14**, 805 (1989).

[12] V. V. Afanasjev, E. M. Dianov, and V. N. Serkin, *IEEE J. Quantum Electron.* **25**, 2656 (1989).

[13] M. Lisak, A. Höök, and D. Anderson, *J. Opt. Soc. Am. B* **7**, 810 (1990).

[14] L. Wang and C. C. Yang, *Opt. Lett.* **15**, 474 (1990).

[15] B. J. Hong, C. C. Yang, and L. Wang, *J. Opt. Soc. Am. B* **8**, 464 (1991).

[16] Yu. S. Kivshar, *Opt. Lett.* **17**, 1322 (1992).

[17] A. V. Buryak, Yu. S. Kivshar, and D. F. Parker, *Phys. Lett. A* **215**, 57 (1996).

[18] S. Trillo, S. Wabnitz, E. M. Wright, and G. Stegeman, *Opt. Lett.* **13**, 871 (1988).

[19] Yu. S. Kivshar and S. K. Turitsyn, *Opt. Lett.* **18**, 337 (1993).

[20] M. Haeltreman and A. P. Sheppard, *Phys. Rev. E* **49**, 4512 (1994).

[21] M. Haelterman and A. P. Sheppard, *Phys. Lett. A* **185**, 265 (1994).

[22] H. T. Tran and R.A. Sammut, *Opt. Commun.* **119**, 583 (1995).

[23] R. Radhakrishnan and M. Lakshmanan, *J. Phys. A* **28**, 2683 (1995).

[24] A. P. Sheppard and Yu. S. Kivshar, *Phys. Rev. E* **55**, 4773 (1997).

[25] E. Seve, G. Millot, and S. Wabnitz, *Opt. Lett.* **23**, 1829 (1998).

[26] Y. Inoue, J. Plasma Phys. **16**, 439 (1976).

[27] M. Shalaby and A. J. Barthelemy, *IEEE J. Quantum Electron.* **28**, 2736 (1992).

[28] M. I. Carvalho, S. R. Singh, D. N. Christodoulides, and R. I. Joseph, *Phys. Rev. E* **53**, R53 (1996).

[29] Z. Chen, M. Segev, T. H. Coskun, D. N. Christodoulides, Yu. S. Kivshar, and V. V. Afanasjev, *Opt. Lett.* **21**, 1821 (1996).

[30] L. Keqing, Q. Shixing, Z. Wei, Z. Yanpeng, and W. Zgensen, *Opt. Commun.* **209**, 437 (2002).

[31] G. P. Agrawal, *Applications of Nonlinear Fiber Optics* (Academic Press, San Diego, 2001), Chap. 2.

[32] J. T. Manassah, *Opt. Lett.* **15**, 670 (1990).

[33] J. T. Manassah, *Opt. Lett.* **16**, 587 (1991).

[34] R. De La Fuente, A. Barthelemy, and C. Froehly, *Opt. Lett.* **16**, 793 (1991).

[35] R. De La Fuente and A. Barthelemy, *IEEE J. Quantum Electron.* **28**, 547 (1992).

[36] G. A. Askar'yan, *Sov. Phys. JETP* **15**, 1088 (1962).

[37] A. W. Snyder and D. J. Love, *Optical Waveguide Theory* (Chapman and Hall, London, 1983).

[38] J. U. Kang, G. I. Stegeman, and J. S. Aitchison, *Opt. Lett.* **20**, 2069 (1995).

[39] M. Morin, G. Duree, G. Salamo, and M. Segev, *Opt. Lett.* **20**, 2066 (1995).

[40] M. Shih, Z. Chen, M. Mitchell, M. Segev, H. Lee, R. S. Feigelson, and J. P. Wilde, *J. Opt. Soc. Am. B* **14**, 3091 (1997).

[41] S. Lan, E. DelRe, Z. Chen, M. Shih, and M. Segev, *Opt. Lett.* **24**, 475 (1999).

[42] S. Lan, M. Shih, G. Mizell, J. A. Giordmaine, Z. Chen, C. Anastassiou, J. Martin, and M. Segev, *Opt. Lett.* **24**, 1145 (1999).

[43] S. Lan, C. Anastassiou, M. Segev, M. Shih, J. A. Giordmaine, and. G. Mizell, *Appl. Phys. Lett.* **77**, 2101 (2000).

[44] E. Fazio, M. Zitelli, M. Bertolotti, A. Carrera, N. G. Sanvito, and G. Chiaretti, *Opt. Commun.* **185**, 331 (2000).

[45] M. Klotz, M. Crosser, A. Guo, M. Henry, M. Segev, and G.L. Wood, *Appl. Phys. Lett.* **79**, 1423 (2001).

[46] A. Guo, M. Henry, G. J. Salamo, M. Segev, and G.L. Wood, *Appl. Phys. Lett.* **26**, 1274 (2001).

[47] S. Lan, J.A. Giordmaine, M. Segev, and D. Rytz, *Opt. Lett.* **27**, 737 (2002).

[48] G. E. Torres-Cisneros, J. J. Sanchez-Mondragon, and V. A. Vysloukh, *Opt. Lett.* **18**, 1299 (1993).

[49] B. Luther-Davies and X. Yang, *Opt. Lett.* **17**, 496 (1992; *Opt. Lett.* **17**, 1775 (1992).

[50] V. G. Makhankov, N. V. Makhaldiani, O. K. Pashaev, *Phys. Lett. A* **81**, 161 (1981).

[51] R. De La Fuente and A. Barthelemy, *Opt. Commun.* **88**, 419 (1992).

[52] M. Florjanczyk and R. Tremblay, *Phys. Lett. A* **141**, 34 (1989).

[53] F. T. Hioe, *Phys. Lett. A* **234**, 351 (1997); *Phys. Rev. E* **56**, 2373 (1997); *Phys. Rev. E* **58**, 6700 (1998).

[54] A. Hasegawa, *Opt. Lett.* **5**, 416 (1980).

[55] G. P. Agrawal, *Fiber-Optic Communication Systems*, 3rd ed. (Wiley, New York, 2002), Chap. 9.

[56] L. A. Bergman, A. J. Mendez, and L. S. Lome, *SPIE Crit. Rev.* **CR 62**, 210 (1996).

[57] C. Yeh and L. Bergman, *J. Appl. Phys.* **80**, 3174 (1996); Phys. Rev. E **57**, 2398 (1998).

[58] L. Bergman, J. Morookian, and C. Yeh, *J. Lightwave Technol.* **16**, 1577 (1998).

[59] C. Yeh, L. Berman, J. Morookian, and S. Monacos, Phys. Rev. E **57**, 6135 (1998).

[60] C. Yeh and L. Bergman, *Phys. Rev. E* **60**, 2306 (1999).

[61] E.A. Ostrovskaya, Yu.S. Kivshar, D. Mihalache, and L.-C. Crasovan, IEEE Selected Topics Quantum Electron. **8**, 591 (2002).

[62] F. T. Hioe, *Phys. Rev. Lett.* **82**, 1152 (1999).

[63] K. Nakkeeran, *Phys. Rev. E* **62**, 1313 (2000).

[64] Y. Nogami and C. S. Warke, *Phys. Lett. A* **59**, 251 (1974).

[65] V. M. Petnikova, V. V. Shuvalov, and V. A. Vysloukh, *Phys. Rev. E* **60**, 1009 (1999).

[66] Q-Han Park, H. J. Shin, and J. Kim, *Phys. Lett. A* **263**, 91 (1999).

[67] A. A. Sukhorukov and N. N. Akhmediev, *Phys. Rev. Lett.* **83**, 4736 (1999).

[68] N. N. Akhmediev, V. M. Eleonskii, N. E. Kulagin, and L. P. Shilnikov, *Pis'ma Zh. Tekh. Fiz.* **15**, 19 (1989) [*Sov. Tech. Phys. Lett.* **15**, 587 (1989)].

[69] V. M. Eleonskii, V. G. Korolev, N. E. Kulagin, and L. P. Shilnikov, *Zh. Eksp. Teor. Fiz.* **99**, 1113 (1991) [*Sov. Phys. JETP* **72**, 619 (1991)].

[70] D. E. Pelinovsky and Yu. S. Kivshar, *Phys. Rev. E* **62**, 8663 (2000).

[71] M. Grillakis, J. Shatah, and W. Strauss, *J. Funct. Anal.* **74**, 160 (1987).

[72] M. Grillakis, *Comm. Pure Appl. Math.* **41**, 747 (1988).

[73] M. Grillakis, *Comm. Pure Appl. Math.* **43**, 299 (1990).

[74] M. Grillakis, J. Shatah, and W. Strauss, *J. Funct. Anal.* **94**, 308 (1990).

[75] Yu. S. Kivshar, *J. Opt. Soc. Am. B* **7**, 2204 (1990).

[76] V. K. Mesentsev and S. K. Turitsyn, *Opt. Lett.* **17**, 1497 (1992).

[77] X. D. Cao and C. J. McKinstrie, *J. Opt. Soc. Am. B* **10**, 1202 (1993).

[78] D. C. Hutchings, J. S. Aitchison, B. S. Wherrett, G. T. Kennedy, and W. Sibbett, *Opt. Lett.* **20**, 991 (1995).

[79] J. S. Aitchison, D. C. Hutchings, J. U. Kang, G. I. Stegeman, and A. Villeneuve, *IEEE J. Quantum Electron.* **33**, 341 (1997).

[80] C. M. De Sterke and J. E. Sipe, *J. Opt. Soc. Am. A* **7**, 636 (1990).

[81] C. M. De Sterke and J. E. Sipe, *Opt. Lett.* **16**, 202 (1991).

[82] N. N. Akhmediev and A. Ankiewicz, *Solitons: Nonlinear Pulses and Beams* (Chapman and Hall, London, 1997), Chap. 7.

[83] C. C. Shang and H. Hsu, *IEEE J. Quantum Electron.* **QE-23**, 177 (1987).

[84] J. Frey, R. Frey, C. Flytzanis, and R. Triboulet, *J. Opt. Soc. Am. B* **9**, 132 (1992).

[85] D. S. Hutchings, J. S. Aitchison, and J. M. Arnold, *J. Opt. Soc. Am. B* **14**, 869 (1997).

[86] A. Villeneuve, J. U. Kang, J. S. Aitchison, and G. I. Stegeman, *Appl. Phys. Lett.* **67**, 760 (1995).

[87] D. C. Hutchings and B. S. Wherrett, *Phys. Rev. B* **52**, 8150 (1995).

[88] Y. Chen and J. Atai, *Opt. Lett.* **19**, 1287 (1994).

[89] P. B. Lundquist and D. R. Andersen, *J. Opt. Soc. Am. B* **14**, 87 (1997).

[90] P. B. Lundquist, D. R. Andersen, and Yu. S. Kivshar, *Phys. Rev. E* **57**, 3551 (1998).

[91] R. A. Sammut, A. V. Buryak, and Yu. S. Kivshar, *Opt. Lett.* **22**, 1385 (1997); *J. Opt. Soc. Am. B* **15**, 1488 (1998).

[92] D. Wang, R. Barille, and G. Rivoire, *J. Opt. Soc. Am. B* **15**, 2731 (1998).

[93] K. J. Blow, N. J. Doran, and D. Wood, *Opt. Lett.* **12**, 202 (1987).

[94] D. Anderson, Yu. S. Kivshar, and M. Lisak, *Physica Scripta* **43**, 273 (1991).

[95] N. N. Akhmediev and J. M. Soto-Crespo, *Phys. Rev. E* **49**, 5742 (1994).

[96] N. N. Akhmediev, A. Buryak, and J. M. Soto-Crespo, *Opt. Commun.* **112**, 278 (1994).

[97] N. N. Akhmediev, A. V. Buryak, J. M. Soto-Crespo, and D. R. Andersen, *J. Opt. Soc. Am. B* **12**, 434 (1995).

[98] J. M. Soto-Crespo, N. N. Akhmediev, and A. Ankiewicz, *J. Opt. Soc. Am. B* **12**, 1100 (1995); *Phys. Rev. E* **51**, 3547 (1995).

[99] J. U. Kang, G. I. Stegeman, J. S. Aitchison, and N. N. Akhmediev, *Phys. Rev. Lett.* **76**, 3699 (1996).

[100] H. G. Winful, *Opt. Lett.* **11**, 33 (1986).

[101] M. V. Tratnik and J. E. Sipe, *Phys. Rev. A* **38**, 2011 (1988).

[102] D. Mihalache, D. Mazilu, and L. Torner, *Phys. Rev. Lett.* **81**, 4353 (1998).

[103] D. Mihalache, D. Mazilu, and L.-C. Crasovan, *Phys. Rev. E* **60**, 7504 (1999).

[104] E. A. Ostrovskaya, N. N. Akhmediev, G. I. Stegeman, J. U. Kang, and J. S. Aitchison, *J. Opt. Soc. Am. B* **14**, 880 (1997).

[105] W. H. Press, S. A. Teukolsky, W. T. Vetterling, and B. P. Flannery, *Numerical Recipes in C: The Art of Scientific Computing* (Cambridge University Press, Cambridge, UK, 1992).

[106] L. D. Landau and E. M. Lifshitz, *Quantum Mechanics: Nonrelativisitc Theory* (Pergamon Press, London, 1977).

[107] J. Yang, *Physica D* **108**, 92 (1997).

[108] R. J. Dowling, *Phys. Rev. A* **42**, 5553 (1990).

[109] S. G. Evangelides, L. F. Mollenauer, J. P. Gordon, and N. S. Bergano, *J. Lightwave Technol.* **10**, 28 (1992).

[110] M. Born and E. Wolf, *Principles of Optics* (Cambridge University Press, New York, 1999).

[111] J. U. Kang, G. I. Stegeman, and J. S. Aitchison, *Opt. Lett.* **21**, 189 (1996).

[112] J. U. Kang, J. S. Aitchison, G. I. Stegeman, and N. N. Akhmediev, *Opt. Quantum Electron.* **30**, 649 (1998).

[113] J. S. Aitchison, D. C. Hutchings, J. M. Arnold, J. U. Kang, G. I. Stegeman, E. A. Ostrovskaya, and N. N. Akhmediev, *J. Opt. Soc. Am. B* **14**, 3032 (1997).

[114] L. Friedrich, R. Malendevich, G. I. Stegeman, J. M. Soto-Crespo, N. N. Akhmediev, and J. S. Aitchison, *Opt. Commun.* **186**, 335 (2000).

[115] R. R. Malendevich, L. Friedrich, G. I. Stegeman, J. M. Soto-Crespo, N. N. Akhmediev, and J. S. Aitchison, *J. Opt. Soc. Am. B* **19**, 695 (2002).

[116] D. N. Christodoulides, S. R. Singh, M. I. Calvalho, and M. Segev, *Appl. Phys. Lett.* **68**, 1763 (1996).

[117] M. Mitchell, M. Segev, and D. N. Christodoulides, *Phys. Rev. Lett.* **80**, 4657 (1998).

[118] A. W. Snyder and Yu. S. Kivshar, *J. Opt. Soc. Am. B* **14**, 3025 (1997).

[119] V. Kutuzov, V. M. Petnikova, V. V. Shuvalov, and V. A. Vysloukh, *Phys. Rev. E* **57**, 6056 (1998).

[120] E. Ostrovskaya, Yu. S. Kivshar, D. Skryabin, and W. J. Firth, *Phys. Rev. Lett.* **83**, 296 (1999).

[121] A. W. Snyder, S. J. Hewlett, and D. J. Mitchell, *Phys. Rev. Lett.* **72**, 1012 (1994).

[122] Y. Silberberg and Y. Barad, *Opt. Lett.* **20**, 246 (1995).

[123] J. Yang and D.J. Benney, *Stud. Appl. Math.* **96**, 111 (1996).

[124] J. Yang, *Physica D* **108**, 92 (1997).

[125] Y. Barad and Y. Silberberg, *Phys. Rev. Lett.* **78**, 3290 (1997).

[126] P. Kochaert and M. Haelterman, *J. Opt. Soc. Am. B* **16**, 732 (1999).

[127] A. C. Yew, B. Sandstede, and C. K. R. T. Jones, *Phys. Rev. E* **61**, 5886 (2000).

[128] C. Cambournac, T. Sylvestre, H. Maillotte, B. Vanderlinden, Ph. Emplit, and M. Haelterman, *Phys. Rev. Lett.* **89**, 083901 (2002).

[129] Z. H. Musslimani, M. Segev, D. N. Christodoulides, and M. Soljačić, *Phys. Rev. Lett.* **84**, 1164 (2000).

[130] J. N. Malmberg, A. H. Carlsson, D. Anderson, M. Lisak, E. A. Ostrovskaya, and Yu. S. Kivshar, *Opt. Lett.* **25**, 643 (2000).

[131] T. Carmon, R. Uzdin, C. Pigier, Z. H. Musslimani, M. Segev, and A. Nepomnyashchy, *Phys. Rev. Lett.* **87**, 14309 (2001).

[132] R.W. Boyd, *Nonlinear Optics* (Academic Press, San Diego, 1992).

[133] D. Anderson, *Phys. Rev. A* **27**, 3135 (1983).

[134] K. Hayata and M. Koshiba, *Opt. Lett.* **19**, 1717 (1994).

[135] A. V. Buryak, Yu. S. Kivshar, M. Shih, and M. Segev, *Phys. Rev. Lett.* **82**, 81 (1999).

[136] L. Gagnon and C. Paré, *J. Opt. Soc. Am. A* **8**, 601 (1991).

[137] J. N. Malmberg, A. H. Carlsson, D. Anderson, M. Lisak, E. A. Ostrovskaya, T. A. Alexander, and Yu. S. Kivshar, *Opt. Lett.* **25**, 643 (2000).

[138] J. J. García-Ripoll, V. Pérez-García, E.A. Ostrovskaya, and Yu. S. Kivshar, *Phys. Rev. Lett.* **85**, 82 (2000).

[139] A. S. Desyatnikov, D. Neshev, E. A. Ostrovskaya, Yu. S. Kivshar, G. McCarthy, W. Krolikowski, and B. Luther-Davies, *J. Opt. Soc. Am. B* **19**, 586 (2002).

[140] A. S. Desyatnikov, Yu. S. Kivshar, K. Motzek, F. Kaiser, C. Weilnau, and C. Denz, *Opt. Lett.* **27**, 634 (2002).

[141] K. Motzek, F. Kaiser, C. Weilnau, C. Denz, G. McCarthy, W. Krolikowski, A. S. Desyatnikov, and Yu. S. Kivshar, *Opt. Commun.* **209**, 501 (2002).

[142] V. Tikhonenko, J. Christou, and B. Luther-Davies, *Phys. Rev. Lett.* **76**, 2698 (1996); *J. Opt. Soc. Am. B* **12**, 2046 (1995).

[143] W. J. Firth and D. V. Skryabin, *Phys. Rev. Lett.* **79**, 2450 (1997).

[144] D. V. Skryabin and W. J. Firth, *Phys. Rev. E* **58**, 3916 (1998).

[145] A. Desyatnikov, D. Neshev, E. Ostrovskaya, Yu. S. Kivshar, W. Krolikowski, B. Luther-Davies, J. J. García-Ripoll, and V. Pérez-García, *Opt. Lett.* **26**, 435 (2001).

[146] A. Barthelemy, C. Froehly, and M. Shalaby, *SPIE Proc.* **2041**, 104 (1994).

[147] M. Soljačić, S. Sears, and M. Segev, *Phys. Rev. Lett.* **81**, 4851 (1998).

[148] A. A. Zozulya and D.Z. Anderson, *Phys. Rev. A* **51**, 1520 (1995).

[149] A. A. Zozulya, D.Z. Anderson, A.V. Mamaev, and M. Saffman, *Phys. Rev. A* **57**, 522 (1998).

[150] W. Krolikowski, M. Saffman, B. Luther-Davies, and C. Denz, *Phys. Rev. Lett.* **80**, 3240 (1998).

[151] D. Neshev, G. McCarthy, W. Krolikowski, E. A. Ostrovskaya, and Yu. S. Kivshar, G. F. Calvo and F. Agullo-Lopez, *Opt. Lett.* **26**, 1185 (2001).

[152] W. Krolikowski, E. A. Ostrovskaya, C. Weinau, M. Geisser, G. McCarthy, Yu. S. Kivshar, C. Denz, and B. Luther-Davies, *Phys. Rev. Lett.* **85**, 1424 (2000).

[153] T. Carmon, C. Anastassiou, S. Lan, D. Kip, Z. H. Musslimani, M. Segev, and D. N. Christodoulides, *Opt. Lett.* **25**, 1113 (2000).

[154] M. D. Iturbe-Castillo, P. A. Marquez-Aguilar, J. J. Sanchez-Mondragon, S. Stepanov, and V. Vysloukh, *Appl. Phys. Lett.* **64**, 408 (1994).

[155] M. Segev, G. C. Valley, B. Crosignani, P. DiPorto, A.Yariv, *Phys. Rev. Lett.* **73**, 3211 (1994).

[156] M. Shih, M. Segev, G.C. Valley, G. Salamo, B. Crosignani, P. DiPorto, *Electron. Lett.* **31**, 826 (1995).

[157] V. E. Zakharov and A.M. Rubenchik, *Zh. Eksp. Teor. Fiz.* **65**, 997 (1973) [*Sov. Phys. JETP* **38**, 494 (1974)].

[158] A. V. Mamaev, M. Saffman, A. A. Zozulya, *Phys. Rev. Lett.* **76**, 2262 (1996).

[159] V. Tikhonenko, J. Christou, B. Luther-Davies, and Yu. S. Kivshar, *Opt. Lett.* **21**, 1129 (1996).

[160] R. A. Fierst, D.-M. Baboiu, B. Lawrence, W. E. Torruellas, G. I. Stegeman, S. Trillo, and S. Wabnitz, *Phys. Rev. Lett.* **78**, 2756 (1997).

[161] X. Liu, K. Beckwitt, and F. Wise, *Phys. Rev. Lett.* **85**, 1871 (2000).

[162] C. Anastassiou, M. Soljačić, M. Segev, E. D. Eugenieva, D. N. Christodoulides, D. Kip, Z. H. Muslimani, and J. P. Torres, *Phys. Rev. Lett.* **85**, 4888 (2000).

[163] H. Fang, R. Malendevich, R. Schiek, and G. I. Stegeman, *Opt. Lett.* **25**, 1786 (2000).

[164] D.V. Skryabin and W.J. Firth, *Phys. Rev. Lett.* **81**, 3379 (1998).

[165] D.V. Skryabin and W. Firth, *Phys. Rev. E* **60**, 1019 (1999).

[166] Z. H. Musslimani, M. Segev, A. Nepomnyashchy, and Yu. S. Kivshar, *Phys. Rev. E* **60**, R1170 (1999).

[167] Z. H. Musslimani and J. Yang, *Opt. Lett.* **26**, 1981 (2001).

[168] D. Neshev, W. Krolikowski, D.E. Pelinovsky, G. McCarthy, and Yu.S. Kivshar, *Phys. Rev. Lett.* **87**, 103903 (2001).

[169] M. Haelterman and A.P. Sheppard, *Opt. Lett.* **19**, 96 (1994).

[170] M. Haelterman and A.P. Sheppard, Chaos, *Solitons and Fractals* **4**, 1731 (1994).

[171] M. Haelterman and A.P. Sheppard, *Phys. Rev. E* **49**, 3389 (1994).

[172] M. Haelterman, *Opt. Commun.* **111**, 86 (1994).

[173] B. A. Malomed, *Phys. Rev. E* **50**, 1565 (1994).

[174] M. Haelterman and M. Badolo, *Opt. Lett.* **20**, 2285 (1995).

[175] Y. Louis, A.P. Sheppard, and M. Haelterman, *Opt. Commun.* **141**, 167 (1997).

[176] B. A. Malomed, A. A. Nepomnyashchy, and M. I. Tribelsky, *Phys. Rev. A* **42**, 7244 (1990).

[177] I. Aranson and L. Tsimring, *Phys. Rev. Lett.* **75**, 3273 (1995).

[178] Yu. S. Kivshar, *Phys. Rev. Lett.* **70**, 3055 (1993).

[179] Yu.S. Kivshar, M. Haelterman, and A.P. Sheppard, *Phys. Rev. E* **50**, 3161 (1994).

[180] S. Pitois, G. Millot, and S. Wabnitz, *Phys. Rev. Lett.* **81**, 1409 (1998).

[181] S. Pitois, G. Millot, P. Grelu, and M. Haelterman, Phys. Rev. E **60**, 994 (1999).

[182] Yu. S. Kivshar and B. Luther-Davies, *Phys. Rep.* **298**, 81 (1998).

[183] L. M. Pismen, *Phys. Rev. Lett.* **72**, 2557 (1994).

[184] I. Velchev, A. Dreischuh, D. Neshev, and S. Dinev, *Opt. Commun.* **130**, 385 (1995).

[185] M. Axenides and L. Perivalaropoulos, *Phys. Rev. D* **56**, 1972 (1997).

[186] K. T. Gahagan and G. A. Swartzlander, Jr., *Opt. Lett.* **21**, 827 (1996).

[187] H. T. Tran and R. A. Sammut, *Phys. Rev. A* **52**, 3170 (1995).

[188] W. Krolikowski, N. N. Akhmediev, and B. Luther-Davies, *Opt. Lett.* **21**, 782 (1996).

[189] E.A. Ostrovskaya and Yu.S. Kivshar, *Opt. Lett.* **23**, 1268 (1998).

[190] A. P. Sheppard and M. Haelterman, *Opt. Lett.* **19**, 859 (1993).

[191] B. V. Gisin and A. A. Hardy, *Phys. Rev. A* **48**, 3466 (1993).

Chapter 10

Parametric Solitons

Optical solitons discussed so far are based on the cubic nonlinearities governed by the third-order susceptibility $\chi^{(3)}$ and occurring when an optical field changes the refractive index of the medium and modifies its own propagation properties. *Parametric solitons* constitute a separate class of solitons in which two or more optical fields become coupled inside a nonlinear medium such that each field propagates as a soliton. When the coupling is provided by the quadratic nonlinearities governed by the second-order susceptibility $\chi^{(2)}$, such parametric solitons are called *quadratic solitons*. This chapter is devoted to the $\chi^{(2)}$-supported parametric solitons. Section 10.1 presents the general concepts behind the two- and three-wave parametric interaction and the underlying nonlinear equations. Section 10.2 describes the properties of $(1 + 1)$-dimensional two-wave quadratic solitons forming inside planar waveguides and focuses on issues related to stability, interaction, and walk-off. It also includes the relevant experimental results and a brief discussion of dark solitons. Section 10.3 is devoted to the $(2 + 1)$-dimensional quadratic solitons. Multifrequency parametric solitons are the focus of Section 10.4, while Section 10.5 is devoted to parametric solitons in quasi-phase-matched (QPM) structures. Several other types of parametric solitons are discussed in Section 10.6, including vortex solitons, temporal solitons, optical bullets, and those supported by the competing (quadratic and cubic) nonlinearities.

10.1 Parametric Interaction

The history behind quadratic solitons can be traced back to 1967, the year in which Ostrovskii [1] discovered that self-induced changes in the phase front of an optical beam, usually associated with Kerr nonlinearity, might occur in a noncentrosymmetric nonlinear crystal when several frequency components are present simultaneously and interact parametrically. Around 1975, Karamzin and Sukhorukov predicted theoretically the existence of two-wave quadratic solitons in both the waveguide and bulk geometries [2, 3]. Soon after, more general, three-wave parametric solitons were investigated theoretically [4, 5].

These early theoretical developments, reported mostly in the Russian literature, are well documented in Sukhorukov's book [6], published in 1988 (in Russian). However, it was not until 1996 that the experimental observation of quadratic solitons was reported. The main reason behind such a delay was the lack of a high-quality material. Moreover, the advantages offered by quadratic solitons for practical applications were not obvious. The situation changed in the 1990s with the rediscovery of the self-action effect in $\chi^{(2)}$ media and with the appearance of high-damage optical materials with long enough propagation lengths [7]. Within a few years, quadratic solitons were observed in materials such as planar LiNbO$_3$ waveguides and bulk KTP crystals. Several reviews have discussed the progress realized during the 1990s [8]–[12]. With the advent of quasi-phase matching and the expected advances in the growth and engineering of $\chi^{(2)}$ materials, parametric solitons are likely to find practical applications.

The theory of quadratic solitons is based on the interaction of three optical fields inside a nonlinear medium in which $\chi^{(2)} \neq 0$. Such a parametric process is known as *three-wave mixing*. If a single pump beam at the frequency ω is launched, parametric interaction can occur provided the pump wave becomes phase matched with its second harmonic at the frequency 2ω. This process is known as *second-harmonic generation* (SHG) and constitutes a special case of three-wave mixing. The phase-matching condition for the SHG process can be satisfied in an anisotropic dielectric medium in two different ways, known as type I and type II in SHG theory [13]–[15]. In an anisotropic crystal, two *different* values of $k(\omega)$ can be found for any wave-vector direction $\mathbf{k}/k$; these correspond to the *ordinary* and *extraordinary* waves with different polarization states and phase velocities. The direction of wave vector $\mathbf{k}$ coincides with the direction of the Poynting vector $\mathbf{S}$ only for the ordinary wave.

We first focus on the general case of three-wave mixing and consider parametric interaction between three continuous-wave (CW) waves with the electric fields $E_j = \Re[A_j \exp(i\mathbf{k}_j \cdot \mathbf{r} - i\omega_j z)]$, where $j = 1$–3 and Re stands for the real part. The three frequencies satisfy the energy-conservation condition $\omega_1 + \omega_2 = \omega_3$. The phase-matching condition is assumed to be nearly satisfied, and we allow for a small mismatch Δk among the three wave vectors; i.e., $\Delta k = k_1(\omega_1) + k_2(\omega_2) - k_3(\omega_3)$. In general, the three waves do not propagate along the same direction, and the beams walk off from each other as they propagate inside the crystal. We choose the z axis along the direction of $\mathbf{k}_1$ and the x axis in the plane defined by $\mathbf{k}_1$ and the direction of the energy walk-off. If all three wave vectors point along the same direction, they have the same phase velocity and exhibit no walk-off.

The theory of $\chi^{(2)}$-mediated three-wave mixing is available in several books devoted to nonlinear optics [13]–[15]. The starting point is the Maxwell wave equation written in MKS units as

$$\nabla \times \nabla \times \mathbf{E} + \frac{1}{c^2} \frac{\partial^2 \mathbf{E}}{\partial t^2} = -\frac{1}{\varepsilon_0 c^2} \frac{\partial^2 \mathbf{P}}{\partial t^2}, \tag{10.1.1}$$

where ε_0 is the vacuum permittivity and c is the speed of light in a vacuum. The induced polarization is written in in the frequency domain as

$$\tilde{\mathbf{P}}(\mathbf{r}, \omega) = \varepsilon_0 \chi^{(1)} \tilde{\mathbf{E}} + \varepsilon_0 \chi^{(2)} \tilde{\mathbf{E}}\tilde{\mathbf{E}} + \cdots, \tag{10.1.2}$$

where a tilde denotes the Fourier transform. Using the slowly varying envelope and paraxial approximations, one can derive the following set of three coupled equations describing the parametric interaction of three waves under type II phase matching:

$$2ik_1\frac{\partial A_1}{\partial z} + \nabla_\perp^2 A_1 + \frac{2\omega_1^2}{\varepsilon_0 c^2}\chi^{(2)}A_3A_2^*e^{-i\Delta kz} = 0, \qquad (10.1.3)$$

$$2ik_2\left(\frac{\partial A_2}{\partial z} - \rho_\omega\frac{\partial A_2}{\partial x}\right) + \nabla_\perp^2 A_2 + \frac{2\omega_2^2}{\varepsilon_0 c^2}\chi^{(2)}A_3A_1^*e^{-i\Delta kz} = 0, \qquad (10.1.4)$$

$$2ik_3\left(\frac{\partial A_3}{\partial z} - \rho_{2\omega}\frac{\partial A_3}{\partial x}\right) + \nabla_\perp^2 A_3 + \frac{2\omega_3^2}{\varepsilon_0 c^2}\chi^{(2)}A_1A_2e^{i\Delta kz} = 0, \qquad (10.1.5)$$

where $\nabla_\perp^2$ is the transverse Laplacian operator, ρ_ω and $\rho_{2\omega}$ are the walk-off parameters, and $\chi^{(2)}$ is an element of the second-order susceptibility tensor. These equations describe the case in which spatial walk-off of all waves occurs in the same plane. Formally, this is true only for uniaxial crystals, but this restriction can easily be relaxed.

In the case of type I SHG, only a single beam at the pump frequency ω_1 is incident on the nonlinear crystal, and a new optical field at the frequency $2\omega_1$ is generated during the SHG process. We can adapt Eqs. (10.1.3)–(10.1.5) to this case with minor modifications. More specifically, we set $\omega_3 = 2\omega_1$, $A_1 = A_2$, and $\rho_\omega = 0$ in these equations. The first two equations then become identical, and one of them can be dropped. The type I SHG process is thus governed by the following set of of two coupled equations:

$$2ik_1\frac{\partial A_1}{\partial z} + \nabla_\perp^2 A_1 + 2k_1dA_3A_1^*e^{-i\Delta kz} = 0, \qquad (10.1.6)$$

$$2ik_3\left(\frac{\partial A_3}{\partial z} - \rho_{2\omega}\frac{\partial A_3}{\partial x}\right) + \nabla_\perp^2 A_3 + 8k_1dA_1^2e^{i\Delta kz} = 0, \qquad (10.1.7)$$

where $d = (\omega_1/\varepsilon_0 n_1 c)\chi^{(2)}$ is the nonlinear parameter and $\Delta k = 2k_1 - k_3$ is the phase-mismatch parameter. As discussed in Chapter 1, one can generalize these equations to the case of optical pulses by adding a dispersion term in the form $\beta_2(\omega_j)(\partial^2 A_j/\partial t^2)$, where $j = 1$ or 3 and β_2 is the group-velocity dispersion (GVD) parameter [16]–[36]. The GVD parameter is generally different for the pump and SHG beams and can be positive or negative.

As in earlier chapters, it is useful to normalize Eqs. (10.1.6) and (10.1.7). However, we cannot use the same scaling as earlier because of the phase mismatch Δk. We first introduce the normalized fields v and w as

$$A_1 = (\beta/d)v\exp(i\beta z), \qquad A_3 = (\beta/2d)w\exp[i(2\beta + \Delta k)z], \qquad (10.1.8)$$

where β takes into account the shift in the phase velocity induced by the nonlinear interaction. If we then normalize the spatial coordinates using

$$x' = (2k_1\beta)^{1/2}x, \qquad y' = (2k_1\beta)^{1/2}y, \qquad z' = \beta z, \qquad (10.1.9)$$

we arrive at the following set of two dimensionless coupled equations for the normalized field envelopes v and w [16]:

$$i\frac{\partial v}{\partial z'} + \nabla_\perp^2 v + r\frac{\partial^2 v}{\partial t^2} - v + wv^* = 0, \qquad (10.1.10)$$

$$i\sigma\frac{\partial w}{\partial z'} - i\delta\frac{\partial w}{\partial x'} + s\nabla_\perp^2 w + s\frac{\partial^2 w}{\partial t^2} - \alpha w + \frac{v^2}{2} = 0, \qquad (10.1.11)$$

where $\alpha = \sigma(2 + \Delta k/\beta)$ and $\sigma = k_3/k_1 \approx 2$. The parameter α is the effective phase-mismatch parameter and plays in important role in the following theory of quadratic solitons. The parameter $\delta = (2k_1/\beta)^{1/2}\rho_{2\omega}$ accounts for the walk-off effects. We have added the GVD terms and introduced the parameters r and s that take positive or negative values, depending on the nature of the dispersion (normal versus anomalous). In the case of spatial solitons excited using CW beams, $r = s = 0$. Equations (10.1.10) and (10.1.11) model the type I SHG process under quite general conditions and are used in the following sections for studying the properties of different kinds of quadratic solitons. For notational simplicity, we drop the prime over spatial variables in the remainder of this chapter.

10.2 Waveguide Geometry

This section considers the case in which SHG occurs inside a planar waveguide. The thickness of the waveguide is assumed to be small enough for it to support a single mode in the y direction at both the fundamental and second-harmonic frequencies. Using the method of separation of variables, we assume $A_j(x,y,z) = F_j(y)B_j(x,z)$ in Eqs. (10.1.6) and (10.1.7) for $j = 1$ and 3, multiply them with $F_j^*(y)$, and integrate over y to eliminate the y dependence completely. If we adopt the normalization scheme used earlier, the normalized coupled equations take the following form:

$$i\frac{\partial v}{\partial z} + r\frac{\partial^2 v}{\partial x^2} - v + wv^* = 0, \qquad (10.2.1)$$

$$i\sigma\frac{\partial w}{\partial z} - i\delta\frac{\partial w}{\partial x} + s\frac{\partial^2 w}{\partial x^2} - \alpha w + \frac{v^2}{2} = 0. \qquad (10.2.2)$$

These equations are written in such a way that they can be used for both spatial and temporal solitons. In the case of spatial solitons, we choose $r = s = 1$. In contrast, r and s can take negative values for temporal solitons, provided x is interpreted as a normalized time variable. The input pulse in this case is assumed to be confined in both transverse dimensions so that it does not diffract during propagation.

10.2.1 One-Dimensional Quadratic Solitons

It is easy to show why one should expect to find soliton-like solutions of Eqs. (10.2.1) and (10.2.2). Consider the limit $\alpha \gg 1$, valid for large positive values of the phase mismatch Δk. Equation (10.2.2) then has the approximate solution $w \approx v^2/2\alpha$. Substituting this solution in Eq. (10.2.1), we recover the standard NLS equation

$$i\frac{\partial v}{\partial z} + r\frac{\partial^2 v}{\partial x^2} - v + (2\alpha)^{-1}|v|^2 v = 0, \qquad (10.2.3)$$

except for the factor of $(2\alpha)^{-1}$ appearing in the nonlinear term. Even this factor can be removed by renormalizing v to $v' = (2\alpha)^{-1/2}v$. As seen in earlier chapters, the NLS equation supports stable bright solitons for $r = +1$ and dark solitons for $r = -1$.

The limit of large phase mismatch ($\alpha \gg 1$) is known as the *cascading limit* because in this limit two second-order $\chi^{(2)}$ effects are cascaded to produce an effective third-order, Kerr-like nonlinearity. The NLS equation (10.2.3) can be derived more rigorously for a material with quadratic nonlinearity by using the method of multiple scales [37]–[40]. It shows clearly the possibility of the existence of NLS-like solitons in $\chi^{(2)}$ media. Of course, solitons are expected to form only after the pump beam has propagated long enough to produce an SHG beam that interacts with the pump such that both fields propagate without changing their intensity distribution.

A quadratic soliton has two components, similar to the case of vector solitons of Chapter 9, except that these components have very different wavelengths and they are coupled parametrically. To find the intensity distribution of the two components, we can solve Eqs. (10.2.1) and (10.2.2) in the asymptotic limit. To simplify the analysis, we neglect the walk-off effects and set $\delta = 0$ in Eq. (10.2.2). Setting the z derivatives to zero in Eqs. (10.2.1) and (10.2.2), we obtain the following set of two coupled ordinary differential equations:

$$r\frac{d^2v}{dx^2} - v + wv^* = 0, \tag{10.2.4}$$

$$s\frac{d^2w}{dx^2} - \alpha w + \frac{v^2}{2} = 0. \tag{10.2.5}$$

These equations can be solved in the form of an asymptotic series whose terms decrease in magnitude for $\alpha \gg 1$ [17]. The solutions are different for $r = \pm 1$ because the two cases correspond to bright and dark quadratic solitons [41]. In the case of bright solitons ($r = +1$), the solution is given by

$$v(x) = 2\sqrt{\alpha}\,\text{sech}\,x\left(1 + \frac{2s}{\alpha}\tanh^2 x + \cdots\right), \tag{10.2.6}$$

$$w(x) = 2\,\text{sech}^2 x\left[1 + \frac{2s}{\alpha}(4 - 5\,\text{sech}^2 x) + \cdots\right]. \tag{10.2.7}$$

In the case of dark solitons ($r = -1$), the solution takes the form

$$v(\zeta) = \sqrt{2\alpha}\left[\tanh\zeta + \frac{s}{\alpha}\text{sech}^2\zeta(\zeta - \tanh\zeta) + \cdots\right], \tag{10.2.8}$$

$$w(\zeta) = \tanh^2\zeta + \frac{s}{\alpha}\text{sech}^2\zeta(2\zeta\tanh\zeta - 4 + 5\,\text{sech}^2\zeta) + \cdots, \tag{10.2.9}$$

where $\zeta = x/\sqrt{2}$. The leading term in the asymptotic series corresponds to the solution of the NLS equation (10.2.3).

Since the properties of Kerr solitons of Eq. (10.2.3) are well known, the existence of the asymptotic solutions suggests that for $\alpha \gg 1$ the SHG process in $\chi^{(2)}$ should evolve in such a way that both the pump and second-harmonic fields form a self-supporting pair of bright solitons for $r = +1$ and of dark solitons for $r = -1$. However, before reaching this conclusion we need to ensure that these solutions are stable against perturbations. Instability can result if a quadratic soliton emits radiation through its interaction with the linear waves. In the case of dark solitons, the solution can also become unstable because of a *parametric* modulation instability [16, 42, 43].

10.2.2 Bright Quadratic Solitons

In this section we focus on *spatial bright* quadratic solitons and set $r = s = +1$ in Eqs. (10.2.1) and (10.2.2). Such solitons display many interesting features that are also found for other types of non-Kerr nonlinearities. We show that an entire family of such solitons exists for different values of the parameter α and that the two-dimensional soliton can have a single hump or multiple humps. We discuss the properties of both types of quadratic solitons together with the experimental results.

Single-Hump Soliton Family

Bright quadratic solitons exist for Eqs. (10.2.1) and (10.2.2) when $r = 1$ and $\alpha > 0$ in the form of single-hump spatial profiles for $v(x)$ and $w(x)$ that maintain their shapes along the nonlinear medium. A family of such coupled parametric solitons was discovered in 1994 by solving Eqs. (10.2.1) and (10.2.2) numerically [25, 28]. Different members of the family correspond to different values of the phase-mismatch parameter α. Figure 10.1 shows two examples of such coupled solitons for $\alpha = 0.2$ and $\alpha = 10$. The entire family is characterized by the ratio w_m/v_m, where v_m and w_m are the peak amplitudes of the pump and SHG fields, respectively. Numerically calculated values of this ratio are shown in Figure 10.1 by a solid line for the whole family.

For $\alpha \gg 1$, the pump amplitude $v(x)$ is much larger than $w(x)$ because the SHG process is heavily phase mismatched. This case corresponds to the asymptotic solution given in Eqs. (10.2.6) and (10.2.7). Retaining only the dominant term, the solution for $\alpha \gg 1$ is given by

$$v(x) \approx 2\sqrt{\alpha}\,\mathrm{sech}(x), \qquad w(x) \approx 2\,\mathrm{sech}^2(x). \tag{10.2.10}$$

The peak-amplitude ratio $w_m/v_m = \alpha^{-1/2}$ in this specific case. The dashed line shows the approximation based on the asymptotic theory. As expected, the agreement is quite good for $\alpha \gg 1$. The solid circle in Figure 10.1 corresponds to the analytical solution first found by Karamzin and Sukhorukov [2] in 1974 for $\alpha = 1$ and given by

$$v(x) = \sqrt{2}\,w(x) = (3/\sqrt{2})\mathrm{sech}^2(x/2). \tag{10.2.11}$$

The variational method has been used to find an approximate analytical form of the quadratic solitons for any value of α [41, 44]. Such an approach requires an initial guess for the soliton shape. It provides a reasonable approximation for most characteristics of solitary waves but fails to reproduce the exact form of spatial variations in the tail region of the quadratic soliton. In order to improve the accuracy of the approximate solutions, Sukhorukov has suggested a semianalytical method based on the scaling properties of quadratic solitons [45]. It leads to the following approximate solution of Eqs. (10.2.4) and (10.2.5):

$$v(x) = v_m\,\mathrm{sech}^p(x/p), \qquad w(x) = w_m\,\mathrm{sech}^p(x/p), \tag{10.2.12}$$

where the parameters v_m, w_m, and p are found from the relations

$$v_m^2 = \frac{\alpha w_m^2}{w_m - 1}, \qquad p = \frac{1}{w_m - 1}, \qquad \alpha = \frac{4(w_m - 1)^3}{2 - w_m}. \tag{10.2.13}$$

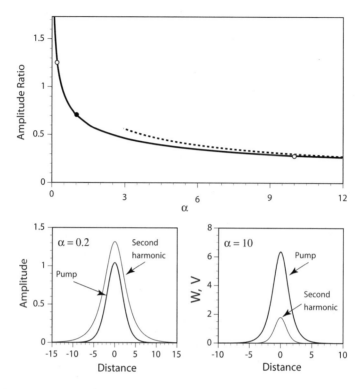

Figure 10.1: Peak ratio w_m/v_m as a function of α and the spatial profiles $v(x)$ and $w(x)$ of the soliton pair for two values of α denoted by open circles. The filled circle corresponds to the exact analytical solution found for $\alpha = 1$. The dashed curve shows the prediction of asymptotic theory. (After Ref. [16]; ©1995 Elsevier.)

For a given value of α, the parameter w_m satisfies a cubic polynomial. Once the amplitude w_m is determined from it, all other parameters are known for the soliton solution in Eq. (10.2.12). It is easy to show that $1 < w_m < 2$ for $0 < \alpha < \infty$. The parameter p can thus vary in the range $1 < p < \infty$. For $\alpha = 1$ and $p = 2$, and the peak values are given by $v_m = 3/\sqrt{2}$ and $w_m = 3/2$. The solution (10.2.12) then reduces to the exact analytical solution given in Eq. (10.2.11). Moreover, the solution (10.2.12) describes correctly the asymptotic solution in Eq. (10.2.10) for $\alpha \gg 1$.

The results shown in Figure 10.1 represent shape-preserving solutions of Eqs. (10.2.1) and (10.2.2). In practice, only a pump beam is launched into the $\chi^{(2)}$ medium. However, quadratic solitons can be generated from a rather broad class of initial conditions. Figure 10.2 shows an example of how the pump and SHG beams develop into a spatial soliton by plotting changes in the peak amplitudes v_m and w_m along the medium length. The incident pump beam is assumed to have a $\text{sech}^2(x)$ intensity profile, and the phase-matching condition is nearly satisfied. Because of diffraction, the pump beam first broadens as it generates the SHG beam. The growth of the SHG beam provides a mechanism through the cascaded nonlinearity for the pump to self-focus. After a few oscillations, during which the two beams diffract and self-focus, the amplitude and the

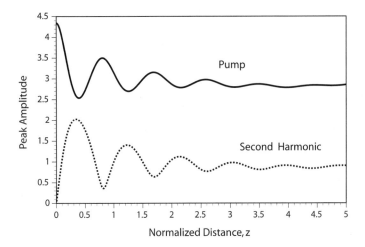

Figure 10.2: Formation of the quadratic soliton when a pump beam with $\mathrm{sech}^2(x)$ shape is launched. The evolution of the peak amplitudes for the pump and harmonics is shown along the medium length. (After Ref. [16]; ©1995 Elsevier.)

width of both beams stop changing, and a pair of bright solitons is formed. This kind of behavior is possible only because of the existence of a continuous family of stable quadratic solitons.

The stability issue is of paramount importance for quadratic solitons. Linear stability of bright quadratic solitons associated with Eqs. (10.2.1) and (10.2.2) was first studied in 1995 using asymptotic bifurcation analysis (near the marginal stability point) of the corresponding linear problem [46]. The instability region was found in Ref. [47] using a Hamiltonian-based approach. The Vakhitov–Kolokolov stability criterion developed in Chapter 2 applies to quadratic solitons as well [46]. This criterion shows that the stability of parametric solitons depends on the numerical values of the parameters α and σ. In the cascading limit ($\alpha \gg 1$), quadratic solitons are always stable, but they become unstable as $\alpha \to 0$. For example, the quadratic soliton shown in Figure 10.1 for $\alpha = 0.2$ is unstable in the spatial soliton regime ($\sigma = 2$). However, parametric solitons become stable even for $\alpha \sim 1$ when σ is smaller than a critical value σ_{cr}. For a fixed value of σ, instability occurs for $\alpha < \alpha_{cr}$. In the spatial case ($\sigma = 2$), $\alpha_{cr} \approx 0.21$.

Similar to the case of of non-Kerr spatial solitons studied in Chapter 2, stable parametric solitons may exhibit amplitude oscillations that persist over long lengths (see Figure 2 in Ref. [26]), unlike the case shown in Figure 10.2. Such persistent oscillations result from the excitation of a soliton internal mode [48]. Such modes correspond to discrete eigenvalues of the linear problem associated with the parametric solitons and are a special feature of solitary waves in nonintegrable models (see Chapter 2). A detailed analysis shows that long-lasting oscillations, often observed in numerical simulations, occur when the amplitude of the pump wave exceeds a threshold value, and they disappear in the limit of large α [49]. However, such oscillations always persist when the phase-matching condition is exactly satisfied ($\Delta k = 0$).

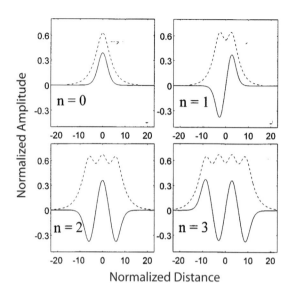

Figure 10.3: Examples of multihump quadratic solitons. The solid and dashed curves show the amplitude profiles for the pump and second harmonic, respectively, and n indicates the number of nodes in the pump field. (After Ref. [32]; ©1997 OSA.)

Multihump Quadratic Solitons

In addition to the localized solutions with a single peak or hump, numerical simulations reveal the existence of bright quadratic solitons with two or more humps for all values of the phase-mismatch parameter α, and the rigorous proof of their existence has been presented [50]. A multihump bright soliton can be thought of as the bound state of several single-hump solitons [29]–[32]. The physical mechanism behind the existence of such solitons is similar to that discussed in Chapter 9 for the solitons of coupled NLS equations and is related to the higher-order modes of an effective soliton-induced waveguide. Several examples of multihump solitons are shown in Figure 10.3 for a fixed value of α. As seen there, the pump amplitude (solid curve) vanishes at several points, similar to the case of higher-order modes associated with a waveguide. Numerical simulations indicate that all multihump quadratic solitons are unstable [50]–[52] and split into stable one-hump solitons or disintegrate completely. The latter scenario occurs for sufficiently small values of α, for which stable single-hump solitons do not exist. These features are similar to those of other coupled-field vector solitons.

Temporal quadratic solitons have some unique features related to the parameters r and s appearing in Eqs. (10.2.4) and 10.2.5). Both of these parameters can be positive or negative depending on the dispersion characteristics of the $\chi^{(2)}$ medium. We choose $r = 1$, corresponding to anomalous GVD at the pump frequency. Parameter s can be negative if GVD is normal at the second-harmonic frequency. In spite of the fact that the effective NLS equation (10.2.3) does not depend in the cascading limit on the value of parameter s, the soliton-like solutions of Eqs. (10.2.4) and 10.2.5) can be quite different for $s = \pm 1$. The case $s = +1$ applies for temporal solitons when the dispersion is anomalous for both the pump and second-harmonic fields. In the other case, $s = -1$, the two fields experience opposite dispersion. It turns out that in this case single-hump quadratic solitons do not exist as a stable entity because they are in resonance with at least one branch of the linear waves [16]. As a result, they loose energy by emitting radiation in the form of dispersive waves and disintegrate.

Although stable single-hump solitons do not exist for $s = -1$ because of such a resonance with linear waves, a numerical analysis of Eqs. (10.2.4) and (10.2.5) shows evidence for the existence of two-hump solitons. Such soliton states are similar to *embedded solitons* discussed in Chapter 2. In the $\chi^{(2)}$ case, each embedded soliton can be considered a radiationless bound state of two single-hump solitons. For such bound states, linear radiation is suppressed in the outside region, but it exists between the solitons in the form of a trapped standing wave; the number of the nodes defines the soliton order. One instance of embedded solitons has been found in an explicit analytical form in the case $r = +1$ and $s = -1$ [24]. This solution of Eqs. (10.2.4) and 10.2.5) exists for $\alpha = 2$ and has the form

$$v(x) = 6\sqrt{2}\tanh(x)\operatorname{sech}(x), \qquad w(x) = 6\operatorname{sech}^2(x). \qquad (10.2.14)$$

It turns out that the solution (10.2.14) is the simplest member of a discrete set of two-soliton embedded bound states existing because of trapped radiation [16]. As is the case for the majority of embedded solitons, discrete solutions of this type are found to be unstable as well.

10.2.3 Walking Quadratic Solitons

Equations (10.2.1) and (10.2.2) contain a walk-off parameter δ that has been taken to be zero so far. In practice, the SHG process in an anisotropic media often leads to *spatial separation* of the SHG beam from the pump because of different propagation directions of the energy and phase fronts. This effect is governed by the δ term in Eq. (10.2.2). The beam walk-off is always present in experiments that employ birefringence or temperature tuning for realizing phase matching [7].

The effect of spatial walk-off on quadratic solitons was studied as early as 1978 in the case of a three-wave parametric interaction inside a planar waveguide [4]. Much later, a detailed analysis of the SHG process in the presence of spatial walk-off (or temporal walk-off in the case of temporal solitons) was carried out [53]–[55], and the existence of the so-called *walking solitons* was predicted. Such solitons form when the pump and SHG beams lock together in such a way that they travel together at a certain angle to the original pump direction as they propagate inside the $\chi^{(2)}$ medium. Walking solitons require a nontrivial phase-front tilt (defined as the transverse derivative of the phase) that is analogous to the chirping of temporal solitons. Figure 10.4 shows two examples of walking solitons. Such "chirped" quadratic solitons are stable only below a certain value of the walk-off parameter [55]. Generalizations of such solitons to the case of type II SHG and three-wave parametric interaction have also been explored in the literature [36].

All quadratic solitons, including mutihump solitons, can be generalized to include the effect of walk-off [56]. Although a linear stability analysis reveals that chirped

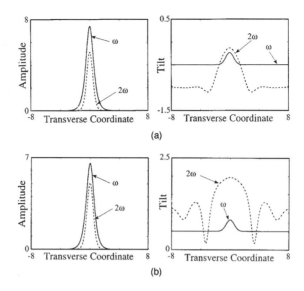

Figure 10.4: Examples of the walking solitons for two different values of the walk-off parameter. The amplitude (left column)and the wave-front tilt (right column) of solitons are shown in each case. (After Ref. [54]; ©1998 OSA.)

walking solitons are always unstable, the instability growth rate can be very small for certain values of parameters. Moreover, such two-hump solitons exhibit spontaneous symmetry breaking that can be useful for all-optical switching [56]. Other decay scenarios explored numerically include soliton splitting and the transformation of a two-hump soliton into a single hump because of the annihilation of the other hump.

The concept of the so-called *guided-center walking soliton* was introduced by Torner [57], who considered self-trapping of light in a periodic $\chi^{(2)}$ medium with large walk-off and analyzed the formation of solitons in quasi-phase-matched structures (see Section 10.5). In each domain of such a structure, the walk-off can be far too large to permit soliton formation, but an average guided-center soliton can still form such that its parameters vary slowly in a periodic manner.

10.2.4   Experimental Observations

In contrast to the case of optical fibers, where propagation distances can be in hundreds of dispersion lengths, spatial soliton experiments are typically limited to propagation over distances of less than five diffraction lengths. This distance is sufficient to form a self-trapped beam, but it is too short to follow soliton collisions to their completion. For the formation of quadratic solitons at reasonable pump power levels, it is necessary to be close to satisfying the phase-matching condition. For this reason, the nonlinear materials used for observing quadratic solitons are the same as those used for SHG.

Spatial quadratic solitons in waveguide geometry were first observed in a 1996 experiment [58] in which a 5-cm-long LiNbO$_3$ waveguide was used and phase matching was realized by controlling the temperature of the waveguide [59]. A relatively con-

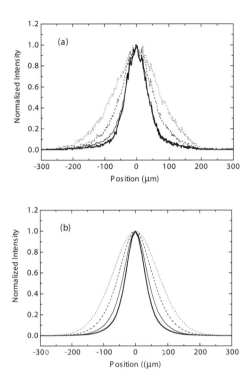

Figure 10.5: (a) Output-pump intensity profiles at a temperature of 335°C. The pump beam narrows and forms a spatial soliton as its peak power is increased to 0.32, 0.76, 1.5, and 1.9 kW. (b) Results of numerical simulations under the same conditions. (After Ref. [59].)

stant temperature over most of the waveguide resulted in an SHG conversion efficiency of more than 50%. At input pump powers in excess of 1 kW (from a 1320-nm, mode-locked, Q-switched Nd:YAG laser), self-trapping of the pump beam was observed. It was induced by a weak second-harmonic beam in the positive phase-mismatch region. The SHG process was weak enough to be in the cascading limit, and it acted as an effective third-order nonlinear process. Figure 10.5 shows the observed spatial profiles of the pump beam at the waveguide output at peak powers in the range 0.3–1.9 kW together with the results of numerical simulations. The pump beam diffracts at low power levels, as expected. However, at high peak powers the cascaded self-focusing nonlinearity counteracts diffraction and leads to the formation of a spatial soliton at power levels above 1.5 kW. The soliton formation was observed even close to the phase-matching condition, but with large pump depletion.

Walking quadratic solitons have also been observed in an experiment making use of type I phase matching. For a TM-polarized pump beam, the group and phase velocities are always *collinear* inside a LiNbO$_3$ waveguide, but the TE-polarized second-harmonic beam propagates at some angle to the x axis. The walking solitons are formed when the pump and second-harmonic beams are locked together and have the same group-velocity direction, in between the direction of propagation and the direction of walk-off. This important feature of quadratic walking solitons was observed in a 1999 experiment [60] by studying the SHG process under different phase-matching conditions at different pump powers. More specifically, the nonlinear translation of the output position of the pump beam was observed as a function of input pump power, thereby demonstrating the mutual locking between the pump and SHG beams.

10.2.5 Soliton Collisions

Since the quadratic solitons are described by a set of two or three equations that are not integrable by the inverse scattering transform, their collisions are not elastic (except in the cascading limit). As a result, strong inelastic effects such as emission of radiation and phase sensitivity are expected to occur, similar to the case of other non-Kerr solitons. Indeed, numerical simulations of the underlying equations reveal that such collisions depend strongly on the initial relative phase θ between the two colliding solitons [61]–[64]. In the cascading limit in which the effective NLS equation is valid (i.e., for large positive phase mismatch), collision behavior is similar to that of Kerr solitons. For smaller values of α, the behavior is quite different, depending on the value of θ. When $\theta = 0$, two solitons of equal amplitude either fuse together or pass through each other almost elastically, depending on the initial conditions. In contrast, they repel each other for $\theta = \pi/2$ [61]. For intermediate values of the relative phase, strongly inelastic symmetry-breaking behavior is observed. More specifically, two identical solitons with equal amplitudes emerge after collision with different amplitudes [61]. When the solitons merge into a single soliton, persistent oscillations in the soliton amplitude are observed; these oscillations are associated with the excitation of an internal mode of the soliton created as a result of the collision [62].

Collisions of quadratic solitons were observed in a 1997 experiment using a LiNbO$_3$ planar waveguide and type I phase matching [65]. The results showed that in the cascading regime, quadratic solitons interacted in an almost elastic fashion, as expected from the NLS equation [61]. However, inelastic behavior was observed outside the cascading regime. In the experiment, two pump beams were launched into a LiNbO$_3$ waveguide in both the parallel and cross configurations. Figure 10.6 shows the results of their interaction in the parallel case; the relative phase between the two launched beams was controlled by two glass wedges. For two in-phase pump beams, two solitons fuse together, as shown in part (a). In contrast, they passed right through each other when they were out of phase, as seen in part (c). For other values of the relative phase difference, collision was inelastic, with an energy exchange between the two solitons [see parts (b) and (d)]. The observed behavior is in agreement with the numerical simulations and with the general picture of soliton interaction in nonintegrable nonlinear models. In fact, the overall behavior resembles closely that of solitons in saturable Kerr media.

When the phase mismatch is changed from positive to negative (i.e., $\alpha < 2\sigma$), the soliton interaction dynamics also change, because the solitons are far from the NLS limit. When the relative phase difference between the two colliding solitons vanishes, they still fuse, but the change from an attractive ($\theta = 0$) to a repulsive ($\theta = \pi/2$) force between the colliding solitons is more gradual and less sensitive to phase variations [61]. Small values of the walk-off parameter do not change these results. However,

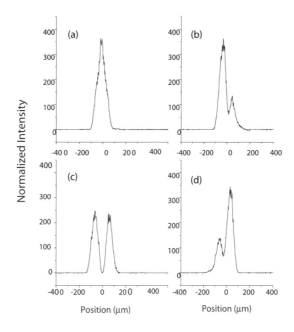

Figure 10.6: Output spatial profiles when the two pump beams are launched in parallel and undergo collision inside the nonlinear waveguide. The relative phase difference between the two beams is (a) 0, (b) $\pi/2$, (c) π, and (d) $3\pi/2$. (After Ref. [65]; ©1997 OSA.)

new features occur for large values [64]. For example, for the same values of soliton parameters, the character of soliton interaction changes from attractive to repulsive as the walk-off parameter $\rho_{2\omega}$ crosses a critical value of 0.53 [64]. If the walk-off angle is small, the two in-phase solitons fuse. However, for a relatively large walk-off ($\rho_{2\omega} = 0.8$), two solitons survive the collision with some energy exchange. This behavior can be understood if we note that the relative phase between the solitons changes during collision in the presence of walk-off.

10.2.6 Quadratic Dark Solitons

In the cascading limit in which the SHG process is well described by an effective NLS equation, one can expect the dark solitons to form for $r = -1$, $s = +1$, because these values correspond to a defocusing-type cubic nonlinearity. The numerical results confirm this expectation [18] and indicate that stable (radiationless) dark quadratic solitons exist for $r = -s = -1$ for Eqs. (10.2.4) and (10.2.5). In fact, a continuous family of quadratic dark solitons exists for $0 < \alpha < \infty$. In the interval $0 < \alpha < 8$, such solitons exhibit oscillatory tails. Two examples of quadratic dark solitons are shown in Figure 10.7 for $\alpha = 1$ and 10.

In the cascading limit ($\alpha \gg 1$), the solution can be written in the form of an asymptotic series given in Eqs. (10.2.8) and (10.2.9). The asymptotic expansion fails to reproduce the oscillatory tails for $\alpha < 8$. The entire family of dark solitons can be char-

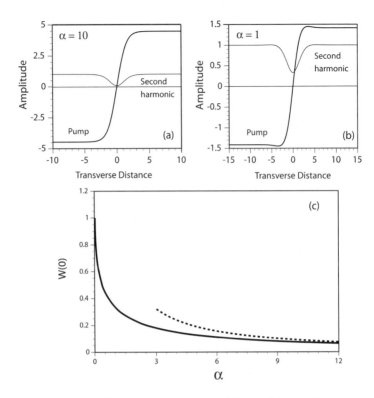

Figure 10.7: Amplitude profiles of the pump and second-harmonic beams for two quadratic dark solitons for (a) $\alpha = 10$ and (b) $\alpha = 10$. (c) $w(0)$ at the dip position as a function of α. The dashed curve shows the cascading limit. (After Ref. [16]; ©1995 Elsevier.)

acterized by the minimum amplitude of the second harmonic $w(0)$ at the location of the dip. The dependence of $w(0)$ on α is shown in Figure 10.7(c). For large values of α, it approaches the asymptotic dashed curve $w(0) \approx 1/\alpha$ obtained in the cascading limit. It should be stressed that quadratic dark solitons have a modulationly stable background only in a certain range of the parameters α and σ. For $\sigma \sim 1$, the soliton background is modulationly stable for $2 < \alpha < \infty$.

Because of the existence of the radiationless oscillating tails, a dark soliton can trap another to form a bound state with two intensity dips [17]. Such twin-hole dark solitons can also form for other non-Kerr solitons, but the $\chi^{(2)}$-based multihole dark solitons constitute the first example of a continuous family of the *stable bound states* of dark solitons. As $\alpha \to 8$, the distance between the neighboring dark solitons increases to infinity, and such bound states cease to exist for $\sigma > 8$. Another family of quadratic dark solitons has been discovered for $r = -s = +1$, but it exists only for $0 < \alpha < 2$ [31]. These dark solitons also possess radiationless oscillating tails and thus can form bound states.

Similar to the case of bright quadratic solitons, radiationless dark solitons do not exist when $r = s = -1$ in Eqs. (10.2.1) and (10.2.2). Only discrete sets of embedded solitons, in the form of radiationless bound states of two or more solitons, have been found [17]. An exact analytic solution exists for $\alpha = 1$ and is given by [66]

$$v(x) = \sqrt{2}\,w(x) = \sqrt{2}\left[1 - \frac{3}{2}\text{sech}^2\left(\frac{x}{2}\right)\right].$$ (10.2.15)

It represents a two-soliton bound state that exists as an embedded dark soliton at a fixed value of α. All such bound states are unstable because of a *parametric* modulation instability. This instability can be suppressed in a medium with two competing nonlinearities [67], and only in this case do dark solitons become stable.

10.2.7 Three-Wave Quadratic Solitons

As discussed in Section 10.1, the most general parametric interaction inside a $\chi^{(2)}$ medium involves the mixing of three waves governed by the set of three coupled equations (10.1.3)–(10.1.5). The same set of equations can be used in the case of type II SHG in which two orthogonally polarized pump beams produce a second-harmonic beam. Starting from the 1975 analysis of three-wave mixing [3], this model was studied in detail in 1981 in the case of perfect phase matching [5]. Since then, several studies have found the families of parametric solitons in which the three waves form bright or dark solitons that are coupled in such a way that they mutually support each other to preserve their shapes [68]–[76].

The three-wave solitons formed in the case of type II SHG behave similarly to their two-wave counterparts. However, some of their properties exhibit a more complicated behavior because of the existence of an additional parameter—the power imbalance between the two polarization components of the fundamental beam. This power imbalance allows for new effects, such as polarization switching [73]. Also, because of the existence of two independent parameters, the linear stability analysis of three-wave solitons goes beyond the standard Vakhitov–Kolokolov stability criterion. Similar to the case of two incoherently coupled NLS equations (see Chapter 9), the marginal stability condition is governed by the following more general determinant criterion:

$$\frac{\partial P_v}{\partial \beta_v}\frac{\partial P_u}{\partial \beta_u} - \frac{\partial P_v}{\partial \beta_u}\frac{\partial P_u}{\partial \beta_v} = 0,$$ (10.2.16)

where β_v and β_u are the propagation constants and P_v and P_u are the powers for the two orthogonally polarized pump beams.

10.3 Quadratic Solitons in Bulk Media

As discussed in Chapters 6 and 7, multidimensional solitons are generally unstable in a bulk Kerr medium, because the cubic NLS equation does not permit stable solutions in $(2 + 1)$ or higher dimensions. The instability manifests itself through a catastrophic collapse as the beam width shrinks to zero at a finite propagation distance. However, several different physical mechanisms can suppress or even eliminate such collapse-type instabilities [77]. The beam coupling induced through parametric interaction plays an important role among such instability-suppression mechanisms. Indeed, in a $\chi^{(2)}$ medium, parametric interaction leads to stable quadratic solitons even in higher dimensions [5, 78].

Historically, two-dimensional quadratic solitons were discovered first for the type I SHG process [3] and then for the three-wave mixing process in the special case of perfect phase matching [5]. The possibility of stable noncollapsing solitons of higher dimensions (e.g., light bullets) was noted as early as 1981. Hayata and Koshiba rediscovered in 1993 the exact analytical solution of Karamzin and Sukhorukov [2] in the one-dimensional case [21]. They also applied a Hartree-like approach for constructing an approximate multidimensional solitary wave of radial symmetry for a specific value of the phase-matching parameter. Numerical simulations [33] have revealed the formation of two-dimensional quadratic solitons at a nonzero value of the phase-mismatch parameter even in the presence of beam walk-off [79, 80]. Later studies focused on the collisions and steering of such solitons [71]–[73].

A two-parameter family of stable two-dimensional quadratic solitons was discovered in 1995 numerically [34, 81]; the two parameters were the phase mismatch and the input pump intensity. The approximate solution of Ref. 20 [21] turned out to be just a single member of this soliton family. An approximate form of two-dimensional quadratic solitons can be found using a variational approach [41]. Such solitons can also form in the presence of competing cubic and quadratic nonlinearities [82] or in a self-defocusing Kerr medium [83, 84]. They can even exist in the presence of walk-off effects (walking solitons) or dispersive effects, leading to the formation of light bullets [86]–[87].

10.3.1   Two-Dimensional Spatial Solitons

In the case of a type I SHG process, the two-dimensional spatial quadratic solitons are the shape-preserving solutions of Eqs. (10.1.10) and (10.1.11) with $r = s = 0$. Such solutions can be found by setting the z derivatives to zero. For simplicity, we focus on the circularly symmetric solitons and use the notation $v(x,y,z) \equiv V(r)$ and $w(x,y,z) \equiv W(r)$, where $r \equiv \sqrt{x^2 + y^2}$ is the radial coordinate. If we neglect the walk-off effects by setting $\delta = 0$ and write the Laplacian operator in the cylindrical coordinates, the radial profile of quadratic solitons is obtained by solving the following two coupled ordinary differential equations:

$$\frac{d^2V}{dr^2} + \frac{1}{r}\frac{dV}{dr} - V + WV = 0, \tag{10.3.1}$$

$$\frac{d^2W}{dr^2} + \frac{1}{r}\frac{dW}{dr} - \alpha W + \frac{V^2}{2} = 0. \tag{10.3.2}$$

These equations can only be solved numerically in general. However, considerable insight can be gained by recasting them as Newtonian equations of motion for a particle moving in the effective potential $U(W,V) = \frac{1}{2}(V^2W - \alpha W^2 - V^2)$. In this mechanical analogy, the first-order derivative in Eqs. (10.3.1) and (10.3.2) plays the role of "anisotropic" dissipation. The solutions of these equations correspond to special (separatrix) trajectories in the four-dimensional phase space $(V, W, dV/dr, dW/dr)$. These trajectories start at the point $(V_m, W_m, 0, 0)$ at $r = 0$ and approach asymptotically the

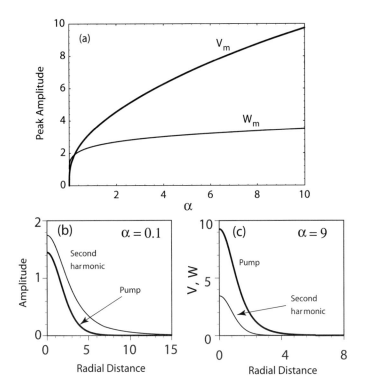

Figure 10.8: (a) Peak amplitudes of the pump and second-harmonic beams as a function of the phase-mismatch parameter α. Beam profiles for (b) $\alpha = 0.1$ and (c) $\alpha = 9$ are also shown. (After Ref. [34]; ©1995 APS.)

point $(0,0,0,0)$ for $r \to \infty$. The separatrix trajectories and the corresponding soliton profiles have been found numerically by employing the shooting technique [34, 47]. The results are summarized in Figure 10.8 by plotting the peak amplitudes V_m and W_m as a function of the phase-mismatch parameter α. The radial profiles are also shown for two members of this soliton family.

Approximate analytical expressions for the beam profiles can be obtained for any α using the variational method with the Lagrangian density

$$\mathcal{L} = \tfrac{1}{2}[(\nabla_\perp W)^2 + (\nabla_\perp V)^2 + W^2 + \alpha V^2 - VW^2]. \tag{10.3.3}$$

To apply the variational approach, we need to assume a specific shape for the two radial profiles. A Gaussian form is often used for this purpose [41]:

$$W(r) = Ae^{-\kappa r^2}, \qquad V(r) = Be^{-\gamma r^2}, \tag{10.3.4}$$

where A, B, κ, and γ are functions of α. Substituting this ansatz into Eq. (10.3.3), we obtain a set of four equations for A, B, κ, and γ from the variational problem. These equations can easily be solved. The solution for κ is obtained from the cubic polynomial

$$32\kappa^3 + 2\alpha\kappa - \alpha = 0. \tag{10.3.5}$$

The other parameters are related to κ as

$$\gamma = 4\kappa^2, \quad B = (1+2\kappa)^2, \quad A = \sqrt{\alpha/8}(1+2\kappa)^2/\kappa. \qquad (10.3.6)$$

Once Eq. (10.3.5) is solved to find κ, all soliton parameters are known for any value of α. The variational results are in good agreement with the numerical solutions [41].

Two-dimensional quadratic solitons can also form in the presence of spatial walk-off (or relative group-velocity mismatch in the case of temporal solitons). Mihalache et al. found numerically a two-parameter family of such solitary waves with a nontrivial phase-front tilt [85, 87]. A new feature is that the intensity profile of such solitons is not circular but *elliptical*; the asymmetry appears to be related to the presence of walk-off. Similar to the case of one-dimensional walking solitons [53], two-dimensional walking solitons propagate at a certain angle to the direction selected by the walk-off effect. In contrast to one-dimensional walking solitons, this family of walking solitons requires the solution of a set of two partial differential equations—an evidently complicated task. The numerical results indicate a nonmonotonic dependence of the soliton power on the propagation constant. This feature indicates that such walking solitons may become unstable. The instability domain depends on two parameters, the phase mismatch α and the walk-off parameter δ. A rigorous stability analysis of this class of solitary waves still remains to be addressed.

Another extension of two-dimensional quadratic solitons consists of including the dispersive effects for pump pulses and studying the formation of light bullets. The variational method has been used to analyze such quadratic solitons in the form of light bullets [88]–[90] by solving Eqs. (10.1.10) and (10.1.11) approximately. Sukhorukov has developed another analytical approach, based on the scaling properties of quadratic solitons [45]; it provides an even better approximation for the radial profiles of two-dimensional quadratic solitons.

10.3.2   Experimental Results

The formation of stable spatial solitons in a bulk nonlinear medium requires some form of saturating nonlinearity. It was shown experimentally in 1995 that $\chi^{(2)}$-induced parametric coupling can provide saturation and thus lead to the formation of stable quadratic solitons by balancing simultaneously the effects for diffraction and walk-off [91]–[93]. In these experiments, Gaussian-shaped pulses of 35-ps duration, obtained from a 1064-nm Nd:YAG laser operating at a repetition rate of 10 Hz, were focused onto a 1-cm-thick KTP crystal to form a 20- or 40-μm spot size, corresponding to a diffraction length of about 0.5 and 2 mm, respectively. For the 20-μm beam diameter, the crystal length was long enough (20 diffraction lengths) for soliton formation. Figure 10.9 shows the output pump-beam profiles under such conditions at two different peak intensities. At low intensities, the beam diffracts as expected in the absence of nonlinear effects. However, above a threshold intensity level, the output beam stops diffracting and can even be narrower than the input beam (right column). A clean, cylindrically symmetric beam profile was observed for both the pump and SHG beams. Numerical simulations show how the beam reshapes inside the crystal during the SHG process. The experiment employed type II phase matching. The energy of pump pulses was depleted by more than 50% close to perfect phase matching.

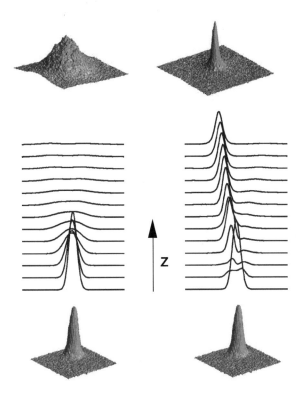

Figure 10.9: Input and output pump-beam profiles for peak intensities of 100 (left) and 10 GW/cm^2 (right) observed during SHG inside a 1-cm-long KTP crystal. The results of numerical simulations are shown in the middle. Spatial walk-off is clearly seen at high intensities. (After Ref. [91]; ©1995 APS.)

Type- II phase matching uses two orthogonally polarized pumps and involves parametric interaction among three waves (two pump beams and one SHG beam) governed by Eqs. (10.1.3)–(10.1.5). In these equations, ρ_ω and $\rho_{2\omega}$ are, respectively, the walk-off angles of the pump and SHG beams propagating as extraordinary beams. These angles had, respectively, values of $0.19°$ and $0.28°$ for the KTP crustal used. The nonlinear coupling coefficient was calculated to be 6 cm^{-1} at an input intensity of 1 GW/cm^2. Linear as well as nonlinear absorption can be neglected at the two wavelengths involved in the SHG process for a 1-cm-long crystal. In the strong coupling regime, the three-wave mixing process is quite different from the case of cubic nonlinearity governed by the cubic NLS equation. As seen in Figure 10.9, the numerical predictions are in excellent agreement with the experimental measurements of the beam waist, both qualitatively and quantitatively.

Figure 10.10(a) shows how the pump-beam width changes with the peak intensity close to phase matching. Once the intensity exceeds 10 GW/cm^2, the beam waist becomes smaller than its input value of 20 μm and gets locked to a size of 12.5 μm for intensity levels exceeding 80 GW/cm^2, showing clearly the formation of a quadratic

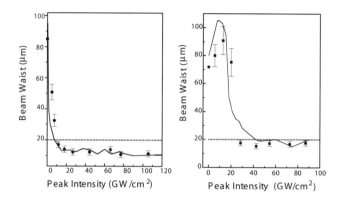

Figure 10.10: (a) Measured beam width as a function of peak intensity close to phase matching for an input pump beam with 20-μm waist. (b) Same as (a) but with negative phase mismatch ($\Delta kL = -5\pi$) resulting in self-defocusing at low input powers. (After Ref. [91]; ©1995 APS.)

soliton. The same evolution pattern is observed on the self-focusing side of phase matching ($\Delta k > 0$), except that the threshold value of the pump intensity is reduced. In contrast, a very different behavior is observed for *negative* values of phase matching ($\Delta k < 0$). Figure 10.10(b) shows changes in the output beam waist with pump intensity for $\Delta kL = -5\pi$. The beam waist first increases to values as high as 100 μm because of the self-defocusing occurring under such conditions. However, it begins to decrease rapidly when the intensity exceeds 15 GW/cm^2, and a quadratic soliton of about 16-μm diameter is formed at intensity levels exceeding 30 GW/cm^2. This behavior is attributed to waveguiding induced by the parametric gain. At large input peak intensities, the growth of second-harmonic field depletes the two pump fields. Eventually, energy from the SHG beam is transferred to pump through down-conversion. The resulting parametric gain guides the two pump fields, which in turn trap the second-harmonic field.

The vectorial nature of the type II SHG process in a KTP crystal can be used for all-optical switching [71]–[73]. If the pump intensities I_o and I_e (for the ordinary and extraordinary waves, respectively) are not the same at the input, a phenomenon known as *soliton dragging* can shift the position of the quadratic soliton, and this shift can be used for all-optical switching. For example, if the ordinary beam is more intense than the extraordinary, the soliton propagation direction is "dragged" toward the ordinary pump beam. Figure 10.11 shows the numerical predictions for the transmission through an aperture placed at the output of a 1-cm-long KTP crystal. The power transmitted through the aperture drops from 45% to below 5% as the imbalance fraction, defined as $(I_e - I_o)/I_o$, turns from negative to positive. This change is due to the lateral displacement of the quadratic soliton. A sharp steering is expected as a function of pump-intensity imbalance because the soliton position shifts by as much as 60 μm close to a specific value of the imbalance fraction. This behavior was observed experimentally by introducing a half-wave plate in the path of the input pump beam for controlling the imbalance fraction. When the input was biased toward the ordinary polarization, the quadratic soliton propagated along a direction perpendicular to the in-

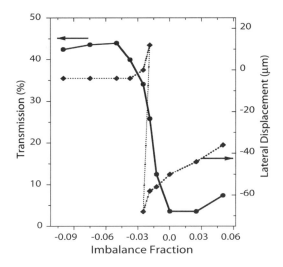

Figure 10.11: Fraction of the power transmitted through a narrow aperture as a function of the intensity imbalance between the two orthogonally polarized pump beams. The shift in the position of the pump-beam waist is shown by a dotted line. (After Ref. [93]; ©1996 AIP.)

put face of the crystal. With the polarization shifted toward the extraordinary side, the soliton propagated at an angle close to the walk-off angle. For a certain position of the half-wave plate, two solitons were observed simultaneously at the output of the crystal.

Most of the experiments on quadratic solitons have required input intensities in excess of 10 GW/cm^2 because the second-order nonlinearities are typically $\sim$1 pm/V. The development of the quasi-phase-matching (QPM) technique for parametric processes has resulted in bulk LiNbO$_3$ samples with effective nonlinearities of about 17 pm/V. Such devices were used in a 1999 experiment in which the use of type I phase matching resulted in the formation of quadratic solitons at pump intensities of only 1 GW/cm^2 at a wavelength of 1064 nm [94]. This geometry has the added advantage that *no walk-off* occurs between the pump and second-harmonic beams.

In another experiment [95], the formation of quadratic solitons was observed in a more general three-wavelength parametric process. A 15-mm-long lithium triborate (LBO) crystal was used as a parametric oscillator. It was pumped by 1.5-ps pulses at a wavelength of 527 nm. As the signal and idler waves grew from quantum noise, their frequencies ω_s and ω_c were set by the phase-matching condition and the energy conservation requirement $2\omega_p = \omega_s + \omega_c$, where ω_p is the pump frequency. When the pump-pulse energy exceeded 1 mJ at an input pump-beam diameter of 57 μm, the output beam reduced to a spot size of only 12 μm. From the numerical simulations based on Eqs. (10.1.3)–(10.1.5) and the experimental results, one can infer the formation of three-wave parametric solitons resulting from the mutual trapping of three beams.

In a related experiment [96], a KTP crystal was used to form a degenerate parametric amplifier for which $\omega_s = \omega_c = \omega_p/2$. This case has been studied theoretically in Ref. [73]. A weak seeding beam at the signal frequency was used in this experiment to initiate the three-wave parametric mixing. The effects of detuning from the

Figure 10.12: Photographs showing the formation of an array of quadratic solitons through the onset of transverse modulation instability. (a) Elliptical input beam, (b) output beam at 48 GW/cm^2, and (c) output beam at 57 GW/cm^2. (After Ref. [101]; ©1997 APS.)

exact phase-matching condition were investigated in detail. An interesting result was that the power of the quadratic soliton associated with the output signal was constant over five orders of magnitude of the input signal. This result is a consequence of a high parametric gain combined with the eigenmode properties of solitons, determined in this case solely by the intensity of the pump beam.

10.3.3 Transverse Modulation Instability

As discussed in earlier chapters, the formation of optical solitons has a strong connection with the phenomenon of modulation instability [77]. Transverse modulation instability is associated with the breakup of an optical beam in both transverse dimensions, forming multiple spatial filaments. For quadratic solitons, such an instability was first studied theoretically [97]–[100]. It has also been observed experimentally using a 1-cm-long KTP crystal [101]. The pump beam used in this SHG experiment had an highly elliptical shape, shown in Figure 10.12(a). When the input pump intensity approached a critical value, the output beam evolved into a single soliton beam [102]. At an input intensity level of 48 GW/cm^2, which exceeded this critical value by a factor of more than 4, an array of four solitons, seen in Figure 10.12(b), was observed such that each soliton had a nearly circular shape [101]. The number of solitons increased to six when the input intensity was increased to 57 GW/cm^2. The generation of such an array of quadratic solitons can be understood using a simple physical argument. The transverse instability breaks the beam into multiple filaments, which in turn undergo self-focusing to form multiple solitons. A part of the energy is radiated away during this evolution, but each two-dimensional soliton is a stable entity.

The situation is somewhat different in a slab waveguide, because the size of the beams in one transverse dimension is fixed by the waveguide. The input pump beam again breaks up into multiple filaments through modulation instability. However, the filaments do not grow to form quadratic solitons because they are unable to change their size in the direction fixed by the waveguide. For samples of practical length (a few diffraction lengths), no solitons are formed. In fact, strong interaction between the multiple filaments complicates the interpretation of the observed output considerably.

In a 2000 experiment [103], a pump beam, in the form of a train of ultrashort pulses, was launched into a BBO crystal cut for type I phase matching, a configuration similar to that used in an earlier experiment [104]. The crystal was either 17 or 25 mm long. Because of the presence of the group-velocity dispersion, a *spatiotemporal* instability was observed to lead to quadratic solitons and their intensity-dependent filamentation. The physics of soliton formation is somewhat similar to that observed for KTP crystals [101]. The main difference is that the quadratic solitons were also localized in the time domain, resembling light bullets.

Quadratic solitons can also be generated from other types of instabilities [77]. For example, vortices embedded into an input pump beam (with sufficient intensity to produce quadratic solitons) can undergo a different form of transverse azimuthal instability that leads to the formation of multiple quadratic solitons [105]. Typically, three quadratic solitons were formed in the experiment when the ring-like vortex beam carrying a phase singularity decayed because of the instability.

10.3.4 Soliton Scattering and Fusion

The set of equations (10.1.10) and (10.1.11) is not invariant under a Galilean transformation. However, when $\sigma = 2$ (i.e., for spatial solitons), these equations become invariant under such a transformation. It is then possible to introduce quadratic solitons moving at a constant velocity through the following gauge transformation:

$$V(\mathbf{r}, z) = V_s(\mathbf{r} - \mathbf{C}z) \exp[i(\mathbf{C} \cdot \mathbf{r})/2 - iC^2 z/4], \tag{10.3.7}$$

$$W(\mathbf{r}, z) = W_s(\mathbf{r} - \mathbf{C}z) \exp[i(\mathbf{C} \cdot \mathbf{r}) + i\delta C_x z - iC^2 z/2], \tag{10.3.8}$$

where the "velocity" vector $\mathbf{C} = (C_x, C_y)$ characterizes the rate of beam displacement in the transverse x–y plane.

Using the preceding transformation, one can investigate different types of collisions among two-dimensional quadratic solitons by solving Eqs. (10.1.10) and (10.1.11) numerically. Such collisions have been studied extensively in the one-dimensional geometry [61]–[64]. Similar to the case of Kerr solitons, collisions depend strongly on the initial relative phase between the two quadratic solitons. A similar behavior was observed for the head-on collisions of two-dimensional quadratic solitons [34]–[72]. When the relative phase of two colliding solitons is zero, the solitons attract each other and eventually fuse into a single two-dimensional soliton of larger amplitude. The amplitude of this "fused" soliton oscillates, indicating the excitation of an *internal mode* [48]. However, when the two colliding solitons are significantly out of phase ($\pi/4 < \Delta\phi < 7\pi/4$), the interaction between them becomes repulsive, and both solitons survive the collision after exchanging some energy.

The experimental observation of stable two-dimensional solitons in saturating nonlinear media initiated a complete study of the interaction for quadratic solitons, and a theory of nonplanar (i.e., fully three-dimensional) collisions of the $\chi^{(2)}$ solitons was developed [106]. The analytical model, capable of describing soliton scattering, spiraling, and fusion, transforms Eqs. (10.1.10) and (10.1.11) into a conservative mechanical system governed by the Lagrangian

$$\mathcal{L} = \tfrac{1}{2} M_R \dot{R}^2 + \tfrac{1}{2} M_\psi \dot{\psi}^2 - U(R, \psi), \tag{10.3.9}$$

where $R \equiv \sqrt{X^2 + Y^2}$ is the relative distance between the two interacting soliton beams, X and Y being the spacing between solitons in the x and y directions, and $\psi = \phi_1 - \phi_2$ is the relative phase between them. The effective masses depend on the intensity profiles of the two beams and can be calculated using

$$M_R \equiv \pi \int\!\!\int_{-\infty}^{\infty} (|v(r)|^2 + 2\sigma|w(r)|^2) r\,dr, \qquad M_\psi = -2\frac{\partial M_R}{\partial \beta}, \tag{10.3.10}$$

where $r = \sqrt{x^2 + y^2}$. The integral in the preceding equation is calculated numerically over the intensity profiles of a radially symmetric quadratic soliton [34]. The effective potential energy governing the attraction or repulsion between the two solitons is defined as

$$U(R, \psi) = \frac{M_R s^2 C^2}{4R^2} + U_1(R)\cos\psi + U_2(R)\cos(2\psi), \tag{10.3.11}$$

where U_1 and U_2 depend on the extent of overlap between the two solitons. The impact parameter s is related to the distance R_0 between the trajectories of noninteracting solitons, and $C \equiv \dot{R}_0$ is the relative velocity between the solitons prior to their interaction.

When the spacing between the two interacting solitons is large, their interaction is governed by the overlapping tails. Far from the beam center, the soliton amplitudes can be found approximately using an asymptotic technique and are given by

$$v(r) \sim \exp(-\sqrt{\beta}r)/\sqrt{r}, \qquad w(r) \sim \exp[-\sqrt{\sigma(2\beta + \Delta)}r]/\sqrt{r}. \tag{10.3.12}$$

Using these results, U_1 and U_2 are found to be

$$U_1(R) = -A\exp(-\sqrt{\beta}R)/\sqrt{R} \qquad U_2(R) = -B\exp[-\sqrt{\sigma(\Delta + 2\beta)}R]/\sqrt{R}, \tag{10.3.13}$$

where the positive constants A and B are determined numerically.

The mechanical model described by the Lagrangian (10.3.9) can be used for predicting the outcome of soliton collisions. As shown in Figure 10.13, the nature of interaction forces depends strongly on the relative phase ψ [107]. In the case of out-of-phase collisions ($\psi = \pi$), the "centrifugal force," governed by the first term in Eq. (10.3.11), and the direct interaction force, governed by the second term, $U_1(R)\cos\psi$, are both *repulsive*. As a result, the two solitons cannot come closer. In fact, they repel each other and are reflected away, as seen in part (a), after spiraling each other. The interaction scenario is very different for two in-phase solitons ($\psi = 0$). This case is shown in part (b). When s is relatively small, the two solitons fuse together. This behavior of quadratic solitons has been observed experimentally [108]. Two quadratic solitons with similar thresholds were generated inside a 2-cm-long KTP crystal by launching two pump beams such that they intersected inside the crystal. The crystal was long enough for the two solitons to from and collide. As expected from the theory of nonintegrable systems, both quasi-elastic and inelastic (fusion) collisions could be observed by controlling the angle at which the beams intersected.

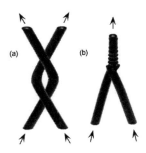

Figure 10.13: Three-dimensional view of two interacting quadratic solitons. (a) Soliton repulsion with spiraling for $\psi = \pi$ and $s = 3.6$; (b) soliton attraction with fusion for $\psi = 0$ and $s = 11$. (After Ref. [107]; ©1999 OSA.)

10.4   Multifrequency Parametric Solitons

The cascading of two or more second-order nonlinear effects in $\chi^{(2)}$ materials offers opportunities for applications such as all-optical processing and optical communications. A parametric process with a single phase-matched interaction results in two-step cascading [7]. For example, the SHG process associated with type I phase matching includes the generation of the second harmonic ($\omega + \omega = 2\omega$) and the regeneration of the pump field through the down-conversion parametric process ($2\omega - \omega = \omega$). These two processes are governed by one phase-matched interaction, and they differ only in the direction of power conversion. In some situations, it is possible to phase match several parametric processes simultaneously. The simplest case, *double phase matching*, has been studied extensively [109]–[113]. It is also attractive for applications such as all-optical transistors, enhanced nonlinearity-induced phase shifts, and polarization switching.

10.4.1   Multistep Cascading

In the simplest case of multistep cascading, a pump beam of frequency ω entering a $\chi^{(2)}$ medium first generates the second-harmonic wave at 2ω via the SHG process, which, in turn, generates higher-order harmonics through sum-frequency mixing process. For example, the pump and SHG beams can generate the third harmonic ($\omega + 2\omega = 3\omega$) or even the fourth harmonic ($2\omega + 2\omega = 4\omega$), if these processes can be phase matched at the same time the SHG process is phase matched [114]. When both such processes are nearly phase matched, they can lead to a large nonlinear phase shift for the pump wave [111]. Such multistep cascading can create a novel type of three-wave spatial soliton in which all three waves preserve their shapes inside the $\chi^{(2)}$ medium.

Consider a cascaded process in which a single pump generates both the second and third harmonics simultaneously. Following the approach of Section 10.1, we can derive a set of three equations similar to Eqs. (10.1.3)–(10.1.5). If we neglect the walkoff effects and focus on the waveguide geometry, these equations can be written in the form

$$2ik_1\frac{\partial A_1}{\partial z} + \frac{\partial^2 A_1}{\partial x^2} + 2k_1 d_1 A_3 A_2^* e^{-i\Delta k_3 z} + 2k_2 d_2 A_2 A_1^* e^{-i\Delta k_2 z} = 0, \quad (10.4.1)$$

$$4ik_1\frac{\partial A_2}{\partial z} + \frac{\partial^2 A_2}{\partial x^2} + 8k_1 d_1 A_3 A_1^* e^{-i\Delta k_3 z} + 4k_1 d_2 A_1^2 e^{i\Delta k_2 z} = 0, \quad (10.4.2)$$

$$6ik_1\frac{\partial A_3}{\partial z} + \frac{\partial^2 A_3}{\partial x^2} + 18k_1 d_1 A_2 A_1 e^{i\Delta k_3 z} = 0, \quad (10.4.3)$$

where d_1 and d_2 are related to the elements of the $\chi^{(2)}$ tensor. In these equations, A_1, A_2, and A_3 are the field envelopes of the pump, second harmonic, and third harmonic, respectively, $\Delta k_2 = 2k_1 - k_2$ is the wave-vector mismatch for the SHG process, and $\Delta k_3 = k_1 + k_2 - k_3$ is the wave-vector mismatch for the sum-frequency mixing process.

As before, it is useful to introduce the normalized field envelopes as

$$A_1 = \frac{\beta w e^{i\beta z}}{2\sqrt{d_1 d_2}}, \quad A_2 = \frac{\beta v}{d_2} e^{2i\beta z + i\Delta k_2 z}, \quad A_3 = \frac{\beta u}{2d_1} e^{3i\beta z + i\Delta k z}, \quad (10.4.4)$$

where $\Delta k = \Delta k_2 + \Delta k_3$. Renormalizing the variables z and x as $z \to z/\beta$ and $x \to x/\sqrt{2\beta k_1}$, we obtain the following set of three coupled equations:

$$i\frac{\partial w}{\partial z} + \frac{\partial^2 w}{\partial x^2} - w + w^* v + v^* u = 0, \quad (10.4.5)$$

$$2i\frac{\partial v}{\partial z} + \frac{\partial^2 v}{\partial x^2} - \alpha v + \frac{1}{2}w^2 + w^* u = 0 \quad (10.4.6)$$

$$3i\frac{\partial u}{\partial z} + \frac{\partial^2 u}{\partial x^2} - \alpha_1 u + \chi v w = 0, \quad (10.4.7)$$

where

$$\alpha = 2(2\beta + \Delta k_2)/\beta, \quad \alpha_1 = 3(3\beta + \Delta k)/\beta \quad (10.4.8)$$

are two dimensionless parameters that characterize the phase mismatch between the two parametric processes involved. The entire nonlinear interaction is governed by a single dimensionless material parameter $\chi = (3d_1/d_2)^2 \sim 1$, whose exact value depends on the type of phase matching used.

Equations (10.4.5)–(10.4.7) describe the parametric interaction of three waves through a two-step cascading process in the absence of walk-off effects. Their solutions provide a two-parameter family of quadratic solitons, α and α_1 playing the role of two parameters. Similar to the case of nondegenerate three-wave mixing, these equations have an exact analytical solution for $\alpha = \alpha_1 = 1$. To find it, we make the substitution

$$w = w_0 \operatorname{sech}^2(\eta x), \quad v = v_0 \operatorname{sech}^2(\eta x), \quad u = u_0 \operatorname{sech}^2(\eta x), \quad (10.4.9)$$

and obtain three algebraic equations for the four unknown parameters. These can be solved for $\eta = 1/2$, for which the amplitude v_0 is the solution of the quadratic equation $4\chi v_0^2 + 6v_0 = 9$. The amplitudes w_0 and u_0 are then found using

$$w_0^2 = 9v_0/(3 + 4\chi v_0), \quad u_0 = (2/3)\chi w_0 v_0. \quad (10.4.10)$$

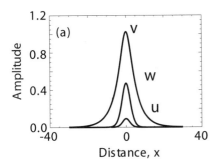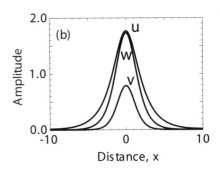

Figure 10.14: Examples of three-wave parametric solitons: (a) $\alpha = 0.05$ and $\alpha_1 = 5$; (b) $\alpha = 5$ and $\alpha_1 = 0.35$. (After Ref. [112]; ©1999 OSA.)

Two different solutions exist for the two different solutions of the quadratic equation, given by $-3(1 \pm \sqrt{5})/(4\chi)$.

In general, the parametric solitons of Eqs. (10.4.5)–(10.4.7) can be found only numerically. Figure 10.14 shows two examples of such solitons for two different sets of the mismatch parameters α and α_1. The case $\alpha_1 \gg 1$, shown in part (a), corresponds to an unmatched sum-frequency mixing process, resulting in a small amplitude of the third harmonic (u vanishes as $\alpha_1 \to \infty$). In the case shown in part (b), α is much larger than α_1, and u exceeds the second-harmonic amplitude v.

10.4.2 Parametric Soliton-Induced Waveguides

When a single pump beam is launched such that its two orthogonally polarized components (ordinary and extraordinary waves in a birefringent crystal) create two orthogonally polarized second-harmonic waves, a different kind of multistep cascaded parametric process can occur that involves only two frequencies, ω and 2ω. If we denote the orthogonally polarized components of the pump as A and B and those of the second harmonic as S and T, then, the multistep cascading process consists of the following steps. First, pump wave A generates S via a type I SHG process (denoted as AA-S). Then, S and A can mix to produce B through a difference-frequency mixing process, denoted as SA-B. Finally, the initial pump wave A is regenerated by the process SB-A or via the route AB-S followed by SA-A. Two principal second-order processes, AA-S and AB-S, correspond to two different components of the $\chi^{(2)}$ tensor. Different types of such multistep cascaded processes are summarized in Table 10.1.

To discuss some of the unique properties of multistep cascading, we consider how it can be employed for soliton-induced waveguiding effects in quadratic media. For this purpose, we consider the principal process (c) in Table 10.1 in the waveguide geometry. It is described by the following set of three coupled equations:

$$2ik_1 \frac{\partial A}{\partial z} + \frac{\partial^2 A}{\partial x^2} + 2k_1 d_1 SA^* e^{-i\Delta k_1 z} = 0, \qquad (10.4.11)$$

$$2ik_1 \frac{\partial B}{\partial z} + \frac{\partial^2 B}{\partial x^2} + 2k_1 d_1 SB^* e^{-i\Delta k_2 z} = 0, \qquad (10.4.12)$$

$$4ik_1 \frac{\partial S}{\partial z} + \frac{\partial^2 S}{\partial x^2} + 4k_1 d_1 A^2 e^{i\Delta k_1 z} + 4k_2 d_2 B^2 e^{i\Delta k_2 z} = 0, \qquad (10.4.13)$$

where Δk_1 and Δk_2 are the mismatch parameters for the parametric processes AA-S and BB-S, respectively.

As before, we introduce the normalized envelopes using the relations

$$A = \frac{u\, e^{i\beta z - i\Delta k_1 z/2}}{4k_1 x_0^2 d_1}, \quad B = \frac{v\, e^{i\beta z - i\Delta k_2 z/2}}{4k_1 x_0^2 \sqrt{d_1 d_2}}, \quad S = \frac{w\, e^{2i\beta z}}{2k_1 x_0^2 d_2}, \qquad (10.4.14)$$

where $z_0 = (\beta - \Delta k_1/2)^{-1}$ and $x_0 = (z_0/2k_1)^{1/2}$ normalize the longitudinal and transverse coordinates z and x, respectively. The resulting normalized equations take the form

$$i\frac{\partial u}{\partial z} + \frac{\partial^2 u}{\partial x^2} - u + u^* w = 0, \qquad (10.4.15)$$

$$i\frac{\partial v}{\partial z} + \frac{\partial^2 v}{\partial x^2} - \alpha_1 v + \chi v^* w = 0, \qquad (10.4.16)$$

$$2i\frac{\partial w}{\partial z} + \frac{\partial^2 w}{\partial x^2} - \alpha w + \frac{1}{2}(u^2 + v^2) = 0, \qquad (10.4.17)$$

where $\chi = \chi_2/\chi_1$ and the two phase-mismatch parameters are defined as

$$\alpha_1 = \frac{\beta - \Delta k_2/2}{\beta - \Delta k_1/2}, \quad \alpha = \frac{4\beta}{\beta - \Delta k_1/2}. \qquad (10.4.18)$$

We now look for the shape-preserving solutions of Eqs. (10.4.15)–(10.4.17) in the asymptotic limit $z \to \infty$. First of all, we notice that when $v = 0$ (or $u = 0$), these equations reduce to Eqs. (10.2.1) and (10.2.2) for the type I SHG process discussed in Section 10.2 (if we set $r = s = 1$). Thus, in this limit, the soliton solution is given by Eqs. (10.2.6) and (10.2.7). More specifically, we again obtain a family of quadratic solitons for different values of the phase-mismatch parameter α, as seen in Figure 10.1. The solution is known analytically for $\alpha = 1$.

When all three field are nonzero, we need to solve Eqs. (10.4.15)–(10.4.17) numerically after setting the z derivatives to zero. However, if we assume that v is real and much smaller than the other two fields, Eq. (10.4.16) for v can be written as the following eigenvalue problem for the effective waveguide created by the second-harmonic

Table 10.1 Several types of multistep cascaded processes

Type	Principal Process	Equivalent Processes
(a)	(AA-S, AB-S)	(BB-S, AB-S); (AA-T, AB-T); (BB-T, AB-T)
(b)	(AA-S, AB-T)	(BB-S, AB-T); (AA-T, AB-S); (BB-T, AB-S)
(c)	(AA-S, BB-S)	(AA-T, BB-T)
(d)	(AA-S, AA-T)	(BB-S, BB-T)

field $w_0(x)$, where w_0 is the solution obtained for $v = 0$:

$$\frac{d^2v}{dx^2} + [\chi\, w_0(x) - \alpha_1]v = 0.$$ (10.4.19)

Physically speaking, the additional parametric process allows one to propagate a probe beam of orthogonal polarization at the pump frequency as a guided mode of an *effective waveguide* created by a quadratic soliton generated using a pump beam of certain polarization. This type of waveguide is different from that studied for Kerr-like solitons because it is coupled *parametrically* to the guided modes. As a result, the physical picture of the guided modes is valid, rigorously speaking, only in the case of stationary phase-matched beams.

To find the guided modes of the parametric waveguide created by a quadratic soliton, we have to solve Eq. (10.4.19) using the exact solution $w_0(x)$ that can only be found numerically. We can solve the problem analytically only approximately if we use the analytic solution given in Eq. (10.2.12). The guided modes exist only for certain discrete values of the parameter α_1 given by

$$\alpha_1^{(n)} = (s - n)^2/p^2, \qquad s = -\tfrac{1}{2} + \tfrac{1}{2}(1 + 4w_m\chi p^2)^{1/2},$$ (10.4.20)

where the integer n stands for the mode order ($n = 0, 1, \ldots$) and the localized solutions are possible, provided $n < s$. The spatial profile of the guided modes is of the form

$$v_n(x) = V\mathrm{sech}^{s-n}(x/p)H(-n, 2s - n + 1, s - n + 1; \zeta/2),$$ (10.4.21)

where $\zeta = 1 - \tanh(x/p)$ and H is the hypergeometric function. The mode amplitude V cannot be determined within the framework of the linear analysis. Thus, a two-wave parametric soliton creates a multimode waveguide; a larger number of guided modes exists for small values of α.

Because a quadratic soliton creates an induced waveguide that is parametrically coupled to its guided modes of the orthogonal polarization, the dynamics of the guided modes may differ drastically from those of conventional waveguides based on Kerr-type nonlinearities. Figure 10.15 shows two examples of the evolution of such guided modes. In the first example, a weak probe at the pump frequency is amplified via the parametric interaction inside the induced waveguide. The mode exchanges its power in a periodic fashion with the orthogonally polarized pump field as it propagates. This process is accompanied by only a weak deformation of the induced waveguide, as seen by the dotted curve in part (a). As seen in parts (b) and (c), this behavior can be interpreted as a power exchange between the two guided modes of orthogonal polarizations in a waveguide created by the second-harmonic field. In the second example, the pump keeps its power constant as it propagates inside the nonlinear medium, as seen in Figure 10.15(d).

When none of the fields in Eqs. (10.4.15)–(10.4.17) are small, the spatial intensity profiles for the three-component soliton can be found numerically for arbitrary values of the phase-mismatch parameters. However, for certain specific values of these parameters, they can be solved in an explicit analytical form. For example, when $\alpha_1 = 1/4$, two families of three-component quadratic solitons have been found for $\alpha \geq 1$, the

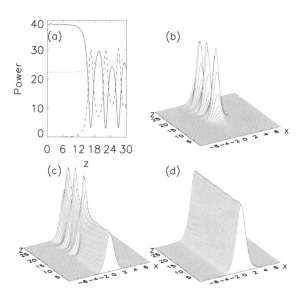

Figure 10.15: (a) Changes in beam powers for pump u (solid), second harmonic w (dotted), and probe v (dashed). Evolution of v and w is shown in (b) and (c), respectively. (d) Stationary propagation of the pump for $\chi = 1$. In all cases, $\alpha = 4$ and $v = 0.1$ initially. (After Ref. [113]; ©1999 APS.)

bifurcation occurring at $\alpha = 1$. These can be identified as the zeroth- and first-order guided modes of the induced waveguide. The zeroth-order guided mode occurs for $\chi = 1/3$ with the spatial profiles

$$u(x) = (3/\sqrt{2})\mathrm{sech}^2(x/2), \quad v(x) = c_2\,\mathrm{sech}(x/2), \quad w(x) = (3/2)\,\mathrm{sech}^2(x/2),$$
$$(10.4.22)$$

where $c_2 = \sqrt{3(\alpha - 1)}$. The first-order guided mode occurs for $\chi = 1$. Its spatial profiles are given by $u(x) = c_1 \mathrm{sech}^2(x/2)$, $v(x) = c_2 \mathrm{sech}^2(x/2)\sinh(x/2)$, and $w(x) = (3/2)\,\mathrm{sech}^2(x/2)$, where $c_1 = (9/2 + c_2^2)^{1/2}$.

10.5   Quasi-Phase Matching

Although quadratic solitons have been observed using KTP crystals as well as LiNbO$_3$ waveguides, the efficiency of the SHG process was relatively low, and the experiments required the use of Q-switched mode-locked pulses with high peak powers. The low efficiency was partly due to the limitations imposed by the conventional phase-matching techniques based on birefringence and temperature tuning.

During the 1990s the technique of quasi-phase matching (QPM) attracted considerable attention for enhancing the efficiency of the SHG and other parametric processes [115]. The QPM technique relies on the periodic modulation of the nonlinear susceptibility. The periodic nature of the modulation creates a $\chi^{(2)}$ grating that helps to compensate the wave-vector mismatch between the the pump and second-harmonic

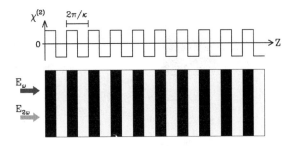

Figure 10.16: Schematic illustration of a QPM structure in which $\chi^{(2)}$ is modulated periodically along the device length. (After Ref. [120]; ©1997 APS.)

waves. With the QPM technique, phase matching becomes possible at ambient temperatures without incurring the walk-off effects. Moreover, one can exploit new materials with strong nonlinearities that cannot be phase-matched with angle or temperature tuning. The concept of QPM has been known since 1962 [116], but it was only during the 1990s that the experimental difficulties were overcome, using techniques such as domain inversion in ferroelectric materials [117], proton exchange [118], and etching with cladding [119]. In this section, we describe the properties of quadratic solitons generated in QPM structures.

10.5.1 Quasi-Phase-Matching Solitons

Consider the interaction of a pump beam and its second harmonic inside a QPM slab waveguide, as shown in Figure 10.16. The parametric interaction is still governed by Eqs. (10.1.6) and (10.1.7), provided d is assumed to vary in a periodic fashion along the z axis and the walk-off parameter ρ is set to zero. Limiting diffraction to only one spatial dimension and introducing the normalization procedure of Section 10.1, these equations can be written in the form

$$i\frac{\partial V}{\partial z} + \frac{1}{2}\frac{\partial^2 V}{\partial x^2} + d(z)WV^*e^{-i\beta z} = 0, \qquad (10.5.1)$$

$$i\frac{\partial W}{\partial z} + \frac{1}{4}\frac{\partial^2 W}{\partial x^2} + d(z)V^2 e^{i\beta z} = 0, \qquad (10.5.2)$$

where V and W represent the normalized amplitudes for the pump and SHG beams. The phase-mismatch parameter $\beta = \Delta k(|k_1|x_0^2)$, where $\Delta k = 2k_1 - k_3$. The variable x is normalized using the input pump width x_0, while z is measured in units of the diffraction length $l_d = |k_1|x_0^2$. The nonlinear parameter d is a periodic function of z and can be expanded in a Fourier series as

$$d(z) = \sum_{n=-\infty}^{\infty} d_n e^{in\kappa z}, \qquad (10.5.3)$$

where $\kappa = 2\pi/l_p$, l_p being the period of the $\chi^{(2)}$ grating.

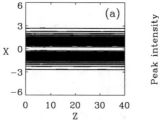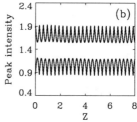

Figure 10.17: (a) Intensity contours for the pump beam in the z–x plane for intensities in the range 0.1–1.92. (b) Peak intensities for the pump and second harmonic for $\tilde{\beta} = 0$ and $\kappa = 10$. (After Ref. [120]; ©1997 APS.)

In many physical applications the QPM grating can be well approximated by the square shape depicted in Figure 10.16. The Fourier series in Eq. (10.5.3) then contains only odd harmonics such that $d_{2n} = 0$ and $d_{2n+1} = 2d_0/[i\pi(2n+1)]$. For some specific value of n (say, m), the SHG process is phase matched by the nonlinear grating such that $m\kappa \approx \beta$. If κ is sufficiently large, the dynamics are well described by the averaged equations. Physically, $m\kappa \gg 1$ implies that the coherence length $l_c = 2\pi/\Delta k$ is much smaller than the diffraction length l_d because $\beta = 2\pi l_d/l_c$.

We use the averaging technique [120] and consider the most efficient QPM of first order and assume $m = 1$. If the phase matching is not perfect, the residual mismatch $\tilde{\beta} = \beta - \kappa$ should be included in the analysis. If we expand V and W in a Fourier series and retain only the dominant zeroth-order terms v_0 and w_0, we obtain the following set of coupled equations [120]:

$$i\frac{\partial v_0}{\partial z} + \frac{1}{2}\frac{\partial^2 v_0}{\partial x^2} - i\chi w_0 v_0^* + \gamma(|v_0|^2 - |w_0|^2)v_0 = 0, \qquad (10.5.4)$$

$$i\frac{\partial w_0}{\partial z} + \frac{1}{4}\frac{\partial^2 w_0}{\partial x^2} - \tilde{\beta} w_0 + i\chi v_0^2 - 2\gamma|v_0|^2 w_0 = 0, \qquad (10.5.5)$$

where both the quadratic and cubic nonlinearity coefficients are known in an explicit form as $\chi = 2/\pi$ and $\gamma = (1 - 8/\pi^2)/\kappa$. Notice the $\pi/2$ phase shift in the quadratic terms (the factor of i) and the opposite signs for the self- (SPM) and cross-phase modulation (XPM) terms in Eq. (10.5.4).

Soliton solutions of Eqs. (10.5.4) and (10.5.5) can be found numerically, and the evolution of peak intensities and spatial profiles can be studied [120]. The results show that the properties of the quadratic soliton obtained for $\gamma = 0$ are not affected much by the SPM and XPM terms governed by the effective cubic nonlinearity γ. The accuracy of Eqs. (10.5.4) and (10.5.5) has also been tested by solving Eqs. (10.5.1) and (10.5.2) directly. The spatial field profiles of a QPM soliton found using Eqs. (10.5.4) and (10.5.5) were used as the initial conditions for solving Eqs. (10.5.1) and (10.5.2). Figure 10.17 shows the numerical results for $\tilde{\beta} = 0$ and $\kappa = 10$. The soliton propagated undistorted along the z axis, although the peak intensities oscillated with the period π/κ.

The average quadratic soliton in a QPM nonlinear medium can be regarded as a *spatial analog* of the *guided-center temporal soliton* found for pulse propagation in

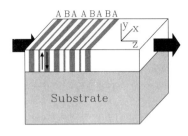 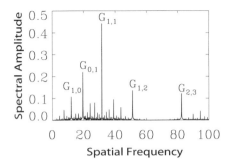

Figure 10.18: (a) Slab waveguide with quasi-periodic QPM superlattice structure composed of building blocks A and B. (b) Numerically calculated amplitude spectrum of $d(z)$. (After Ref. [122]; ©1999 APS.)

optical fibers with periodic amplification and dispersion management [121]. However, unlike the guided-center soliton, the periodic modulation of the quadratic nonlinearity does not alter the existing nonlinearity but induces higher-order nonlinearities.

10.5.2 Quasi-Periodic Parametric Solitons

Solitons are usually thought to be *coherent* localized modes of nonlinear systems with particle-like dynamics that are quite different from the irregular and stochastic behavior observed for chaotic systems. However, when the nonlinearity is sufficiently strong, it can act as an effective *phase-locking mechanism* by producing a large frequency shift for different components with a random phase. In essence, it can introduce a long-range order into an incoherent wave packet, thus enabling the formation of a localized structure for quasi-periodic modes. We discuss in this section that such solitons can exist in the form of quasi-periodic nonlinear localized modes. As an example of this phenomenon, we consider the SHG process in a *Fibonacci superlattice* [122]. Numerical results show the possibility of localized waves whose envelope amplitude varies in a quasi-periodic fashion while maintaining a stable, well-defined spatial shape. We refer to such an object as a *quasi-periodic soliton*.

Consider a QPM structure in which $\chi^{(2)}$ does not vary periodically but follows a quasi-periodic sequence. Such optical superlattices are one-dimensional analogs of quasi-crystals [123]. They are usually designed for studying the Anderson localization in the linear regime of wave propagation. For example, transmission through a quasi-periodic multilayer stack of SiO_2 and TiO_2 thin films has been observed to be strongly suppressed through Anderson localization [124]. For the SHG process, a nonlinear quasi-periodic superlattice of $LiTaO_3$ was formed in which two antiparallel ferroelectric domains were arranged in a Fibonacci sequence [125]. Such devices can produce SHG with energy conversion efficiencies in the range 5–20%. They can also be used for efficient third-harmonic generation [126].

Figure 10.18 shows schematically a quasi-periodic QPM structure. It uses two types of layers (marked A and B) of different lengths l_A and l_B that are ordered in a Fibonacci sequence. In each pair, two layers have $\chi^{(2)}$ with opposite signs but equal

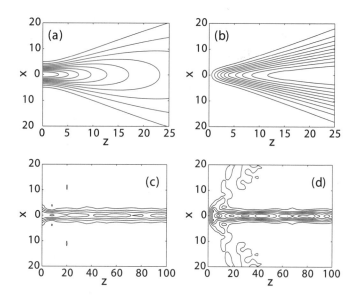

Figure 10.19: Intensity contours for the pump (a) and SHG (b) beams at low input pump powers [$V(0) = 0.25$]. Parts (c) and (d) show that a quasi-periodic soliton forms at a large pump power with $V(0) = 5$. (After Ref. [122]; ©1999 APS.)

magnitudes, realized by using positive and negative ferroelectric domains. The lengths l_A and l_B are chosen such that $l_A = l(1 + \eta)$ and $l_B = l(1 - \tau\eta)$, where η and τ are two constants. In one set of numerical simulations, they were chosen such that $\eta = 2(\tau - 1)/(1 + \tau^2) = 0.34$, where $\tau = (1 + \sqrt{5})/2$ was the so-called *golden ratio*.

Equations (10.5.1) and (10.5.2) can still be used for describing the SHG process in such a structure, but the grating function $d(z)$ is no longer periodic. However, because it follows the Fibonacci sequence, it can be expanded in a double Fourier series as

$$d(z) = \sum_m \sum_n d_{mn} \exp(iG_{m,n}z), \qquad G_{m,n} = (m + n\tau)(2\pi/D), \qquad (10.5.6)$$

where $D = \tau l_A + l_B$. Since the integers m and n take both positive and negative values, the spectrum d_{mn} consists of a large number of spatial frequencies found by adding and subtracting the multiple of the basic wave numbers $\kappa_1 = 2\pi/D$ and $\kappa_2 = 2\pi\tau/D$. These spectral components fill the whole Fourier space densely because κ_1 and κ_2 are incommensurate. Figure 10.18 shows the numerically calculated Fourier spectrum d_{mn} using $l = 0.1$. The most intense peaks are marked by $G_{m,n}$ and correspond to relatively small values of $|m|$ and $|n|$.

To analyze the SHG process in such a quasi-periodic QPM grating, Eqs. (10.5.1) and (10.5.2) were solved numerically with a Gaussian shape for the pump beam at the input end. Figure 10.19 shows the evolution of the pump and SHG beams at low (upper row) and high pump powers (bottom row). At low pump powers, the nonlinear effects are relatively weak. As a result, both beams spread without forming a quadratic soliton. At high pump powers, the nonlinearity becomes so strong that it leads to self-focusing

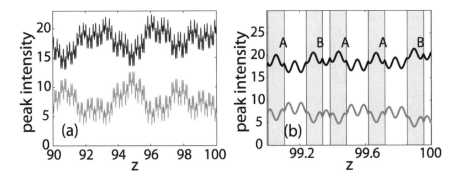

Figure 10.20: (a) oscillations of the peak intensity for the pump and SHG beams. (b) Close up view in the range $z = 99$ to 100. The Fibonacci building blocks A and B are also shown. (After Ref. [122]; ©1999 APS.)

and mutual self-trapping of the two fields, resulting in the formation of a quadratic soliton, despite the continuous scattering from the quasi-periodic grating.

It is important to note that such quadratic solitons do not preserve their shape perfectly. As seen in Figure 10.19, both the pump and SHG beams are self-trapped, but their peak intensities and widths vary in a quasi-periodic fashion. It turns out that, after initial transients have died out, the peak intensities of both beams oscillate out of phase but follow the quasi-periodic nature of $d(z)$. This is illustrated in Figure 10.20. Part (b) shows the asymptotic regime in more detail together with the structure of the QPM grating. The spectrum of amplitude oscillations is quite similar to that of the QPM grating shown in Figure 10.18.

The numerical results show that the quasiperiodic solitons can be generated over a broad range of the phase-mismatch parameter β. The amplitude and the width of solitons depend on the effective mismatch governed by the the separation between β and the nearest strong peak $G_{m,n}$ in the Fibonacci spectrum shown in Figure 10.18. Thus, low-amplitude broad solitons are excited for β values in between the spectral peaks, but high-amplitude narrow solitons are excited when β is close to a strong peak. The important point is that self-trapping occurs such that the quasi-periodicity of the nonlinear grating is preserved in variations of the amplitude of the two components of a QPM soliton.

10.6   Related Concepts

10.6.1   Competing Nonlinearities

The nonlinear response of any $\chi^{(2)}$ material always includes the contribution of the next-order (cubic or $\chi^{(3)}$) nonlinearity, which, under certain conditions, can become important enough to compete strongly with the $\chi^{(2)}$ nonlinearity. The influence of cubic nonlinearity on the SHG and other second-order parametric processes has been studied since the early 1980s and leads to such effects as distortion of the second-harmonic spectrum and saturation of the SHG conversion efficiency [127]–[129].

Several physical mechanisms exist that can lead to a competition between two different kinds of nonlinearities. First of all, any $\chi^{(2)}$ material has an *inherent* $\chi^{(3)}$ nonlinearity that can become important at high powers or when the pump and its second harmonic are not closely phase matched. A second mechanism is related to the QPM technique. As discussed in Section 10.5, the SHG process in a QPM medium is always influenced by the effects of *induced cubic nonlinearity* resulting from an incoherent coupling among the waves generated through diffraction from the spatial (QPM) grating [120].

Another mechanism that can lead to the cubic nonlinear terms in the conventional model of $\chi^{(2)}$-based parametric processes is due to an incoherent coupling of the two main interacting waves with other modes that may exist and may be excited by the higher-order cascading effects. This is the situation, for example, when the SHG occurs in a waveguide that supports a single mode at the pump frequency ω but two modes at 2ω. Similar situation may occur in the case of multistep cascading when the influence of other second-order processes involving the sum- and difference-frequency mixing is taken into account [112, 113]. If only one of the processes is nearly phase matched, the others can be treated in the cascading limit, which always leads to cubic nonlinear terms. In short, any mismatched parametric coupling among optical fields induces an effective cubic nonlinearity in the dynamical equations for parametrically coupled harmonics. Clearly, the competition between the quadratic and cubic nonlinearities is a general physical phenomenon and occurs in many physical situations. The dynamic evolution of parametric solitons in nonlinear media with competing quadratic and cubic nonlinearities has attracted considerable attention in recent years [130]–[136].

10.6.2 Parametric Vortex Solitons

As discussed in Chapter 8, vortex solitons with a central hole can form in a cubic nonlinear medium so that the beam shape is in the form of a "donut." It is natural to consider the existence and stability of such solitary waves of circular symmetry in a medium with quadratic nonlinearity. In the case of a cubic medium, vortex solitons can have circular symmetry without any angular dependence [137], or they may follow a spiral path [138]. Both kinds of parametric vortex solitons are expected to exist in quadratic media. For such solitons, both the pump and SHG beams have a donut shape with a vortex-like structure.

To find solutions of Eqs. (10.1.10) and (10.1.11) with a nontrivial angular dependence, one should look for solutions of radial symmetry in the form

$$v(x,y,z) = V(r)\exp[i(\kappa z + l\phi)], \tag{10.6.1}$$

$$w(x,y,z) = W(r)\exp[2i(\kappa z + l\phi)], \tag{10.6.2}$$

where $r = \sqrt{x^2 + y^2}$ is the radial coordinate, ϕ is the polar angle, and the integer l is a measure of the angular momentum associated with the quadratic soliton. The case $l = 0$ corresponds to the two-dimensional quadratic solitons discussed earlier in Section 10.3. The localized solutions with $l \neq 0$ can exist only if both V and W vanish as $r \to 0$. Such radially symmetric ring-like solitary waves are referred to as *donut solitons* [139]. They have a finite orbital angular momentum characterized by l and are usually unstable and break into multiple filaments that fly out tangentially from the initial ring, behaving just like free particles. A similar scenario occurs for the vortex solitons forming in a cubic medium with saturable nonlinearity and observed experimentally using Rb vapors [139]–[141].

Parametric vortex solitons have been analyzed numerically and observed experimentally [142]–[146]. In the numerical simulations, Eqs. (10.1.10) and (10.1.11) were solved using an input pump beam with a phase dislocation such that

$$v(r, z = 0) = Ar^{|l|}e^{il\phi}e^{-r^2/w^2}. \tag{10.6.3}$$

The SHG process led to breaking of the pump beam into into multiple spatial solitons after formation of an unstable parametric vortex soliton [142]. Numerical simulations show that all such vortex solitons are *azimuthally unstable* and decay into a set of stable two-dimensional quadratic solitons [144]. Depending on the nature of the initial perturbation, donut vortex solitons with $l = 2$ and 3 decayed into four or five individual quadratic solitons.

Generalization of the vortex-like quadratic solitons is also possible in the case of three-wave parametric interaction occurring for type II phase matching [145, 146]. In this case, the vortex or the donut soliton has three components, corresponding to the three waves interacting parametrically, and each component can have a finite orbital angular momentum governed by the integer l. As in the case of two-wave mixing, all such solitons are unstable, and they decay into multiple three-wave quadratic solitons of circular symmetry. It has been suggested that such vortex solitons can be used to make optical devices that process information by mixing the orbital angular momentum (also called the *topological charge*) of the three optical beams [146].

Quadratic vortex solitons with a nonvanishing background represent an extension of the vortex dark solitons forming in a $\chi^{(3)}$ medium (see Chapter 8) to a $\chi^{(2)}$ medium. They were first analyzed in 1998 in the form of two parametrically coupled vortex modes [67]. The excitation of such solitons requires a pump field with a 2π phase twist at the beam center, resulting in a 4π phase twist for the second-harmonic field. Two types of such parametric vortices are possible [136]. In the case of a *halo vortex*, two coupled vortex cores are surrounded by a bright ring of its second-harmonic field. In the other case of a *ring vortex*, the second-harmonic field guides a ring-like structure at the pump frequency. The important feature of all such parametric vortices is the requirement that a $\chi^{(3)}$ contribution with a *defocusing* kind of nonlinearity must exist for stabilizing the otherwise unstable background beams at the pump and SHG frequencies. However, such solitons are hard to observe experimentally because the cubic nonlinearity usually leads to self-focusing in realistic $\chi^{(2)}$ materials. Nevertheless, they were created in an experiment [147] in which the combination of a finite beam size and the transverse walk-off helped to eliminate the parametric modulation instability. A stable quadratic vortex soliton was observed to form such that both the pump and SHG beams had phase singularities at their centers.

10.6.3   Parametric Optical Bullets

As discussed in Chapter 7, *optical bullets* is the name given to three-dimensional solitons that maintain their shape in both the temporal and spatial domains. In the case of
a $\chi^{(3)}$ medium, optical bullets represent the radially symmetric solutions of a $(3+1)$-
dimensional NLS equation with either the Kerr or non-Kerr (transiting or saturable)
nonlinearities [148]–[150]. Generalization of the concept of light bullets to the case of
parametric solitons in $\chi^{(2)}$ media is far from being straightforward. The main difficulty
is related to the dependence of the group-velocity dispersion (GVD) on the carrier frequency of a pulse. As a result, the GVD terms for the pump and SHG pulses do not
have the same magnitude. This feature makes the multidimensional solitons *radially
asymmetric*, and the asymmetry depends on the ratio $\delta_\omega \equiv \beta_2(\omega)/\beta_2(2\omega)$ of the GVD
coefficients.

The possibility of forming stable spatiotemporal $\chi^{(2)}$ solitons was discussed as
early as 1981 in the special case of $\delta_\omega = 1$ and perfect phase matching [5]. Several
other specific cases were discussed during the 1990s [21, 78]. A more general formulation of this problem was developed later using a variational technique [88]–[90]. This
approach can be used to find approximate profiles of the $\chi^{(2)}$-based optical bullets. In
the special case of $\delta_\omega = 1$, the numerical results show a reasonably good agreement
with those obtained from the variational method. The stability of such optical bullets
has also been studied using both the Vakhitov–Kolokolov criterion and direct numerical simulations [88, 89]. The $\chi^{(2)}$ optical bullets are found to be always stable if both
the fundamental and SHG pulses propagate in the anomalous-dispersion regime. When
only the pump pulse experiences anomalous dispersion, $\chi^{(2)}$ bullets behave in a quasi-
stable manner if the parameter $|\delta|$ is small enough [88]. The more realistic case $|\delta| \gg 0$
has also been analyzed theoretically [86, 151].

The experimental observation of $\chi^{(2)}$ bullets has been hampered by the lack of suitable materials with a large enough nonlinearity and large values of the GVD parameters with the correct sign at both the pump and SHG frequencies. A clever approach
involving spatially tilted phase fronts was used in 1998 to demonstrate the formation of
spatiotemporal $\chi^{(2)}$ solitons in one spatial dimension [152]. This approach can induce
large, effective (positive or negative) GVD and was used to demonstrate spatiotemporal
self-trapping over one dispersion length. Even in this case, a long nonlinear crystal and
ultrashort pulses of <50-fs duration were required for the formation of spatiotemporal
$\chi^{(2)}$ solitons. Moreover, the third-order dispersive, refractive, and absorptive effects influence the pulse evolution and need to be properly accounted for when comparing the
experimental data with numerical simulations. The experiment was performed in the
cascading limit in which a large positive phase mismatch reduced the SHG component
of the quadratic soliton to such a small amplitude that it was only necessary to control
the GVD for the pump pulse [152].

In two 1999 experiments, self-trapping was realized in both *space* and *time* using
a 1-cm-long LiNbO$_3$ crystal cut for type I phase matching [153] and a 2.5-cm-long
barium metaborate (Ba$_2$BO$_4$, or BBO) crystal [104]. Wavefront tilting was used in
the cascading limit to control the GVD in one spatial dimension, and the diffraction
length in that dimension was made equal to the dispersion and nonlinear lengths by
adjusting the input-beam diameter and the peak intensity. The spot size in the second

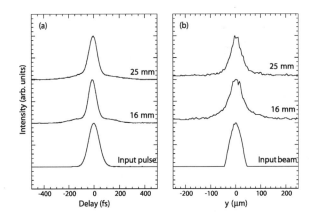

Figure 10.21: Temporal (a) and spatial (b) profiles of a spatiotemporal soliton observed using BBO crystals of different lengths at a pump peak intensity of 8 GW/cm^2. (After Ref. [104]; ©2000 APS.)

transverse beam dimension was made so large that virtually no diffraction occurred over the crystal length. The evolution of the temporal and spatial profiles is shown in Figure 10.21, where a slight asymmetry is also evident. These results provide the first example of spatiotemporal quadratic solitons or parametric optical bullets, but diffraction is limited to only one spatial dimension.

10.6.4 Discrete Quadratic Solitons

The molecular-beam epitaxial (MBE) technique can be used to make molecular crystalline multilayer structures, analogous to inorganic superlattices and quantum-well structures [154]–[157]. Even organic multilayered structures of high optical quality can be grown with this technique [156]. Surface nonlinearities in such materials can lead to new linear and nonlinear optical effects. For example, novel localized states known as *Fermi-resonance interface modes* have been predicted [158]–[161]. Such states can lead to the formation of a new type of soliton called the *Fermi-resonance interface soliton* [162, 163]. The existence of these states can enhance the quadratic and cubic nonlinear susceptibilities [159]. In fact, the Fermi resonance in molecular systems is quite similar to SHG in optics, and the interface solitons are analogous to the quadratic solitons discussed in this chapter. For example, the exact two-wave soliton solution in Ref. [162] constitutes a rediscovery of the quadratic-soliton solution known in nonlinear optics since 1974 [2]. Similarly, the variational formulation developed in Ref. [44] is similar to that used in nonlinear optics [41]. The most interesting feature of the Fermi-resonance modes localized to an interface is the discreteness of the nonlinear lattice model [164]–[168].

In this section we consider a nonlinear lattice each of whose elements consists of a fundamental mode resonantly interacting with its second harmonic [165]–[169]. This kind of *nonlinear $\chi^{(2)}$ lattice* describes an array of weakly interacting but identical optical waveguides. Mathematically, the nonlinear process is described by a set of equations obtained by discretizing Eqs. (10.2.1) and (10.2.2) and given by

$$i\frac{dv_n}{dz} + \eta_v(v_{n+1} + v_{n-1}) + v_n^* w_n = 0, \tag{10.6.4}$$

$$i\frac{dw_n}{dz} + \eta_w(w_{n+1} + w_{n-1}) + \frac{1}{2}v_n^2 = 0, \tag{10.6.5}$$

where we set $r = s = 1$ and $\delta = 0$ (no walk-off). The parameters η_w and η_v determine the strength of the coupling between the fields at neighboring waveguides. The integer n varies in the range 1–N, with N being the total number of waveguides. Equations (10.6.4) and (10.6.5) can be considered a generalization of the discrete NLS equation given in Chapter 11. As discussed there, the discrete NLS equation applies to an array of nonlinear waveguides with cubic nonlinearity [170], the simplest example being of two coupled waveguides in a directional coupler [171].

Equations (10.6.4) and (10.6.5) can be solved exactly only for $N = 1$ and become nonintegrable even for $N = 2$. In fact, the $N = 2$ case applies to the so-called $\chi^{(2)}$ dimer and is known to display chaotic dynamics [165]. However, the stationary (z-independent) soliton-like modes of the $\chi^{(2)}$ dimer can easily be found and their stability can be analyzed using a standard technique. Moreover, in the cascading limit (large values of phase mismatch), Eqs. (10.6.4) and (10.6.5) with $N = 2$ reduce to the integrable model of the nonlinear coupler [171]. Bang et al. [165] used this nearly integrable limit to show how the stationary modes develop a gradual transition to chaotic dynamics.

The case of larger N is more complicated and less well studied, although stationary solutions of Eqs. (10.6.4) and (10.6.5) are known for $N = 4$ and 6 [164]. When $N \gg 1$, these equations have been shown to support strongly localized, self-trapped states of odd and even configurations that involve only a few neighboring sites [166]–[168]. Such states are referred to as *discrete quadratic solitons*. In the small-amplitude limit, such solitons exhibit behavior similar to that observed for the quadratic solitons of the continuum model, in the sense that they not only can move through the lattice but can also fuse during a collision.

The nonlinear localized modes can also occur in a photonic crystal with quadratic nonlinearities in which the linear refractive index varies periodically [172]. The $\chi^{(2)}$ nonlinearity was introduced through thin layers embedded periodically into the linear structure. Different kinds of nonlinear modes in such structures were analyzed. The case of a single nonlinear layer in a linear periodic structure was studied in detail. Some of the properties of the two-frequency nonlinear localized modes were found to resemble closely the properties of quadratic solitons excited in a homogeneous medium [173].

References

[1] L. A. Ostrovskii, *Pisma Zh. Eksp. Teor. Fiz.* **5**, 331 (1967).

[2] Yu. N. Karamzin and A. P. Sukhorukov, *Pisma Zh. Eksp. Teor. Fiz.* **20**, 730 (1974) [*JETP Lett.* **20**, 338 (1975)]

[3] Yu. N. Karamzin and A. P. Sukhorukov, *Zh. Eksp. Teor. Fiz.* **68**, 834 (1975) [*Sov. Phys. JETP* **41**, 414 (1976)].

[4] Yu. N. Karamzin, A. P. Sukhorukov, and T. S. Filipchuk, *Moscow Univ. Phys. Bull.* **33**, 73 (1978).

[5] A. A. Kanashov and A. M. Rubenchik, *Physica D* **4**, 122 (1981).

[6] A. P. Sukhorukov, *Nonlinear Wave Interactions in Optics and Radiophysics* (Nauka, Moscow, 1988) (in Russian).

[7] G. I. Stegeman, D. J. Hagan, and L. Torner, *Opt. Quantum Electron.* **28**, 1691 (1996).

[8] L. Torner, in *Beam Shaping and Control with Nonlinear Optics*, F. Kajzer and R. Reinish, Eds. (Plenum, New, York, 1998).

[9] Yu. S. Kivshar, in *Advanced Photonics with Second-order Optically Nonlinear Processes*, A. Boardman et al., Eds. (Kluwer, Dordrecht, 1999).

[10] C. Etrich, F. Lederer, B. A. Malomed, T. Peschel, and U. Peschel, in *Progress in Optics*, Vol. 41, E. Wolf, Ed. (Elsevier Science, Amsterdam, 2000), pp. 483–568.

[11] W. Torruellas, Yu. S. Kivshar, and G. I. Stegeman, in *Spatial Solitons*, S. Trillo and W. Torruellas, Eds. (Springer, Berlin, 2001), pp. 127–168.

[12] A. V. Buryak, P. Di Trapani, D. V. Skryabin, and S. Trillo, *Phys. Rep.* **370**, 63 (2002).

[13] Y. R. Shen, *The Principles of Nonlinear Optics* (Wiley, New York, 1984).

[14] P. N. Butcher and D. Cotter, *The Elements of Nonlinear Optics* (Cambridge University Press, Cambridge, UK, 1992).

[15] R. W. Boyd, *Nonlinear Optics* (Academic Press, San Diego, 1992).

[16] A. V. Buryak and Yu. S. Kivshar *Phys. Lett. A* **197**, 407 (1995).

[17] A. V. Buryak and Yu. S. Kivshar, *Phys. Rev. A* **51**, R41 (1995)

[18] A. V. Buryak and Yu. S. Kivshar, *Opt. Lett.* **20**, 834 (1995).

[19] A. V. Buryak and Yu. S. Kivshar, *Opt. Lett.* **20**, 1080 (1995).

[20] O. Bang, *J. Opt. Soc. Am. B* **14**, 51 (1997).

[21] K. Hayata and M. Koshiba, *Phys. Rev. Lett.* **71**, 3275 (1993).

[22] R. Schiek, *J. Opt. Soc. Am. B* **10**, 1848 (1993).

[23] M. J. Werner and P.D. Drummond, *J. Opt. Soc. Am. B* **10**, 2390 (1993).

[24] M. J. Werner and P.D. Drummond, *Opt. Lett.* **19**, 613 (1994).

[25] A. V. Buryak and Yu. S. Kivshar, *Opt. Lett.* **19**, 1612 (1994).

[26] L. Torner, C. R. Menyuk, and G. I. Stegeman, *Opt. Lett.* **19**, 1615 (1994).

[27] D. Ferro and S. Trillo, *Phys. Rev. E* **51**, 4994 (1995).

[28] L. Torner, Opt. Comm. **114**, 136 (1995).

[29] A.D. Boardman, K. Xie, and A. Sangarpaul, *Phys. Rev. A* **52**, 4099 (1995).

[30] D. Mihalache, F. Lederer, D. Mazilu, and L.-C. Crasovan, *Opt. Eng.* **35**, 1616 (1996).

[31] H. He, M. J. Werner, and P. D. Drummond, *Phys. Rev. E* **54**, 896 (1996).

[32] M. Haelterman, S. Trillo, and P. Ferro, *Opt. Lett.* **22**, 84 (1997).

[33] L. Torner, C.R. Menyuk, W. E. Torruellas, and G. I. Stegeman, *Opt. Lett.* **20**, 13 (1995).

[34] A. V. Buryak, Yu. S Kivshar, and V.V. Steblina, *Phys. Rev. A* **52**, 1670 (1995).

[35] A. V. Buryak, Yu. S. Kivshar, and S. Trillo, *Phys. Rev. Lett.* **77**, 5210 (1996).

[36] D. Mihalache, D. Mazilu, L.-C. Crasovan, and L. Torner, *Phys. Rev. E* **56**, R6294 (1997).

[37] Q. Guo, *Quantum Opt.* **5**, 133 (1993).

[38] A. G. Kalocsai and J. W. Haus, *Opt. Commun.* **97**, 239 (1993).

[39] A. G. Kalocsai and J. W. Haus, *Phys. Rev. A* **49**, 574 (1994).

[40] A. G. Kalocsai and J. W. Haus, *Phys. Rev. E* **52**, 3166 (1995).

[41] V. V. Steblina, Yu. S. Kivshar, M. Lisak, and B. A. Malomed, *Opt. Commun.* **118**, 345 (1995).

[42] S. Trillo and P. Ferro, *Opt. Lett.* **20**, 438 (1995).

[43] H. He, P.D. Drummond, and B.A. Malomed, *Opt. Commun.* **123**, 394 (1996).

[44] V. M. Agranovich, S. A. Darmanyan, O. A. Dubovsky, A. M. Kamchatnov, E. I. Ogievetsky, T. Neidlinger, and P. Reineker, *Phys. Rev. B* 53, 15451 (1996).

[45] A. A. Sukhorukov, *Phys. Rev. E* **61**, 4530 (2000).

[46] D. E. Pelinovsky, A. V. Buryak, and Yu. S. Kivshar, *Phys. Rev. Lett.* **75**, 591 (1995).

[47] L. Torner, D. Mihalache, D. Mazilu, and N.N. Akhmediev, *Opt. Lett.* **20**, 2183 (1995).

[48] C. Etrich, U. Peschel, F. Lederer, B. A. Malomed, and Yu. S. Kivshar, *Phys. Rev. E* **54**, 4321 (1996).

[49] D. E. Pelinovsky, J. E. Sipe, and J. Yang, *Phys. Rev. E* **59**, 7250 (1999).

[50] A. C. Yew, A.R. Champneys, and P.J. McKenna, *J. Nonlinear Sci.* **9**, 33 (1999).

[51] C. Etrich, U. Peschel, F. Lederer, D. Mihalache, and D. Mazilu, *Opt. Quantum Electron.* **30**, 881 (1998).

[52] A. C. Yew, B. Sandstede, and C. K. R. T. Jones, *Phys. Rev. E* **61**, 5886 (2000).

[53] L. Torner, D. Mazilu, and D. Mihalache, *Phys. Rev. Lett.* **77**, 2455 (1996).

[54] L. Torner, D. Mihalache, D. Mazilu, and N.N. Akhmediev, *J. Opt. Soc. Am. B* **15**, 1476 (1998).

[55] C. Etrich, U. Peschel, F. Lederer, and B. A. Malomed, *Phys. Rev. E* **55**, 6155 (1997).

[56] C. Etrich, U. Peschel, F. Lederer, D. Mihalache, and D. Mazilu, *Opt. Quantum Electron.* **30**, 881 (1998).

[57] L. Torner, *Opt. Lett.* **23**, 1256 (1998).

[58] R. Schiek, Y. Baek, and G. I. Stegeman, *Phys. Rev. E* **53**, 1138 (1996).

[59] Y. Baek, "Cascaded Second-Order Nonlinearities in Lithium Niobate Waveguides," Ph.D. Thesis (Department of Physics, University of Central Florida, Orlando, FL, 1997).

[60] R. Schiek, Y. Baek, G. I. Stegeman, and W. Sohler, *Opt. Lett.* **24**, 83 (1999).

[61] D.-M. Baboiu, G. I. Stegeman, and L. Torner, *Opt. Lett.* **20**, 2282 (1995).

[62] C. Etrich, U. Peschel, F. Lederer, and B. A. Malomed, *Phys. Rev. A* **52**, 3444 (1995).

[63] C. B. Clausen, P.L. Christiansen, and L. Torner, *Opt. Commun.* **136**, 185 (1997).

[64] D.-M. Baboiu and G. I. Stegeman, *J. Opt. Soc. Am. B* **14**, 3143 (1997).

[65] Y. Baek, R. Schiek, G. I. Stegeman, I. Baumann, and W. Sohler, *Opt. Lett.* **22**, 1550 (1997).

[66] K. Hayata and M. Koshiba, *Phys. Rev. A* **50**, 675 (1994).

[67] T. J. Alexander, A. V. Buryak, and Yu. S. Kivshar, *Opt. Lett.* **28**, 670 (1998).

[68] B. A. Malomed, D. Anderson, and M. Lisak, *Opt. Commun.* **126**, 251 (1996).

[69] H. T. Tran, *Opt. Commun.* **118**, 581 (1995).

[70] U. Peschel, C. Etrich, F. Lederer, and B. A. Malomed, *Phys. Rev. E* **55**, 7704 (1997).

[71] G. Leo, G. Assanto, and W. E. Torruellas, *Opt. Lett.* **22**, 7 (1997).

[72] G. Leo, G. Assanto, and W. E. Torruellas, *Opt. Commun.* **134**, 223 (1997).

[73] G. Leo and G. Assanto, *Opt. Lett.* **22**, 1391 (1997).

[74] B. S. Azimov, A. P. Sukhorukov, and D. V. Trukhov, *Izv. Akad. Nauk USSR* **51**, 229 (1987) [*Bull. Acad. Sci. USSR Phys.* **51**(2), 19 (1987)].

[75] A. D. Capobianco, B. Costantini, C. De Angelis, D. Modotto, A. Lauteri Palma, G. F. Nalesso, and C. G. Someda, *Opt. Quantum Electron.* **30**, 483 (1998).

[76] A. V. Buryak, Yu. S. Kivshar, and S. Trillo, *J. Opt. Soc. Am. B* **14**, 3110 (1997).

[77]   Yu. S. Kivshar and D. E. Pelinovsky, *Phys. Rep.* **331**, 117 (2000).

[78]   L. Bergé, V. K. Mezentsev, J. J. Rasmussen, and J. Wyller, *Phys. Rev. A* **52**, R28 (1995).

[79]   L. Torner, W. E. Torruellas, G. I. Stegeman, and C. R. Menyuk, *Opt. Lett.* **20**, 1952 (1995).

[80]   L. Torner and E. M. Wright, *J. Opt. Soc. Am. B* **13**, 864 (1996).

[81]   L. Torner, D. Mihalache, D. Mazilu, E. M. Wright, W. E. Torruellas, and G. I. Stegeman, *Opt. Commun.* **121**, 149 (1995).

[82]   L. Bergé, O. Bang, J. J. Rasmussen, and V. K. Mezentsev, *Phys. Rev. E* **55**, 3555 (1997).

[83]   O. Bang, Yu. S. Kivshar, and A. V. Buryak, *Opt. Lett.* **22**, 1680 (1997).

[84]   O. Bang, Yu. S. Kivshar, A. V. Buryak, A. De Rossi, and S. Trillo, *Phys. Rev. E* **58**, 5057 (1998).

[85]   D. Mihalache, D. Mazilu, L.-C. Crasovan, and L. Torner, *Opt. Commun.* **137**, 113 (1997).

[86]   D. Mihalache, D. Mazilu, J. Dörring, and L. Torner, *Opt. Commun.* **159**, 129 (1999).

[87]   D. Mihalache, D. Mazilu, B. A. Malomed, and L. Torner, *Opt. Commun.* **169**, 341 (1999).

[88]   B. A. Malomed, P. Drummond, H. He, A. Berntson, D. Anderson, and M. Lisak, *Phys. Rev. E* **56**, 4725 (1997).

[89]   D. V. Skryabin and W.J. Firth, *Opt. Commun.* **148**, 79 (1998).

[90]   D. V. Skryabin and W.J. Firth, *Phys. Rev. Lett.* **81**, 3379 (1998).

[91]   W. E. Torruellas, Z. Wang, D. J. Hagan, E. W. Van Stryland, and G. I. Stegeman, *Phys. Rev. Lett.* **74**, 5036 (1995).

[92]   W. E. Torruellas, Z. Wang, L. Torner, and G. I. Stegeman, *Opt. Lett.* **20**, 1949 (1995).

[93]   W. E. Torruellas, G. Assanto, B. L. Lawrence, R. A. Fuerst, and G. I. Stegeman, *Appl. Phys. Lett.* **68**, 1449 (1996).

[94]   B. Bourliaguet, V. Couderc, A. Barthélémy, G. W. Ross, P. G. R. Smith, D. C. Hanna, and C. De Angelis, *Opt. Lett.* **24**, 1410 (1999).

[95]   P. Di Trapani, G. Valiulis, W. Chinaglia, and A. Andreoni, *Phys. Rev. Lett.* **80**, 265 (1998).

[96]   M. T. G. Canva, R. A. Fuerst, D. Baboiu, G. I. Stegeman, and G. Assanto, *Opt. Lett.* **22**, 1683 (1997).

[97]   A. De Rossi, S. Trillo, A. V. Buryak, and Yu. S. Kivshar, *Opt. Lett.* **22**, 868 (1997).

[98]   A. De Rossi, S. Trillo, A. V. Buryak, Yu. S. Kivshar, *Phys. Rev. E* **56**, R4959 (1997).

[99]   D.-M. Baboiu and G. I. Stegeman, *Opt. Lett.* **23**, 31 (1998).

[100] D. V. Skryabin, *Phys. Rev. E* **60**, 7511 (1999).

[101] R. A. Fuerst, D. M. Baboiu, B. Lawrence, W. E. Torruellas, G. I. Stegeman, S. Trillo, S. Wabnitz, *Phys. Rev. Lett.* **78**, 2756 (1997).

[102] R. A. Fuerst, B. L. Lawrence, W. E. Torruellas, and G. I. Stegeman, *Opt. Lett.* **22**, 19 (1997).

[103] X. Liu, K. Beckwitt, and F. Wise, *Phys. Rev. Lett.* **85**, 1871 (2000).

[104] X. Liu, K. Beckwitt, and F. Wise, *Phys. Rev. E* **62**, 1328 (2000).

[105] D. V. Petrov, L. Torner, J. Martorell, R. Vilaseca, J. P. Torres, and C. Cojocaru *Opt. Lett.* **23**, 1444 (1998).

[106] V. V. Steblina, Yu. S. Kivshar, and A. V. Buryak, *Opt. Lett.* **23**, 156 (1997).

[107] V. V. Steblina and A. V. Buryak, *J. Opt. Soc. Am. B* **16**, 245 (1999).

[108] B. Costantini, C. De Angelis, A. Barthelemy, B. Bourliaguet, and V. Kermene, *Opt. Lett.* **23**, 1376 (1998).

[109] G. Assanto, I. Torelli, and S. Trillo, *Opt. Lett.* **19**, 1720 (1994).

[110] A. D. Boardman, P. Bontemps, and K. Xie, *Opt. Quantum Electron.* **30**, 891 (1998).

[111] K. Koynov and S. Saltiel, *Opt. Commun.* **152**, 96 (1998).

[112] Yu. S. Kivshar, T. J. Alexander, and S. M. Saltiel, *Opt. Lett.* **24**, 759 (1999).

[113] Yu. S. Kivshar, A. A. Sukhorukov, and S. M. Saltiel, *Phys. Rev. E* **60**, R5056 (1999).

[114] S. A. Akhmanov, A. N. Dubrovik, S. M. Saltiel, I. V. Tomov, and V. G. Tunkin, *Pis'ma Zh. Eksp. Teor. Fiz.* **20**, 264 (1974) [*JETP Lett.* **20**, 117 (1974)].

[115] M. M. Fejer, G. A. Magel, D. H. Jundt, and R. L. Byer, *IEEE J. Quantum Electron.* **28**, 2631 (1992).

[116] J. A. Armstrong, N. Bloembergen, J. Ducuing, and P. S. Pershan, *Phys. Rev.* **127**, 1918 (1962); P. A. Franken and J. F. Ward, *Rev. Mod. Phys.* **35**, 23 (1963).

[117] E. J. Lim, M. M. Fejer, and R. L. Byer, *Electron. Lett.* **25**, 174 (1989).

[118] K. Mizuuchi, K. Yamamoto, and T. Taniuchi, *Appl. Phys. Lett.* **58**, 2732 (1991).

[119] T. Fujimura, T. Suhara, and H. Nishihara, *Electron. Lett.* **27**, 1207 (1991).

[120] C. B. Clausen, O. Bang, and Yu. S. Kivshar, *Phys. Rev. Lett.* **78**, 4749 (1997).

[121] G. P. Agrawal, *Applications of Nonlinear Fiber Optics* (Academic Press, San Diego, 2001).

[122] C. B. Clausen, Yu. S. Kivshar, O. Bang, and P. L. Christiansen, *Phys. Rev. Lett.* **83**, 4740 (1999).

[123] D. Schechtman, I. Blech, D. Gratias, and J. W. Cahn, *Phys. Rev. Lett.* **53**, 1951 (1984).

[124] W. Gellermann, M. Kohmoto, B. Sutherland, and P. C. Taylor, *Phys. Rev. Lett.* **72**, 63 (1994).

[125] S. Zhu, Y. Zhu, Y. Qin, H. Wang, C. Ge, and N. Ming, *Phys. Rev. Lett.* **78**, 26752 (1997).

[126] S. Zhu, Y. Zhu, and N. Ming, *Science* **278**, 843 (1997); Y. Zhu *et al.*, *Appl. Phys. Lett.* **73**, 432 (1998).

[127] L. S. Telegin and A. S. Chirkin, *Sov. J. Quantum Electron.* **12**, 1354 (1982).

[128] S. Trillo and S. Wabnitz, *Opt. Lett.* **17**, 1572 (1992).

[129] A. Kobyakov, F. Lederer, O. Bang, and Yu. S. Kivshar, *Opt. Lett.* **23**, 506 (1998).

[130] M. V. Komissarova and A. P. Sukhorukov, *Bull. Russian Acad. Sci. Phys.* **56**, 1995 (1992).

[131] M. A. Karpierz, *Opt. Lett.* **20**, 1677 (1995).

[132] A. V. Buryak, Yu. S. Kivshar, and S. Trillo, *Opt. Lett.* **20**, 1961 (1995).

[133] S. Trillo, A. V. Buryak, and Yu. S. Kivshar, *Opt. Commun.* **122**, 200 (1996).

[134] O. Bang, L. Bergé, and J. J. Rasmussen, *Opt. Commun.* **146**, 231 (1998).

[135] A. De Rossi, G. Assanto, S. Trillo, and W. E. Torruellas, *Opt. Commun.* **150**, 390 (1998).

[136] T. J. Alexander, Yu. S. Kivshar, A. V. Buryak, and R.A. Sammut, *Phys. Rev. E* **61**, 2042 (2000).

[137] Z. K. Yankauskas, *Izv. Vuzov Radiofiz.* **9**, 412 (1966) [*Sov. Radiophys.* **9**, 261 (1966)]

[138] V. I. Kruglov and R.A. Vlasov, *Phys. Lett. A* **111**, 401 (1985).

[139] W. J. Firth and D. V. Skryabin, *Phys. Rev. Lett.* **79**, 2450 (1997).

[140] V. Tikhonenko, J. Christou, and B. Luther-Davies, *J. Opt. Soc. Am. B* **12**, 2046 (1995).

[141] V. Tikhonenko, J. Christou, and B. Luther-Davies, *Phys. Rev. Lett.* **76**, 2698 (1996).

[142] L. Torner and D. V. Petrov, *Electron. Lett.* **33**, 608 (1997).

[143] L. Torner and D. V. Petrov, *J. Opt. Soc. Am. B* **14**, 2017 (1997).

[144] J. P. Torres, J. M. Soto-Crespo, L. Torner, and D. V. Petrov, *J. Opt. Soc. Am. B* **15**, 625 (1998).

[145] J. P. Torres, J. M. Soto-Crespo, L. Torner, and D. V. Petrov, *Opt. Commun.* **149**, 77 (1998).

[146] L. Torner, J. P. Torres, D. V. Petrov, and J. M. Soto-Crespo, *Opt. Quantum Electron.* **30**, 809 (1998).

[147] P. Di Trapani, W. Chinaglia, S. Minardi, A. Piskarskas, and G. Valiulis, *Phys. Rev. Lett.* **84**, 3843 (2000).

[148] Y. Silberberg, *Opt. Lett.* **15**, 1282 (1990).

[149] D. E. Edmundson and R.H. Enns, *Opt. Lett.* **17**, 596 (1992).

[150] D. E. Edmundson and R.H. Enns, *Opt. Lett.* **18**, 1609 (1993).

[151] D. Mihalache, D. Mazilu, B. Maloned, and L. Torner, *Opt. Commun.* **152**, 365 (1998).

[152] P. Di Trapani, D. Caironi, G. Valiulis, A. Dubietis, R. Danielius, and A. Piskarskas, *Phys. Rev. Lett.* **81**, 570 (1998).

[153] X. Liu, L.J. Qian, and F.W. Wise, *Phys. Rev. Lett.* **82**, 4631 (1999).

[154] F. F. So, S. R. Forrest, Y. Q. Shi, W. H. Steier, *Appl. Phys. Lett.* **56**, 674 (1990).

[155] Y. Imanishi, S. Hattori, A. Kakuta, and S. Numata, *Phys. Rev. Lett.* **71**, 2098 (1993).

[156] T. Nanaka, Y. Mori, N. Nagai, Y. Nakagawa, N. Saeda, T. Nakahaki, and A. Ishitani, *Thin Solid Films* **239**, 214 (1994).

[157] V. Bulovic and S. R. Forrest, *Chem. Phys. Lett.* **238**, 88 (1995).

[158] V. M. Agranovich and O. A. Dubovsky, *Chem. Phys. Lett.* **210**, 458 (1993).

[159] V. M. Agranovich, O. A. Dubovsky, and A. M. Kamchatnov, *J. Chem. Phys.* **28**, 13607 (1994).

[160] V. M. Agranovich, P. Reineker, and V. I. Yudson, *Syntetic Metals* **64**, 147 (1994).

[161] V. M. Agranovich, O. A. Dubovsky, and A. M. Kamchatnov, *Chem. Phys. Lett.* **198**, 245 (1995).

[162] V. M. Agranovich and A. M. Kamchatnov, *Pis'ma Zh. Eksp. Teor. Fiz.* **59**, 397 (1994) [*JETP Lett.* **59**, 424 (1994)].

[163] V. M. Agranovich, S. A. Darmanyan, A. M. Kamchatnov, T. A. Leskova, and A.D. Boardman, *Phys. Rev. E* **55**, 1894 (1997).

[164] S.A. Dubovsky and A. V. Orlov, *Phys. Solid State* **38**, 675 (1996); *Phys. Solid State* **38**, 1067 (1996).

[165] O. Bang, P. L. Christiansen, and C. B. Clausen, *Phys. Rev. E* **56**, 7257 (1997).

[166] S. Darmanyan, A. Kobyakov, and F. Lederer, *Phys. Rev. E* **57**, 2344 (1998).

[167] S. Darmanyan, A. Kamchatnov, and F. Lederer, *Phys. Rev. E* **58**, R4120 (1998).

[168] T. Peschel, U. Peschel, and F. Lederer, *Phys. Rev. E* **57**, 1127 (1998).

[169] A. Kobyakov, S. Darmanyan, T. Pertsch, and F. Lederer, *J. Opt. Soc. Am. B* **16**, 1737 (1999).

[170] A. B. Aceves, C. De Angelis, T. Peschel, R. Muschall, F. Lederer, S. Trillo, and S. Wabnitz, *Phys. Rev. E* **53**, 1172 (1996).

[171] S. M. Jensen, *IEEE J. Quantum Electron.* **18**, 1580 (1982).

[172] A. A. Sukhorukov, Yu. S. Kivshar, O. Bang, and C. M. Soukoulis *Phys. Rev. E* **63**, 016615 (2001).

[173] A. A. Sukhorukov, Yu. S. Kivshar, and O. Bang, *Phys. Rev. E* **59**, R41 (1999).

Chapter 11

Discrete Solitons

A periodic array of optical waveguides creates a novel kind of device in which new types of spatial solitons can be generated and studied experimentally. The properties of spatially localized modes in a waveguide array are usually analyzed in the framework of a set of coupled-mode equations, each equation representing the soliton amplitude in a specific waveguide but coupled to the neighboring waveguides. We consider such a coupled set of equations in Section 11.1, referred to as the *discrete* nonlinear Schrödinger (NLS) equation, and discuss its localized solutions, known as *discrete solitons*. In Section 11.2 we follow a different approach and model the waveguide array as a periodic structure formed by a sequence of thin-film nonlinear waveguides embedded in an otherwise linear dielectric medium. This model goes beyond the approximations inherent in the use of the discrete NLS equation and thus allows us to study several interesting properties of more realistic waveguide arrays. Section 11.3 focuses on the modulation instability occurring in such nonlinear arrays, while Section 11.4 describes the bright and dark spatial solitons of different types associated with this instability. Experimental results on the generation and steering of spatial solitons in waveguide arrays are summarized in Section 11.5. In Section 11.6 we present two generalizations of the concept of discrete solitons. First, we discuss the concept of two-dimensional discrete soliton networks that may prove useful for signal-processing operations such as routing and time gating. Second, we describe the recent ideas and the experimental results on optically induced waveguide arrays.

11.1 Discrete NLS Equation

Discrete spatial solitons were first introduced in 1988 by Christodoulides and Joseph [1], who studied theoretically the spatially localized modes of a periodic optical structure (created by using an array of optical waveguides) based on the analogy with the localized modes of a discrete lattice [2, 3]. Although discrete spatial solitons were studied extensively after 1988 from a theoretical viewpoint [4]–[9], it was only after 1998 that they were observed experimentally using arrays of single-mode nonlinear optical waveguides [10]–[15].

Figure 11.1: A polymer-based linear waveguide array (before applying the polymer cladding) consisting of 75 single-mode waveguides. (After Ref. [15]; ©2002 APS.)

11.1.1 Nonlinear Waveguide Arrays

The standard theoretical approach for studying discrete spatial solitons is based on analyzing the solutions of an effective discrete NLS equation [9]. A similar approach in solid-state physics is known as the *tight-binding approximation*. In the context of an optical waveguide array, this approximation corresponds to the assumption that the fundamental modes in all waveguides are only *weakly coupled*. The concept of weak coupling appears in other contexts, such as the study of nonlinear dynamics of a Bose–Einstein condensate in optical lattices [16].

Figure 11.1 shows an example of a waveguide array made using an organic polymer as the nonlinear material. The theoretical model for describing the dynamics of such a periodic array of nonlinear optical waveguides is based on the generalization of the theory behind a two-core directional coupler to the case of N evanescently coupled waveguides [17]. The main assumption of the model is that the evolution of the slowly varying field envelopes associated with the individual guided modes of a homogeneous array of single-mode waveguides can be described by a set of coupled equations that takes into account only the nearest-neighbor interaction resulting from the weakly overlapping guided modes. In the case of an ideal, infinite-size array (no losses) with only the Kerr-type nonlinearity, this set of equations takes the form [1]

$$i\frac{dA_n}{dz} + \beta A_n + C(A_{n-1} + A_{n+1}) + \gamma|A_n|^2 A_n = 0, \qquad (11.1.1)$$

where A_n is the mode amplitude in the nth waveguide, β is the linear propagation constant, C is the coupling coefficient, and $\gamma = (\omega_0 n_2)/(cA_{\text{eff}})$ is the nonlinear parameter introduced in Chapter 1. Here, ω_0 is the optical frequency associated with the modes, n_2 is the Kerr coefficient, and A_{eff} is the effective area of the waveguide modes. No second-order derivatives appear in Eq. (11.1.1) because all dispersive and diffractive effects are ignored in each waveguide. This set of infinite equations is referred to as the *discrete NLS equation*. It describes the physical situation reasonably well when a CW beam is launched into a large array in which each waveguide supports a single mode and confines it in both transverse dimensions (no diffraction).

For a finite-size array consisting of N waveguides, the subscript n in Eq. (11.1.1) varies in the range 1–N with the boundary condition $A_0 = A_{N+1} = 0$. When $N = 2$, Eq. (11.1.1) describes the well-known case of a nonlinear directional coupler, which is known to be completely integrable [18, 19]. However, Eq. (11.1.1) is not integrable analytically, even for $N = 3$, and exhibits self-trapping as well as chaotic dynamics [20, 21]. It is easy to show that Eq. (11.1.1) possesses the following two constants of motion for all values of N:

$$P = \sum_{n=1}^{N} |A_n|^2, \qquad H = \sum_{n=1}^{N} \left(\beta |A_n|^2 + C|A_n - A_{n-1}|^2 - \frac{\gamma}{2} |A_n|^4 \right). \qquad (11.1.2)$$

These are related to two conserved quantities—the total power P and the Hamiltonian H or the total system energy [3].

11.1.2   Discrete Diffraction

Even though no diffraction occurs within each waveguide, an optical beam can still spread over the whole array because of the coupling among the waveguides. It is common to refer to this spreading as the *interwaveguide diffraction* or *discrete diffraction*. The linear terms in Eq. (11.1.1) govern the diffractive properties of the whole array. This equation can be solved analytically when the nonlinear term is neglected. The solution is in the form of a plane wave $A_n = A_0 \exp(ink_x d + ik_z z)$, where d represents the center-to-center spacing between the waveguides in the x direction. Substituting this solution in Eq. (11.1.1) with $\gamma = 0$, we obtain the following dispersion relation between k_x and k_z [22]:

$$k_z = \beta + 2C \cos(k_x d). \qquad (11.1.3)$$

Following the standard definition of the group-velocity dispersion (GVD) for optical pulses [23], we can introduce a diffraction parameter for the entire array using $\mathcal{D} = \partial^2 k_z / \partial k_x^2$. Equation (11.1.3) then yields

$$\mathcal{D} = -2Cd^2 \cos(k_x d). \qquad (11.1.4)$$

An important property of the dispersion relation (11.1.3) is its periodicity in k_x. In fact, because of the periodicity and continuity of the dispersion relation, there exists a maximum angle, $\theta_{\max}$, for the diffraction of a plane wave inside the array. Using the standard terminology from the field of solid-state physics, the Brillouin zone is formed in the range $|k_x d| < \pi$. The eigenmodes for a periodic system are known as the *Floquet–Bloch waves*. Such waves are encountered in optics in other contexts such as the propagation of light in fiber Bragg gratings or a corrugated planar waveguide [24]–[26]. They have also been used in the context of two-dimensional optical lattice [27]. Generally speaking, the Floquet–Bloch treatment is not based on the tight-binding approximation and allows a more general study of the properties of periodic media. Nevertheless, as in solid-state physics, both approaches lead to the same qualitative results [28].

The most important feature of Eq. (11.1.4) in the context of waveguide arrays is the change in the sign of the diffraction parameter $\mathcal{D}$ in the outer parts of the Brillouin zone. The positive values of $\mathcal{D}$ in the range $\pi/2 < |k_x d| \leq \pi$ are analogous to optical pulses experiencing anomalous dispersion. In this region, diffractive properties are anomalous in the sense that they are opposite to those experienced in nature. Moreover, diffraction completely disappears around the two points $k_x = \pm\pi/2d$. A similar feature occurs for photonic crystals [29].

Similar to an optical beam propagating inside a bulk medium, the diffraction within a waveguide array spreads the beam from one waveguide to another at low powers, for which the cubic term in Eq. (11.1.1) can be neglected. This infinite set of equations is analytically integrable in this linear case [30]. When only one waveguide is excited initially such that $A_n = 0$ at $z = 0$ for all $n \neq 0$, the solution is given by

$$A_n(z) = A_0(i)^n J_n(2Cz) \exp(i\beta z),\tag{11.1.5}$$

where J_n is the Bessel function of order n. Physically, as the CW beam propagates along the waveguides, its power spreads into many nearby waveguides in a symmetric fashion such that the intensities are distributed as $J_n^2(2Cz)$ at any z. At a distance such that $2Cz \approx 2.405$, J_0 vanishes and all the power disappears from the original waveguide. A part of the power reappears with further propagation, as dictated by the the zeroth-order Bessel function.

11.1.3 Discrete Spatial Solitons

When the input intensity becomes large enough that the Kerr term cannot be neglected, Eq. (11.1.1) cannot be solved analytically in general. However, when the intensity varies slowly over adjacent waveguides (i.e, for weak coupling), the discrete set of ordinary differential equations in Eq. (11.1.1) can be converted to a single continuous NLS equation [8], which can be solved and which supports one-dimensional spatial solitons, discussed in Chapter 2. Indeed, the numerical solutions based on Eq. (11.1.1) reveal that at high enough intensity levels, the field distribution inside the array is well described by

$$A_n(z) = A_0 \operatorname{sech}(X_n/X_0) \exp(i\beta z + 2iCz),\tag{11.1.6}$$

where $X_n = nd$ is the location of the nth waveguide and X_0 is the width of the discrete soliton [1, 7]. Of course, X_0 is relatively large because the soliton tails are spread over a large number of waveguides.

In the general case of a moderate to strong coupling, numerical solutions of Eq. (11.1.1) can be found in the form of nonlinear modes localized over only a few wave-guides [31, 32]. At each power level, there is one solution centered on a single wave-guide and another centered in between the two neighboring waveguides. Figure 11.2 shows an example of the two types of such discrete solitons. It turns out that the soliton centered inside a single waveguide is stable because it corresponds to the minimum of the Hamiltonian [5]. If the discrete soliton is forced to move sideways, it has to jump from one waveguide to the next, passing from a stable to an unstable configuration. The difference between the Hamiltonians in the two cases—the so-called Peierls–Nabarro potential—accounts for the resistance that the soliton has to overcome during trans-verse propagation [33]. This potential increases as the input power level increases. As a result, the soliton becomes localized to a single waveguide and is effectively decoupled from the rest of the array. Moving discrete solitons can form for certain input parameters, but they have a more complicated phase structure [31, 34].

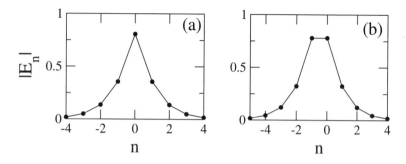

Figure 11.2: Two examples of discrete solitons found numerically by solving the discrete NLS equation. Only the soliton shown in the panel (a) is stable.

Numerical simulations show that discrete solitons share a few basic properties with standard spatial solitons but differ in others. For example, just like spatial solitons, discrete solitons do not necessarily always propagate along the length of waveguides. When a discrete soliton is launched with a linear phase gradient across the waveguides, it propagates at an angle to the direction of the waveguides. However, as the soliton power is increased, it may change its direction of propagation. Such properties of discrete solitons have been suggested for soliton-based optical switching and beam steering using waveguide arrays [35]–[40].

The discrete NLS equation (11.1.1) can also be used for describing the dynamics of discrete temporal solitons in fiber arrays, provided it is generalized to include the dispersive effects, governed by the parameter β_2 [23], for each individual fiber of the array by adding a term of the form $\beta_2(\partial^2 A_n/\partial t^2)$. The resulting set of N coupled but continuous NLS equations is known to display complex dynamics, even in the simple case of $N = 3$ [41]. However, for all values of N this set of equations possesses specific solutions that represent stable soliton-like pulses localized not only in time but also in the spatial direction perpendicular to the direction of propagation [42]–[45]. Such multidimensional spatiotemporal solitons can form even in the case of a Kerr nonlinearity, because the collapse is prevented by the discreteness of the waveguide array. This mechanism of avoiding the beam collapse was first found in the context of a (1+1)-dimensional generalized NLS equation with power-law nonlinearity [46, 47].

11.2   General Theory

The theory of discrete solitons based on the discrete NLS equation does not properly account for many of the experimental results for periodic structures of more complicated geometries. When the weak-coupling approximation is not valid, the applicability of the tight-binding approach and the resulting discrete NLS equation become questionable. For example, the recent experimental results [48] on the measurement of the bandgap structure and the Floquet–Bloch modes of waveguide arrays cannot be explained within the framework of the discrete NLS equation, because it describes

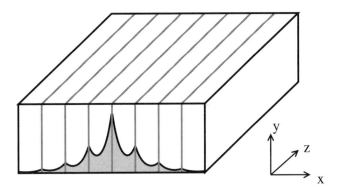

Figure 11.3: An array of thin-film nonlinear waveguides embedded in a linear slab waveguide. Gray shading shows the spatial profile of a nonlinear localized mode.

only a single band. This section focuses on an improved analytic model that takes into account the bandgap structure of periodic arrays.

11.2.1 Improved Analytical Model

An essential feature of wave propagation in any periodic structure (that follows from the Floquet–Bloch theory) is the existence of one or more forbidden bandgaps in the transmission spectrum. Any incident light is totally reflected when its frequency lies in one of these bandgaps. The nonlinear effects can change this behavior because they can lead to the formation of localized states inside such bandgaps. As discussed in Chapter 5, such a localized state is sometimes called a *gap soliton*. The discrete NLS equation, derived in the tight-binding approximation, describes only one transmission band surrounded by two semi-infinite bandgaps. On the other hand, the coupled-mode theory developed for gap solitons describes only the modes localized in an isolated narrow gap (see Chapter 5 and Refs. [49]–[52]). As a result, it does not allow one to consider simultaneously gap solitons and conventional spatial solitons.

The inclusion of the complete bandgap structure of the transmission spectrum is very important for the stability analysis of nonlinear localized modes [53]. Such an analysis is especially important for the nonlinear localized modes in photonic crystals (see Ref. [54] and Chapter 12). For this reason, we consider a periodic array of thin-film nonlinear waveguides embedded in an otherwise linear dielectric medium [55]–[57]. Such a structure, shown schematically in Figure 11.3, can be regarded as a nonlinear analog of the so-called Dirac comb lattice [58]. The linear effects of the periodicity and the bandgap spectrum are taken into account explicitly; the nonlinear effects enter through the boundary conditions at the interfaces and can be studied analytically.

Consider an electromagnetic wave propagating along the Z direction of such a structure created by a periodic array of thin-film nonlinear waveguides (see Figure 11.3). Assuming that the field structure in the Y direction is defined by the linear guided mode $W(Y;X)$ of the slab waveguide, we write the total field as $\mathcal{E}(X,Y,Z) = W(Y;X)A(X,Z)$. The evolution of the field envelope $A(X,Z)$ along the Z axis is then governed by the paraxial equation

$$i\frac{\partial A}{\partial Z} + \frac{1}{2k_0}\frac{\partial^2 A}{\partial X^2} + \varepsilon(X)k_0 A + g(X)|A|^2 A = 0, \tag{11.2.1}$$

where $k_0 = \omega_0/c$ for a CW beam at frequency ω_0. The linear propagation of the guided wave is governed by the dielectric constant $\varepsilon(X)$, whereas $g(X)$ characterizes the Kerr-type nonlinear response of thin layers. We assume that both $\varepsilon(X)$ and $g(X)$ are periodic functions of X. The periodicity of $g(X)$ describes the nonlinear, thin-film, multilayer structure seen in Figure 11.3 [59]–[61]. On the other hand, the periodicity of $\varepsilon(X)$ describes an impurity band in a deep photonic bandgap [62].

To reduce the number of physical parameters, we normalize Eq. (11.2.1) using

$$A(X,Z) = A_0\psi(x,z)\exp(i\bar{\varepsilon}k_0 Z), \quad x = X/d, \quad z = Z/(2kd^2), \tag{11.2.2}$$

where $\bar{\varepsilon}$ represents the mean value of $\varepsilon(X)$, d is a characteristic transverse scale (usually array spacing), and A_0 is a typical value of the field amplitude. The normalized NLS equation then takes the form

$$i\frac{\partial \psi}{\partial z} + \frac{\partial^2 \psi}{\partial x^2} + \mathcal{F}(I;x)\psi = 0, \tag{11.2.3}$$

where $I \equiv |\psi|^2$ is the normalized intensity and the real function

$$\mathcal{F}(I;x) = (2k_0 d^2)\{[\varepsilon(X) - \bar{\varepsilon}]k_0 + g(X)|A_0|^2 I\} \tag{11.2.4}$$

describes both the *nonlinear* and *periodic* properties of the layered medium.

In the absence of losses, Eq. (11.2.3) describes a Hamiltonian system, and its spatially localized solutions conserve the total power, defined as

$$P = \int_{-\infty}^{\infty} |\psi(x,z)|^2 \, dx. \tag{11.2.5}$$

It is also important to note that Eq. (11.2.3) describes the beam evolution in the *paraxial approximation* and is valid for waves propagating mainly along the z direction [63]). More precisely, it is valid as long as the length scale over which the CW beam changes its shape is much longer than its width in the transverse direction x. This leads to the condition that the dielectric constant must be weakly modulated, or $|\varepsilon(X) - \bar{\varepsilon}| \ll |\bar{\varepsilon}|$.

We look for shape-preserving localized solutions of Eq. (11.2.3) in the form

$$\psi(x,z) = u(x;\beta)\exp(i\beta z), \tag{11.2.6}$$

where β is the propagation constant. Using Eq. (11.2.6) in Eq. (11.2.3), $u(x;\beta)$ is found to satisfy the following ordinary nonlinear differential equation:

$$\frac{d^2 u}{dx^2} + \mathcal{F}(I;x)u = \beta u. \tag{11.2.7}$$

If there is no energy flow along the transverse direction x, the function $u(x)$ is real (up to a constant phase that can be removed by a coordinate shift $z \to z - z_0$). This is always the case for spatially localized solutions for which $u(x \to \pm\infty) = 0$.

To simplify the analysis further, we assume that the linear periodicity is associated only with the presence of an array of thin-film waveguides and use the following form for $\mathcal{F}(I;x)$:

$$\mathcal{F}(I;x) = \sum_{n=-\infty}^{+\infty} (\alpha + \gamma I)\delta(x - nh), \tag{11.2.8}$$

where h is the spacing between the neighboring thin films (the lattice period) and n is an integer. The total response of each thin film is approximated by a delta function, and the real parameters α and γ describe the *linear* and *nonlinear* properties of the layer, respectively. Without any loss of generality, the nonlinear coefficient γ can be normalized to unity such that $\gamma = +1$ corresponds to a *self-focusing* nonlinearity and $\gamma = -1$ to a *self-defocusing* nonlinearity. The linear coefficient ($\alpha > 0$) defines the low-intensity response and characterizes the coupling strength among the waveguides. This model based on Eq. (11.2.8) was first analyzed in Refs. [56] and [57], and it can be regarded as a nonlinear analog of the Dirac comb lattice [58].

11.2.2 Dispersion Characteristics

We can convert Eq. (11.2.7) into a discrete NLS equation by representing the eigenmodes associated with this equation in the form [63]

$$u(x) = a_n e^{-\mu(x-nh)} + b_n e^{+\mu(x-nh)}, \tag{11.2.9}$$

where $nh \le x \le (n+1)h$ and $\mu = \sqrt{\beta}$. The coefficients a_n and b_n can be written in terms of the amplitudes, $u_n \equiv u(hn)$ and $u_{n+1} \equiv u(nh+h)$, at the two ends of the nth waveguide as

$$a_n = \frac{u_n e^{\mu h} - u_{n+1}}{2\sinh(\mu h)}, \qquad b_n = u_n - a_n. \tag{11.2.10}$$

When we substitute Eqs. (11.2.9) and (11.2.10) into Eqs. (11.2.7) and (11.2.8), we find that the normalized amplitudes $U_n = |\xi\gamma|^{1/2}u_n$ satisfy the following discrete equation:

$$\eta U_n + (U_{n-1} + U_{n+1}) + \chi|U_n|^2 U_n = 0, \tag{11.2.11}$$

where $\chi = \text{sign}(\xi\gamma)$ and the parameters ξ and η are defined as

$$\xi = \sinh(\mu h)/\mu, \qquad \eta = -2\cosh(\mu h) + \alpha\xi. \tag{11.2.12}$$

Both of these parameters depend on β through $\mu = \sqrt{\beta}$.

The linear solutions of Eq. (11.2.11) for $\chi = 0$ can be written in the form $U_n = U_0 \exp(iKn)$ and yield the dispersion relation $\eta = -2\cos K$. Therefore, plane-wave-like solutions with real K can exist only for $|\eta| \le 2$. On the other hand, nonlinear localized modes with exponentially decaying tails can appear only for $|\eta| > 2$ (when K is imaginary). This condition defines the bandgap structure of the eigenvalue spectrum. The dependence of η and ξ on the propagation constant β is shown in Figure 11.4, where the transmission bands are indicated by the gray shading. The last semi-infinite bandgap corresponds to total internal reflection (IR). Other bandgaps are due to the resonant Bragg reflection (BR) from the periodic structure.

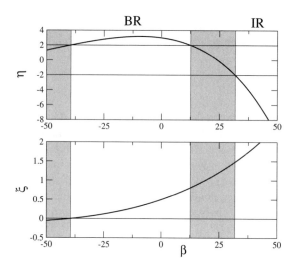

Figure 11.4: Dependence of η and ξ on the propagation constant β for a linear periodic array with $h = 0.5$ and $\alpha = 10$. Gray shading marks the transmission bands. The bandgaps marked BR and IR correspond to Bragg and internal reflections, respectively. (After Ref. [56]; ©2002 APS.)

A *linear* stability analysis of the localized modes can be performed by writing the solution in the form

$$\psi(x,z) = \left[u(x) + v(x)e^{i\Gamma z} + w^*(x)e^{-i\Gamma^* z} \right] e^{i\beta z} \qquad (11.2.13)$$

and considering the evolution of small-amplitude perturbations v and w. As usual, by using this form of solution in Eq. (11.2.3) we obtain a linear eigenvalue problem for $v(x)$ and $w(x)$. In general, the solutions of this eigenvalue problem fall into one of the following categories: (i) *internal modes* with real eigenvalues that describe periodic oscillations ("breathing") of the localized state, (ii) *instability modes* that correspond to purely imaginary eigenvalues with $\text{Im}(\Gamma) < 0$, and (iii) *oscillatory unstable modes* that appear when the eigenvalues are complex and $\text{Im}(\Gamma) < 0$. Additionally, there can exist decaying modes when $\text{Im}(\Gamma) > 0$. However, it follows from the structure of the eigenvalue equations that the eigenvalue spectrum is invariant with respect to the transformation $\Gamma \to \pm\Gamma^*$ when the amplitudes u_n are real (up to an arbitrary constant phase). In this case, the exponentially growing and decaying modes always coexist, and the latter do not significantly affect the dynamics. It is important to note that the functions $\eta(\beta)$ and $\xi(\beta)$ fully characterize the existence and stability of all localized and extended solutions of Eq. (11.2.3) with the response function (11.2.8).

11.2.3 Analytical Approximations

The theoretical model based on Eqs. (11.2.3) and (11.2.8) can be solved analytically in several approximations. In this section we consider three such cases, known as the *tight-binding approximation*, the *coupled-mode approach*, and the *two-component model*.

Tight-Binding Approximation

When each thin-film waveguide of the periodic structure supports a single mode that weakly overlaps with the modes of the two neighboring waveguides, the modes are only weakly coupled via a small change of the refractive index in the waveguides. The properties of discrete solitons can then be analyzed in the framework of the tight-binding approximation discussed earlier [64].

To employ this approximation, we first analyze the properties of a single thin-film waveguide in the linear regime and solve Eq. (11.2.7) with $\mathcal{F}(I;x) = \alpha\delta(x)$ to find the spatial profile of the linear guided mode. The solution is given by

$$u_s(x) = \exp\left(-\alpha|x|/2\right), \tag{11.2.14}$$

the corresponding value of the propagation constant being $\beta_s = \alpha^2/4$. We then consider the interaction among the waveguides, assuming that the total field can be written as a superposition of slightly perturbed waves localized at the isolated waveguides. Specifically, we assume that the propagation constant remains close to its unperturbed value β_s and neglect small variations in the spatial profiles of the localized modes. Thus, we seek a general solution of Eqs. (11.2.3) and (11.2.8) in the form

$$\psi(x,z) = \sum_{n=-\infty}^{+\infty} \psi_n(z)u_s(x-nh). \tag{11.2.15}$$

In this approximation, the wave evolution is characterized by the amplitude function $\psi_n(z)$ only.

To find the corresponding evolution equation, we substitute Eq. (11.2.15) into Eqs. (11.2.3) and (11.2.8), multiply the resulting equation by $u_s(x-mh)$, and integrate over the transverse profile of the mode. Following the original assumption of weakly interacting waveguides (valid for $\alpha h \gg 1$ and $|\gamma||\psi_n(z)|^2 \ll \alpha$), we use the approximation $|\psi(nh,z)|^2 \simeq |\psi_n(z)|^2$ and neglect the overlap integrals, $\int_{-\infty}^{\infty} u_s(x-nh)u_s(x-mh)\,dx$, for $|n-m| > 1$. We then arrive at the following set of coupled discrete equations for the field amplitudes at the nonlinear layers:

$$i\frac{d\psi_n}{dz} + \beta_s\psi_n + \frac{\alpha^2}{2}e^{-\alpha h/2}(\psi_{n-1} + \psi_{n+1}) + \frac{\gamma\alpha}{2}|\psi_n|^2\psi_n = 0. \tag{11.2.16}$$

The stationary solutions of Eq. (11.2.16) can be written in the form of Eq. (11.2.11) if we use $\psi_n = |\xi\gamma|^{-1/2}U_n e^{i\beta z}$ and define the parameter η and ξ as

$$\eta = -\frac{2}{\alpha^2}\left(\beta - \frac{\alpha^2}{4}\right)e^{\alpha h/2}, \qquad \xi = \frac{1}{\alpha}e^{\alpha h/2}. \tag{11.2.17}$$

These relations represent a series expansion of the original dispersion relation in Eq. (11.2.12) near the edges of the first transmission band for $\alpha h \gg 1$.

Coupled-Mode Theory

We now consider the opposite limit, $\alpha h \ll 1$, in which the first BR gap is narrow (see Figure 11.4). Mathematically, we assume that $|\beta_1 - \beta_2|$ is small compared with both $|\beta_1|$ and $|\beta_2|$, which represent the propagation constants at the edges of the bandgap and are obtained by using the condition $\eta(\beta) = 2$ in Eq. (11.2.12). It follows from this equation that $\beta_2 = -(\pi/h)^2$ and $\beta_1 \simeq \beta_2 + 2\alpha/h + O[(\alpha h)^{3/2}]$. To find solutions close to the BR gap, we write the total field in the form

$$\psi(x,z) = a_1(x,z)u_b(x;\beta_1) + a_2(x,z)u_b(x;\beta_2), \tag{11.2.18}$$

where a_j with $j = 1, 2$ are unknown nonlinear amplitudes and $u_b(x;\beta_j)$ are the linear Bloch functions satisfying Eqs. (11.2.7) and (11.2.8) when $\gamma = 0$. These Bloch functions can be found in an explicit form and are given by

$$u_b(x;\beta_2) = \sin(x\pi/h), \quad u_b(x+nh;\beta_1) = (-1)^n \sin[\sqrt{|\beta_1|}(h/2-x)], \tag{11.2.19}$$

where $0 \le x \le h$. Note that the field amplitudes at the layers are $u_n \sim (-1)^n a_1(nh)$, since $u_b(nh;\beta_2) \equiv 0$.

To find the evolution equations for the amplitudes a_1 and a_2, we substitute Eq. (11.2.18) into the original Eqs. (11.2.3) and (11.2.8). Next, we use the fact that the gap is narrow and that close to its edges the Bloch functions are weakly modulated; i.e., $|\partial a_j/\partial x| \ll |a_j/h|$ for $j = 1, 2$. This assumption allows us to keep only the lowest-order terms. If we multiply the resulting equation by $u_b(x;\beta_j)$ and integrate it over one period, the coupled-mode equations take the form

$$i\frac{\partial a_1}{\partial z} + \beta_1 a_1 + \frac{2\pi}{h}\frac{\partial a_2}{\partial x} + \frac{2\gamma}{h}|a_1|^2 a_1 = 0, \tag{11.2.20}$$

$$i\frac{\partial a_2}{\partial z} + \beta_2 a_2 - \frac{2\pi}{h}\frac{\partial a_1}{\partial x} = 0. \tag{11.2.21}$$

Equations (11.2.20) and (11.2.21) allow a direct comparison between the solutions of the coupled-mode theory and the solutions obtained assuming $a_j(x,z) = b_j(x)e^{i\beta z}$, where b_j does not change with z and $j = 1$ or 2. Moreover, since the functions a_j are weakly modulated, their spatial derivatives can be approximated by the finite differences close to the bandgap edges, where $\beta \simeq \beta_j$. After simple algebra, we obtain a discrete equation for the field amplitudes at the layers that has the form of the discrete NLS equation (11.2.11) with the parameters η and ξ given by

$$\eta(\beta) = 2 - \frac{h^4}{4\pi^2}(\beta - \beta_1)(\beta - \beta_2), \qquad \xi(\beta) = \frac{h^3}{2\pi^2}(\beta - \beta_2). \tag{11.2.22}$$

Similar to the case of the tight-binding approximation, these dispersion relations can be found from the general result in Eq. (11.2.12) through a series expansion near the band-edge value β_2.

Two-Component Model

The principal limitation of the tight-binding approximation and the coupled-mode theory is that both of them are valid only in narrow regions around the bandgap edges.

Indeed, these two approaches are applicable when the dimensionless parameter αh is either small or large. In the intermediate case, one must rely on the numerical solutions with the exact dispersion relations (11.2.12). However, it is useful to consider a simplified model that can describe, at least qualitatively, the situation close to the Bragg-reflection gap ($\beta \simeq \beta_1$) and is valid for $\alpha h \simeq 1$. To achieve this goal, we extend the tight-binding approximation and the corresponding discrete NLS equation to this new regime.

Equation (11.2.16) can be considered a rough discretization of the original model in Eq. (11.2.3) such that only one node per period is located at $x = nh$. A natural generalization is to include an additional *linear node* located between the nonlinear layers at the positions $x = (n + \frac{1}{2})h$. Such a generalization corresponds to a two-component superlattice. Since the refractive indices at the node position are now different, we obtain the following set of coupled discrete equations:

$$i\frac{d\psi_n}{dz} + \beta_1 \psi_n + \rho_1(\psi_{n-1/2} + \psi_{n+1/2}) + \tilde{\gamma}|\psi_n|^2 \psi_n = 0, \qquad (11.2.23)$$

$$i\frac{d\psi_{n+1/2}}{dz} + \beta_2 \psi_{n+1/2} + \rho_2(\psi_n + \psi_{n+1}) = 0, \qquad (11.2.24)$$

where the constants ρ_1 and ρ_2 depend on α and h.

As before, the stationary-mode profile of the discrete soliton can be expressed in terms of the amplitudes U_n. These amplitudes satisfy the normalized discrete NLS equation (11.2.11) with the following parameters:

$$\eta(\beta) = 2 - (\rho_1\rho_2)^{-1}(\beta - \beta_1)(\beta - \beta_2), \qquad (11.2.25)$$

$$\xi(\beta) = \tilde{\gamma}(\gamma\rho_1\rho_2)^{-1}(\beta - \beta_2), \qquad (11.2.26)$$

where the parameters $\rho_{1,2}$ and $\beta_{1,2}$ are selected to match the exact dispersion relation [56].

It is easy to conclude that the new model describes a system with a semi-infinite IR gap and a BR gap of a finite width. Although the new dispersion relations in Eqs. (11.2.25) and (11.2.26) look similar to those found for the coupled-mode theory and given in Eqs. (11.2.22), the coupled-mode theory is valid only near a single isolated Bragg-reflection bandgap. For this reason, the two-component discrete model based on Eqs. (11.2.23) and (11.2.24) represents an important generalization of the discrete NLS theory and has a wider applicability than the coupled-mode theory.

The three approximate models discussed in this section allow one to gain considerable physical understanding, because they can be solved analytically in several limiting cases. We use them in the next section for analyzing important properties of the extended and localized modes associated with an array of nonlinear waveguides.

11.3 Modulation Instability

Before discussing the properties of the discrete solitons associated with Eqs. (11.2.3) and (11.2.8), it is useful to consider the stability of the plane-wave solution of these equations. As in earlier chapters, the plane-wave solution becomes unstable under some condition through the onset of the modulation instability.

11.3.1   Plane-Wave Dispersion Relation

The plane-wave solution of Eqs. (11.2.3) and (11.2.8) has a constant intensity I_0 at all nonlinear layers and belongs to the first transmission band in Figure 11.4. This solution can be found by solving Eq. (11.2.11) with $U_n = U_0 \exp(iKn)$, where the wave number K is in the first Brillouin zone ($|K| \leq \pi$). Using Eq. (11.2.11), we find the dispersion relation

$$K(\beta) = \mp \cos^{-1} \tfrac{1}{2}[\eta(\beta) + \alpha\gamma\xi I_0], \qquad (11.3.1)$$

where we replaced α in Eq. (11.2.12) with $\alpha + \gamma I_0$ to take into account the layer nonlinearity. Since the transmission bands corresponds to the condition $|\eta| < 2$, the band structure shifts as intensity increases. Indeed, from the dispersion relation we find the following relation between the propagation constant and the intensity:

$$I_0(\beta) = -\frac{2\cos K + \eta(\beta)}{\gamma\xi(\beta)}. \qquad (11.3.2)$$

In the first transmission band, $\beta > -(\pi/h)^2$, ensuring that $\xi(\beta) > 0$ (see Figure 11.4). It follows from Eq. (11.2.12) that the propagation constant β increases at higher intensities in a self-focusing medium ($\gamma > 0$) and decreases in a self-defocusing medium ($\gamma < 0$).

To describe the stability properties of this plane-wave solution, we consider the evolution of weak perturbations [65] by using Eq. (11.2.13) in Eq. (11.2.3) and linearizing the resulting equations for v and w. From the periodicity of the steady-state solution and the Bloch theorem, it follows that the eigenmodes of the linear eigenvalue problem would also be periodic. If we use the solution in the form

$$v(x+h) = v(x)e^{i(q+K)}, \qquad w(x+h) = w(x)e^{i(q-K)}, \qquad (11.3.3)$$

the eigenvalues Γ are found from the solvability condition

$$[\eta(\beta + \Gamma) + 2\gamma\xi(\beta + \Gamma)I_0 + 2\cos(q+K)] \times [\eta(\beta - \Gamma)$$
$$+ 2\gamma\xi(\beta - \Gamma)I_0 + 2\cos(q-K)] = \gamma^2\xi(\beta + \Gamma)\xi(\beta - \Gamma)I_0^2, (11.3.4)$$

the eigenvalues Γ are determined using the requirement that all spatial modulation frequencies q are real. The eigenvalue spectrum consists of bands, and thus the instability growth rate can change continuously only from zero to some maximum value. Moreover, since the spectrum is symmetric under $\Gamma \to \pm\Gamma^*$, it is sufficient to consider only the solutions with $\text{Re}(\Gamma) \geq 0$.

11.3.2   Staggered and Unstaggered Modes

The stability condition in Eq. (11.3.4) should be solved numerically to find the eigenvalues Γ. In what follows, we focus on two special cases, $K = 0$ and π, that correspond to two edges of the Brillouin zone and consider the function $Q(\Gamma) = \cos[q(\Gamma)]$. The steady-state solutions for $K = \pi$ and $K = 0$ are referred to as the *staggered* and *unstaggered* modes, respectively. The mode amplitude has the same phase ($U_n > 0$) for an unstaggered mode, but the phase alternates between 0 and π for staggered modes.

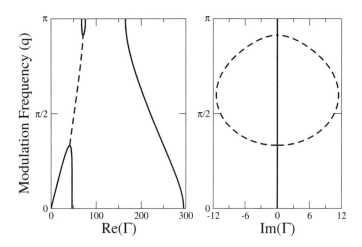

Figure 11.5: Modulation frequency q versus real (left) and imaginary (right) parts of the eigen-values Γ for a self-focusing medium with $K = \pi$, $\beta = 7$, $\alpha = 3$, $h = 0.5$, and $\gamma = +1$. Dashed lines show the unstable parts. (After Ref. [56]; ©2002 APS.)

A steady-state solution is stable for real Γ and becomes unstable as soon as Γ be-comes complex. Assuming that the imaginary part of Γ is relatively small, we expand Q in a Taylor series as

$$Q(\text{Re}\,\Gamma + i\text{Im}\,\Gamma) = Q(\text{Re}\,\Gamma) + Q'(\text{Re}\,\Gamma)(i\text{Im}\,\Gamma) + \tfrac{1}{2}Q''(\text{Re}\,\Gamma)(i\text{Im}\,\Gamma)^2 + \cdots, \quad (11.3.5)$$

where the prime denotes differentiation with respect to the argument. Since Q should remain real, the second term in this series expansion should vanish. Thus, we conclude that complex eigenvalues can appear only at the critical points, where $dQ/d\Gamma = 0$.

This interesting result shows that instability can be predicted by using the function $Q(\Gamma)$ on the real axis only and then extending the solution to the complex plane at the critical points (if they are present). The spatial modulation frequency $q = \cos^{-1}(Q)$ is real in the interval $-1 \le Q \le 1$. As a result, modulation instability appears only at the critical points, where $|Q| = 1$ or $dq/d\Gamma = 0$. Figure 11.5 shows $\text{Re}(\Gamma)$ and $\text{Im}(\Gamma)$ as a function of q. The dashed line shows the region in which modulation instability occurs because $\text{Im}(\Gamma)$ becomes nonzero.

Modulation instability of the nonlinear Bloch waves in a periodic medium has also been studied in the context of the Bose–Einstein condensates forming in opti-cal lattices [65]. The dynamics of such condensates is described in the mean-field approximation by the Gross–Pitaevskii equation, which is mathematically equivalent to Eq. (11.2.3) with $\mathcal{F}(I;x) = v(x) + \gamma I$. It was found that the unstaggered modes are always *modulationly unstable* in a self-focusing medium ($\chi = \text{sign}(\gamma) = +1$) but that they are *stable* in a self-defocusing medium ($\chi = -1$). The same conclusion holds for discrete solitons. In fact, since the unstaggered waves are the fundamental nonlinear modes in a periodic potential, oscillatory instabilities cannot occur, and modulation instability can only correspond to purely imaginary eigenvalues Γ. Such an instability should appear at the critical points, where $dQ/d\Gamma = 0$. Two critical points are found to

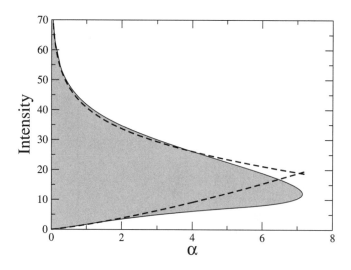

Figure 11.6: modulation instability of staggered Bloch-wave modes (gray shading) in a self-focusing medium ($\gamma = 1$). The intensity range over which instability occurs is shown as a function of α for $h = 0.5$. Dashed lines show the analytical approximations in the low- and high-intensity limits. (After Ref. [56]; ©2002 APS.)

occur for $Q = 1$ and $Q = \eta + 3$. Therefore, the range of the unstable modulation frequencies is $0 < q < \cos^{-1}(\eta + 3)$ for $-4 \le \eta < -2$ and $0 < q \le \pi$ for $\eta < -4$. Note that at small intensities ($\eta \simeq -2$) the modulation instability in a self-focusing medium corresponds to long-wave modulations.

In a self-defocusing medium, even the staggered Bloch waves ($\chi = -1$) become modulationly unstable [65]. This happens because the staggered waves experience effectively normal diffraction [13]. Such waves exist for $\eta > 2$. Similar to the case of unstaggered waves in a self-focusing medium, we identify the range of unstable frequencies corresponding to purely imaginary eigenvalues (Re $\Gamma = 0$) to be $0 < q < \cos^{-1}(3 - \eta)$ for $2 < \eta \le 4$ and $0 < q \le \pi$ for $\eta > 4$.

Finally, we analyze the stability of staggered Bloch waves in a self-focusing medium ($\chi = +1$). Since such modes exist for $\eta < 2$, the domain $1 < Q(\Gamma) < 3 - \eta$ for Re$(\Gamma) = 0$ does not correspond to physically possible modulation frequencies. However, the oscillatory instabilities (i.e., those with complex Γ) can appear due to resonances between the modes that belong to different bands. As shown in Figure 11.6, such instabilities appear in a certain range of the beam intensity and disappear when the parameter α exceeds a certain value (the instability exists for $\alpha h < 3.57$). Approximate analytic expressions for the boundaries of the unstable region have been found. The lower boundary is given by

$$\gamma I_0^{\min} \simeq \alpha + 2\sqrt{2h}\alpha^{3/2}/\pi + O(\alpha^2).$$ (11.3.6)

The upper boundary follows the relation

$$\alpha \simeq 4\gamma I_0^{\max} \exp(-\gamma I_0^{\max} h/4).$$ (11.3.7)

These analytical estimates are shown with the dashed lines in Figure 11.6.

It is interesting to compare these results with those obtained in the framework of the coupled-mode theory (see Section 11.2.3), valid for a narrow bandgap (i.e., for small α and small I_0). Although the nonlinear coupling coefficient in Eq. (11.2.20) is different than that of the coupled-mode theory for shallow gratings [66], the key stability result remains the same—the oscillatory instability appears above a certain critical intensity proportional to the bandgap width. In our case, the bandgap width is approximately $2\alpha/h$, resulting in a good agreement with the results of the coupled-mode theory. However, the coupled-mode equations (11.2.20) and (11.2.21) cannot predict the stability region at high intensities, simply because they become invalid in this region.

In the limit of large α, the Bloch-wave dynamics can be studied in the tight-binding approximation (see Sec. 11.2.3). The effective discrete NLS equation (11.2.16) predicts the stability of the staggered modes in a self-focusing medium [67]–[69]. Numerical and analytical results confirm that the steady-state solutions are indeed stable in the corresponding parameter regime.

The two-component discrete model introduced in Section 11.2.3 predicts the existence of oscillatory instabilities of the Bloch waves in a certain region of β. More importantly, Eqs. (11.2.23) and (11.2.24) predict the key pattern of oscillatory modulation instabilities: (i) Instability appears only for a finite range of intensities when the grating depth (α) is below a critical value, and (ii) it is completely suppressed for large α. Thus, unlike the discrete NLS equation, the two-component discrete model predicts qualitatively all major features of the modulation instability occurring in a periodic medium.

The preceding analysis shows that the staggered Bloch waves ($K = \pi$) are always stable in a self-focusing medium with respect to low-frequency modulations. However, at larger intensities, unstable frequencies appear closer to an edge of the Brillouin zone, and as they shift toward the $q = \pi$ edge, they experience the largest instability growth rate. The modulation instability in this region manifests itself through period-doubling modulations.

11.4 Bright and Dark Solitons

As discussed earlier, nonlinear localized modes in the form of discrete bright solitons can exist only inside the bandgaps ($|\eta| > 2$). Moreover, such solutions can exist only if the nonlinear and diffraction parameters have opposite signs, i.e., when $\eta\chi < -2$. It follows from Eq. (11.2.12) that $\beta > -(\pi/h)^2$ and $\xi > 0$ in the IR bandgap as well as the first BR bandgap (see Figure 11.4). The self-focusing nonlinearity ($\chi = +1$) can support bright solitons in the IR region, where $\eta < -2$ (the conventional waveguiding regime). Even in the case of self-defocusing ($\chi = +1$), bright solitons can exist in the first BR bandgap owing to the fact that the sign of the effective diffraction is inverted for $\eta > 2$. In this case, the mode forms in the so-called *antiwaveguiding regime*. In this section we consider both regimes.

11.4.1   Odd and Even Bright Solitons

As shown in Figure 11.2, two types of bright discrete solitons can exist in a waveguide array. The soliton centered inside a nonlinear waveguide is called *odd* because the field reverses its phase on the two sides of this waveguide (and also because an odd number of waveguides is involved). The other kind of soliton, centered between two neighboring waveguides, is called *even*. If the two waveguides are numbered as $n = 0$ and $n = 1$, the symmetry properties are such that $U_{|n|} = \chi^s U_{-|n|-s}$, where $s = 0$ for the odd soliton and $s = 1$ for the even soliton [70]. The mode profile is "unstaggered" in both cases (i.e., $U_n > 0$) if $\eta < -2$. On the other hand, Eq. (11.2.11) possesses the symmetry

$$U_n \to (-1)^n U_n, \quad \text{when} \quad \eta \to -\eta, \quad \chi \to -\chi. \tag{11.4.1}$$

It implies that the solutions become "staggered" for $\eta > 2$. Because of this symmetry, it is sufficient to find soliton solutions of Eq. (11.2.11) for $\eta < -2$ and $\chi = +1$.

The approximate solutions based on the discrete NLS equation give accurate results either in the case of highly localized modes, for which $|\eta| \gg 2$ [9, 71] or in the continuous limit, valid for $|\eta| \simeq 2$ [6]. To describe the soliton profiles for arbitrary values of η, we use a new approach based on the physical properties of localized solutions. The approach is based on the following observation. The nonlinear localized modes are similar to the impurity states for large values of η, as evident from the sharp central peak (or two peaks) in Figure 11.2. On the other hand, the tails of all localized modes are always quite smooth. We can thus construct an approximate solution for any value of η by matching the soliton tail with the central impurity node.

Consider first the soliton tail. Because of its smooth nature, its profile can be well approximated in the continuum limit of the discrete NLS equation. In a simple but effective approach, we match the discrete solution at the beginning of the tail, which we define as the *zero concavity point*, and then ensure that the asymptotic behavior of the tail corresponds to the *linear limit*. In the continuum limit valid for large n, Eq. (11.2.11) can be converted into the following approximate equation for the soliton tails:

$$\frac{\lambda}{\rho^2} \frac{d^2 U}{dn^2} = \lambda U - U^3, \tag{11.4.2}$$

where $\lambda = -(\eta + 2) > 0$ and $\rho = \cosh^{-1}(1 + \lambda/2)$. Using the boundary condition that $U(n) \to 0$ as $|n| \to \infty$, we obtain the simple solution

$$U(n; n_s) = \sqrt{\lambda} \, \text{sech}[\rho(n + n_s)], \tag{11.4.3}$$

where the parameter n_s takes into account the shift of the soliton. In the limit $\lambda \to 0$, $\rho = \sqrt{\lambda}$, and we recover the standard result expected for the continuous NLS equation [1].

We now construct the central part of the soliton. The odd soliton will be centered inside the central $n = 0$ node, while the even soliton has two peaks, located on each side of the $n = 0$ node. Since the soliton profile is symmetric, we need to calculate it only for $n \geq 0$. We assume that we can use Eq. (11.4.3) for all values of $n \geq 1$; i.e., $U_n \approx U(n; n_s)$ for $n \geq 1$. For $n = 0$ and 1, we solve the original discrete equations (11.2.11) at these

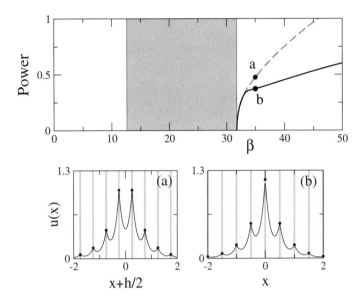

Figure 11.7: Total power versus propagation constant for odd (solid) and even (dashed) solitons in a self-focusing ($\gamma = 1$) medium. All even solitons are unstable. Gray shading marks the transmission band. Intensity profiles of the localized modes corresponding to the marked points (a) and (b) are also shown. Black dots represent analytical approximations for the node amplitudes. The parameters used are the same as those in Figure 11.4. (After Ref. [56]; ©2002 APS.)

two nodes. The amplitudes U_0 and U_1 are then found to satisfy

$$(2-s)U_1 + U_0^3 = (2-s+\lambda)U_0, \tag{11.4.4}$$

$$U_0 + U_2 + U_1^3 = (2+\lambda)U_1, \tag{11.4.5}$$

where $s = 0$ and 1 for odd and even modes, respectively. These equations can be solved numerically using U_1, U_2 and n_s from Eq. (11.4.3), treating n_s as an unknown parameter. For all $\lambda > 0$, a solution exists that joins smoothly with the solution obtained in the continuous limit (as $\lambda \to 0$). Figures 11.7 and 11.8 show the $P(\beta)$ curves for odd (solid line) and even (dashed line) modes in the cases of self-focusing and self-defocusing nonlinearity, respectively. In each case, an example of the odd and even discrete solitons is also shown.

A comparison of the analytical approximation with the exact numerical solution of the original Eq. (11.2.11) shows that the error for the peak amplitude U_0 does not exceed 1.5% for odd and 0.8% for even solitons. The mode profiles are also adequately represented by the analytical approximation, as seen clearly in Figs. 11.7 and 11.8. Thus, the matching procedure allows one to obtain reasonably accurate approximate analytical solutions in the whole parameter range, including the extreme limits, in which $\lambda \to 0$ or $\lambda \to +\infty$.

The stability analysis reveals that even modes are *always unstable* with respect to a translational shift along the x axis. On the other hand, odd modes are always stable in the self-focusing regime (see Figure 11.7) but can exhibit *oscillatory instabilities* in

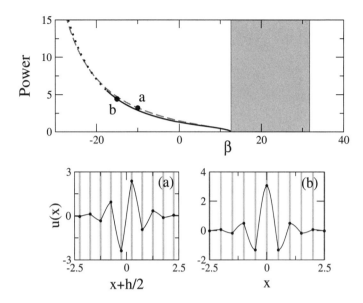

Figure 11.8: Same as in Figure 11.7, except the nonlinear medium is of the self-defocusing type ($\gamma = -1$). (After Ref. [56]; ©2002 APS.)

the self-defocusing case when the power exceeds a certain value (see Figure 11.8). At this point, an eigenmode of the linearized problem resonates with the bandgap edge, and the quantity $\beta - \text{Re}(\Gamma)$ moves inside the band while at the same time Γ develops a nonzero imaginary part. This behavior is shown by an example in Figure 11.9, where a real eigenvalue Γ turns into a complex quantity at the bifurcation point located near $\beta = -18$. This instability scenario is similar to that identified earlier in Chapter 5 for gap solitons [72]. It also occurs for the modes localized at a single nonlinear layer in a linear periodic structure [53]. The latter example demonstrates a deep similarity between the periodic systems with localized and distributed nonlinearities.

11.4.2 Twisted Modes

The periodic modulation of the linear refractive index can lead to a bound state of two discrete solitons [73]. The so-called *twisted* mode is an example of such a bound state. Its existence was first predicted in 1998 using the discrete NLS equation [74]–[76]. It represents a combination of two out-of-phase bright discrete solitons [73]. Such solitons do not have a counterpart in the continuous case. In fact, they can exist only when the discreteness effects are strong, i.e., when the parameter $|\eta|$ exceeds a certain value η_{cr}.

The properties of a twisted mode depend on the spatial separation between the modes forming the bound state. Consider a bound state of two lowest-order solitons. Both of these solitons can be even modes with zero nodes ($m = 0$) in between the peaks or odd soliton with one node ($m = 1$) in the middle such that $U_0 = 0$. The symmetry properties for such twisted solitons dictate that $U_{|n|+m} = -\chi^{m+1} U_{-|n|-1}$. In view of the

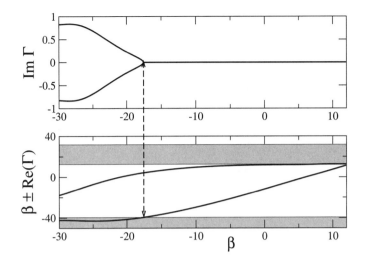

Figure 11.9: Example of an oscillatory instability for the odd soliton induced by a resonance at the bandgap edge. Parameter values correspond to those of Figure 11.7. (After Ref. [56]; ©2002 APS.)

symmetry seen in Eq. (11.4.1), we only have to construct the solution for $\chi = +1$. We assume that the soliton tails for $n > m$ can be approximated as

$$U_n = U(n - m; n_s),\qquad(11.4.6)$$

where $U(n; n_s)$ is given by Eq. (11.4.3). The matching conditions for U_m and n_s are obtained from Eq. (11.2.11) and are given by

$$-(3 - m + \lambda)U_m + U_{m+1} + U_m^3 = 0,\qquad(11.4.7)$$

$$-(2 + \lambda)U_{m+1} + U_m + U_{m+2} + U_{m+1}^3 = 0,\qquad(11.4.8)$$

where, as before, $\lambda = -(\eta + 2) > 0$.

One can find the solution of Eqs. (11.4.7) starting with a highly localized mode described earlier in the limit $\lambda \gg 1$ [74] and then gradually decrease the parameter λ. It turns out that the solution exists for $\lambda > \lambda_{cr} > 0$, but it ceases to exist when $n_s(\lambda_{cr}) = 0$. Substituting this condition into Eq. (11.4.7), the critical value of η is found to be approximately $\eta_{cr} \simeq 3.32$ for $m = 0$ and $\eta_{cr} \simeq 2.95$ for $m = 1$. These analytical values agree with the numerical results to within 1%. Moreover, the approximate solution describes quite accurately the profiles of the twisted modes. Figures 11.10 and 11.11 show the $P(\beta)$ curves for odd (solid line) and even (dashed line) twisted modes for the cases of self-focusing and self-defocusing nonlinearity, respectively. In each case, an example of the odd and even discrete solitons is also shown. As seen there, the relative error for the peak amplitude at any node is less than 0.5%.

In the self-focusing regime, stability properties of the twisted solitons in the IR bandgap are similar to those found earlier in the framework of the discrete NLS equation [74]–[76]. In the example shown in Figure 11.10, the solitons are stable for large

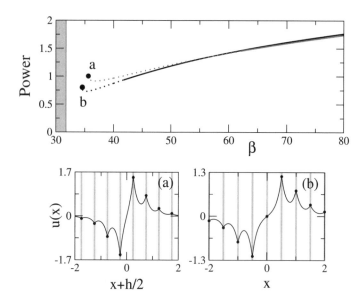

Figure 11.10: Soliton power versus propagation constant for odd (solid) and even (dashed) twisted solitons in a self-focusing ($\gamma = 1$) medium. Gray shading marks the transmission band. Intensity profiles of the localized modes corresponding to the marked points (a) and (b) are also shown. The notation and the parameters used are the same as in Figure 11.4. (After Ref. [56]; ©2002 APS.)

values of the propagation constant β, but they exhibit an oscillatory instability close to the boundary of the existence region. Quite importantly, the stability region is much wider in the case of odd twisted modes because of a larger separation between the individual solitons of the bound state.

The characteristics of the twisted modes in the BR bandgap are quite different from the preceding case of the IR gap. First, the value of η is limited from above ($2 < \eta < \eta_{max}$) and, therefore, some families of the twisted modes with $m < m_{cr}$ may not exist. For example, for the medium parameters corresponding to Figure 11.4, $\eta_{cr} < \eta_{max} \simeq 3.18$ when $m = 1$ but $\eta_{cr} < \eta_{cr}$ when $m = 0$. Under these conditions, even twisted modes with $m = 0$ cannot exist in the BR regime. In contrast, not only do the odd twisted modes with $m = 1$ exist, but they are stable over a wide parameter region, as seen in see Figure 11.11. In general, the oscillatory instabilities for IR and BR twisted modes are associated with an exponential increase in amplitude modulations and emission of radiation waves.

11.4.3 Discrete Dark Solitons

Similar to the case of the continuous NLS equation with self-defocusing nonlinearity discussed in Chapter 4 or the discrete NLS equation considered in Section 11.1 [78]–[80], the model based on Eq. (11.2.11) supports *dark* discrete solitons in the form of a localized dip on the Bloch-wave background. The new feature is that dark discrete

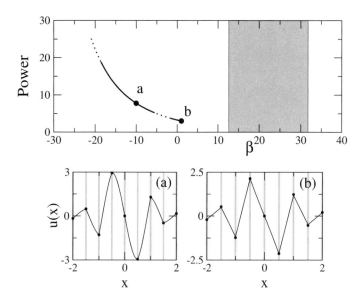

Figure 11.11: Same as in Figure 11.10, except the nonlinear medium is of the self-defocusing type ($\gamma = -1$). (After Ref. [56]; ©2002 APS.)

solitons can exist in a periodic medium for both signs of the nonlinearity (i.e., even in a self-focusing medium). This is a direct manifestation of the anomalous diffraction that can occur in optical media with a periodically varying refractive index. To be specific, we consider the case of a background corresponding to the Bloch-wave solution obtained in Section 11.3 for $K = 0$ and $K = \pi$. Then dark solitons can appear at the bandgap edge, where $\eta = 2\chi = 2\,\mathrm{sign}(\gamma)$, since in this case the nonlinear and dispersion terms have the same signs [77].

Similar to the case of bright discrete solitons, two types of dark discrete solitons can be identified: *odd* solitons are centered inside a nonlinear thin-film waveguide, whereas *even* ones are centered between the two neighboring waveguides [79]. Both kinds satisfy the symmetry condition

$$U_{|n|+s} = -(-\chi)^{s+1}U_{-|n|-1}, \qquad (11.4.9)$$

where $s = 0$ for even modes but 1 for odd modes. The Bloch-wave background is unstaggered for $\chi = -1$ and staggered for $\chi = +1$; the corresponding solutions can be constructed from Eq. (11.2.11) with the help of the symmetry transformation $U_n \to (-1)^n U_n$. However, the stability properties of these two types of background states can be quite different. As discussed in Section 11.3, the staggered background can become unstable in a self-focusing medium ($\chi = +1$), while the unstaggered background is always stable if $\chi = -1$.

To find the approximate analytical solutions for dark solitons, we consider the case $\chi = -1$; solutions for $\chi = +1$ can be obtained by applying the symmetry transformation in Eq. (11.4.1)]. In this case, the solution of Eq. (11.2.11) close to the background level is found to be $(U_\infty - U_n) \simeq e^{\rho n}$, where $U_\infty = \sqrt{\lambda}$ is the background amplitude,

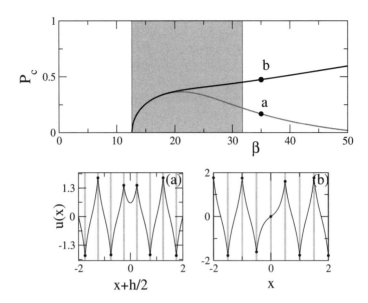

Figure 11.12: Complementary power versus propagation constant for odd (black) and even (gray) dark discrete solitons in a self-focusing ($\gamma = +1$) medium. The notation and the parameters used are the same as in Figure 11.4. (After Ref. [56]; ©2002 APS.)

$\rho = \cosh^{-1}(1 + \lambda)$ is the localization parameter, and $\lambda = \eta + 2 > 0$. The tails of the dark soliton are then approximately governed by the following continuous equation, valid for large n:

$$\lambda U + \frac{2\lambda}{\rho^2}\frac{d^2 U}{dn^2} - U^3 = 0. \tag{11.4.10}$$

This equation can be solved easily to find

$$U(n; n_s) = \sqrt{\lambda}\tanh[\rho(n + n_s)/2]. \tag{11.4.11}$$

In the limit $\lambda \to 0$, $\rho \to \sqrt{\lambda}$, and we recover the result expected for conventional continuous dark solitons.

Similar to the case of bright discrete solitons, we find the complete solution by matching the soliton tails with the central part of the dark soliton obtained by solving Eq. (11.2.11). The matching conditions are found to be

$$(\lambda - 3 + s)U_0 + (1 - s)U_1 - U_0^3 = 0, \tag{11.4.12}$$

$$(\lambda - 2)U_1 + U_0 + U_2 - U_1^3 = 0. \tag{11.4.13}$$

They can be used to determine the shift parameter n_s and the amplitudes U_0 and U_1. Of course, $U_0 = 0$ for odd modes, as expected from their symmetry properties.

Two types of dark spatial soliton are shown for both staggered and unstaggered Bloch-wave backgrounds in Figure 11.12 for the self-focusing case and in Figure 11.13 for the self-focusing case. In each case, the family of dark solitons is quantified by

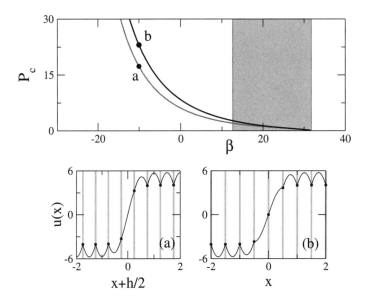

Figure 11.13: Same as in Figure 11.12 but for the self-defocusing nonlinearity ($\gamma = -1$). (After Ref. [56]; ©2002 APS.)

using the concept of the *complementary power*, introduced in Chapter 4 as

$$P_c = \lim_{n \to +\infty} \int_{-nh}^{+nh} \left(|u(x+2nh)|^2 - |u(x)|^2 \right) dx, \qquad (11.4.14)$$

where n is integer. The $P_c(\beta)$ curves shown in Figs. 11.12 and 11.13 characterize the family of odd (black curve) and even (gray curve) dark solitons. An example of each type of soliton is also shown in these figures. The analytical and numerical results for the soliton amplitude at the nonlinear layers agree quite well. In fact, the error does not exceed 2% for odd and 1% for even modes. The soliton shapes are similar to those found in the context of the superflow dynamics on a periodic potential [81].

Inherent discreteness of nonlinear systems can qualitatively alter their dynamical behavior compared with their continuous counterparts, a feature that also holds for dark solitons. In particular, discreteness can induce new soliton-formation mechanisms, resulting in the existence of shock waves with two finite backgrounds as well as asymmetric dark solitons [82]. Such dark solitons exhibit nontrivial intrinsic phase dynamics in the sense that their background oscillates at two different frequencies, and the transition region is characterized by a combination of these frequencies. A new family of symmetric dark solitons without the phase jump π at the soliton center was also found. Such nonlinear localized modes are stable only in the limit of strong discreteness, and thus their existence is due to the effective pinning mechanisms of the lattice.

Numerical simulations show that, similar to the case of bright discrete solitons, dark discrete solitons of the even type are unstable with respect to asymmetric perturbations. In contrast, odd modes can propagate in a stable (or weakly unstable) manner. Dark solitons also exhibit an oscillatory instability close to the continuum limit, but a detailed analysis of the discreteness-induced oscillatory instability has been performed only for dark solitons of the discrete NLS equation [79] and for unbounded standing waves [83].

11.5   Experimental Results

As discussed earlier, the propagation of light in an array of coupled waveguides permits control over diffraction, not normally possible in a bulk medium or in a single waveguide. The reason behind this possibility is easily understood by noting that the spreading of light in an array is governed by the extent of coupling between the neighboring waveguides. As a result, the expansion rate, or "discrete diffraction," depends on the strength of the coupling and can easily be controlled. Even the sign of the diffraction can be chosen for a given waveguide array by controlling the input conditions. Such properties are a direct manifestation of the discreteness of the system and have been observed in several experiments. Most of the experiments have been performed in waveguide arrays made of the same AlGaAs material used for observing spatial solitons in a single slab waveguide [10]. Only recently an array of single-mode polymer waveguides was used for observing discrete solitons [15].

11.5.1   Self-Focusing Regime

In the case of the AlGaAs-based experiments, an array of 4-μm-wide single-mode waveguides was created using the technique of reactive-ion etching [10]. The array typically consisted of 40–60 waveguides with the composition $Al_{0.18}Ga_{0.82}As$, and they were separated by cladding layers with 24% Al content. The coupling between the neighboring waveguides was controlled by varying the spacing among the waveguides in the range 2–7 μm. The inset in Figure 11.14 shows the array design schematically. A synchronously pumped optical parametric oscillator was used for observing discrete solitons. It emitted 100- to 200-fs-wide optical pulses whose center wavelength was tunable in the 1.5-μm region. This wavelength region lies below the half-bandgap of AlGaAs and helps in reducing the effects of two-photon absorption. The short pulses were needed to reach the required peak power levels; the maximum available peak power was 1.5 kW.

In the 1998 experimental study of Eisenberg et al. [10], the input beam was focused onto a single waveguide of a 6-mm-long array, and the intensity distribution at the output end was recorded using the setup shown in Figure 11.14. The resulting output intensity distribution is shown in Figure 11.15 for three power levels: 70, 320, and 500 W . At the lowest input power level, the propagation was linear, and the beam spread over more than 30 waveguides (top row). The coupling length was estimated to be about 1.4 mm from this intensity distribution. As the input power was increased, the output beam spread over fewer waveguides (middle row) until it was mostly confined to five waveguides around the input waveguide at a power level of 500 W (bottom row).

When light is launched into a single waveguide, the discrete soliton that eventually forms propagates along the array. However, discrete solitons do not have to propagate

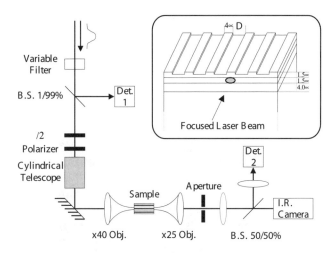

Figure 11.14: Experimental setup for observing discrete solitons. The inset shows the sample design schematically. (After Ref. [10]; ©1998 APS.)

in the direction of the waveguides. When an input beam with a linear phase tilt is coupled into several waveguides, the transverse momentum leads to transverse motion of the discrete soliton across the waveguides. Moreover, such a discrete soliton behaves differently when it is launched in different directions. Even without the phase tilt, a discrete soliton can shift considerably with small changes in input-beam position [11]. Figure 11.16 shows the experimental results as the input-beam position is varied over 15 μm.

According to the theory, there are two types of soliton that propagate along the waveguides, one centered on a single waveguide and one centered between two waveguides. In the experimental results of Morandotti et al. [11] shown in Figure 11.16, solitons were steered sideways due to the asymmetry in the input coupling. When the input distribution was centered on a waveguide or exactly between two waveguides, the

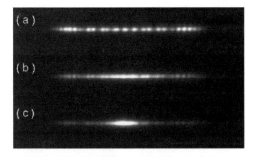

Figure 11.15: Images at the output facet of a 6-mm-long waveguide array with 4-μm spacing for input peak power of (a) 70 W, (b) 320 W, and (c) 500 W. A discrete soliton is formed in the last case. (After Ref. [10]; ©1998 APS.)

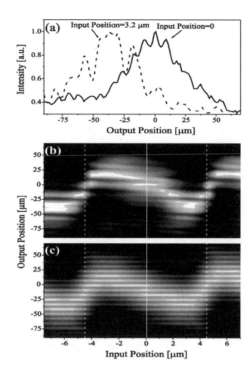

Figure 11.16: Steering of a discrete soliton by an internally induced velocity. (a) Output intensity distributions for two input-beam positions at a fixed input peak power of 1500 W; (b) output distributions for various input-beam positions; (c) the results of numerical simulations under the same conditions. The solid white line marks the center of the input waveguide, while dashed lines are centered between the two waveguides. (After Ref. [11]; ©1999 APS.)

soliton propagated in the waveguide direction without any transverse motion. However, when the input condition was not symmetric, a transverse motion was observed because of the phase tilt that was induced through the Kerr effect. The periodic nature of the output location is clearly seen in Figure 11.16.

The formation of discrete solitons in waveguide arrays because of self-focusing is similar to the the case of self-focusing in a single planar waveguide. However, because an array possesses a much more complicated bandgap structure, self-focusing can be observed to occur in several different bandgaps, similar to the case of gap solitons in fiber gratings. In one experiment, Mandelik et al. measured the entire bandgap structure of waveguide arrays and observed the excitation of array modes belonging to high-frequency bands [48]. Such discrete solitons are called the Floquet–Bloch solitons or, simply, the *discrete* gap solitons, because they represent a generalization of the gap solitons to the discrete case.

The complete bandgap structure of a waveguide array cannot be described by the discrete NLS equations, and an improved analytical model should be used for the rigorous analysis. However, in some cases the discrete gap solitons can be described by a generalization of the two-component discrete model that can account for many of their properties in a simple manner. Based on these results, the novel concept of array engineering has been suggested [84]. The basic idea consists of using a *binary* waveguide array composed of alternating wide and narrow waveguides. Propagation of optical waves in such a structure can be approximately described by the discrete NLS equation, with the main difference being that the propagation constant β is not constant but takes two values. Such an approach accounts for the structure of the first bandgap in the transmission spectrum, and it can serve as a good approximation to the more completed analytical model discussed earlier.

11.5.2 Diffraction Management

Diffraction in a waveguide array can be controlled through the choice of input conditions, and under some conditions even the sign of diffraction can be reversed. This anomalous diffraction seems a counterintuitive phenomenon because it implies that high spatial frequency components have a smaller transverse velocity compared with the low-frequency components. Experimentally, this happens when the input beam has a phase tilt so that adjacent waveguides are excited out of phase. As far as linear optics is concerned, both normal and anomalous diffraction lead to broadening of a finite-width optical beam. However, similar to the case of temporal dispersion, a sign change in diffraction can force a self-focusing Kerr medium to exhibit self-defocusing .

Such anomalous behavior was indeed observed in a 2001 experiment [14] using a specially designed sample with 61 waveguides, each 3 μm wide and separated by 3 μm from the neighboring waveguides. The experiment allowed one to observe the self-focusing and self-defocusing behavior in the same array of waveguides using slightly different initial conditions. For normal dispersion, i.e., when the input beam was normal to the input facet, it broadened at low powers and shrank to form a discrete bright soliton, as expected for the self-focusing nonlinearity (see Figure 11.15). However, when the input beam was injected at an angle of $2.6 \pm 0.3°$, it experienced anomalous diffraction and broadened even more at high input powers because of self-defocusing produced by diffraction reversal. In this case, the optical fields in adjacent waveguides had a π phase difference.

The anomalous diffractive properties of waveguide arrays can be used to produce structures with designed diffraction and to fabricate waveguide arrays with reduced, cancelled, or even reversed diffraction [13]. The experimental results obtained with such waveguides are in agreement with the predictions based on the discrete NLS equation and the coupled-mode theory. Figure 11.17 shows the basic idea. Instead of a tilted input beam, the beam is launched normal to the waveguides, but the array is tilted from the normal to the input facet. The tilt angle α determines the transverse wave number k_x through $\theta = k_x d$, where d is the distance between the centers of two adjacent waveguides. Since θ can vary from 0 to π, in the experimental configuration the maximum value of π is equivalent to a tilt angle of $\alpha = \sin^{-1}(\theta/k_x d) \approx 1.5°$.

Figure 11.17 shows the intensity distribution of the output beam for a 6-mm-long sample with $d = 9$ μm (coupling length about 2 mm) . The 21-μm-wide input Gaussian beam excited mostly three waveguides. The beam expanded to 42 μm when propagating in the untilted array, whereas it expanded to 54 μm in a continuous slab waveguide with no array. For $\theta = \pi/2$, where diffraction should vanish, the output field broad-

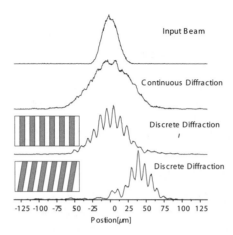

Figure 11.17: Diffraction management in waveguide arrays. A 21-μm-wide input beam spreads to 54 μm under continuous diffraction and to 42 μm in the case of an array with parallel waveguides. Tilting of waveguides suppresses diffraction and the beam narrows down to 29 μm. The residual broadening is due to the higher-order diffractive effects. (After Ref. [13]; ©2000 APS.)

ened to only 29 μm. It also exhibited an asymmetric profile, attributed to the residual third-order diffraction. These results are analogous to the case of an optical pulse propagating inside an optical fiber. In fact, the output shape for $\theta = \pi/2$ resembles the temporal shape of an optical pulse expected when $\beta_2 = 0$ (no group-velocity dispersion) so that it experiences only third-order dispersion.

Motivated by the experimental observation of anomalous diffraction in a waveguide array, a new theoretical model has been proposed for describing the propagation of an optical beam in diffraction-managed nonlinear waveguide arrays [90, 91]. This model supports discrete solitons whose beam width and peak amplitude evolve periodically. A nonlocal integral equation governs the slow evolution of the soliton amplitude. This new discrete equation permits a novel type of discrete spatial soliton such that it breathes (changes its width in a periodic fashion) during propagation and its phase profile has a nonlinear chirp. Similar to the case of dispersion-managed solitons discussed in Chapter 3, the discrete soliton regains its initial power and shape at the end of each diffraction map. This behavior opens the possibility of fabricating a customized waveguide array such that it admits diffraction-managed, spatially confined discrete solitons.

11.5.3 Interaction of Discrete Solitons

Waveguides created in periodically modulated structures possess many interesting properties. For example it is possible to guide waves through a low-index core with no radiation losses, a feature that cannot be realized using conventional waveguides based on total internal reflection [85, 86]. This type of localization can occur only in the Bragg-reflection bandgap, where the nature of diffraction is effectively inverted. Such a phenomenon was found to occur in a nonuniform waveguide array [87] when the

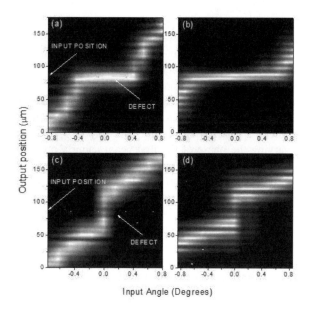

Figure 11.18: Generation of nonlinear defect modes and discrete solitons centered at the defect location (shown with a wavy line). The experimental (left column) and and numerical (right column) results are shown for the attractive (top row) and repulsive (bottom row) cases. (After Ref. [88]; ©2001 OSA.)

width of a single waveguide was reduced compared with that of all other waveguides. This "defect" reduces the effective mode index in that specific waveguide. At low powers, the defect waveguide supports linear guided waves in the Bragg-reflection regime such that their profile resembles that of gap solitons.

Defects can also be created by decreasing or increasing the spacing between a pair of waveguides. Such an approach was used in a 2002 experiment [88] to investigate the interaction among discrete solitons with defects, and it was found that the outcome strongly depends on the inclination angle of an input beam. At small angles the soliton is reflected from a defect because of its small velocity. On the other hand, transmission is observed for large angles. Even more interesting behavior is observed when the input beam is centered at the defect, as shown in Figure 11.18. An attractive defect created by reducing the waveguide spacing increases the Peierls–Nabarro potential, and large inclination angles are required to make the soliton escape from the defect site [see parts (a) and (b)]. In contrast, smaller angles are sufficient to move the soliton away from a repulsive defect [see parts (c) and (d)]. The latter phenomenon can be potentially useful for optical switching applications because the output position is very sensitive to the sign of the inclination angle. Such sensitivity is due to an instability that is similar in nature to the symmetry-breaking instability of "even" localized modes.

Soliton tunneling through a wide gap between two uniform waveguide arrays has also been observed experimentally [88]. Such a gap creates an effectively repulsive defect; however there are some differences compared to the case of narrow defects. It was found that moderately localized discrete solitons can get trapped at the edge of the waveguide boundary. At larger input powers, highly localized solitons are formed, and they get reflected from the boundary. In both cases, solitons tunnel through the gap when the inclination angle of the input beam is increased beyond some critical level.

The experiments on soliton interaction in nonlinear discrete waveguide arrays show radically different behavior for low (but still nonlinear) and high intensities when two or three channels are initially excited and their relative phase is measured [89]. By injecting two parallel copolarized beams into an array of weakly coupled AlGaAs waveguides and varying the phase of the first beam using a precision delay line, it was possible to test the stability of the relative phase interferometrically. For the smallest input power, each beam showed significant diffraction in the absence of the second beam. As the incident power was increased, the beams collapsed due to self-focusing. At an average power of 6 μW, narrow soliton-like beams had widths comparable to the input beam widths. For moderate input powers and a phase difference close to zero, the two beams fused into one strongly localized beam located in the channel between the two incident beams. When the two input beams had a π relative phase difference, two output beams with separation larger than the inputs were obtained. However, for other values of the relative phase difference, the output did not have a fixed location. Rather, its position changed spatially in a periodic manner (scanning) among the three positions such that most of the energy was localized in just a few channels.

11.5.4   Bloch Oscillations

Discrete systems such as semiconductor superlattices and waveguide arrays share a lot of interesting features. One of the most remarkable phenomena is the occurrence of Bloch oscillations [92, 93], when a charge carrier in an ideal crystal exhibits periodic oscillations in the presence of a uniform electric field. Physically speaking, the carrier is accelerated by the electric field until its momentum satisfies the Bragg condition associated with the periodic potential and is thus reflected backward. A similar effect is known to occur in semiconductor superlattices, where an oscillating current is generated when a static electric field is applied, in contrast to the dc flow observed in bulk materials [94]. Because of the fundamental relevance of discreteness in nature, one expects to find similar effects in other systems of quite different origin. In fact, Bloch oscillations have been predicted to occur in molecular chains [95] and periodically spaced curved optical waveguides [96], and they have been observed experimentally for cold atoms in optical lattices [97, 98]. It was suggested in 1998 that waveguide arrays with a spatially varying refractive index in individual waveguides constitute an ideal environment to observe optical Bloch oscillations [99].

Morandotti et al. [12] observed optical Bloch oscillations in a waveguide array in which the effective index increased linearly within each individual waveguide and played the role of an effective dc electric field in semiconductor superlattices. In the linear limit (small input powers), the output profiles for different propagation lengths indicated a periodic transverse motion of the optical field and a complete recovery of the initial excitation, in excellent agreement with the basic theory of Bloch oscillations. In contrast, a broad beam moves across the array without changing its shape considerably. For larger input powers, the nonlinear effects led to a loss of recovery of the input field and inhibited Bloch oscillations because of symmetry breaking and power-induced beam spreading. In another experiment, optical Bloch oscillations were observed by using an array of polymer waveguides in which the linear variation of the refractive index was realized by applying a temperature gradient [100].

The experimentally observed effects induced by the nonlinearity are well reproduced in numerical simulations. They can be explained by the nonlinear phase shift that changes the linear nature of the phase relations and makes the recovery incomplete. As a result, the field distribution starts to disperse. In general, the onset of the nonlinearity tends to randomize the field distribution. The strong coherence disappears, and the light starts to behave more and more in a conventional way. Similar effects are observed in quantum-mechanical systems at densities high enough that scattering suppresses coherent recovery and particles behave as classical objects. Such power-induced changes in the field distribution may be useful for applications related to signal steering and switching because the beam can be scanned almost over the entire array [12].

It was predicted theoretically in 2002 that *hybrid* discrete solitons can exist in waveguide arrays exhibiting a linear variation of the effective index [101]. Such solitons can be considered as the bound state of a conventional discrete soliton and the satellite tail of a linear Wannier–Stark state. There exist different soliton families, which are characterized by the symmetry (odd or even) and the spatial separation between the two soliton components in a bound state. Such a multiplicity of states can be considered as a nonlinear analog of the linear Wannier–Stark ladder of eigenvalues.

Linear stability analysis and direct numerical simulations reveal that the hybrid solitons with strong satellite content are unstable. It is quite remarkable that even when the soliton decays due to the development of an instability, its power remains rather localized. This happens because, as the soliton broadens, the effect of nonlinearity decreases, and the field evolves as a linear superposition of Wannier–Stark states, which are localized. The excitation dynamics of hybrid solitons is also unusual. A realistic input beam cannot match the soliton profile exactly, and therefore radiation waves are generated as well. However, since the linear waves are localized, the radiation cannot escape, in contrast to homogeneous waveguide arrays, where radiation waves can propagate freely. Therefore, stable hybrid solitons can be excited only if they can withstand repeated interactions with the radiation waves [101].

11.6 Related Concepts

The concept of discrete optical solitons can be extended into several directions, and we discuss two of them in this section. The first one, two-dimensional networks of nonlinear waveguides operating using discrete solitons, may be useful for designing signal-processing circuits based on waveguide arrays. The other generalization is associated with discrete solitons excited in optically induced waveguide arrays.

11.6.1 Discrete-Soliton Networks

The signal-processing applications of discrete solitons are limited by the one-dimensional topology of waveguide arrays. The concept of a *discrete-soliton network* intro-

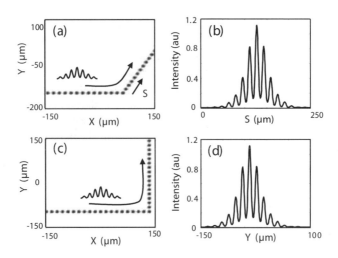

Figure 11.19: (a) A discrete soliton propagating along a 120° bend; (b) soliton intensity after traversing the bend; (c) same as (a) for a 90° bend; (d) soliton intensity after the bend along the vertical branch. (After Ref. [102]; ©2001 APS.)

duced in 2001 solves this problem to a large extent [102]–[104]. In such networks, optical beams are launched into a two-dimensional array of waveguides that act like optical wires along which discrete solitons can travel. The main limitation results from the reflections occurring at the junctions where two or more waveguides cross. Through an appropriate design, reflection losses at such sharp bends in a two-dimensional wave-guide array can be almost eliminated.

Consider a two-dimensional network of nonlinear waveguides involving identical elements. Each branch is composed of regularly spaced waveguides separated by a distance d. Every waveguide is designed to support a single mode at the operating wavelength. A moderately confined bright discrete soliton (extending over five to seven sites) is excited in one of the branches of the network. Figure 11.19(a) shows such a soliton propagating along the branch. As shown in Section 11.5, the envelope profile of such a soliton is approximately given by $u(n) = u_0 \operatorname{sech}(nd/x_0) \exp(i\alpha nD)$, where α describes the phase tilt necessary to make the soliton move along the array. By imposing a linear-phase chirp (tilt) on the beam profile, a discrete soliton can be set in motion along any branch of the network. The important question is how the soliton changes when it traverses a sharp bend along the branch.

Numerical simulations have been used to answer this question for a silica-waveguide network [102]–[104]. The circular waveguides had a core radius of 5.3 μm and were spaced 15.9 μm apart. They were formed by increasing the refractive index by 0.003 from the background value of 1.5. Figure 11.19(a) shows a discrete soliton propagating along a branch with a 120° bend. It slides along the horizontal branch and after passing the intersection moves to the upper branch. Figure 11.19(b) shows that such a cross-ing can occurs with almost no change in the soliton intensity profile or speed. Even more importantly, the bend-induced losses are extremely low (less than 0.7%). In other words, the array behaves like a *soliton wire*. Such low levels of losses at $120°$ junctions suggest that discrete solitons can be used as information carriers in honeycomb-like networks. Similar results were obtained for a $90°$ bend, as shown in parts (c) and (d). In this case, the energy loss increased to 5% because of the sharpness of the bend, but the soliton shape and intensity distribution remained practically intact. This behavior is reminiscent of the waves propagating along sharp bends in photonic crystals [105].

Reflection losses occurring along very sharp bends in a two-dimensional discrete-soliton network can be almost eliminated by modifying the guiding properties of the waveguide corner [103]. The analytical and numerical calculations show that this can be achieved by introducing a defect site at the bend corner. Moreover, by using vector or incoherent interactions at network junctions, soliton signals can be routed along specific pathways; i.e., discrete solitons can be navigated anywhere within a two-dimensional network of nonlinear waveguides. The possibility of realizing useful operations such as blocking, routing, logic functions, and time gating is discussed in Ref. [102]. By appropriately engineering the intersection site, the switching efficiency of such junctions can be improved further [106].

11.6.2 Optically Induced Waveguide Arrays

In a waveguide array, a low-intensity optical beam focused onto a specific waveguide spreads to its neighbors via discrete diffraction, such that the intensity is concentrated mainly in the outer lobes. When the nonlinearity is sufficiently high (e.g., at high intensities if the nonlinearity is Kerr type), the nonlinear effects suppress the coupling between adjacent waveguides. The combination of discrete diffraction and nonlinearity leads to discrete solitons. In a recent theoretical study [107], it was shown that this concept can be extended to the situation in which the waveguide array is not a physical entity before the input light is launched but is induced optically using the photorefractive nonlinearity. The photorefractive nonlinearity is utilized to create a waveguide array in real time (in one or even two dimensions) by interfering pairs of plane waves. This concept offers considerable flexibility, in the sense that the same photorefractive waveguide array can exhibit self-focusing or self-defocusing, depending on the polarity of the external bias, with adjustable lattice parameters.

The creation of an optically induced waveguide array requires that each waveguide br as uniform as possible. Therefore, the coupling between the interfering plane waves (used to form the waveguide array) must be eliminated so that the interference pattern does not vary in the propagation direction. The soliton-forming signal beam, on the other hand, must experience the highest possible nonlinearity. To achieve these two, seemingly conflicting objectives, one should choose a photorefractive crystal with a strong electro-optic anisotropy, polarize the interfering waves in a non-electro-optic direction, and at the same time polarize the signal beam in the crystalline orientation that yields the highest possible nonlinearity. In this arrangement, two (or more) interfering plane waves polarized perpendicular to the crystalline c axis will propagate almost linearly, while the signal beam polarized along the c axis will experience both a periodic potential and a significant (screening) nonlinearity.

In the experimental realization of such a scheme [108], a 6-mm-long SBN crystal was employed, because it offers the most appropriate values of the electro-optic param-

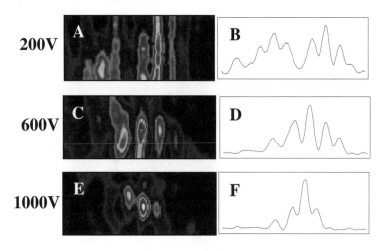

Figure 11.20: (a) Formation of a discrete soliton in an optically induced waveguide array as the self-focusing nonlinearity is enhanced by increasing the external voltage from 200 to 1000 V. Crystal output is shown on the left, with the corresponding on-axis intensity profile on the right. (After Ref. [108]; ©2002 APS.)

eters ($r_{33} \sim 1340$ pm/V and $r_{13} \sim 67$ pm/V). The one-dimensional optically induced array was created by interfering "ordinarily polarized" plane waves through a Mach–Zehnder interferometer. The signal beam was an "extraordinary polarized" beam and it was coupled into a single waveguide. The external voltage sets the photorefractive screening nonlinearity: it increases (with a nonlinear intensity dependence) the index contrast, creates a waveguide array, and also leads to localization of the signal beam. At the same time, the nonzero electro-optic coefficient for the interfering waves allows a sufficiently high electric field to be applied to create a two-dimensional waveguide array by breaking up the one-dimensional interference pattern through the transverse modulation instability.

Figure 11.20 shows the experimental results of Ref. [108]. It depicts the transition of the signal beam from discrete diffraction to discrete soliton as the self-focusing nonlinearity is enhanced by increasing the external voltage from 200 to 1000 V. The array has a waveguide spacing of 8.8 μm, and the laser power of 200 mW is divided between the array-creating beam and the signal beam with an intensity ratio of 5:1. At low intensities, the signal beam experiences linear discrete diffraction (top row). Even though it initially excites a single waveguide, two intensity lobes (separated by three waveguides) appear at the output end. At a 600-V level, the nonlinear effects set in, and the output becomes single-lobed even though the beam still diffracts considerably (middle row). In the highly nonlinear regime, a highly localized discrete soliton is formed, as seen in the bottom row of Figure 11.20. For a grating spacing of 7.8 μm, the output indicated the presence of the in-phase and staggered bright solitons. The staggered discrete soliton was formed at the edge of the first Brillouin zone. In this case, the central lobe should be out of phase with its neighbors; this feature was verified by observing constructive interference in the center and destructive interference at the outer lobes. A similar approach can be employed to create two-dimensional discrete solitons.

References

[1] D. N. Christodoulides and R. I. Joseph, *Opt. Lett.* **13**, 794 (1988).

[2] A. C. Scott and L. Macneil, *Phys. Lett. A* **98**, 87 (1983).

[3] J. C. Eilbeck, P. S. Lomdahl, and A. C. Scott, *Physica D* **14**, 318 (1985).

[4] Yu. S. Kivshar, *Opt. Lett.* **18**, 1147 (1993).

[5] W. Krolikówski and Yu. S. Kivshar, *J. Opt. Soc. Am. B* **13**, 876 (1996).

[6] A. B. Aceves, C. De Angelis, T. Peschel, R. Muschall, F. Lederer, S. Trillo, and S. Wabnitz, *Phys. Rev. E* **53**, 1172 (1996).

[7] S. Darmanyan, A. Kobyakov, E. Schmidt, and F. Lederer, *Phys. Rev. E* **57**, 3520 (1998).

[8] F. Lederer and J.S. Aitchison, in *Optical Solitons: Theorertical Challenges and Industrial Perspectives*, V. E. Zakharov and S. Wabnitz, Eds. (EDP Sciences, Les Ulis, 1999), pp. 349–365.

[9] F. Lederer, S. Darmanyan, and A. Kobyakov, in *Spatial Solitons*, S. Trillo and W. Torruellas, Eds. (Springer, Berlin, 2001), pp. 267–290.

[10] H. S. Eisenberg, Y. Silberberg, R. Morandotti, A. R. Boyd, and J. S. Aitchison, *Phys. Rev. Lett.* **81**, 3383 (1998).

[11] R. Morandotti, U. Peschel, J. S. Aitchison, H. S. Eisenberg, and Y. Silberberg, *Phys. Rev. Lett.* **83**, 2726 (1999).

[12] R. Morandotti, U. Peschel, J. S. Aitchison, H. S. Eisenberg, and Y. Silberberg, *Phys. Rev. Lett.* **83**, 4756 (1999).

[13] H. S. Eisenberg, Y. Silberberg, R. Morandotti, and J. S. Aitchison, *Phys. Rev. Lett.* **85**, 1863 (2000).

[14] R. Morandotti, H. S. Eisenberg, Y. Silberberg, M. Sorel, and J. S. Aitchison, *Phys. Rev. Lett.* **86**, 3296 (2001).

[15] T. Pertsch, T. Zentgraf, U. Peschel, A. Bräuer, and F. Lederer, *Phys. Rev. Lett.* **88**, 93901 (2002).

[16] A. Trombettoni and A. Smerzi, *Phys. Rev. Lett.* **86**, 2353 (2001).

[17] G. P. Agrawal, *Applications of Nonlinear Fiber Optics* (Academic Press, San Diego, 2001), Chap. 2.

[18] S. M. Jensen, *IEEE J. Quantum Electron.* **18**, 1580 (1982).

[19] A. A. Mayer, *Sov. J. Quantum Electron.* **14**, 101 (1984).

[20] M. I. Molina and G. P. Tsironis, *Phys. Rev. A* **46**, 1124 (1992).

[21] N. Finlayson, K. J. Blow, L. J. Bernstein, and K. W. DeLong, *Phys. Rev. A* **48**, 3863 (1993).

[22] S. Somekh, E. Garmire, A. Yariv, H. L. Garvin, and R. G. Hunsperger, *Appl. Phys. Lett.* **22**, 46 (1973).

[23] G. P. Agrawal, *Nonlinear Fiber Optics*, 3rd ed. (Academic Press, San Diego, 2001), Chap. 1.

[24] A. Yariv and A. Gover, *Appl. Phys. Lett.* **26**, 537 (1975).

[25] P. St. J. Russell, *Appl. Phys. B* **39**, 231 (1986).

[26] P. St. J. Russell, *Phys. Rev. A* **33**, 3232 (1986).

[27] J. Feng and N. B. Ming, *Phys. Rev. A* **40**, 7047 (1989).

[28] P. St. J. Russell, *Opt. Commun.* **48**, 71 (1983).

[29] H. Kosaka, T. Kawashima, A. Tomita, M. Notomi, T. Tamamura, T. Sato, and S. Kawakami, *Appl. Phys. Lett.* **74**, 11 (1999).

[30] A. Yariv, *Optical Electronics* (Saunders College Publishing, Philadelphia, 1991), pp. 519–524.

[31] D. B. Duncan, J. C. Eilbeck, H. Feddersen, and J. A. D. Wattis, *Physica D* **68**, 1 (1993).

[32] A. Lahiri, S. Panda, and T. K. Roy, *Phys. Rev. Lett.* **84**, 3570 (2000).

[33] Yu. S. Kivshar and D. K. Campbell, *Phys. Rev. E* **48**, 3077 (1993).

[34] S. Flach and K. Kladko, *Physica D* **127**, 61 (1999).

[35] W. Krolikowski, U. Trutschel, M. Cronin-Golomb, C. Schmidt-Hattenberger, *Opt. Lett.* **19**, 321 (1994).

[36] R. Muschall, C. Schmidt-Hattenberger, and F. Lederer, *Opt. Lett.* **19**, 323 (1994).

[37] A. B. Aceves, C. De Angelis, S. Trillo, and S. Wabnitz, *Opt. Lett.* **19**, 332 (1994).

[38] M. Matsumoto, S. Katayama, and A. Hasegawa, *Opt. Lett.* **20**, 1758 (1995).

[39] T. Peschel, R. Muschall, and F. Lederer, *Opt. Commun.* **136**, 16 (1997).

[40] D. Cai, A. R. Bishop, and N. Grønbech-Jensen, *Phys. Rev. E* **56**, 7246 (1997).

[41] R. M. Abrarov, P. L. Christiansen, S. A. Darmanyan, A. C. Scott, and M. P. Soerensen, *Phys. Lett. A* **171**, 298 (1992).

[42] A. B. Aceves, C. De Angelis, A. M. Rubenchik, and S. K. Turitsyn, *Opt. Lett.* **19**, 329 (1994).

[43] A. B. Aceves, C. De Angelis, G. G. Luther, and A. M. Rubenchik, *Opt. Lett.* **19**, 1186 (1994).

[44] A. B. Aceves, G. G. Luther, C. De Angelis, and A. M. Rubenchik, and S. K. Turitsyn, *Phys. Rev. Lett.* **75**, 73 (1995).

[45] E. W. Laedke, K. H. Spatschek, S. K. Turitsyn, and V. K. Mesentsev, *Phys. Rev. E* **52**, 5549 (1995).

[46] O. Bang, J. J. Rasmussen, and P. L. Christiansen, *Nonlinearity* **7**, 205 (1994).

[47] E. W. Laedke, K. H. Spatschek, and S. K. Turitsyn, *Phys. Rev. Lett.* **73**, 1055 (1994).

[48] D. Mandelik, H. S. Eidenberg, Y. Silberberg, R. Morandotti, and J. S. Aitchison, *Proc. Nonlinear Guided Waves and Their Applications* (Optical Society of America, Washington, DC, 2002).

[49] Yu. I. Voloshchenko, Yu. N. Ryzhov, and V. E. Sotin, *Zh. Tekh. Fiz.* **51**, 902 (1981) [*Sov. Phys. Tech. Phys.* **26**, 541 (1981)].

[50] W. Chen and D. L. Mills, *Phys. Rev. Lett.* **58**, 160 (1987).

[51] D. N. Christodoulides and R. I. Joseph, *Phys. Rev. Lett.* **62**, 1746 (1989).

[52] C. M. de Sterke and J. E. Sipe, in *Progress in Optics*, Vol. 33, E. Wolf, Ed. (North-Holland, Amsterdam, 1994), pp. 203–260.

[53] A. A. Sukhorukov and Yu. S. Kivshar, *Phys. Rev. Lett.* **87**, 083901 (2001).

[54] S. F. Mingaleev and Yu. S. Kivshar, *Phys. Rev. Lett.* **86**, 5474 (2001).

[55] I. V. Gerasimchuk and A. S. Kovalev, *Fiz. Nizk. Temp.* **26**, 799 (2000) [*Low Temp. Phys.* **26**, 586 (2000)].

[56] A. A. Sukhorukov and Yu. S. Kivshar, *Phys. Rev. E* **65**, 036609 (2002).

[57] A. A. Sukhorukov and Yu. S. Kivshar, *J. Opt. Soc. Am. B* **19**, 772 (2002).

[58] J. P. Dowling and C. M. Bowden, *Phys. Rev. A* **46**, 612 (1992); I. Alvarado-Rodriguez, P. Halevi, and A. S. Sánchez, *Phys. Rev. E* **63**, 056613 (2001).

[59] H. Grebel and W. Zhong, *Opt. Lett.* **18**, 1123 (1993).

[60] R. F. Nabiev, P. Yeh, and D. Botez, *Opt. Lett.* **18**, 1612 (1993).

[61] M. D. Tocci, M. J. Bloemer, M. Scalora, J. P. Dowling, and C. M. Bowden, *Appl. Phys. Lett.* **66**, 2324 (1995).

[62] S. Lan, S. Nishikawa, and O. Wada, *Appl. Phys. Lett.* **78**, 2101 (2001).

[63] A. A. Sukhorukov, Yu. S. Kivshar, O. Bang, and C. M. Soukoulis, *Phys. Rev. E* **63**, 016615 (2001).

[64] S. Mookherjea and A. Yariv, *Opt. Exp.* **9**, 91 (2001).

[65] J. C. Bronski, L. D. Carr, B. Deconinck, J. N. Kutz, and K. Promislow, *Phys. Rev. E* **63**, 036612 (2001).

[66] C. M. de Sterke, *J. Opt. Soc. Am. B* **15**, 2660 (1998).

[67] Yu. S. Kivshar and M. Peyrard, *Phys. Rev. A* **46**, 3198 (1992).

[68] Yu. S. Kivshar and M. Salerno, *Phys. Rev. E* **49**, 3543 (1994).

[69] S. Darmanyan, I. Relke, and F. Lederer, *Phys. Rev. E* **55**, 7662 (1997).

[70] Yu. S. Kivshar and D. K. Campbell, *Phys. Rev. E* **48**, 3077 (1993).

[71] B. Malomed and M. I. Weinstein, *Phys. Lett. A* **220**, 91 (1996).

[72] I. V. Barashenkov, D. E. Pelinovsky, and E. V. Zemlyanaya, *Phys. Rev. Lett.* **80**, 5117 (1998).

[73] Yu. S. Kivshar, A. R. Champneys, D. Cai, and A. R. Bishop, *Phys. Rev. B* **58**, 5423 (1998).

[74] S. Darmanyan, A. Kobyakov, and F. Lederer, *Zh. Eksp. Teor. Fiz.* **86**, 1253 (1998) [*JETP* **86**, 682 (1998)].

[75] P. G. Kevrekidis, A. R. Bishop, and K. Ø. Rasmussen, *Phys. Rev. E* **63**, 036603 (2001).

[76] T. Kapitula, P. G. Kevrekidis, and B. A. Malomed, *Phys. Rev. E* **63**, 036604 (2001).

[77] Yu. S. Kivshar and B. Luther-Davies, *Phys. Rep.* **298**, 81 (1998).

[78] Yu. S. Kivshar, W. Królikowski, and O. A. Chubykalo, *Phys. Rev. E* **50**, 5020 (1994).

[79] M. Johansson and Yu. S. Kivshar, *Phys. Rev. Lett.* **82**, 85 (1999).

[80] V. V. Konotop and S. Takeno, *Phys. Rev. E* **60**, 1001 (1999).

[81] F. Barra, P. Gaspard, and S. Rica, *Phys. Rev. E* **61**, 5852 (2000).

[82] S. Darmanyan, A. Kobyakov, and F. Lederer, *Zh. Eksp. Teor. Fiz.* **120**, 486 (2001) [*JETP* **93**, 429 (2001)].

[83] A. M. Morgante, M. Johansson, G. Kopidakis, and S. Aubry, *Phys. Rev. Lett.* **85**, 550 (2000).

[84] A. A. Sukhorukov and Yu. S. Kivshar, in *Proc. Nonlinear Guided Waves and Their Applications* (Optical Society of America, Washington, DC, 2002).

[85] P. Yeh and A. Yariv, *Opt. Commun.* **19**, 427 (1976).

[86] A. Y. Cho, A. Yariv, and P. Yeh, *Appl. Phys. Lett.* **30**, 471 (1977).

[87] U. Peschel, R. Morandotti, J. S. Aitchison, H. S. Eisenberg, and Y. Silberberg, *Appl. Phys. Lett.* **75**, 1348 (1999).

[88] R. Morandotti, H. S. Eisenberg, D. Mandelik, Y. Silberberg, D. Modotto, M. Sorel, and J. S. Aitchison, in *OSA Trends in Optics and Photonics*, vol. 74 (Optical Society of America, Washington DC, 2002), p. 239.

[89] J. Meier, G. Stegeman, H. S. Eisenberg, Y. Silberberg, R. Morandotti, and J. S. Aitchison, *Proc. Nonlinear Guided Waves and their Applications* (Optical Society of America, Washington DC, 2002)

[90] M. J. Ablowitz and Z. H. Musslimani, *Phys. Rev. Lett.* **87**, 254102 (2001).

[91] U. Peschel and F. Lederer, *J. Opt. Soc. Am. B* **19**, 544 (2002).

[92] F. Bloch, *Z. Phys.* **52**, 555 (1928).

[93] C. Zener, *Proc. R. Soc. London A* **154**, 523 (1932).

[94] C. Waschke, H. Roskos, R. Schwendler, K. Leo, H. Kurz, and K. Kohler, *Phys. Rev. Lett.* **70**, 3319 (1993).

[95] D. Cai, A. Bishop, N. Gronbech-Jensen, and M. Salerno, *Phys. Rev. Lett.* **74**, 1186 (1995)

[96] G. Lenz, I. Talanina, C.M. de Sterke, *Phys. Rev. Lett.* **83**, 963 (1999).

[97] M. Dahan, E. Peik, J. Reichel, Y. Castin, and C. Salomon, *Phys. Rev. Lett.* **76**, 4508 (1996)

[98] Q. Niu, X.-G. Zhao, G. A. Georgakis, and M. G. Raizen, *Phys. Rev. Lett.* **76**, 4504 (1996)

[99] U. Peschel, T. Pertsch, and F. Lederer, *Opt. Lett.* **23**, 1701 (1998).

[100] T. Pertsch, P. Dannberg, W. Elflein, A. Bräuer, and F. Lederer, *Phys. Rev. Lett.* **83**, 4752 (1999).

[101] T. Pertsch, U. Peschel, and F. Lederer, *Phys. Rev. E* **67** (2003).

[102] D. N. Christodoulides and E. D. Eugenieva, *Phys. Rev. Lett.* **87**, 233901 (2001).

[103] D. N. Christodoulides and E. D. Eugenieva, *Opt. Lett.* **26**, 1876 (2001); *Opt. Lett.* **27**, 369 (2002).

[104] E. D. Eugenieva, N. K. Efremidis, and D. N. Christodoulides, *Opt. Photon. News* **12** (12) 57 (2001).

[105] J. D. Joannopoulos, P. R. Villeneuve, and S. Fan, *Nature* **386**, 143 (1997).

[106] E. D. Eugenieva, N. K. Efremidis, and D. N. Christodoulides, *Opt. Lett.* **26**, 1978 (2001).

[107] N. K. Efremidis, S. Sears, D. N. Christodoulides, J. W. Fleischer, and M. Segev, *Phys. Rev. E* **66**, 046602 (2002).

[108] J. W. Fleischer, T. Carmon, M. Segev, N. K. Efremidis, and D. N. Chrsidoulides, *Phys. Rev. Lett.* **90** (2003).

Chapter 12

Solitons in Photonic Crystals

Photonic crystals can be viewed as an optical analog of semiconductors, in the sense that they modify the propagation characteristics of light just as an atomic lattice modifies the properties of electrons through a bandgap structure. Photonic crystals with embedded nonlinear impurities or made of a nonlinear material are called *nonlinear photonic crystals*, and they create an ideal environment for the generation and observation of localized modes in the form of solitons. Such solitons can be viewed as an extension of the concepts of the discrete and gap solitons to two (and even three) spatial dimensions. Moreover, in the case of short pulses propagating inside photonic-crystal waveguides and circuits, the concept of such solitons can be further extended to the temporal domain. For this reason, the study of nonlinear localized modes in photonic crystals unifies several fundamental concepts of soliton physics and nonlinear optics. This chapter focuses on solitons forming inside photonic crystals. Section 12.1 considers a simple model of a two-dimensional photonic crystal in the form of a lattice of dielectric rods, and it shows that an isolated defect can support a linear localized mode, while an array of such defects creates a waveguide. Section 12.2 presents several approaches to study nonlinear effects in photonic-crystal waveguides based on certain simplified nonlinear models. Additionally, this section introduces a set of discrete equations capable of describing such waveguides in photonic crystals. The properties of quasi-one-dimensional spatial solitons (nonlinear localized modes) in photonic waveguides are discussed in Section 12.3, while Section 12.4 summarizes the results on two-dimensional solitons in photonic crystals. Section 12.5 is devoted to the properties of quadratic photonic crystals.

12.1 Linear Characteristics

Before discussing the nonlinear effects, such as soliton formation, it is important to understand how a photonic crystal modifies the propagation of light even in the absence of nonlinear effects. An important property of photonic crystals is the existence of a complete bandgap: Incident light whose frequency falls within this bandgap cannot propagate through the photonic crystal. In this section we consider a simple photonic-

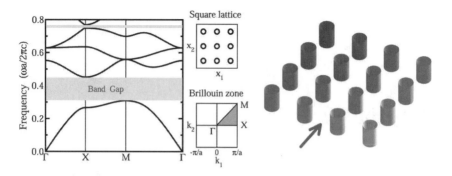

Figure 12.1: Band structure of a photonic crystal consisting of a square lattice of dielectric rods (right) exhibiting a bandgap (shaded). The two insets show the square lattice and the corresponding Brillouin zone; the irreducible zone is shown shaded.

crystal structure to discuss these features. Although we restrict the following discussion to the simpler two-dimensional case appropriate for a planar photonic-bandgap structure, many of the results can be extended to the more general three-dimensional case allowing an absolute bandgap.

12.1.1 Photonic Bandgap

Many types of two-dimensional (2D) and three-dimensional (3D) periodic dielectric structures are known to possess a complete bandgap [1]–[4]. As a simple example, we consider a 2D photonic crystal, shown schematically in Figure 12.1 and created by a periodic lattice of infinitely long dielectric rods aligned vertically (parallel to the z axis). Optical wave propagation in such a structure is characterized by a dielectric constant $\varepsilon(\mathbf{r}) \equiv \varepsilon(x,y)$. Photonic crystals of this type can possess a complete bandgap when the electric field $\mathbf{E}$ is polarized along the z axis and propagates in the (x,y) plane. The field evolution is then governed by the scalar wave equation

$$\nabla^2 E(\mathbf{r},t) - \frac{1}{c^2} \frac{\partial^2}{\partial t^2} [\varepsilon(\mathbf{r})E] = 0, \tag{12.1.1}$$

where $\nabla^2 \equiv \partial^2/\partial x^2 + \partial^2/\partial y^2$ is the Laplacian operator. As usual, we introduce the slowly varying pulse envelope $A(\mathbf{r},t)$ by writing the electric field in the form

$$E(\mathbf{r},t) = \mathrm{Re}[A(\mathbf{r},t)e^{-i\omega t}]. \tag{12.1.2}$$

Assuming that A varies slowly with t, Eq. (12.1.1) reduces to

$$\left[\nabla^2 + \varepsilon(\mathbf{r}) \left(\frac{\omega}{c} \right)^2 \right] A(\mathbf{r},t) \approx -2i\varepsilon(\mathbf{r}) \frac{\omega}{c^2} \frac{\partial A}{\partial t}. \tag{12.1.3}$$

In the continuous-wave (CW) case, A does not depend on t. Equation (12.1.3) then describes an eigenvalue problem that can be solved by using the well-known plane-wave method [5].

In the case of a perfect photonic crystal, the dielectric constant $\varepsilon(\mathbf{r}) \equiv \varepsilon_p(\mathbf{r})$ is a periodic function defined as

$$\varepsilon_p(\mathbf{r} + \mathbf{s}_{ij}) = \varepsilon_p(\mathbf{r}), \qquad \mathbf{s}_{ij} = i\,\mathbf{a}_1 + j\,\mathbf{a}_2, \tag{12.1.4}$$

where i and j are arbitrary integers and $\mathbf{a}_1$ and $\mathbf{a}_2$ are the two lattice vectors along the x and y directions. Figure 12.1 shows that band structure of such a 2D photonic crystal using the notation of solid-state physics [6, 7]. The cylindrical rods are assumed to form a square lattice with the distance a between two neighboring rods, so that $\mathbf{a}_1 = a\mathbf{r}_1$ and $\mathbf{a}_2 = a\mathbf{r}_2$, $\mathbf{r}_1$ and $\mathbf{r}_2$ being the unit vectors of the lattice.

Figure 12.1 shows the band structure of such a square lattice assuming that each rod has a radius $r_0 = 0.18a$ and that the dielectric constant $\varepsilon_0 = 11.56$. The band structure is shown using the irreducible triangular region forming the Brillouin zone (shaded region in Figure 12.1), with its three corners marked as Γ, M, and X. This 2D photonic crystal exhibits a complete bandgap in the frequency range extending from $\omega(a/2\pi c) = 0.303$ to 0.444. It has been used for studying the bound states [7], transmission of light through sharp bends [8], waveguide branches [9], waveguide intersections [10], channel-drop filters [11], nonlinear localized modes [12]–[14], and discrete spatial solitons [15]. This type of photonic crystal has also been fabricated using porous silicon with $a = 0.57$ μm and was found to exhibit a complete bandgap at 1.55 μm [16].

12.1.2 Defect Modes

One of the most intriguing properties of a photonic crystal is the emergence of localized modes that may appear within its photonic bandgap when a defect is embedded in an otherwise perfect photonic crystal. The simplest way to create a defect in a 2D photonic crystal is to introduce a "defect rod" with radius r_d and dielectric constant ε_d. Writing the total dielectric constant as

$$\varepsilon(\mathbf{r}) = \varepsilon_p(\mathbf{r}) + \varepsilon_d(\mathbf{r}), \tag{12.1.5}$$

Eq. (12.1.3) takes the form

$$\left[\nabla^2 + \left(\frac{\omega}{c}\right)^2 \varepsilon_p(\mathbf{r})\right] A(\mathbf{r},t) = -\hat{\mathcal{L}}A(\mathbf{r},t), \tag{12.1.6}$$

where the operator

$$\hat{\mathcal{L}} = \left(\frac{\omega}{c}\right)^2 \varepsilon_d(\mathbf{r}) + 2i\varepsilon(\mathbf{r})\frac{\omega}{c^2}\frac{\partial}{\partial t} \tag{12.1.7}$$

is introduced for convenience.

Equation (12.1.6) can be solved using the Green function approach, and the solution is given by

$$A(\mathbf{r},t) = \int_{-\infty}^{\infty} G(\mathbf{r},\mathbf{r}')\hat{\mathcal{L}}A(\mathbf{r}',t)\,d^2\mathbf{r}', \tag{12.1.8}$$

where the Green function $G(\mathbf{r},\mathbf{r}')$ satisfies

$$\left[\nabla^2 + \left(\frac{\omega}{c}\right)^2 \varepsilon_p(\mathbf{r})\right] G(\mathbf{r},\mathbf{r}';\omega) = -\delta(\mathbf{r} - \mathbf{r}'). \tag{12.1.9}$$

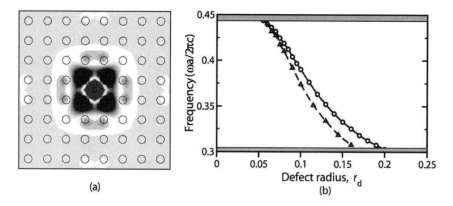

Figure 12.2: (a) A localized mode supported by a single defect rod with radius $r_d = 0.1a$ in a 2D array of rods with $r_0 = 0.18a$ and $\varepsilon_0 = 11.56$. Rod locations are indicated by circles. (b) Frequency of the defect mode as a function of rod radius r_d. The dashed line shows the approximate solution. (After Ref. [18]; ©2001 Kluwer.)

The properties of the Green function associated with a perfect 2D photonic crystal are described in Ref. [5]. The two most noteworthy among them are the symmetric and periodic properties, expressed as

$$G(\mathbf{r}, \mathbf{r}') = G(\mathbf{r}', \mathbf{r}), \qquad G(\mathbf{r} + \mathbf{s}_{ij}, \mathbf{r}' + \mathbf{s}_{ij}) = G(\mathbf{r}, \mathbf{r}'), \qquad (12.1.10)$$

where $\mathbf{s}_{ij}$ is defined in Eq. (12.1.4). The Green function can be calculated using the Fourier transform

$$G(\mathbf{r}, \mathbf{r}'; \omega) = \int_{-\infty}^{\infty} G(\mathbf{r}, \mathbf{r}', t) e^{i\omega t} dt. \qquad (12.1.11)$$

The time-dependent Green function then satisfies the equation

$$\left[\nabla^2 - \varepsilon_p(\mathbf{r}) \frac{\partial^2}{\partial t^2} \right] G(\mathbf{r}, \mathbf{r}', t) = -\delta(t) \delta(\mathbf{r} - \mathbf{r}'), \qquad (12.1.12)$$

which can be solved using the finite-difference time-domain (FDTD) method [17].

Once the Green function is known, we can find the defect modes by solving Eq. (12.1.8) directly. As an example, consider again the 2D photonic crystal shown in Figure 12.1. Figure 12.2(a) shows the defect mode created by introducing a single defect rod with a smaller radius ($r_d = 0.1a$) but the same dielectric constant into the 2D array of rods with $r_0 = 0.18a$ and $\varepsilon_0 = 11.56$. The electric-field contours show that light at a specific frequency lying inside the photonic bandgap remains confined to the defect site. Figure 12.2(b) shows the dependence of the frequency associated with the defect mode as a function of r_d. The solid curve is obtained by solving Eq. (12.1.8) numerically, while the dashed line shows an approximate solution. Although a direct numerical solution of the integral equation (12.1.8) is possible in the case of one or a few defect rods, the required computing resources become overly prohibitive as the number of defect rods increases. In fact, the use of approximate numerical techniques becomes necessary for investigating a row of defects, forming a waveguide, and bending or crossing of several such waveguides.

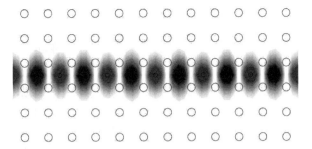

Figure 12.3: Electric-field contours for a linear guided mode in a waveguide created by an array of defect rods. Rod positions are indicated by circles.

12.1.3 Waveguides in Photonic Crystals

One of the most promising applications of the photonic-bandgap (PBG) structures is the possibility of creating a novel type of optical waveguide. In conventional waveguides, such as optical fibers, light is confined by *total internal reflection* occurring because of different refractive indices for the core and the cladding. A weakness of such waveguides is that they cannot be bent sharply, because much of the light escapes from the core region if the bend radius is not large compared with the wavelength. This is a serious obstacle for creating "integrated optical circuits," in which the space required for large-radius bends is often unavailable.

The waveguides based on PBG materials employ a different physical mechanism. More specifically, light is guided by a line of coupled defects, each defect supporting a localized mode whose frequency lies inside the photonic bandgap. The simplest PBG waveguide can be created by a straight line of defect rods, as shown in Figure 12.3. Instead of a single localized state of an isolated defect, a waveguide supports propagating states (guided modes) with the frequencies in a narrow band located inside the bandgap of a perfect crystal. Such guided modes have a periodic profile along the waveguide, but they decay exponentially in the transverse direction. The photonic-crystal waveguides operate in a manner similar to resonant cavities, because the guiding mode is forbidden from propagating in the bulk. Because of this property, when a bend is created in such a waveguide, light remains trapped within the waveguide, no matter how sharp the bend is. Of course, a part of energy may be reflected for very sharp bends. However, as was predicted numerically [6] and also demonstrated using microwave [8] and optical [19] experiments, it is possible to realize high transmission efficiencies for nearly all frequencies supported by a PBG waveguide.

12.2 Effective Discrete Models

12.2.1 General Overview

A straightforward way to analyze the properties of photonic crystals and photonic-crystal waveguides is to employ direct numerical simulations, although an analytic approach, if possible, can provide more physical insight. It turns out that the basic properties of a PBG waveguide composed of weakly interacting (or weakly coupled) defect modes can be described by applying the concepts of solid-state physics based on the so-called tight-binding approximation. Using this approach, it was shown in a 2002 study [20] that *spatiotemporal discrete solitons* can propagate undistorted along a chain of coupled defects that are embedded in a photonic-crystal structure and form a nonlinear high-Q microcavity. Such states are possible as a result of the balance between the effects of discrete dispersion and material nonlinearity. Moreover, such self-localized entities are capable of exhibiting very low group velocities, depending on the coupling strength among neighboring microcavities; in principle, they can even remain immobile. Assuming that the defect modes are weakly coupled, a discrete NLS equation governs such discrete solitons.

In another study [21], a similar discrete equation was obtained by employing the Fourier-transform technique. It allowed a theoretical and numerical description of the coupled defects in photonic-bandgap crystals using a set of coupled ordinary differential. The actual configuration of the defects (chain, lattice, bend, or anything else) enters the equations as a linear coupling between neighboring defects. The results based on this approach were compared in Ref. [21] with the results obtained using the numerical methods, the transfer-matrix method and the FTDT method, and a good agreement was found. The nonlinear pulse propagation in coupled-resonator optical waveguides (CROWs) has also been studied [22, 23] using a continuous analog of the effective discrete equations (see Ref. [24]).

Numerical solutions do not always provide much physical insight. Moreover, the effective discrete equations derived in Refs. [20, 21] are applicable only for describing the arrays of weakly coupled defect modes. For this reason, we derive in this section the effective discrete equations using the Green-function approach discussed earlier. It turns out that many properties of photonic-crystal waveguides and circuits, including the transmission spectra of waveguides with sharp bends [25], can be accurately described by such discrete equations. These equations are somewhat analogous to the Kirchhoff equations for electric circuits, but, in contrast to electronics, both diffraction and interference become important in photonic crystals, and one must include the effects of long-range interaction.

12.2.2   Green-Function Approach

The starting point is Eq. (12.1.8). The operator $\hat{\mathcal{L}}$ appearing in this equation consists of two terms, as seen in Eq. (12.1.7) . If we consider the CW case and neglect the time-derivative term, the problem of light propagation in a PBG waveguide reduces to solving the following integral equation [12]:

$$A(\mathbf{r}, \omega) = \left(\frac{\omega}{c}\right)^2 \int_{-\infty}^{\infty} G(\mathbf{r}, \mathbf{r}', \omega) \delta\varepsilon(\mathbf{r}') A(\mathbf{r}', \omega) d^2\mathbf{r}', \qquad (12.2.1)$$

where $\delta\varepsilon(\mathbf{r}')$ denotes the dielectric perturbation induced by defects forming the waveguide. The frequency dependence of the Green function G is shown explicitly, to emphasize that the results depend on the frequency of the optical field $A(\mathbf{r}, \omega)$.

A single defect rod is described by $\delta\varepsilon(\mathbf{r}) = \varepsilon_d f(\mathbf{r})$, where $f(\mathbf{r})$ denotes the location of the defect. This defect can support several localized modes, all of which can be found by solving the eigenvalue problem

$$A_l(\mathbf{r}, \omega) = \left(\frac{\omega_l}{c}\right)^2 \int_{-\infty}^{\infty} G(\mathbf{r}, \mathbf{r}', \omega) \varepsilon_d f(\mathbf{r}') A_l(\mathbf{r}', \omega) d^2\mathbf{r}', \tag{12.2.2}$$

where ω_l is the frequency (a discrete eigenvalue) of the lth eigenmode and $A_l(\mathbf{r})$ is the corresponding optical field.

As the number of defect rods increases in PBG waveguide circuits, the numerical solution of the integral equation (12.2.1) becomes hard to obtain and requires extensive computer resources [6]–[11]. The discrete model simplifies this task by considering the modes of a single defect that are coupled to the modes of the nearest neighbors. This approach is similar to that of Chapter 11, used in the context of discrete solitons associated with waveguide arrays. Mathematically, the optical field is expanded as a linear combination of the localized modes $A_l(\mathbf{r})$ supported by the isolated defects:

$$A(\mathbf{r}, \omega) = \sum_l \sum_n \psi_n^{(l)}(\omega) A_l(\mathbf{r} - \mathbf{r}_n), \tag{12.2.3}$$

where $\mathbf{r}_n$ denotes the location of the nth defect.

Substituting Eq. (12.2.3) into Eq. (12.2.1), multiplying it by $A_{l'}^*(\mathbf{r} - \mathbf{r}_{n'})$ and integrating over $\mathbf{r}$, we obtain the following set of discrete equations for the amplitude $\psi_n^{(l)}$ of the lth eigenmode at the nth defect rod and its coupling to the neighboring defect rods:

$$\sum_l \sum_n \lambda_{l,n}^{l',n'} \psi_n^{(l)} = \sum_l \sum_n \sum_m \varepsilon_d \mu_{l,n,m}^{l',n'}(\omega) \psi_n^{(l)}, \tag{12.2.4}$$

where

$$\lambda_{l,n}^{l',n'} = \int_{-\infty}^{\infty} A_l(\mathbf{r} - \mathbf{r}_n) A_{l'}^*(\mathbf{r} - \mathbf{r}_{n'}) d^2\mathbf{r}, \tag{12.2.5}$$

$$\mu_{l,n,m}^{l',n'}(\omega) = \left(\frac{\omega}{c}\right)^2 \int_{-\infty}^{\infty} d^2\mathbf{r} A_{l'}^*(\mathbf{r} - \mathbf{r}_{n'}) \times$$
$$\int_{-\infty}^{\infty} G(\mathbf{r}, \mathbf{r}', \omega) f(\mathbf{r}' - \mathbf{r}_m) A_l(\mathbf{r}' - \mathbf{r}_n) d^2\mathbf{r}'. \tag{12.2.6}$$

It should be emphasized that the discrete equations (12.2.4)–(12.2.6) are derived by using only the approximation inherent in the expansion (12.2.3). A comparison of their solutions with the direct numerical solutions of Eq. (12.2.1) shows that this approximation is quite accurate and can be used for many physical problems.

The effective discrete equations (12.2.4) are still quite complicated. In some cases they can be simplified further while remaining reasonably accurate. A good example is provided by photonic-crystal waveguides that are created by using widely separated defect rods. Such waveguides are also known as CROWs or coupled-cavity waveguides [26]–[28]. Because of widely separated defect rods, the localized modes at any defect are only weakly coupled to the modes of neighboring defects. As discussed in Chapter 11, this situation can be described quite accurately by using the tight-binding approximation [29]. Its use implies that both $\lambda_{l,n}^{l',n'}$ and $\mu_{l,n,m}^{l',n'}$ vanish for $|n' - n| > 1$ and $|n' - m| > 1$. The most important feature of the CROW circuits is that their bends are reflectionless throughout the entire band [28]. This is in a sharp contrast with the conventional photonic-crystal waveguides created by removing or inserting closely spaced defect rods, for which 100% transmission through a waveguide bend is known to occur only at certain resonant frequencies [6]. In spite of this advantage, the CROW structures have a very narrow guiding band, and, as a result, they produce 100% transmission only in a narrow frequency band.

A major simplification of Eq. (12.2.4) occurs when the indirect coupling between the remote defect modes, caused by the slowly decaying Green function, is assumed to have a limited range; i.e., $\mu_{l,n,m}^{l',n'} = 0$ for $|n' - n| > N$, where the number N of effectively coupled defects lies between five and ten. This type of interaction, neglected in the tight-binding approximation, is important for understanding the transmission properties of photonic-crystal waveguides. With its use, one can even neglect the direct overlap among the nearest-neighbor eigenmodes and use the approximation

$$\lambda_{l,n}^{l',n'} = \delta_{l,l'}\,\delta_{n,n'}, \qquad \mu_{l,n,m}^{l',n'} = 0 \ \text{ for } \ n \neq m. \tag{12.2.7}$$

Taking this overlap into account leads to only minor corrections.

Assuming that the defects support only a *single* eigenmode ($l = 1$), the coefficients in Eqs. (12.2.5) and (12.2.6) can be calculated reasonably accurately, even with the approximation that the electric field remains constant inside the defect rods, i.e., $A_1(\mathbf{r}) \sim f(\mathbf{r})$. This approximation corresponds to averaging of the electric field in the integral equation (12.2.1) over the cross section of defect rods [12, 30]. The resulting approximate discrete equations for the amplitudes $E_n(\omega) \equiv \psi_n^1(\omega)$ of the eigenmode excited at the defect sites can be written in a matrix form,

$$\sum_m M_{nm}(\omega)E_m(\omega) = 0, \tag{12.2.8}$$

where the matrix M depends on the dielectric constant of the defects as

$$M_{nm}(\omega) = \varepsilon_d(E_m)J_{nm}(\omega) - \delta_{nm}. \tag{12.2.9}$$

The coupling constant $J_{n,m}(\omega) \equiv \mu_{1,m,n}^{1,n}(\omega)$ can be calculated in the approximation $A_1(\mathbf{r}) \sim f(\mathbf{r})$ using

$$J_{nm}(\omega) = \left(\frac{\omega}{c}\right)^2 \int_{r_d} G(\mathbf{r}_n, \mathbf{r}_m + \mathbf{r}', \omega)\, d^2\mathbf{r}', \tag{12.2.10}$$

where the integration is over the defect area. Note that matrix M is completely determined by the Green function of the 2D photonic crystal without any defects [12].

To check the accuracy of the discrete model, we apply it to the simple case of a single defect located at point $\mathbf{r}_0$. In this case, the solution of Eq. (12.2.9) requires

$$M_{00}(\omega) = \varepsilon_d J_{00}(\omega) - 1 = 0. \tag{12.2.11}$$

Its solution $J_{00}(\omega_d) = 1/\varepsilon_d$ defines the frequency ω_d of the defect mode. When the defect is created by removing a single rod, this condition leads to the frequency $\omega_d = 0.391 \times 2\pi c/a$, which differs by only 1% from the value $\omega_d = 0.387 \times 2\pi c/a$ calculated using the full numerical code [31].

The preceding analysis can be extended to include the nonlinearities of the defect rods by assuming that the dielectric constant ε_d depends on the field intensity at the site of the defect. In the case of a Kerr-type nonlinearity, we assume $\varepsilon_d(E_m) = \varepsilon_{d0} + \varepsilon_2 |E_m|2$ in Eq. (12.2.9). We can also extend the analysis to the case of optical pulses by using Eq. (12.1.7). Assuming again that the optical field inside a defect rod is almost constant, we obtain, for the amplitudes of the optical field inside the defect rods, the following time-dependent *discrete nonlinear equation*:

$$i\sigma \frac{\partial}{\partial t} E_n - E_n + \sum_m J_{n-m}(\omega)(\varepsilon_{d0} + |E_m|^2)E_m = 0. \qquad (12.2.12)$$

The nonlinear parameter ε_2 has been eliminated by renormalizing the field. As before, the parameter σ and the coupling constants $J_{n-m} \equiv J_{nm}$ are determined from the Green function $G(\mathbf{r}, \mathbf{r}', \omega)$ of the perfect photonic crystal. We have introduced the simpler notation J_{n-m} because J_{nm} in Eq. (12.2.10) depends only on the difference $n - m$.

12.3 Bandgap Solitons

One of the important physical concepts associated with nonlinearity is related to self-trapping and *localization*. In the linear case, the concept of localization in a periodic medium is always associated with the disorder that breaks translational invariance. However, it has become clear during recent years that localization can occur in the absence of any disorder if the nonlinearity leads to the formation of *intrinsic localized modes* [32, 33]. A rigorous proof of the existence of such nonlinear modes exists for a broad class of Hamiltonian-type, coupled-oscillator, nonlinear lattices [34, 35]. Approximate analytical solutions can be found in many other cases and demonstrate the general nature of the concept of *nonlinear* localized modes.

Although nonlinear localized modes can easily be identified in numerical simulations of molecular dynamics using several different physical models [32], it is only recently that they were observed experimentally in mixed-valence transition-metal complexes [36], quasi-one-dimensional antiferromagnetic chains [37], and arrays of Josephson junctions [38, 39]. Importantly, similar types of spatially localized nonlinear modes have been observed experimentally in *macroscopic* mechanical [40] and guided-wave optical [41] systems. From the standpoint of practical applications, the existence of nonlinear localized modes in optical systems is most promising, because it can lead to all-optical switching devices. Nonlinear localized modes in photonic crystals are especially attractive, because 3D photonic crystals for visible light have been successfully fabricated and are being pursued for creating tunable bandgap switches and optical transistors operating entirely with light.

12.3.1 Nonlinear Photonic Crystals

To realize the potential of photonic-crystal waveguides, it is crucial that their transmission properties be tunable. Several approaches have been suggested for this purpose. For instance, it has been found both numerically [42] and experimentally [43] that the transmission spectrum of straight and sharply bent waveguides in *quasi-periodic* photonic crystals contains a rich structure such that perfect transmission occurs near some specific frequencies. A *channel-dropping filter* is also useful in this context. It consists of two parallel waveguides coupled by point defects between them and allows frequency-selective power transfer between the two waveguides by creating resonant defect states of different symmetry [11]. However, such devices do not allow *dynamical tunability* because of their frequency-selective nature. Dynamic tuning can be realized using waveguides formed inside *nonlinear photonic crystals*. Such waveguides are created by inserting an additional row of rods made using a nonlinear material characterized by second-order [44] or third-order [12] nonlinear susceptibility.

In this section we discuss the properties of nonlinear photonic crystals in the framework of the discrete model developed in Section 12.2. For definiteness, we assume that the nonlinear defect rods are embedded in the photonic crystal along a selected direction $\mathbf{s}_{ij}$; i.e., they are located at points $\mathbf{r}_m = \mathbf{r}_0 + m\mathbf{s}_{ij}$. By changing the radius r_d of these defect rods and their location $\mathbf{r}_0$ in the crystal, one can create nonlinear waveguides with quite different properties.

The discrete model requires the solution of Eq. (12.2.12), in which the coefficients J_{n-m} depend on the Green function $G(\mathbf{r}, \mathbf{r}', \omega)$. As mentioned earlier, the Green function is a long-range function and decays only slowly as one moves away from a specific site [12]. This can be seen from Figure 12.4, where a typical spatial profile of the Green function is shown. Along the $\mathbf{s}_{01}$ and $\mathbf{s}_{10}$ directions, the coupling coefficients in Eq. (12.2.10) can be approximated by an exponential function as

$$|J_n(\omega)| \approx \begin{cases} J_0(\omega) & \text{for } n = 0, \\ J_*(\omega)\, e^{-\alpha(\omega)|n|} & \text{for } |n| \geq 1, \end{cases} \qquad (12.3.1)$$

where the characteristic decay rate $\alpha(\omega)$ can be as small as 0.85, depending on the values of ω, $\mathbf{r}_0$, and r_d. It can be even smaller for other types of photonic crystals.

With this form of J_n, Eq. (12.2.12) is a nontrivial long-range generalization of the 2D discrete nonlinear Schrödinger (NLS) equation of Chapter 11 [45]–[47]. It allows us to make use of an analogy between the problem under consideration and a class of the NLS equations that describe nonlinear excitations in quasi-one-dimensional molecular chains with long-range (e.g., dipole–dipole) interaction between the particles and local on-site nonlinearities [48, 49]. The nonlocal interparticle interaction introduces some new features to the properties of nonlinear localized modes (such as bistability in their transmission spectrum). The coupling coefficients $J_n(\omega)$ can be either unstaggered [$J_n(\omega) = |J_n(\omega)|$] or staggered and oscillate from site to site such that $J_n(\omega) = (-1)^n |J_n(\omega)|$. We therefore expect that the nonlocal nature of both linear and nonlinear terms in Eq. (12.2.12) would introduce new features into the properties of nonlinear localized modes excited in a photonic-crystal waveguide.

12.3.2 Staggered and Unstaggered Modes

Similar to the case of discrete solitons discussed in Chapter 11, the nonlinear localized modes inside a PBG waveguide can be classified as being staggered or unstaggered. In this section we discuss the properties of the two types of such modes.

As can be seen from the structure of the Green function in Figure 12.4, $J_n(\omega)$ varies monotonically for the waveguide oriented in the $\mathbf{s}_{01}$ direction with $\mathbf{r}_0 = \mathbf{a}_1/2$.

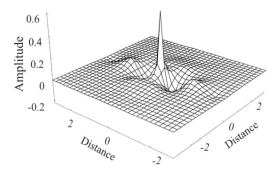

Figure 12.4: The Green function $G(\mathbf{r}_0, \mathbf{r}_0 + \mathbf{r}', \omega)$ for the photonic crystal shown in Figure 12.1, with $\mathbf{r}_0 = \mathbf{a}_1/2$ and $\omega = 0.33 \times 2\pi c/a$. (After Ref. [12]; ©2000 APS.)

Figure 12.5(a) shows the bandgap structure in this case using the parameter values $\varepsilon_0 = \varepsilon_d = 11.56$, $r_0 = 0.18a$, and $r_d = 0.1a$. The solid line within the bandgap shows the frequency of the linearly guided mode. The normalized frequency $\omega(a/2\pi c)$ at points A and B take the values 0.378 and 0.412, respectively. Since this frequency takes its minimum value at $k = 0$, the corresponding nonlinear mode is expected to be unstaggered.

Numerical solutions of Eq. (12.2.12) show that nonlinearity can lead to the existence of *guided modes* that are localized in the direction perpendicular to the waveguide by the defect rods and by the nonlinearity-induced self-trapping effect in the direction of the waveguide. Figure 12.5(b) shows the results for the waveguide oriented along the s_{01} direction by plotting the mode power $Q = \sum_n |E_n|^2$ as a function of the frequency of the nonlinear mode that corresponds to a spatial soliton. Such nonlinear modes exist with frequencies below the frequency of the linear guided mode of the waveguide, i.e., below the frequency at point A in Figure 12.5(a). The bell-shaped profile of the non-

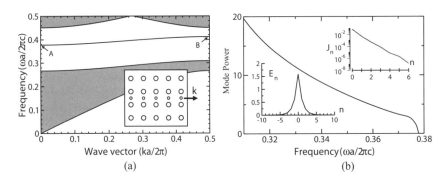

Figure 12.5: (a) Dispersion relation (solid line) for a PBG waveguide oriented along the s_{10} direction. The gray areas show the band structure of the perfect 2D photonic crystal. (b) Mode power $Q(\omega)$ of the nonlinear mode. The right inset shows the dependence of $J_n(\omega)$ on n for $\omega = 0.37(2\pi c/a)$, while the left inset shows the soliton amplitude. (After Ref. [12]; ©2000 APS.)

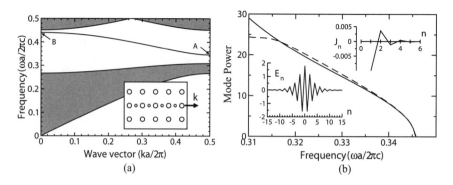

Figure 12.6: (a) Same as in Figure 12.5 except the waveguide is oriented along the s_{10} direction as shown in the inset. (b) Mode power $Q(\omega)$ of the nonlinear mode. The dashed line shows the case of a completely nonlinear photonic crystal. The right inset shows the dependence $J_n(\omega)$ for $n \geq 1$ ($J_0 = 0.045$) at $\omega = 0.33 \times 2\pi c/a$, while the left inset shows the soliton amplitude. (After Ref. [12]; ©2000 APS.)

linear modes indicates that they are indeed unstaggered. The mode power Q is closely related to the energy stored within the nonlinear mode in the 2D photonic crystal.

From the structure of the Green function in Figure 12.4, the case of staggered coupling corresponds to the waveguide oriented in the s_{10} direction with $r_0 = a_1/2$. Figure 12.6(a) shows the band structure for such a waveguide with the same parameter values used for Figure 12.5. The normalized frequency $\omega(a/2\pi c)$ at points A and B now takes the values 0.346 and 0.440, respectively. In this case, this frequency of the linearly guided mode takes its minimum at $k = \pi/a$. Accordingly, the staggered guided mode localized along the direction of the waveguide is expected to exist, with the frequency below the lowest frequency of the linear guided mode occurring at point A. Figure 12.6(b) shows the mode power for such modes as a function of their frequency. The amplitude profile of such a 2D nonlinear localized mode is shown in the left inset.

These results are obtained for linear photonic crystals with nonlinear waveguides created by a row of defect rods. The same analysis can also be carried out in the general case of a *completely nonlinear* photonic crystal in which rods of different size are made of the same nonlinear material. It turns out that the results are nearly the same, with only minor differences, as long as the nonlinearity is relatively weak. As an example, the dashed line in Figure 12.6(b) shows the mode power Q for a completely nonlinear photonic crystal. A direct comparison shows that the mode power is nearly the same in the two cases for $Q < 20$.

12.3.3   Soliton Stability

Let us consider a PBG waveguide created by a row of defect rods located at the points $r_0 = (a_1 + a_2)/2$ along a straight line in either the s_{10} or s_{01} direction. Since the coupling coefficients J_n decay slowly with the site number n, the effective interaction decays on a length scale longer than the two cases considered before. The numerical results for this case are shown in Figure 12.7 using the same parameter values as for

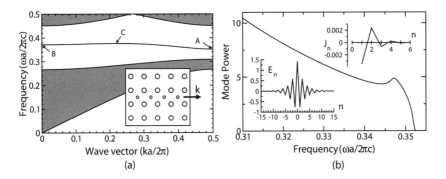

Figure 12.7: (a) Same as in Figure 12.5 except the waveguide is oriented as shown in the inset. (b) Mode power $Q(\omega)$ of nonlinear modes. The left and right insets show an example of the mode amplitude and the coupling coefficients, respectively. (After Ref. [12]; ©2000 APS.)

Figures 12.5 and 12.6. Figure 12.7(a) shows the band structure for the waveguide given in the inset. The normalized frequency $\omega(a/2\pi c)$ at points A, B, and C takes the values $0.352, 0.371$, and 0.376, respectively. Again, the frequency of the linearly guided mode takes its minimum value at point A, where $k = \pi/a$. Accordingly, a nonlinear staggered mode, localized along the direction of the waveguide, exists at a frequency below the lowest frequency of the linear guided mode at point A. Figure 12.7(b) shows the mode power for such modes as a function of their frequency. The left and right insets show, respectively, an example of the mode profile and the coupling coefficients $J_n(\omega)$ for $n \geq 1$ ($J_0 = 0.068$) for $\omega = 0.345 \times 2\pi c/a$.

A comparison of Figs. 12.6 and 12.7 reveals a new feature of the mode power. Similar to the case of long-range dispersive interaction [48, 49], the mode power Q varies in a *nonmonotonic* fashion for this type of nonlinear PBG waveguide. More specifically, $Q(\omega)$ *increases* in the frequency interval $0.344 < \omega a/(2\pi c) < 0.347$. Figure 12.8 shows this region as shaded and marked as "instability region." The nonlinear localized modes in this frequency interval are unstable and eventually decay or transform into the modes of higher or lower frequency [50]. Moreover, such a nonlinear waveguide exhibits *optical bistability*; i.e, two stable nonlinear localized modes of different widths coexist for the same value of mode power. Since the mode power is closely related to the mode energy, the mode energy is also a nonmonotonic function of ω and exhibits bistability. This behavior is a direct manifestation of the nonlocal nature of the effective (linear and nonlinear) interaction between the defect rod sites.

Such nonlinear PBG waveguides support *antisymmetric localized modes* in addition to the symmetric modes. Figure 12.8 shows the $Q(\omega)$ curves for the symmetric and antisymmetric branches, together with an example of the intensity distribution for each type of localized mode. These numerical results show that the power $Q(\omega)$ of the antisymmetric modes always exceeds that of symmetric ones (for all values of ω and all types of waveguides). Thus, antisymmetric modes are expected to be unstable and to transform into a lower-energy symmetric mode. The difference between the powers for the two types of modes determines the Peierls–Nabarro potential. One can see from Figure 12.8 that this barrier is negligible for $0.347 < \omega a/(2\pi c) < 0.352$, and

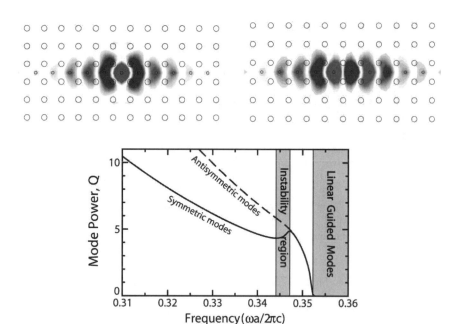

Figure 12.8: Examples of symmetric and antisymmetric localized modes. Mode Power Q as a function of its frequency is shown for the two kinds of modes; the bistable region is shown shaded and is marked unstable (After Ref. [18]; ©2001 Kluwer.)

thus such modes should be mobile. However, the Peierls–Nabarro barrier becomes sufficiently large for nonlinear modes in the frequency range $\omega < 0.344 \times 2\pi c/a$ that such modes become immobile. Hence, this bistability phenomenon may be useful [49] for *switching* between the immobile localized modes (used for energy storage) and the mobile localized modes (used for energy transport).

12.4   Two-Dimensional Photonic Crystals

A low-intensity optical beam cannot propagate through a photonic crystal if its frequency falls into the bandgap. However, in the case of a 2D periodic medium made using a Kerr-type nonlinear material, high-intensity light can propagate even when its frequency lies inside the bandgap. The propagation occurs in the form of solitary waves that are referred to as *2D gap solitons* that are found to be *stable* [51, 52]. This conclusion is based on the use of coupled-mode equations that are valid only for a *weakly* modulated dielectric constant $\varepsilon(\mathbf{r})$. In real photonic crystals, the modulation of $\varepsilon(\mathbf{r})$ is *comparable to its average value*. Thus, the results of Ref. [52] cannot always be used to discuss the properties of localized modes in realistic photonic crystals.

More specifically, the coupled-mode equations are valid if and only if the bandgap Δ is vanishingly small. The parameter Δ is related to a small parameter in the multiscale asymptotic expansions [53]. If we apply this analysis to describe nonlinear modes in

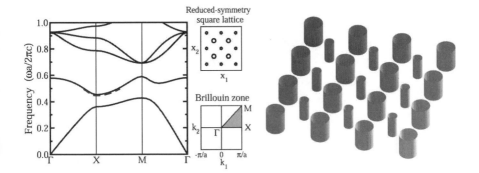

Figure 12.9: Bandgap structure of a reduced-symmetry photonic crystal (shown on the right) with $r_0 = 0.1a$, $r_d = 0.05a$, and $\varepsilon = 11.4$ for both types of rods. Solid lines are calculated using the full nonlinear model, whereas the dashed line is found from the discrete model. The top inset shows a cross-sectional view of the 2D photonic crystal. The bottom inset shows the corresponding Brillouin zone. (After Ref. [15]; ©2001 APS.)

a wider-gap photonic crystal, we obtain a $(2+1)$-dimensional NLS equation that does not possess any stable localized solutions. Moreover, the 2D localized modes of the coupled-mode equations are expected to exhibit an *oscillatory instability* that occurs for a broad class of Thirring-like coupled-mode equations [54]. Thus, if nonlinear localized modes do exist in realistic PBG materials, their stability should be associated with *different physical mechanisms* not accounted for by simplified continuum models.

In this section we follow Ref. [15] and study the properties of nonlinear localized modes in a 2D photonic crystal shown in Figure 12.9. It is composed of two types of circular rods: Rods of radius r_0 made from a linear dielectric material are placed at the corners of a square lattice with lattice spacing a, while rods of radius r_d made from a nonlinear dielectric material are placed at the center of each unit cell. Such photonic crystals of *reduced symmetry* have attracted attention because they possess much larger absolute bandgaps [55]. The band structure of this photonic crystal is shown in Figure 12.9. As seen there, it possesses two bandgaps, the first of which extends from $\omega = 0.426 \times 2\pi c/a$ to $\omega = 0.453 \times 2\pi c/a$.

The reduced-symmetry "diatomic" photonic crystal shown in Figure 12.9 can be viewed as a square lattice of "nonlinear defect rods" of small radius r_d ($r_d < r_0$) embedded in the ordinary single-rod photonic crystal formed by a square lattice of rods of larger radius, r_0, in air. The positions of the defect rods can then be described by the vectors $\mathbf{r}_{n,m} = n\mathbf{a}_1 + m\mathbf{a}_2$, where $\mathbf{a}_1$ and $\mathbf{a}_2$ are the primitive lattice vectors of the 2D photonic crystal. In contrast to the photonic-crystal waveguides discussed in the previous section, the nonlinear defect rods are characterized by two integer indices n and m. However, it is straightforward to extend Eq. (12.2.12) to obtain the following approximate 2D discrete nonlinear equation:

$$i\sigma\frac{\partial}{\partial t}E_{n,m} - E_{n,m} + \sum_{k,l} J_{n-k,m-l}(\omega)(\varepsilon_{d0} + |E_{k,l}|^2)E_{k,l} = 0, \qquad (12.4.1)$$

where $E_{n,m}(t,\omega) \equiv E(\mathbf{r}_{n,m},t,\omega)$ is the amplitude of the electric field inside the defect

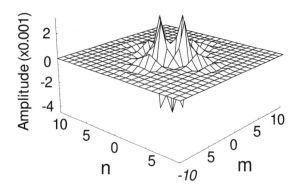

Figure 12.10: Coupling coefficients $J_{n,m}(\omega)$ for the photonic crystal depicted in Figure 12.9 ($J_{0,0} = 0.039$ is not shown). The frequency $\omega = 0.4456$ falls into the first bandgap. (After Ref. [15]; ©2001 APS.)

rod located at the nmth site.

The accuracy of this equation can be checked by solving it in the linear limit. However, as seen in Figure 12.10, the coupling coefficients $J_{n,m}(\omega)$ are long-range functions of n and m. Thus, one should take into account the interaction among at least 10 neighbors to obtain accurate results. The band structure shown in Figure 12.9 indicates that the frequencies of the linear modes (depicted by a dashed line, with a minimum at $\omega = 0.446 \times 2\pi c/a$) calculated from Eq. (12.4.1) are in a good agreement with those calculated directly from Eq. (12.1.3). The nonlinear modes described by Eq. (12.4.1) can be found numerically by using the Newton–Raphson iteration scheme. The results reveal the existence of a *continuous family* of such modes. A typical example of the nonlinear localized mode is shown in Figure 12.11.

Figure 12.12 shows the dependence of the mode power

$$Q(\omega) = \sum_{n,m} |E_{n,m}|^2 \qquad (12.4.2)$$

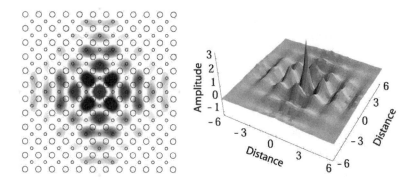

Figure 12.11: Top view (left) and 3D view (right) of a nonlinear localized mode in the first bandgap of 2D photonic crystal. (After Ref. [15]; ©2001 APS.)

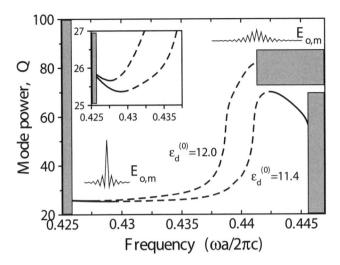

Figure 12.12: Mode power Q as a function of mode frequency ω for two different values of ε_{d0} for the nonlinear localized modes existing in the 2D photonic crystal of Figure 12.9. Dashed lines indicate unstable modes. The insets show typical profiles of stable modes and an enlarged portion of the $Q(\omega)$ curves. Gray areas show the lower and upper bands of delocalized modes surrounding the bandgap. (After Ref. [15]; ©2001 APS.)

on the frequency ω for the photonic crystal shown in Figure 12.9. As discussed in Chapter 2, this dependence represents an important characteristic of the nonlinear localized modes, because it allows one to determine their stability by means of the Vakhitov–Kolokolov stability criterion. More specifically, $dQ/d\omega > 0$ for unstable modes. This criterion has also been extended to 2D NLS models [56].

In the case of the 2D discrete cubic NLS equation, only high-amplitude localized modes are stable, and no stable modes exist in the continuum limit [56]–[58]. The high-amplitude modes are also stable (see inset in Figure 12.12), but they are not accessible under realistic conditions. To excite such modes, one needs to increase the refractive index at the mode center to more than two times. Thus, for realistic conditions and relatively small values of $\chi^{(3)}$, only low-amplitude localized modes can be excited in an experiment. However, such modes are always unstable; they either collapse or spread out [45]. They can be stabilized by some external forces (e.g., interaction with boundaries or disorder [59]), but in this case the excitations are pinned and cannot be used for energy or signal transfer.

In a sharp contrast to the 2D discrete NLS models, the low-amplitude localized modes of Eq. (12.4.1) can be stabilized, owing to *nonlinear long-range dispersion* inherent in photonic crystals. It is important to note that such stabilization does not occur in models with only *linear long-range* dispersion [45]. To gain a better insight into the stabilization mechanism, one can solve Eq. (12.2.12) for exponentially decaying coupling coefficients $J_{n,m}$. The results show that the most important factor that determines stability of the low-amplitude localized modes is the ratio of the coefficients at the local nonlinearity ($\sim J_{0,0}$) and the nonlinear dispersion ($\sim J_{0,1}$). If the coupling coefficients

$J_{n,m}$ decrease with distances n and m rapidly, the low-amplitude modes of Eq. (12.4.1) with $\varepsilon_{d0} = 11.4$ are essentially stable for $J_{0,0}/J_{0,1} \leq 13$. This estimate is usually lower in our case because the stabilization is favored by the long-range interactions.

It should be mentioned that the stabilization of low-amplitude 2D localized modes is not inherent to all types of nonlinear photonic crystals. On the contrary, photonic crystals must be *carefully designed* to support *stable low-amplitude nonlinear modes*. For example, for the photonic crystal considered in this section, such modes are stable for $11 < \varepsilon_{d0} < 12$ but become unstable for $\varepsilon_{d0} \geq 12$ (see Figure 12.12). The stability of these modes can also be controlled by varying the photonic-crystal parameters. Thus, experimental observation of the nonlinear localized modes would require not only the use of photonic materials with a relatively large nonlinear refractive index (such as Galas/Alas periodic structures or polymer PBG crystals [60]–[62]), but also a fine adjustment of the parameters of the photonic crystal. The latter can be achieved, in principle, by employing a surface-coupling technique [63] that is able to provide coupling to specific points of the dispersion curve and opens up a straightforward way to access nonlinear effects.

12.5   Future Perspectives

The field of nonlinear photonic crystals is growing rapidly and is expected to evolve in the near future with the ongoing research. In this section we indicate the directions in which photonic crystals are likely to make an impact.

Recent research has indicated that the use of quadratic (or $\chi^{(2)}$) nonlinearities can lead to strong, ultrafast, self-phase modulation through the cascading effect when the fundamental wave and its second harmonic are nearly phase matched. As discussed in Chapter 10, the cascaded nonlinearities allow for many important $\chi^{(3)}$-like effects to occur at much lower input powers [64]. This feature has led to considerable interest in the application of $\chi^{(2)}$ materials for all-optical signal processing.

The important area where photonic crystals promise to play a crucial role for the cascading effects is frequency mixing and harmonic generation [65]-[67]. The importance of periodic structures is well known for the quasi-phase-matching (QPM) technique. A traditional QPM technique relies on one-dimensional periodic modulation of the second-order susceptibility for compensating the mismatch between the wave vectors of the collinear fundamental and second-harmonic waves. The use of nonlinear photonic crystals allows one to extend this concept into higher dimensions, such that $\chi^{(2)}$ is modulated in two or three dimensions [68]. In a 2000 experiment, a two-dimensional QPM nonlinear structure with hexagonal symmetry was created using lithium niobate as a nonlinear material [69]. Such a nonlinear photonic crystal permits efficient (>60%) second-harmonic generation using multiple reciprocal lattice vectors of the periodic lattice. More importantly, such two-dimensional, $\chi^{(2)}$-based, nonlinear structures can provide simultaneous phase matching at several wavelengths [70, 71], thus opening a road for the experimental verification and practical implementation of the theoretical concepts based on parametric multi-step cascading [72, 73].

Figure 12.13 shows an expanded view of the two-dimensional, quadratically nonlinear, photonic crystal fabricated by Broderick et al. [69]. Each hexagon is a region

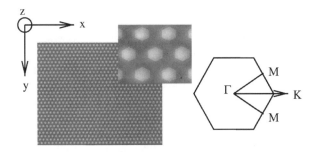

Figure 12.13: The hexagonally poled lithium niobate crystal and its first Brillouin zone. The 18.05–μm period of the crystal is uniform over the whole sample. (After Ref. [69]; ©2000 APS.)

of domain-inverted material; the total inverted area comprises 30% of the total sample area. Periodic poling was accomplished by applying an electric field via liquid electrodes on the opposite faces at room temperature. This quadratic photonic crystal had a period suitable for noncollinear frequency doubling of 1536-nm radiation, and it permitted efficient generation of the second harmonic through QPM. Since the second harmonic could be simultaneously phase matched in different directions using multiple reciprocal lattice vectors, the output consisted of multiple coherent beams.

Rigorously speaking, such structures do not fit the classical definition of a photonic crystal [1], since they do not possess a bandgap in the limit of small intensities. However, the frequency bandgap is not crucial for observing interesting effects such as harmonic generation in such nonlinear photonic crystals. In fact, quadratic nonlinear crystals seem to be the most suitable for observing numerous effects based on phase-matched parametric interaction. The fabrication technique is extremely versatile and allows for the fabrication of a broad range of two-dimensional quadratic crystals, including quasi-crystals.

Many other frontiers related to the physics of photonic crystals and, in particular, the soliton aspects of photonic crystals remain to be explored. The use of nonlinear photonic crystals in all-optical devices and circuits is under active research from the fundamental as well as application viewpoints. The photonic-crystal concept offers enormous potential for the field of nonlinear optics. Many of the well-known nonlinear phenomena, such as optical bistability and all-optical switching, studied earlier using nonlinear guided-wave optics, find their unique and unexpected manifestations in these novel materials [74]–[77].

Photonic crystals seem to be an ideal material where many properties of discrete optical solitons can be engineered in a simple way. For example, the slow group velocity of light in photonic-crystal circuits can dramatically increase the accumulated nonlinear phase shifts required for an efficient performance of an all-optical switch [77], and this should lead to a decrease in the size of many photonic devices operating at much smaller powers. These advantages could be employed to design very small all-optical logical gates using readily available materials and perhaps combining several thousands of such devices on a chip of a few square centimeters.

References

[1] J. D. Joannoupoulos, R. B. Meade, and J. N. Winn, *Photonic Crystals: Molding the Flow of Light* (Princeton University Press, Princeton, NJ, 1995).

[2] T. F. Krauss and R. M. De la Rue, *Prog. Quantum Electron.* **23**, 51 (1999).

[3] K. Sakoda, *Optical Properties of Photonic Crystals* (Springer, Berlin, 2001).

[4] S. G. Johnson and J. D. Joannopoulos, *Photonic Crystals: The Road from Theory to Practice* (Kluwer Academic, Boston, 2002).

[5] A. A. Maradudin and A. R. McGurn, in *Photonic Bandgaps and Localization*, NATO Series B, Vol. 308, C. M. Soukoulis, Ed. (Plenum Press, New York, 1993).

[6] A. Mekis, J. C. Chen, I. Kurland, S. Fan, P. R. Villeneuve, and J. D. Joannopoulos, *Phys. Rev. Lett.* **77**, 3787 (1996).

[7] A. Mekis, S. Fan, and J. D. Joannopoulos, *Phys. Rev. B* **58**, 4809 (1998).

[8] S.-Y. Lin, E. Chow, V. Hietala, P. R. Villeneuve, and J. D. Joannopoulos, *Science* **282**, 274 (1998).

[9] S. Fan, S. G. Johnson, J. D. Joannopoulos, C. Manolatou, and H. A. Haus, *J. Opt. Soc. Am. B* **18**, 162 (2001).

[10] S. G. Johnson, C. Manolatou, S. Fan, P. R. Villeneuve, J. D. Joannopoulos, and H. A. Haus, *Opt. Lett.* **23**, 1855 (1998).

[11] S. Fan, P. R. Villeneuve, J. D. Joannopoulos, and H. A. Haus, *Phys. Rev. Lett.* **80**, 960 (1998).

[12] S. F. Mingaleev, Yu. S. Kivshar, and R. A. Sammut, *Phys. Rev. E* **62**, 5777 (2000).

[13] A. R. McGurn, *Phys. Lett. A* **251**, 322 (1999).

[14] A. R. McGurn, *Phys. Lett. A* **260**, 314 (1999).

[15] S. F. Mingaleev and Yu. S. Kivshar, *Phys. Rev. Lett.* **86**, 5474-5477 (2001).

[16] T. Zijlstra, E. van der Drift, M. J. A. de Dood, E. Snoeks, and A. Polman, *J. Vac. Sci. Technol. B* **17**, 2734-2739 (1999).

[17] A. J. Ward and J. B. Pendry, Phys. Rev. B **58**, 7252 (1998).

[18] S. F. Mingaleev, Yu. S. Kivshar, and R. A. Sammut, in *Soliton-driven Photonics*, A. D. Boardman and A. P. Sukhorukov, Eds. (Kluwer, Dordrecht, Netherlands, 2001), pp. 487–504.

[19] M. Tokushima, H. Kosaka, A. Tomita, and H. Yamada, *Appl. Phys. Lett.* **76**, 952 (2000).

[20] D. N. Christodoulides and N. K. Efremidis, *Opt. Lett.* **27**, 568 (2002).

[21] A. L. Reynolds, U. Peschel, F. Lederer, P. J. Roberts, T. F. Krauss, and P. J. I. de Maagt, *IEEE Trans. Microwave Theory and Tech.* **49**, 1860 (2001).

[22] S. Mookherjea, D. S. Cohen, and A. Yariv, *Opt. Lett.* **27**, 933 (2002).

[23] S. Mookherjea and A. Yariv, *IEEE J. Sel. Topics Quantum Electron.* **8**, 448 (2002).

[24] N. A. R. Bhat and J. E. Sipe, *Phys. Rev. E* **64**, 056604 (2001).

[25] S. F. Mingaleev and Yu. S. Kivshar, *Opt. Lett.* **27**, 231 (2002).

[26] A. Yariv, Y. Xu, R. K. Lee, and A. Scherer, *Opt. Lett.* **24**, 711 (1999).

[27] Y. Xu, R. K. Lee, and A. Yariv, *J. Opt. Soc. Am. B* **17**, 387 (2000).

[28] M. Bayindir, B. Temelkuran, and E. Ozbay, *Phys. Rev. B* **61**, R11855 (2000).

[29] E. Lidorikis, M. M. Sigalas, E. Economou, and C. M. Soukoulis, *Phys. Rev. Lett.* **81**, 1405 (1998).

[30] A. R. McGurn, *Phys. Rev. B* **53**, 7059 (1996).

[31] S. G. Johnson and J. D. Joannopoulos, *Opt. Exp.* **8**, 173 (2001).

[32] S. Flach and C. R. Willis, *Phys. Rep.* **295**, 181 (1998).

[33] O. M. Braun and Yu. S. Kivshar, *Phys. Rep.* **306**, 1 (1998).

[34] R. S. MacKay and S. Aubry, *Nonlinearity* **7**, 1623 (1994).

[35] S. Aubry, *Physica D* **103**, 201 (1997).

[36] B. I. Swanson, J. A. Brozik, S. P. Love, G. F. Strouse, A. P. Shreve, A. R. Bishop, W. Z. Wang, and M. I. Salkola, *Phys. Rev. Lett.* **82**, 3288 (1999).

[37] U. T. Schwarz, L. Q. English, and A. J. Sievers, *Phys. Rev. Lett.* **83**, 223 (1999).

[38] E. Trias, J. J. Mazo, and T. P. Orlando, *Phys. Rev. Lett.* **84**, 741 (2000).

[39] P. Binder, D. Abraimov, A. V. Ustinov, S. Flach, and Y. Zolotaryuk, *Phys. Rev. Lett.* **84**, 745 (2000).

[40] F. M. Russel, Y. Zolotaryuk, and J. C. Eilbeck, *Phys. Rev. B* **55**, 6304 (1997).

[41] H. S. Eisenberg, Y. Silberberg, R. Marandotti, A. R. Boyd, and J.S. Aitchison, *Phys. Rev. Lett.* **81**, 3383 (1998).

[42] S. S. M. Cheng, L. M. Li, C. T. Chan, and Z. Q. Zhang, *Phys. Rev. B* **59**, 4091 (1999).

[43] C. Jin, B. Cheng, B. Mau, Z. Li, D. Zhnag, S. Ban, B. Sun, *Appl. Phys. Lett.* **75**, 1848 (1999).

[44] A. A. Sukhorukov, Yu. S. Kivshar, and O. Bang, *Phys. Rev. E* **60**, R41 (1999).

[45] V. K. Mezentsev, S. L. Musher, I. V. Ryzhenkova, and S. K. Turitsyn, *JETP Lett.* **60**, 829 (1994).

[46] S. Flach, K. Kladko, and R. S. MacKay, *Phys. Rev. Lett.* **78**, 1207 (1997).

[47] P. L. Christiansen, Yu. B. Gaididei, M. Johansson, and K. Ø. Rasmussen, V. K. Mezentsev, and J. J. Rasmussen, *Phys. Rev. B* **57**, 11303 (1998).

[48] Y. B. Gaididei, S. F. Mingaleev, P. L. Christiansen, and K. Ø. Rasmussen, *Phys. Rev. E* **55**, 6141 (1997).

[49] M. Johansson, Y. B. Gaididei, P. L. Christiansen, and K. Ø. Rasmussen, *Phys. Rev. E* **57**, 4739 (1998).

[50] D. E. Pelinovsky, V. V. Afanasjev, and Yu. S. Kivshar, *Phys. Rev. E* **53**, 1940 (1996).

[51] S. John and N. Aközbek, *Phys. Rev. Lett.* **71**, 1168 (1993).

[52] S. John and N. Aközbek, *Phys. Rev. E* **57**, 2287 (1998).

[53] Yu. S. Kivshar, O. A. Chubykalo, O. V. Usatenko, and D. V. Grinyoff, *Int. J. Mod. Phys. B* **9**, 2963 (1995).

[54] I. V. Barashenkov, D. E. Pelinovsky, and E. V. Zemlyanaya, *Phys. Rev. Lett.* **80**, 5117 (1998).

[55] C. M. Anderson and K. P. Giapis, *Phys. Rev. Lett.* **77**, 2949 (1996); *Phys. Rev. B* **56**, 7313 (1997).

[56] E. W. Laedke *et al.*, *JETP Lett.* **62**, 677 (1995).

[57] E. W. Laedke, K. H. Spatschek, S. K. Turitsyn, and V. K. Mezentsev, *Phys. Rev. E* **52**, 5549 (1995).

[58] Yu. B. Gaididei, P. L. Christiansen, K. Ø. Rasmussen, and M. Johansson, *Phys. Rev. B* **55**, R13365 (1997).

[59] Yu. B. Gaididei, D. Hendriksen, P. L. Christiansen, and K. Ø. Rasmussen, *Phys. Rev. B* **58**, 3075 (1998).

[60] P. Millar, M. De La Rue, T. F. Krauss, J. S. Aitchison, N. G. R. Broderick, and D. J. Richardson, *Opt. Lett.* **24**, 685 (1999).

[61] A. S. Helmy, D. C. Hutchings, T. C. Kleckner, J. H. Marsh, A. C. Bryce, J. M. Arnold, C. R. Stanley, J. S. Aitchison, C. T. A. Brown, K. Moutzouris, and M. Ebrahimzadeh, *Opt. Lett.* **25**, 1370 (2000).

[62] S. Shoji and S. Kawata, *Appl. Phys. Lett.* **76**, 2668 (2000).

[63] V. N. Astratov, D. M. Whittaker, I. S. Culshaw, R. M. Stevenson, M. S. Skolnick, T. F. Krauss, and R. M. De La Rue, *Phys. Rev. B* **60**, R16255 (1999).

[64] Yu. S. Kivshar, in *Advanced Photonics with Second-order Optically Nonlinear Processes*, A. D. Boardman, L. Pavlov, and S. Tanev, Eds. (Kluwer, , Netherlands, 1998), pp. 451–475.

[65] B. Xu and N.-B. Ming, *Phys. Rev. Lett.* **71**, 1003 (1993).

[66] G. D'Aguanno, M. Centini, C. Sibilia, M. Bertolotti, M. Scalora, M. J. Bloemer, and C. M. Bowden, *Opt. Lett.* **24**, 1663 (1999).

[67] B. Shi, Z. M. Jiang, and X. Wang, *Opt. Lett.* **26**, 1194 (2001).

[68] V. Berger, *Phys. Rev. Lett.* **81**, 4136 (1998).

[69] N. G. R. Broderick, G. W. Ross, H. L. Offerhaus, D. J. Richardson, and D.C. Hanna, *Phys. Rev. Lett.* **84**, 4345 (2000).

[70] S. Saltiel and Yu. S. Kivshar, *Opt. Lett.* **25**, 1204 (2000).

[71] M. De Sterke, S. M. Saltiel, and Yu. S. Kivshar, *Opt. Lett.* **26**, 539 (2001).

[72] Yu. S. Kivshar, A. A. Sukhorukov, and S. Saltiel, *Phys. Rev. E* **60**, R5056 (1999).

[73] A. Chowdhury, C. Staus, B. F. Boland, T. F. Kuech, and L. McCaughan, *Opt. Lett.* **26**, 1353 (2001).

[74] M. Scalora, J. P. Dowling, C. M. Bowden, and M. J. Bloemer, *Phys. Rev. Lett.* **73**, 1368 (1994).

[75] P. Tran, *Opt. Lett.* **21**, 1138 (1996).

[76] S. F. Mingaleev and Yu. S. Kivshar, *J. Opt. Soc. Am. B* **19**, 2241 (2002).

[77] M. Soljacic, S. G. Johnson, S. Fan, M. Ibanescu, E. Ippen, and J. D. Joannopoulos, *J. Opt. Soc. Am. B* **19**, 2052 (2002).

Chapter 13

Incoherent Solitons

One of the main features of optical solitons discussed so far is their coherent nature. Indeed, the amplitude, phase, and frequency associated with an optical soliton are well-behaved deterministic quantities. However, starting in 1996 the soliton concept could be extended to cover a more general class of self-trapped beams that are partially coherent or even totally incoherent. Such spatial solitons are called *incoherent solitons*, and they are found to exhibit some unique features that have no counterparts in the coherent regime. In this chapter, we focus on incoherent solitons and discuss how they are different from their coherent counterparts. After providing a brief historical perspective in Section 13.1, we present in Section 13.2 four different theoretical approaches used for their description. Section 13.3 focuses on the properties of bright incoherent solitons and discusses some of the experimental results on self-trapping of incoherent beams. The effect of partial coherence on modulation instability is also discussed in this section. The physics of dark and vortex-type incoherent solitons is considered in Section 13.4 together with the experimental results.

13.1 Historical Perspective

Traditionally, optical solitons have been observed using only coherent sources of light, such as lasers. The possibility of self-trapping of a pulse with a randomly varying phase was first discussed by Hasegawa in the 1970s in the context of plasma physics [1] while studying the average dynamics of quasi-particles forming the plasma through the Vlasov equation; this approach is somewhat similar to the ray approximation used in optics. During the 1990s several studies considered how the coherence properties of partially incoherent beams are modified in nonlinear Kerr media without focusing on optical solitons [2]–[8].

In 1996, Mitchell et al. recognized that the noninstantaneous nature of nonlinearity is the necessary condition for generating incoherent solitons. They used a partially (spatially) incoherent light beam and showed experimentally, for the first time, that such a beam can be self-trapped inside a photorefractive crystal [9]. In this experiment, a laser beam was sent through a rotating diffuser that changed the phase across the wave front in a random manner every 1 μs or so, thus making the beam spatially incoherent. The beam was then launched into a slow-responding photorefractive crystal. Under appropriate conditions, the beam size was observed to become constant in the form of a narrow filament. Such a partially coherent, self-trapped beam can be interpreted to form an incoherent spatial soliton because the diffractive effects are countered by the medium nonlinearity, similar to the case of coherent spatial solitons. In a subsequent experiment [10], a white light beam from an incandescent light bulb was used to form an incoherent soliton, even though such a beam was incoherent, both temporally and spatially.

Soon after the first experimental observation of the incoherent soliton, two different theories were developed for explaining its formation. In the *coherent density theory* [11] an angular-spectrum approach was used to find an analytical solution in the case of a saturable logarithmic nonlinearity [12]. In the *modal theory* [13], as the name suggests, partially coherent light was decomposed into individual modes. By virtue of its simplicity, the modal approach has became the method of choice for finding incoherent solitons, their range of existence, and their correlation properties. In 1998, a third theory, based on the propagation of the *mutual coherence function*, was proposed [14]. Historically, the underlying technique behind this method is the oldest and is based on an equation derived in 1974 by Pasmanik [15]. More recently, a fourth method, based on the Wigner transform, has been suggested as an alternative statistical theory for describing the dynamics of partially coherent fields propagating inside a nonlinear medium [16] (see also Ref. [17]). Even though, at first sight, these theoretical approaches appear to be quite dissimilar, they are, in fact, formally equivalent to one another [18, 19] and describe just different representations of the same propagation process, each having its own strength and weakness.

In addition to these exact theories, a more simplified, ray-optics approach was suggested in 1998 [20]. The ray-optics formulation of incoherent solitons almost fully coincides with that developed for random-phase solitons in plasma physics [1]. However, ray optics can provide only simple and intuitive information about incoherent solitons, because all phase information is lost. For example, it fails to describe the coherence properties of a partially coherent soliton, since it views such an entity as a bundle of completely uncorrelated rays.

To understand the concept of incoherent solitons, one needs first to know what is meant by incoherent or partially coherent light [21]. A partially coherent beam results from an optical field whose amplitude and phase fluctuate with time, even for a continuous-wave (CW) beam. As a result, the beam intensity at any moment consists of many tiny bright and dark "patches" (the so-called speckle pattern), whose distribution varies across the entire cross section randomly in time [22]. Consider how such a spatially incoherent beam is perceived by a slow photodetector (e.g., a human eye). When this detector responds much more slowly than the characteristic time of amplitude and phase fluctuations, all the detector "sees" is the time-averaged intensity, which may appear to be quite uniform, in spite of the underlying speckle pattern. However, such a beam diffracts much more than a coherent beam of the same width because each tiny speckle contributes to the diffraction of the time-averaged beam envelope. In the limiting case in which the speckles are much smaller than the beam size, diffraction is dominated by the degree of coherence rather than by the diameter of the beam.

Instantaneous nonlinearities cannot trap an incoherent beam and form a soliton out of it. If an incoherent beam is launched into a self-focusing nonlinear medium that responds almost instantaneously (e.g., a Kerr medium), each speckle forms a self-induced waveguide that captures a small fraction of the beam. These tiny induced waveguides intersect and cross each other in a random manner during beam propagation. The net effect is that the beam breaks up into small fragments, and self-trapping of the beam envelope does not occur. Thus, only a slow-responding nonlinear medium can support incoherent solitons. This is exactly how the idea of incoherent solitons first occurred [23]. More explicitly, the slow response of the photorefractive nonlinearity hinted that rapidly varying spatial information carried by an optical beam will be averaged out and that the nonlinear medium would respond only to the time-averaged intensity of the beam.

To sum up, the noninstantaneous nature of the underlying nonlinearity is the *necessary condition* for generation of incoherent solitons. However, several other conditions should be satisfied for self-trapping of an incoherent beam to occur [22]. First, the response time of the nonlinear medium should be much longer than the fluctuation time across the incoherent beam. Second, the multimode speckle pattern should be able to induce a multimode waveguide via the nonlinearity. Third, as with all solitons, self-trapping requires self-consistency, i.e., the multimode beam must be able to guide itself inside the self-induced waveguide.

13.2 Theoretical Methods

As mentioned in Section 13.1, four different methods were developed during the brief period of one year or so to explain the experimentally observed features of incoherent solitons. In this section, we discuss the four methods and identify their strengths and weaknesses.

13.2.1 Coherent Density Theory

The coherent density theory was the first analytical method developed for describing the experimentally observed self-trapping of partially coherent optical fields in noninstantaneous nonlinear media [11]. With this method, the intensity as well as the correlation statistics of a partially coherent beam can be studied along the medium length using a modified version of the Van Cittert–Zernike theorem [21]. Physically speaking, the incoherent input beam is decomposed into many coherent fragments, all of which are assumed to be mutually incoherent with respect to one another. The initial relative weight of each fragment is determined from the angular spectrum of the source. Each fragment propagates in a coherent fashion inside the nonlinear material, and the total intensity of the partially coherent field is found by superimposing the intensities of these coherent parts.

To describe the angular-spectrum method, well known in linear optics [21], we first apply it to solve a linear diffraction problem, namely, the propagation of a partially coherent beam in the absence of nonlinearities. In the paraxial approximation, we need to solve Eq. (1.2.12) with $n_2 = 0$. The resulting equation is

$$2i\beta_0 \frac{\partial A}{\partial z} + \left(\frac{\partial^2 A}{\partial x^2} + \frac{\partial^2 A}{\partial y^2} \right) = 0. \tag{13.2.1}$$

In the angular-spectrum method, this linear equation is solved in the Fourier domain using the two-dimension Fourier transform

$$A(\mathbf{r}, z) = \int_{-\infty}^{\infty} a(\mathbf{p}, z) \exp(i\beta_0 \mathbf{p} \cdot \mathbf{r}) d^2\mathbf{p}, \tag{13.2.2}$$

where $\mathbf{p} = (p_x, p_y)$ and $\mathbf{r} = (x, y)$ are two-dimensional vectors. Physically, $a(\mathbf{p}, z)$ is called the *angular spectrum* because it represents the amplitude of plane waves propagating in different directions. The components p_x and p_y represent direction cosines of a plane wave with respect to the propagation axis z.

Using Eq. (13.2.2) in Eq. (13.2.1), the angular spectrum at any distance z is related to the input value at $z = 0$ as

$$a(\mathbf{p}, z) = a(\mathbf{p}, 0) \exp(-i\beta_0 \mathbf{p}^2 z/2). \tag{13.2.3}$$

Substituting this relation in Eq. (13.2.2), we obtain the angular spectrum representation of the optical field in the form

$$A(\mathbf{r}, z) = \int_{-\infty}^{\infty} a(\mathbf{p}, 0) \exp[i\beta_0 (\mathbf{p} \cdot \mathbf{r} - i\mathbf{p}^2 z/2)] d^2\mathbf{p}. \tag{13.2.4}$$

Let us write the input optical field as $A(\mathbf{r}, 0) = m(\mathbf{r})\phi_0(\mathbf{r})$, where $m(\mathbf{r})$ is a complex modulation function and $\phi_0(\mathbf{r})$ is the field before modulation that describes all spatial statistical properties of a partially coherent source. If we assume that source fluctuations follow a stationary random process, the autocorrelation function of $\phi_0(x, y)$ satisfies

$$\langle \phi_0(\mathbf{r}) \phi_0^*(\mathbf{r}') \rangle = R(\mathbf{r} - \mathbf{r}'). \tag{13.2.5}$$

From this relation, the autocorrelation of the angular spectrum is found to satisfy

$$\langle \Phi_0(\mathbf{p}) \Phi_0^*(\mathbf{p}') \rangle = (\beta_0/2\pi)^2 G(\mathbf{p}) \delta^2(\mathbf{p} - \mathbf{p}'), \tag{13.2.6}$$

where $\Phi_0(\mathbf{p})$ and $G(\mathbf{p})$ are the Fourier transforms of $\phi_0(\mathbf{r})$ and $R(\mathbf{r})$, respectively. Physically, $G(\mathbf{p})$ represents the angular power spectrum of the partially coherent source.

We can now calculate the intensity I at any point z using Eq. (13.2.2) with $I = \langle |A|^2 \rangle$. Noting from the convolution theorem that a is related to Φ_0 as

$$a(\mathbf{p}, 0) = \int_{-\infty}^{\infty} M(\mathbf{p} - \mathbf{p}') \Phi_0(\mathbf{p}') d^2\mathbf{p}', \tag{13.2.7}$$

where $M(\mathbf{p})$ is the Fourier transform of the spatial modulation function $m(\mathbf{r})$, the final result is given by

$$I(\mathbf{r}, z) = \int_{-\infty}^{\infty} |f(\mathbf{p}, \mathbf{r}, z)|^2 d^2\mathbf{p}, \tag{13.2.8}$$

where the transfer function $f(\mathbf{p}, \mathbf{r}, z)$ is defined as [22]

$$f(\mathbf{p}, \mathbf{r}, z) = \frac{\beta_0^2 \sqrt{G(\mathbf{p})}}{(2\pi)^2} \int \int_{-\infty}^{\infty} M(\mathbf{p}_1) \exp\{i\beta_0[\mathbf{p}_1 \cdot (\mathbf{r} - \mathbf{p}z) - i\mathbf{p}_1^2 z/2]\} \, d^2\mathbf{p}_1. \quad (13.2.9)$$

The auxiliary function f has a simple meaning. It shows that the diffraction behavior of a partially incoherent beam can be described by considering all angular components, finding the intensity $|f|^2$ of a specific component, and then adding the intensity contributions of all the components involved. The function f is the *coherent density function* associated with a partially coherent source. It is easy to show from Eq. (13.2.9) that the coherent density function f is a solution of the following modified paraxial equation:

$$i\left(\frac{\partial f}{\partial z} + \mathbf{p} \cdot \nabla f\right) + \frac{1}{2\beta_0} \nabla^2 f = 0, \quad (13.2.10)$$

where ∇ is the two-dimensional gradient operator. This equation should be solved with the initial condition that $f(\mathbf{p}, \mathbf{r}, 0) = \sqrt{G(\mathbf{p})} \, m(\mathbf{r})$.

Equation (13.2.10) is derived for a linear medium with $n_2 = 0$. However, it can easily be generalized to include nonlinear effects as long as the nonlinear contribution depends only on the average intensity I. This is the case for a slow-responding nonlinear medium. Using the nonlinear term from Eq. (1.2.12), the coherent density in such a nonlinear medium is found to evolve as [11]

$$i\left(\frac{\partial f}{\partial z} + \mathbf{p} \cdot \nabla f\right) + \frac{1}{2\beta_0} \nabla^2 f + k_0 n_{nl}(I) f = 0, \quad (13.2.11)$$

where $n_{nl}(I)$ represents the functional dependence of the refractive index on the average intensity. The coherent density theory is useful for studying different types of dynamics of partially incoherent beams, including the beam collapse [24] and incoherent spatial solitons forming in isotropic saturable media [25].

13.2.2 Mutual Coherence Function

The coherent density theory does not deal directly with the degree of coherence associated with a partially incoherent beam. A theory that deals with that is well known and makes use of the mutual coherence function, defined as [21]

$$\Gamma(\mathbf{r}_1, \mathbf{r}_2, z; t_1, t_2) = \langle A(\mathbf{r}_1, z; t_1) A^*(\mathbf{r}_2, z; t_2) \rangle. \quad (13.2.12)$$

In most cases of physical interest, the underlying stochastic process is stationary, and the coherence function Γ depends only on the time difference $\tau = t_1 - t_2$. Moreover, when temporal coherence effects are not of interest, one can isolate the spatial coherence effects by setting $t_1 = t_2$. The spatial coherence function is then defined as

$$J_{12}(\mathbf{r}_1, \mathbf{r}_2, z) \equiv \Gamma(\mathbf{r}_1, \mathbf{r}_2, z; t, t) = \langle A(\mathbf{r}_1, z, t) A^*(\mathbf{r}_2, z, t) \rangle. \quad (13.2.13)$$

Associated with J_{12} is the spatial degree of coherence, defined as

$$\mu_{12} = J_{12} / \sqrt{J_{11} J_{22}}, \quad (13.2.14)$$

where $J_{11} = I(\mathbf{r}_1, z)$ and $J_{22} = I(\mathbf{r}_2, z)$ are the intensities at the points $\mathbf{r}_1$ and $\mathbf{r}_2$, respectively. This quantity governs all spatial coherence properties of the beam.

In the case of CW beams propagating inside a nonlinear medium, $A(\mathbf{r}_1, t)$ satisfies Eq. (1.2.12), rewritten here as

$$i\frac{\partial A}{\partial z} + \frac{1}{2\beta_0}\nabla^2 A + k_0 n_{nl}(I)A = 0. \tag{13.2.15}$$

It was shown in 1974 that the standard coherence theory can be extended to the case of nonlinear media [15]. In fact, using Eqs. (13.2.13) and (13.2.15) it is easy to show that the mutual coherence function J_{12} satisfies the following four-dimensional equation, provided the intensity appearing in the nonlinear term is the average intensity:

$$i\frac{\partial J_{12}}{\partial z} + \frac{1}{2\beta_0}(\nabla_1^2 - \nabla_2^2)J_{12} + k_0[n_{nl}(I_1) - n_{nl}(I_2)]J_{12} = 0, \tag{13.2.16}$$

This equation is the nonlinear version of the linear equation governing propagation of the mutual coherence function in the paraxial approximation [21]. Notice that it also requires the nonlinear medium to be slow responding.

The mutual coherence function can be related to the coherent density function $f(\mathbf{p}, \mathbf{r}, z)$ introduced earlier as [26]

$$J_{12}(\mathbf{r}_1, \mathbf{r}_2; z) = \int_{-\infty}^{\infty} f(\mathbf{p}, \mathbf{r}_1, z)f^*(\mathbf{p}, \mathbf{r}_2, z)\exp[i\beta_0\mathbf{p}\cdot(\mathbf{r}_1 - \mathbf{r}_2)]\,d^2\mathbf{p}. \tag{13.2.17}$$

Noting that the intensity at $\mathbf{r}_j$ can be obtained from Eq. (13.2.17) as

$$I_j = I(\mathbf{r}_j, z) \equiv J_{jj} = \int_{-\infty}^{\infty} |f(\mathbf{p}, \mathbf{r}, z)|^2\,d^2 p, \tag{13.2.18}$$

which is the same as Eq. (13.2.2). Equation (13.2.17) represents a modified version of the Van Cittert–Zernike theorem [27]. The equivalence of the coherent density approach and the mutual coherence method can also be established [18] by showing that Eqs. (13.2.11) and (13.2.17) lead to Eq. (13.2.16).

Even though Eq. (13.2.16) is valid for any form of the nonlinear function $n_{nl}(I)$, an explicit form of the coherence function for partially coherent solitons can be obtained in only a few cases, such as a nonlinear medium with logarithmic nonlinearity [28]. This case has been studied extensively [29]–[31] and is discussed in Section 13.3.

13.2.3 Modal Theory

The third approach for describing the self-trapping of a partially incoherent beam makes use of the concept of optical modes and is easier to understand intuitively [13]. It is well suited for identifying the coherence properties of incoherent solitons as well as the range of parameters for which they can exist. Moreover, the modal theory reduces to a set of incoherently coupled NLS equations similar to those studied in Chapter 9. This feature makes it easy to solve them analytically in a number of cases of practical interest [32]–[38].

The basic idea behind the modal approach is well known in the optical coherence theory (see Section 2.5 of Ref. [21]). To explain the modal theory in a simple way, we assume that the slowly varying envelope of the partially incoherent beam can be written in terms of an orthonormal set of modes $u_m(\mathbf{r},z)$ using

$$A(\mathbf{r},z) = \sum_m c_m u_m(\mathbf{r},z), \qquad (13.2.19)$$

where the modal coefficients c_m are random variables that are uncorrelated with one another, i.e, $\langle c_m c_n^* \rangle = \lambda_m \delta_{nm}$, where λ_m (the modal occupancy) is a real positive quantity. This kind of representation is known as a Karhunen–Loéve expansion [21, 39]. For this expansion to be valid, the functions $u_m(\mathbf{r},z)$ should remain orthogonal during propagation provided that this was the case at $z = 0$, i.e.,

$$\int_{-\infty}^{\infty} u_m(\mathbf{r},z)\, u_n^*(\mathbf{r},z)\, dx\, dy = \delta_{nm}. \qquad (13.2.20)$$

Using Eq. (13.2.19) in Eq. (13.2.15), each eigenfunction u_m is found to evolve as

$$i\frac{\partial u_m}{\partial z} + \frac{1}{2\beta_0}\nabla^2 u_m + k_0 n_{\mathrm{nl}}(I)u_m = 0, \qquad (13.2.21)$$

where $I = \sum_m \langle |c_m|^2 \rangle |u_m|^2$. It can be shown that, irrespective of the nature of the eigenfunctions involved, the set u_m remains orthonormal during propagation.

The mutual coherence function can also be expanded in terms of the same orthonormal eigenfunctions as

$$J_{12}(\mathbf{r}_1,\mathbf{r}_2;z) = \sum_m \langle |c_m|^2 \rangle u_m(\mathbf{r}_1,z)u_m^*(\mathbf{r}_2,z), \qquad (13.2.22)$$

where the relation $\langle c_m c_n^* \rangle = \langle |c_m|^2 \rangle \delta_{nm}$ was used. We can differentiate this equation with respect to z and obtain Eq. (13.2.16), which is the basis of the mutual coherence method. The main advantage of the modal approach is that it allows us to describe the evolution of a partially incoherent beam by a set of incoherently coupled NLS equations, given in Eq. (13.2.21), rather than solving the much more complicated Eq. (13.2.16).

As discussed in Chapter 9, the set of incoherently coupled NLS equations is exactly integrable in the one-dimensional case, under certain conditions, for a slow-responding Kerr medium for which $n_{\mathrm{nl}}(I) = n_2 I$, and explicit analytical solutions have been found in this case [32]–[38]. Such models truncate the infinite set of NLS equations because only a finite number of modes contribute in the expansion (13.2.19). They are nonetheless quite useful for providing physical insight. For example, it was found that a change in the number of modes from $N = 2$ to $N = 3$ introduces several new features. In particular, such incoherent solitons can have modes with an asymmetric shape, and the shape can change during collisions even though the soliton maintains a stationary structure [40]–[42]. A similar behavior is observed for a saturable medium, even though the incoherent soliton splits into multiple beams after the collision [42].

13.2.4 Wigner Transform Method

An alternative statistical theory for describing the dynamics of partially coherent fields propagating in a nonlinear medium has recently been developed [16, 17]. This approach is based on the Wigner function [43], first introduced in statistical quantum mechanics in 1932. Since then, it has been successfully applied in connection with weak plasma turbulence [44] and nonstationary and relativistic plasma [45]. More recently, the Wigner transform method has been used to analyze the longitudinal dynamics of charged-particle beams in accelerators [46] and to study the dynamics of Bose–Einstein condensates in the presence of a chaotic external potential [47].

The starting point is again the NLS equation (13.2.15), which we rewrite in a more convenient form as

$$i\frac{\partial A}{\partial z} + \frac{d}{2}\nabla^2 A + \gamma G(I)A = 0, \tag{13.2.23}$$

where the function $G(I)$ with $I = \langle|A|^2\rangle$ characterizes the nonlinear properties of the medium and the parameter d takes into account the diffractive or dispersive (in the temporal case) properties of the medium. As discussed in Chapter 7, both the diffractive and dispersive effects can be included simultaneously by adding an additional term in the Laplacian and treating $\mathbf{r}$ as a vector with three components.

The Wigner transform is defined as [43]

$$\rho(\mathbf{p},\mathbf{r},z) = \frac{1}{(2\pi)^D}\int_{-\infty}^{+\infty}\langle A^*(\mathbf{r}+\mathbf{s}/2,z)A(\mathbf{r}-\mathbf{s}/2,z)\rangle e^{i\mathbf{p}\cdot\mathbf{s}}d^D\mathbf{s}, \tag{13.2.24}$$

where r and s are D-dimensional vectors. It is easy to show that the average intensity is related to the Wigner function ρ as

$$\langle A^*(\mathbf{r},t)A(\mathbf{r},t)\rangle = \int_{-\infty}^{+\infty}\rho(\mathbf{p},\mathbf{r},z)d^D\mathbf{p}. \tag{13.2.25}$$

Comparing this relation with Eq. (13.2.8), one can see that the Wigner function is similar in nature to the coherent density function.

Applying the Wigner transform to Eq. (13.2.23), one obtains the following Wigner–Moyal equation governing the evolution of the Wigner function:

$$\frac{\partial\rho}{\partial z} + d\mathbf{p}\cdot\nabla_r\mathbf{r} + 2\gamma G(I)\sin\left(\frac{1}{2}\overleftarrow{\nabla}_r\cdot\overrightarrow{\nabla}_p\right)\rho = 0, \tag{13.2.26}$$

where a subscript denotes the variable used for the gradient operator. The two gradient operators within the sine function act on the left or right, as indicated by the arrow on top. The nonlinear term is quite complicated in this equation because it involves spatial derivatives of all orders. It can be simplified considerably in the ray-optics approximation, in which $\Delta\mathbf{p}\Delta\mathbf{r} \gg 2\pi$, where $\Delta\mathbf{p}$ is the width of the Wigner spectrum and $\Delta\mathbf{r}$ is the width of the medium response function. In this approximation, we can use the approximation $\sin(x) \approx x$, and the Wigner–Moyal equation is transformed into a much simpler Vlasov-like equation:

$$\frac{\partial\rho}{\partial z} + d\mathbf{p}\cdot\nabla_r\mathbf{r} + \gamma\nabla_r G(I)\cdot\nabla_p\rho = 0, \tag{13.2.27}$$

This equation represents a continuity equation, implying the conservation of the number of quasi-particles in the phase space. Using the Liouville theorem, we can obtain the canonical Hamilton equations of motion for a quasi-particle of mass d or, equivalently, the ray equations of the geometrical-optics approximation, with $\mathbf{r}$ and $\mathbf{p}$ playing the role of the canonical variables. Equation (13.2.27) is similar to the radiative transfer equation used in Refs. [48] and [49].

The Wigner transform method provides an equivalent way to treat the propagation of partially coherent fields in nonlinear optical media. It can be shown that it is equivalent to the other three methods discussed in this section [19]. To recover all the information, we first solve Eq. (13.2.26) and then take the inverse Fourier transform of ρ, using

$$J_{12}(\mathbf{s},\mathbf{r},z) \equiv J_{12}(\mathbf{r}_1,\mathbf{r}_2,z) = \int_{-\infty}^{\infty} \rho(\mathbf{p},\mathbf{s},z)e^{-i\mathbf{p}\cdot\mathbf{s}}d^2\mathbf{p}, \qquad (13.2.28)$$

where $\mathbf{r} = \frac{1}{2}(\mathbf{r}_1 - \mathbf{r}_2)$ and $\mathbf{s} = \mathbf{r}_2 - \mathbf{r}_1$. The main advantage of this method is that the complexity of the problem is considerably reduced, because the Wigner–Moyal equation is real and thus the number of equations to be solved is reduced by the factor of 2.

13.3   Bright Incoherent Solitons

In this section we use two specific forms of the slow-responding nonlinearity that allow us to obtain an analytic form of bright incoherent solitons and use it to discuss the coherence properties of such solitons. More specifically, we consider the Kerr nonlinearity and a logarithmically saturable nonlinearity of the form $n_{nl}(I) = n_2 \ln(1 + I/I_s)$, where I_s is the saturation intensity. The case of logarithmic nonlinearity was first solved in 1997 using the coherent density approach [12]; since then, other methods have also been used [29]–[31].

13.3.1   Properties of Partially Coherent Solitons

We first use the modal theory in the case of a logarithmically saturable nonlinearity and discuss the coherence properties of the resulting partially incoherent solitons. Assuming $I \gg I_s$ and focusing on the one-dimensional case to simplify the following discussion, Eq. (13.2.21) for the coherent modes becomes

$$i\frac{\partial u_m}{\partial z} + \frac{1}{2\beta_0}\frac{\partial^2 u_m}{\partial x^2} + k_0 n_2 \ln\left(\frac{I}{I_s}\right) u_m = 0. \qquad (13.3.1)$$

We assume a Gaussian form for the intensity profile of the input beam and use

$$I(x,0) = I_s b_0 \exp(-x^2/w_0^2), \qquad (13.3.2)$$

where w_0 is the input-beam width and b_0 is a measure of the input peak intensity, in units of the saturation intensity. We also normalize Eq. (13.3.1) using

$$x' = x/w_0, \qquad z' = z/L_d, \qquad L_d = \beta_0 w_0^2, \qquad (13.3.3)$$

where L_d is the diffraction length. The modal equation then becomes

$$i\frac{\partial u_m}{\partial z} + \frac{1}{2}\frac{\partial^2 u_m}{\partial x^2} + \frac{\alpha^2}{2}(\ln b_0 - x^2)u_m = 0, \tag{13.3.4}$$

where we have dropped the primes over x and z for notational simplicity and the dimensionless parameter α is defined as

$$\alpha = w_0\sqrt{2n_2k_0\beta_0}. \tag{13.3.5}$$

For a beam to become an incoherent soliton, its intensity should not change during propagation; i.e., $I(x,z) = I(x,0)$ for all z. This is possible if the parameters b_0 and w_0 in Eq. (13.3.2) are constants. We assume this to be the case. Equation (13.3.4) can now be solved by assuming a solution in the form $u_m(x,z) = U_m(x)\exp(i\lambda_m z)$, resulting in the eigenvalue equation

$$\frac{d^2U_m}{dx^2} + \alpha^2(\ln b_0 - x^2)U_m = 2\lambda_m U_m. \tag{13.3.6}$$

This is the standard harmonic-oscillator equation and has solutions in the form of Hermite–Gauss functions. The eigenfunctions and eigenvalues are thus given by

$$U_m(x) = H_m(\sqrt{\alpha}x)e^{-\alpha x^2/2}, \qquad \lambda_m = \tfrac{1}{2}[\alpha^2\ln(b_0) - (2m+1)\alpha]. \tag{13.3.7}$$

The intensity at any point can now be constructed by using the relation

$$I(x,z) = \sum_m \langle|c_m|^2\rangle|u_m(x,z)|^2, \tag{13.3.8}$$

where c_m is the expansion coefficient in Eq. (13.2.19). Physically, $\langle|c_m|^2\rangle$ represents the mode-occupation probability. It turns out that Eq. (13.3.8) leads to the Gaussian form in Eq. (13.3.2) when $\langle|c_m|^2\rangle|$ follows the Poisson distribution [33]

$$\langle|c_m|^2\rangle = b_1\frac{q^m e^{-q}}{m!}, \tag{13.3.9}$$

where b_1 and q are related to the beam parameters as

$$b_1 = b_0 e^{q/2}\sqrt{1-q^2}, \qquad q = (\alpha-1)/(\alpha+1). \tag{13.3.10}$$

Since $q > 0$ is required, the bright incoherent soliton can exist for the logarithmically saturable nonlinearity only if α exceeds 1. From Eq. (13.3.4), the nonlinear parameter n_2 must exceed a threshold value parameter $(2k_0L_d)^{-1}$ for solitons to exist. In the completely coherent case ($q = 0$), in which only the fundamental mode is excited, $\alpha = 1$ is required. The completely incoherent case corresponds to $q = 1$ and occurs for very large values of α.

The coherence properties of the partially coherent soliton can be found from Eq. (13.2.22) using Eqs. (13.3.7)–(13.3.9). The degree of coherence is obtained from Eq. (13.2.14) and is found to be

$$\mu_{12}(d) = \exp\left[-\frac{qd^2}{(1-q)^2}\right], \tag{13.3.11}$$

where d is the distance between two spatial points at a given distance z. If we define the correlation length l_c (sometimes called the *coherence radius*) as the value of d for which $\mu_{12} = 1/e$, this length is given by

$$l_c/w_0 = (1-q)/\sqrt{q} = 2/\sqrt{\alpha^2 - 1}. \tag{13.3.12}$$

In the completely coherent case ($q = 0$), $l_c \to \infty$, as expected. The completely incoherent case corresponds to $q = 1$, for which $l_c \to 0$. For intermediate values of q, the soliton is only partially coherent, and Eq. (13.3.11) provides its degree of coherence. As an example, the correlation length equals the beam width ($l_c = w_0$) for $\alpha = \sqrt{5}$ or $q \approx 0.38$.

As a second example of partially coherent solitons, we consider a slow-responding Kerr medium and replace $\ln(I/I_s)$ in Eq. (13.3.1) with I. If we assume that the soliton shape remains the same as in the coherent case, use $I = I_0 = \mathrm{sech}^2(x/w_0)$, and employ the normalization scheme of Eq. (13.3.3), Eq. (13.3.1) can be written as

$$i\frac{\partial u_m}{\partial z} + \frac{1}{2}\frac{\partial^2 u_m}{\partial x^2} + \frac{\alpha^2}{2}\mathrm{sech}^2(x)u_m = 0, \tag{13.3.13}$$

where the parameter α is now defined as $\alpha^2 = k_0 L_d n_2 I_0$.

The coherent modes are obtained by assuming a solution of Eq. (13.3.13) in the form $u_m(x,z) = U_m(x)\exp(i\lambda_m z)$ and solving the resulting eigenvalue equation. With the transformation $y = \tanh(x)$, the eigenvalue equation takes the form [35]

$$(1-y^2)\frac{d^2 U_m}{dx^2} - 2y\frac{dU_m}{dx} + \left(\alpha^2 - \frac{2\lambda_m}{1-y^2}\right)U_m, \tag{13.3.14}$$

and has the solution $U_m(x) = P_n^m(\tanh x)$, where $P_n^m(x)$ is the associated Legendre function of the first kind and the integers m and n are defined as $\lambda_m = m^2/2$ and $\alpha^2 = n(n+1)$. Although n can be any positive integer, m is restricted to be in the range 1–n. For $n = m = 1$, only one coherent mode is excited, and we recover the fully coherent case. For other values of n, we obtain a family of partially coherent solitons containing a superposition of modes. Equation (13.3.8) represents the self-consistency condition and can be used to find the expansion coefficients. The degree of coherence can then be found using the procedure outlined earlier.

The simplest case corresponds to $n = 2$, for which the partially coherent soliton consists of just two coherent modes, corresponding to $m = 1$ and 2. The coherence properties of such solitons can be found for any value of n, and the analytic expressions for the modes and the coherence function are given in Ref. [35] for n as large as six. As an example, Figure 13.1 shows for $n = 2$ and 3 how the normalized spatial coherence function $\mu_{12}(x,d)$ changes with the spacing d as x is varied. These two-dimensional plots indicate the complex nature of the coherence properties of a Kerr-type partially coherent soliton; the plots becomes more and more complicated as n increases.

The modal analysis can easily be extended to two transverse dimensions and shows that incoherent solitons can exist in the form of a Gaussian beam with an elliptical spot [30]. Other methods discussed in Section 13.2 can also be used and lead to the same conclusions [29]–[31]. The theory based on the mutual coherence function has

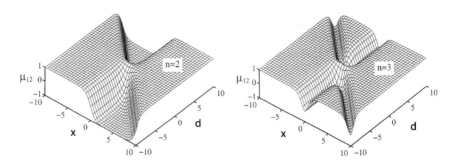

Figure 13.1: Spatial coherence function $\mu_{12}(x,d)$ plotted as a function of x and d for $n = 2$ and 3 for Kerr-type partially coherent solitons. (After Ref. [35]; ©1999 APS.)

been used to find the so-called *twisted incoherent solitons*, whose phase is position dependent and vanishes only in the fully coherent limit [50]. This theory can also be used to prove the validity of a superposition principle that states that any linear combination of partially coherent solitons will also propagate as a soliton as long as they have the same intensity profile [51].

13.3.2 Modulation Instability of Partially Coherent Beams

Modulation instability of partially coherent beams was first analyzed using the coherent density theory [52]. The theory was later extended to the two-dimensional case to discuss a symmetry-breaking instability that leads to pattern formation [53]. Here, we apply the Wigner transform method and consider the modulation instability of a partially coherent CW beam propagating inside a slow-responding Kerr medium using Eq. (13.2.23) with $G(I) = I = |A|^2$. First, we recall the fully coherent case discussed in Section 1.5.1. For a coherent CW beam of input intensity I_0 at $z = 0$, $A(\mathbf{r}, z) = A_0 \exp(i\gamma I_0 z)$ is the solution of Eq. (13.2.23). Such a beam experiences modulation instability when $d\gamma > 0$ and $K < K_c$, where K is the spatial frequency of the perturbation and $K_c = (4\gamma I_0/d)^{1/2}$ is the cutoff value. The instability growth rate g in the coherent case is given by

$$g(K) = \text{Im}(\Omega) = \tfrac{1}{2} dK \sqrt{K_c^2 - K^2}. \tag{13.3.15}$$

In the case of a partially coherent CW beam, we solve Eq. (13.2.26) assuming that the Wigner function can be written in the form

$$\rho(p, x, z) = \rho_0(p) + \rho_1 \exp[i(Kx - \Omega z)], \tag{13.3.16}$$

where $\rho_0(p)$ is the steady-state distribution function corresponding to a plane wave with the complex amplitude

$$A(\mathbf{r}, z) = \sqrt{I_0} \exp[i\gamma I_0 z + i\phi(x)]. \tag{13.3.17}$$

The coherence properties are governed by the randomly varying phase $\phi(x)$. By definition, the input intensity satisfies the relation $I_0 = \int_{-\infty}^{+\infty} \rho_0(p) \, dp$. The quantity ρ_1 represents a small perturbation. Linearizing Eq. (13.2.26) in the perturbation ρ_1, we obtain the following dispersion relation from the linearized Wigner–Moyal equation [17]:

$$1 + \frac{\gamma}{d} \int_{-\infty}^{+\infty} \frac{\rho_0(p + K/2) - \rho_0(p - K/2)}{K(p - \Omega/\beta K)} \, dp = 0. \tag{13.3.18}$$

The instability gain is obtained by solving this equation for Ω and using $g \equiv \text{Im}(\Omega)$. In fact, using $\rho_0(p) = I_0 \delta(p)$ in Eq. (13.3.18), it is easy to show that Eq. (13.3.15) is recovered in the limit of a fully coherent coherent wave.

The dispersion relation resulting from the linearized Vlasov-like equation (13.2.27) has the form

$$1 + \frac{\gamma}{d} \int_{-\infty}^{+\infty} \frac{d\rho_0/dp}{(p - \Omega/\beta K)} \, dp = 0, \tag{13.3.19}$$

which can also be directly obtained from Eq. (13.3.18) in the limit of small K. Relation (13.3.19) is similar to the dispersion relation obtained for electron plasma waves. In general, the integrals in Eqs. (13.3.18) and (13.3.19) can be represented as the sum of a principal value and a residue contribution, the latter leads to a Landau-type damping of the perturbation, similar to that occurring for electron plasma. This stabilizing effect is not an ordinary dissipative damping. Rather it represents an energy-conserving, self-action effect within a partially coherent field, causing a redistribution of the Wigner spectrum because of the interaction among its different parts that counteracts the modulation instability. Similar phenomena occur in connection with nonlinear propagation of electron plasma waves interacting with intense electromagnetic radiation [44]–[48], nonlinear interaction between random-phase photons and sound waves in electron-positron plasmas [49], and longitudinal dynamics of charged-particle beams in accelerators [46].

To illustrate the partially coherent case, let us assume that the optical field in Eq. (13.3.17) satisfies the following autocorrelation function:

$$\langle A^*(x + y/2, z)A(x - y/2, z)\rangle = \exp(-p_0|y|), \tag{13.3.20}$$

where p_0^{-1} is the correlation length. The corresponding Wigner function is found to exhibit a Lorentzian shape

$$\rho_0(p) = \frac{I_0}{\pi} \frac{p_0}{p^2 + p_0^2}. \tag{13.3.21}$$

Using this form in the dispersion relation (13.3.18) then yields the following expression for the instability gain:

$$g(K) = \frac{1}{2}dK \left(\sqrt{K_c^2 - K^2} - 2p_0 \right). \tag{13.3.22}$$

This result is similar to that obtained in Ref. [52] using a different approach.

A comparison of Eq. (13.3.22) with Eq. (13.3.15) shows that the stabilizing effect of the Landau damping is due to the finite bandwidth p_0 of the Lorentzian spectrum (or a finite correlation length of field fluctuations). In fact, if the width of the Lorentzian spectrum satisfies the condition $p_0 > p_c$, where $p_c = K_c/2 = (\gamma I_0/d)^{1/2}$, the modulation instability is *completely suppressed* for all wave numbers K. In other words, when the Landau damping induced by the partial incoherence of a CW beam is strong enough, it can overcome the coherent growth associated with the modulation instability. This effect has been observed experimentally [54].

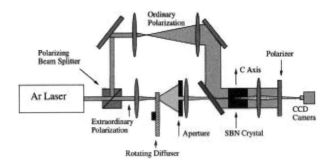

Figure 13.2: Experimental setup used for observing self-trapping of incoherent light, resulting in the formation of a spatial incoherent soliton. (After Ref. [9]; ©1996 APS.)

13.3.3 Experimental Results

The first observation of the self-trapping of a spatially incoherent optical beam used the slow-responding nonlinearity associated with photorefractive crystals [9]. Under appropriate conditions, a CW beam propagated as a bright soliton of 30-μm width, in spite of its partially coherent nature. Since the correlation length across the beam intensity profile was much shorter than the beam width, the self-trapped beam could be considered to form a bright incoherent soliton.

The setup used in this 1996 experiment is shown in Figure 13.2 [9]. It used a 488-nm argon laser beam that was split into two orthogonally polarized parts using a polarizing beam splitter. The ordinarily polarized beam was expanded and used as a background for illuminating a photorefractive SBN crystal uniformly. This background beam generates a certain level of electrons in the conduction band that optimizes the photorefractive self-focusing. The extraordinarily polarized beam was sent through a *rotating diffuser* for producing a partially spatially incoherent beam [55]. The diffuser was rotating with a period much shorter than the response time of the photorefractive crystal. The beam was then sent through the SBN crystal, oriented such that the beam polarization was parallel to the crystal c axis. Self-focusing was observed to occur with the application of an external voltage of appropriate magnitude and polarity. The applied electric field gives rise to a space charge field that has a large component along the c axis and enhances the nonlinearity.

Figure 13.3 shows the horizontal and vertical beam profiles of the input beam, the diffracted output beam at zero voltage, and the self-trapped beam at 550 V. Since the nonlinear medium responds only to the *time-averaged* intensity, it does not "see" rapidly varying speckles. However, when the diffuser does not rotate, the speckle pattern stops changing in time, and the beam breaks up into randomly located multiple filaments. When the diffuser is again rotated much faster than the response time of the nonlinear medium, the filaments disappear and a single self-trapped beam reappears. This behavior confirms that the observed self-trapped beam is indeed an incoherent soliton. Even though the self-trapped beam is composed of many randomly changing coherent components, the time-averaged intensity corresponds to a smooth single beam that induces a single waveguide (via the photorefractive effect) and guides itself

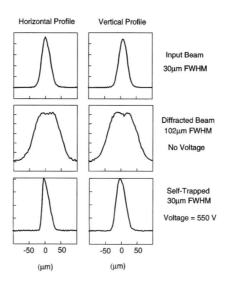

Figure 13.3: Experimental results on the generation of partially coherent solitons. Shown are the horizontal and vertical profiles of the input beam, the diffracted output beam at zero voltage, and the self-trapped output beam formed at 550 V. (After Ref. [9]; ©1996 APS.)

inside it in a self-consistent manner. At any given instant, however, the guided beam is a speckled beam.

One of the most attractive properties of solitons is their ability to create a waveguide for guiding a probe beam. In the context of incoherent solitons, this property has been used to demonstrate the image transmission through a slowly responding self-focusing medium. In one experiment [56], a partially coherent soliton was used to form a multimode waveguide in a photorefractive crystal, and the modes of that waveguide were used to transmit an incoherent image through the nonlinear medium. The transmission of images through such a self-focusing medium can be understood by noting that the multimode waveguide induced by the incoherent soliton can act as a graded-index lens. This method works well only for finite propagation distances as long as intermodal dispersion among the guided modes of the soliton-induced waveguide does not destroy the image.

As discussed in Section 13.3.2, modulation instability has a higher threshold for a partially coherent beam and can be observed only when the input power exceeds a certain threshold value that depends on the degree of coherence [57]–[59]. Figure 13.4 shows examples of the experimentally observed spatial patterns at the output of the photorefractive medium when a wide CW beam is launched at the input end [59]. The correlation length across the beam is only 13 μm, and the beam intensity is equal to the saturation intensity of the nonlinear medium. The strength of nonlinearity is controlled by an external voltage applied to the photorefractive crystal. Photograph A shows the beam profile in the absence of the nonlinearity (uniform intensity). When the nonlinearity is increased, first the threshold for one-dimensional modulation instability is reached in part B. Here a mixed state can appear in which the order and disorder

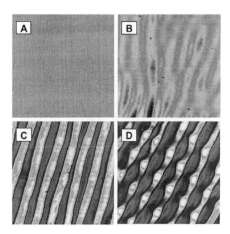

Figure 13.4: Photographs showing modulation instability of a partially coherent beam. (A) Linear propagation; (B) intensity pattern close to instability threshold; (C) stripe formation well above threshold. (D) At much higher values of the nonlinearity, even stripes become unstable and form a two-dimensional pattern. (After Ref. [59]; ©2002 OSA.)

coexist, in the sense that only certain parts of the beam develop filaments, intensity in other parts remaining homogeneous. This behavior indicates that the nonlinear interaction undergoes an order–disorder phase transition. Part C in Figure 13.4 corresponds to a value of the nonlinearity significantly above the instability threshold. In this case the filaments (soliton stripes) have been formed everywhere. When the nonlinearity is further increased, a second transition occurs, shown in part D: The stripes become unstable and start to break into an ordered array of two-dimensional spots corresponding to two-dimensional incoherent solitons. In all plots, the coherence length is much shorter than the distance between any two adjacent stripes or filaments.

The clustering of incoherent solitons has also been observed [60]. The clustering process is initiated by noise-driven modulation instability, which leads to the formation of soliton-like self-trapped filaments. These soliton filaments tend to attract one another, eventually forming clusters. The incoherence of the wave front (which can be varied in a controlled manner) along with the noninstantaneous nature of the nonlinearity give rise to attractive forces between the soliton filaments. The experimental results were in agreement with the theoretical predictions and were confirmed using numerical simulations.

The suppression of modulation instability for partially coherent beams has also been used for forming stable one-dimensional soliton stripes. Normally such stripes are unstable in the coherent case in the other spatial dimension. However, consider a stripe that is fully coherent in the self-trapping dimension yet partially incoherent in the other transverse dimension. It turns out that such soliton stripes become transversely stable when the correlation length is below a threshold value [61]. It was found experimentally that a transversely stable one-dimensional bright Kerr soliton can exist in a bulk medium when the transverse instability of a soliton stripe is completely eliminated by making it sufficiently incoherent along the transverse dimension [54]. An experimental setup similar to that shown in Figure 13.2 was used. The only difference was that a cylindrical lens was used to focus the beam in only one direction, say, the x direction, creating a highly elliptical beam in the shape of a stripe. The nonlinearity was turned on by applying an electric field of 2.7 kV/cm, and the beam formed a stable soliton stripe. When the beam was fully coherent, the soliton suffered from the transverse instability and broke up into filaments. When the beam was made incoherent in the y direction with a speckle size of about 5 μm, the instability was eliminated, and the one-dimensional soliton stripe became stable. For the soliton stripe to be stable, the degree of coherence in the y dimension must be such that the nonlinearity is below the threshold of transverse instability.

13.4 Dark and Vortex Solitons

With the discovery of bright incoherent solitons, it may seem natural that a slow-responding nonlinear medium would also support dark incoherent solitons. However, this issue is not so simple if we recall from Chapter 4 that dark solitons require a transverse phase shift at the center of the dark stripe [62]. If the phase is initially uniform across the dark region of an input beam, the beam does not form a dark soliton but splits into two diverging grey solitons. Recall also from Chapter 8 that two-dimensional coherent vortex solitons require a helical phase structure.

13.4.1 Structure of Dark Incoherent Solitons

Extending the concept of dark solitons to partially coherent beams raises an important question [22]. If dark incoherent solitons were to exist, what would be their phase structure, which was found to be irrelevant for bright incoherent beams? This question was answered in 1998 theoretically using the coherent density approach [26]. Numerical simulations revealed that, when an incoherent beam with a central dark stripe is launched into a slow-responding self-defocusing nonlinear medium, the beam undergoes considerable evolution but eventually becomes self-trapped, with only small oscillations around the dark-soliton solution. Surprisingly, this dark incoherent soliton requires an initial transverse phase jump and is always gray (intensity does not drop to zero at the stripe center). These theoretical results suggested that dark incoherent solitons should exist, and indeed their existence was confirmed soon after in several experiments [63]–[66].

To emphasize the difference between coherent and incoherent light propagation, it is useful to compare the diffraction of a notch (dark stripe) in a linear medium with and without a π phase jump at the center. Consider first the fully coherent case. In the presence of the phase hump, the stripe maintains its zero amplitude at the center for all propagation distances, but in its absence, diffraction fills the central notch and produces a finite amplitude at the center. Therefore, we expect that the nonlinear propagation of a notch-bearing coherent beam would also show a marked difference between uniform-phase and π-phase-jump cases. This is exactly what occurs in practice. In the presence of the phase shift, a single dark soliton forms, but in its absence, the beam splits and forms an even number of gray solitons [62].

Now consider the linear diffraction of a partially coherent beam with a central dark stripe. It turns out that the diffraction pattern becomes indistinguishable after a relatively short distance for the uniform-phase and π-phase-jump cases. In both cases, the notch diminishes rapidly, irrespective of the phase at its center [64]. This happens because the independent speckles that make up the beam diffract in an uncorrelated manner and diffuse the phase information all over the beam. It is therefore quite surprising that the same partially incoherent beam, when propagating inside a nonlinear medium, "remembers" its initial phase imprint, and evolves accordingly into a single dark soliton or into two gray solitons [26]. It appears that a *phase-memory effect* exists, in the case of a slow-responding nonlinear medium, that vanishes in the absence of nonlinearities [64].

The modal theory of incoherent dark spatial solitons helps to understand some of the new features of incoherent dark solitons or, generally speaking, solitons on a background [33] (see also Refs. [67, 68]). It shows that the underlying modal structure of incoherent dark solitons is quite different compared with the case of incoherent bright solitons. More specifically, dark solitons consist not only of the bound modes (as bright solitons do) but also of the *radiation modes*. This feature explains why incoherent dark solitons are always gray and why a transverse π phase flip can facilitate their observation.

The main idea behind the modal analysis is to study the evolution of an input beam whose intensity varies as

$$I(x,0) = I_0[1 - d_s^2 \text{sech}^2 x] \tag{13.4.1}$$

and thus exhibits a central notch. The parameter d_s governs the depth of the notch and is ultimately related to the grayness of the dark soliton. Using Eq. (13.4.1) with $n_{\text{nl}}(I) = -I$ in Eq. (13.2.21) and assuming a solution in the form $u_m(z,z) = U_m(x)\exp(i\lambda_m z)$, we obtain the following eigenvalue equation for the modes:

$$\frac{\partial^2 u_m}{\partial x^2} - 2\beta_0 k_0 I_0[1 - d_s^2 \text{sech}^2 x]u_m = \lambda_m u_m. \tag{13.4.2}$$

This equation supports both the bound and radiation modes. The radiation modes need to be included to ensure that the intensity is finite as $|x| \rightarrow \infty$. The intensity profile at a distance z is found from this superposition of the bound and radiation modes and has the form [22]

$$I = A^2 \text{sech}^2 x + \int_0^\infty D(Q)[Q^2 + \tanh^2 x]dQ, \tag{13.4.3}$$

where $D(Q)$ represents the contribution of radiation modes. The first term in Eq. (13.4.3) arises from bound modes. The self-consistency condition for the formation of incoherent solitons requires that the intensity given by Eq. (13.4.3) should be identical to that in Eq. (13.4.1). This can occur only if

$$I_0 = \int_0^\infty D(Q)(Q^2 + 1)dQ, \tag{13.4.4}$$

$$A^2 = \int_0^\infty D(Q)[1 - d_s^2(Q^2 + 1)]dQ. \tag{13.4.5}$$

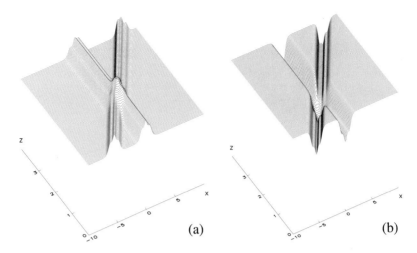

(a)   (b)

Figure 13.5: Interaction of two incoherent solitons on a multi-component background in the case of (a) self-focusing (bright solitons) and (b) self-defocusing (dark solitons). (After Ref. [68]; ©2001 Elsevier.)

The analytical solution given in Eqs. (13.4.4) and (13.4.5) shows that incoherent dark solitons indeed exist, but they involve an infinite number of radiation modes with amplitudes $D(Q)$. The form of $D(Q)$ depends on the properties of the input beam. It is evident from Eqs. (13.4.4) and (13.4.5) that the distribution function $D(Q)$ is by no means unique, and many self-consistent solutions are possible, depending on the specific form of $D(Q)$. A similar approach can be employed to study incoherent bright solitons on a finite background, called *antidark solitons* because they have a higher intensity than the background [68, 69]. The background itself can be stabilized against the development of modulation instability, as discussed earlier.

Two examples of the interaction of two incoherent solitons are shown in Figure 13.5 for self-focusing (bright solitons) and self-defocusing (dark solitons) cases. Since each incoherent soliton consists of several coherent modes, its shape changes after collision. In particular, a symmetric soliton can acquire an asymmetrical shape after the collision. Additionally, in contrast with the scalar solitons, such multicomponent solitons experience larger lateral translational shifts during a collision because of multiple contributions from all constituent components.

13.4.2   Experimental Results

Similar to the bright-soliton case, the experiments on dark solitons employed a rotating diffuser to generate an input beam that was only partially coherent spatially [63]. The experimental setup was similar to that shown in Figure 13.2, with the only difference that, after the rotating diffuser, the beam was reflected from a phase mask (a mirror with a $\lambda/4$ step at the beam center). This phase shift produced a dark notch on a broad partially coherent background. The notch-bearing beam was launched into a photorefractive crystal. Self-trapping of the dark notch was observed at a certain bias

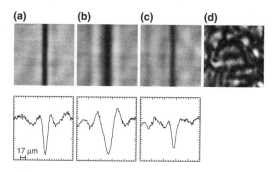

Figure 13.6: Self-trapping of a dark stripe and formation of a dark incoherent soliton. Shown are photographs (top row) and beam profiles (bottom row) of (a) the input beam, (b) the diffracted output beam without nonlinearities, and (c) the self-trapped output beam. The output beam for a coherent beam (stationary diffuser) exhibits a speckle pattern (d), as expected. (After Ref. [63]; ©1998 AAAS.)

field that introduced the necessary nonlinearity so that diffraction was balanced with self-defocusing.

The experimental results showing the formation of a dark incoherent soliton are shown in Figure 13.6. In agreement with the prediction of Ref. [26], the observed incoherent dark soliton was always gray. Thus, unlike coherent dark solitons, which can be either black or gray, dark incoherent solitons are always gray. Another difference between the incoherent and coherent dark (or bright) solitons is the nature of the temporal response of the nonlinearity. Coherent spatial solitons can form in either instantaneous or noninstantaneous nonlinear media, but incoherent spatial solitons require a nonlinear response that is much slower than the time scale of fluctuations associated with a partially coherent beam. For example, Figure 13.6(d) shows what happens when the rotation of the diffuser is stopped. The self-defocusing medium responds to the stationary speckle pattern by fragmenting the beam. Since self-trapping does not occur in this case, no dark soliton is observed to form. Self-trapping of a dark notch depends critically on the spatial degree of coherence. As the degree of coherence decreases, the self-trapped notch becomes grayer and a higher nonlinearity is needed for soliton formation [63].

The effect of the initial phase distribution at the center of the dark stripe of a partially coherent beam has been investigated in several experimental and theoretical studies [26, 64]. The results reveal that the initial phase shift at the stripe center is crucial for the evolution of dark incoherent solitons. If the phase jumps by π at the center of the input beam, a single gray incoherent soliton emerges. In contrast, if the phase is continuous across the input dark stripe, then two gray incoherent solitons emerge but they separate from each other with propagation [64]. This behavior is very similar to that occurring for coherent dark solitons [62].

Similar to the case of bright solitons, a dark soliton also changes the refractive index of the medium and induces a graded-index waveguide. An interesting question is whether a coherent light beam can be guided through the waveguide created by an incoherent dark soliton. The answer turned out to be affirmative in a 1999 experiment [65].

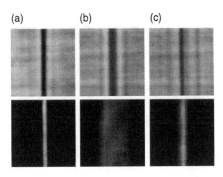

Figure 13.7: Photographs showing the guiding of a probe beam (bottom row) by an incoherent dark soliton (top row): input (a) and output beam intensities without (b) and with (c) the nonlinear effects. (After Ref. [65]; ©1999 OSA.)

Figure 13.7 shows the experimental results. The input dark beam (a) has a coherence distance of 15 μm (estimated from the average speckle size). It forms an incoherent dark soliton (c) of 18-μm (FWHM) width at a bias field of 950 V/cm. In the absence of nonlinearity, the probe beam (bottom row) diffracts from 20 μm to ~68 μm [Figure 13.7(b)], as expected from diffraction theory. However, once the dark incoherent soliton has formed, as seen in part (c), the probe beam stops diffracting because it is guided by the soliton-induced waveguide. In this experiment, the incoherent soliton beam had an average intensity of ~4.5 mW/cm^2, while the intensity of the probe beam reached a level of 50 mW/cm^2. At the output, nearly 80% of the input power of the probe beam was guided into this waveguide channel (after accounting for Fresnel reflections and crystal absorption). The nature of the waveguide created depends on the initial phase shift imposed on the incoherent beam. In the one-dimensional case, incoherent solitons produce either a straight or a Y-junction planar waveguide, depending on the phase shift. In the two-dimensional case, such solitons would produce a cylindrical waveguide. These results point to the possibility of controlling high-power laser beams with low-power incoherent light sources.

Another interesting observation turns outs to be that the spatial degree of coherence of a bright, partially coherent signal beam can be affected through its interaction with a dark coherent (or incoherent) spatial soliton [66]. Physically speaking, during the nonlinear interaction of two such beams, a part of the incoherent bright beam is trapped within the dark (or gray) notch of the controlling soliton, thus forming a sharp intensity spike. In this region, the correlation length increases by at least two orders of magnitude. In other words, incoherent light can be effectively cooled (its entropy reduced) at any arbitrarily chosen point by using a dark spatial soliton. This is the only passive system that is known to exhibit an increase in both the local intensity and the local coherence simultaneously.

The concept of incoherent dark solitons can be readily extended to two transverse dimensions, resulting in the formation of incoherent vortex solitons. The experimental approach is similar to that used for coherent beams (see Chapter 8). More specifically, one needs to use an input beam with phase dislocations [62]. In one experiment, the

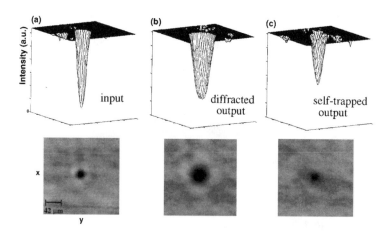

Figure 13.8: Self-trapping of an optical vortex on a partially coherent beam. Both the the three-dimensional intensity plots (top row) and the actual photographs (bottom row) are shown for (a) the input beam, (b) the diffracted output beam, and (c) the self-trapped output beam. (After Ref. [63]; ©1998 AAAS.)

input vortex beam was generated by using a helicoidal phase mask in combination with the rotating diffuser, thereby creating an optical vortex of unit topological charge that was nested in a broad spatially incoherent beam. When this incoherent beam was propagated through a photorefractive crystal, it developed into a self-trapped incoherent vortex soliton [63]. The experimental results are shown in Figure 13.8, where both the diffracted output in the absence of nonlinearities (b) and the self-trapped output (c) are shown together with the input (a). Similar to the one-dimensional case, the central intensity dip does not extend to zero, and the vortex is of gray type. Also, as the degree of coherence of the background beam decreases, the self-trapped vortex becomes grayer, and a larger nonlinearity is needed for its formation.

One more interesting feature of incoherent self-trapping is that a bright spatial soliton can be guided by a partially coherent background. Although dark solitons require a background, this is not so for bright solitons. Over the years, the idea of a bright soliton that can exist on top of a nonvanishing background (the antidark soliton) has received considerable attention [70]–[75]. The stability of antidark solitons turns out to be a critical issue, and stable anti-dark solitons in the coherent case form only when higher-order effects or nontrivial nonlinearities are involved [74]. The instability is triggered by the modulation instability of the coherent background beam needed to confine the bright part of the antidark soliton. However, as discussed earlier in this section, such a modulation instability can be suppressed for an incoherent background beam. The existence of a threshold for incoherent modulation instability is a feature unique of the partially coherent nature of light, and such a threshold does not exist in the coherent regime. It is thus natural to expect the formation of stable antidark solitons by controlling the coherence of the background beam. This is indeed what was observed in a 2000 experiment [69]. Both the numerical simulations and the experiments indicate that these incoherent antidark solitons propagate in a stable fashion, provided that spatial coherence of their background is reduced below the threshold of modulation instability.

References

[1] A. Hasegawa, *Phys. Fluids* **18**, 77 (1975); *Phys. Fluids* **20**, 2155 (1977).

[2] B. Gross and J. T. Manassah, *Opt. Lett.* **16**, 1835 (1991); *Opt. Lett.* **17**, 166 (1992).

[3] M. T. de Araujo, H. R. da Cruz, and A. S. Gouveia-Neto, *J. Opt. Soc. Am. B* **8**, 2094 (1991).

[4] H. R. da Cruz, J. M. Hickmann, and A. S. Gouveia-Neto, *Phys. Rev. A* **45**, 8268 (1992).

[5] J. N. Elgin, *Opt. Lett.* **18**, 10 (1993); *Phys. Rev. A* **47**, 4331 (1993).

[6] S. B. Cavalcanti, G. P. Agrawal, and M. Yu, *Phys. Rev. A* **51**, 4086 (1995).

[7] J. Garnier, L. Videau, C. Gouédard, and A. Migus, *J. Opt. Soc. Am. B* **15**, 2773 (1998).

[8] S. B. Cavalcanti, *New J. Phys.* **4**, 19 (2002).

[9] M. Mitchell, Z. Chen, M. Shih, and M. Segev, *Phys. Rev. Lett.* **77**, 490 (1996).

[10] M. Mitchell and M. Segev, *Nature* **387**, 880 (1997).

[11] D. N. Christodoulides, T. H. Coskun, M. Mitchell, and M. Segev, *Phys. Rev. Lett.* **78**, 646 (1997).

[12] D. N. Christodoulides, T. H. Coskun, and R. I. Joseph, *Opt. Lett.* **22**, 1080 (1997).

[13] M. Mitchell, M. Segev, T. Coskun, and D. N. Christodoulides, *Phys. Rev. Lett.* **79**, 4990 (1997).

[14] V. V. Shkunov and D. Z. Anderson, *Phys. Rev. Lett.* **81**, 2683 (1998).

[15] G. A. Pasmanik, *Sov. Phys. JETP* **39**, 234 (1974).

[16] B. Hall, M. Lisak, and D. Anderson, R. Fedele, and V. E. Semenov, *Phys. Rev. E* **65**, 035602 (2002).

[17] L. Helczynski, D. Anderson, R. Fedele, B. Hall, and M. Lisak, *IEEE J. Sel. Topics Quantum Electron.* **8**, 408 (2002).

[18] D. N. Christodoulides, E. D. Eugenieva, T. H. Coskun, M. Segev, and M. Mitchell, *Phys. Rev. E* **63**, 035601 (2001).

[19] M. Lisak, L. Helczynski, and D. Anderson (private communication).

[20] A. W. Snyder and D. J. Mitchell, *Phys. Rev. Lett.* **80**, 1422 (1998).

[21] L. Mandel and E. Wolf, *Optical Coherence and Quantum Optics* (Cambridge University Press, New York, 1995).

[22] M. Segev and D. N. Christodoulides, in *Spatial Solitons*, S. Trillo and W. Torruellas, Eds. (Springer, Berlin, 2001), pp. 87–126.

[23] M. Segev and G. I. Stegeman, *Phys. Today* **51**, 42 (1998).

[24] O. Bang, D. Edmundson, and W. Krolikowski, *Phys. Rev. Lett.* **83**, 5479 (1999).

[25] E. D. Eugenieva, D. N. Christodoulides, and M. Segev, *Opt. Lett.* **25**, 972 (2000).

[26] T. H. Coskun, D. N. Christodoulides, M. Mitchell, Z. Chen, and M. Segev, *Opt. Lett.* **23**, 418 (1998).

[27] B. Ya. Zel'dovich, N. F. Pilipetsky, and V. V. Shkunov, *Principles of Phase Conjugation* (Springer, Berlin, 1985).

[28] I. Bialynicki-Birula and J. Mycielski, *Physica Scripta* **20**, 539 (1979).

[29] A. W. Snyder and J. Mitchell, *Opt. Lett.* **22**, 16 (1997).

[30] D. N. Christodoulides, T. H. Coskun, M. MItchell, and M. Segev, *Phys. Rev. Lett.* **80**, 2310 (1998).

[31] W. Krolikowski, D. Eddmundson, and O. Bang, *Phys. Rev. E* **61**, 3122 (2000).

[32] Y. Nogami and C. S. Warke, *Phys. Lett. A* **59**, 251 (1976).

[33] D. N. Christodoulides, T. H. Coskun, M. MItchell, Z. Chen, and M. Segev, *Phys. Rev. Lett.* **80**, 5113 (1998).

[34] N. N. Akhmediev, W. Krolikowski, and A. W. Snyder, *Phys. Rev. Lett.* **81**, 4632 (1998).

[35] M. I. Carvalho, T. H. Coskun, D. N. Christodoulides, M. Mitchell, and M. Segev, *Phys. Rev. E* **59**, 1193 (1999).

[36] A. A. Sukhorukov and N. N. Akhmediev, *Phys. Rev. Lett.* **83**, 4736 (1999).

[37] A. Ankiewicz, W. Krolikowski, and N. N. Akhmediev, *Phys. Rev. E* **59**, 6079 (1999).

[38] T. Kanna and M. Lakshmanan, *Phys. Rev. Lett.* **86**, 5043 (2001).

[39] M. M. Loeve, *Probability Theory* (Van Nostrand, New York, 1955).

[40] N. N. Akhmediav, W. Krolikowski, and A.W. Snyder, *Phys. Rev. Lett.* **81**, 4632 (1998).

[41] W. Krolikowski, N. N. Akhmediev, and B. Luther-Davies, *Phys. Rev. E* **59**, 4654 (1999).

[42] N. M. Litchinitser, W. Krolikowski, N. N. Akhmediev, and G. P. Agrawal, *Phys. Rev. E* **60**, 2377 (1999).

[43] E. Wigner, Phys. Rev. **40**, 749 (1932).

[44] A. A. Venedov, *Theory of Turbulent Plasma* (Nauka, Moscow, 1965) [Israel Program for Scientific Translations, Jerusalem, 1966].

[45] N. L. Tsintsadze and J. T. Mendonça, *Phys. Plasmas* **5**, 3609 (1998).

[46] R. Fedele, D. Anderson, and M. Lisak, *Physica Scripta* **T84**, 27 (2000).

[47] S. A. Gardiner, D. Jaksch, R. Dum, J. I. Cirac, and P. Zoller, *Phys. Rev. A* **62**, 023612 (2000).

[48] R. Bingham, J. T. Mendonça, and J. M. Dawson, *Phys. Rev. Lett.* **78**, 247 (1997).

[49] P. K. Shukla and L. Stenflo, *Phys. Plasmas* **5**, 1554 (1998).

[50] S. A. Ponomarenko, *Phys. Rev. E* **64**, 038818 (2001).

[51] S. A. Ponomarenko, *Phys. Rev. E* **65**, 055601 (2002).

[52] M. Soljacic, M. Segev, T. Coskun, D. N. Christodoulides, and A. Vishwanath, *Phys. Rev. Lett.* **84**, 467 (2000).

[53] S. M. Sears, M. Soljacic, D. N. Christodoulides, and M. Segev, *Phys. Rev. E* **65**, 036620 (2002).

[54] C. Anastassiou, M. Soljacic, M. Segev, E. D. Eugenieva, D. N. Christodoulides, D. Kip, Z. H. Musslimani, and J. P. Torres, *Phys. Rev. Lett.* **85**, 4888 (2000).

[55] J. W. Goodman, *Statistical Optics* (Wiley, New York, 1985).

[56] D. Kip, C. Anastassiou, E. Eugenieve, D. N. Christodoulides, and M. Segev, *Opt. Lett.* **26**, 524 (2001).

[57] D. Kip, M. Soljacic, M. Segev, E. Eugenieve, and D. N. Christodoulides, *Science* **290**, 495 (2000).

[58] J. Klinger, H. Martin, and Z. Chen, *Opt. Lett.* **26**, 271 (2001).

[59] D. Kip, M. Soljacic, M. Segev, S. M. Seras, and D. N. Chrisdoulides, *J. Opt. Soc. Am. B* **19**, 502 (2002).

[60] Z. Chen, S. M. Sears, H. Martin, D. N. Christodoulides, and M. Segev, *Proc. Nat. Acad. sci. USA* **99**, 5223 (2002).

[61] J. P. Torres, C. Anastassiou, M. Segev, M. Soljacic, and D. N. Christodoulides, *Phys. Rev. E* **65**, 015601 (2001).

[62] Yu. S. Kivshar and B. Luther-Davies, *Phys. Rep.* **298**, 81 (1998).

[63] Z. Chen, M. Mitchell, M. Segev, T. H. Coskun, and D. N. Christodoulides, *Science* **280**, 889 (1998).

[64] T. Coskun, D. N. Christodoulides, Z. Chen, and M. Segev, *Phys. Rev. E* **59**, R4777 (1999).

[65] Z. Chen, M. Segev, D. N. Christodoulides, and R.S. Feigelson, *Opt. Lett.* **24**, 1160 (1999).

[66] T. H. Coskun, A. G. Grandpierre, D. N. Christodoulides, and M. Segev, *Opt. Lett.* **25**, 826 (2000).

[67] A. A. Sukhorukov and N. N. Akhmediev, *Phys. Rev. E* **61**, 5893 (2000).

[68] A. A. Sukhorukov, A. Ankiewicz, and N. N. Akhmediev, *Opt. Commun.* **195**, 293 (2001).

[69] T. H. Coskun, D. N. Christodoulides, Y.-R. Kim, Z. Chen, M. Soljacic, and M. Segev, *Phys. Rev. Lett.* **84**, 2374 (2000).

[70] Yu. S. Kivshar, *Phys. Rev. A* **43**, 1677 (1991).

[71] Yu. S. Kivshar and V. V. Afanasjev, *Phys. Rev. A* **44**, R1446 (1991).

[72] L. Gagnon, *J. Opt. Soc. Am. B* **10**, 469 (1993).

[73] N. Belanger and P.A. Bélanger, *Opt. Commun.* **124**, 301 (1996).

[74] Yu. S. Kivshar, V. V. Afansjev, and A. W. Snyder, *Opt. Commun.* **126**, 348 (1996).

[75] D. J. Frantzeskakis, K. Hizanidis, B. A. Malomed, and C. Polymilis, *Phys. Lett. A* **248**, 203 (1998).

Chapter 14

Related Concepts

Several important concepts are closely connected with the optical solitons described in preceding chapters. Moreover, in many cases the physics of optical solitons is quite useful for getting a deeper insight into the novel phenomena described by similar nonlinear models. This chapter aims to present several such related concepts. In Section 14.1 we discuss reorientation nonlinearities in liquid crystals and the resulting nonlocal solitons. Section 14.2 focuses on the physics of optically induced waveguides in photosensitive materials. We describe the growth and interaction of self-written filaments and compare them with conventional optical solitons. Section 14.3 is devoted to the physics of dissipative and cavity solitons. We discuss the formation, propagation, and interaction of such solitons within an optical resonator for both cubic (Kerr-like) and quadratic nonlinearities. The last two sections take us away from optics but emphasize the relevance of optical solitons to other domains of nonlinear physics. Section 14.4 focuses on the spatiotemporal self-focusing of spin waves in magnetic films, resulting in the formation of spin-wave or magnetic solitons in ferrite films. In Section 14.5 we consider the rapidly developing field of coherent matter waves and nonlinear atom optics. In particular, we describe the nonlinear dynamics of the Bose–Einstein condensates and explore the close connection between self-focusing of light in nonlinear optics and the dynamics of matter-wave solitons.

14.1   Self-Focusing and Solitons in Liquid Crystals

Liquid crystals are fascinating materials with many unique properties and applications. What is often not appreciated is that they also constitute an important nonlinear medium and have numerous applications in nonlinear optics [1]–[3]. In this section we discuss the physical origin of optical nonlinearities in liquid crystals and their use for making nonlinear planar waveguides capable of supporting spatial solitons. We also discuss the specific features of these solitons associated with the nonlocal response of the reorientation nonlinearities.

Figure 14.1: Schematic illustration of a liquid-crystal planar waveguide.

14.1.1 Reorientation Nonlinearities

The nonlinear effects in liquid crystals arise mainly from thermal and reorientation processes. While the thermal effects are similar to those observed in other materials, the reorientation effect is a unique characteristic of the liquid-crystalline phase. The cubic, Kerr-like nonlinearity induced by the reorientation effect in the nematic phase of liquid crystals is responsible for numerous nonlinear effects that are not observed in other materials. Not only does the reorientation nonlinearity induce extremely large intensity-dependent changes in the refractive index at relatively low power levels, but such changes can be modified by external optical or electrical fields. Moreover, the nonlinearity depends on light polarization but is independent of light wavelength within a wide range. The nonlinear optics of liquid crystals has been of interest for many years, and the experimental and theoretical studies on self-focusing in such materials date back to early 1990s [4]–[7].

Liquid crystals are composed of anisotropic molecules with a rod-like shape. They behave in a fluid-like fashion but exhibit a *long-range order* that is characteristic of all crystals. The liquid-crystalline phase is observed in some range of temperature for pure compounds and mixtures (thermotropic liquid crystals), for solutions (lyotropic liquid crystals), and for polymers. Several different types of long-range order is observed in thermotropic liquid crystals, and they are classified as smectics, nematics, and cholesterics, depending on the nature of the long-range interaction among their molecules. The simplest type of order is observed in nematic liquid crystals, in which the position of the molecules is arbitrary but they are all oriented in nearly the same direction. Figure 14.1 shows this situation schematically. At a given temperature, the molecules fluctuate around the mean direction, denoted by the unit vector **n** and called *the director*. The orientation order is described by the parameter

$$S = \tfrac{1}{2}\langle 3\cos^2\Theta - 1 \rangle, \tag{14.1.1}$$

where Θ is the local angle that each molecule makes with respect to the director and the angle brackets denote averaging over both time and space. For crystals, the order parameter $S = 1$, while $S = 0$ for isotropic liquids. For nematics, $0.4 < S < 0.7$, but $S \approx 0.9$ for smectics.

The anisotropy of liquid crystals manifest itself in its various properties, such as electrical permittivity, magnetic permeability, conductivity, and optical birefringence. As a result, an external electric field **E** induces an electrical dipole with moment **p** that is not parallel to **E**. Consequently, the torque $\mathbf{p} \times \mathbf{E}$ tends to rotate the molecules into alignment with the applied electric field. This reorientation does not depend on the sign of the electric field and occurs for time-varying fields as well, including optical fields. A similar behavior is observed for magnetic fields, but magnetic anisotropy is usually much lower than the electrical one. At optical frequencies, the interaction with the magnetic field can be neglected, and the interaction between light and a liquid crystal is described by the electrical dipole.

The rotation induced by the electrical dipole is opposed by the elastic forces that maintain the long-range order within a liquid-crystalline cell. The orientation of each molecule is determined by these two opposing forces. Because the birefringence of liquid crystal is connected with the orientation of molecules, changes in orientation cause the rotation of the optical birefringence axis. Physically speaking, the light incident on a liquid crystal modifies the electric permittivity tensor, leading to the *reorientation nonlinearity*. Because the anisotropy for a liquid crystal is relatively large, the reorientation nonlinearity can create large changes in the refractive index at relatively low intensity levels ($\sim 1 \text{ kW/cm}^2$).

The reorientation nonlinearity can be calculated by minimizing the total free-energy density, which includes the deformation energy, the energy of interaction with the external field, and the effects of boundaries. The key variable that governs the orientation problem is the angle θ between the director $\mathbf{n}$ and the axis along which the input light is polarized (see Figure 14.1). The magnitude of the reorientation nonlinearity depends on the initial orientation $\mathbf{n}$, and therefore on the liquid-crystal configuration. For studying the self-focusing phenomena and spatial solitons, one can place the liquid crystal either in a capillary or between two plates (as shown in Figure 14.1). However, if transversal dimensions of the liquid-crystal film are much larger than the wavelength and the size of the input beam, then liquid crystal can be treated as a bulk medium. Such a configuration is appropriate for observing two-dimensional solitons. When the film thickness is comparable to the wavelength, the situation is similar to a planar waveguide. In this case, large refractive index changes ($\Delta n \sim 0.1$) can be induced at intensity levels of 10 mW/μm. This large nonlinearity can be further enhanced by two orders of magnitudes in the presence of organic dyes in a liquid-crystal mixture [8], but the effect depends strongly on the light wavelength.

14.1.2   Spatial Solitons in Liquid Crystals

In *planar waveguides*, the liquid-crystal layer can play the role of a core or a cladding layer. Moreover, the liquid crystal can be oriented in several different ways. The nonlinear effects experienced by an optical field propagating inside the waveguide depend strongly on both the orientation of liquid-crystal molecules and the state of polarization. In the configuration shown in Figure 14.1, both the director $\mathbf{n}$ and the electric field vector $\mathbf{E}$ are in the x–z plane. The orientation of liquid-crystal molecules is governed by the angle θ and the director $\mathbf{n} = (\cos\theta, 0, \sin\theta)$. The field-induced rotation of the molecule is then described by the Euler–Lagrange equation

$$\nabla^2\theta + \frac{\varepsilon_0\Delta\varepsilon}{4K}\left[(E_xE_z^* + E_x^*E_z)\cos(2\theta) + (|E_x|^2 - |E_z|^2)\sin(2\theta)\right] = 0, \quad (14.1.2)$$

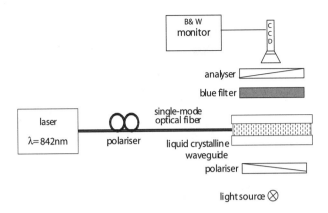

Figure 14.2: Experimental setup for observing the self-focusing phenomenon in a liquid-crystal cell. (After Ref. [11]; ©1998 Gordon & Breach.)

where K is the elastic constant, ε_0 is the electrical permittivity, and $\Delta\varepsilon$ is a measure of electrical anisotropy. This equation should be solved with the boundary conditions at $x = 0$ and $x = d$, in combination with the Maxwell equations for the optical field. Because of the complexity of such coupled equations, the solution of the problem often requires the use of numerical methods.

The physical mechanism behind self-focusing in a liquid-crystal waveguide can be understood as follows. As the guided field reorients liquid-crystal molecules, the refractive index of the medium increases most where the field is most intense, which in turn modifies the guided field itself [9, 10]). Due to the large anisotropy of liquid crystals, the guided mode changes its profile significantly, resulting in self-focusing. The nonlinear change in the refractive index is bistable and begins above a threshold value of the light intensity. Optical bistability is caused by a large nonlinearity and the threshold character of the molecular reorientation. The reorientation effect depends not only on the local value of the electric field but also on the entire field profile across the waveguide. The nonlocal nature of the nonlinear reorientational effect is a source of the feedback, which is necessary for optical bistability. When the electric field tends to reorient liquid-crystal molecules positioned at an angle $\pi/2$, the reorientation starts above a threshold value of the electric field. This phenomenon is called *the Freedericksz threshold effect*.

The self-focusing phenomena has been observed in liquid-crystal waveguides [11] using the setup shown schematically in Figure 14.2. A 10-μm-thick layer of nematic liquid crystal between two glass plates formed the waveguide. Light at a wavelength of 842 nm from a semiconductor laser was launched into the liquid-crystal waveguide. Propagation of the light beam inside the liquid-crystal waveguide was observed through the scattered light detected by a CCD camera. Figure 14.3 shows the photographs obtained. At low values of light power ($P \ll 20$ mW), the beam diffracted, as expected for a linear medium. At a power level of 20 mW, the self-focusing was unstable for the TE-polarized beam, and the beam diffracted even more. However, a TM-polarized beam with a power of 30 mW became self-trapped and appeared to form a spatial

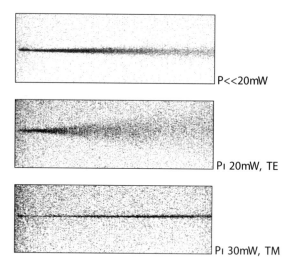

Figure 14.3: Experimental results on self-focusing observed in a liquid-crystal waveguide. The optical beam diffracts at low powers (top) as well as when it is TE-polarized (middle) but becomes self-trapped for TM polarization (bottom). The length of all photographs is only 0.6 mm, but it correspond to three diffraction lengths. (After Ref. [11]; ©1998 Gordon & Breach.)

soliton. The birefringence measurements show that liquid-crystal molecules in the beam area rotate with a spatial period equal to the birefringence length [11]. This observation was found to be in agreement with theoretical predictions, and it proves that the reorientation is the source of self-trapping.

When the transversal size of a liquid-crystal cell is larger than the beam size, it can be treated as a *bulk medium*. The size of the cell cannot be very large because of difficulties in retaining homogenous orientation in thick layers. Although the light beam is concentrated in a small area, the reorientation has a nonlocal character and can appear far from the light beam. Therefore, the shape of the cell and the orientation on the boundaries are important. In general, the beam propagation in a bulk nonlinear medium in the scalar and paraxial approximations is governed by Eq. (1.2.12). The main difference in the case of liquid crystals is that the nonlinear change in the refractive index depends on the angle θ, which depends on the field intensity I. Moreover, in order to eliminate the threshold nature of the nonlinear response, an external voltage is applied to the cell. Therefore, the nonlinear wave equation for the wave envelope A can be written as

$$2ik\frac{\partial A}{\partial z} + \nabla^2 A + k_0^2 n_a^2 [\sin^2\theta - \sin^2\theta_0]A = 0, \qquad (14.1.3)$$

where ∇^2 is the transverse Laplacian, k_0 is the vacuum wave number, $n_a^2 = n_\parallel^2 - n_\perp^2$ is the optical anisotropy, $k^2 = k_0^2(n_\perp^2 + n_a^2\sin^2\theta_0)$, and θ_0 is the tilt in the absence of a light beam but in the presence of an external quasi-static field.

Equation (14.1.3) neglects the polarization effects and is valid for only small reorientation angles. However, the results obtained using it are in good agreement with the experiments [6, 7]. When the external electric field E is applied along the x axis, the reorientation dynamics is described by the following equation [2]:

$$K\nabla^2\theta + \frac{1}{2}\Delta\varepsilon_{RF}E^2\sin(2\theta) + \frac{1}{4}\varepsilon_0 n_a^2|A|^2\sin(2\theta) = 0, \tag{14.1.4}$$

where $\Delta\varepsilon_{RF}$ is the low-frequency anisotropy.

Two-dimensional self-focusing of a light beam in bulk nematics has been observed using both capillaries and planar cells. In one set of experiments [4, 5], the nematic liquid crystal was inserted in a glass capillary with an inner diameter of 1.5 mm. The optical beam from an argon laser was focused to a spot of size of around 50 μm. The beam was observed by looking at the scattered light with the microscope (see Figure 14.2). The polarization direction was perpendicular to the director of the liquid crystal, and thus the reorientation was achieved above the Freedericksz threshold. As the beam power was increased, first the focal spots were observed (for intensities of $\sim$0.6 kW/cm^2). Then the beam began to exhibit undulations at an intensity of $\sim$1.9 kW/cm^2 and finally broke up into two distinct filaments when $I = 2$ kW/cm^2. Because of the threshold nature of the nonlinearity, spatial solitons were not observed. These experimental results are in agreement with the theory of liquid crystals [6, 7].

In another set of experiments [12, 13], light from an argon laser was launched into the capillary cell through an optical fiber. This allowed the researchers to reduce the input beam size in the range 4–10 μm. In a capillary of diameter 250 μm, a nematic liquid crystal was doped with a small amount of the anthraquinone dye to enhance the reorientation resulting from the Janossy effect [8]. The nonlinear self-focusing was observed at a power level of a few milliwatts, together with the appearance of a "self-waveguiding" structure.

In a separate set of experiments [14]–[16], stable spatial solitons were observed using a planar liquid-crystal cell. Figure 14.4 shows the experimental results for a 75-mm-thick nematic cell observed using a 2-mW argon-laser beam. To eliminate the threshold nature of the nonlinear response, an external voltage was applied to the cell. The optical beam from an argon laser was collimated to a waist of <2.5 μm before launching it into the liquid-crystal cell. Without the external electric field, the CW beam exhibited instabilities and often broke up into several parts. The use of an external electric field oriented the molecules in the field direction, resulting in an optical nonlinearity without a threshold. As a consequence, the optical beam was stably self-trapped, and an optical soliton could be formed. The reorientation origin of this effect was confirmed by using a weak collinear He–Ne probe that was guided by the soliton only when it was copolarized with the argon-laser beam.

The two-dimensional spatial solitons could be formed even by using spatially incoherent beams [15]. The soliton-induced waveguides were able to confine a weaker signal of the same polarization, even in the presence of significant angular misalignment, a feature that allows one to steer the signal angularly by tilting the soliton-forming beam [17]. These phenomena are wavelength independent due to the nonresonant nature of the nonlinearity, and they can be observed at low powers using light-emitting diodes or white-light sources.

Beam propagation inside twisted nematic liquid-crystal waveguides has also been analyzed [18]. The reorientation nonlinearity in such waveguides is large enough for

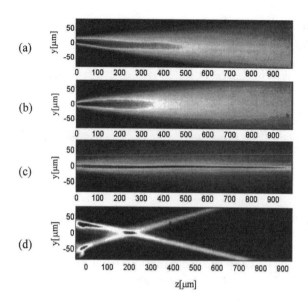

Figure 14.4: Self-focusing in a 75-mm-thick nematic cell observed using a 2-mW argon-laser beam. (a) Diffraction of a y-polarized beam; (b) diffraction of an x-polarized beam with no bias voltage; (c) spatial soliton formation for an x-polarized beam with bias voltage; (d) collision of two solitons launched at nearly opposite angles. (After Ref. [16]; ©2002 APS.)

observing spatial solitons at a few milliwatts of input power. Not only does the reorientation nonlinearity induces self-focusing, but it can also change the direction of beam propagation.

14.1.3 Nonlocal Solitons

Because of a spatially nonlocal response of a liquid-crystalline medium, the spatial solitons generated inside nematic liquid crystals represent a novel class of self-trapped optical beams that exist only in nonlocal nonlinear media. Writing the angle distribution in Eqs. (14.1.3) and (14.1.4) in the form $\theta \approx \theta_0 + \phi$ and linearizing in ϕ, we obtain the following set of two normalized coupled equations [19]:

$$i\frac{\partial u}{\partial z} + \nabla^2 u + \phi u = 0, \tag{14.1.5}$$

$$\nabla^2 \phi - \alpha^2 \phi + |u|^2 = 0. \tag{14.1.6}$$

These equations resemble those governing two-frequency parametric solitons in quadratic media (see Chapter 10), with the main difference that the second field ϕ is real. A direct link between the parametric and nonlocal solitons can be established, as discussed in Ref. [20].

If the $\nabla^2 \phi$ term in Eq. (14.1.6) is negligible, e.g., for large values of α, we can use the solution $\phi = |u|^2/\alpha^2$ in Eq. (14.1.5) and obtain the standard NLS equation. The resulting Kerr-type soliton depends on the intensity only locally. When the $\nabla^2 \phi$ term

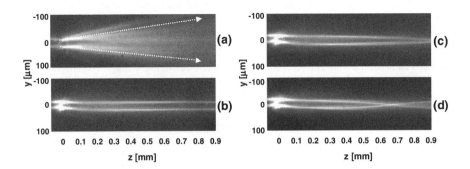

Figure 14.5: Attraction of two identical solitons in a liquid-crystal cell. The photos show the propagation in the y–z plane of two Gaussian beams with 10-μm waist and 28-μm separation: (a) linear behavior; (b) weak attraction at a power of 2.8 mW; (c) stronger attraction at 3.6 mW; (d) crossing and interlacing at 4.5 mW. (After Ref. [32]; ©2002 OSA.)

cannot be neglected, the solution is expected to have a nonlocal character because it depends on a range of field intensities over a certain spatial domain. Such solutions correspond to the so-called *nonlocal solitons*. They remain stable even in a bulk Kerr medium (in contrast with the standard Kerr solitons) and exhibit long-range interaction; i.e., they can interact even at separations exceeding their transverse extension.

Equations (14.1.5) and (14.1.6) represent a special case of the NLS equation with a nonlocal nonlinear response with the general form

$$i\frac{\partial u}{\partial z} + \nabla^2 u + u \int V(\mathbf{r} - \mathbf{r}')|u(\mathbf{r}')|^2 d\mathbf{r}' = 0, \qquad (14.1.7)$$

where the response function $V(\mathbf{R})$ governs the nonlocal nature of the medium response. In the case of a nematic liquid crystal, $V(\mathbf{R}) = \exp(-|\alpha\mathbf{R}|)/|\mathbf{R}|$. An equation similar to Eq. (14.1.7) appears in the context of plasmas [21]–[23] and Bose–Einstein condensates consisting of bosonic particles with long-range dipole interaction [24, 25]. In the limit of weak nonlocality, Eq. (14.1.7) can be reduced to a perturbed NLS equation containing a nonlinear-dispersion term [23]; its solutions for bright and dark solitons can be found in an explicit form [26]. The most interesting case is that of strong nonlocality. Nonlocality tends to suppress the modulation instability [27] in a way similar to a saturable nonlinearity. More importantly, it can completely eliminate the beam collapse in a Kerr medium [28, 29]. Additionally, the nonlocal nature of the nonlinear response in Eq. (14.1.7) changes the character of the soliton interaction because it depends on the type of the response function and can become long-ranged, in the sense that two solitons that are far apart experience an interaction force. This specific feature of nonlocal solitons is known to lead to the formation of multisoliton bound states even in the two-dimensional geometry, as first discussed by Mironov et al. in 1981 [30] (see also Ref. [31]).

As a typical example of the soliton interaction in nonlocal media, Figure 14.5 shows the experimental results for two spatial solitons excited with a 514-nm argon-laser beam inside nematic liquid crystals [32]. Two Gaussian beams with 10-μm waist and 28-μm separation were launched to form two spatial solitons. In spite of their relatively large separation, the refractive-index perturbation created by one beam can diffuse and affect the other beam, causing the two solitons to attract each other. In part (d) of Figure 14.5 such an attraction leads to complete interlacing when the Gaussian beams are launched in parallel and are separated by 1 μm at the input interface. Notice that the distance at which two solitons collide decreases as the input power is increased.

14.2   Self-Written Waveguides

Optical waveguides that form when an optical beam is self-trapped inside a nonlinear medium are of transient nature, in the sense that they disappear when the input beam propagating as a spatial soliton is turned off [33]. Many applications use self-induced waveguides for guiding other beams of different polarizations or wavelengths. These applications would benefit if soliton-induced changes in the refractive index were to persist even after the beam producing them was turned off, resulting in the "freezing" of a soliton-induced waveguide. This is possible only if a *photosensitive material* is used [34]. This section is devoted to the photosensitivity phenomenon and its applications in the context of optical solitons.

14.2.1   Photosensitive Materials

It was discovered in 1978 that when an argon-laser beam was launched into a germanium-doped optical fiber, the reflected signal increased from its nominal value of 4% to more than 60% within a few minutes [35]. This observation ultimately led to the development of fiber Bragg gratings, which are used routinely for numerous applications [36]. It turned out that the refractive index of germanium-doped glasses can be increased permanently by a small amount ($\sim 10^{-4}$) by illuminating it with ultraviolet light at wavelengths close to 250 nm [37]–[39]. This phenomenon is known as *photosensitivity*. Although heating accelerates the aging of photosensitivity-induced changes in the refractive index, they can be considered to be effectively permanent for practical purposes.

The phenomenon of self-writing has been observed in a number of photosensitive optical materials [40]–[53]. The list includes the writing of waveguide tapers in ultraviolet-cured epoxy [41], the self-trapping and self-focusing observed in photopolymers such as a liquid diacrylate polymer [42, 43], self-writing in planar chalcogenide glasses [44], the interaction of self-written fibers in a photopolymerizable resin [45]–[47], micro-fabrication through single- and two-photon polymerization [48]–[52], and self-writing in undoped glasses by femtosecond infrared laser pulses [53]. Although microscopic details of the index-change mechanisms vary significantly among different materials, simple phenomenological models can be adapted for a broad range of optical materials.

The physics of self-writing in all cases is very similar to the physics of spatial solitons and is based on the self-action of light. Consider a Gaussian beam incident on a uniform planar waveguide (see Figure 14.6) made of a photosensitive material. This

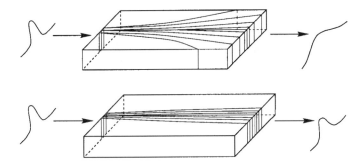

Figure 14.6: An optical beam incident on a photosensitive nonlinear material initially diffracts (top) but the refractive index begins to change in response to its intensity. Over time, the beam is guided by the waveguide (bottom) that it has self-written. (After Ref. [34]; ©2001 Taylor & Francis.)

photosensitivity produces the largest changes in refractive index at the location where the optical intensity is higher. The beam initially diffracts in the plane of the waveguide (top panel in Figure 14.6), but then it forms a region of raised refractive index along the propagation axis. This reduces the beam diffraction, and the beam begins to become confined to the channel it has created. If the writing beam is left on for some time, the refractive index distribution eventually evolves into a fairly uniform channel waveguide along the propagation axis (bottom panel in Figure 14.6). This is a self-written waveguide because the beam of light that creates the waveguide is subsequently guided by it. Such a self-writing process is not restricted to the planar waveguide geometry, and waveguides can also be formed in bulk photosensitive materials.

14.2.2 Theoretical Models

To understand the self-writing process, one should employ an appropriate model that describes the growth of self-written waveguides in a photosensitive material. The photosensitive process in both glasses and polymers occurs slowly relative to the transit time of light inside the material. For this reason, the propagation of a CW beam is governed by the following "stationary" wave equation for the field envelope [34, 54]:

$$i\frac{\partial A}{\partial z} + \frac{1}{2\beta_0}\nabla^2 A + k_0\Delta n(I,t)A + \frac{i\alpha}{2}A = 0, \tag{14.2.1}$$

where Δn represents the change in the refractive index because of the absorption by the medium governed by the attenuation coefficient α. The term containing Δn is time-dependent and governs the changes in the field evolution through the refractive-index changes resulting from photosensitivity. Note that the optical field itself depends on time implicitly.

This equation should be supplemented with an appropriate "material equation" describing how the nonlinear index change $\Delta n(I,t)$ varies with time and the field intensity $I = |A|^2$. In a simple phenomenological model of the self-writing process, the induced

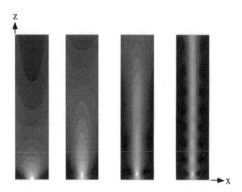

Figure 14.7: Guiding of a Gaussian beam as it creates a self-written waveguide inside a photopolymer. Intensity contours in the x–z plane are shown at $t/t_0 = 0$, 5, 10, and 15. The panel size is $100\lambda_0 \times 10\lambda_0$. (After Ref. [42]; ©1996 OSA.)

refractive index is assumed to change with time as [37]

$$\frac{\partial \Delta n}{\partial t} = B_p I^p, \tag{14.2.2}$$

where $p = 1$ or 2, depending on whether the index growth is governed by a one-photon or a two-photon absorption process. The material parameter B_p depends on p, properties of the photosensitive material, and the wavelength λ_0 of the writing beam [55]. Its values are typically positive ($A > 0$) because the refractive index increases with intensity in most materials [56]. Equation (14.2.2) holds for every point in space locally, but the index change depends on the temporal history of the illumination at that position.

Although the simple model based on Eq. (14.2.2) serves well when the index changes are relatively small, it is necessary to incorporate the saturation effects when index changes saturate to a maximum value Δn_s. A better model for the evolution of the refractive index uses [34]

$$\frac{\partial \Delta n}{\partial \tau} = B_p I^p \left(1 - \frac{\Delta n}{\Delta n_s} \right). \tag{14.2.3}$$

This model has been used for saturation in both photopolymers [42] and photosensitive glasses [57] and provides good agreement with the experimental data in both cases. It also applies to the formation of fiber Bragg gratings [36]. In the case $\Delta n_s \rightarrow \infty$ (no saturation), Eqs. (14.2.1) and (14.2.3) possess interesting self-similar solutions [58]. However, in general, they should be analyzed numerically [59]–[61].

In one set of numerical simulations [42], Eqs. (14.2.1) and (14.2.3) were solved using the split-step (or beam-propagation) method with the transparent boundary conditions. Figure 14.7 shows how a Gaussian beam propagating inside a photopolymer liquid became self-trapped after a time interval of $15t_0$, where t_0 represents the time needed to attain the critical exposure at the location of the beam center; t_0 depends on the material used and is typically $\sim$1 ms. In these simulations, the input beam width was $2\lambda_0$ and the propagation distance equals $500\lambda_0$, although only the first $100\lambda_0$ is

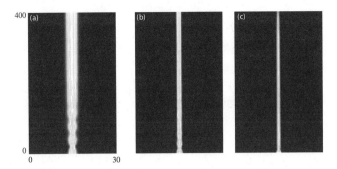

Figure 14.8: Final refractive index profile of the self-written waveguide for three normalized values of the intensity threshold I_{th}/I_0: (a) 0, (b) 0.01, and (c) 0.025. The other relevant parameters are $\Delta n_s = 0.02$, $t_0 = 1$ ms, and $w_0 = 1$ μm. (After Ref. [47]; ©2002 OSA.)

shown in Figure 14.7. Because the absorption length is only $103\lambda_0$, self-trapping stops after this distance. Note also that the horizontal scale in Figure 14.7 is magnified by a factor of 10 to improve the transverse resolution. The leftmost panel illustrates the initial state of beam propagation before any index changes are induced. Because the beam "writes" a waveguide by increasing the index more where it is more intense, diffractive effects are countered more and more by the waveguide. Almost complete self-trapping of the optical beam is apparent in the last panel in Figure 14.7. Oscillations in the beam diameter occur because during the index buildup phase, the waveguide becomes so strong that it begins to support more than one mode.

The photopolymerization process in a photosensitive resin is somewhat more complicated, for two reasons. First, the polymerization reaction is initially delayed [42]. Second, for a given exposure time, the polymerization occurs only in the regions where optical intensity exceeds a certain *threshold* value. The threshold nature of the process can be included by modifying Eq. (14.2.3) as follows [47]:

$$\frac{\partial \Delta n}{\partial \tau} = B_1 \left(1 - \frac{\Delta n}{\Delta n_s}\right) \begin{cases} I - I_{th}, & I \geq I_{th}, \\ 0, & I < I_{th}, \end{cases} \qquad (14.2.4)$$

where $p = 1$ was used assuming a one-photon absorption process.

The use of Eq. (14.2.4) provides an excellent qualitative agreement with the experimental data [47]. Figure 14.8 shows the results of numerical simulations in the one-dimensional geometry using an input Gaussian beam with amplitude $A(x,z) = \sqrt{I_0}\exp(-x^2/x_0^2)$ for three different values of I_{th}. The $I_{th} = 0$ case in part (a) shows that the self-written waveguide in this case is not uniform, exhibits width oscillations, and becomes broader away from the input face, features also observed experimentally in photosensitive glasses [57]. In contrast, the waveguide is quite uniform and its width is nearly constant when the threshold is taken into account, as seen in parts (b) and (c). This is exactly the behavior observed experimentally in the case of photopolymerizable resins [45].

Equations (14.2.1) and (14.2.4) possess stationary solutions (at $\alpha = 0$) that describe the properties of self-written waveguides in a self-consistent manner. Because of the saturation of the refractive index after long exposure to incident light, the final index profile has the following simple form:

$$\Delta n(x) = \begin{cases} \Delta n_s, & |x| < d/2, \\ 0, & |x| \geq d/2, \end{cases} \tag{14.2.5}$$

where d is the waveguide width. The optical modes of this waveguide should satisfy the self-consistency relation given in Eq. (14.2.4); i.e., $|A(x,z)|^2 < I_{th}$ for $|x| > d/2$ and $|A(x,z)|^2 > I_{th}$ for $|x| < d/2$. These relations uniquely define the optical field profile when the asymmetric modes are not excited and the waveguide supports only a single symmetric mode. For $d < 2\pi(2\beta_0 k_0 \Delta n_s)^{-1/2}$, the guided field is given by

$$E(x,z) = \sqrt{I_{th}} e^{i\kappa z} \begin{cases} \cos(qx)/\cos(qd/2), & |x| < d/2, \\ \exp[-Q(|x| - d/2)], & |x| \geq d/2, \end{cases} \tag{14.2.6}$$

where κ is the propagation constant of the mode and

$$Q = \sqrt{2\beta_0 \kappa}, \qquad q = \sqrt{2\beta_0(k_0 \Delta n_s - \kappa)}. \tag{14.2.7}$$

The propagation constant κ should be chosen to satisfy the continuity of the solution $E(x,z)$ at $|x| = d/2$. This condition leads to the standard eigenvalue equation $\tan(qd/2) = Q/q$. It turns out that a waveguide supporting a single symmetric mode can be formed if $I_{th} < I_0 < 7 I_{th}$. These analytical results explain the stabilization of the waveguide width as the threshold level is increased in numerical simulations.

14.2.3   Experimental Results

In a 1992 experiment [40], the self-writing process in ion-implanted $Bi_4Ge_3O_{12}$ glass was observed to produce an optically written waveguide. Independently, it was observed that permanent waveguide tapers can be formed in a commercially available epoxy (containing acrylic acid and hydroxypropyl methacrylate) because of permanent refractive-index changes induced in response to 532-nm radiation [41]. In this experiment a CW beam created the taper, which guided the beam through the epoxy. Because the index changes were long-lasting, tapers written in this way could subsequently be used at other wavelengths.

Several experiments have focused on the self-writing process occurring in photosensitive glasses. In a 1998 experiment [57], self-trapping of light was observed in a self-written channel created inside a germanosilicate glass waveguide through a two-photon absorption process. In this experiment, three layers of silica were deposited on the 11-mm-long silicon substrate. The 3-mm-thick photosensitive middle layer, containing 8% germanium, was sandwiched between two undoped silica layers whose refractive index was lower by 0.022. The structure was grown using the hollow-cathode, plasma-enhanced chemical vapor deposition (CVD) process that produced low-loss, nonporous silica glass with high intrinsic photosensitivity [62]. When irradiated with 488-nm light, the refractive index could be increased by 10^{-3}, which reduced the width of the propagating beam by a factor of 13.

The strong red luminescence observed during the writing process gave an indication of the beam intensity throughout the slab, and it was used to study the evolution of

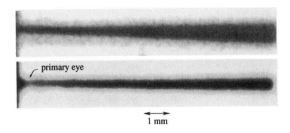

1 mm

Figure 14.9: Photographs showing luminescence distribution in the region inside (bottom) and outside (top) of the self-written channel in a photosensitive silica waveguide. (After Ref. [57]; ©1998 APS.)

the self-written waveguide. Figure 14.9 shows two photographs obtained using this luminescence. The top photograph shows the normal beam diffraction in the region away from the self-induced waveguiding channel. The bottom photograph shows how the beam spreading is reduced within the channel. Initially, the beam diffracts even in this region. However, after some time, the index change grows large enough to counteract the initial diffraction, and the beam is focused into a waist, developing the so-called "primary eye" [59]. Comparisons with numerical simulations indicate the qualitative agreement between the experiment and theory discussed earlier assuming that the absorption occurred through a two-photon process ($p = 2$) in the experiment [57].

The one-photon absorption process has also been used for writing permanent channel waveguides in photopolymers [46, 51]). In one experiment [46], optical waveguides were created in a photopolymerizing resin mixture using a multimode optical fiber for guiding the light. The self-written channel was formed by the selective photopolymerization of the higher-refractive-index monomer by an argon-laser beam. A continuous, straight waveguide can be grown by this method. This technique allows for the formation of three-dimensional optical circuits because it enables regrowth after passing through thick transparent glass plates. It is likely to enable automation of optical interconnections and packaging and could also be potentially useful for optical networks.

The photopolymerizing resins can be used to form multiple waveguides [45], and two such waveguides can be employed for studying soliton collisions. Experimentally, both waveguides were created at the same time by splitting a single laser beam from a He–Cd laser ($\lambda = 441.6$ nm) into two beams (using a beam splitter), which were then focused onto the photopolymerizing resin solution. The optical axes of two beams were made to intersect inside the solution so that the two beams are guided by the self-induced waveguides and eventually collide with each other inside the photosensitive material [47]. The collision process was observed using a CCD camera from a side of the sample cell. Figure 14.10(a) shows that when two self-written waveguides collide with each other, they can merge to form a single waveguide. The waveguide after merging grows along the direction that bisects the angle between the two waveguides that formed before the collision, because the input powers of two beams was equal (0.1 mW) and balance each other. In this case, the input beam width was only 0.96 μm. Similar to the case of two interacting solitons [33], the merging of self-growing waveg-

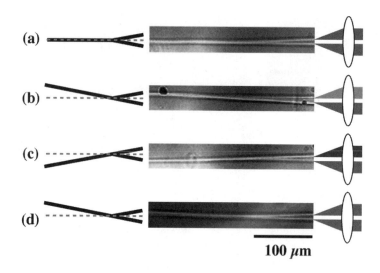

Figure 14.10: Experimental data showing the collisions between two self-written waveguides. The power of the lower beam is 0.1 mW; the power of the upper beam for the four parts is (a) 0.1 mW, (b) 0.07 mW, and (c) 0.13 mW, and (d) 0.1 mW. Only in case (d) is the upper beam delayed. The black lines on the left amplify the waveguide pattern, for clarity. (After Ref. [47]; ©2002 OSA.)

uides depends strongly on the collision angle between the waveguides [45]. In the resin used, the merging does not occur when the collision angle is larger than 9°.

Figure 14.10 shows the experimental results for a collision angle of 6.4° so that the growing waveguides always merge after collision. As seen there, the power ratio of two input beams can change the growth direction of the waveguide after merging. The power of one input beam is fixed at 0.1 mW, while that of the other beam is changed to 0.07 mW in (b) and 0.13 mW in (c). The merged waveguide follows the optical beam with the high input power, in contrast with the symmetric case shown in part (a) where the new waveguide is formed in the center. A similar effect is observed when one beam is delayed, as shown in part (d). In this case, the resulting waveguide follows the direction of the beam that enters the medium first. These features are in sharp contrast with the interaction of spatial solitons of unequal amplitudes [33]. The theoretical model based on Eqs. (14.2.1) and (14.2.4) has been used for studying the interaction between two self-written waveguides [47]. The numerical results agree well with the experimental observations shown in Figure 14.10. They also confirm that the two waveguides merge only when the collision angle is below a threshold value.

One may wonder what would happen if the input beam with a central dark region were sent though a photosensitive material. In one experiment, a phase mask was used to change the phase by π in the beam center [63], and the beam was sent though a thin film of a polymethyl methacrylate polymer doped with an organic dye (known as PMMA/DCM). It was found that a stable structure similar to a spatial dark soliton was formed, because the nonlinear photobleaching of the dye-doped polymer led to a *decrease* in the value of the local refractive index [64]. Such a dark soliton-like

Figure 14.11: Numerical simulation showing that a pipe structure in the center of a photosensitive glass can be created using a donut-shaped writing beam. Successive frames show the narrowing of the pipe induced through diffractive broadening of the writing beam. (After Ref. [65]; ©2002 OSA.)

structure trapped a probe beam and directed it in the form of a channel waveguide. The intensity of CW radiation required to build it was below 1 kW/cm^2. This intensity is comparable to that used for forming dark spatial solitons in photorefractive materials. One possible application of these permanent light-induced waveguides is in networks of optical interconnects between linear arrays of optical transmitters and receivers.

It was found in a 2002 experiment that illuminating the bulk Nd-doped BK-7 glass with a 488-nm laser beam *reduces* the refractive index by an amount $\sim 10^{-4}$ [65]. This effect has been used to demonstrate that, similar to an earlier work [63], it is possible to use self-writing to enhance the divergence of a Gaussian beam [65]. The simple model based on Eqs. (14.2.1) and (14.2.3) can be modified to reproduce numerically a decrease in the refractive index, and it is in good agreement with the experimental data. Numerical simulations were also performed in the case in which the input writing beam has the donut shape similar to that of the first-order Laguerre–Gauss mode. Figure 14.11 shows the changes in the refractive index at three different times. The low-index regions (white shaded) spread outwards over time as the beam begins to diffract more. However, this beam spreading also helps in narrowing the central high-index region, thereby creating a uniform channel waveguide in the center of the unexposed material with a relatively higher refractive index. The propagation characteristics of the channel formed were also investigated numerically by sending a Gaussian beam of a longer wavelength through it. It was found that this beam could be guided, showing that the pipe structure formed through index reduction can be used to guide light of different wavelengths [65].

14.3 Dissipative and Cavity Solitons

The types of optical solitons discussed so far are associated with an optical beam (or pulse) that propagates through the nonlinear medium once. If the nonlinear medium is placed inside an optical resonator (or cavity), the same beam passes through the nonlinear medium many times because of the feedback occurring at the cavity mirrors. Although the beam may experience substantial losses at the cavity mirrors, such losses can be balanced by an external beam. Spatial localized structures forming in such externally driven optical cavities are called *cavity solitons*. While cavity solitons share many

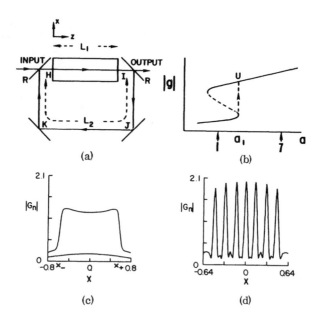

Figure 14.12: Formation of solitons in an optically bistable ring cavity: (a) schematic; (b) plane-wave bistability; (c) beginning of a soliton train after 23 roundtrips (lower curve shows the input Gaussian beam after a single pass); (d) steady-state train of seven solitons after 200 roundtrips. (After Ref. [68]; ©1983 APS.)

properties with standard spatial solitons, optical feedback also introduces several new features. For example, such solitons may exist even when the same nonlinear medium does not support any spatial solitons in the absence of the cavity. Since all cavities have losses and may even provide gain if the nonlinear medium is suitably pumped, such solitons fall in the general category of *dissipative solitons*. In this section, we discuss the basic physics of nondiffracting localized states forming in a nonlinear system with gain and loss.

Different types of localized structures can exist in driven and lossy nonlinear systems. Examples include lasers (vortices), lasers with a saturable absorber (bright solitons), parametric oscillators (phase solitons), and driven nonlinear resonators (bright and dark solitons) [66]. In all cases, solitons exist as self-trapped domains of one field state surrounded by another state of the field. Because cavity solitons result from a balance between the gain and the loss, one of the main differences between solitons in conservative and dissipative systems is the following: Whereas conservative solitons form *continuous families* of localized solutions, dissipative solitons are associated with certain *discrete values* of the parameters that satisfy the energy balance condition. Temporal cavity solitons can be created by using optical pulses, and they can be used to store images or information. The ability to control and manipulate cavity solitons externally is useful for many applications. For example, such solitons can be used as natural "bits" for parallel processing of optical information when they employ semiconductor microresonators.

14.3.1 History and Basic Physics

The simplest cavity that can support spatially localized structures and is easy to analyze theoretically is a ring cavity containing a self-focusing Kerr medium, as shown in Figure 14.12(a). The internal optical field propagates around the cavity and combines coherently with the driving beam at the input end. This nonlinear system is known to exhibit *optical bistability*, a phenomenon in which two outputs are possible for a given input under certain conditions. Part (b) of Figure 14.12 shows the standard *S*-shaped bistability curve for the output power as a function of input power. The middle branch of this curve with the negative slope is unstable, resulting in bistability with hysteresis. This curve corresponds to a continuous-wave (CW) beam whose amplitude is constant in the transverse dimensions (a plane wave).

The earliest analytical and numerical studies of the transverse spatial effects in a bistable cavity were performed by using a plane-wave or Gaussian-shape input beam [67]–[69]. Even when only one spatial dimension is included, the intracavity field envelope $A_n(x,z)$ exhibits interesting dynamics as the roundtrip number n is increased. When the input intensity is ramped up to exceed the upper switching threshold, the beam center switches, and a switching wave moves out, switching up most of the beam, as shown in Figure 14.12(c). The interface between the "on" and "off" domains spawns what would now be termed a *modulation instability* of the on region as it breaks up into a set of distinct peaks, as shown in Figure 14.12(d). These peaks in the modulated structure were interpreted by Moloney et al. [68] as an array of spatial solitons circulating inside the cavity. However, the number of peaks was nearly proportional to the cavity width, whereas proper cavity solitons are created and removed individually [70].

During the 1990s, a closely related concept of *diffractive autosolitons* was introduced from the study of switching waves in optical bistability [71]–[75]. Switching waves between the coexisting stable states are known in many fields (such as reaction-diffusion systems) and, in particular, they can exist as the self-trapped domains of one field state surrounded by another state of the field. Purely diffusive switching waves have monotonic profiles, but diffraction leads to *oscillating tails* or ripples. Having oscillating tails, such a switching wave can trap another similar wave, resulting in the formation of various types of bound states; several examples of such bound states are shown in Figure 14.13. In higher dimensions, one switching wave might bend around and close on itself, forming a stable island of one phase surrounded by the other, similar to a diffractive autosoliton. This approach provides another physical interpretation of cavity solitons: They can be treated as *self-trapped switching waves* in gain–loss systems with multiple stable equilibrium states.

In the 1983 study of cavity solitons [68], an input beam launched with intensities below the switching threshold did not form solitons although multiple peaks appeared above the threshold because of modulation instability. Later, it was shown that it was possible to switch individual solitons independently [71, 76], e.g., by using a pump beam with a spatially varying amplitude. Numerical simulations showed the possibility of switching on and off 20 memory bits with an out-of-phase address pulse [77]. These early studies are said to belong to the "Stone Age" in the short history of cavity solitons [70]—the transition to the "Modern Age" occurring around 1990 [78]. Recent advances include experimental observations, stability analysis, and engineering

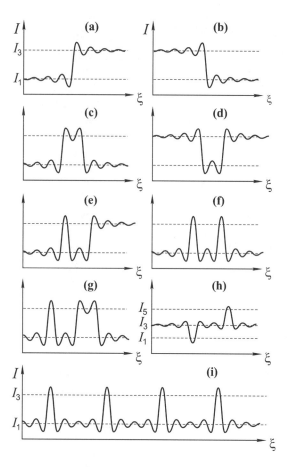

Figure 14.13: Transverse profiles of dissipative localized structures in a bistable system: (a, b) "left" and "right" switching waves; (c, d) positive and negative dissipative solitons; (e) switching wave coupled to a soliton; (f, g) symmetric and asymmetric two-soliton structures; (h) a bound state of positive and negative solitons; and (i) a periodic array of dissipative solitons. (After Ref. [66]; ©2002 Springer.)

applications of cavity solitons [70, 79].

14.3.2   Kerr and Kerr-like Media

Unlike the earlier numerical studies of localized structures forming inside a bistable ring cavity [68, 69], the modern approach is based on the so-called *mean-field cavity model*. In the framework of this approach, propagation around the cavity is replaced by a single partial differential equation with both driving and loss terms. The mean-field cavity model was first introduced in 1987 in the context of spatial pattern formation [80] and is based on the perturbed NLS equation and its modifications.

The mean-field model can be understood with a help of simple physics. The conventional NLS equation describes an infinite nonlinear medium. In a cavity, the propagation occurs between mirrors placed around the nonlinear medium, which confine the beam to a finite slab of material. Real mirrors and materials are lossy, but the loss can be compensated by "feeding" the confined beam in a cavity with an input field. Thus, an appropriate NLS-like model should include both the external-field and loss terms. After averaging over a cavity roundtrip and proper normalization, one obtains the following perturbed NLS equation:

$$i\frac{\partial u}{\partial t} + \nabla^2 u + |u|^2 u = i\varepsilon[-(1+i\theta)u + u_{in}], \tag{14.3.1}$$

where ε is a small perturbation parameter, u_{in} represents the external field, and θ is related to the detuning of the field frequency from the cavity resonance. An important change from the usual NLS equation is that the propagation coordinate z is replaced by evolution in time t, related to the roundtrip time inside the cavity: Changes in t indicate how $u(x,t)$ evolves over multiple roundtrips. The replacement of z with t makes physical sense because the solitons are trapped inside a cavity of finite length.

The perturbation terms include cavity losses ($\varepsilon > 0$) and the frequency mismatch or the detuning θ between the cavity field u and the driving field u_{in}. Time is scaled to the cavity response time. Equation (14.3.1) applies in the case of a low-loss or high-finesse cavity and assumes that only one longitudinal mode is excited [70, 81]. It can be generalized easily to include the nonlinearities associated with a two-level atomic medium, which becomes Kerr-like far from an atomic resonance [82, 83]. When the external field is in resonance with the atomic system, the medium acts like a saturable absorber with no nonlinear contribution to the refractive index.

An NLS equation, such as Eq. (14.3.1), with complex coefficients is usually called the *Ginzburg–Landau equation* and has been studied extensively because it describes the physics of a vast variety of phenomena, ranging from second-order phase transitions, superconductivity, superfluidity, and Bose–Einstein condensation to liquid crystals and strings in field theory [84]. In the temporal domain, the second term in Eq. (14.3.1) accounts for dispersion rather than diffraction and is appropriate for *fiber cavities*, where soliton-like structures in synchronously pumped fiber loops have been found and analyzed [85, 86].

For $\varepsilon = 1$, Eq. (14.3.1) is similar to that introduced in Ref. [80]. In this case, it has exact solutions when the driving field is homogeneous and has the form $u_s = u_{in}/[1 + i(\theta - |u_s|^2)]$. For $\theta < \sqrt{3}$, this implicit solution is single-valued but three solutions exist for $\theta > \sqrt{3}$, indicating the existence of optical bistability. It turns out that a solution is unstable if $|u_s| > 1$, an instability that leads to spontaneous pattern formation [80]. The basic spatial entity whose repetitions form the pattern is related to a cavity soliton. The shape s of this soliton can be found numerically by solving Eq. (14.3.1) with the ansatz $u = u_s[1 + s(r, \phi)]$, where r and ϕ are the cylindrical coordinates. Figure 14.14 shows an example of a two-dimensional cavity soliton found numerically in the radially symmetric case. It consists of a bright peak on a flat background u_s, with a few weak diffraction rings. The period and the damping rate of these rings are set by the tails of an isolated cavity soliton and the fact that these tails are governed by a pair of the generalized Bessel functions [87]. The rings become much more pronounced as

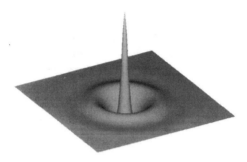

Figure 14.14: Spatial structure of a cavity soliton shown by plotting $u(x,y)$ for $\theta = 1.2$ and $|u_s|^2 = 0.9$. (After Ref. [81]; ©2002 OSA.)

$|u_s| \to 1$. To compare the cavity solitons with the periodic solutions leading to patterns, we show in Figure 14.15 the branches for both homogeneous (spatially uniform) solutions and the two-dimensional cavity solitons. The cavity-soliton branch appears at a *subcritical bifurcation point*. Hence, a cavity soliton can be interpreted as a single spot of the hexagonal pattern.

Stability of cavity solitons is a crucial issue and should be examined carefully. A linear stability analysis shows that the lower branch of the bistability loop is always unstable. The upper branch may become stable in some parameter domain. The shaded area in Figure 14.15 shows the stability domain. The onset of instability for large values of θ is related to a Hopf bifurcation but not to the collapse, as was the case for the cubic NLS equation (without cavity). More specifically, a pair of complex-conjugate eigenvalues crosses the imaginary axis for $|u_{in}|^2 < 1$. Numerical simulations confirm the linear stability analysis. A perturbed cavity soliton exhibits damped oscillations in the stable domain, which become undamped as the stability boundary is crossed. For large θ, the cavity soliton either decays or becomes very narrow, suggestive of the collapse, much as the two-dimensional Kerr solitons do.

More recent studies have confirmed the absolute stability of cavity solitons supported by Eq. (14.3.1) for small cavity detunings [81]. However, these studies also show that, in parts of the domain in which a cavity soliton is stable radially, it can become unstable azimuthally. The character of the radial instability is explained by a Hopf bifurcation [88], and this instability leads to oscillations of a cavity soliton whose amplitude increases with the background intensity. The oscillating soliton is rather robust, and it neither collapses nor decays, even well above the Hopf bifurcation threshold. The azimuthal instability depends on the mode number m for the mode, whose phase varies as $\exp(\pm im\varphi)$. It turns out that typically $m = 5$ or 6 for the unstable mode. The resultant dynamics lead to the formation of an expanding pattern, which maintains a fivefold or sixfold symmetry as it grows for $m = 5$ and 6, respectively. In both cases, the emerging pattern is dynamical, in the sense that each spot oscillates with a location-dependent phase. This behavior of cavity solitons clearly indicates that cavity solitons are qualitatively different from ordinary spatial solitons of conservative (or even weakly perturbed) systems.

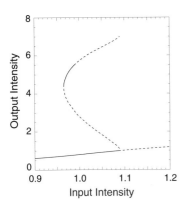
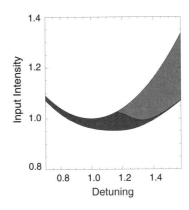

Figure 14.15: (left) Branches showing the homogeneous solution and the cavity-soliton solution for $\theta = 1.3$. (right) Stability domain of the two-dimensional cavity soliton (shaded region) in the parameter space. (After Ref. [70]; ©2001 Springer.)

14.3.3 Semiconductor Microresonators

A semiconductor microresonator, such as shown in Figure 14.16, is commonly used for vertical-cavity surface-emitting lasers, also called VCSELs [89]. It consists of a thin central region that is sandwiched between two distributed Bragg reflectors (DBRs). The central region is made of multiple quantum-well (MQW) GaAs/AlGaAs layers. Such a structure is quite useful for cavity solitons because of its microscopic size; the cavity length is approximately 1 μm, even though the transverse size can exceed 1 cm. Such short-length, wide-area microresonators support only one longitudinal mode but an enormous number of transverse modes. This situation allows for a very large number of cavity solitons to coexist. The DBR mirrors used to form the resonator are of the plane-mirror type, resulting in frequency degeneracy for all transverse modes, and thus allowing arbitrary field patterns to exist inside the resonator.

The phenomenological model used for a driven, wide-area semiconductor microresonator [90]–[92] is somewhat different from that based on Eq. (14.3.1). It describes the optical field u inside the resonator using the mean-field approach, just as Eq. (14.3.1) does. However, nonlinear absorption and refractive-index changes, induced by the intracavity field in the vicinity of the MQW band edge, are assumed to be proportional to the carrier density N. The rate equation for N includes the effects of nonresonant pumping P, carrier recombination, and carrier diffusion. The resulting equations describing the spatiotemporal dynamics of u and N have the following normalized form:

$$\frac{\partial u}{\partial t} = u_{\text{in}} - \sqrt{T}\{[1 + C\alpha''(1 - N)] + i(\theta - C\alpha'N - \nabla^2)\}u, \qquad (14.3.2)$$

$$\frac{\partial N}{\partial t} = P - \gamma[N - |u|^2(1 - N) - d\nabla^2 N], \qquad (14.3.3)$$

where the complex parameter $\alpha = \alpha' + i\alpha''$ governs both the absorptive and refractive nonlinearities, C takes into account saturable absorption (scaled to the DBR transmission T and assumed to be small), θ represents detuning of the driving field from a

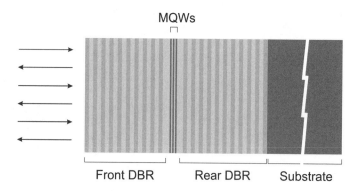

Figure 14.16: Schematic of a semiconductor microresonator consisting of multiple quantum wells sandwiched between two distributed Bragg reflectors. (After Ref. [100]; ©2002 IEEE.)

resonator resonance, γ is the ratio of the photon lifetime to the carrier lifetime, and d is related to the diffusion coefficient. The cavity is driven by an external field u_{in}. This model is quite successful in describing cavity solitons. Sometimes it becomes important to include the thermal effects, but the analysis becomes much more complicated because it requires an additional equation for the temperature distribution [93].

Linear effects within a resonator spread the input beam (CW or pulsed) because of diffraction and dispersion. The material nonlinearity that can balance this linear spreading can come from several sources. The nonlinear changes in the resonator finesse (because of absorption saturation) is one possibility. The other possibility is related to the change in the resonator optical length because of nonlinear changes in the refractive index. Both of these constitute what are called the *longitudinal nonlinear effects* [94]. The transverse effect of the nonlinear refractive index can be self-focusing or self-defocusing, favoring the formation of bright and dark solitons, respectively. Absorption (or gain) saturation can also lead to the transverse effects through the phenomenon of spatial-hole burning. The longitudinal and transverse effects can work in opposition or they can cooperate, depending on the device parameters. The two main external control parameters are the driving-field intensity $|u_{in}|^2$ and the resonator detuning θ. At input intensity levels below the resonance threshold for the whole beam, the light is accumulated in isolated spots where the intensity is high enough locally to reach the resonance condition, resulting in the formation of a bright–dark pattern.

Figure 14.17 shows the setup for a 2001 experiment performed using semiconductor microresonators consisting of multiple GaAs/AlGaAs layers [95], sandwiched between the two high-reflectivity (≥ 0.995) DBRs (see Figure 14.16). The microresonator structures were grown on a GaAs substrate by using a molecular-beam-epitaxy technique that makes it possible to fabricate high-quality MQW structures with negligible thickness variations in the radial direction. The best sample had <0.3 nm/mm variation in the resonance wavelength over the entire sample cross section. A tunable Ti:sapphire laser or a single-mode laser diode, operating in CW regime, was used for injecting a 50-μm-diameter beam into the sample (Fresnel number > 100). The thermal effects mitigated by using an acousto-optic modulator.

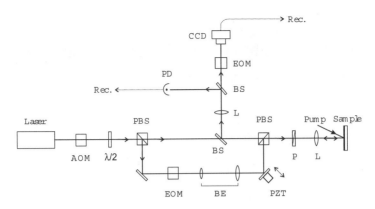

Figure 14.17: Experimental setup for producing localized structures in semiconductor microresonators. AOM: acousto-optic modulator; $\lambda/2$: half-wave plate; PBS: polarization beam splitters; EOM: electro-optical modulator; BE: beam expander; PZT: piezo-electric transducer; P: polarizer; L: lense; BS: beam splitter; PD: photodiode; CCD: camera. (After Ref. [95]; ©2001 Springer.)

Part of the laser light is split away from the driving beam and then superimposed with the main beam using a Mach–Zehnder interferometer arrangement, to serve as the address beam. The address beam is sharply focused and directed to a specific location on the illuminated sample. The switching light is opened only for a few nanoseconds using an electro-optic modulator. In the case of incoherent switching, the polarization of the address beam was orthogonal to that of the main beam to avoid interference. In the case of coherent switching, the two polarizations were parallel, and the phase of the switching beam was changed for switching the soliton on and off. One of the interferometer mirrors could be moved by a piezoelectric transducer to control the phase difference between the driving and the address beams. Optical pumping of the MQW resonator was done by using a multimode laser diode (or a single-mode Ti:sapphire laser).

To find the most stable localized structures, one can play with the nonlinear response (absorptive versus dispersive) by changing the driving-field wavelength, with the resonator detuning or with the carrier population created through pumping. When working $\sim$30 nm below the band edge, spontaneous formation of hexagonal patterns was observed. The hexagon period scaled with the detuning as $\theta^{-1/2}$ [96], indicating that the tilted-wave mechanism [97] was responsible for the formation of hexagons [98]. Moreover, hexagons with a dark spot were converted into bright-spot hexagons when the driving-beam intensity was increased. At this intensity level, individual spots of these patterns could not be switched on or off independently, as expected for a strongly correlated spot structure.

When the driving-beam intensity was increased further, the bright spots in such hexagonal patterns could be switched independently by using orthogonally polarized (incoherent) optical pulses [96]. Thus, the hexagonal pattern was not coherent at high intensity levels. In fact, individual spots acted independently even when they were so densely packed that the distance between any two was close to the spot size. These

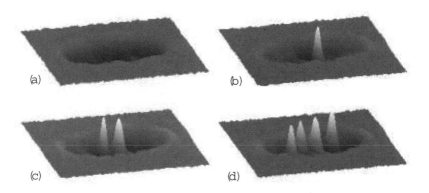

Figure 14.18: Switched beam observed in reflection at four input intensity levels. Switching of the beam to the upper branch of the bistability curve in part (a) is followed by formation of one or more localized structures in parts (b) to (d) as intensity is increased. (Courtesy C. O. Weiss.)

experimental findings can be understood using the model based on Eqs. (14.3.2) and (14.3.3) under the experimental conditions [99, 100]. Indeed, it was found that at high intensity levels, the resonator has a rather wide freedom to arrange self-consistently its field structure, thus allowing for a large number of possible stable patterns among which the system can choose.

When the wavelength of the driving field is tuned close to the band edge, the semiconductor microresonator supports individual localized structures [101]. Figure 14.18 shows four examples of such structures observed in reflection (a dip indicates a peak in transmission). Part (a) shows the switched domain of the optical beam on the upper branch of the bistability curve. Parts (b) to (d) show how one, two, and four peaks form spontaneously as the input intensity is increased. The noteworthy point is that that the shape and size of the bright spots are independent of the shape and intensity of the driving beam. However, there was no experimental evidence that the spots were switchable objects; they can be regarded as *precursors of cavity solitons*.

Nonlinearity of the MQW structure near the band edge is predominantly absorptive. Therefore, to the first-order approximation, the refractive part of the nonlinearity in Eq. (14.3.3) can be neglected. Numerical simulations shown in Figure 14.19 confirm the existence of both bright and dark cavity solitons. This figure shows the existence domains for bright and dark solitons in the parameter space formed using the detuning and the incident intensity. The domain of optical bistability is also shown by dashed lines. The cavity spatial solitons of Figure 14.19 should be contrasted with the conventional spatial solitons in a bulk nonlinear material: The latter cannot be supported by a saturable absorber. At excitation above the bandgap, bright solitons form in a similar manner [102]. The numerical analysis also shows that an increase in the pump intensity leads to shrinking of the existence domain of cavity solitons. Moreover, solitons form at lower intensity levels. When pump intensity approaches the transparency point of the semiconductor material, the cavity solitons cease to form [103]. However, they reappear above the transparency point. In the experiment, the contribution of the imaginary part of the complex nonlinearity was quite strong near the band edge. Since

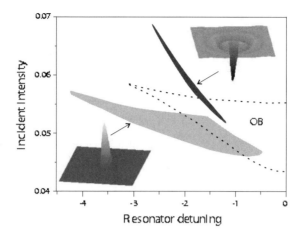

Figure 14.19: Numerically calculated domains for the existence of bistability (dashed line), bright solitons (gray area), and dark solitons (black area). Insets show examples of bright and dark solitons. Parameters are $C = 20$, $T = 0.005$, $\alpha'' = 1$, and $d = 0.01$. (Courtesy C. O. Weiss.)

the transparency point is quite close to the lasing threshold, inversion without lasing is difficult to realize.

The ability to switch cavity solitons on and off and to control their location and motion by applying optical pulses suggests that such solitons should be useful as pixels for making reconfigurable arrays and all-optical processing units. Such an ability was demonstrated in a 2002 experiment by pumping a semiconductor microresonator electrically above the transparency point but slightly below the lasing threshold [104]. The device had a large diameter (150 μm) to ensure that its boundaries did not affect the individual control of pixels. The injected field was tunable in the range 960–980

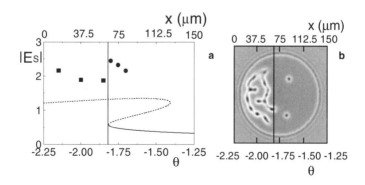

Figure 14.20: Intracavity field as a function of detuning θ (left) and pattern formation in the region where no stable state exists (right). Dashed lines show the unstable part, and the vertical line separates the stable and unstable regions. Circles and squares show the maximum intensity in the two regions, respectively. Black spots in the stable region show cavity solitons. (After Ref. [104]; ©2002 Nature.)

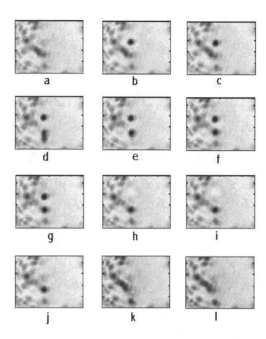

Figure 14.21: Controlled manipulation of cavity solitons using a writing beam. Different panels show how two cavity solitons can be created at different locations by turning on the writing beam and then erased by the same writing beam with a π phase shift. (After Ref. [104]; ©2002 Nature.)

nm, and its intensity was varied using an acousto-optic modulator or a polarizer. The detuning parameter θ varied along the sample width because of a gradient in the cavity length. Figure 14.20 shows the calculated intracavity field as a function of the detuning parameter θ. Only the solid part of the curve is stable. Since no stable steady-state exists for $\theta < -1.81$, one expects pattern formation, because of the onset of the modulation instability, only in the left part of the sample. This is exactly what is seen in the numerical simulations shown on the right and what was observed experimentally. In the homogeneous region, two cavity solitons were turned on by applying two 6-ns pulses at different times.

Figure 14.21 shows how such cavity solitons can be turned on and off by using a writing beam with 50-μW average power and a well-controlled phase. The 12 frames show the intensity distribution of the output field over a $60 \times 60 \ \mu m^2$ region in the sample center under different writing conditions. The holding beam is always on (power 8 mW), and all other parameters are kept constant. Frame (a) shows the initial situation with no cavity solitons. A single cavity soliton (10-μm spot size) is turned on in frame (b). As seen in frame (c), the soliton persists even after the writing beam is blocked. Frames (d) to (f) show how the writing beam creates a second cavity soliton at a different location. In frames (g) to (i), one of the spots is erased by changing the phase of the writing beam by π. Frames (j) to (l) show that the second spot can also be erased using the same approach. These results clearly indicate that cavity solitons can serve as pixels that can be written, erased, and manipulated as objects independent of each other and of the boundary.

14.3.4 Parametric Cavity Solitons

We have seen in Chapter 10 that parametric solitons can exist in the case of a quadratic nonlinearity as a coupled structure of the fundamental and second harmonics. It is natural to expect that similar solitons can exist in planar resonators in the form of *parametric cavity solitons*. Resonators with quadratically nonlinear media are used routinely for frequency up- and down-conversion [105, 106]. A driving field at the fundamental frequency in such a cavity can generate the second-harmonic while forming two-frequency parametric solitons inside a quadratically nonlinear medium, some of which may turn out to be stable [107]–[112]. A further generalization considers the case of vectorial cavity solitons in nondegenerate optical parametric oscillators [113, 114].

In a planar resonator containing a quadratically nonlinear medium, the situation is similar to the case of an optical parametric oscillator (OPO). The frequencies of the incident fundamental field and the generated second harmonic are often close to a cavity resonance. One can then employ a well-established modal theory whose use simplifies the analysis considerably compared with the approach based on forward- and backward-propagating fields. The appropriately scaled evolution equations for the transmitted fields A and B at the fundamental and second harmonics can be written in the form [107, 111]

$$i\frac{\partial A}{\partial t} + \nabla^2 A + (i\gamma_A + \Delta_A)A + A^*B = A_{\text{in}}, \tag{14.3.4}$$

$$i\frac{\partial B}{\partial t} + \alpha\nabla^2 B + (i\gamma_B + \Delta_B)B + A^2 = 0, \tag{14.3.5}$$

where ∇^2 is the transverse Laplacian, γ_j and Δ_j are, respectively, the cavity decay rate and frequency detunings for the two fields ($j = A, B$), and α is the ratio of the diffraction lengths at the fundamental and second-harmonic frequencies. The fields are scaled in terms of the effective nonlinear coefficients arising from the second-order susceptibility and the overlap integrals entering into the modal theory. The absolute value of the overlap integrals depends critically on the phase mismatch between the fundamental and the second harmonics. Large values of detuning for the second harmonics produce focusing (for $\Delta_B < 0$) or defocusing (when $\Delta_B > 0$). This can be seen by neglecting the derivatives in Eq. (14.3.5) for large Δ_B and substituting the resulting B in Eq. (14.3.4).

We are interested in bright cavity solitons that form on a finite background. Considerable insight can be obtained from the plane-wave solutions of Eqs. (14.3.4) and (14.3.5) obtained by setting all derivatives to zero. The resulting cubic polynomial for the fundamental-field amplitude has three real solutions in a certain parameter range set by the conditions [107]

$$\frac{|\Delta_B|(|\Delta_A| - \sqrt{3})}{\sqrt{3}|\Delta_A| + 1} > \gamma_B, \qquad \Delta_A\Delta_B > 0, \tag{14.3.6}$$

where $\gamma_A = 1$ was assumed. Using a linear stability analysis with spatially homogeneous spatially modulated perturbations, one can show that these conditions correspond

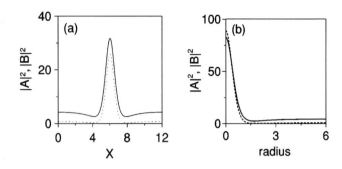

Figure 14.22: Amplitude profiles of (a) one-dimensional and (b) two-dimensional cavity solitons for $\Delta_A = -3$, $\Delta_B = -5$, $\gamma_A = 1$, $\gamma_B = 0.5$, and $A_{in} = 5$. (After Ref. [111]; ©2001 APS.)

to the standard optical bistability behavior. When one considers spatially modulated perturbations, the lower branch is found to be stable, but the upper branch is unstable to such perturbations.

The localized structure of a parametric cavity solitons can be calculated numerically by solving Eqs. (14.3.4) and (14.3.5). Such solitons exist both in one- and two-dimensional geometries. Figure 14.22 shows the soliton shape in the two cases. The soliton is assumed to be radially symmetric in the two-dimensional case, for which the peak intensity is also much higher. This can be understood in the cubic limit. If diffraction of the second-harmonic wave can be neglected, the process of up- and down-conversion generates a phase shift of the fundamental wave that is similar to the one produced by a cubic nonlinearity. In the presence of a cubic nonlinearity, the two-dimensional beams tend to collapse, whereas one-dimensional field distributions are stable. Consequently, in the presence of a quadratic nonlinearity, much more energy is accumulated at the center of a two-dimensional cavity soliton. Note that the collapse is easily eliminated by the combined action of losses and quadratic nonlinearity. In both one and two dimensions, the parameter domain where cavity solitons exist is quite limited.

The stability of parametric cavity solitons can be checked numerically using the split-step or beam-propagation method [107]. At the point where the localized soliton structure exists, the plane waves become modulationally unstable, with either a finite or infinite period. In the latter case, this occurs at a limit point (the same as homogeneously stable). The branch of stable solitons ends where the background destabilizes, i.e., at the same point in the parameter space where such solitons form. In effect, such solitons are the substitute for the plane-wave hysteresis, even outside the range of optical bistability.

Parametric cavity solitons can exist in more complicated settings. For example, they were found to form in the vectorial case in which two orthogonally polarized fundamental fields combine to generate a second harmonic of either polarization through the type II phase-matching process inside a high-finesse Fabry–Perot cavity containing a quadratic nonlinear material. The model is similar to the OPO case, but one should

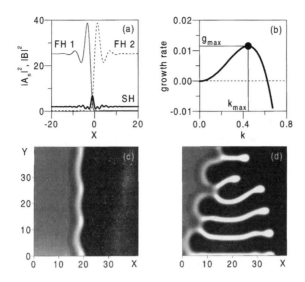

Figure 14.23: (a) Parametric cavity soliton formed when two orthogonally polarized fundamental harmonics (FH 1 and FH 2) create a polarization front. Parameters used are $\Delta_A = 1$, $\Delta_B = 1.5$, and $\gamma_A = \gamma_B = 1$. (b) Gain of transverse modulation instability for the one-dimensional soliton. (c) Snake-like instability of the polarization front at $T = 320$. (d) Decay of the polarization front at $T = 560$. (After Ref. [113]; ©1998 APS.)

include the two polarization components for the driving field acting as pumps [115]:

$$i\frac{\partial A_j}{\partial t} + \nabla^2 A_j + (i\gamma_A + \Delta_A)A_j + A_{3-j}^* B = A_{j\text{in}}, \tag{14.3.7}$$

$$i\frac{\partial B}{\partial t} + \alpha\nabla^2 B + (i\gamma_B + \Delta_B)B + A_1 A_2 = 0, \tag{14.3.8}$$

where the notation is the same as used earlier except that $j = 1, 2$ to account for the two polarization components. As usual, the instabilities of stationary plane-wave solutions serve as a point of departure for the identification of transverse localized structures. The most interesting structures in this model are associated with the symmetry-breaking effects and occur when the two input field components have the same intensity.

There are two distinct but mathematically identical situations, depending on whether $A_1 > A_2$ or $A_1 < A_2$. Because these states can coexist, the transition between them can lead to the formation of a one-dimensional topological soliton, similar to the domains of different magnetization in a ferromagnetic material. Figure 14.23(a) shows this soliton by plotting the intensities for the two pump components (solid and dashed curves) and the second-harmonic field (thick solid curve). Recalling that the two pumps are orthogonally polarized, it is easy to see that the central region corresponds to a *polarization front*. With only one transverse degree of freedom, (e.g., in a thin-film waveguide resonator), this soliton is stable, and it stays in the middle because of its symmetry with respect to the two pump components. However, when the second transverse dimension is included, the polarization front turns out to be modulationally unstable with respect to periodic perturbations, which grow exponentially in time. The growth rate $g(k)$ of these perturbations is plotted in part (b) as a function of the modulation wave number k. For $k = 0$ the gain is zero because the structure is stable in the one-dimensional case. The exponential growth results mainly in a snake-like instability of the entire front, as shown in part (c). The polarization front eventually decays, resulting in a complicated moving structure such as shown in part (d).

The filamentary structures seen in Figure 14.23(d) possess many interesting properties, the most important being that they can be stable in two transverse dimensions [113]. It can be interpreted as a stable, growing stripe, identical to the one-dimensional soliton, with two-dimensional moving heads at both ends. If such solitons move, their interaction and collision behavior becomes a critical issue. Numerical simulations on collision experiments indicate that such moving solitons are quite robust [113]. A central collision of two moving solitons results in an unconventional final state with the following features. Prior to the actual collision, the two solitons halt and form a localized state consisting of two truncated resting solitons. After an off-axis collision they do not fuse or penetrate each other as observed in conservative systems; rather, they simply try to avoid close contact. Interaction with a moving truncated soliton induces complete decay of unstable solitons starting at the collision site; moving solitons are emitted, alternating between the two sites. A tree-like structure eventually develops and starts to cover the whole plane with a roll pattern. The so-called *walking cavity solitons* have also been studied in the context of a degenerate OPO operating below threshold [112]. Such solitons exist in the presence of nonvanishing walk-off, and they resemble the walking quadratic solitons discussed in Chapter 10.

14.4 Magnetic Solitons

The concept of spin waves as the dynamic excitations of magnetic media was introduced by Bloch [116], who studied how small perturbations of the magnetic momentum propagate as waves through a magnetic medium. Since the spin waves are almost entirely determined by magnetic dipole interactions, they are usually called *dipolar magnetostatic spin waves*. The frequency ω of a spin wave depends on the orientation of its wave vector $\mathbf{q}$ because of the anisotropic nature of the magnetic dipole interaction. For large values of $\mathbf{q}$, the magnetic spin-wave exchange interactions cannot be neglected, and the resulting spin waves are referred to as *dipole-exchange spin waves*. This section considers solitons associated with spin waves under such conditions.

14.4.1 Nonlinear Spin Waves

Spin waves provide the basis for describing the spatial and temporal evolution of the magnetization distribution in a magnetic medium, under the general assumption that the length of the magnetization vector is constant locally. This condition is fulfilled when the magnetic sample is magnetized to saturation by an external bias magnetic field, and it forms a single domain state. The dynamics of the magnetization vector is then described by a nonlinear Landau–Lifshitz torque equation [117]. In a magnetic film of finite thickness, the spin-wave spectrum is modified because the translational invariance is broken in the vicinity of the film surface. As a result, the dispersion relation of spin waves, $\omega(\mathbf{q})$, depends on the normal modes of the film and acquires dispersive properties. The net result is that such a magnetic film acts like a nonlinear dispersive medium for the propagation of *nonlinear spin waves* and can support solitons known as *magnetic solitons*.

From the experimental point of view, nonlinear spin waves at microwave frequencies can be observed in monocrystalline ferrite films; an example is the yttrium-iron garnet (YIG) film made of the magnetic material $Y_3Fe_5O_{12}$, with a very low ferromagnetic resonance line width [118]–[120]. Moreover, the nonlinear and dispersive characteristics of spin waves can be controlled by changing the magnitude and orientation of the bias magnetic field. A wide variety of nonlinear phenomena, including self-focusing of spin-wave packets and the formation, propagation, and collision of bright and dark magnetic solitons, have been observed in ferrite films at moderate microwave power levels (<1 W).

Particularly interesting are the spin-wave processes in films that are magnetized in the film plane, because they show a strong intrinsic anisotropy for waves having different relative orientations of their in-plane wave vector q_z and the saturation magnetization. The ratio between the parameters describing diffraction and dispersion is much smaller in tangentially magnetized magnetic films than that found in optical systems. This feature makes it easy to study the diffractive and dispersive effects simultaneously, resulting in the formation of spatiotemporal localized structures, or *magnetic bullets* [121]. When the magnetization m in the direction perpendicular to the z axis increases, the magnitude of the magnetization along the z axis, M_z, is reduced to

$$M_z = M_s \sqrt{1 - (m/M_s)^2} \approx M_s(1 - U^2), \qquad (14.4.1)$$

where U is the dimensionless amplitude of the variable magnetization. If the spin-wave packet is narrow, we can expand the nonlinear dispersion relation $\omega(q_z, |U|^2)$ near the point (ω_0, q_0), corresponding to the carrier wave. Similar to the case of light waves, this procedure leads to the following $(2+1)$-dimensional NLS equation:

$$i\left(\frac{\partial U}{\partial t} + V_g \frac{\partial U}{\partial z}\right) + \frac{D}{2}\frac{\partial^2 U}{\partial z^2} + S\frac{\partial^2 U}{\partial y^2} - N|U|^2 U = -i\omega_r U, \qquad (14.4.2)$$

where $V_g = (\partial\omega/\partial q_z)$ is the group velocity, D and S are, respectively, the dispersion and diffraction parameters, N is the nonlinear coefficient, and ω_r is the dissipation parameter, proportional to the ferromagnetic resonance line width. Note that, in contrast with the optical case, the propagation variable is time t. The magnetization vector lies in the y–z plane, which coincides with the film plane.

The coefficients V_g, D, S, and N in Eq. (14.4.2) can be calculated for different directions of the spin-wave propagation [118]. A more rigorous calculation of the nonlinear coefficients uses the Hamiltonian formalism [122]. The $(2+1)$-dimensional NLS equation (two in-plane coordinates plus time), Eq. (14.4.2), is used extensively for describing the nonlinear effects associated with the dipolar spin waves. The link to the optical problems of self-trapping and self-focusing follows from the similarity of the governing NLS equations. The dissipation contribution, often negligible in optical problems, plays an important role in the case of magnetic films [123].

14.4.2   Bright and Dark Magnetic Solitons

For dipolar magnetostatic spin waves propagating along a bias magnetic field in a tangentially magnetized YIG film, the nonlinear coefficient N is negative in Eq. (14.4.2), while both the dispersion (D) and the diffraction (S) parameters are positive [124]. Such spin waves fulfill the criterion for modulation instability [125] in both in-plane directions because $SN < 0$ and $DN < 0$. As a result, temporal modulations (leading to the formation of temporal solitons) as well as spatial modulations (leading to the formation of spatial solitons) can occur. Both of these effects have been observed in YIG films acting either separately or together in the form of spatiotemporal self-focusing [126]–[131].

Modulation instability in magnetic films was first observed in a 1988 experiment [126]. Using pulsed excitation, spin waves were found to propagate in the form of envelope solitons. It was established later that the modulation instability can intensify weak spin waves, excited at a frequency close to the carrier frequency of the pump spin wave [132]. This phenomenon of induced modulation instability has been observed in ferromagnetic films [133]. Both modulation instability and soliton formation are strongly affected by dissipation inside the magnetic medium. The dissipation not only changes the shape of the threshold curve but also determines the threshold level. The calculated threshold curve for the formation of magnetic solitons from input rectangular pulses in the presence of weak dissipation [123] agrees well with that measured in YIG films.

If the YIG film is relatively narrow, a spin-wave packet (the equivalent of an optical pulse) is trapped in the transverse direction, and the film becomes an effective waveguide, somewhat resembling an optical waveguide. A spin-wave packet excited at the input of this waveguide will propagate with the spin-wave group velocity while being affected by both dispersion and nonlinearity. Accordingly, we can expect the formation of *bright magnetic solitons* when the dispersion and nonlinearity have the opposite signs. The observation of such a magnetic envelope soliton dates back to 1983 [134]. This experiment generated considerable interest in the study of magnetic solitons described by a $(1 + 1)$-dimensional NLS equation [135]–[140].

Even more interesting effects are possible for wide YIG films, in which spatiotemporal evolution of spin-wave packets is observed to occur in the two in-plane dimensions [131]. The spatiotemporal self-focusing of dipolar spin waves eventually leads to the formation of two-dimensional wave packets—magnetic bullets—with a YIG film. Such magnetic bullets are similar to the light bullets forming in an optical nonlinear medium [141] and discussed in Chapter 7. Because the magnetic bullets are spin-wave packets that maintain their size in both in-plane directions (y and z), they can propagate a considerable distance without changing spatial size, but they constantly lose energy due to dissipation. For this reason, such magnetic bullets always begin to diverge when their amplitude is no longer sufficient for self-focusing to occur.

Figure 14.24 shows the experimentally measured distribution of the magnetic intensity in a spin-wave packet formed at an input power level of $P_{in} = 460$ mW. The shape of the wave packet is shown at five different delay times after launching the packet at $t = 0$. The bottom traces show the cross section of the wave packets at the level where the intensity has dropped by 50%. The position of the input antenna is also shown at

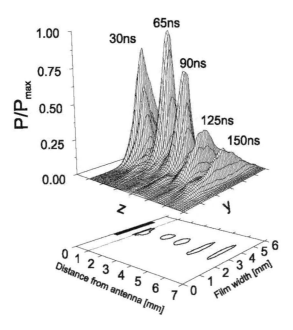

Figure 14.24: Two-dimensional distributions of a dipolar spin-wave packet propagating inside a YIG sample at five different delay times. The input power was 460 mW. The cross sections of the propagating wave packets at half-maximum power are also shown on the (y, z) plane. (After Ref. [131]; ©1998 APS.)

$z = 0$. As seen clearly, the input wave packet exhibits *spatiotemporal self-focusing*, and its intensity is larger at the focal point situated near $z = 2.5$ mm ($t = 60$ ns), where the packet width along the y axis is also a minimum. The focusing effect is further illustrated in Figure 14.25, where the experimentally measured widths of the wave packet in the y and z directions are shown as a function of the propagation time.

The results shown in Figure 14.25 can be understood physically as follows. For $t < 40$ ns, the spin-wave packet generated by the microwave field of the antenna is entering the region of the film accessible by the scanning technique. Therefore, the visible size of the wave packet is linearly increasing with time. For t in the range 40–45 ns, a rapid collapse-like self-focusing of the packet is observed, and the packet width decreases rapidly in both in-plane directions. The collapse is subsequently stabilized by dissipation. As a result, in the time interval $50 < t < 100$ ns, both widths of the propagating packet are almost constant. This region indicates that a quasi-stable magnetic bullet has formed inside the YIG film. For $t > 100$ ns, the transverse size L_y of the packet starts to increase rapidly because of diffraction while L_z remains relatively constant. The increase in L_y is due to the reduction of the peak power below the self-focusing threshold. The reduction in the peak power of the wave packet with time is shown in Figure 14.25(c). The focal point near $t = 60$ ns is clearly defined by the peak, and it corresponds to the minimum transverse size of the spin-wave packet. The results of numerical simulations, shown by solid lines in Figure 14.25, indicate quali-

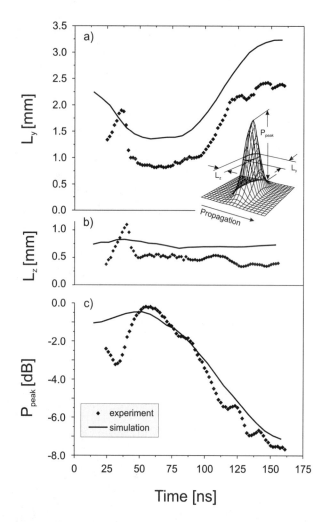

Figure 14.25: Widths of the wave packet in the (a) y and (b) z directions as a function of the propagation (delay) time. The normalized peak power of the wave packet shown in part (c) under the same conditions. Results of numerical simulations are shown for comparison as solid lines. (After Ref. [131]; ©1998 APS.)

tative agreement with the experimental data. No agreement is expected for $t < 40$ ns because numerical simulations assume that the entire field enters the sample at $t = 0$.

Microwave magnetic *dark solitons* were first observed in 1993 by propagating spin waves perpendicular to the direction of a bias magnetic field in a tangentially magnetized YIG film [142]. In such a geometry, the dispersion and nonlinear coefficients have the same sign [143], making it possible to generate dark solitons. The experimental results revealed an unusual feature: The number of dark solitons changed from even to odd with increasing input power. This observation can be explained by noting that the localized spin wave acquires *an induced spatial phase shift* that is inversely proportional to the spin-wave group velocity [144]. Such a phase shift is negligible for large group velocities, such as those occurring in fibers. However, the induced phase is not small in solids, and its effect becomes important in the case of spin waves. Based on this general concept, it is found that an arbitrary small phase shift across the initial pulse can change the character of the soliton generation and affect the number of dark solitons formed at the output end. By 1998, it was possible to generate a train of magnetic bright [145] and dark [146] solitons in a ring geometry.

14.4.3 Soliton and Bullet Collisions

Collisions of magnetic solitons were first studied in a 1994 experiment [147]. After a head-on collision of two such solitons, it was found that the soliton shape remained essentially unchanged. Since the experiment used narrow strips of magnetic films (YIG waveguides) that were 0.5–1 mm wide but more than 40 mm long, the $(1 + 1)$-dimensional NLS equation could be used for understanding the experimental results. The spin-wave envelope solitons were found to be stable, and they practically retain their shapes after collisions, even though they were propagating in a medium with relatively large dissipation (about 500 times larger than that in optical fibers). The situation is more complex in the case of two-dimensional magnetic solitons. Head-on collisions between two such self-focused spin-wave packets were observed in a 1999 experiment [148] using a space- and time-resolved Brillouin scattering technique. It was found that although quasi-one-dimensional magnetic solitons formed in narrow film strips retained their shape after collision, the two-dimensional wave packets—magnetic bullets—formed in wide YIG films were destroyed during a collision.

Let us consider first the case of $(1 + 1)$-dimensional magnetic solitons. Figure 14.26 shows the experimentally measured intensity profiles of two counterpropagating spin-wave packets in a YIG film at four different times covering their head-on collision. The lower part in each case shows the region over which intensity exceeds 50% of the peak value. Two microstrip antennas, from which the two counterpropagating spin-wave packets were launched, were oriented along the y axis and were situated at $z = 0$ and $z = 8$ mm, respectively. The radiation efficiencies of the left and right input antennas were slightly different. For this reason, even though the input microwave pulses supplied to both antennas were identical, the peak intensity of the wave packet propagating from the left was about 20% lower than that of the other. In this YIG film, the spin-wave packets became quasi-one-dimensional after some propagation time. As seen in part (b), the two wave packets become elongated along the y axis such that they occupy almost the full width of the waveguide. At this point quasi-one-dimensional magnetic solitons appear to have been formed from the initially two-dimensional spin-wave packets. During collision in part (c), the two spin-wave packet first merge and then separate. As seen in part (d), soliton profiles after the collision are almost the same as before the collision, i.e., quasi-one-dimensional magnetic solitons really behave like standard NLS solitons and retain their shapes after collisions, in spite of the considerable dissipation [note the change in the vertical scale in part (d)].

In sharp contrast to the previous case, the observed behavior is quite different in a wide film sample, in which the transverse size of the spin wave is not restricted by the film surface. As expected, strong self-focusing of spin-wave packets takes place in

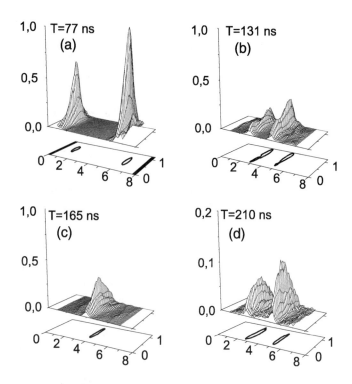

Figure 14.26: Formation and collision of quasi-one-dimensional magnetic solitons in a narrow YIG film. In each case, the upper frame shows the intensity distribution while the lower frame shows the cross-sectional area of the wave packet. (After Ref. [148]; ©1999 APS.)

two dimensions, and the initially elliptical cross sections of the wave packet become much narrower and almost circular, a well-known feature of wave packets approaching collapse. The collapse, however, is stopped by the dissipation, and quasi-stable two-dimensional magnetic bullets are formed [121]. When two such magnetic bullets collide inside the film, the increase in the intensity in the colliding region leads to a catastrophic self-focusing. As a result, the equilibrium between the self-focusing and dissipation is broken, and both spin-wave bullets are destroyed in the collision process.

The qualitative differences between the collision properties of quasi-one-dimensional magnetic solitons in a waveguide and those of two-dimensional magnetic bullets in a wide film are shown in Figure 14.27, where the transverse (L_y) and longitudinal (L_z) widths of propagating spin-wave packets are plotted as a function of the propagation time. In the case of a waveguide (left column), after an initial 100-ns period of soliton formation, during which the shape of the spin-wave packet becomes elliptical and elongated along the y direction $(L_y/L_z \approx 2.5)$, the spatial sizes of the resulting magnetic soliton remain practically constant afterwards and are not significantly affected, even after the collision with another soliton occurring close to 165 ns (dotted vertical line). In contrast, the behavior of two-dimensional spin-wave packets in a wide film (right column) is quite different. Strong two-dimensional self-focusing leads to the for-

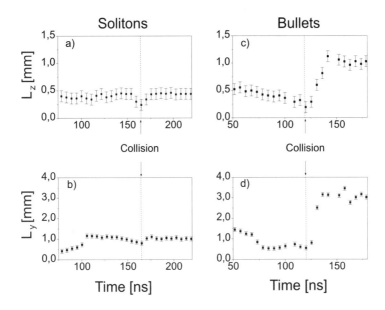

Figure 14.27: Widths L_z and L_y of two spin wave packets as a function of time when the two packets propagate and collide inside a waveguide (left column) and a wide film (right column). The time of collision is shown by vertical dotted lines. (After Ref. [148]; ©1999 APS.)

mation of magnetic bullets of nearly circular shape ($L_y/L_z \approx 1$). The collision of two such bullets at 120 ns leads to a dramatic increase in the size of the bullets along both in-plane directions and eventually to their destruction after 140 ns.

14.5 Bose–Einstein Condensates

The phenomenon known as *Bose–Einstein condensation* (BEC) was actually predicted in 1924 for systems whose particles obey the Bose statistics and whose total particle number is conserved. It was shown that there exists a critical temperature below which a finite fraction of all particles *condenses* into the same quantum state. Since 1995, the BEC phenomenon has been observed using several different types of atoms, confined by a magnetic trap and cooled down to extremely low temperatures [149]–[152].

From a mathematical point of view, the dynamics of BEC wave function can be described by an effective mean-field equation known as the *Gross–Pitaevskii (GP) equation* [153]. This is a classical nonlinear equation that takes into account the effects of particle interaction through an effective mean field. Because of the similarities between the GP equation in the BEC theory and the NLS equation in nonlinear optics, many of the phenomena predicted and observed in nonlinear optics are expected to occur for the BEC macroscopic quantum states, even though the underlying physics can be quite dissimilar. In this section, we discuss the theoretical background and the experimental observations of dark and bright solitons and vortices in BECs.

14.5.1   Gross–Pitaevskii Equation

The complete theoretical description of a BEC requires a quantum many-body approach [153]. The many-body Hamiltonian describing N interacting bosons is expressed through the boson field operators $\hat{\Phi}(\mathbf{r})$ and $\hat{\Phi}^{\dagger}(\mathbf{r})$ that, respectively, annihilate and create a particle at the position $\mathbf{r}$. A mean-field approach is commonly used for the interacting systems to overcome the problem of solving exactly the full many-body Schrödinger equation. Apart from the convenience of avoiding heavy numerical work, mean-field theories allow one to understand the behavior of a system in terms of a set of parameters that have a clear physical meaning. This is particularly true in the case of trapped bosons. Actually, most of the experimental results show that the mean-field approach is very effective in providing both qualitative and quantitative predictions for the static and dynamic properties of the trapped ultracold gases.

The basic idea behind the mean-field description of a dilute Bose gas was formulated in 1947 by Bogoliubov [154]. The key point consists in separating out the condensate contribution from the bosonic field operator. If we introduce a macroscopic wave function $\Psi(\mathbf{r})$, defined as the expectation value of the field operator, $\Psi(\mathbf{r},t) = <\hat{\Phi}(\mathbf{r},t)>$, its modulus fixes the condensate density, $n(\mathbf{r},t) = |\Psi(\mathbf{r},t)|^2$, and it possesses a well-defined phase. The macroscopic wave function $\Psi(\mathbf{r},t)$ has the meaning of the order parameter. The equation for $\Psi(\mathbf{r},t)$ is similar to the Schrödinger equation for a single-particle quantum state, but it also includes the effect of interparticle interactions through a mean-field nonlinear term.

To be more specific, we consider the macroscopic dynamics of a condensed atomic cloud in a three-dimensional, external parabolic potential created by a magnetic trap. The BEC dynamics are described by the following GP equation:

$$i\hbar\frac{\partial\Psi}{\partial t} = -\frac{\hbar^2}{2m}\nabla^2\Psi + V(\mathbf{r})\Psi + U_0|\Psi|^2\Psi, \tag{14.5.1}$$

where $\Psi(\mathbf{r},t)$ is the macroscopic wave function of the condensate, $V(\mathbf{r})$ is a parabolic trapping potential, and the parameter $U_0 = 4\pi\hbar^2(a/m)$ characterizes the two-particle interaction proportional to the s-wave scattering length a. When $a > 0$, the interaction between the particles in the condensate is *repulsive*, but it becomes *attractive* for $a < 0$. In fact, the scattering length a can be continuously tuned from positive to negative values by varying the external magnetic field near the so-called Feshbach resonances [155].

To simplify Eq. (14.5.1), it is useful to consider the case of a highly anisotropic (cigar-shaped) trap of the axial symmetry such that

$$V(\mathbf{r}) = \frac{1}{2}m\omega_{\perp}^2(r_{\perp}^2 + \lambda x^2), \tag{14.5.2}$$

where $r_{\perp} = \sqrt{y^2 + z^2}$, $\omega_{\perp}$ is the trap frequency in the transverse plane, and $\lambda = \omega_x^2/\omega_{\perp}^2$. Introducing normalized variables as $t' = t(\omega_{\perp}\sqrt{\lambda}/2)$, $\mathbf{r}' = \mathbf{r}(m\omega_{\perp}\sqrt{\lambda}/\hbar)^{1/2}$, and $\Psi' = \Psi(2U_0\sqrt{\lambda}/\hbar\omega_{\perp})^{1/2}$, we obtain the following dimensionless equation:

$$i\frac{\partial\Psi}{\partial t} + \nabla^2\Psi - [\lambda^{-1}(y^2 + z^2) + x^2]\Psi + \sigma|\Psi|^2\Psi = 0, \tag{14.5.3}$$

where $\sigma = \text{sgn}(a) = \pm 1$, depending on the sign of the s-wave scattering length and we have dropped the primes over the variables for notational simplicity.

If we assume that the nonlinear interaction is weak relative to the trapping potential force in the transverse dimensions ($\lambda \ll 1$), it follows from Eq. (14.5.3) that the transverse size of the condensate is much smaller than its length, and the condensate has a cigar-like shape. When $\lambda \ll 1$, the condensate cloud is confined in the transverse direction, and its transverse structure is defined mostly by the harmonic trapping potential [156]. Therefore, we can look for solutions of Eq. (14.5.3) in the form

$$\Psi(r,x,t) = \Phi(r_\perp)\psi(x,t)\exp(-2i\gamma t), \tag{14.5.4}$$

where $\Phi(r_\perp)$ is a solution of the following auxiliary problem for the two-dimensional radially symmetric quantum harmonic oscillator:

$$\nabla_\perp^2 \Phi + 2\gamma\Phi - (r_\perp^2/\lambda)\Phi = 0. \tag{14.5.5}$$

The ground state of this harmonic oscillator is a Gaussian-shape wave function $\Phi_0(r_\perp) = C\exp(-\gamma r_\perp^2/2)$, where $\gamma = 1/\sqrt{\lambda}$. Using the normalization condition $\int \Phi_0^2 d\mathbf{r}_\perp = 1$, we find $C^2 = \gamma/\pi$.

After substituting the factorized solution into Eq. (14.5.3), multiplying it by Φ^*, and integrating over the transverse cross section of the cigar-shaped condensate, we finally obtain the following one-dimensional form of the GP equation:

$$i\frac{\partial \psi}{\partial t} + \frac{\partial^2 \psi}{\partial x^2} - x^2\psi + \sigma|\psi|^2\psi = 0. \tag{14.5.6}$$

The particle number N is given by $N = (\hbar\omega/2U_0\sqrt{\lambda})Q$, where

$$Q = \int_{-\infty}^{\infty} |\psi(x,t)|^2\,dx \tag{14.5.7}$$

is the integral of motion for the normalized GP equation (14.5.6).

The case when the transverse size of the condensate varies along its length can be analyzed using a variational method with the transverse width acting as a free parameter. Minimization of the the system Lagrangian then allows one to derive an effective NLS equation with a generalized nonlinearity [157]. To the lowest order, this equation can be reduced to a cubic-quintic NLS equation, analyzed in Chapter 2. This approach allows one to describe many features of a realistic three-dimensional condensate with attractive ($\sigma = -1$) interaction, such as instability and collapse, in a relatively simple way. In particular, it predicts collapse for $N > N_c$, where $N_c = 2a_r/3|a|$ and $a_r = \sqrt{\hbar/m\omega_\perp}$.

14.5.2 Nonlinear Modes in a Parabolic Trap

Equation (14.5.6) describes the *longitudinal* profile of the condensate in a highly anisotropic trap along the weak confinement direction. In the linear limit, i.e., when formally $\sigma \to 0$, Eq. (14.5.6) becomes the well-known equation for a one-dimensional quantum harmonic oscillator. Its stationary solutions have the form

$$\psi(x,t) = \phi(x)e^{-i\Omega t} \tag{14.5.8}$$

and exist only for discrete values of Ω such that $\Omega = 2n + 1$, where $n = 0, 1, 2, \ldots$. The stationary wave function in the nth quantum state is given in terms of Hermite–Gauss polynomials as

$$\phi_n(x) = c_n H_n(x) e^{-x^2/2}, \qquad H_n(x) = (-1)^n e^{x^2/2} \frac{d^n(e^{-x^2/2})}{dx^n}, \qquad (14.5.9)$$

where $c_n = (2^n n! \sqrt{\pi})^{-1/2}$.

In general, localized solutions of Eq. (14.5.6) for $\sigma \neq 0$ can be found only numerically [158]. All such solutions can be characterized by the dependence of the invariant Q in Eq. (14.5.7), which is similar to the beam power in optics, on the normalized chemical potential (energy per particle in the condensate), which is similar to the beam propagation constant. In some limiting cases, we can employ different approximate methods to find the localized solutions and their characteristics in an analytical form.

To describe the effects of *weak nonlinearity*, we use the perturbation theory based on the expansion of the general solution of Eq. (14.5.6) in the infinite set of the eigenfunctions (14.5.9). A similar approach has been used earlier in the theory of the dispersion-managed temporal solitons [159]. To apply such a perturbation theory, we look for solutions of Eq. (14.5.6) in the form [see Eq. (14.5.8)]

$$\psi(x,t) = e^{-i\Omega t} \sum_{n=0}^{\infty} B_n \phi_n(x), \qquad (14.5.10)$$

where $\phi_n(x)$ are the harmonic-oscillator eigenfunctions given in Eq. (14.5.9). Inserting the expansion (14.5.10) into Eq. (14.5.6), multiplying by ϕ_n and integrating over x, we obtain the following set of algebraic equations for the coefficients:

$$(\Omega - \Omega_m) B_m - \sigma \sum_{n,l,k} V_{m,n,l,k} B_n B_l B_k = 0, \qquad (14.5.11)$$

where the coupling coefficients are given by

$$V_{m,n,l,k} = \int_{-\infty}^{+\infty} \phi_m(x) \phi_n(x) \phi_l(x) \phi_k(x) dx. \qquad (14.5.12)$$

Equation (14.5.11) can be rewritten in the traditional form $\delta(H + \Omega Q) = 0$, where H is the Hamiltonian associated with Eq. (14.5.6), and this allows us to develop a perturbation theory for small nonlinearities. For example, consider the ground-state mode, for which $n = 0$. Assuming $B_0 \gg B_m$ for $m \neq 0$ and the case of the symmetric solution $\phi(x) = \phi(-x)$, we find $\Omega \approx \Omega_0 + \sigma V_{0,0,0,0}|B_0|^2$, and

$$B_{2k} = \frac{\sigma V_{2k,0,0,0}}{(\Omega - \Omega_{2k})} |B_0|^2 B_0, \qquad B_{2k+1} = 0. \qquad (14.5.13)$$

Using the expansion of the invariant Q for small nonlinearities, we find

$$Q = \sum_k |B_{2k}|^2 \approx -\sigma a_0 (\Omega - \Omega_0)[1 + b_0(\Omega - \Omega_0)^2], \qquad (14.5.14)$$

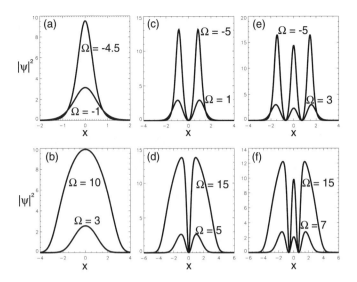

Figure 14.28: Condensate density $|\psi(x)|^2$ of the first three nonlinear modes for several values of Ω in the case of $\sigma = +1$ (upper row) and $\sigma = -1$ (lower row). (After Ref. [158]; ©2001 Elsevier.)

where the coefficients are given by

$$a_0 = \sqrt{2\pi}, \qquad b_0 = \sum_{k=1}^{\infty} \frac{(2k)!}{(4k)^2 (k! \, 2^{2k})^2}. \qquad (14.5.15)$$

Higher-order modes can be considered in a similar way, and the results are similar to Eq. (14.5.14) except that a_0 and b_0 are replaced with a_n and b_n, respectively, both of which depend on the mode order n.

In the opposite limit, in which the nonlinear potential terms are large in comparison with the kinetic linear term given by the second-order derivative, we can use two different approximations for finding the localized modes. For $\sigma = +1$ and large *negative* Ω, the localized modes are described by the stationary solutions of the NLS equation because we can neglect the trapping potential. Using the one-soliton solution of the cubic NLS equation,

$$\phi_s(x) = \sqrt{-2\Omega} \, \mathrm{sech}(x\sqrt{-\Omega}), \qquad (14.5.16)$$

we obtain the dependence $Q(\Omega)$, which coincides with the soliton invariant $Q_s = 4\sqrt{-\Omega}$. For $\sigma = -1$ and large *positive* Ω, the ground-state solution can be obtained by using the so-called Thomas–Fermi approximation based on neglecting the kinetic term. In this case, $\phi_{\mathrm{TF}}(x) \approx \sqrt{\Omega - x^2}$.

In general, to find the stationary states of the condensate in a trap, i.e., the nonlinear modes, one should solve Eq. (14.5.6) numerically, looking for the solutions in the form (14.5.8). Figure 14.28 shows the first three nonlinear modes of Eq. (14.5.6) by plotting the BEC density $|\psi(x)|^2$ for several values of the dimensionless parameter Ω using $\sigma = +1$ (top row) and $\sigma = -1$ (bottom row). For $\Omega \to 1$, i.e., in the limit of the harmonic oscillator ground-state mode, the solution is close to a Gaussian shape in both cases. When Ω deviates from 1, the solution shape is defined by the type of nonlinearity. For attraction ($\sigma = +1$), the shape approaches that of a sech-type soliton, whereas for repulsion ($\sigma = -1$) the solution flattens and is better described by the Thomas–Fermi approximation.

Figure 14.29 shows the dependence of the invariant Q on the parameter Ω for $\sigma = \pm 1$, corresponding to two different signs of the scattering length. The dashed curves show the predictions based on the NLS equation in the $\sigma = 1$ case. For the zeroth-order mode, the dashed curve almost coincides with the corresponding solid curve. This feature implies that for such a narrow localized state, the effect of a parabolic potential is negligible, and the condensate ground state becomes localized mostly due to attractive interparticle interaction. Even for higher-order modes, the soliton solution of the NLS equation agrees with the solid curves in the asymptotic region of *negative* Ω, ($\Omega < -5$). In contrast, for large *positive* Ω the effect of trapping potential is crucial, and the solution of the Thomas–Fermi approximation, $\phi_{TF}(x) = \sqrt{\Omega - x^2}$, determines the asymptotic behavior not only for the fundamental lowest-order mode but also for all the higher-order modes. In this limit, $Q_{TF} \sim \frac{4}{3}\Omega^{3/2}$, as shown by the dashed-dotted curves in Figure 14.29

As mentioned before, in the linear limit ($\sigma \to 0$), Eq. (14.5.6) possesses a discrete set of localized modes described by the Hermite–Gauss polynomials. Such modes can be readily calculated using perturbation theory in the weakly nonlinear case. All of them should also exist for the strongly nonlinear problem since they describe an analytical continuation of the Hermite–Gaussian linear modes to a set of nonlinear stationary states [160]. The results of the numerical analysis presented in Figures 14.28(a-f) confirm this expectation. In the limit $\sigma \to 0$, nonlinear modes transform into the corresponding eigenfunctions of a linear harmonic oscillator.

It is clear from Figure 14.28 that the nonlinearity has a different effect for the negative and positive scattering lengths. First of all, in the case of the attractive interaction ($\sigma = +1$), the ground state transforms into a self-trapped state—a bright soliton—which is only slightly modified by the parabolic potential, whereas for the repulsive interaction ($\sigma = -1$) the localization is due solely to the potential trapping. Consequently, for the negative scattering length (attraction), the higher-order modes transform into *multisoliton states* consisting of an array of solitary waves with alternating phases. This is also confirmed by Figure 14.29, where all the branches of the higher-order modes approach asymptotically the soliton dependencies $Q_n \sim (n+1)Q_s$, where n is the mode number ($n = 0, 1, \ldots$). From the physical point of view, the higher-order stationary modes exist because of a balance between the *repulsion* among out-of-phase bright NLS solitons and the *attraction* imposed by the trapping potential. The analysis of the global stability of such higher-order multihump soliton modes is still an open problem, although there are some indications that multihump soliton states can be stable, at least in some nonlinear models [161]. In the case of a positive scattering length ($\sigma = -1$), the higher-order modes transform into a sequence of dark solitons or *kinks* [162], such that the first-order mode corresponds to a single dark soliton, the second-order mode to a pair of dark solitons, etc. Again, these stationary solutions result from a balance: The repulsion among dark solitons is exactly compensated by the attractive force of the trapping potential. Similar physics applies in higher dimensions.

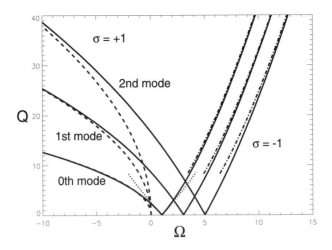

Figure 14.29: Invariant $Q(\Omega)$ for the first three nonlinear modes in the case of $\sigma = 1$ (left) and $\sigma = -1$ (right). Dashed curves for $\sigma = +1$ show the NLS limit. Dashed-dotted curves for $\sigma = -1$ are obtained in the Thomas–Fermi approximation. Dotted lines show the results of perturbation theory. (After Ref. [158]; ©2001 Elsevier.)

14.5.3 Dark Solitons

To study the dynamics of dark solitons, we can employ the perturbation theory used for optical dark solitons in Chapter 4 after modifying it to include the trapping potential [163]. Similar results can be obtained by other methods, such as the boundary layer theory developed in Ref. [164]. Using the perturbation theory, we obtain an equation of motion for a dark (or gray) soliton propagating through an effectively one-dimensional BEC cloud, assuming only that the background density and velocity vary slowly on the soliton scale. Denoting the center of the dark soliton by x_0, this equation has the form

$$\frac{d^2 x_0}{dt^2} = -\frac{1}{2} \frac{dV(x_0)}{dx_0}, \qquad (14.5.17)$$

where $V(x)$ is the external potential that enters the GP equation. In some studies, this equation is used without the factor of $\frac{1}{2}$ [165]. It turns out that the factor of $\frac{1}{2}$ disappears if one assumes that the soliton does not move relative to the background [166].

In a harmonic trap, Eq. (14.5.17) implies that the dark soliton oscillates with a frequency that is smaller by a factor of $\sqrt{2}$ compared with the trap frequency. This result can also be obtained for small oscillations by solving the Bogolubov equations for a motionless soliton in a trap. Equation (14.5.17) holds for an arbitrary potential as long as it varies slowly on the soliton length scale. The accuracy of Eq. (14.5.17) has been verified by solving the GP equation numerically for a wide range of potentials [164]; an example is shown in Figure 14.30, where the evolution of the dark soliton is shown in a nearly parabolic trap. The dark soliton tunnels through a small dip in the background cloud caused by a small hump in the condensate potential and continues to oscillate back and forth in a periodic manner. Since one can generate microwells

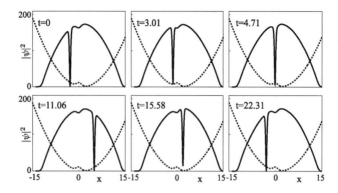

Figure 14.30: Condensate density $|\psi|^2$ showing oscillations of a dark soliton through a static Thomas–Fermi cloud in a weakly perturbed external potential. (After Ref. [164]; ©2000 APS.)

or barriers in a trap using lasers, it should be possible to realize this type of potential experimentally.

It is important to estimate the effects of dissipative losses that can become an important factor in any experiment and may lead to soliton decay. In a finite trap, coupling to the background condensate modes does not create dissipation. However, dissipation can come from corrections to the mean-field theory, in particular, from collisions with uncondensed atoms of the thermal cloud as well as three-body collisions. In most experimental situations, with 99% of the particles in the condensate, the dark-soliton decay time is estimated to be about 1 s [167, 168].

Matter-wave dark solitons have been observed in several experiments. In a 2000 experiment [169], the technique of macroscopic quantum-phase engineering was used to imprint a specific spatial phase distribution on the BEC cloud. A macroscopic quantum state was designed and produced by optically imprinting a phase pattern onto a BEC of sodium atoms, and matter-wave interferometry with spatially resolved imaging was used to analyze the resultant phase distribution. An appropriate phase imprint created dark solitons. The subsequent evolution of such a dark soliton was investigated both experimentally and theoretically. To observe soliton propagation, the BEC density was measured using absorption imaging. The condensate was released from the magnetic trap 1 ms before being imaged, but the expansion of the released BEC during that time was negligible. Figure 14.31 shows an example of the experimental results. For a phase imprint of 0.5π, two dark solitons were formed, and they moved in opposite directions at the speed of sound within the experimental uncertainty. For a larger phase imprint of 1.5π, two dark solitons separated at slower speeds, in agreement with numerical simulations. An even larger phase imprint generated many solitons, as shown in part C.

In another experiment [170], dark solitons were observed in a cigar-shaped BEC formed using a dilute vapor of Rb atoms. Several excited states in the form of dark-soliton modes were produced by imprinting a local phase onto the BEC wave function. By monitoring the evolution of the density profile, it was found that the density dip travelled at a smaller velocity than the speed of sound in the trapped condensate. By

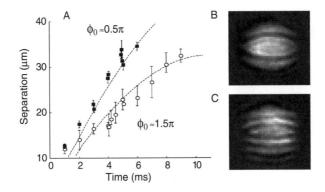

Figure 14.31: (A) Measured separation between two dark solitons as a function of time for phase imprints of 0.5π (squares) and 1.5π (circles). The dashed lines show the results of numerical simulations. Condensate 6 ms after a phase imprint of (B) 0.5π and (C) 2π. (After Ref. [169]; ©2000 AAAP.)

comparing the data with the numerical solutions of the GP equation under the experimental conditions, this density dip could be identified as a moving dark soliton.

Similar to the case of optical solitons, dark-soliton stripes are subject to transverse modulation instability and can decay into pairs of vortices with opposite topological charge [171]. To observe this instability, a 2001 experiment used a two-component BEC in which the dark soliton exists in one of the condensate components, whereas the second component fills the hole [172]. Such filled dark solitons were stable for hundreds of milliseconds. The filling was selectively removed, making the soliton more susceptible to dynamical instabilities. For a condensate in a spherically symmetric potential, these instabilities caused a dark soliton to decay into stable vortex rings, as predicted by the theory [171].

The connection between the quantized vortices and dark solitons in a waveguide-like trap geometry has also been studied [173]. By changing the size of the transverse confinement, it was found that the quasi-one-dimensional regime, where dark-soliton stripes are stable, leads to two- or three-dimensional confinement, where dark-soliton stripes are subject to a transverse instability known as the "snake instability." In the regime of stronger confinement, dark solitons decay into single deformed vortices with solitonic properties, rather than vortex pairs as associated with the "snake" metaphor. Further increase of the condensate size in the transverse dimensions leads to the production of two and then three vortices, in agreement with the Bogoliubov stability analysis. The decay of a stationary dark soliton into a single solitonic vortex is predicted to be experimentally observable in a three-dimensional harmonically confined BEC.

14.5.4   Vortices

As was already mentioned, many effects described in optics are expected to find their counterparts in the physics of BEC, including the dynamics of vortices [174, 175]. Historically, quantum vortices in trapped atomic gases were first observed in 1999, using two-component condensates [176]. The possibility of trapping more than one BEC component arises from the hyperfine atomic structure. Atoms in internal states with different total angular momentum may coexist in the BEC fraction, and it is possible to induce transitions between their different states. To form a vortex soliton, a phase gradient was imprinted in one of the BEC components, which caused it to rotate. The system was stabilized at a configuration in which the nonrotating component was localized at the center of the trap acting as an effective potential on the rotating component, which resided in the outer region.

The main properties of a two-component BEC can be described using a system of two incoherently coupled GP equations, similar to the two coupled NLS equations that describe optical vector solitons, except for the presence of a trapping potential that prevents the condensate with repulsive interaction from spreading. The two-component GP equation has solutions with remarkable properties [175]. In a situation where the two components overlap considerably, the creation of a vortex in just one of the components is not dynamically stable. The reason is that the angular momentum can be transferred after some time from one component to the other one, initiating a cyclic process. If one only monitors the density profile of the two atomic species, it may seem that the vortex disappears and eventually reappears in a periodic manner.

If a nearly two-dimensional trap is made to rotate, the situation changes qualitatively. In the rotating frame the Coriolis force manifests itself through an additional term in the Hamiltonian, $-\Omega L_z$, where L_z is a component of the angular momentum operator $\mathbf{L}$ and Ω is the angular frequency [177]. This additional force produces the centrifugal barrier proportional to $L_z^2 z$, where $L_z = m\hbar$ is the angular momentum per particle. Thus, it is always energetically costly to have a high angular momentum. However, for nonzero values of Ω, it may be energetically favorable to develop small positive values of L_z. If Ω is sufficiently large, solutions with L_z greater than $\hbar$ may have the lowest energy; a vortex is created in this situation.

For Ω greater than a critical frequency $\Omega_c = 0.22\omega_0$, where ω_0 is the trap frequency, the lowest energy corresponds to the state with $L_z = \hbar$. As the frequency increases, states with increasingly high angular momentum become the effective ground state in the rotating frame, thereby creating vortices with a topological charge $m > 1$. Such configurations are unstable and decompose into several single-charge vortices [178]. The interaction between vortices is generally believed to be repulsive. Thus, at rotating frequencies high enough to generate many stable vortices, vortices tend to move apart and drift toward the borders of the condensate, where they would disappear if their existence were not favored by the rotation. The result is that, at very high angular frequencies, vortices tend to form a regular array, as also observed experimentally. Figure 14.32 shows examples of vortices for a rotating Rb-vapor BEC. The array formation is akin to what has been long known for type II superconductors, where it is the presence of a magnetic field that forms a triangular vortex crystal—the so-called Abrikosov lattice [179].

In another experiment [180], the formation of a regular vortex array was observed in a [87]Rb BEC as the number of stable vortices raises from 0 to 4 by increasing the angular frequency. The formation of a triangular vortex lattice with as many as 130 vortices was observed in an experiment where BEC was obtained using sodium atoms

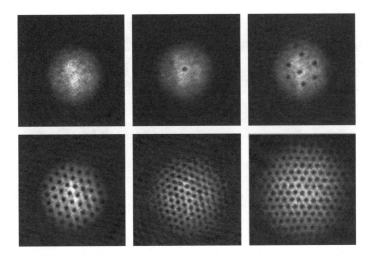

Figure 14.32: Generation of vortices and vortex lattices in a rotating ^{87}Rb condensate. (Courtesy P. Engels).

[181]. The optimum configuration in terms of size and regularity is achieved after 500 ms. For times shorter than that, regular order is not completely established, and a blurry structure is formed because of the misalignment of some vortices with respect to the rotation axis. For times much longer than 500 ms, inelastic collisions induce atom losses and a decrease in the number of vortices.

Several experiments have studied the nucleation of vortices in a BEC stirred by a laser beam. In one experiment [182], vortices were generated in a BEC cloud stirred by a laser beam and observed with time-of-flight absorption imaging. Depending on the stirrer size, either discrete resonances or a broad response was visible as the stir frequency was varied. Stirring beams that were small compared to the condensate size generated vortices below the critical rotation frequency for the nucleation of surface modes, suggesting a local mechanism of vortex generation. In addition, it was observed that the centrifugal distortion of the condensate induced by a rotating vortex lattice led to bending of the vortex lines.

14.5.5 Bright Solitons

Similar to the dispersive broadening of optical pulses, the kinetic energy of condensed atoms cause localized matter-wave clouds to spread as they propagate. Solitons may be formed when the nonlinear attractive interaction produces self-focusing of the matter-wave packet that compensates for spreading. The BEC dynamics are described by the GP equation, which in the case of attractive interactions has the same form as the NLS equation for light waves propagating in a medium with Kerr nonlinearity. In this sense, bright matter-wave solitons in one dimension are expected to be similar to optical solitons in optical fibers. Unlike dark solitons in a repulsive BEC, which survive only for short times after the condensate is released from the trap, bright atomic solitons do

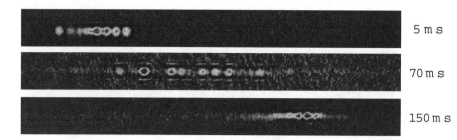

Figure 14.33: Generation of a train of bright solitons in a BEC of Li atoms. The three images show a soliton train near the two turning points (top and bottom) and near the center of oscillation (middle). The number of solitons varies from image to image because of shot-to-shot variations and a slow loss of soliton signal with time. (After Ref. [183]; ©2002 Nature.)

not rely on the trap and can propagate over much longer distances, and they can be used for forming non-spreading atom-laser output beams [183, 184].

Since the cubic nonlinearity resulting from the mean-field interaction among atoms is not limited by the saturation effects, the dynamics of a BEC with attractive interaction in a two- or three-dimensional confinement geometry displays the collapse phenomenon [185, 186]. However, when the spatial confinement is highly anisotropic, the condensate stability can be achieved in a quasi-one-dimensional case. The degree of transverse confinement necessary to achieve soliton stability has been investigated theoretically [158, 156, 187]. The main requirement is that the condensate occupation number N is limited to values such that $N \ll a_r/|a|$, where $a_r = (\hbar/m\omega_\perp)^{1/2}$ is the radial scale length, m is the atomic mass, and a is the s-wave scattering length.

The formation of a train of bright solitons of ^{7}Li atoms was first observed in a 2002 experiment [183] using a quasi-one-dimensional optical trap and magnetically tuning the interatomic interaction within the stable BEC from repulsive to attractive. In this experiment, the condensate was created on a side of the optical potential by axially displacing the focus of the infrared beam relative to the centers of the magnetic trap and the box potential formed by the end caps. The end caps prevented the condensate from moving under the influence of the infrared potential until the end caps are switched off and the condensate is set in motion. The condensate is allowed to evolve for a set period of time before an image is taken. The BEC spreads for repulsive interaction, but multiple bright solitons are formed in the case of attractive interaction. Typically, four solitons were created from an initially stationary condensate via modulation instability. The train propagates in the parabolic potential for many oscillatory cycles, and the solitons eventually disappear because of loss of atoms observed for times exceeding 3 s. Figure 14.33 shows three images of such a soliton train near the two turning points (top and bottom) and near the center of potential (middle).

The alternating phase structure of the soliton train can be inferred from the relative motion of solitons. Noninteracting solitons, simultaneously released from different points in a harmonic potential, would be expected to pass through one another. But this is not the case in experiments, as can be seen from Figure 14.33, where the spacing between the solitons increases near the center of oscillation while they bunch together at the end points. This behavior indicates the presence of a short-range repulsive force between the neighboring solitons because of their alternating phases.

In another experiment on bright matter-wave solitons, an ultracold ^{7}Li gas was used to create the BEC cloud [184]. The effective interaction between atoms was tuned from *repulsive* to *attractive* using a Feshbach resonance, and the propagation of bright solitons over a macroscopic distance of 1.1 mm was observed. In the experiment, at all times the soliton width remained equal to the resolution limit of the imaging system used, and no decay of the soliton was observed in a 10-ms time interval the BEC remained in the detection region. However, a substantial fraction of atoms (>65%) remained in a noncondensed state, creating a pedestal around the soliton.

The remarkable similarities between matter-wave solitons and optical solitons in fibers emphasize the intimate connection between nonlinear atom optics and fiber optics. However, many issues remain to be addressed, including the dynamical process of soliton formation and a comprehensive study of soliton interactions and collisions. An "atomic soliton laser" based on bright matter-wave solitons may prove useful for precision measurement applications such as atom interferometry.

References

[1] N. V. Tabiryan, A. V. Sukhov, and B. Ya. Zeldovich, *Mol. Cryst. Liq. Cryst.* **136**, 1 (1986).

[2] I. C. Khoo, in *Progress in Optics*, Vol. 36, E. Wolf, Ed. (Elsevier, Amsterdam, 1988), pp. 105–161.

[3] I. C. Khoo and N. T. Wu, *Optics and Nonlinear Optics of Liquid Crystals* (World Scientific, Singapore, 1993).

[4] E. Braun, L. P. Faucheux, A. Libchaber, D. W. McLaughlin, D. J. Muraki, and M. J. Shelley, *Europhys. Lett.* **23**, 239 (1993).

[5] E. Braun, L. P. Faucheux, and A. Libchaber, *Phys. Rev. A* **48**, 611 (1993).

[6] D. W. McLaughlin, D. J. Muraki, M. J. Shelley, and X. Wang, *Physica D* **88**, 55 (1995).

[7] D. W. McLaughlin, D. J. Muraki, and M. J. Shelley, *Physica D* **97**, 471 (1996).

[8] I. Janossy, *Phys. Rev. E* **49**, 2957 (1994).

[9] M. A. Karpierz, A. W. Domański, M. Sierakowski, M. Swillo, and T. R. Woliński, *Acta Phys. Polonica A* **95**, 783 (1999).

[10] M. A. Karpierz, in *Soliton-Driven Photonics*, A. D. Boardman and A. P. Sukhorukov, Eds. (Kluwer, Dordrecht, Netherlands, 2001), pp. 41-57.

[11] M. A. Karpierz, M. Sierakowski, M. Swillo, and T. Woliński, *Mol. Cryst. Liq. Cryst.* **320**, 157 (1998).

[12] M. Warenghem, J. F. Henninot, and G. Abbate, *Mol. Cryst. Liq. Cryst.* **320**, 207 (1998).

[13] M. Warenghem, J. F. Henninot, and G. Abbate, *Opt. Exp.* **2**, 483 (1998).

[14] M. Peccianti, A. De Rossi, G. Assanto, A. De Luca, C. Umeton, and I. C. Khoo, *Appl. Phys. Lett.* **77**, 7 (2000).

[15] M. Peccianti and G. Assanto, *Opt. Lett.* **26**, 1791 (2001).

[16] M. Peccianti and G. Assanto, *Phys. Rev. E* **65**, 035803 (2002).

[17] M. Peccianti and G. Assanto, *Opt. Lett.* **26**, 1690 (2001).

[18] M. A. Karpierz, M. Sierakowski, and T. Woliński, *Mol. Cryst. Liq. Cryst.* **375**, 313 (2002).

[19] C. Conti, M. Peccianti, and G. Assanto, *Proc. Nonlinear Guided Waves and Their Applications* (Optical Society of America, Washington, DC, 2002).

[20] I. V. Shadrivov and A. A. Zharov, *J. Opt. Soc. Am. B* **19**, 596 (2002).

[21] A. G. Litvak and A. M. Sergeev, *JETP Lett.* **27**, 517 (1978) [*Pis'ma Zh. Eksp. Teor. Fiz.* **27**, 548 (1978)].

[22] H. L. Pesceli and J. J. Rasmussen, *Plasma Phys.* **22**, 421 (1980).

[23] T. A. Davydova and A. I. Fishchuk, *Ukr. J. Phys.* **40**, 487 (1995).

[24] V. M. Pérez-García, V. V. Konotop, and J. J. García-Ripoll, *Phys. Rev. E* **62**, 4300 (2000).

[25] H. Pu, W. Zhang, and P. Meystre, *Phys. Rev. Lett.* **87**, 140405 (2001).

[26] W. Królikowski and O. Bang, *Phys. Rev. E* **63**, 016610 (2000).

[27] W. Królikowski, O. Bang, J. J. Rasmussen, and J. Wyller, *Phys. Rev. E* **64**, 016612 (2001).

[28] S. K. Turitsyn, *Teor. Mat. Fiz.* **64**, 226 (1985).

[29] N. N. Rosanov, A. G. Vladimirov, D. V. Skryabin, and W. J. Firth, *Phys. Lett. A* **293**, 45 (2002).

[30] V. A. Mironov, A. M. Sergeev, and E. M. Sher, *Sov. Phys. Dokl.* **26**, 861 (1981) [*Dokl. Akad. Nauk SSSR* **260**, 325 (1981)].

[31] G. L. Alfimov, V. M. Eleonsky, and N. V. Mitskevich, *Sov. Phys. JETP* **76**, 563 (1993) [*Zh. Eksp. Teor. Fiz.* **103**, 1152 (1993)].

[32] M. Peccianti, C. Conti, and G. Assanto, *Proc. Nonlinear Guided Waves and Their Applications* (Optical Society of America, Washington, DC, 2002).

[33] G. I. Stegeman and M. Segev, *Science* **286**, 1518 (1999).

[34] T. M. Monro, C. M. De Sterke, and L. Poladian, *J. Mod. Opt.* **48**, 191 (2001).

[35] K. O. Hill, Y. Fujii, D. C. Johnson, and B. S. Kawasaki, *Appl. Phys. Lett.* **32**, 647 (1978).

[36] R. Kashyap, *Fiber Bragg Gratings* (Academic Press, San Diego, 1999).

[37] D. K. W. Lam and B. K. Garside, *Appl. Opt.* **20**, 440 (1981).

[38] G. Meltz, W. W. Morey, and W. H. Glenn, *Opt. Lett.* **14**, 823 (1989).

[39] B. Malo, K. A. Vineberg, F. Bilodeau, J. Albert, D. C. Johnson, and K. O. Hill, *Opt. Lett.* **15**, 953 (1990).

[40] W. S. Brocklesby *et al.*, *Opt. Materials* **1**, 177 (1992).

[41] S. J. Frisken, *Opt. Lett.* **18**, 1035 (1993).

[42] A. S. Kewitsch and A. Yariv, *Opt. Lett.* **21**, 24 (1996).

[43] A. S. Kewitsch and A. Yariv, *Appl. Phys. Lett.* **68**, 455 (1996).

[44] C. Meneghini and A. Villeneuve, *J. Opt. Soc. Am. B* **15**, 2946 (1998).

[45] S. Shoji and S. Kawata, *Appl. Phys. Lett.* **75**, 737 (1999).

[46] M. Kagami, T. Yamashita, and H. Ito, *Appl. Phys. Lett.* **79**, 1079 (2001).

[47] S. Shoji, S. Kawata, A. A. Sukhorukov, and Yu. S. Kivshar, *Opt. Lett.* **27**, 185 (2002).

[48] S. Maruo and K. Ikuta, *Appl. Phys. Lett.* **76**, 2656 (2000).

[49] S. Maruo, O. Nakamura, and S. Kawata, *Opt. Lett.* **22**, 132 (1997).

[50] B. H. Cumpston, S. P. Ananthavel, S. Barlow, D. L. Dyer, J. E. Ehrlich, L. L. Erskine, A. A. Heikal, S. M. Kuebler, I.-Y. Sandy Lee, D. McCord-Maughon, J. Qin, H. Rckel, M. Rumi, X. Wu, S. R. Marder, and J. W. Perry, *Nature* **398**, 51 (1999).

[51] S. Shoji and S. Kawata, *Appl. Phys. Lett.* **76**, 2668 (2000).

[52] S. Kawata, H.-B. Sun, T. Tanaka, and K. Takada, *Nature* **412**, 697 (2001).

[53] D. Homoelle, S. Wielandy, A.L. Gaeta, N. F. Borrelli, and C. Smith, *Opt. Lett.* **24**, 1311 (1999).

[54] T. M. Monro, C. M. de Sterke, and L. Poladian, *Opt. Commun.* **119**, 523 (1995).

[55] V. Mizrahi, S. LaRochelle, G. I. Stegeman, and J. E. Sipe, *Phys. Rev. A* **43**, 433 (1991).

[56] C. M. de Sterke, S. An, and J.E. Sipe, *Opt. Commun.* **83**, 315 (1991).

[57] T. M. Monro, D. Moss, M. Bazylenko, C. M. de Sterke, and L. Poladian, *Phys. Rev. Lett.* **80**, 4072 (1998).

[58] T. M. Monro, P. D. Miller, L. Poladian, and C. M. de Sterke, *Opt. Lett.* **23**, 268 (1998).

[59] T. M. Monro, C. M. de Sterke, and L. Poladian, *J. Opt. Soc. Am. B* **13**, 2824 (1996).

[60] T. M. Monro, L. Poladian, and C. M. de Sterke, *Phys. Rev. E* **57**, 1104 (1998).

[61] T. M. Monro, C. M. de Sterke, and L. Poladian, *J. Opt. Soc. Am. B* **16**, 1680 (1999).

[62] M. Bazylenko, M. Gross, P. L. Chu, and D. Moss, *Electron. Lett.* **32**, 1198 (1996).

[63] S. Sarkisov, M. Curley, A. Wilkosz, and V. Grymalsky, *Opt. Commun.* **161**, 132 (1999).

[64] S. Sarkisov, A. Taylor, P. Venkateswarlu, and A. Wilkosz, *Opt. Commun.* **145**, 265 (1998).

[65] A. M. Ljungström and T. M. Monro, *Opt. Exp.* **10**, 230 (2002).

[66] N. N. Rosanov, *Spatial Hysteresis and Optical Patterns* (Springer, Berlin, 2002).

[67] N. N. Rosanov and V. E. Semenov, *Opt. Spectrosc.* **48**, 59 (1980).

[68] D. W. McLaughlin, J. V. Moloney, and A. C. Newell, *Phys. Rev. Lett.* **51**, 75 (1983).

[69] J. V. Moloney, *Phys. Rev. Lett.* **53**, 556 (1984).

[70] W. J. Firth and G. K. Harkness, in *Spatial Solitons*, S. Trillo and W. Torruellas, Eds. (Springer, Berlin, 2001), pp. 343–358.

[71] N. N. Rosanov and G. V. Khodova, *Opt. Spectrosc.* **65**, 1399 (1988).

[72] N. N. Rosanov and G. V. Khodova, *J. Opt. Soc. Am. B* **7**, 1057 (1990).

[73] N. N. Rosanov, *Proc. SPIE* **1840**, 130 (1991).

[74] S. V. Fedorov, G. V. Khodova, and N. N. Rosanov, *Proc. SPIE* **1840**, 208 (1991).

[75] N. N. Rosanov, in *Progress in Optics*, Vol. 35, E. Wolf, Ed. (Elsevier, Amsterdam, 1996), pp. 1–60.

[76] G. S. McDonald and W. J. Firth, *J. Opt. Soc. Am. B* **7**, 1328 (1990).

[77] G. S. McDonald and W. J. Firth, *J. Opt. Soc. Am. B* **10**, 1081 (1993).

[78] N. B. Abraham and W. J. Firth, *J. Opt. Soc. Am. B* **7**, 951 (1990).

[79] C. O. Weiss, G. Slekys, V. B. Taranenko, K. Staliunas, and R. Kuszelewicz, in *Spatial Solitons*, S. Trillo and W. Torruellas, Eds. (Springer, Berlin, 2001), pp. 395–416.

[80] L. A. Lugiato and R. Lefever, *Phys. Rev. Lett.* **58**, 2209 (1987).

[81] W. J. Firth, G. K. Harkness, A. Lord, J. M. McSloy, D. Gomila, and P. Colet, *J. Opt. Soc. Am. B* **19**, 747 (2002).

[82] M. Tlidi, P. Mandel, and R. Lefever, R. *Phys. Rev. Lett.* **73**, 640 (1994).

[83] W. J. Firth and A.J. Scroggie, *Phys. Rev. Lett.* **76**, 1623 (1996).

[84] I. S. Aranson and L. Kramer, *Rev. Mod. Phys.* **74**, 99 (2002).

[85] S. Wabnitz, *Opt. Lett.* **18**, 601 (1993).

[86] G. Steinmeyer, A. Schwache, and F. Mitschke, *Phys. Rev. E* **53**, 5399 (1996).

[87] W. J. Firth and A. Lord, *J. Mod. Opt.* **43**, 1071 (1996).

[88] D.V. Skryabin, *Phys. Rev. E* **64**, 056601 (2001).

[89] T. E. Sale, *Vertical Cavity Surface Emitting Lasers* (Wiley, New York, 1995).

[90] D. Michaelis, U. Peschel, and F. Lederer, *Phys. Rev. A* **56**, R3366 (1997).

[91] L. Spinelli, G. Tissoni, M. Brambilla, F. Prati, and L. A. Lugiato, *Phys. Rev. A* **58**, 2542 (1998).

[92] L. A. Lugiato, L. Spinelli, G. Tissoni, and M. Brambilla, *J. Opt. B* **1** 43 (1999).

[93] G. Tissoni, L. Spinelli, L. A. Lugiato, M. Brambilla, I. M. Perrini, and T. Maggipinto, *Opt. Exp.* **19**, 1009 (2002).

[94] K. Staliunas and V. J. Sanchez-Morcillo, *Opt. Commun.* **139**, 306 (1997).

[95] V. B. Taranenko, I. Ganne, R. Kuszelewicz, and C. O. Weiss, *Appl. Phys. B* **72**, 377 (2001).

[96] V. B. Taranenko, I. Ganne, R. Kuszelewicz, and C. O. Weiss, *Phys. Rev. A* **61**, 063818 (2000).

[97] P. K. Jacobsen, J. V. Moloney, A. C. Newell, and R. Indik, *Phys. Rev. A* **45**, 8129 (1992).

[98] W. J. Firth and A. J. Scroggie, *Europhys. Lett.* **26**, 521 (1994).

[99] V. B. Taranenko, C. O. Weiss, and B. Schäpers, *Phys. Rev. A* **65**, 013812 (2002).

[100] V. B. Taranenko and C. O. Weiss, *IEEE J. Sel. Topics Quantum Electron.* **8**, 488 (2002).

[101] V. B. Taranenko and C. O. Weiss, webpage: arXiv.org/abs/nlin.PS/0206029 (2002).

[102] V. B. Taranenko, C. O. Weiss, and W. Stolz, *J. Opt. Soc. Am. B* **19**, 8129 (1992).

[103] V. B. Taranenko, C. O. Weiss, and W. Stolz, *Opt. Lett.* **26**, 1574 (2001).

[104] S. Barland, J. R. Tredicce, M. Brambilla, L. A. Lugiato, S. Balle, M. Giudici, T. Maggipinto, L. Spinelli, G. Tissoni, T. Knödl, M. Miller, and R. Jäger, *Nature* **419**, 699 (2002).

[105] R. L. Byer and A. Piskarskas, Eds., *J. Opt. Soc. Am. B* **10**, 1655 (1993); special issue on parametric oscillation and amplification.

[106] G.-L. Oppo, M. Brambilla, D. Camesasca, A. Gatti, and L. A. Lugiato, *J. Mod. Opt.* **41**, 1151 (1994).

[107] C. Etrich, U. Peschel, and F. Lederer, *Phys. Rev. Lett.* **79**, 2454 (1997).

[108] F. Lederer, C. Etrich, U. Peschel, and D. Michaelis, *Chaos, Solitons, Fractals* **10**, 895 (1999).

[109] D. V. Skryabin, *Phys. Rev. E* **60**, R3508 (1999).

[110] G.-L. Oppo, A. J. Scroggie, and W. J. Firth, *Phys. Rev. A* **63**, 066209 (2001).

[111] S. Fedorov, D. Michaelis, U. Peschel, C. Etrich, D. V. Skryabin, N. Rosanov, and F. Lederer, *Phys. Rev. A* **64**, 036610 (2001).

[112] D. V. Skryabin and A. R. Champneys, *Phys. Rev. E* **63**, 066610 (2001).

[113] U. Peschel, D. Michaelis, C. Etrich, and F. Lederer, *Phys. Rev. E* **58**, R2745 (1998).

[114] G. J. de Valcárcel, E. Roldán, and K. Staliunas, *Opt. Commun.* **181**, 207 (2000).

[115] G.-L. Oppo, M. Brambilla, and L. A. Lugiato, *Phys. Rev. A* **49**, 2028 (1994).

[116] F. Bloch, Z. Phys. **61**, 206 (1930).

[117] B. Lax and K. J. Button, *Microwave Ferrites and Ferrimagnetics* (McGraw-Hill, New York, 1962).

[118] P. E. Wigen, Ed., *Nonlinear Phenomena and Chaos in Magnetic Materials* (World Scientific, Singapore, 1994).

[119] M. G. Cottam, Ed., *Linear and Nonlinear Spin Waves in Magnetic Films and Superlattices* (World Scientific, Singapore, 1994).

[120] S. O. Demokritov, B. Hillebrands, and AN. Slavin, *Phys. Rep.* **348**, 441 (2001).

[121] O. Büttner, M. Bauer, S. O. Demokritov, B. Hillebrands, Yu. S. Kivshar, V. Grimalsky, Yu. Rapoport, and A. N. Slavin, *Phys. Rev. B* **61**, 11576 (2000).

[122] A. N. Slavin and I. V. Rojdestvenski, *IEEE Trans. Magn.* **30**, 37 (1994).

[123] A. N. Slavin, *Phys. Rev. Lett.* **77**, 4644 (1996).

[124] A. N. Slavin, B. A. Kalinikos, and N. G. Kovshikov, in *Nonlinear Phenomena and Chaos in Magnetic Materials* P. E. Wigen, Ed. (World Scientific, Singapore, 1994), Chap. 9.

[125] M. J. Lighthill, *J. Inst. Math. Appl.* **1**, 269 (1965).

[126] B. A. Kalinikos, N. G. Kovshikov, and A. N. Slavin, *Sov. Phys. JETP* **67**, 303 (1988).

[127] O. von Geisau, U. Netzelmann, S. M. Rezende, and J. Pelzl, *IEEE Trans. Magn.* **26**, 1471 (1990).

[128] M. Chen, M. A. Tsankov, J. M. Nash, and C. E. Patton, *Phys. Rev. B* **49**, 12 773 (1994).

[129] J. W. Boyle, S. A. Nikitov, A. D. Boardman, J. G. Booth, and K. Booth, *Phys. Rev. B* **53**, 12 173 (1996).

[130] M. Bauer, C. Mathieu, S. O. Demokritov, B. Hillebrands, P. A. Kolodin, S. Sure, H. Dötsch, V. Grimalsky, Yu. Rapoport, and A. N. Slavin, *Phys. Rev. B* **56**, R8483 (1997).

[131] M. Bauer, O. Büttner, S. O. Demokritov, B. Hillebrands, V. Grimalsky, Yu. Rapoport, and A. N. Slavin, *Phys. Rev. Lett.* **81**, 2582 (1998).

[132] B. A. Kalinikos, N. G. Kovshikov, M. P. Kostylev, and H. Benner, *JETP Lett.* **64**, 171 (1996).

[133] V. E. Demidov, *JETP Lett.* **68**, 869 (1998) [Pis'ma Zh. Eksp. Teor. Fiz. **68**, 828 (1998)].

[134] B. A. Kalinikos, N. G. Kovshikov, and A. N. Slavin, *JETP Lett.* **38**, 413 (1983).

[135] A. D. Boardman, G. S. Cooper, A. A. Maradudin, and T.P. Shen, *Phys. Rev. B* **34**, 8273 (1986).

[136] A. D. Boardman, S. A. Nikitov, K. Xie, and H. Mehta, *J. Magn. Magn. Mat.* **145**, 357 (1995).

[137] J. M. Nash, C. E. Patton, and P. Kabos, *Phys. Rev. B* **51**, 15079 (1995).

[138] N. G. Kovshikov, B. A. Kalinikos, C. E. Patton, E. S. Wright, and J. M. Nash, *Phys. Rev. B* **54**, 15210 (1996).

[139] H. Xia, P. Kabos, C. E. Patton, and H. E. Ensle, *Phys. Rev. B* **55**, 15018 (1997).

[140] H. Xia, P. Kabos, R. A. Staudinger, and C. E. Patton, *Phys. Rev. B* **58**, 2708 (1998).

[141] Y. Silberberg, *Opt. Lett.* **15**, 1282 (1990).

[142] M. Chen, M. A. Tsankov, J. M. Nash, and C. E. Patton, *Phys. Rev. Lett.* **70**, 1707 (1993).

[143] A. K. Zvezdin and A. F. Popkov, *Zh. Eksp. Teor. Fiz.* **84**, 606 (1983) [*Sov. Phys. JETP* **57**, 350 (1983)].

[144] A. N. Slavin, Yu. S. Kivshar, E. A. Ostrovskaya, and H. Benner, *Phys. Rev. Lett.* **82**, 2583 (1999).

[145] B. A. Kalinikos, N. G. Kovshikov, and C. E. Patton, *Phys. Rev. Lett.* **80**, 4301 (1998).

[146] B. A. Kalinikos, N. G. Kovshikov, and C. E. Patton, *JETP Lett.* **68**, 243 (1998) [Pis'ma Zh. Eksp. Teor. Fiz. **68**, 229 (1998)].

[147] B. A. Kalinikos and N. G. Kovshikov, *JETP Lett.* **60**, 305 (1994) [Pis'ma Zh. Eksp. Teor. Fiz. **60**, 290 (1994)].

[148] O. Büttner, M. Bauer, S. O. Demokritov, B. Hillebrands, M. P. Kostylev, B. A. Kalinikos, and A. N. Slavin, *Phys. Rev. Lett.* **82**, 4320 (1999).

[149] M. N. Anderson *et al.*, *Science* **269**, 198 (1995).

[150] C. C. Bradley, C. A. Sackett, J. J. Tollett, and R. G. Hulet, *Phys. Rev. Lett.* **75**, 1687 (1995).

[151] K. B. Davis, M. -O. Mewes, M. R. Andrews, N. J. van Druten, D. S. Durfee, D. M. Kurn, and W. Ketterle, *Phys. Rev. Lett.* **75**, 3969 (1995).

[152] D. G. Fried, T. C. Killian, L. Willmann, D. Landhuis, S. C. Moss, D. Kleppner, and T. J. Greytak, *Phys. Rev. Lett.* **81**, 3811 (1998).

[153] F. Dalfovo, S. Giorgini, L. P. Pitaevskii, and S. Stringari, *Rev. Mod. Phys.* **71**, 463 (1999).

[154] N. N. Bogoliubov, *J. Phys. (Moscow)* **11**, 23 (1947).

[155] S. Inouye, M. R. Andrews, J. Stenger, H.-J. Miesner, D. M. Stamper-Kurn, and W. Ketterle *Nature* **392**, 151 (1998).

[156] V. M. Pérez-García, H. Michinel, and H. Herrero, *Phys. Rev. A* **57**, 3837 (1998).

[157] L. Salasnich, A. Parola, and L. Reatto, *Phys. Rev. A* **65**, 043614 (2002).

[158] Yu. S. Kivshar, T. J. Alexander, and S. K. Turitsyn, *Phys. Lett. A* **278**, 225 (2001).

[159] S. K. Turitsyn and V. K. Mezentsev, *JETP Lett.* **64**, 616 (1998).

[160] V. I. Yukalov, E. P. Yukalov, and V. S. Bagnato, *Phys. Rev. A* **56**, 4845 (1997).

[161] E. A. Ostrovskaya, Yu. S. Kivshar, D. V. Skryabin, and W. J. Firth, *Phys. Rev. Lett.* **83**, 296 (1999).

[162] Yu. Kivshar and B. Luther-Davies, *Phys. Rep.* **298**, 81 (1998).

[163] D. J. Frantzesakakis, G. Theocharis, F. K. Diakonos, P. Schmelcher, and Yu. S. Kivshar, *Phys. Rev. A* **63**, Nov. (2002)

[164] T. Busch and J. R. Anglin, *Phys. Rev. Lett.* **84**, 2298 (2000).

[165] W. P. Reinhardt and C. W. Clark, *J. Phys. B* **30**, L785 (1997).

[166] S. A. Morgan, R. J. Ballagh, and K. Burnett, *Phys. Rev. A* **55**, 4338 (1997).

[167] A. E. Muryshev, H.B. van Linden van den Heuvell, and G. V. Shlyapnikov, *Phys. Rev. A* **60**, R2665 (1999).

[168] P. O. Fedichev, A. E. Muryshev, and G. V. Shlyapnikov, *Phys. Rev. A* **60**, 3220 (1999).

[169] J. Denschlag, J. E. Simsarian, D. L. Feder, C. W. Clark, L. A. Collins, J. Cubizolles, L. Deng, E. W. Hagley, K. Helmerson, W. P. Reinhardt, S. L. Rolston, B. I. Schneider, and W. D. Phillips, *Science* **287**, 97 (2000).

[170] S. Burger, K. Bongs, S. Dettmer, W. Ertmer, K. Sengstock, A. Sanpera, G. V. Shlyapnikov, and M. Lewenstein, *Phys. Rev. Lett.* **83**, 5198 (1999).

[171] D. L. Feder, M. S. Pindzola, L. A. Collins, B. I. Schneider, and C. W. Clark, *Phys. Rev. A* **62**, 053606 (2000).

[172] B. P. Anderson, P. C. Haljan, C. A. Regal, D. L. Feder, L. A. Collins, C. W. Clark, and E. A. Cornell, *Phys. Rev. Lett.* **86**, 2926 (2001).

[173] J. Brand and W. P. Reinhardt, *Phys. Rev. A* **65**, 043612 (2002).

[174] J. E. Williams and M. J. Holland, *Nature* **401**, 568 (1999).

[175] J. J. García-Ripoll and V. M. Pérez-García, *Phys. Rev. Lett.* **84**, 4267 (2000).

[176] M. R. Matthews, B .P. Anderson, P. C. Haljan, D. S. Hall, C. E. Wieman, and E. A. Cornell, *Phys. Rev. Lett.* **83**, 2498 (1999).

[177] A. L. Fetter and A. A. Svidzinsky, *J. Phys.: Condens. Matter* **13**, R135 (2001).

[178] D. A. Butts and D. S. Rokhsar, *Nature* **397**, 327 (1999).

[179] A. A. Abrikosov, *J. Exp. Theor. Phys.* **5**, 1174 (1957) [Zh. Eksp. Teor. Fiz. **32**, 1442 (1957).

[180] K. W. Madison, F. Chevy, W. Wohlleben, and J. Dalibard, *Phys. Rev. Lett.* **84**, 806 (2000).

[181] J. R. Abo-Shaeer, C. Raman, J. M. Vogels, and W. Ketterle, *Science* **292**, 476 (2001).

[182] C. Raman, J. R. Abo-Shaeer, J. M. Vogels, K. Xu, and W. Ketterle, *Phys. Rev. Lett.* **87**, 210402 (2001).

[183] K. E. Strecker, G. B. Partridge, A. G. Truscott, and R. G. Hulet, *Nature* **417**, 150 (2002).

[184] L. Khaykovich, F. Schreck, G. Ferrari, T. Bourdel, J. Cubizolles, L. D. Carr, Y. Castin, and C. Salomon, *Science* **296**, 1290 (2002).

[185] J. M. Gerton, D. Strekalov, I. Prodan, and R. G. Hulet, *Nature* **408**, 692 (2000).

[186] E. A. Donley, N. R. Claussen, S. L. Cornish, J. L. Roberts, E. A. Cornell, and C. E. Wieman, *Nature* **412**, 295 (2001).

[187] L. D. Carr, M. A. Leung, and W. P. J. Reinhardt, *Phys. Rev. B* **33**, 3983 (2000).

Index